EMERY'S Elements of Medical Genetics

EMERY'S

Elements of Medical Genetics

14th EDITION

Peter D. Turnpenny
BSc, MB, ChB, FRCP, FRCPCH
Consultant Clinical Geneticist
Royal Devon and Exeter Hospital
and
Honorary Senior Clinical Lecturer
Peninsula Medical School
Exeter, United Kingdom

Sian Ellard
BSc, PhD, FRCPath
Consultant Clinical Molecular Geneticist
Royal Devon and Exeter Hospital
and
Professor of Human Molecular Genetics
Peninsula Medical School
Exeter, United Kingdom

ELSEVIER
CHURCHILL
LIVINGSTONE

1600 John F. Kennedy Blvd.
Ste 1800
Philadelphia, PA 19103-2899

EMERY'S ELEMENTS OF MEDICAL GENETICS ISBN: 978-0-7020-4043-6

Notices

Knowledge and best practice in this field are constantly changing. As new research and experience broaden our understanding, changes in research methods, professional practices, or medical treatment may become necessary.

Practitioners and researchers must always rely on their own experience and knowledge in evaluating and using any information, methods, compounds, or experiments described herein. In using such information or methods they should be mindful of their own safety and the safety of others, including parties for whom they have a professional responsibility.

With respect to any drug or pharmaceutical products identified, readers are advised to check the most current information provided (i) on procedures featured or (ii) by the manufacturer of each product to be administered, to verify the recommended dose or formula, the method and duration of administration, and contraindications. It is the responsibility of practitioners, relying on their own experience and knowledge of their patients, to make diagnoses, to determine dosages and the best treatment for each individual patient, and to take all appropriate safety precautions.

To the fullest extent of the law, neither the Publisher nor the authors, contributors, or editors assume any liability for any injury and/or damage to persons or property as a matter of products liability, negligence or otherwise, or from any use or operation of any methods, products, instructions, or ideas contained in the material herein.

ISBN: 978-0-7020-4043-6

Publishing Director: Anne Lenehan
Developmental Editor: Andrew Hall
Publishing Services Manager: Anne Altepeter
Project Manager: Cindy Thoms
Senior Designer: Ellen Zanolle

Printed in Spain

Last digit is the print number: 9 8 7 6 5 4 3 2 1

To our fathers—
sources of encouragement and support
who would have been proud of this work

Preface

Alan E.H. Emery
Emeritus Professor of Human Genetics & Honorary Fellow
University of Edinburgh

"A man ought to read just as inclination leads him; for what he reads as a task will do him little good."
Dr. Samuel Johnson

Advances and breakthroughs in genetic science are continually in the news, attracting great interest because of the potential, not only for diagnosing and eventually treating disease, but also for what we learn about humankind through these advances. In addition, almost every new breakthrough raises a fresh ethical, social, and moral debate about the uses to which genetic science will be put, particularly in reproductive medicine and issues relating to identity and privacy. Increasingly, today's medical graduates, and mature postgraduates, must be equipped to integrate genetic knowledge and science appropriately into all areas of medicine, for the task cannot be left solely to clinical geneticists, who remain small in number; indeed, in many countries there is either no structured training program in clinical genetics or the specialty is not recognized at all.

Since the publication of the thirteenth edition of *Emery's Elements of Medical Genetics* there has been a huge surge forward in our knowledge and understanding of the human genome as the technology of *microarray comparative genomic hybridization* has been extensively applied, both in research and clinical service settings. We know so much more about the normal variability of the human genome as the extent of *copy number variants* (of DNA) has become clearer, though we are still trying to unravel the possible significance of these in relation to health and disease. And as we write this there is great excitement about the next technological revolution that is underway, namely *next generation sequencing*. Already there are dramatic examples of gene discovery in mendelian conditions through analysis of the whole *exome* of very small numbers of patients with clear phenotypes. There is also more realistic anticipation than before that breakthroughs will be made in the treatment of genetic disease, which will take a variety of different forms. Whilst discovery and knowledge proceed apace, however, the foundation for those who aspire to be good clinical practitioners in this field lies in a thorough grasp of the basics of medical genetics, which must include the ability to counsel patients and families with sensitivity and explain difficult concepts in simple language.

In this fourteenth edition of *Emery's Elements of Medical Genetics* we have tried to simplify some of the language and reduce redundant text where possible, to make way for some new, updated material. Several chapters have undergone significant revisions, and the range of illustrations has increased. We have listened to those colleagues (a small number!) who identified one or two errors in the last edition and also suggested ideas for improvement. Once again, we have sought to provide a balance between a basic, comprehensive text and one that is as up to date as possible, still aiming at medical undergraduates and those across both medical and non-medical disciplines who simply want to "taste and see." The basic layout of the book has not changed because it seems to work well, and for that we remain in debt to our predecessors in this project, namely Alan Emery, Bob Mueller, and Ian Young.

Peter D. Turnpenny and Sian Ellard
Exeter, United Kingdom
November 2010

Acknowledgments

As with the previous two editions, we are very grateful to those of our patients who were asked for consent to publish their photographs for the first time; again, not one refused, which was enormously helpful. In preparation of this edition we thank colleagues who cast a critical but very constructive eye over particular chapters, which led to some very necessary changes to the text. These were Dr. Paul Kerr (Consultant Hematologist, Royal Devon and Exeter Hospital, Exeter) and Dr. Claire Bethune (Consultant Immunologist, Derriford Hospital, Plymouth). Dr. Rachel Freathy (Sir Henry Wellcome Postdoctoral Fellow, Peninsula Medical School, Exeter) provided new insights and assisted with revision of the chapters describing polygenic inheritance and common disorders. We thank those at Elsevier who communicated very fully and promptly throughout the revision, and were patient with delays on our part. We again thank those at our respective homes who had to put up with a season of early mornings and late nights, without which the revision would not have been possible.

Contents

EMERY'S Elements of Medical Genetics

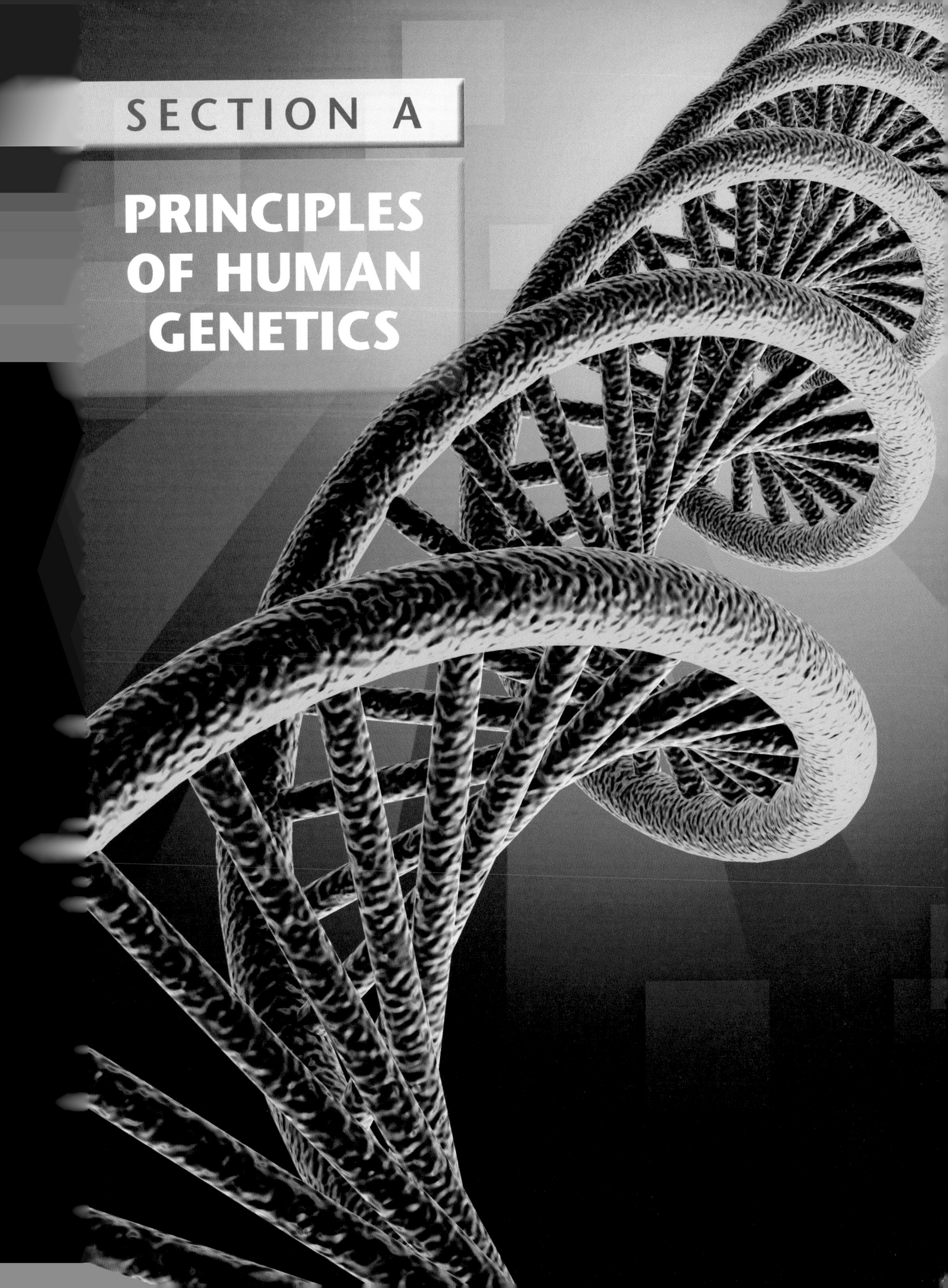
SECTION A
PRINCIPLES OF HUMAN GENETICS

CHAPTER 1

The History and Impact of Genetics in Medicine

It's just a little trick, but there is a long story connected with it which it would take too long to tell.

GREGOR MENDEL, IN CONVERSATION WITH C. W. EICHLING

It has not escaped our notice that the specific pairing we have postulated immediately suggests a possible copying mechanism for the genetic material.

WATSON & CRICK (APRIL 1953)

Presenting historical truth is at least as challenging as the pursuit of scientific truth and our view of human endeavors down the ages is heavily biased in favor of winners—those who have conquered on military, political, or, indeed, scientific battlefields. The history of genetics in relation to medicine is one of breathtaking discovery from which patients and families already benefit hugely, but in the future success will be measured by ongoing progress in translating discoveries into both treatment and prevention of disease. As this takes place, we should not neglect looking back with awe at what our forebears achieved with scarce resources and sheer determination, sometimes aided by serendipity, in order to lay the foundations of this dynamic science. A holistic approach to science can be compared with driving a car: without your eyes on the road ahead, you will crash and make no progress; however, the competent driver will glance in the rear and side mirrors regularly to maintain control.

Gregor Mendel and the Laws of Inheritance

Early Beginnings

Developments in genetics during the twentieth century have been truly spectacular. In 1900 Mendel's principles were awaiting rediscovery, chromosomes were barely visible, and the science of molecular genetics did not exist. By contrast, at the time of writing this text in 2010, chromosomes can be rapidly analyzed to an extraordinary level of sophistication by microarray techniques and the sequence of the entire human genome has been published. Some 13,000 human genes with known sequence are listed and nearly 6500 genetic diseases or **phenotypes** have been described, of which the molecular genetic basis is known in approximately 2650.

Few would deny that genetics is of major importance in almost every medical discipline. Recent discoveries impinge not just on rare genetic diseases and syndromes, but also on many of the common disorders of adult life that may be predisposed by genetic variation, such as cardiovascular disease, psychiatric illness, and cancer, not to mention influences on obesity, athletic performance, musical ability, and longevity. Consequently a fundamental grounding in genetics should be an integral component of any undergraduate medical curriculum.

To put these exciting developments into context, we start with an overview of some of the most notable milestones in the history of medical genetics. The importance of understanding its role in medicine is then illustrated by reviewing the overall impact of genetic factors in causing disease. Finally, new developments of major importance are discussed.

It is not known precisely when *Homo sapiens* first appeared on this planet, but according to current scientific consensus based on the finding of fossilized human bones in Ethiopia, man was roaming East Africa about 200,000 years ago. It is reasonable to suppose that our early ancestors were as curious as ourselves about matters of inheritance and, just as today, they would have experienced the birth of babies with all manner of physical defects. Engravings in Chaldea in Babylonia (modern-day Iraq) dating back at least 6000 years show pedigrees documenting the transmission of certain characteristics of the horse's mane. However, any early attempts to unravel the mysteries of genetics would have been severely hampered by a total lack of knowledge and understanding of basic processes such as conception and reproduction.

Early Greek philosophers and physicians such as Aristotle and Hippocrates concluded, with typical masculine modesty, that important human characteristics were determined by semen, using menstrual blood as a culture

FIGURE 1.1 Gregor Mendel. (Reproduced with permission from BMJ Books.)

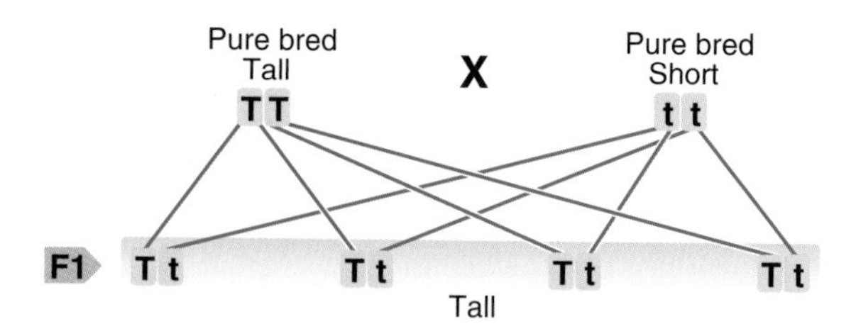

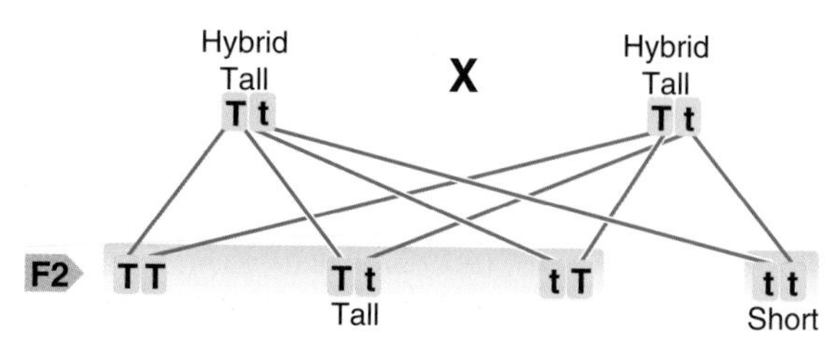

FIGURE 1.2 An illustration of one of Mendel's breeding experiments and how he correctly interpreted the results.

medium and the uterus as an incubator. Semen was thought to be produced by the whole body; hence bald-headed fathers would beget bald-headed sons. These ideas prevailed until the seventeenth century, when Dutch scientists such as Leeuwenhoek and de Graaf recognized the existence of sperm and ova, thus explaining how the female could also transmit characteristics to her offspring.

The blossoming of the scientific revolution in the 18th and 19th centuries saw a revival of interest in heredity by both scientists and physicians, among whom two particular names stand out. Pierre de Maupertuis, a French naturalist, studied hereditary traits such as extra digits (polydactyly) and lack of pigmentation (albinism), and showed from pedigree studies that these two conditions were inherited in different ways. Joseph Adams (1756–1818), a British doctor, also recognized that different mechanisms of inheritance existed and published *A Treatise on the Supposed Hereditary Properties of Diseases*, which was intended as a basis for genetic counseling.

Our present understanding of human genetics owes much to the work of the Austrian monk Gregor Mendel (1822–1884; Figure 1.1) who, in 1865, presented the results of his breeding experiments on garden peas to the Natural History Society of Brünn in Bohemia (now Brno in the Czech Republic). Shortly after, Mendel's observations were published by that association in the Transactions of the Society, where they remained largely unnoticed until 1900, some 16 years after his death, when their importance was first recognized. In essence, Mendel's work can be considered as the discovery of genes and how they are inherited. The term **gene** was first coined in 1909 by a Danish botanist, Johannsen, and was derived from the term 'pangen' introduced by De Vries. This term was itself a derivative of the word 'pangenesis,' coined by Darwin in 1868. In acknowledgement of Mendel's enormous contribution, the term **mendelian** is now part of scientific vocabulary, applied both to the different patterns of inheritance shown by single-gene characteristics and to disorders found to be the result of defects in a single gene.

In his breeding experiments, Mendel studied contrasting characters in the garden pea, using for each experiment varieties that differed in only one characteristic. For example, he noted that when strains bred for a feature such as tallness were crossed with plants bred to be short all of the offspring in the first filial or F1 generation were tall. If plants in this F1 generation were interbred, this led to both tall and short plants in a ratio of 3:1 (Figure 1.2). Characteristics that were manifest in the F1 hybrids were referred to as **dominant**, whereas those that reappeared in the F2 generation were described as being **recessive**. On reanalysis it has been suggested that Mendel's results were 'too good to be true' in that the segregation ratios he derived were suspiciously closer to the value of 3:1 than the laws of statistics would predict. One possible explanation is that he may have published only those results that best agreed with his preconceived single-gene hypothesis. Whatever the truth of the matter, events have shown that Mendel's interpretation of his results was entirely correct.

Mendel's proposal was that the plant characteristics being studied were each controlled by a pair of factors, one of which was inherited from each parent. The pure-bred plants, with two identical genes, used in the initial cross would now be referred to as **homozygous**. The hybrid F1 plants, each of which has one gene for tallness and one for shortness, would be referred to as **heterozygous**. The genes responsible for these contrasting characteristics are referred to as **allelomorphs**, or **alleles** for short.

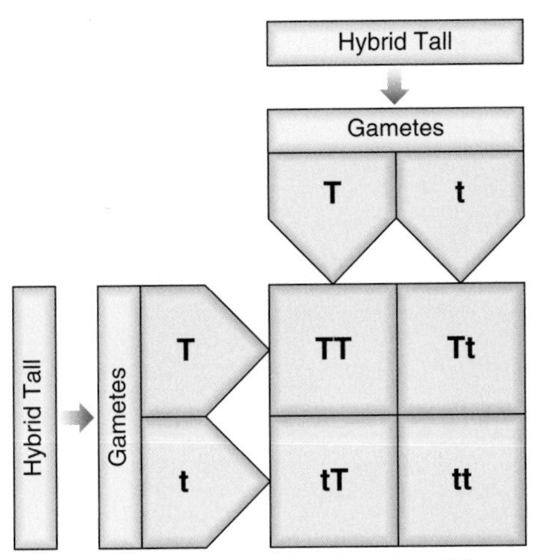

FIGURE 1.3 A Punnett square showing the different ways in which genes can segregate and combine in the second filial cross from Figure 1.2. Construction of a Punnett square provides a simple method for showing the possible gamete combinations in different matings.

An alternative method for determining **genotypes** in offspring involves the construction of what is known as a Punnett square (Figure 1.3). This is used further in Chapter 8 when considering how genes segregate in large populations.

On the basis of Mendel's plant experiments, three main principles were established. These are known as the laws of uniformity, segregation, and independent assortment.

The Law of Uniformity

The *law of uniformity* refers to the fact that when two homozygotes with different alleles are crossed, all of the offspring in the F1 generation are identical and heterozygous. In other words, the characteristics do not blend, as had been believed previously, and can reappear in later generations.

The Law of Segregation

The *law of segregation* refers to the observation that each person possesses two genes for a particular characteristic, only one of which can be transmitted at any one time. Rare exceptions to this rule can occur when two allelic genes fail to separate because of chromosome non-disjunction at the first meiotic division (p. 43).

The Law of Independent Assortment

The *law of independent assortment* refers to the fact that members of different gene pairs segregate to offspring independently of one another. In reality, this is not always true, as genes that are close together on the same chromosome tend to be inherited together, because they are 'linked' (p. 136). There are a number of other ways by which the laws of mendelian inheritance are breached but, overall, they remain foundational to our understanding of the science.

The Chromosomal Basis of Inheritance

As interest in mendelian inheritance grew, there was much speculation as to how it actually occurred. At that time it was also known that each cell contains a nucleus within which there are several threadlike structures known as **chromosomes**, so called because of their affinity for certain stains (*chroma* = color, *soma* = body). These chromosomes had been observed since the second half of the nineteenth century after development of cytologic staining techniques. Human mitotic figures were observed from the late 1880s, and it was in 1902 that Walter Sutton, an American medical student, and Theodour Boveri, a German biologist, independently proposed that chromosomes could be the bearers of heredity (Figure 1.4). Subsequently, Thomas Morgan transformed Sutton's chromosome theory into the theory of the gene, and Alfons Janssens observed the formation of chiasmata between homologous chromosomes at meiosis. During the late 1920s and 1930s, Cyril Darlington helped to clarify chromosome mechanics by the use of tulips collected on expeditions to Persia. It was during the 1920s that the term **genome** entered the scientific vocabulary, being the fusion of *genom* (German for 'gene') and *ome* from 'chromosome'.

When the connection between mendelian inheritance and chromosomes was first made, it was thought that the normal chromosome number in humans might be 48, although various papers had come up with a range of figures. The number 48 was settled on largely as a result of a paper in 1921 from Theophilus Painter, an American cytologist who had been a student of Boveri. In fact, Painter himself had some preparations clearly showing 46 chromosomes, even though he finally settled on 48. These discrepancies were probably from the poor quality of the material at that time; even into the early 1950s, cytologists were counting 48 chromosomes. It was not until 1956 that the correct number of 46 was established by Tjio and Levan, 3 years after the correct structure of DNA had been proposed. Within a few years, it was shown that some disorders in humans could be caused by loss or gain of a whole chromosome as well as by an abnormality in a single gene. Chromosome disorders are discussed at length in Chapter 18. Some chromosome aberrations, such as translocations, can run in families (p. 44), and are sometimes said to be segregating in a mendelian fashion.

DNA as the Basis of Inheritance

Whilst James Watson and Francis Crick are justifiably credited with discovering the structure of DNA in 1953, they were attracted to working on it only because of its key role as the genetic material, as established in the 1940s. Formerly many believed that hereditary characteristics were transmitted by proteins, until it was appreciated that their molecular structure was far too cumbersome. Nucleic acids were actually discovered in 1849. In 1928 Fred Griffith, working on two strains of *Streptococcus*, realized

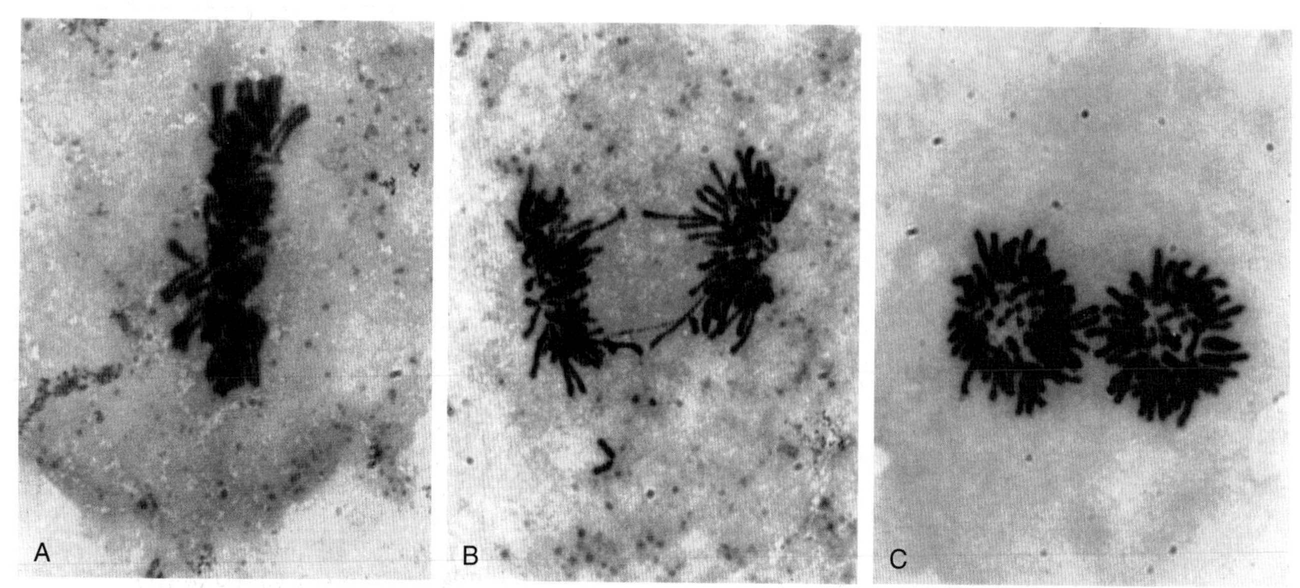

FIGURE 1.4 Chromosomes dividing into two daughter cells at different stages of cell division. **A,** Metaphase; **B,** anaphase; **C,** telophase. The behavior of chromosomes in cell division (mitosis) is described at length in Chapter 3. (Photographs courtesy Dr. K. Ocraft, City Hospital, Nottingham.)

that characteristics of one strain could be conferred on the other by something that he called the **transforming principle**. In 1944, at the Rockefeller Institute in New York, Oswald Avery, Maclyn McCarty, and Colin MacLeod identified DNA as the genetic material while working on the pneumococcus (*Streptococcus pneumoniae*). Even then, many in the scientific community were skeptical; DNA was only a simple molecule with lots of repetition of four nucleic acids—very boring! The genius of Watson and Crick, at Cambridge, was to hit on a structure for DNA that would explain the very essence of biological reproduction, and their elegant double helix has stood the test of time. Crucial to their discovery was the x-ray crystallography work of Maurice Wilkins and Rosalind Franklin at King's College, London.

This was merely the beginning, for it was necessary to discover the process whereby DNA, in discrete units called genes, issues instructions for the precise assembly of proteins, the building blocks of tissues. The sequence of bases in DNA, and the sequence of amino acids in protein, the **genetic code**, was unravelled in some elegant biochemical experiments in the 1960s and it became possible to predict the base change in DNA that led to the amino-acid change in the protein. Further experiments, involving Francis Crick, Paul Zamecnik, and Mahlon Hoagland, identified the molecule transfer RNA (tRNA) (p. 20), which directs genetic instructions via amino acids to intracellular ribosomes, where protein chains are produced. Confirmation of these discoveries came with DNA sequencing methods and the advent of recombinant DNA techniques. Interestingly, however, the first genetic trait to be characterized at the molecular level had already been identified in 1957 by laborious sequencing of the purified proteins. This was sickle-cell anemia, in which the mutation affects the amino-acid sequence of the blood protein hemoglobin.

The Fruit Fly

Before returning to historical developments in human genetics, it is worth a brief diversion to consider the merits of an unlikely creature, which has proved to be of great value in genetic research. The fruit fly, *Drosophila*, possesses several distinct advantages for the study of genetics:

1. It can be bred easily in a laboratory.
2. It reproduces rapidly and prolifically at a rate of 20 to 25 generations per annum.
3. It has a number of easily recognized characteristics, such as *curly wings* and a *yellow body*, which follow mendelian inheritance.
4. *Drosophila melanogaster*, the species studied most frequently, has only four pairs of chromosomes, each of which has a distinct appearance so that they can be identified easily.
5. The chromosomes in the salivary glands of *Drosophila* larvae are among the largest known in nature, being at least 100 times bigger than those in other body cells.

In view of these unique properties, fruit flies were used extensively in early breeding experiments. Today their study is still proving of great value in fields such as developmental biology, where knowledge of gene homology throughout the animal kingdom has enabled scientists to identify families of genes that are important in human embryogenesis (see Chapter 6). When considering major scientific achievements in the history of genetics, it is notable that sequencing of the 180 million base pairs of the

Drosophila melanogaster genome was completed toward the end of 1999.

The Origins of Medical Genetics

In addition to the previously mentioned Pierre de Maupertuis and Joseph Adams, whose curiosity was aroused by polydactyly and albinism, there were other pioneers. John Dalton, of atomic theory fame, observed that some conditions, notably color blindness and hemophilia, show what is now referred to as sex- or X-linked inheritance, and to this day color blindness is still occasionally referred to as **daltonism**. Inevitably, these founders of human and medical genetics could only speculate on the nature of hereditary mechanisms.

In 1900 Mendel's work resurfaced. His papers were quoted almost simultaneously by three European botanists—De Vries (Holland), Correns (Germany), and Von Tschermak (Austria)—and this marked the real beginning of medical genetics, providing an enormous impetus for the study of inherited disease. Credit for the first recognition of a single-gene trait is shared by William Bateson and Archibald Garrod, who together proposed that alkaptonuria was a rare recessive disorder. In this relatively benign condition, urine turns dark on standing or on exposure to alkali because of the patient's inability to metabolize homogentisic acid (p. 171). Young children show skin discoloration in the napkin (diaper) area and affected adults may develop arthritis in large joints. Realizing that this was an inherited disorder involving a chemical process, Garrod coined the term **inborn error of metabolism** in 1908. However, his work was largely ignored until the mid-twentieth century, when the advent of electrophoresis and chromatography revolutionized biochemistry. Several hundred such disorders have now been identified, giving rise to the field of study known as **biochemical genetics** (see Chapter 11). The history of alkaptonuria neatly straddles almost the entire twentieth century, starting with Garrod's original observations of recessive inheritance in 1902 and culminating in cloning of the relevant gene on chromosome 3 in 1996.

During the course of the twentieth century, it gradually became clear that hereditary factors were implicated in many conditions and that different genetic mechanisms were involved. Traditionally, hereditary conditions have been considered under the headings of **single gene, chromosomal,** and **multifactorial**. Increasingly, it is becoming clear that the interplay of different genes (**polygenic inheritance**) is important in disease, and that a further category—**acquired somatic genetic disease**—should also be included.

Single-Gene Disorders

In addition to alkaptonuria, Garrod suggested that albinism and cystinuria could also show recessive inheritance. Soon other examples followed, leading to an explosion in knowledge and disease delineation. By 1966 almost 1500 single-gene disorders or traits had been identified, prompting the publication by an American physician, Victor McKusick (Figure 1.5), of a catalog of all known single-gene conditions. By 1998, when the 12th edition of this catalog was published, it contained more than 8500 entries (Figure 1.6). The growth of 'McKusick's Catalog' has been exponential and is now available electronically as *Online Mendelian Inheritance in Man* (OMIM) (see Appendix). By 2010 OMIM contained a total of almost 20,000 entries.

FIGURE 1.5 Victor McKusick in 1994, whose studies and catalogs have been so important to medical genetics.

Chromosome Abnormalities

Improved techniques for studying chromosomes led to the demonstration in 1959 that the presence of an additional number 21 chromosome (*trisomy 21*) results in

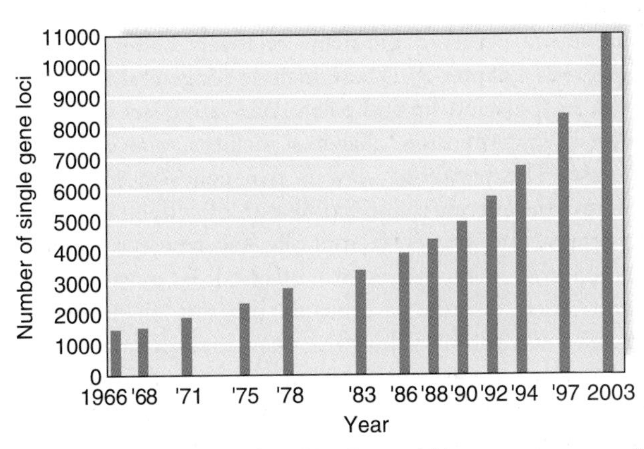

FIGURE 1.6 Histogram showing the rapid increase in recognition of conditions and characteristics (traits) showing single-gene inheritance. (Adapted from McKusick, 1998, and OMIM—see Appendix.)

Down syndrome. Other similar discoveries followed rapidly—Klinefelter and Turner syndromes—also in 1959. The identification of chromosome abnormalities was further aided by the development of banding techniques in 1970 (p. 33). These enabled reliable identification of individual chromosomes and helped confirm that loss or gain of even a very small segment of a chromosome can have devastating effects on human development (see Chapter 18).

Later it was shown that several rare conditions featuring learning difficulties and abnormal physical features are due to loss of such a tiny amount of chromosome material that no abnormality can be detected using even the most high-powered light microscope. These conditions are referred to as microdeletion syndromes (p. 280) and can be diagnosed using a technique known as **FISH** (**fluorescent in-situ hybridization**), which combines conventional chromosome analysis (**cytogenetics**) with newer DNA diagnostic technology (**molecular genetics**) (p. 34). Already, however, the latest technique of microarray **CGH** (**comparative genomic hybridization**) is revolutionizing clinical genetics through the detection of subtle genomic imbalances (p. 36).

Multifactorial Disorders

Francis Galton, a cousin of Charles Darwin, had a longstanding interest in human characteristics such as stature, physique, and intelligence. Much of his research was based on the study of identical twins, in whom it was realized that differences in these parameters must be largely the result of environmental influences. Galton introduced to genetics the concept of the **regression coefficient** as a means of estimating the degree of resemblance between various relatives. This concept was later extended to incorporate Mendel's discovery of genes, to try to explain how parameters such as height and skin color could be determined by the interaction of many genes, each exerting a small additive effect. This is in contrast to single-gene characteristics in which the action of one gene is exerted independently, in a non-additive fashion.

This model of **quantitative inheritance** is now widely accepted and has been adapted to explain the pattern of inheritance observed for many relatively common conditions (see Chapter 9). These include congenital malformations such as cleft lip and palate, and late-onset conditions such as hypertension, diabetes mellitus, and Alzheimer disease. The prevailing view is that genes at several loci interact to generate a susceptibility to the effects of adverse environmental trigger factors. Recent research has confirmed that many genes are involved in most of these adult-onset disorders, although progress in identifying specific susceptibility loci has been disappointingly slow. It has also emerged that in some conditions, such as type I diabetes mellitus, different genes can exert major or minor effects in determining susceptibility (p. 233). Overall, **multifactorial** or **polygenic** conditions are now known to make a major contribution to chronic illness in adult life (see Chapter 15).

Acquired Somatic Genetic Disease

Not all genetic errors are present from conception. Many billions of cell divisions (mitoses) occur in the course of an average human lifetime. During each **mitosis,** there is an opportunity for both single-gene mutations to occur, because of DNA copy errors, and for numerical chromosome abnormalities to arise as a result of errors in chromosome separation. Accumulating somatic mutations and chromosome abnormalities are now known to play a major role in causing cancer (see Chapter 14), and they probably also explain the rising incidence with age of many other serious illnesses, as well as the aging process itself. It is therefore necessary to appreciate that not all disease with a genetic basis is hereditary.

Before considering the impact of hereditary disease, it is helpful to introduce a few definitions.

Incidence

Incidence refers to the rate at which new cases occur. Thus, if the birth incidence of a particular condition equals 1 in 1000, then on average 1 in every 1000 newborn infants is affected.

Prevalence

This refers to the proportion of a population affected at any one time. The prevalence of a genetic disease is usually less than its birth incidence, either because life expectancy is reduced or because the condition shows a delayed age of onset.

Frequency

Frequency is a general term that lacks scientific specificity, although the word is often taken as being synonymous with incidence when calculating gene 'frequencies' (see Chapter 8).

Congenital

Congenital means that a condition is present at birth. Thus, cleft palate represents an example of a congenital **malformation**. Not all genetic disorders are congenital in terms of age of onset (e.g., Huntington disease), nor are all congenital abnormalities genetic in origin (e.g., fetal disruptions, as discussed in Chapter 16).

The Impact of Genetic Disease

During the twentieth century, improvements in all areas of medicine, most notably public health and therapeutics, resulted in changing patterns of disease, with increasing recognition of the role of genetic factors at all ages. For some parameters, such as perinatal mortality, the actual numbers of cases with exclusively genetic causes have probably remained constant but their **relative** contribution to overall figures has increased as other causes, such as infection, have declined. For other conditions, such as the chronic diseases of adult life, the overall contribution of genetics has almost certainly increased as greater life expectancy has provided more opportunity for adverse genetic and environmental interaction to manifest itself, for example in Alzheimer disease, macular degeneration, cardiomyopathy, and diabetes mellitus.

Consider the impact of genetic factors in disease at different ages from the following observations.

Spontaneous Miscarriages

A chromosome abnormality is present in 40% to 50% of all recognized first-trimester pregnancy loss. Approximately 1 in 6 of all pregnancies results in spontaneous miscarriage, thus around 5% to 7% of all recognized conceptions are chromosomally abnormal (p. 273). This value would be much higher if unrecognized pregnancies could also be included, and it is likely that a significant proportion of miscarriages with normal chromosomes do in fact have catastrophic submicroscopic genetic errors.

Newborn Infants

Of all neonates, 2% to 3% have at least one major congenital abnormality, of which at least 50% are caused exclusively or partially by genetic factors (see Chapter 16). The incidences of chromosome abnormalities and single-gene disorders in neonates are approximately 1 in 200 and 1 in 100, respectively.

Childhood

Genetic disorders account for 50% of all childhood blindness, 50% of all childhood deafness, and 50% of all cases of severe learning difficulty. In developed countries, genetic disorders and congenital malformations together also account for 30% of all childhood hospital admissions and 40% to 50% of all childhood deaths.

Adult Life

Approximately 1% of all malignancy is caused by single-gene inheritance, and between 5% and 10% of common cancers such as those of the breast, colon, and ovary have a strong hereditary component. By the age of 25 years, 5% of the population will have a disorder in which genetic factors play an important role. Taking into account the genetic contribution to cancer and cardiovascular diseases, such as coronary artery occlusion and hypertension, it has been estimated that more than 50% of the older adult population in developed countries will have a genetically determined medical problem.

Major New Developments

The study of genetics and its role in causing human disease is now widely acknowledged as being among the most exciting and influential areas of medical research. Since 1962 when Francis Crick, James Watson, and Maurice Wilkins gained acclaim for their elucidation of the structure of DNA, the Nobel Prize for Medicine and/or Physiology has been won on 22 occasions by scientists working in human and molecular genetics or related fields (Table 1.1), and for the first time in 2009 two such prizes were awarded in a single year. These pioneering studies have spawned a thriving molecular technology industry with applications as diverse as the development of genetically modified disease-resistant crops, the use of genetically engineered animals to produce therapeutic drugs, and the possible introduction of DNA-based vaccines for conditions such as malaria. Pharmaceutical companies are investing heavily in the DNA-based **pharmacogenomics**—drug therapy tailored to personal genetic makeup.

The Human Genome Project

With DNA technology rapidly progressing, a group of visionary scientists in the United States persuaded Congress in 1988 to fund a coordinated international program to sequence the entire human genome. The program would run from 1990 to 2005 and US$3 billion were initially allocated to the project. Some 5% of the budget was allocated to study the ethical and social implications of the new knowledge in recognition of the enormous potential to influence public health policies, screening programs, and personal choice. The project was likened to the Apollo moon mission in terms of its complexity, although in practical terms the long-term benefits are likely to be much more tangible. The draft DNA sequence of 3 billion base pairs was completed successfully in 2000 and the complete sequence was published ahead of schedule in October 2004. Before the closing stages of the project, it was thought that there might be approximately 100,000 coding genes that provide the blueprint for human life. It has come as a surprise to many that the number is much lower, with current estimates at slightly more than 25,000. However, many genes have the capacity to perform multiple functions, which in some cases is challenging traditional concepts of disease classification. The immediate benefits of the sequence data are being realized in research that is leading to better diagnosis and counseling for families with a genetic disease. A number of large, long-term, population-based studies are under way in the wake of the successful Human Genome Project, including, for example, UK Biobank, which aims to recruit 500,000 individuals ages 40 to 69 to study the progression of common disease, lifestyle, and genetic susceptibility.

The technique of microarray CGH is one of the most significant developments in the investigation of genetically determined disease since the discovery of chromosomes, but whole-genome sequencing is likely to be the future of genetic testing once rapid and affordable technologies are developed—but with these developments will come additional ethical challenges centered around their use and application. In the longer term an improved understanding of how genes are expressed will hopefully lead to the development of new strategies for the prevention and treatment of both single-gene and polygenic disorders.

Table 1.1 Genetic Discoveries That Have Led to the Award of the Nobel Prize for Medicine and/or Physiology and/or Chemistry, 1962–2009

Year	Prize Winners	Discovery
1962	Francis Crick James Watson Maurice Wilkins	The molecular structure of DNA
1965	François Jacob Jacques Monod André Lwoff	Genetic regulation
1966	Peyton Rous	Oncogenic viruses
1968	Robert Holley Gobind Khorana Marshall Nireberg	Deciphering of the genetic code
1975	David Baltimore Renato Dulbecco Howard Temin	Interaction between tumor viruses and nuclear DNA
1978	Werner Arber Daniel Nathans Hamilton Smith	Restriction endonucleases
1980	Baruj Benacerraf Jean Dausset George Snell	Genetic control of immunologic responses
1983	Barbara McClintock	Mobile genes (transposons)
1985	Michael Brown Joseph Goldstein	Cell receptors in familial hypercholesterolemia
1987	Susumu Tonegawa	Genetic aspects of antibodies
1989	Michael Bishop Harold Varmus	Study of oncogenes
1993	Richard Roberts Phillip Sharp	'Split genes'
1995	Edward Lewis Christiane Nüsslein-Volhard Eric Wieschaus	Homeotic and other developmental genes
1997	Stanley Prusiner	Prions
1999	Günter Blobel	Protein transport signaling
2000	Arvid Carlsson Paul Greengard Eric Kandel	Signal transduction in the nervous system
2001	Leland Hartwell Timothy Hunt Paul Nurse	Regulators of the cell cycle
2002	Sydney Brenner Robert Horritz John Sulston	Genetic regulation in development and programmed cell death (apoptosis)
2006	Andrew Fire Craig Mello	RNA interference
2007	Mario Capecchi Martin Evans Oliver Smithies	Gene modification by the use of embryonic stem cells
2009	Elizabeth Blackburn Carol Greider Jack Szostak	The role of telomerase in protecting chromosome telomeres (Medicine prize)
	Venkatraman Ramakrishnan Thomas A. Steitz Ada E. Yonath	Structure and function of the ribosome (Chemistry prize)

Gene Therapy

Most genetic disease is resistant to conventional treatment so that the prospect of successfully modifying the genetic code in a patient's cells is extremely attractive. Despite major investment and extensive research, success in humans has so far been limited to a few very rare immunologic disorders. For more common conditions, such as cystic fibrosis, major problems have been encountered, such as targeting the correct cell populations, overcoming the body's natural defense barriers, and identifying suitably non-immunogenic vectors. However, the availability of mouse models for genetic disorders, such as cystic fibrosis (p. 301), Huntington disease (p. 293), and Duchenne muscular dystrophy (p. 307), has greatly enhanced research opportunities, particularly in unraveling the cell biology of these conditions. In recent years there has been increasing optimism for novel drug therapies and stem cell treatment (p. 356), besides the prospects for gene therapy itself (p. 350).

The Internet

The availability of information in genetics has been enhanced greatly by the development of excellent online databases, and a selection of the well established is listed in the Appendix—GenBank, Ensembl, DDBJ. By 2010 there were more than a thousand molecular biology databases, so navigating this ever-growing maze can be daunting—and not just for the novice. This has developed into the exciting growth area of **Bioinformatics**, the science where biology, computer science, and information technology merge into a single discipline that encompasses gene maps, DNA sequences, comparative and functional genomics, and a lot more. Familiarity with interlinking databases is essential for the molecular geneticist, but is increasingly relevant for the keen clinician with an interest in genetics, who will find OMIM a good place to start for an account of all mendelian disorders, together with pertinent clinical details and extensive references. Although it is unlikely that more traditional sources of information, such as this textbook, will become completely obsolete, it is clear that only electronic technology can hope to match the explosive pace of developments in all areas of genetic research.

FURTHER READING

Baird PA, Anderson TW, Newcombe HB, Lowry RB 1988 Genetic disorders in children and young adults: a population study. Am J Hum Genet 42:677–693

A comprehensive study of the incidence of genetic disease in a large Western urban population.

Dunham I, Shimizu N, Roe BA, et al 1999 The DNA sequence of human chromosome 22. Nature 402:489–495

The first report of the complete sequencing of a human chromosome.

Emery AEH, 1989 Portraits in medical genetics—Joseph Adams 1756–1818. J Med Genet 26:116–118

An account of the life of a London doctor who made remarkable observations about hereditary disease in his patients.

Garrod AE 1902 The incidence of alkaptonuria: a study in chemical individuality. Lancet ii:1916–1920

A landmark paper in which Garrod proposed that alkaptonuria could show mendelian inheritance and also noted that 'the mating of first cousins gives exactly the conditions most likely to enable a rare, and usually recessive, character to show itself'.

Orel V 1995 Gregor Mendel: the first geneticist. Oxford: Oxford University Press

A detailed biography of the life and work of the Moravian monk who was described by his abbot as being 'very diligent in the study of the sciences but much less fitted for work as a parish priest'.

Ouellette F 1999 Internet resources for the clinical geneticist. Clin Genet 56:179–185.

A guide to how to access some of the most useful online databases.

Shapiro R 1991 The human blueprint: the race to unlock the secrets of our genetic script. New York: St Martin's Press.

Watson J 1968 The Double Helix. New York: Atheneum.

The story of the discovery of the structure of DNA, through the eyes of Watson himself.

Databases

Online Mendelian Inheritance in Man:
www.ncbi.nlm.nih.gov/omim

For literature:
http://www.ncbi.nlm.nih.gov/PubMed/
http://scholar.google.com/

Genome:
www.ncbi.nlm.nih.gov/omim/GenBank
www.hgmd.cf.ac.uk (human, Cardiff)
www.ensembl.org (human, comparative, European, Cambridge)
http://genome.ucsc.edu (American browser)

ELEMENTS

1 A characteristic manifest in a hybrid (heterozygote) is dominant. A recessive characteristic is expressed only in an individual with two copies of the mutated gene (i.e., a homozygote).

2 Mendel proposed that each individual has two genes for each characteristic: one is inherited from each parent and one is transmitted to each child. Genes at different loci act and segregate independently.

3 Chromosome separation at cell division facilitates gene segregation.

4 Genetic disorders are present in at least 2% of all neonates, account for 50% of childhood blindness, deafness, learning difficulties and deaths, and affect 5% of the population by the age of 25 years.

5 From the rediscovery of Mendel's genetic research on peas, to the full sequencing of the human genome, almost exactly 100 years elapsed.

6 Molecular genetics and molecular biology are at the forefront of medical research, embraced within the new scientific discipline of bioinformatics, and hold promise novel forms of treatment for genetic diseases.

CHAPTER 2

The Cellular and Molecular Basis of Inheritance

The hereditary material is present in the nucleus of the cell, whereas protein synthesis takes place in the cytoplasm. What is the chain of events that leads from the gene to the final product?

This chapter covers basic cellular biology outlining the structure of DNA, the process of DNA replication, the types of DNA sequences, gene structure, the genetic code, the processes of transcription and translation, the various types of mutations, mutagenic agents, and DNA repair.

There is nothing, Sir, too little for so little a creature as man.
It is by studying little things that we attain the great art of having as little misery and as much happiness as possible.
SAMUEL JOHNSON

The Cell

Within each cell of the body, visible with the light microscope, is the **cytoplasm** and a darkly staining body, the **nucleus,** the latter containing the hereditary material in the form of **chromosomes** (Figure 2.1). The phospholipid bilayer of the plasma membrane protects the interior of the cell but remains selectively permeable and has integral proteins involved in recognition and signaling between cells. The nucleus has a darkly staining area, the **nucleolus**. The nucleus is surrounded by a membrane, the **nuclear envelope**, which separates it from the cytoplasm but still allows communication through **nuclear pores**.

The cytoplasm contains the **cytosol**, which is semifluid in consistency, containing both soluble elements and cytoskeletal structural elements. In addition, in the cytoplasm there is a complex arrangement of very fine, highly convoluted, interconnecting channels, the **endoplasmic reticulum**. The endoplasmic reticulum, in association with the **ribosomes**, is involved in the biosynthesis of proteins and lipids. Also situated within the cytoplasm are other even more minute cellular organelles that can be visualized only with an electron microscope. These include the Golgi apparatus, which is responsible for the secretion of cellular products, the **mitochondria**, which are involved in energy production through the oxidative phosphorylation metabolic **pathways**, and the **peroxisomes** (p. 180) and **lysosomes**, both of which are involved in the degradation and disposal of cellular waste material and toxic molecules.

DNA: The Hereditary Material

Composition

Nucleic acid is composed of a long polymer of individual molecules called **nucleotides**. Each nucleotide is composed of a nitrogenous base, a sugar molecule, and a phosphate molecule. The nitrogenous bases fall into two types, **purines** and **pyrimidines**. The purines include adenine and guanine; the pyrimidines include cytosine, thymine and uracil.

There are two different types of nucleic acid, **ribonucleic acid (RNA)**, which contains the five carbon sugar ribose, and **deoxyribonucleic acid (DNA)**, in which the hydroxyl group at the 2 position of the ribose sugar is replaced by a hydrogen (i.e., an oxygen molecule is lost, hence 'deoxy'). DNA and RNA both contain the purine bases adenine and guanine and the pyrimidine cytosine, but thymine occurs only in DNA and uracil is found only in RNA.

RNA is present in the cytoplasm and in particularly high concentrations in the nucleolus of the nucleus. DNA, on the other hand, is found mainly in the chromosomes.

Structure

For genes to be composed of DNA, it is necessary that the latter should have a structure sufficiently versatile to account for the great variety of different genes and yet, at the same time, be able to reproduce itself in such a manner that an identical replica is formed at each cell division. In 1953, Watson and Crick, based on x-ray diffraction studies by themselves and others, proposed a structure for the DNA molecule that fulfilled all the essential requirements. They suggested that the DNA molecule is composed of two chains of nucleotides arranged in a double helix. The backbone of each chain is formed by phosphodiester bonds between the 3′ and 5′ carbons of adjacent sugars, the two chains being held together by hydrogen bonds between the nitrogenous bases, which point in toward the center of the helix. Each DNA chain has a polarity determined by the orientation of the sugar–phosphate backbone. The chain end terminated by the 5′ carbon atom of the sugar molecule is referred to as the **5′ end**, and the end terminated by the

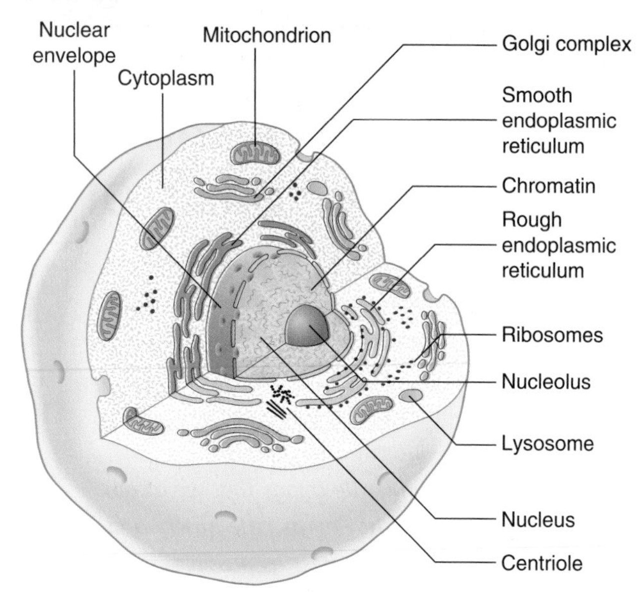

FIGURE 2.1 Diagrammatic representation of an animal cell.

3′ carbon atom is called the **3′ end**. In the DNA duplex, the 5′ end of one strand is opposite the 3′ end of the other, that is, they have opposite orientations and are said to be **antiparallel**.

The arrangement of the bases in the DNA molecule is not random. A purine in one chain always pairs with a pyrimidine in the other chain, with specific pairing of the base pairs: guanine in one chain always pairs with cytosine in the other chain, and adenine always pairs with thymine, so that this base pairing forms complementary strands (Figure 2.2). For their work Watson and Crick, along with Maurice Wilkins, were awarded the Nobel Prize for Medicine or Physiology in 1962 (p. 10).

Replication

The process of **DNA replication** provides an answer to the question of how genetic information is transmitted from one generation to the next. During nuclear division the two strands of the DNA double helix separate through the action of enzyme DNA helicase, each DNA strand directing the synthesis of a complementary DNA strand through specific base pairing, resulting in two daughter DNA duplexes that are identical to the original parent molecule. In this way, when cells divide, the genetic information is conserved and transmitted unchanged to each daughter cell. The process of DNA replication is termed **semiconservative**, because only one strand of each resultant daughter molecule is newly synthesized.

DNA replication, through the action of the enzyme DNA polymerase, takes place at multiple points known as **origins of replication**, forming bifurcated Y-shaped structures known as **replication forks**. The synthesis of both complementary antiparallel DNA strands occurs in the 5′ to 3′ direction. One strand, known as the **leading strand**, is

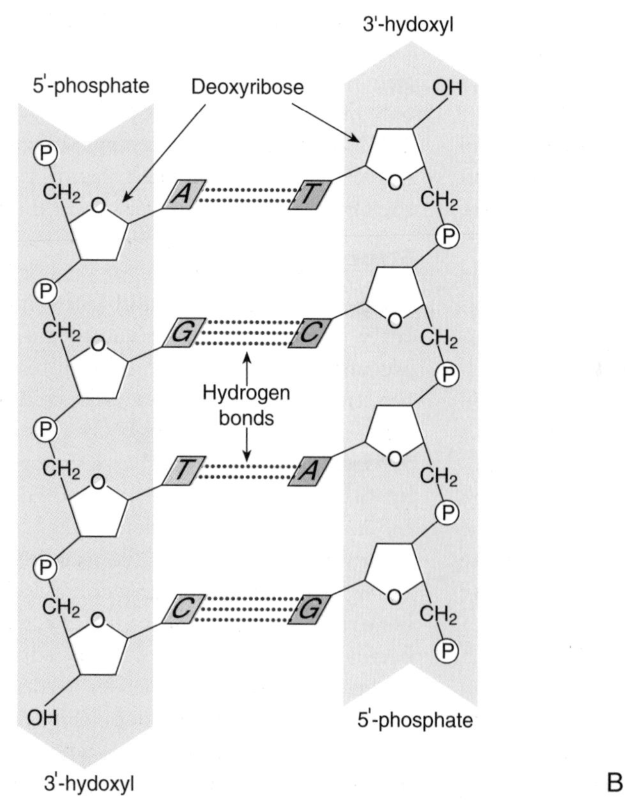

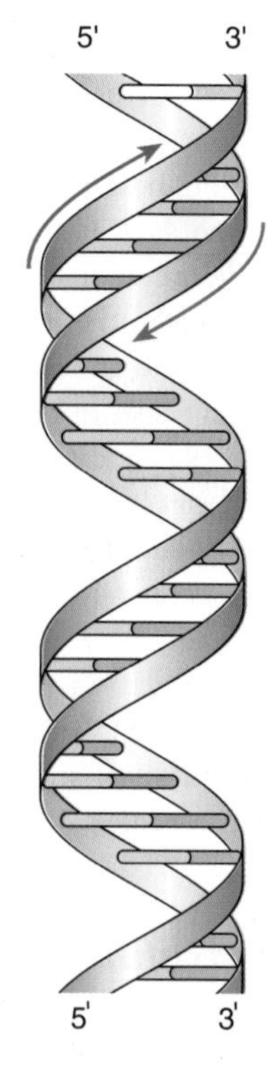

FIGURE 2.2 **DNA double helix. A,** Sugar-phosphate backbone and nucleotide pairing of the DNA double helix (*P*, phosphate; *A*, adenine; *T*, thymine; *G*, guanine; *C*, cytosine). **B,** Representation of the DNA double helix.

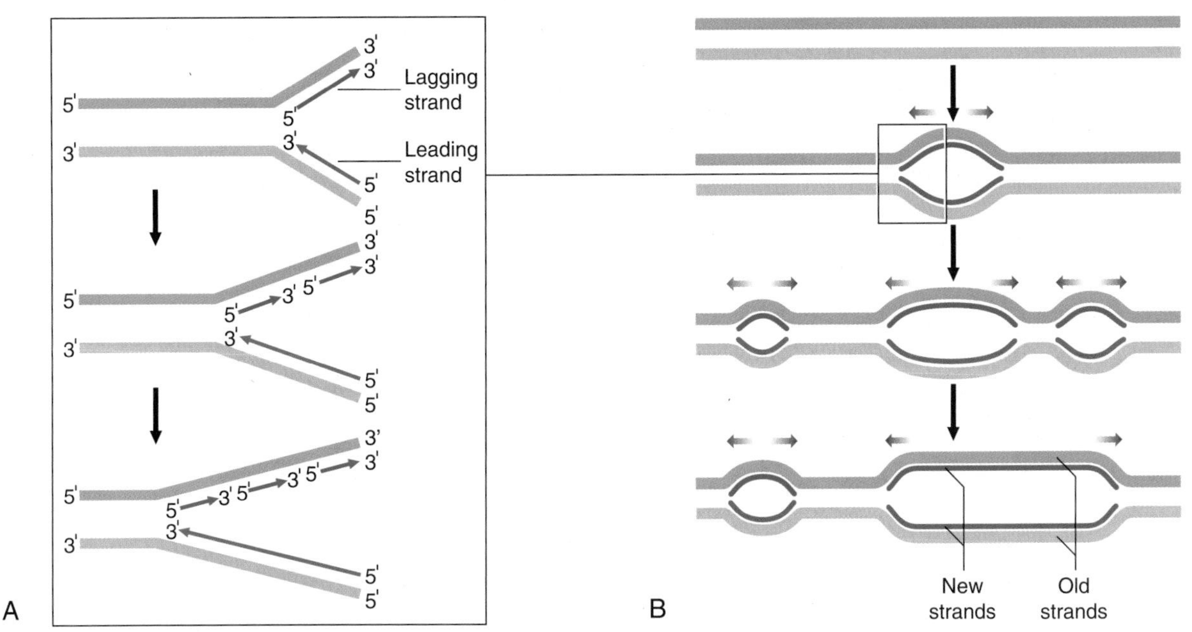

FIGURE 2.3 **DNA replication. A,** Detailed diagram of DNA replication at the site of origin in the replication fork showing asymmetric strand synthesis with the continuous synthesis of the leading strand and the discontinuous synthesis of the lagging strand with ligation of the Okazaki fragments. **B,** Multiple points of origin and semiconservative mode of DNA replication.

synthesized as a continuous process. The other strand, known as the **lagging strand**, is synthesized in pieces called Okazaki fragments, which are then joined together as a continuous strand by the enzyme DNA ligase (Figure 2.3*A*).

DNA replication progresses in both directions from these points of origin, forming bubble-shaped structures, or **replication bubbles** (Figure 2.3*B*). Neighboring replication origins are approximately 50 to 300 kilobases (kb) apart and occur in clusters or **replication units** of 20 to 80 origins of replication. DNA replication in individual replication units takes place at different times in the S phase of the cell cycle (p. 39), adjacent replication units fusing until all the DNA is copied, forming two complete identical daughter molecules.

Chromosome Structure

The idea that each chromosome is composed of a single DNA double helix is an oversimplification. A chromosome is very much wider than the diameter of a DNA double helix. In addition, the amount of DNA in the nucleus of each cell in humans means that the total length of DNA contained in the chromosomes, if fully extended, would be several meters long! In fact, the total length of the human chromosome complement is less than half a millimeter.

The packaging of DNA into chromosomes involves several orders of DNA coiling and folding. In addition to the primary coiling of the DNA double helix, there is secondary coiling around spherical **histone** 'beads', forming what are called **nucleosomes**. There is a tertiary coiling of the nucleosomes to form the **chromatin fibers** that form long loops on a scaffold of non-histone acidic proteins, which are further wound in a tight coil to make up the chromosome as visualized under the light microscope (Figure 2.4), the whole structure making up the so-called **solenoid** model of chromosome structure.

Types of DNA Sequence

DNA, if denatured, will reassociate as a duplex at a rate that is dependent on the proportion of unique and repeat sequences present, the latter occurring more rapidly. Analysis of the results of the kinetics of the reassociation of human DNA have shown that approximately 60% to 70% of the human genome consists of single- or low-copy number DNA sequences. The remainder of the genome, 30% to 40%, consists of either moderately or highly **repetitive** DNA sequences that are not transcribed. This latter portion consists of mainly satellite DNA and interspersed DNA sequences (Box 2.1).

Nuclear Genes

It is estimated that there are between 25,000 and 30,000 genes in the nuclear genome. The distribution of these genes varies greatly between chromosomal regions. For example, heterochromatic and centromeric (p. 32) regions are mostly non-coding, with the highest gene density observed in subtelomeric regions. Chromosomes 19 and 22 are gene rich, whereas 4 and 18 are relatively gene poor. The size of genes also shows great variability: from small genes with single exons to genes with up to 79 exons (e.g., dystrophin, which occupies 2.5 Mb of the genome).

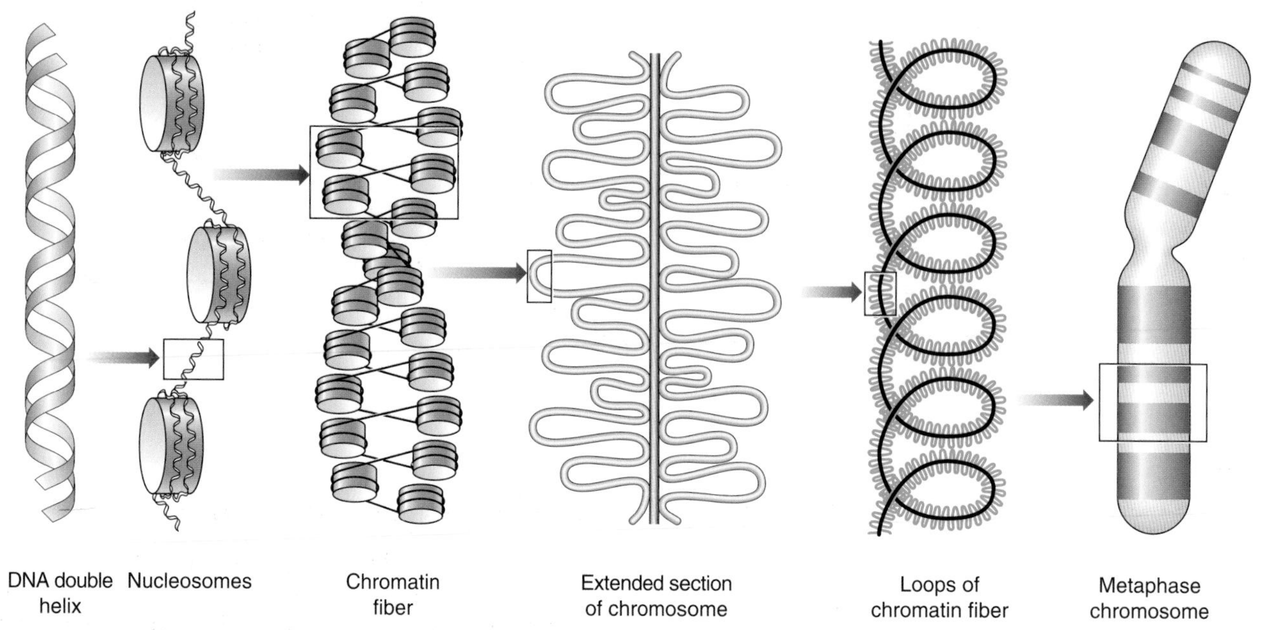

FIGURE 2.4 Simplified diagram of proposed solenoid model of DNA coiling that leads to the visible structure of the chromosome.

Unique Single-Copy Genes

Most human genes are unique single-copy genes coding for polypeptides that are involved in or carry out a variety of cellular functions. These include enzymes, hormones, receptors, and structural and regulatory proteins.

Multigene Families

Many genes have similar functions, having arisen through gene duplication events with subsequent evolutionary divergence making up what are known as **multigene families**. Some are found physically close together in clusters; for example, the α- and β-globin gene clusters on chromosomes 16 and 11 (Figure 2.5), whereas others are widely dispersed throughout the genome occurring on different chromosomes, such as the *HOX* homeobox gene family (p. 87).

Multigene families can be split into two types, **classic gene families** that show a high degree of sequence homology and **gene superfamilies** that have limited sequence homology but are functionally related, having similar structural domains.

Box 2.1 Types of DNA Sequence

Nuclear (~3 × 10⁹ bp)
Genes (~30,000)
Unique single copy
Multigene families
Classic gene families
Gene superfamilies
Extragenic DNA (unique/low copy number or moderate/highly repetitive)
Tandem repeat
Satellite
Minisatellite
  Telomeric
  Hypervariable
Microsatellite
Interspersed
Short interspersed nuclear elements
Long interspersed nuclear elements
Mitochondrial (16.6 kb, 37 genes)
Two rRNA genes
22 tRNA genes

Classic Gene Families

Examples of classic gene families include the numerous copies of genes coding for the various ribosomal RNAs, which are clustered as tandem arrays at the nucleolar organizing regions on the short arms of the five acrocentric chromosomes (p. 32), and the different transfer RNA (p. 20) gene families, which are dispersed in numerous clusters throughout the human genome.

Gene Superfamilies

Examples of gene superfamilies include the HLA (human leukocyte antigen) genes on chromosome 6 (p. 200) and the T-cell receptor genes, which have structural homology with the immunoglobulin (Ig) genes (p. 200). It is thought that these are almost certainly derived from duplication of

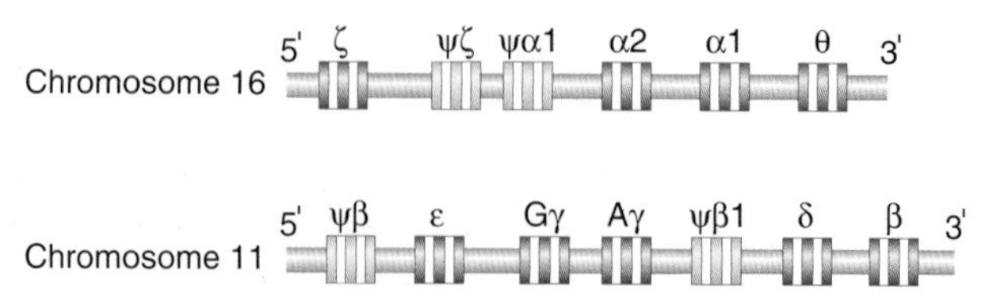

FIGURE 2.5 Representation of the α- and β-globin regions on chromosomes 16 and 11.

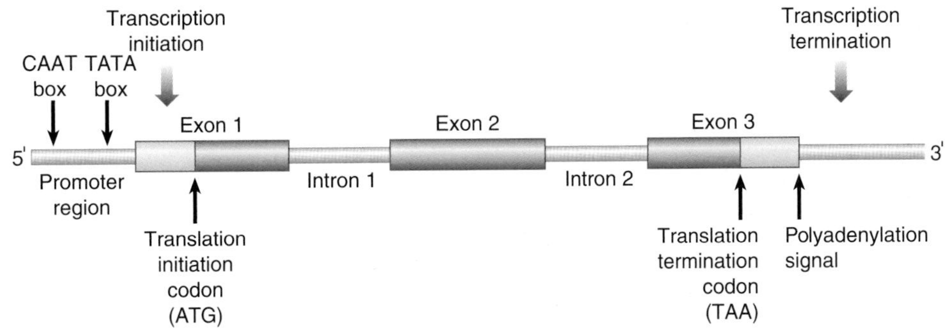

FIGURE 2.6 Representation of a typical human structural gene.

a precursor gene, with subsequent evolutionary divergence forming the Ig superfamily.

Gene Structure

The original concept of a gene as a continuous sequence of DNA coding for a protein was turned on its head in the early 1980s by detailed analysis of the structure of the human β-globin gene. It was revealed that the gene was much longer than necessary to code for the β-globin protein, containing non-coding intervening sequences, or introns, that separate the coding sequences or exons (Figure 2.6). Most human genes contain introns, but the number and size of both introns and exons is extremely variable. Individual introns can be far larger than the coding sequences and some have been found to contain coding sequences for other genes (i.e., genes occurring within genes). Genes in humans do not usually overlap, being separated from each other by an average of 30 kb, although some of the genes in the HLA complex (p. 200) have been shown to be overlapping.

Pseudogenes

Particularly fascinating is the occurrence of genes that closely resemble known structural genes but which, in general, are not functionally expressed: so-called **pseudogenes**. These are thought to have arisen in two main ways: either by genes undergoing duplication events that are rendered silent through the acquisition of mutations in coding or regulatory elements, or as the result of the insertion of complementary DNA sequences, produced by the action of the enzyme **reverse transcriptase** on a naturally occurring messenger RNA transcript, that lack the promoter sequences necessary for expression.

Extragenic DNA

The estimated 25,000 to 30,000 unique single-copy genes in humans represent less than 2% of the genome encoding proteins. The remainder of the human genome is made up of repetitive DNA sequences that are predominantly transcriptionally inactive. It has been described as **junk** DNA, but some regions show evolutionary conservation and may play a role in the regulation of gene expression.

Tandemly Repeated DNA Sequences

Tandemly repeated DNA sequences consist of blocks of tandem repeats of non-coding DNA that can be either highly dispersed or restricted in their location in the genome. Tandemly repeated DNA sequences can be divided into three subgroups: satellite, minisatellite, and microsatellite DNA.

Satellite DNA

Satellite DNA accounts for approximately 10% to 15% of the repetitive DNA sequences of the human genome and consists of very large series of simple or moderately complex, short, tandemly repeated DNA sequences that are transcriptionally inactive and are clustered around the centromeres of certain chromosomes. This class of DNA sequences can be separated on density-gradient centrifugation as a shoulder, or 'satellite', to the main peak of genomic DNA, and has therefore been referred to as satellite DNA.

Minisatellite DNA

Minisatellite DNA consists of two families of tandemly repeated short DNA sequences: telomeric and hypervariable minisatellite DNA sequences that are transcriptionally inactive.

Telomeric DNA. The terminal portion of the telomeres of the chromosomes (p. 32) contains 10 to 15 kb of tandem repeats of a 6-base pair (bp) DNA sequence known as telomeric DNA. The telomeric repeat sequences are necessary for chromosomal integrity in replication and are added to the chromosome by an enzyme known as telomerase (p. 32).

Hypervariable minisatellite DNA. Hypervariable minisatellite DNA is made up of highly polymorphic DNA sequences consisting of short tandem repeats of a common core sequence. The highly variable number of repeat units in different hypervariable minisatellites forms the basis of the DNA fingerprinting technique developed by Professor Sir Alec Jeffreys in 1984 (p. 69).

Microsatellite DNA

Microsatellite DNA consists of tandem single, di-, tri-, and tetra-nucleotide repeat base-pair sequences located throughout the genome. Microsatellite repeats rarely occur within coding sequences but trinucleotide repeats in or near genes are associated with certain inherited disorders (p. 59).

This variation in repeat number is thought to arise by incorrect pairing of the tandem repeats of the two complementary DNA strands during DNA replication, or what is known as **slipped strand mispairing**. Duplications or deletions of longer sequences of tandemly repeated DNA are thought to arise through unequal crossover of non-allelic DNA sequences on chromatids of homologous chromosomes or sister chromatids (p. 32).

Nowadays DNA microsatellites are used for forensic and paternity tests (p. 69). They can also be helpful for gene tracking in families with a genetic disorder but no identified mutation (p. 70).

Highly Repeated Interspersed Repetitive DNA Sequences

Approximately one-third of the human genome is made up of two main classes of short and long repetitive DNA sequences that are interspersed throughout the genome.

Short Interspersed Nuclear Elements. About 5% of the human genome consists of some 750,000 copies of **short interspersed nuclear elements**, or **SINEs**. The most common are DNA sequences of approximately 300 bp that have sequence similarity to a signal recognition particle involved in protein synthesis. They are called **Alu repeats** because they contain an *AluI* restriction enzyme recognition site.

Long Interspersed Nuclear Elements. About 5% of the DNA of the human genome is made up of **long interspersed nuclear elements**, or **LINEs**. The most commonly occurring LINE, known as LINE-1 or an L1 element, consists of more than 100,000 copies of a DNA sequence of up to 6000 bp that encodes a reverse transcriptase.

The function of these interspersed repeat sequences is not clear. Members of the *Alu* repeat family are flanked by short direct repeat sequences and therefore resemble unstable DNA sequences called transposable elements or **transposons**. Transposons, originally identified in maize by Barbara McClintock (p. 10), move spontaneously throughout the genome from one chromosome location to another and appear to be ubiquitous in the plant and animal kingdoms. It is postulated that *Alu* repeats could promote unequal recombination, which could lead to pathogenic mutations (p. 22) or provide selective advantage in evolution by gene duplication. Both *Alu* and LINE-1 repeat elements have been implicated as a cause of mutation in inherited human disease.

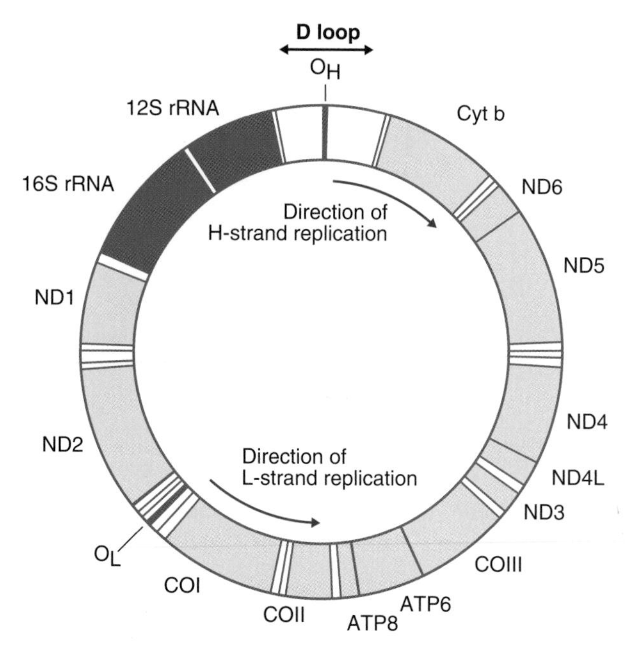

FIGURE 2.7 The human mitochondrial genome. *H* is the heavy strand and *L* the light strand.

Mitochondrial DNA

In addition to nuclear DNA, the several thousand mitochondria of each cell possess their own 16.6 kb circular double-stranded DNA, **mitochondrial DNA** (or **mtDNA**) (Figure 2.7). The mtDNA genome is very compact, containing little repetitive DNA, and codes for 37 genes, which include two types of ribosomal RNA, 22 transfer RNAs (p. 20) and 13 protein subunits for enzymes, such as cytochrome *b* and cytochrome oxidase, which are involved in the energy producing oxidative phosphorylation pathways. The genetic code of the mtDNA differs slightly from that of nuclear DNA.

The mitochondria of the fertilized zygote are inherited almost exclusively from the oocyte, leading to the maternal pattern of inheritance that characterizes many mitochondrial disorders (p. 181).

Transcription

The process whereby genetic information is transmitted from DNA to RNA is called **transcription**. The information stored in the genetic code is transmitted from the DNA of a gene to **messenger RNA**, or **mRNA**. Every base in the mRNA molecule is complementary to a corresponding base in the DNA of the gene, but with uracil replacing thymine in mRNA. mRNA is single stranded, being synthesized by the enzyme RNA polymerase II, which adds the appropriate complementary ribonucleotide to the 3′ end of the RNA chain.

In any particular gene, only one DNA strand of the double helix acts as the so-called **template strand**. The

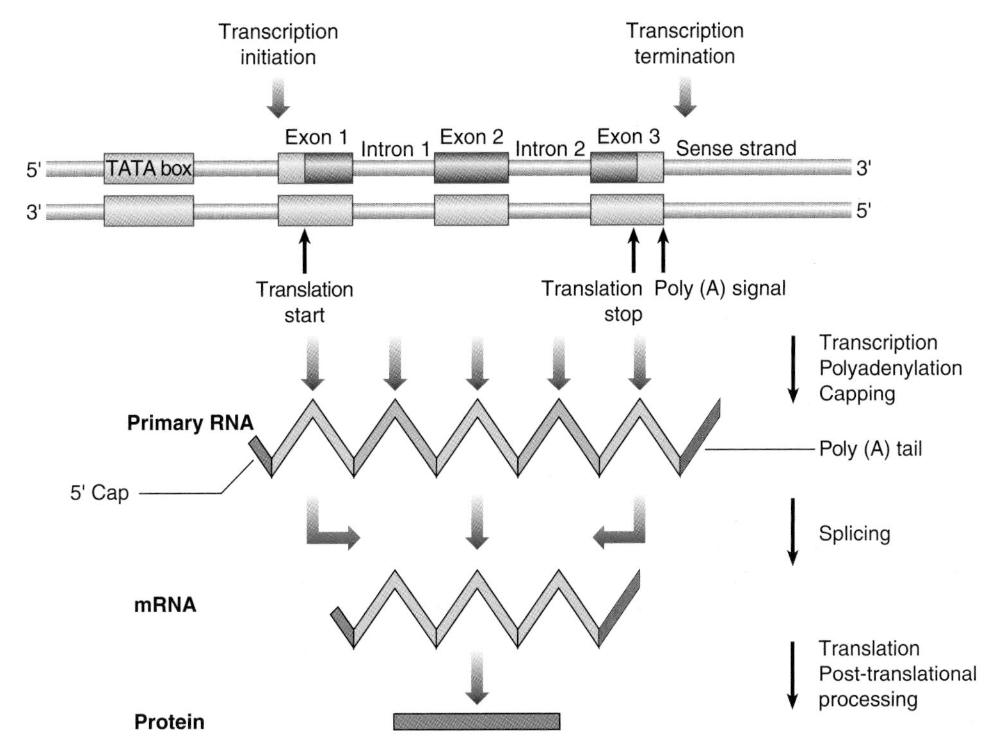

FIGURE 2.8 Transcription, post-transcriptional processing, translation, and post-translational processing.

transcribed mRNA molecule is a copy of the complementary strand, or what is called the **sense strand** of the DNA double helix. The template strand is sometimes called the **antisense strand**. The particular strand of the DNA double helix used for RNA synthesis appears to differ throughout different regions of the genome.

RNA Processing

Before the primary mRNA molecule leaves the nucleus it undergoes a number of modifications, or what is known as **RNA processing**. This involves splicing, capping, and polyadenylation.

mRNA Splicing

During and after transcription, the non-coding introns in the precursor (pre) mRNA are excised, and the non-contiguous coding exons are spliced together to form a shorter mature mRNA before its transportation to the ribosomes in the cytoplasm for translation. The process is known as **mRNA splicing** (Figure 2.8). The boundary between the introns and exons consists of a 5′ donor GT dinucleotide and a 3′ acceptor AG dinucleotide. These, along with surrounding short splicing consensus sequences, another intronic sequence known as the branch site, small nuclear RNA (snRNA) molecules and associated proteins, are necessary for the splicing process.

5′ Capping

The **5′ cap** is thought to facilitate transport of the mRNA to the cytoplasm and attachment to the ribosomes, as well as to protect the RNA transcript from degradation by endogenous cellular exonucleases. After 20 to 30 nucleotides have been transcribed, the nascent mRNA is modified by the addition of a guanine nucleotide to the 5′ end of the molecule by an unusual 5′ to 5′ triphosphate linkage. A methyltransferase enzyme then methylates the N7 position of the guanine, giving the final 5′ cap.

Polyadenylation

Transcription continues until specific nucleotide sequences are transcribed that cause the mRNA to be cleaved and RNA polymerase II to be released from the DNA template. Approximately 200 **adenylate residues**—the so-called **poly(A) tail**—are added to the mRNA, which facilitates nuclear export and translation

Translation

Translation is the transmission of the genetic information from mRNA to protein. Newly processed mRNA is transported from the nucleus to the cytoplasm, where it becomes associated with the **ribosomes**, which are the site of protein synthesis. Ribosomes are made up of two different sized subunits, which consist of four different types of **ribosomal RNA (rRNA)** molecules and a large number of ribosomal specific proteins. Groups of ribosomes associated with the same molecule of mRNA are referred to as **polyribosomes** or **polysomes**. In the ribosomes, the mRNA forms the template for producing the specific sequence of amino acids of a particular **polypeptide**.

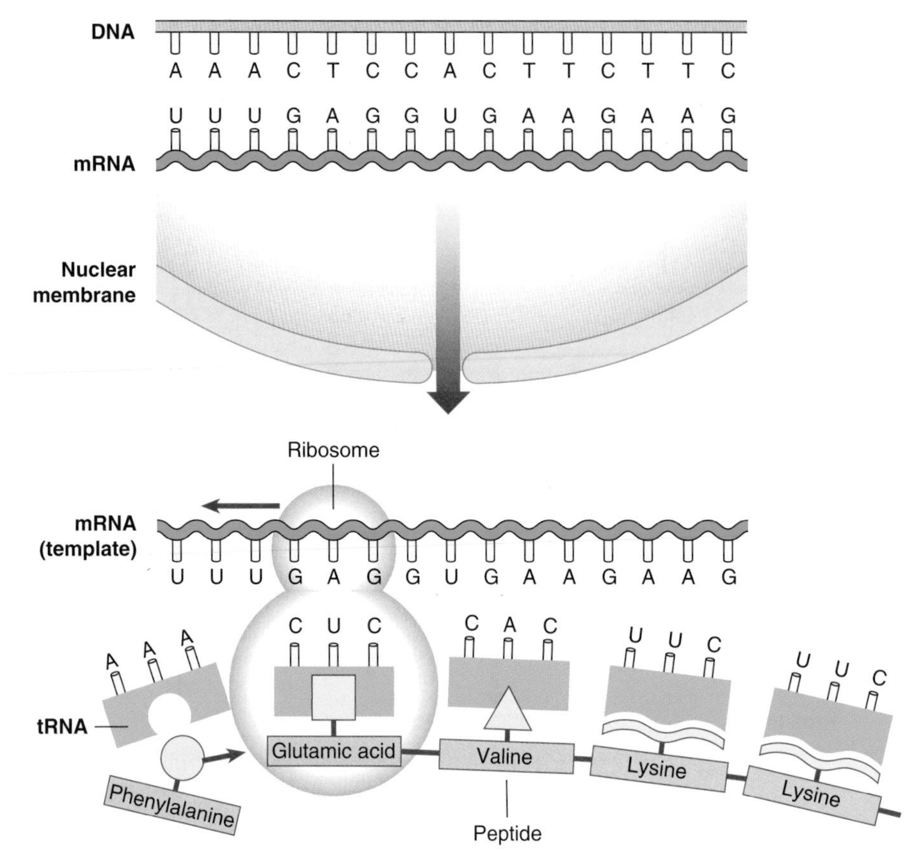

FIGURE 2.9 Representation of the way in which genetic information is translated into protein.

Transfer RNA

In the cytoplasm there is another form of RNA called **transfer RNA**, or **tRNA**. The incorporation of amino acids into a **polypeptide chain** requires the amino acids to be covalently bound by reacting with ATP to the specific tRNA molecule by the activity of the enzyme aminoacyl tRNA synthetase. The ribosome, with its associated rRNAs, moves along the mRNA, the amino acids linking up by the formation of peptide bonds through the action of the enzyme peptidyl transferase to form a polypeptide chain (Figure 2.9).

Post-Translational Modification

Many proteins, before they attain their normal structure or functional activity, undergo **post-translational modification**, which can include chemical modification of amino-acid side chains (e.g., hydroxylation, methylation), the addition of carbohydrate or lipid moieties (e.g., glycosylation), or proteolytic cleavage of polypeptides (e.g., the conversion of proinsulin to insulin).

Thus post-translational modification, along with certain short amino-acid sequences known as **localization sequences** in the newly synthesized proteins, results in transport to specific cellular locations (e.g., the nucleus), or secretion from the cell.

The Genetic Code

Twenty different amino acids are found in proteins; as DNA is composed of four different nitrogenous bases, obviously a single base cannot specify one amino acid. If two bases were to specify one amino acid, there would only be 4^2 or 16 possible combinations. If, however, three bases specified one amino acid then the possible number of combinations of the four bases would be 4^3 or 64. This is more than enough to account for all the 20 known amino acids and is known as the genetic code.

Triplet Codons

The triplet of nucleotide bases in the mRNA that codes for a particular amino acid is called a **codon**. Each triplet codon in sequence codes for a specific amino acid in sequence and so the genetic code is non-overlapping. The order of the triplet codons in a gene is known as the translational **reading frame**. However, some amino acids are coded for by more than one triplet, so the code is said to be **degenerate** (Table 2.1). Each tRNA species for a particular amino acid has a specific trinucleotide sequence called the **anticodon**, which is complementary to the codon of the mRNA. Although there are 64 codons, there are only 30 cytoplasmic tRNAs, the anticodons of a number of the tRNAs recognizing codons that differ at the position of the third base, with

Table 2.1 Genetic Code of the Nuclear and Mitochondrial Genomes

First Base	Second Base: U	C	A	G	Third Base
U	Phenylalanine	Serine	Tyrosine	Cysteine	U
	Phenylalanine	Serine	Tyrosine	Cysteine	C
	Leucine	Serine	Stop	Stop (*Tryptophan*)	A
	Leucine	Serine	Stop	Tryptophan	G
C	Leucine	Proline	Histidine	Arginine	U
	Leucine	Proline	Histidine	Arginine	C
	Leucine	Proline	Glutamine	Arginine	A
	Leucine	Proline	Glutamine	Arginine	G
A	Isoleucine	Threonine	Asparagine	Serine	U
	Isoleucine	Threonine	Asparagine	Serine	C
	Isoleucine (*Methionine*)	Threonine	Lysine	Arginine	A
	Methionine	Threonine	Lysine	Arginine (*Stop*)	G
A	Isoleucine	Threonine	Asparagine	Serine	U
	Isoleucine	Threonine	Asparagine	Serine	C
	Isoleucine (*Methionine*)	Threonine	Lysine	Arginine (*Stop*)	A
	Methionine	Threonine	Lysine	Arginine	G
G	Valine	Alanine	Aspartic acid	Glycine	U
	Valine	Alanine	Aspartic acid	Glycine	C
	Valine	Alanine	Glutamic acid	Glycine	A
	Valine	Alanine	Glutamic acid	Glycine	G

Differences in the mitochondrial genetic code are in *italics*.

guanine being able to pair with uracil as well as cytosine. Termination of translation of the mRNA is signaled by the presence of one of the three **stop** or **termination codons**.

The genetic code of mtDNA differs from that of the nuclear genome. Eight of the 22 tRNAs are able to recognize codons that differ only at the third base of the codon, 14 can recognize pairs of codons that are identical at the first two bases, with either a purine or pyrimidine for the third base, the other four codons acting as stop codons (see Table 2.1).

Regulation of Gene Expression

Many cellular processes, and therefore the genes that are expressed, are common to all cells, for example ribosomal, chromosomal and cytoskeleton proteins, constituting what are called the **housekeeping** genes. Some cells express large quantities of a specific protein in certain tissues or at specific times in development, such as hemoglobin in red blood cells (p. 155). This differential control of gene expression can occur at a variety of stages.

Control of Transcription

The control of transcription can be affected permanently or reversibly by a variety of factors, both environmental (e.g., hormones) and genetic (cell signaling). This occurs through a number of different mechanisms that include signaling molecules that bind to regulatory sequences in the DNA known as **response elements**, intracellular receptors known as **hormone nuclear receptors**, and receptors for specific ligands on the cell surface involved in the process of **signal transduction**.

All of these mechanisms ultimately affect transcription through the binding of the general transcription factors to short specific DNA promoter elements located within 200 bp 5′ or **upstream** of most eukaryotic genes in the so-called core **promoter region** that leads to activation of RNA polymerase (Figure 2.10). Promoters can be broadly

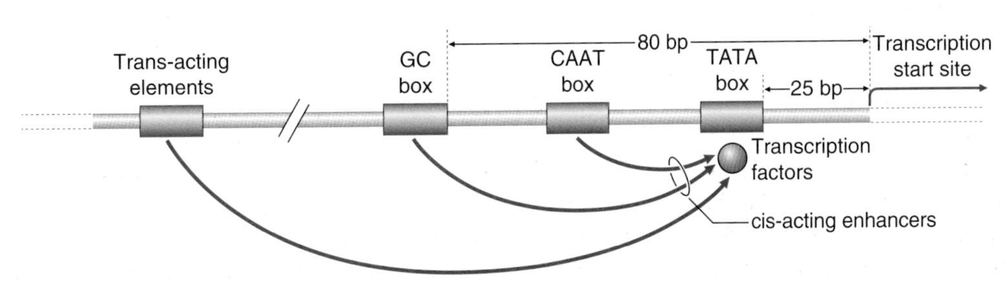

FIGURE 2.10 Diagrammatic representation of the factors that regulate gene expression.

classed into two types, TATA box-containing and GC rich. The TATA box, which is about 25 bp upstream of the transcription start site, is involved in the initiation of transcription at a basal constitutive level and mutations in it can lead to alteration of the transcription start site. The GC box, which is about 80 bp upstream, increases the basal level of transcriptional activity of the TATA box.

The regulatory elements in the promoter region are said to be **cis-acting**, that is, they only affect the expression of the adjacent gene on the same DNA duplex, whereas the transcription factors are said to be **trans-acting**, acting on both copies of a gene on each chromosome being synthesized from genes that are located at a distance. DNA sequences that increase transcriptional activity, such as the GC and CAAT boxes, are known as **enhancers**. There are also negative regulatory elements or **silencers** that inhibit transcription. In addition, there are short sequences of DNA, usually 500 bp to 3 kb in size and known as **boundary elements**, which block or inhibit the influence of regulatory elements of adjacent genes.

Transcription Factors

A number of genes encode proteins involved in the regulation of gene expression. They have DNA-binding activity to short nucleotide sequences, usually mediated through helical protein motifs, and are known as **transcription factors**. These gene regulatory proteins have a transcriptional activation domain and a DNA-binding domain. There are four types of DNA-binding domain, the most common being the **helix–turn–helix**, made up of two α helices connected by a short chain of amino acids that make up the 'turn'. The three other types are the **zinc finger**, **leucine zipper**, or **helix–loop–helix** motifs, so named as a result of specific structural features.

Post-Transcriptional Control of Gene Expression

Regulation of expression of most genes occurs at the level of transcription but can also occur at the levels of RNA processing, RNA transport, mRNA degradation and translation. For example, the G to A variant at position 20,210 in the 3′ untranslated region of the prothrombin gene increases the stability of the mRNA transcript, resulting in higher plasma prothrombin levels.

RNA-Mediated Control of Gene Expression

RNA-mediated silencing was first described in the early 1990s, but it is only recently that its key role in controlling post-transcriptional gene expression has been both recognized and exploited (see Chapter 23). Small interfering RNAs (siRNAs) were discovered in 1998 and are the effector molecules of the RNA interference pathway (RNAi). These short double-stranded RNAs (21 to 23 nucleotides) bind to mRNAs in a sequence-specific manner and result in their degradation via a ribonuclease-containing RNA-induced silencing complex (RISC). MicroRNAs (miRNAs) also bind to mRNAs in a sequence-specific manner. They can either cause endonucleolytic cleavage of the mRNA or act by blocking translation.

Alternative Isoforms

The majority of human genes (at least 74%) undergo **alternative splicing** and therefore encode more than one protein. **Alternative polyadenylation** generates further diversity. Some genes have more than one promoter, and these **alternative** promoters may result in tissue-specific isoforms. Alternative splicing of exons is also seen with individual exons present in only some isoforms. The extent of alternative splicing in humans may be inferred from the finding that the human genome includes only 25,000 to 30,000 genes, far fewer than the original prediction of more than 100,000.

RNA-directed DNA Synthesis

The process of the transfer of the genetic information from DNA to RNA to protein has been called the **central dogma**. It was initially believed that genetic information was transferred only from DNA to RNA and thence translated into protein. However, there is evidence from the study of certain types of virus—retroviruses—that genetic information can occasionally flow in the reverse direction, from RNA to DNA (p. 210). This is referred to as **RNA-directed DNA synthesis**. It has been suggested that regions of DNA in normal cells serve as templates for the synthesis of RNA, which in turn then acts as a template for the synthesis of DNA that later becomes integrated into the nuclear DNA of other cells. Homology between human and retroviral oncogene sequences could reflect this process (p. 211), which could be an important therapeutic approach for the treatment of inherited disease in humans.

Mutations

A **mutation** is defined as a heritable alteration or change in the genetic material. Mutations drive evolution but can also be pathogenic. Mutations can arise through exposure to mutagenic agents (p. 27), but the vast majority occur spontaneously through errors in DNA replication and repair. Sequence variants with no obvious effect upon phenotype may be termed polymorphisms.

Somatic mutations may cause adult-onset disease, such as cancer, but cannot be transmitted to offspring. A mutation in **gonadal tissue** or a **gamete** can be transmitted to future generations unless it affects fertility or survival into adulthood. It is estimated that each individual carries up to six lethal or semilethal recessive mutant alleles that in the homozygous state would have very serious effects. These are conservative estimates and the actual figure could be many times greater. Harmful alleles of all kinds constitute the so-called genetic load of the population.

There are also rare examples of 'back mutation' in patients with recessive disorders. For example, reversion of inherited deleterious mutations has been demonstrated in

Table 2.2 Main Classes, Groups, and Types of Mutation and Effects on Protein Product

Class	Group	Type	Effect on Protein Product
Substitution	Synonymous	Silent*	Same amino acid
	Non-synonymous	Missense*	Altered amino acid—may affect protein function or stability
		Nonsense*	Stop codon—loss of function or expression due to degradation of mRNA
		Splice site	Aberrant splicing—exon skipping or intron retention
		Promoter	Altered gene expression
Deletion	Multiple of 3 (codon)		In-frame deletion of one or more amino acid(s)—may affect protein function or stability
	Not multiple of 3	Frameshift	Likely to result in premature termination with loss of function or expression
	Large deletion	Partial gene deletion	May result in premature termination with loss of function or expression
		Whole gene deletion	Loss of expression
Insertion	Multiple of 3 (codon)		In-frame insertion of one or more amino acid(s)—may affect protein function or stability
	Not multiple of 3	Frameshift	Likely to result in premature termination with loss of function or expression
	Large insertion	Partial gene duplication	May result in premature termination with loss of function or expression
		Whole gene duplication	May have an effect because of increased gene dosage
	Expansion of trinucleotide repeat	Dynamic mutation	Altered gene expression or altered protein stability or function

*Some have been shown to cause aberrant splicing.

phenotypically normal cells present in a small number of patients with Fanconi anemia.

Types of Mutation

Mutations can range from single base substitutions, through insertions and deletions of single or multiple bases to loss or gain of entire chromosomes (Table 2.2). Base substitutions are most prevalent (Table 2.3) and missense mutations account for nearly half of all mutations. A standard nomenclature to describe mutations (Table 2.4) has been agreed on (see http://www.hgvs.org/mutnomen/). Examples of chromosome abnormalities are discussed in Chapter 3.

Substitutions

A **substitution** is the replacement of a single nucleotide by another. These are the most common type of mutation. If the substitution involves replacement by the same type of nucleotide—a pyrimidine for a pyrimidine (C for T or vice versa) or a purine for a purine (A for G or vice versa); this is termed a **transition**. Substitution of a pyrimidine by a purine or vice versa is termed a **transversion.** Transitions occur more frequently than transversions. This may be due to the relatively high frequency of C to T transitions, which is likely to be the result of the nucleotides cytosine and guanine occurring together, or what are known as CpG dinucleotides (p represents the phosphate) frequently

Table 2.3 Frequency of Different Types of Mutation

Type of Mutation	Percentage of Total
Missense or nonsense	56
Splicing	10
Regulatory	2
Small deletions, insertions or indels*	24
Gross deletions or insertions	7
Other (complex rearrangements or repeat variations)	<1

Data from http://www.hgmd.org

*Indels are mutations that involve both an insertion and a deletion of nucleotides.

Table 2.4 Mutation Nomenclature: Examples of CFTR Gene Mutations

Type of Mutation	Nucleotide	Protein Designation	Consequence Description
Missense	c.482G>A	p.Arg117His	Arginine to histidine
Nonsense	c.1756G>T	p.Gly542X	Glycine to stop
Splicing	c.621 + 1G>T		Splice donor site mutation
Deletion (1 bp)	c.1078T	p.Val358TyrfsX11	Frameshift mutation
Deletion (3 bp)	c.1652_1654delCTT	p.Phe508del	In-frame deletion of phenylalanine
Insertion	c.3905_3906insT	p.Leu1258PhefsX7	Frameshift mutation

Mutations can be designated according to the genomic or cDNA (mRNA) sequence and are prefixed by 'g.' or 'c.', respectively. The first base of the start codon (ATG) is c.1. However, for historical reasons this is not always the case, and the first base of the *CFTR* cDNA is actually nucleotide 133.

being methylated in genomic DNA with spontaneous deamination of methylcytosine converting them to thymine. CpG dinucleotides have been termed 'hotspots' for mutation.

Deletions

A **deletion** involves the loss of one or more nucleotides. If this occurs in coding sequences and involves one, two, or more nucleotides that are not a multiple of three, the reading frame will be disrupted. Larger deletions may result in partial or whole gene deletions and may arise through unequal crossover between repeat sequences (e.g., hereditary neuropathy with liability to pressure palsies; see p. 296).

Insertions

An **insertion** involves the addition of one or more nucleotides into a gene. Again, if an insertion occurs in a coding sequence and involves one, two, or more nucleotides that are not a multiple of three, it will disrupt the reading frame. Large insertions can also result from unequal crossover (e.g., hereditary sensory and motor neuropathy type 1a; see p. 296) or the insertion of transposable elements (p. 18).

In 1991, expansion of trinucleotide repeat sequences was identified as a mutational mechanism. A number of single-gene disorders have subsequently been shown to be associated with triplet repeat expansions (Table 2.5). These are described as **dynamic** mutations because the repeat sequence becomes more unstable as it expands in size. The mechanism by which amplification or expansion of the triplet repeat sequence occurs is not clear at present. Triplet repeats below a certain length for each disorder are faithfully and stably transmitted in mitosis and meiosis. Above a certain repeat number for each disorder, they are more likely to be transmitted unstably, usually with an increase or decrease in repeat number. A variety of possible explanations has been offered as to how the increase in triplet repeat number occurs. These include unequal crossover or unequal sister chromatid exchange (see Chapter 18) in non-replicating DNA, and slipped-strand mispairing and polymerase slippage in replicating DNA.

Triplet repeat expansions usually take place over a number of generations within a family, providing an explanation for some unusual aspects of patterns of inheritance as well as possibly being the basis of the previously unexplained phenomenon of anticipation (p. 120).

The exact mechanisms by which repeat expansions cause disease are not known. Unstable trinucleotide repeats may be within coding or non-coding regions of genes and hence vary in their pathogenic mechanisms. Expansion of the CAG repeat in the coding region of the *HD* gene and some *SCA* genes results in a protein with an elongated polyglutamine tract that forms toxic aggregates within certain cells. In fragile X the CGG repeat expansion in the 5′ untranslated region (UTR) results in methylation of promoter sequences and lack of expression of the *FMR1* protein. In myotonic dystrophy (MD) it is thought that a gain-of-function RNA mechanism results from both the CTG expansion in the 3′ UTR of the *DMPK* (type 1 MD) and the CCTG expansion within intron 1 of the *ZNF9* gene. The expanded transcripts bind splice regulatory proteins to form RNA-protein complexes that accumulate in the nuclei of cells. The disruption of these splice regulators causes abnormal developmental processing where embryonic isoforms of the resulting proteins are expressed in adult myotonic dystrophy tissues. The immature proteins then appear to cause the clinical features common to both diseases (p. 295).

The spectrum of repeat expansion mutations also includes a dodecamer repeat expansion upstream from the cystatin B gene that causes progressive myoclonus epilepsy

Table 2.5 Examples of Diseases Arising from Triplet Repeat Expansions

Disease	Repeat Sequence	Normal Range (Repeats)	Pathogenic Range (Repeats)	Repeat Location
Huntington disease (HD)	CAG	9–35	36–100	Coding
Myotonic dystrophy type 1 (DM1)	CTG	5–35	50–4000	3′ UTR
Fragile X site A (FRAXA)	CGG	10–50	200–2000	5′ UTR
Kennedy disease (SBMA)	CAG	13–30	40–62	Coding
Spinocerebellar ataxia 1 (SCA1)	CAG	6–38	39–80	Coding
Spinocerebellar ataxia 2 (SCA2)	CAG	16–30	36–52	Coding
Machado–Joseph disease (MJD, SCA3)	CAG	14–40	60–>85	Coding
Spinocerebellar ataxia 6 (SCA6)	CAG	5–20	21–28	Coding
Spinocerebellar ataxia 7 (SCA7)	CAG	7–19	37–220	Coding
Spinocerebellar ataxia 8 (SCA8)	CTG	16–37	100–>500	3′ UTR
Spinocerebellar ataxia 12 (SCA12)	CAG	9–45	55–78	5′ UTR
Spinocerebellar ataxia 17 (SCA17)	CAG	25–42	47–55	Coding
Dentatorubral-pallidoluysian atrophy (DRPLA)	CAG	7–23	49–>75	Coding
Friedreich ataxia (FA)	GAA	8–33	100–900	Intronic
Fragile X site E (FRAXE)	CCG	6–25	>200	Promoter
Oculopharyngeal muscular dystrophy	GCG	6	8–13	Coding

UTR, untranslated region.

(EPM1) and a pentanucleotide repeat expansion in intron 9 of the *ATXN10* gene shown in families with spinocerebellar ataxia type 10. Spinocerebellar ataxia is an extremely heterogeneous disorder and, in addition to the dynamic mutations shown in Table 2.5, non-repeat expansion mutations have been reported in four additional genes.

Structural Effects of Mutations on the Protein

Mutations can also be subdivided into two main groups according to the effect on the polypeptide sequence of the encoded protein, being either *synonymous* or *non-synonymous*.

Synonymous or Silent Mutations

If a mutation does not alter the polypeptide product of the gene, it is termed a **synonymous** or **silent mutation**. A single base-pair substitution, particularly if it occurs in the third position of a codon because of the degeneracy of the genetic code, will often result in another triplet that codes for the same amino acid with no alteration in the properties of the resulting protein.

Non-Synonymous Mutations

If a mutation leads to an alteration in the encoded polypeptide, it is known as a **non-synonymous mutation**. Non-synonymous mutations are observed to occur less frequently than synonymous mutations. Synonymous mutations are selectively neutral, whereas alteration of the amino-acid sequence of the protein product of a gene is likely to result in abnormal function, which is usually associated with disease, or lethality, which has an obvious selective disadvantage.

Non-synonymous mutations can occur in one of three main ways.

Missense

A single base-pair substitution can result in coding for a different amino acid and the synthesis of an altered protein, a so-called **missense** mutation. If the mutation codes for an amino acid that is chemically dissimilar, for example has a different charge, the structure of the protein will be altered. This is termed a **non-conservative** substitution and can lead to a gross reduction, or even a complete loss, of biological activity. Single base-pair mutations can lead to qualitative rather than quantitative changes in the function of a protein, such that it retains its normal biological activity (e.g., enzyme activity) but differs in characteristics such as its mobility on electrophoresis, its pH optimum, or its stability so that it is more rapidly broken down in vivo. Many of the abnormal hemoglobins (p. 157) are the result of missense mutations.

Some single base-pair substitutions result in the replacement of a different amino acid that is chemically similar, and may have no functional effect. These are termed **conservative** substitutions.

Nonsense

A substitution that leads to the generation of one of the stop codons (see Table 2.1) will result in premature termination of translation of a peptide chain, or what is termed a **nonsense** mutation. In most cases the shortened chain is unlikely to retain normal biological activity, particularly if the termination codon results in the loss of an important functional domain(s) of the protein. mRNA transcripts containing premature termination codons are frequently degraded by a process known as **nonsense-mediated decay**. This is a form of RNA surveillance that is believed to have evolved to protect the body from the possible consequences of truncated proteins interfering with normal function.

Frameshift

If a mutation involves the insertion or deletion of nucleotides that are not a multiple of three, it will disrupt the reading frame and constitute what is known as a **frameshift** mutation. The amino-acid sequence of the protein subsequent to the mutation bears no resemblance to the normal sequence and may have an adverse effect on its function. Most frameshift mutations result in a premature stop codon downstream to the mutation. This may lead to expression of a truncated protein, unless the mRNA is degraded by nonsense-mediated decay.

Mutations in Non-Coding DNA

In general, mutations in non-coding DNA are less likely to have a phenotypic effect. Exceptions include mutations in promoter sequences or other regulatory regions that affect the level of gene expression. With our new knowledge of the role of RNA interference in gene expression, it has become apparent that mutations in miRNA or siRNA binding sites within UTRs can also result in disease.

Splicing Mutations

Mutations of the highly conserved splice donor (GT) and splice acceptor (AG) sites (p. 19) usually result in aberrant splicing. This can result in the loss of coding sequence (exon skipping) or retention of intronic sequence, and may lead to frameshift mutations. **Cryptic splice** sites, which resemble the sequence of an authentic splice site, may be activated when the conserved splice sites are mutated. In addition, base substitutions resulting in apparent silent, missense and nonsense mutations can cause aberrant splicing through mutation of **exon splicing enhancer** sequences. These purine-rich sequences are required for the correct splicing of exons with weak splice-site consensus sequences.

Functional Effects of Mutations on the Protein

Mutations exert their phenotypic effect in one of two ways, through either loss or gain of function.

Loss-of-Function Mutations

Loss-of-function mutations can result in either reduced activity or complete loss of the gene product. The former can be the result of reduced activity or of decreased stability of the gene product and is known as a **hypomorph**, the latter being known as a **null allele** or **amorph**. Loss-of-function mutations involving enzymes are usually inherited in an autosomal or X-linked recessive manner, because the catalytic activity of the product of the normal allele is more than adequate to carry out the reactions of most metabolic pathways.

Haplo-insufficiency

Loss-of-function mutations in the heterozygous state in which half normal levels of the gene product result in phenotypic effects are termed **haplo-insufficiency mutations**. The phenotypic manifestations sensitive to gene dosage are a result of mutations occurring in genes that code for either receptors, or more rarely enzymes, the functions of which are rate limiting; for example, familial hypercholesterolemia (p. 175) and acute intermittent porphyria (p. 179).

In a number of autosomal dominant disorders, the mutational basis of the functional abnormality is the result of haplo-insufficiency in which, not surprisingly, homozygous mutations result in more severe phenotypic effects; examples are angioneurotic edema and familial hypercholesterolemia (p. 175).

Gain-of-Function Mutations

Gain-of-function mutations, as the name suggests, result in either increased levels of gene expression or the development of a new function(s) of the gene product. Increased expression levels from activating point mutations or increased gene dosage are responsible for one type of Charcot-Marie-Tooth disease, hereditary motor, and sensory neuropathy type I (p. 296). The expanded triplet repeat mutations in the Huntington gene cause qualitative changes in the gene product that result in its aggregation in the central nervous system leading to the classic clinical features of the disorder (p. 293).

Mutations that alter the timing or tissue specificity of the expression of a gene can also be considered to be gain-of-function mutations. Examples include the chromosomal rearrangements that result in the combination of sequences from two different genes seen with specific tumors (p. 212). The novel function of the resulting chimeric gene causes the neoplastic process.

Gain-of-function mutations are dominantly inherited and the rare instances of gain-of-function mutations occurring in the homozygous state are often associated with a much more severe phenotype, which is often a prenatally lethal disorder, for example homozygous achondroplasia (p. 93) or Waardenburg syndrome type I (p. 91).

Dominant-Negative Mutations

A **dominant-negative** mutation is one in which a mutant gene in the heterozygous state results in the loss of protein activity or function, as a consequence of the mutant gene product interfering with the function of the normal gene product of the corresponding allele. Dominant-negative mutations are particularly common in proteins that are dimers or multimers, for instance structural proteins such as the collagens, mutations in which can lead to osteogenesis imperfecta.

Genotype-Phenotype Correlation

Many genetic disorders are well recognized as being very variable in severity, or in the particular features manifested by a person with the disorder (p. 112). Developments in molecular genetics increasingly allow identification of the mutational basis of the specific features that occur in a person with a particular inherited disease, or what is known as the phenotype. This has resulted in attempts to correlate the presence of a particular mutation, which is often called the genotype, with the specific features seen in a person with an inherited disorder, this being referred to as **genotype-phenotype correlation**. This can be important in the management of a patient. One example includes the association of mutations in the *BRCA1* gene with the risk of developing ovarian cancer as well as breast cancer (p. 224). Particularly striking examples are mutations in the receptor tyrosine kinase gene *RET* which, depending on their location, can lead to four different syndromes that differ in the functional mechanism and clinical phenotype. Loss-of-function nonsense mutations lead to lack of migration of neural crest–derived cells to form the ganglia of the myenteric plexus of the large bowel, leading to Hirschsprung disease, whereas gain-of-function missense mutations result in familial medullary thyroid carcinoma or one of the two types of multiple endocrine neoplasia type 2 (p. 100). Mutations in the *LMNA* gene are associated with an even broader spectrum of disease (p. 112).

Mutations and Mutagenesis

Naturally occurring mutations are referred to as **spontaneous mutations** and are thought to arise through chance errors in chromosomal division or DNA replication. Environmental agents that cause mutations are known as mutagens. These include natural or artificial ionizing radiation and chemical or physical mutagens.

Radiation

Ionizing radiation includes electromagnetic waves of very short wavelength (x-rays and γ rays) and high-energy particles (α particles, β particles, and neutrons). X-rays, γ rays, and neutrons have great penetrating power, but α particles can penetrate soft tissues to a depth of only a fraction of a millimeter and β particles only up to a few millimeters.

Table 2.6 Approximate Average Doses of Ionizing Radiation from Various Sources to the Gonads of the General Population

Source of Radiation	Average Dose per Year (mSv)	Average Dose per 30 Years (mSv)
Natural		
Cosmic radiation	0.25	7.5
External γ radiation*	1.50	45.0
Internal γ radiation	0.30	9.0
Artificial		
Medical radiology	0.30	9.0
Radioactive fallout	0.01	0.3
Occupational and miscellaneous	0.04	1.2
Total	2.40	72.0

Data from Clarke RH, Southwood TRE 1989 Risks from ionizing radiation. Nature 338:197–198

*Including radon in dwelling.

Dosimetry is the measurement of radiation. The dose of radiation is expressed in relation to the amount received by the gonads because it is the effects of radiation on germ cells rather than somatic cells that are important as far as transmission of mutations to future progeny is concerned. The **gonad dose** of radiation is often expressed as the amount received in 30 years. This period has been chosen because it corresponds roughly to the generation time in humans.

The various sources and average annual doses of the different types of natural and artificial ionizing radiation are listed in Table 2.6. Natural sources of radiation include cosmic rays, external radiation from radioactive materials in certain rocks, and internal radiation from radioactive materials in tissues. Artificial sources include diagnostic and therapeutic radiology, occupational exposure and fallout from nuclear explosions.

The average gonadal dose of ionizing radiation from radioactive fallout resulting from the testing of nuclear weapons is less than that from any of the sources of background radiation. However, the possibility of serious accidents involving nuclear reactors, as occurred at Three Mile Island in the United States in 1979 and at Chernobyl in the Soviet Union in 1986, with widespread effects, must always be borne in mind.

Genetic Effects

Experiments with animals and plants have shown that the number of mutations produced by irradiation is proportional to the dose: the larger the dose, the greater the number of mutations produced. It is believed that there is no threshold below which irradiation has no effect—even the smallest dose of radiation can result in a mutation. The genetic effects of ionizing radiation are also cumulative, so that each time a person is exposed to radiation, the dose received has to be added to the amount of radiation already received. The total number of radiation-induced mutations is directly proportional to the total gonadal dose.

Unfortunately, in humans there is no easy way to demonstrate genetic damage caused by mutagens. Several agencies throughout the world are responsible for defining what is referred to as the maximum permissible dose of radiation. In the United Kingdom, the Radiation Protection Division of the Health Protection Agency advises that occupational exposure should not exceed 15 mSv in a year. To put this into perspective, 1 mSv is roughly 50 times the dose received in a single chest x-ray and 100 times the dose incurred when flying from the United Kingdom to Spain in a jet aircraft!

There is no doubting the potential dangers, both somatic and germline, of exposure to ionizing radiation. In the case of medical radiology, the dose of radiation resulting from a particular procedure has to be weighed against the ultimate beneficial effect to the patient. In the case of occupational exposure to radiation, the answer lies in defining the risks and introducing and enforcing adequate legislation. With regard to the dangers from fallout from nuclear accidents and explosions, the solution would seem obvious.

Chemical Mutagens

In humans, chemical mutagenesis may be more important than radiation in producing genetic damage. Experiments have shown that certain chemicals, such as mustard gas, formaldehyde, benzene, some basic dyes, and food additives, are mutagenic in animals. Exposure to environmental chemicals may result in the formation of DNA adducts, chromosome breaks, or aneuploidy. Consequently all new pharmaceutical products are subject to a battery of mutagenicity tests that include both in vitro and in vivo studies in animals.

DNA Repair

The occurrence of mutations in DNA, if left unrepaired, would have serious consequences for both the individual and subsequent generations. The stability of DNA is dependent upon continuous **DNA repair** by a number of different mechanisms (Table 2.7). Some types of DNA damage can be repaired directly. Examples include the dealkylation of O^6-alkyl guanine or the removal of thymine dimers by photoreactivation in bacteria. The majority of DNA repair mechanisms involve cleavage of the DNA strand by an endonuclease, removal of the damaged region by an exonuclease, insertion of new bases by the enzyme DNA polymerase, and sealing of the break by DNA ligase.

Nucleotide excision repair removes thymine dimers and large chemical adducts. It is a complex process involving more than 30 proteins that remove fragments of approximately 30 nucleotides. Mutations in at least eight of the genes encoding these proteins can cause xeroderma pigmentosum (p. 289), characterized by extreme sensitivity to ultraviolet light and a high frequency of skin cancer. A different set of repair enzymes is used to excise single abnormal bases (**base excision repair**), with mutations in the gene encoding the DNA glycosylase MYH having recently been

Table 2.7 DNA Repair Pathways, Genes, and Associated Disorders

Type of DNA Repair	Mechanism	Genes	Disorders
Base excision repair (BER)	Removal of abnormal bases	*MYH*	Colorectal cancer
Nucleotide excision repair (NER)	Removal of thymine dimers and large chemical adducts	*XP*	Xeroderma pigmentosum
Post-replication repair	Removal of double-strand breaks by homologous recombination or non-homologous end-joining	*NBS* *BLM* *BRCA1/2*	Nijmegen breakage syndrome Bloom syndrome Breast cancer
Mismatch repair (MMR)	Corrects mismatched bases caused by mistakes in DNA replication	*MSH* and *MLH*	Colorectal cancer (HNPCC)

HNPCC, hereditary non-polyposis colorectal cancer.

shown to cause an autosomal recessive form of colorectal cancer (p. 223).

Naturally occurring reactive oxygen species and ionizing radiation induce breakage of DNA strands. Double-strand breaks result in chromosome breaks that can be lethal if not repaired. **Post-replication repair** is required to correct double-strand breaks and usually involves homologous recombination with a sister DNA molecule. Human genes involved in this pathway include *NBS*, *BLM*, and *BRCA1/2*, mutated in Nijmegen breakage syndrome, Bloom syndrome (p. 288), and hereditary breast cancer (p. 224), respectively. Alternatively, the broken ends may be rejoined by non-homologous end-joining, which is an error-prone pathway.

Mismatch repair (MMR) corrects mismatched bases introduced during DNA replication. Cells defective in MMR have very high mutation rates (up to 1000 times higher than normal). Mutations in at least six different MMR genes cause hereditary non-polyposis colorectal cancer (hereditary non-polyposis colorectal cancer; see p. 222).

Although DNA repair pathways have evolved to correct DNA damage and hence protect the cell from the deleterious consequences of mutations, some mutations arise from the cell's attempts to tolerate damage. One example is **translesion DNA synthesis**, in which the DNA replication machinery bypasses sites of DNA damage, allowing normal DNA replication and gene expression to proceed downstream. Human disease may also be caused by defective cellular responses to DNA damage. Cells have complex signaling pathways that allow cell-cycle arrest to provide increased time for DNA repair. If the DNA damage is irreparable, the cell may initiate programmed cell death (**apoptosis**). The ATM protein is involved in sensing DNA damage and has been described as the 'guardian of the genome'. Mutations in the *ATM* gene cause ataxia telangiectasia (see p. 204), characterized by hypersensitivity to radiation and a high risk of cancer.

FURTHER READING

Alberts B, Johnson A, Lewis J, et al 2007 Molecular biology of the cell, 5th ed. London: Garland.

Very accessible, well written, and lavishly illustrated comprehensive text of molecular biology with accompanying problems book and CD-ROM using multimedia review and self-assessment.

Dawkins R 1989 The selfish gene, 3rd ed. Oxford: Oxford University Press.

An interesting, controversial concept.

Fire A, Xu S, Montgomery MK, et al 1998 Potent and specific genetic interference by double-stranded RNA in *Caenorhabditis elegans*. Nature 391:806–811.

Landmark paper describing the discovery of RNAi.

Lewin B 2011 Genes X, 10th ed. Oxford: Oxford University Press.

The tenth edition of this excellent textbook of molecular biology with color diagrams and figures. Hard to improve upon.

Mettler FA, Upton AC 2008 medical effects of ionising radiation, 3rd ed. Philadelphia: Saunders.

Good overview of all aspects of the medical consequences of ionizing radiation.

Schull WJ, Neel JV 1958 Radiation and the sex ratio in man. Sex ratio among children of survivors of atomic bombings suggests induced sex-linked lethal mutations. Science 228:434–438.

The original report of possible evidence of the effects of atomic radiation.

Strachan T, Read AP 2011 Human molecular genetics, 4th ed. London: Garland Science.

An up-to-date, comprehensive textbook of all aspects of molecular and cellular biology as it relates to inherited disease in humans.

Turner JE 1995 Atoms, radiation and radiation protection. Chichester, UK: John Wiley.

Basis of the physics of radiation, applications, and harmful effects.

Watson JD, Crick FHC 1953 Molecular structure of nucleic acids—a structure for deoxyribose nucleic acid. Nature 171:737–738.

The concepts in this paper, presented in just over one page, resulted in the authors receiving the Nobel Prize!

ELEMENTS

1 Genetic information is stored in DNA (deoxyribonucleic acid) as a linear sequence of two types of nucleotide, the purines (adenine [A] and guanine [G]) and the pyrimidines (cytosine [C] and thymine [T]), linked by a sugar–phosphate backbone.

2 A molecule of DNA consists of two antiparallel strands held in a double helix by hydrogen bonds between the complementary G–C and A–T base pairs.

3 DNA replication has multiple sites of origin and is semiconservative, each strand acting as a template for synthesis of a complementary strand.

4 Genes coding for proteins in higher organisms (eukaryotes) consist of coding (exons) and non-coding (introns) sections.

5 Transcription is the synthesis of a single-stranded complementary copy of one strand of a gene that is known as messenger RNA (mRNA). RNA (ribonucleic acid) differs from DNA in containing the sugar ribose and the base uracil instead of thymine.

6 mRNA is processed during transport from the nucleus to the cytoplasm, eliminating the non-coding sections. In the cytoplasm it becomes associated with the ribosomes, where translation (i.e., protein synthesis) occurs.

7 The genetic code is 'universal' and consists of triplets (codons) of nucleotides, each of which codes for an amino acid or termination of peptide chain synthesis. The code is degenerate, as all but two amino acids are specified by more than one codon.

8 The major control of gene expression is at the level of transcription by DNA regulatory sequences in the 5′ flanking promoter region of structural genes in eukaryotes. General and specific transcription factors are also involved in the regulation of genes.

9 Mutations occur both spontaneously and as a result of exposure to mutagenic agents such as ionizing radiation. Mutations are continuously corrected by DNA repair enzymes.

CHAPTER 3

Chromosomes and Cell Division

Let us not take it for granted that life exists more fully in what is commonly thought big than in what is commonly thought small.

VIRGINIA WOOLF

At the molecular or submicroscopic level, DNA can be regarded as the basic template that provides a blueprint for the formation and maintenance of an organism. DNA is packaged into **chromosomes** and at a very simple level these can be considered as being made up of tightly coiled long chains of genes. Unlike DNA, chromosomes can be visualized during cell division using a light microscope, under which they appear as threadlike structures or 'colored bodies'. The word *chromosome* is derived from the Greek *chroma* (= color) and *soma* (= body).

Chromosomes are the factors that distinguish one species from another and that enable the transmission of genetic information from one generation to the next. Their behavior at somatic cell division in mitosis provides a means of ensuring that each daughter cell retains its own complete genetic complement. Similarly, their behavior during gamete formation in meiosis enables each mature ovum and sperm to contain a unique single set of parental genes. Chromosomes are quite literally the vehicles that facilitate reproduction and the maintenance of a species.

The study of chromosomes and cell division is referred to as **cytogenetics.** Before the 1950s it was thought, incorrectly, that each human cell contained 48 chromosomes and that human sex was determined by the number of X chromosomes present at conception. Following the development in 1956 of more reliable techniques for studying human chromosomes, it was realized that the correct chromosome number in humans is 46 (p. 5) and that maleness is determined by the presence of a Y chromosome regardless of the number of X chromosomes present in each cell. It was also realized that abnormalities of chromosome number and structure could seriously disrupt normal growth and development.

Table 3.1 highlights the methodological developments that have taken place during the past 5 decades that underpin our current knowledge of human cytogenetics.

Human Chromosomes

Morphology

At the submicroscopic level, chromosomes consist of an extremely elaborate complex, made up of supercoils of DNA, which has been likened to the tightly coiled network of wiring seen in a solenoid (p. 31). Under the electron microscope chromosomes can be seen to have a rounded and rather irregular morphology (Figure 3.1). However, most of our knowledge of chromosome structure has been gained using light microscopy. Special stains selectively taken up by DNA have enabled each individual chromosome to be identified. These are best seen during cell division, when the chromosomes are maximally contracted and the constituent genes can no longer be transcribed.

At this time each chromosome can be seen to consist of two identical strands known as **chromatids**, or **sister chromatids**, which are the result of DNA replication having taken place during the S (synthesis) phase of the cell cycle (p. 39). These sister chromatids can be seen to be joined at a primary constriction known as the **centromere**. Centromeres consist of several hundred kilobases of repetitive DNA and are responsible for the movement of chromosomes at cell division. Each centromere divides the chromosome into short and long arms, designated p (= petite) and q ('g' = grande), respectively.

The tip of each chromosome arm is known as the **telomere**. Telomeres play a crucial role in sealing the ends of chromosomes and maintaining their structural integrity. Telomeres have been highly conserved throughout evolution and in humans they consist of many tandem repeats of a TTAGGG sequence. During DNA replication, an enzyme known as **telomerase** replaces the 5′ end of the long strand, which would otherwise become progressively shorter until a critical length was reached when the cell could no longer divide and thus became senescent. This is in fact part of the normal cellular aging process, with most cells being unable to undergo more than 50 to 60 divisions. However, in some tumors increased telomerase activity has been implicated as a cause of abnormally prolonged cell survival.

Morphologically chromosomes are classified according to the position of the centromere. If this is located centrally, the chromosome is **metacentric**, if terminal it is **acrocentric**, and if the centromere is in an intermediate position the chromosome is **submetacentric** (Figure 3.2). Acrocentric chromosomes sometimes have stalk-like appendages called **satellites** that form the nucleolus of the resting interphase

Table 3.1 Development of Methodologies for Cytogenetics

Decade	Development	Examples of Application
1950–1960s	Reliable methods for chromosome preparations	Chromosome number determined to be 46 (1956) and Philadelphia chromosome identified as t(9;22) (1960)
1970s	Giemsa chromosome banding	Mapping of *RB1* gene to chromosome 13q14 by identification of deleted chromosomal region in patients with retinoblastoma (1976)
1980s	Fluorescent in-situ hybridization (FISH)	Interphase FISH for rapid detection of Down syndrome (1994) Spectral karyotyping for whole genome chromosome analysis (1996)
1990s	Comparative genomic hybridization (CGH)	Mapping genomic imbalances in solid tumors (1992)
2000s	Array CGH	Analysis of constitutional rearrangements; e.g., identification of ~5 Mb deletion in a patient with CHARGE syndrome that led to identification of the gene (2004)

CHARGE, **c**oloboma of the eye, **h**eart defects, **a**tresia of the choanae, **r**etardation of growth and/or development, **g**enital and/or urinary abnormalities, and **e**ar abnormalities and deafness.

cell and contain multiple repeat copies of the genes for ribosomal RNA.

Classification

Individual chromosomes differ not only in the position of the centromere, but also in their overall length. Based on the three parameters of length, position of the centromere, and the presence or absence of satellites, early pioneers of cytogenetics were able to identify most individual chromosomes, or at least subdivide them into groups labeled A to G on the basis of overall morphology (A, 1–3; B, 4–5; C, 6–12 1 X; D, 13–15; E, 16–18; F, 19–20; G, 21–22 1 Y). In humans the normal cell nucleus contains 46 chromosomes, made up of 22 pairs of **autosomes** and a single pair of sex chromosomes—XX in the female and XY in the male. One member of each of these pairs is derived from each parent. Somatic cells are said to have a **diploid** complement of 46 chromosomes, whereas gametes (ova and sperm) have a **haploid** complement of 23 chromosomes. Members of a pair of chromosomes are known as **homologs**.

The development of chromosome banding (p. 33) enabled very precise recognition of individual chromosomes and the detection of subtle chromosome abnormalities. This technique also revealed that **chromatin**, the combination of DNA and histone proteins that comprise chromosomes, exists in two main forms. **Euchromatin** stains lightly and

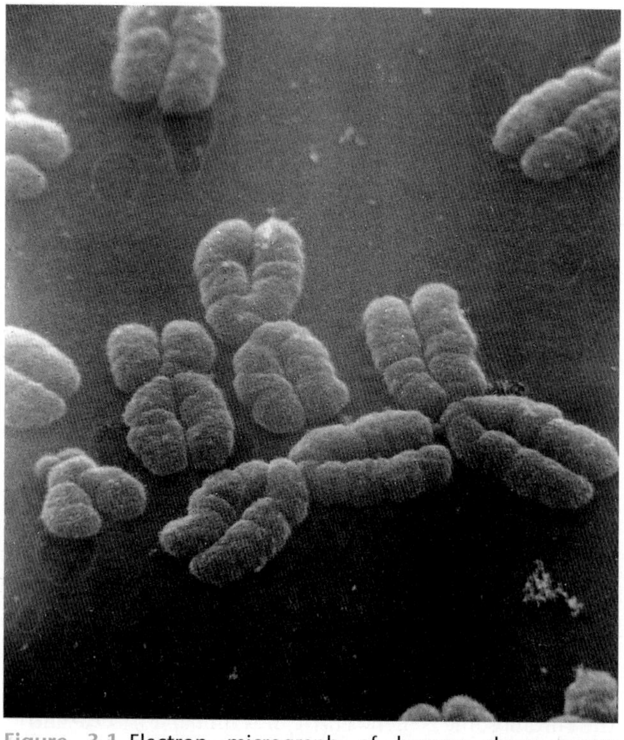

Figure 3.1 Electron micrograph of human chromosomes showing the centromeres and well-defined chromatids. (Courtesy Dr. Christine Harrison. Reproduced from Harrison et al 1983 Cytogenet Cell Genet 35: 21–27; with permission of the publisher, S. Karger, Basel.)

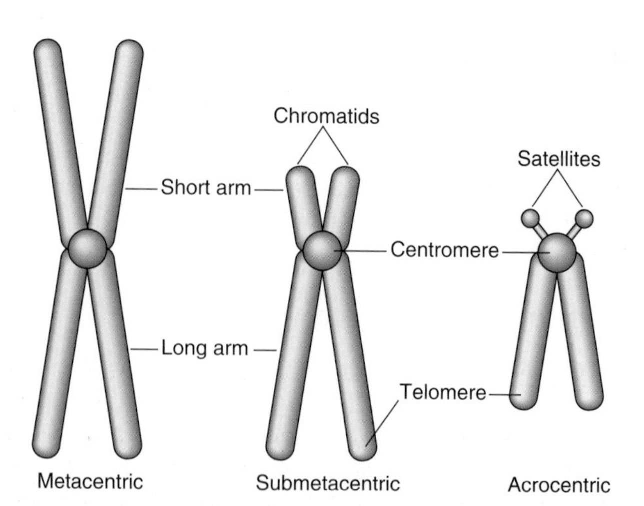

Figure 3.2 Morphologically chromosomes are described as metacentric, submetacentric, or acrocentric, depending on the position of the centromere.

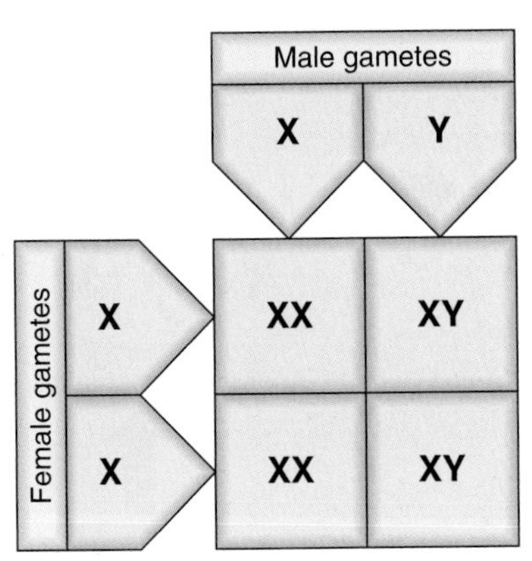

Figure 3.3 Punnett square showing sex chromosome combinations for male and female gametes.

consists of genes that are actively expressed. In contrast, **heterochromatin** stains darkly and is made up largely of inactive, unexpressed, repetitive DNA.

The Sex Chromosomes

The X and Y chromosomes are known as the sex chromosomes because of their crucial role in sex determination. The X chromosome was originally labeled as such because of uncertainty as to its function when it was realized that in some insects this chromosome is present in some gametes but not in others. In these insects the male has only one sex chromosome (X), whereas the female has two (XX). In humans, and in most mammals, both the male and the female have two sex chromosomes—XX in the female and XY in the male. The Y chromosome is much smaller than the X and carries only a few genes of functional importance, most notably the testis-determining factor, known as *SRY* (p. 92). Other genes on the Y chromosome are known to be important in maintaining spermatogenesis.

In the female each ovum carries an X chromosome, whereas in the male each sperm carries either an X or a Y chromosome. As there is a roughly equal chance of either an X-bearing sperm or a Y-bearing sperm fertilizing an ovum, the numbers of male and female conceptions are approximately equal (Figure 3.3). In fact, slightly more male babies are born than females, although during childhood and adult life the sex ratio evens out at 1 : 1.

The process of sex determination is considered in detail later (p. 101).

Methods of Chromosome Analysis

It was generally believed that each cell contained 48 chromosomes until 1956, when Tjio and Levan correctly concluded on the basis of their studies that the normal human somatic cell contains only 46 chromosomes (p. 5). The methods they used, with certain modifications, are now universally employed in cytogenetic laboratories to analyze the chromosome constitution of an individual, which is known as a **karyotype**. This term is also used to describe a photomicrograph of an individual's chromosomes, arranged in a standard manner.

Chromosome Preparation

Any tissue with living nucleated cells that undergo division can be used for studying human chromosomes. Most commonly circulating lymphocytes from peripheral blood are used, although samples for chromosomal analysis can be prepared relatively easily using skin, bone marrow, chorionic villi, or cells from amniotic fluid (amniocytes).

In the case of peripheral (venous) blood, a sample is added to a small volume of nutrient medium containing phytohemagglutinin, which stimulates T lymphocytes to divide. The cells are cultured under sterile conditions at 37°C for about 3 days, during which they divide, and colchicine is then added to each culture. This drug has the extremely useful property of preventing formation of the spindle, thereby arresting cell division during metaphase, the time when the chromosomes are maximally condensed and therefore most visible. Hypotonic saline is then added, which causes the red blood cells to lyze and results in spreading of the chromosomes, which are then fixed, mounted on a slide and stained ready for analysis (Figure 3.4).

Chromosome Banding

Several different staining methods can be used to identify individual chromosomes but G (**Giemsa**) banding is used most commonly. The chromosomes are treated with trypsin, which denatures their protein content, and then stained with a DNA-binding dye—also known as 'Giemsa'—that gives each chromosome a characteristic and reproducible pattern of light and dark bands (Figure 3.5).

G banding generally provides high-quality chromosome analysis with approximately 400 to 500 bands per haploid set. Each of these bands corresponds on average to approximately 6000 to 8000 kilobases (kb) (i.e., 6 to 8 megabases [mb]) of DNA. High-resolution banding of the chromosomes at an earlier stage of mitosis, such as prophase or prometaphase, provides greater sensitivity with up to 800 bands per haploid set, but is much more demanding technically. This involves first inhibiting cell division with an agent such as methotrexate or thymidine. Folic acid or deoxycytidine is added to the culture medium, releasing the cells into mitosis. Colchicine is then added at a specific time interval, when a higher proportion of cells will be in prometaphase and the chromosomes will not be fully contracted, giving a more detailed banding pattern.

Karyotype Analysis

The next stage in chromosome analysis involves first counting the number of chromosomes present in a specified number of cells, sometimes referred to as **metaphase spreads**, followed by careful analysis of the banding pattern of each individual chromosome in selected cells.

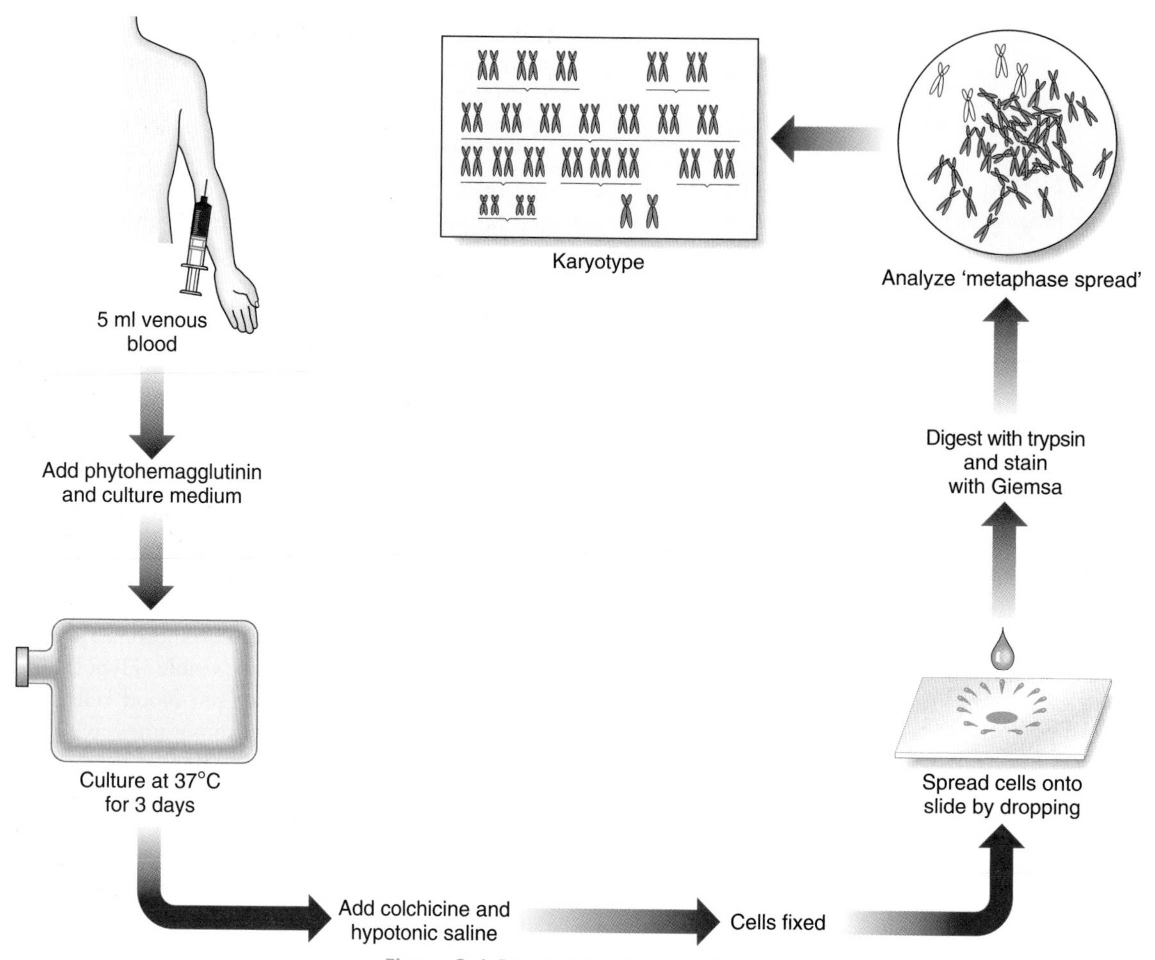

Figure 3.4 Preparation of a karyotype.

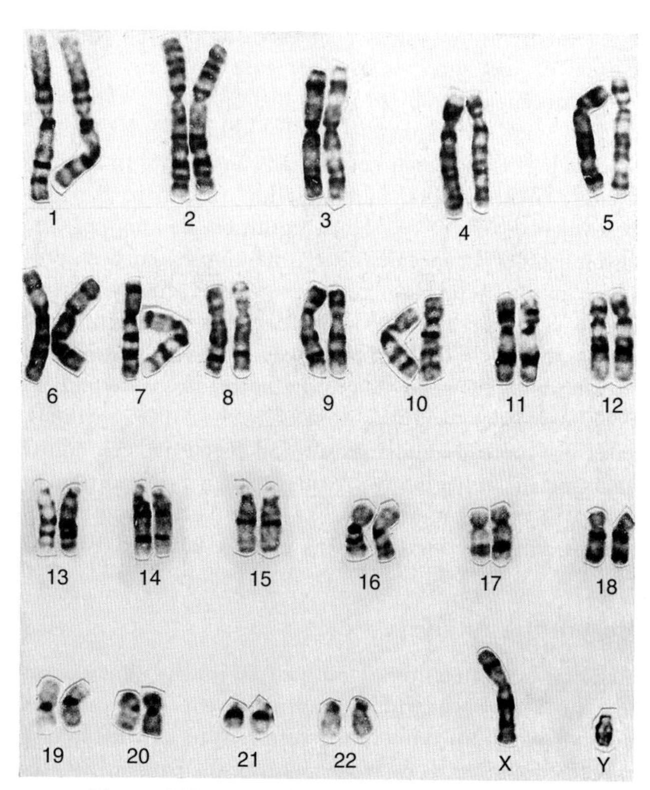

Figure 3.5 A normal G-banded male karyotype.

The banding pattern of each chromosome is specific and can be shown in the form of a stylized ideal karyotype known as an **idiogram** (Figure 3.6). The cytogeneticist analyzes each pair of homologous chromosomes, either directly by looking down the microscope or using an image capture system to photograph the chromosomes and arrange them in the form of a karyogram (Figure 3.7).

Molecular Cytogenetics

Fluorescent In-Situ Hybridization

This diagnostic tool combines conventional cytogenetics with molecular genetic technology. It is based on the unique ability of a portion of single-stranded DNA (i.e., a probe; see p. 35) to anneal with its complementary target sequence on a metaphase chromosome, interphase nucleus or extended chromatin fiber. In **fluorescent in-situ hybridization (FISH)**, the DNA probe is labeled with a fluorochrome which, after hybridization with the patient's sample, allows the region where hybridization has occurred to be visualized using a fluorescence microscope. FISH has been widely used for clinical diagnostic purposes during the past 15 years and there are a number of different types of probes that may be employed.

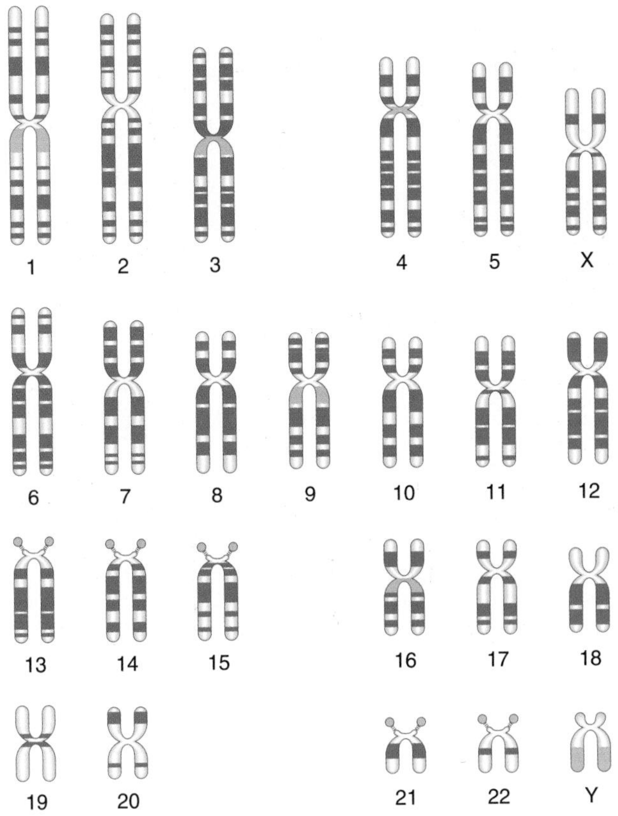

Figure 3.6 An idiogram showing the banding patterns of individual chromosomes as revealed by fluorescent and Giemsa staining.

Different Types of FISH Probe

Centromeric Probes

These consist of repetitive DNA sequences found in and around the centromere of a specific chromosome. They were the original probes used for rapid diagnosis of the common aneuploidy syndromes (trisomies 13, 18, 21; see p. 274) using non-dividing cells in interphase obtained from a prenatal diagnostic sample of chorionic villi. In the present, quantitative fluorescent polymerase chain reaction is more commonly used to detect these trisomies.

Chromosome-specific Unique-sequence Probes

These are specific for a particular single locus. Unique-sequence probes are particularly useful for identifying tiny submicroscopic deletions and duplications (Figure 3.8). The group of disorders referred to as the **microdeletion** syndromes are described in Chapter 18. Another application is the use of an interphase FISH probe to identify *HER2* overexpression in breast tumors to identify patients likely to benefit from Herceptin treatment.

Telomeric Probes

A complete set of telomeric probes was been developed for all 24 chromosomes (i.e., autosomes 1 to 22 plus X and Y). Using these, a method has been devised that enables the simultaneous analysis of the subtelomeric region of every chromosome by means of only one microscope slide per

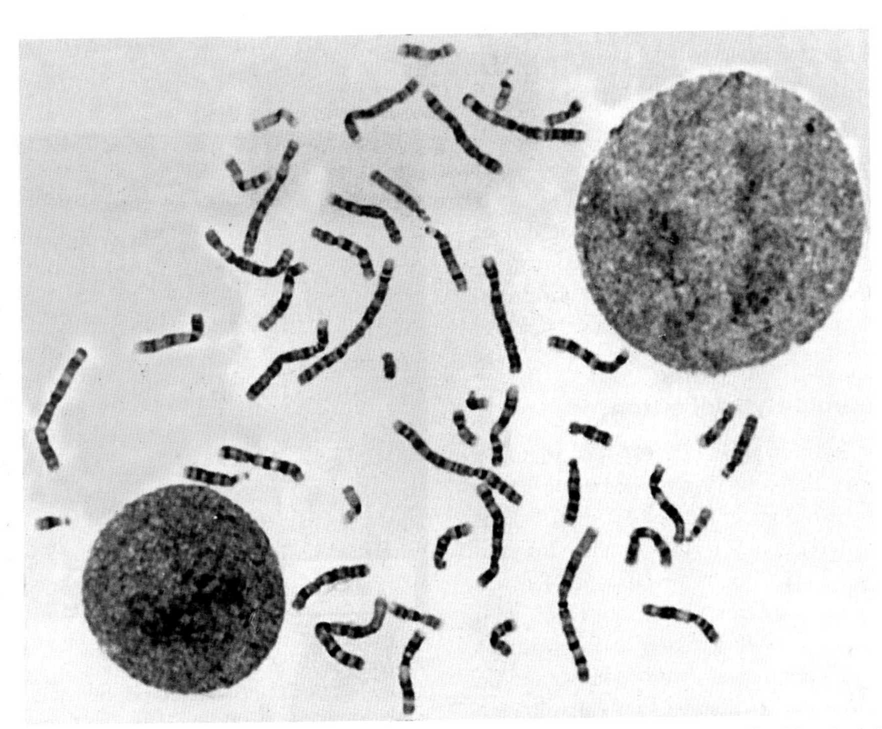

Figure 3.7 A G-banded metaphase spread. (Courtesy Mr. A. Wilkinson, Cytogenetics Unit, City Hospital, Nottingham, UK.)

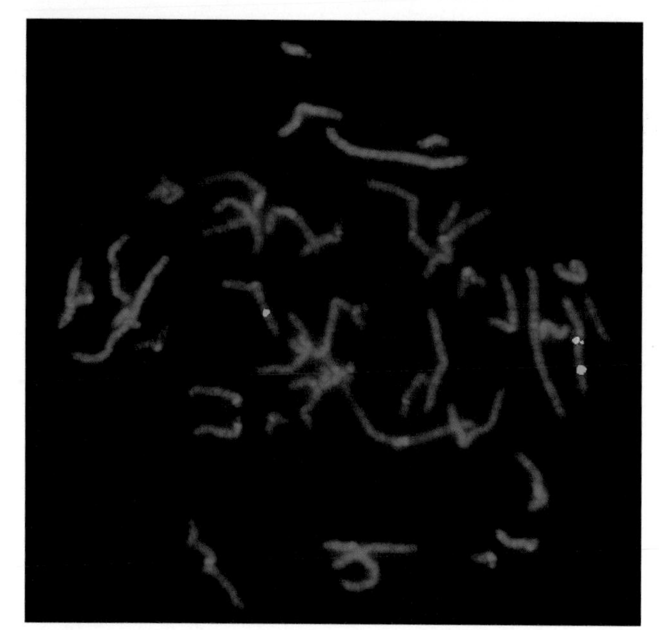

Figure 3.8 Metaphase image of Williams (*ELN*) region probe (Vysis), chromosome band 7q11.23, showing the deletion associated with Williams syndrome. The normal chromosome has signals for the control probe (*green*) and the *ELN* gene probe (*orange*), but the deleted chromosome shows only the control probe signal. (Courtesy Catherine Delmege, Bristol Genetics Laboratory, Southmead Hospital, Bristol, UK.)

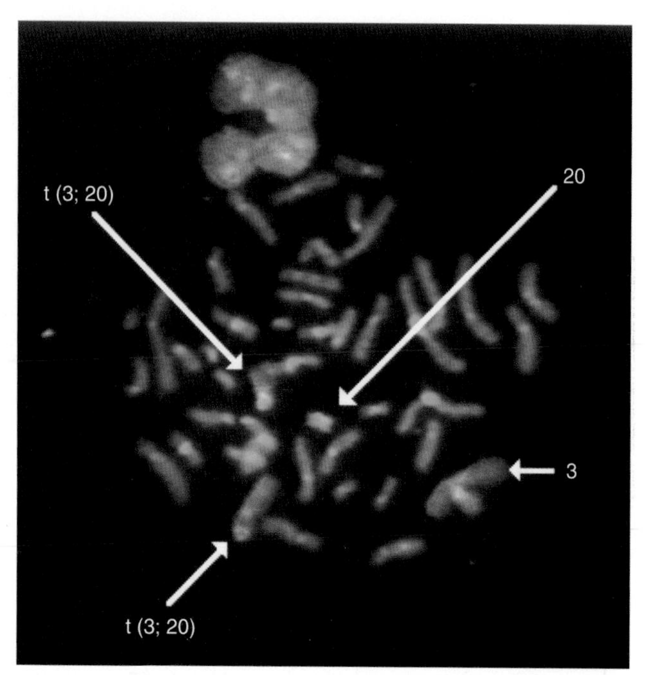

Figure 3.9 Chromosome painting showing a reciprocal translocation involving chromosomes 3 *(red)* and 20 *(green)*.

patient. This proved to be a useful technique for identifying tiny 'cryptic' subtelomeric abnormalities, but has largely been replaced with a quantitative polymerase chain reaction method, multiplex ligation-dependent probe amplifications, that simultaneously measures dosage for all the subtelomeric chromosome regions.

Whole-Chromosome Paint Probes

These consist of a cocktail of probes obtained from different parts of a particular chromosome. When this mixture of probes is used together in a single hybridization, the entire relevant chromosome fluoresces (i.e., is 'painted'). Chromosome painting is extremely useful for characterizing complex rearrangements, such as subtle translocations (Figure 3.9), and for identifying the origin of additional chromosome material, such as small supernumerary markers or rings.

Comparative Genomic Hybridization

Comparative genomic hybridization (CGH) was originally developed to overcome the difficulty of obtaining good-quality metaphase preparations from solid tumors. This technique enabled the detection of regions of allele loss and gene amplification (p. 220). Tumor or 'test' DNA was labeled with a green paint, and control normal DNA with a red paint. The two samples were mixed and hybridized competitively to normal metaphase chromosomes, and an image captured (Figure 3.10). If the test sample contained more DNA from a particular chromosome region than the control sample, that region was identified by an increase in the green to red fluorescence ratio (Figure 3.11). Similarly a deletion in the test sample was identified by a reduction in the green to red fluorescence ratio.

Array CGH

Cytogenetic techniques are traditionally based on microscopic analysis. However, the increasing application of microarray technology is also having a major impact on cytogenetics. Although array CGH is a molecular biology

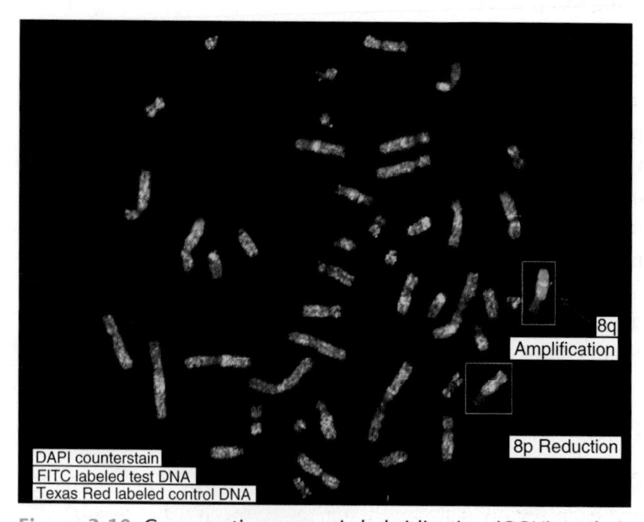

Figure 3.10 Comparative genomic hybridization (CGH) analysis showing areas of gene amplification and reduction (deletion) in tumor DNA. *DAPI*, diamidinophenylindole; *FITC*, fluorescein isothiocyanate. (Courtesy Dr. Peter Lichter, German Cancer Research Center, Heidelberg, and Applied Imaging.)

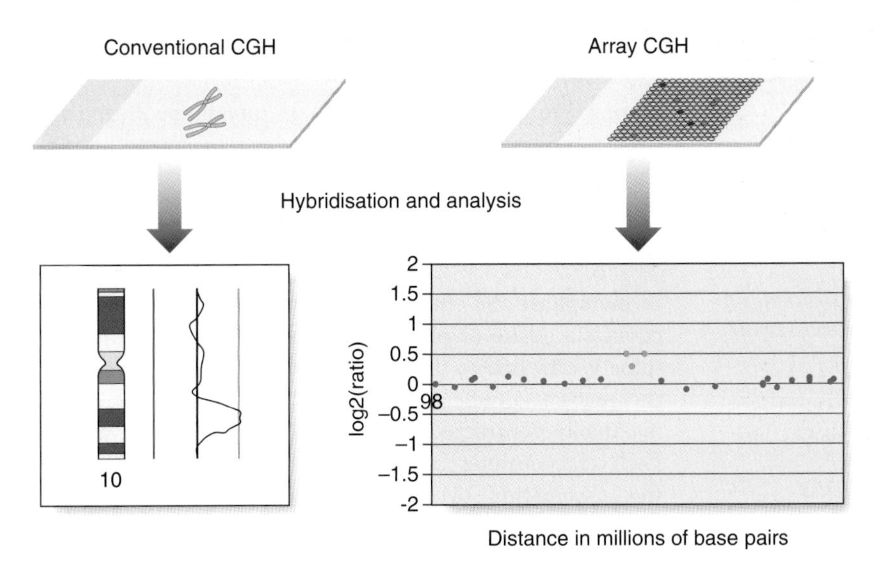

Figure 3.11 Comparison of conventional and array comparative genomic hybridization (CGH). Both techniques involve the hybridization of differentially labeled normal and patient DNA, but the targets of the hybridization are metaphase chromosomes and microarrays, respectively. The results show deletions of chromosome 10q and deletion of three clones on a 1-Mb bacterial artificial chromosome (BAC) array. (Array CGH data courtesy Dr. John Barber, National Genetics Reference Laboratory [Wessex], Salisbury, UK.)

technique, it is introduced in this chapter because it has evolved from metaphase CGH and is being used to investigate chromosome structure.

Array CGH also involves the hybridization of patient and reference DNA, but metaphase chromosomes are replaced as the target by large numbers of DNA sequences bound to glass slides (Figure 3.11). The DNA target sequences have evolved from mapped clones (yeast artificial chromosome [YAC], bacterial artificial chromosome [BAC], or P1-derived artificial chromosome [PAC] or cosmid), to oligonucleotides. They are spotted on to the microscope slides using robotics to create a microarray, in which each DNA target has a unique location. Following hybridization and washing to remove unbound DNA, the relative levels of fluorescence are measured using computer software. Oligonucleotide arrays provide the highest resolution and can include up to 1 million probes.

The application of microarray CGH has extended from cancer cytogenetics to the detection of any type of gain or loss, including the detection of subtelomeric deletions in patients with unexplained intellectual impairment. Array CGH is faster and more sensitive than conventional metaphase analysis for the identification of constitutional rearrangements (with the exception of balanced translocations) and has replaced conventional karyotyping as the first-line test in the investigation of patients with severe developmental delay/learning difficulties and/or congenital abnormalities.

Chromosome Nomenclature

By convention each chromosome arm is divided into regions and each region is subdivided into bands, numbering always from the centromere outwards (Figure 3.12). A given point on a chromosome is designated by the chromosome number, the arm (p or q), the region, and the band (e.g., 15q12). Sometimes the word *region* is omitted, so that 15q12 would be referred to simply as band 12 on the long arm of chromosome 15.

A shorthand notation system exists for the description of chromosome abnormalities (Table 3.2). Normal male and female karyotypes are depicted as 46,XY and 46,XX, respectively. A male with Down syndrome as a result of trisomy 21 would be represented as 47,XY,+21, whereas a female with a deletion of the short arm of one number 5 chromosome (cri du chat syndrome; see p. 281) would be represented as 46,XX,del(5p). A chromosome report

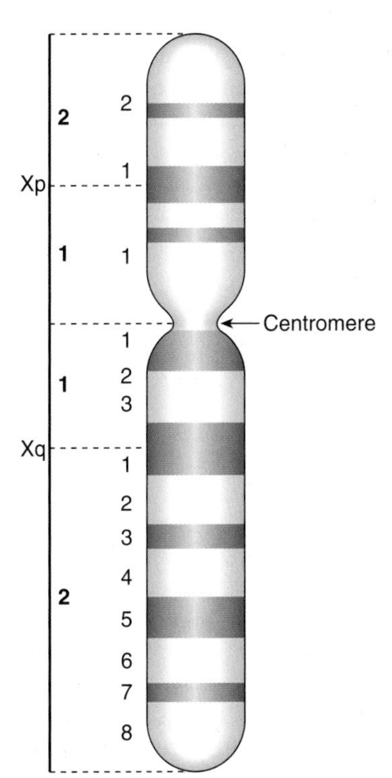

Figure 3.12 X chromosome showing the short and long arms each subdivided into regions and bands.

Table 3.2 Symbols Used in Describing a Karyotype

Term	Explanation	Example
p	Short arm	
q	Long arm	
cen	Centromere	
del	Deletion	46,XX,del(1)(q21)
dup	Duplication	46,XY, dup(13)(q14)
fra	Fragile site	
i	Isochromosome	46,X,i(Xq)
inv	Inversion	46XX,inv(9)(p12q12)
ish	In-situ hybridization	
r	Ring	46;XX,r(21)
t	Translocation	46,XY,t(2;4)(q21;q21)
ter	Terminal or end	Tip of arm; e.g., pter or qter
/	Mosaicism	46,XY/47,XXY
+ or –	Sometimes used after a chromosome arm in text to indicate gain or loss of part of that chromosome	46,XX,5p–

reading 46,XY,t(2;4)(p23;q25) would indicate a male with a reciprocal translocation involving the short arm of chromosome 2 at region 2 band 3 and the long arm of chromosome 4 at region 2 band 5.

Cell Division

Mitosis

At conception the human zygote consists of a single cell. This undergoes rapid division, leading ultimately to the mature human adult consisting of approximately 1×10^{14} cells in total. In most organs and tissues, such as bone marrow and skin, cells continue to divide throughout life. This process of somatic cell division, during which the nucleus also divides, is known as **mitosis**. During mitosis each chromosome divides into two daughter chromosomes, one of which segregates into each daughter cell. Consequently, the number of chromosomes per nucleus remains unchanged.

Prior to a cell entering mitosis, each chromosome consists of two identical sister chromatids as a result of DNA replication having taken place during the S phase of the cell cycle (p. 39). Mitosis is the process whereby each of these pairs of chromatids separates and disperses into separate daughter cells.

Mitosis is a continuous process that usually lasts 1 to 2 hours, but for descriptive purposes it is convenient to distinguish five distinct stages. These are prophase, prometaphase, metaphase, anaphase, and telophase (Figure 3.13).

Prophase

During the initial stage of **prophase,** the chromosomes condense and the mitotic spindle begins to form. Two **centrioles** form in each cell, from which **microtubules** radiate as the centrioles move toward opposite poles of the cell.

Prometaphase

During **prometaphase** the nuclear membrane begins to disintegrate, allowing the chromosomes to spread around the cell. Each chromosome becomes attached at its centromere to a microtubule of the mitotic spindle.

Metaphase

In **metaphase** the chromosomes become aligned along the equatorial plane or plate of the cell, where each chromosome is attached to the centriole by a microtubule forming the mature spindle. At this point the chromosomes are maximally contracted and, therefore, most easily visible. Each chromosome resembles the letter X in shape, as the chromatids of each chromosome have separated

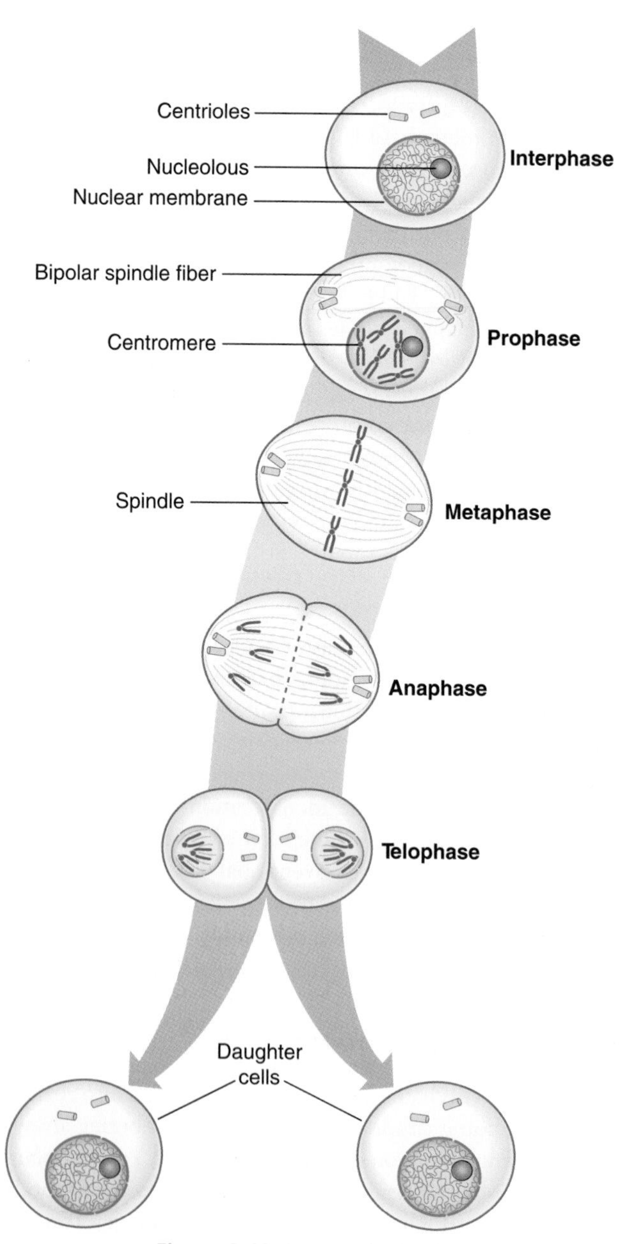

Figure 3.13 Stages of mitosis.

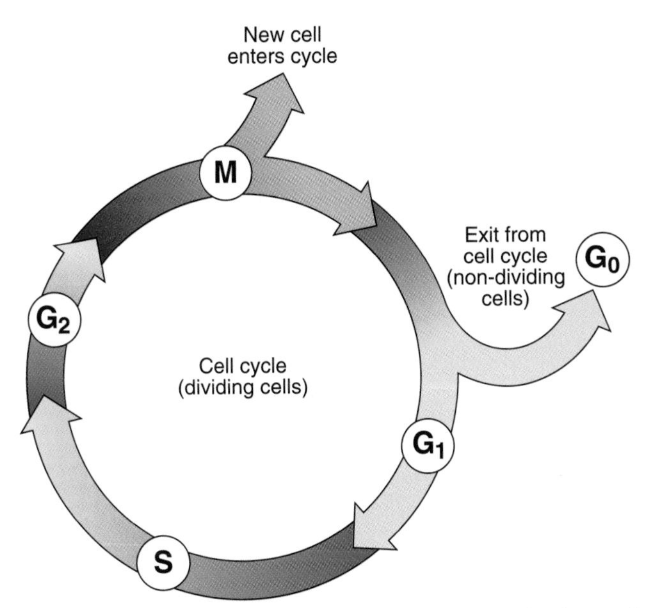

Figure 3.14 Stages of the cell cycle. G_1 and G_2 are the first and second 'resting' stages of interphase. *S* is the stage of DNA replication. *M*, mitosis.

longitudinally but remain attached at the centromere, which has not yet undergone division.

Anaphase

In **anaphase** the centromere of each chromosome divides longitudinally and the two daughter chromatids separate to opposite poles of the cell.

Telophase

By **telophase** the chromatids, which are now independent chromosomes consisting of a single double helix, have separated completely and the two groups of daughter chromosomes each become enveloped in a new nuclear membrane. The cell cytoplasm also separates (cytokinesis), resulting in the formation of two new daughter cells, each of which contains a complete diploid chromosome complement.

The Cell Cycle

The period between successive mitoses is known as the **interphase** of the cell cycle (Figure 3.14). In rapidly dividing cells this lasts for between 16 and 24 hours. Interphase commences with the G_1 (G = gap) phase during which the chromosomes become thin and extended. This phase of the cycle is very variable in length and is responsible for the variation in generation time between different cell populations. Cells that have stopped dividing, such as neurons, usually arrest in this phase and are said to have entered a noncyclic stage known as G_0.

The G_1 phase is followed by the S phase (S = synthesis), when DNA replication occurs and the chromatin of each chromosome is replicated. This results in the formation of two chromatids, giving each chromosome its characteristic X-shaped configuration. The process of DNA replication commences at multiple points on a chromosome (p. 14).

Homologous pairs of chromosomes usually replicate in synchrony. However, one of the X chromosomes is always late in replicating. This is the inactive X chromosome (p. 103) that forms the **sex chromatin** or so-called **Barr body**, which can be visualized during interphase in female somatic cells. This used to be the basis of a rather unsatisfactory means of sex determination based on analysis of cells obtained by scraping the buccal mucosa—a 'buccal smear'.

Interphase is completed by a relatively short G_2 phase during which the chromosomes begin to condense in preparation for the next mitotic division.

Meiosis

Meiosis is the process of nuclear division that occurs during the final stage of gamete formation. Meiosis differs from mitosis in three fundamental ways:

1. Mitosis results in each daughter cell having a diploid chromosome complement (46). During meiosis the diploid count is halved so that each mature gamete receives a haploid complement of 23 chromosomes.
2. Mitosis takes place in somatic cells and during the early cell divisions in gamete formation. Meiosis occurs only at the final division of gamete maturation.
3. Mitosis occurs as a one-step process. Meiosis can be considered as two cell divisions known as meiosis I and meiosis II, each of which can be considered as having prophase, metaphase, anaphase, and telophase stages, as in mitosis (Figure 3.15).

Meiosis I

This is sometimes referred to as the reduction division, because it is during the first meiotic division that the chromosome number is halved.

Prophase I

Chromosomes enter this stage already split longitudinally into two chromatids joined at the centromere. Homologous chromosomes pair and, with the exception of the X and Y chromosomes in male meiosis, exchange of homologous segments occurs between non-sister chromatids; that is, chromatids from each of the pair of homologous chromosomes. This exchange of homologous segments between chromatids occurs as a result of a process known as **crossing over** or **recombination**. The importance of crossing over in linkage analysis and risk calculation is considered later (pp. 136, 345).

During prophase I in the male, pairing occurs between homologous segments of the X and Y chromosomes at the tip of their short arms, with this portion of each chromosome being known as the **pseudoautosomal** region (p. 118).

The prophase stage of meiosis I is relatively lengthy and can be subdivided into five stages.

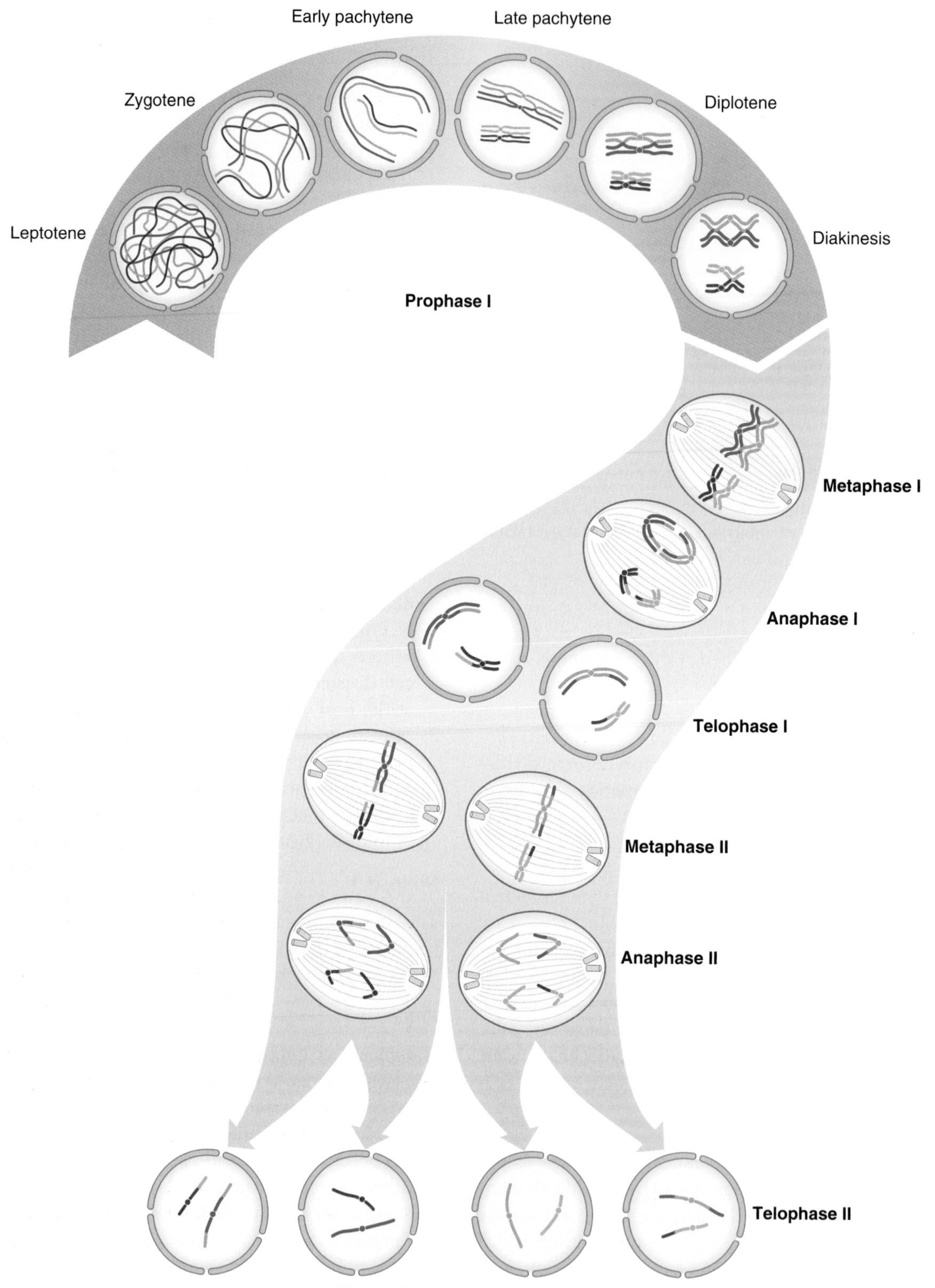

Figure 3.15 Stages of meiosis.

Leptotene. The chromosomes become visible as they start to condense.

Zygotene. Homologous chromosomes align directly opposite each other, a process known as synapsis, and are held together at several points along their length by filamentous structures known as **synaptonemal** complexes.

Pachytene. Each pair of homologous chromosomes, known as a **bivalent**, becomes tightly coiled. Crossing over occurs, during which homologous regions of DNA are exchanged between chromatids.

Diplotene. The homologous recombinant chromosomes now begin to separate but remain attached at the points where crossing over has occurred. These are known as **chiasmata**. On average, small, medium, and large chromosomes have one, two, and three chiasmata, respectively, giving an overall total of approximately 40 recombination events per meiosis per gamete.

Diakinesis. Separation of the homologous chromosome pairs proceeds as the chromosomes become maximally condensed.

Metaphase I

The nuclear membrane disappears and the chromosomes become aligned on the equatorial plane of the cell where they have become attached to the spindle, as in metaphase of mitosis.

Anaphase I

The chromosomes now separate to opposite poles of the cell as the spindle contracts.

Telophase I

Each set of haploid chromosomes has now separated completely to opposite ends of the cell, which cleaves into two new daughter gametes, so-called **secondary spermatocytes** or **oocytes**.

Meiosis II

This is essentially the same as an ordinary mitotic division. Each chromosome, which exists as a pair of chromatids, becomes aligned along the equatorial plane and then splits longitudinally, leading to the formation of two new daughter gametes, known as spermatids or ova.

The Consequences of Meiosis

When considered in terms of reproduction and the maintenance of the species, meiosis achieves two major objectives. First, it facilitates halving of the diploid number of chromosomes so that each child receives half of its chromosome complement from each parent. Second, it provides an extraordinary potential for generating genetic diversity. This is achieved in two ways:

1. When the bivalents separate during prophase of meiosis I, they do so independently of one another. This is consistent with Mendel's third law (p. 5). Consequently each gamete receives a selection of parental chromosomes. The likelihood that any two gametes from an individual will contain exactly the same chromosomes is 1 in 2^{23}, or approximately 1 in 8 million.
2. As a result of crossing over, each chromatid usually contains portions of DNA derived from both parental homologous chromosomes. A large chromosome typically consists of three or more segments of alternating parental origin. The ensuing probability that any two gametes will have an identical genome is therefore infinitesimally small. This dispersion of DNA into different gametes is sometimes referred to as **gene shuffling**.

Table 3.3 Differences in Gametogenesis in Males and Females

	Males	Females
Commences	Puberty	Early embryonic life
Duration	60–65 days	10–50 years
Numbers of mitoses in gamete formation	30–500	20–30
Gamete production per meiosis	4 spermatids	1 ovum + 3 polar bodies
Gamete production	100–200 million per ejaculate	1 ovum per menstrual cycle

Gametogenesis

The process of gametogenesis shows fundamental differences in males and females (Table 3.3). These have quite distinct clinical consequences if errors occur.

Oogenesis

Mature ova develop from oogonia by a complex series of intermediate steps. Oogonia themselves originate from primordial germ cells by a process involving 20 to 30 mitotic divisions that occur during the first few months of embryonic life. By the completion of embryogenesis at 3 months of intrauterine life, the oogonia have begun to mature into primary oocytes that start to undergo meiosis. At birth all of the primary oocytes have entered a phase of maturation arrest, known as **dictyotene**, in which they remain suspended until meiosis I is completed at the time of ovulation, when a single secondary oocyte is formed. This receives most of the cytoplasm. The other daughter cell from the first meiotic division consists largely of a nucleus and is known as a polar body. Meiosis II then commences, during which fertilization can occur. This second meiotic division results in the formation of a further polar body (Figure 3.16).

It is probable that the very lengthy interval between the onset of meiosis and its eventual completion, up to 50 years later, accounts for the well documented increased incidence of chromosome abnormalities in the offspring of older mothers (p. 44). The accumulating effects of 'wear and tear' on the primary oocyte during the dictyotene phase probably damage the cell's spindle formation and repair mechanisms, thereby predisposing to non-disjunction (p. 17).

Spermatogenesis

In contrast, spermatogenesis is a relatively rapid process with an average duration of 60 to 65 days. At puberty spermatogonia, which will already have undergone approximately 30 mitotic divisions, begin to mature into primary spermatocytes which enter meiosis I and emerge as haploid secondary spermatocytes. These then undergo the second meiotic division to form spermatids, which in turn develop without any subsequent cell division into

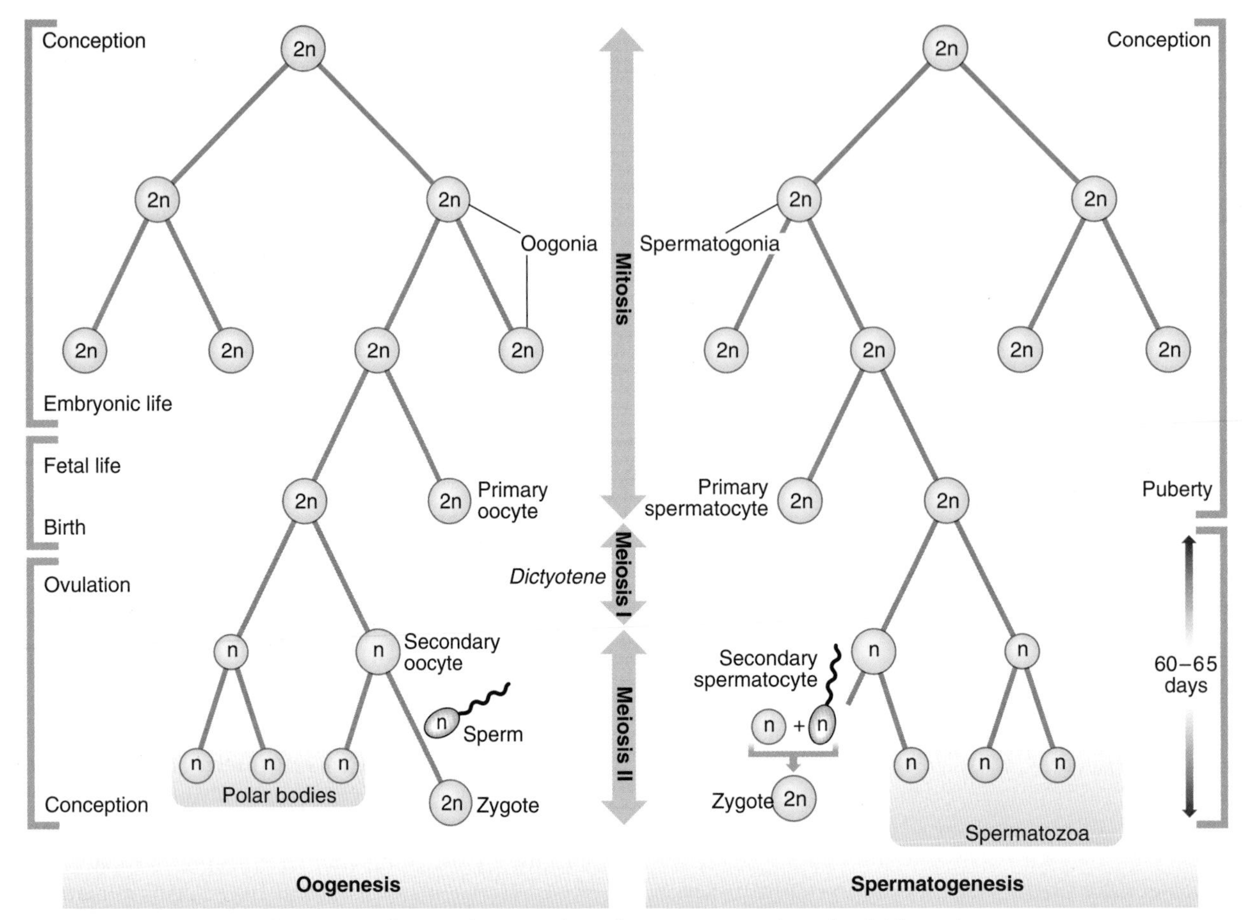

Figure 3.16 Stages of oogenesis and spermatogenesis. *n*, haploid number.

mature spermatozoa, of which 100 to 200 million are present in each ejaculate.

Spermatogenesis is a continuous process involving many mitotic divisions, possibly as many as 20 to 25 per annum, so that mature spermatozoa produced by a man of 50 years or older could well have undergone several hundred mitotic divisions. The observed paternal age effect for new dominant mutations (p. 113) is consistent with the concept that many mutations arise as a consequence of DNA copy errors occurring during mitosis.

Chromosome Abnormalities

Specific disorders caused by chromosome abnormalities are considered in Chapter 18. In this section, discussion is restricted to a review of the different types of abnormality that may occur. These can be divided into numerical and structural, with a third category consisting of different chromosome constitutions in two or more cell lines (Box 3.1).

Numerical Abnormalities

Numerical abnormalities involve the loss or gain of one or more chromosomes, referred to as **aneuploidy**, or the

Box 3.1 Types of Chromosome Abnormality

Numerical
Aneuploidy
Monosomy
Trisomy
Tetrasomy
Polyploidy
Triploidy
Tetraploidy

Structural
Translocations
Reciprocal
Robertsonian
Deletions
Insertions
Inversions
Paracentric
Pericentric
Rings
Isochromosomes

Different Cell Lines (Mixoploidy)
Mosaicism
Chimerism

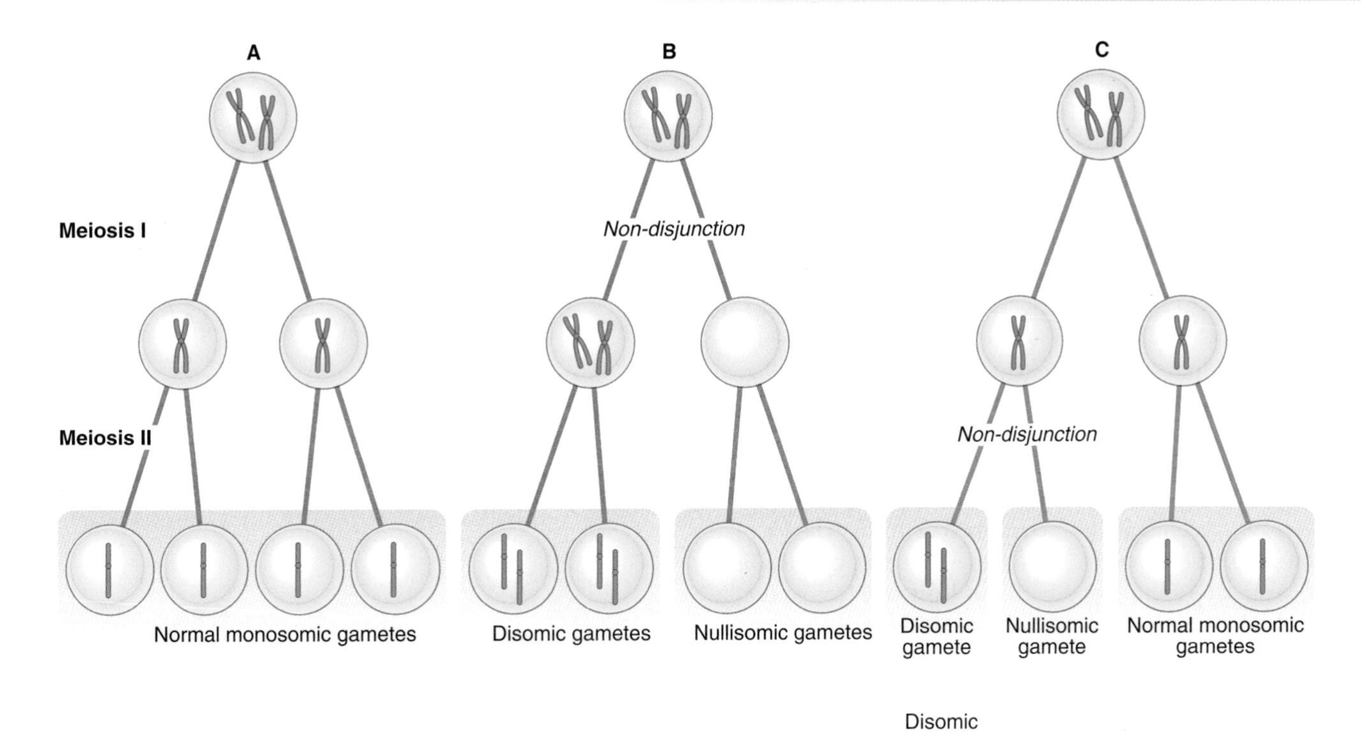

Figure 3.17 Segregation at meiosis of a single pair of chromosomes in, **A,** normal meiosis, **B,** non-disjunction in meiosis I, and, **C,** non-disjunction in meiosis II.

addition of one or more complete haploid complements, known as **polyploidy**. Loss of a single chromosome results in **monosomy**. Gain of one or two homologous chromosomes is referred to as **trisomy** or **tetrasomy**, respectively.

Trisomy

The presence of an extra chromosome is referred to as **trisomy**. Most cases of Down syndrome are due to the presence of an additional number 21 chromosome; hence, Down syndrome is often known as trisomy 21. Other autosomal trisomies compatible with survival to term are Patau syndrome (trisomy 13) (p. 275) and Edwards syndrome (trisomy 18) (p. 275). Most other autosomal trisomies result in early pregnancy loss, with trisomy 16 being a particularly common finding in first-trimester spontaneous miscarriages. The presence of an additional sex chromosome (X or Y) has only mild phenotypic effects (p. 104).

Trisomy 21 is usually caused by failure of separation of one of the pairs of homologous chromosomes during anaphase of maternal meiosis I. This failure of the bivalent to separate is called **non-disjunction**. Less often, trisomy can be caused by non-disjunction occurring during meiosis II when a pair of sister chromatids fails to separate. Either way the gamete receives two homologous chromosomes (**disomy**); if subsequent fertilization occurs, a trisomic conceptus results (Figure 3.17).

The Origin of Non-disjunction

The consequences of non-disjunction in meiosis I and meiosis II differ in the chromosomes found in the gamete. An error in meiosis I leads to the gamete containing both homologs of one chromosome pair. In contrast, non-disjunction in meiosis II results in the gamete receiving two copies of one of the homologs of the chromosome pair. Studies using DNA markers have shown that most children with an autosomal trisomy have inherited their additional chromosome as a result of non-disjunction occurring during one of the maternal meiotic divisions (Table 3.4).

Non-disjunction can also occur during an early mitotic division in the developing zygote. This results in the presence of two or more different cell lines, a phenomenon known as **mosaicism** (p. 50).

The Cause of Non-disjunction

The cause of non-disjunction is uncertain. The most favored explanation is that of an aging effect on the primary oocyte, which can remain in a state of suspended inactivity for up

Table 3.4 Parental Origin of Meiotic Error Leading to Aneuploidy

Chromosome Abnormality	Paternal (%)	Maternal (%)
Trisomy 13	15	85
Trisomy 18	10	90
Trisomy 21	5	95
45,X	80	20
47,XXX	5	95
47,XXY	45	55
47,XYY	100	0

to 50 years (p. 41). This is based on the well-documented association between advancing maternal age and increased incidence of Down syndrome in offspring (see Table 18.4; see p. 275). A maternal age effect has also been noted for trisomies 13 and 18.

It is not known how or why advancing maternal age predisposes to non-disjunction, although research has shown that absence of recombination in prophase of meiosis I predisposes to subsequent non-disjunction. This is not surprising, as the chiasmata that are formed after recombination are responsible for holding each pair of homologous chromosomes together until subsequent separation occurs in diakinesis. Thus failure of chiasmata formation could allow each pair of homologs to separate prematurely and then segregate randomly to daughter cells. In the female, however, recombination occurs before birth whereas the non-disjunctional event occurs any time between 15 and 50 years later. This suggests that at least two factors can be involved in causing non-disjunction: an absence of recombination between homologous chromosomes in the fetal ovary, and an abnormality in spindle formation many years later.

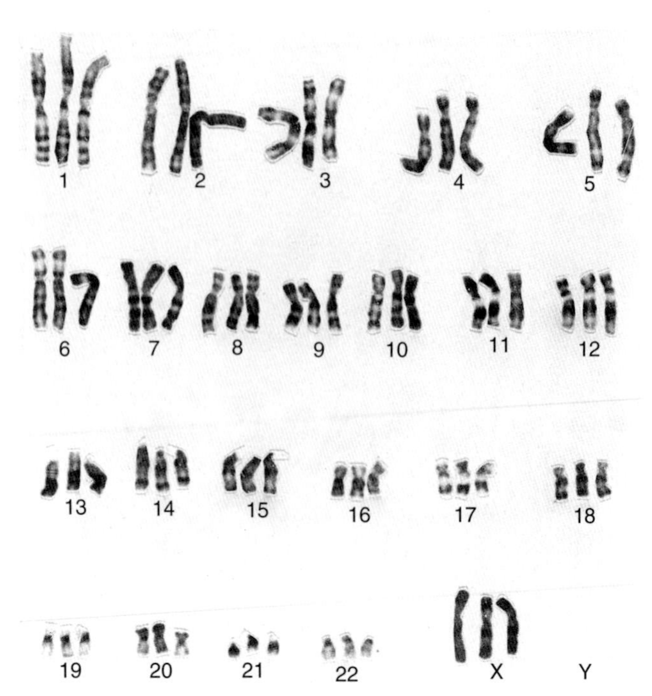

Figure 3.18 Karyotype from products of conception of a spontaneous miscarriage showing triploidy.

Monosomy

The absence of a single chromosome is referred to as **monosomy**. Monosomy for an autosome is almost always incompatible with survival to term. Lack of contribution of an X or a Y chromosome results in a 45,X karyotype, which causes the condition known as Turner syndrome (p. 277).

As with trisomy, monosomy can result from non-disjunction in meiosis. If one gamete receives two copies of a homologous chromosome (**disomy**), the other corresponding daughter gamete will have no copy of the same chromosome (**nullisomy**). Monosomy can also be caused by loss of a chromosome as it moves to the pole of the cell during anaphase, an event known as **anaphase lag**.

Polyploidy

Polyploid cells contain multiples of the haploid number of chromosomes such as 69, **triploidy**, or 92, **tetraploidy**. In humans, triploidy is found relatively often in material grown from spontaneous miscarriages, but survival beyond mid-pregnancy is rare. Only a few triploid live births have been described and all died soon after birth.

Triploidy can be caused by failure of a maturation meiotic division in an ovum or sperm, leading, for example, to retention of a polar body or to the formation of a diploid sperm. Alternatively it can be caused by fertilization of an ovum by two sperm: this is known as **dispermy**. When triploidy results from the presence of an additional set of paternal chromosomes, the placenta is usually swollen with what are known as hydatidiform changes (p. 101). In contrast, when triploidy results from an additional set of maternal chromosomes, the placenta is usually small. Triploidy usually results in early spontaneous miscarriage (Figure 3.18). The differences between triploidy due to an additional set of **paternal** chromosomes or **maternal** chromosomes provide evidence for important 'epigenetic' and 'parent of origin' effects with respect to the human genome. These are discussed in more detail in Chapter 6.

Structural Abnormalities

Structural chromosome rearrangements result from chromosome breakage with subsequent reunion in a different configuration. They can be balanced or unbalanced. In balanced rearrangements the chromosome complement is complete, with no loss or gain of genetic material. Consequently, balanced rearrangements are generally harmless with the exception of rare cases in which one of the breakpoints damages an important functional gene. However, carriers of balanced rearrangements are often at risk of producing children with an unbalanced chromosomal complement.

When a chromosome rearrangement is unbalanced the chromosomal complement contains an incorrect amount of chromosome material and the clinical effects are usually serious.

Translocations

A **translocation** refers to the transfer of genetic material from one chromosome to another. A reciprocal translocation is formed when a break occurs in each of two chromosomes with the segments being exchanged to form two new derivative chromosomes. A Robertsonian translocation is a particular type of reciprocal translocation in which the breakpoints are located at, or close to, the centromeres of two acrocentric chromosomes (Figure 3.19).

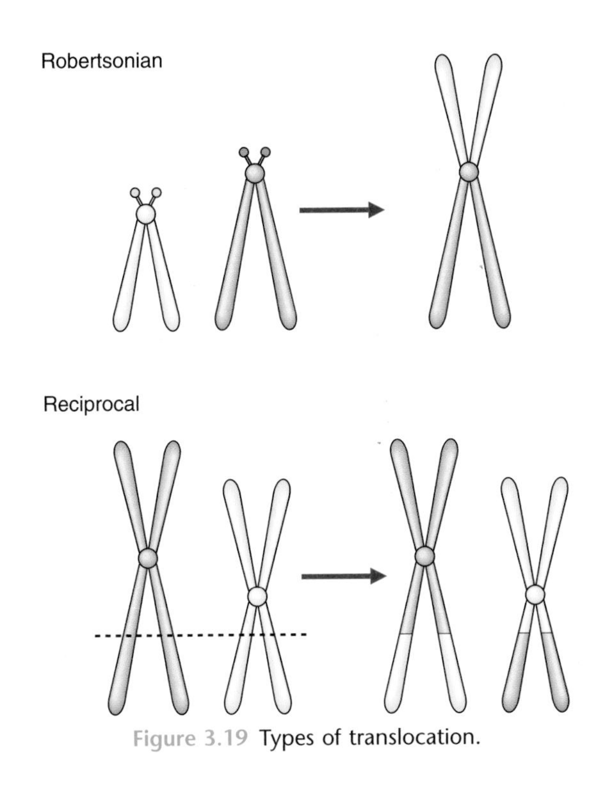

Figure 3.19 Types of translocation.

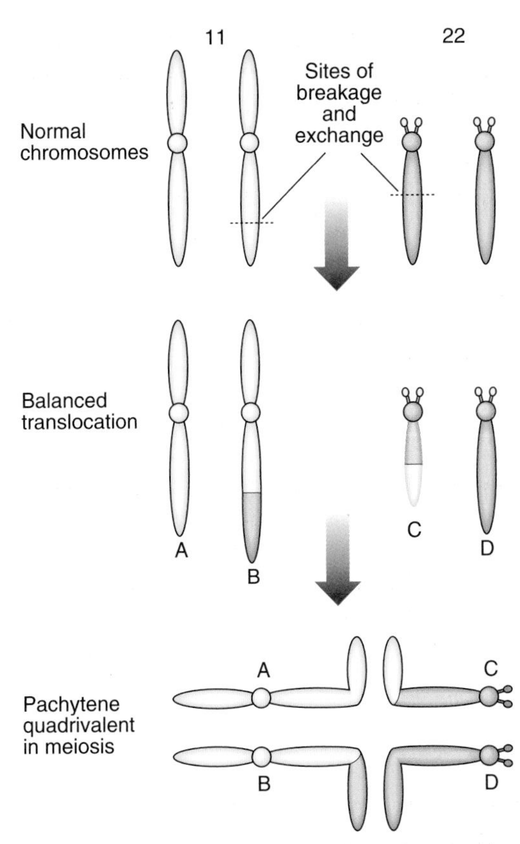

Figure 3.20 How a balanced reciprocal translocation involving chromosomes 11 and 22 leads to the formation of a quadrivalent at pachytene in meiosis I. The quadrivalent is formed to maintain homologous pairing.

Reciprocal Translocations

A reciprocal translocation involves breakage of at least two chromosomes with exchange of the fragments. Usually the chromosome number remains at 46 and, if the exchanged fragments are of roughly equal size, a reciprocal translocation can be identified only by detailed chromosomal banding studies or FISH (see Figure 3.9). In general, reciprocal translocations are unique to a particular family, although, for reasons that are unknown, a particular balanced reciprocal translocation involving the long arms of chromosomes 11 and 22 is relatively common. The overall incidence of reciprocal translocations in the general population is approximately 1 in 500.

Segregation at Meiosis. The importance of balanced reciprocal translocations lies in their behavior at meiosis, when they can segregate to generate significant chromosome imbalance. This can lead to early pregnancy loss or to the birth of an infant with multiple abnormalities. Problems arise at meiosis because the chromosomes involved in the translocation cannot pair normally to form bivalents. Instead they form a cluster known as a **pachytene quadrivalent** (Figure 3.20). The key point to note is that each chromosome aligns with homologous material in the quadrivalent.

2:2 Segregation. When the constituent chromosomes in the quadrivalent separate during the later stages of meiosis I, they can do so in several different ways (Table 3.5). If alternate chromosomes segregate to each gamete, the gamete will carry a normal or balanced haploid complement (Figure 3.21) and with fertilization the embryo will either have normal chromosomes or carry the balanced rearrangement. If, however, adjacent chromosomes segregate together, this will invariably result in the gamete acquiring an unbalanced chromosome complement. For example, in Figure 3.20, if the gamete inherits the normal number 11 chromosome (A) and the derivative number 22 chromosome (C), then fertilization will result in an embryo with monosomy for the distal long arm of chromosome 22 and trisomy for the distal long arm of chromosome 11.

3:1 Segregation. Another possibility is that three chromosomes segregate to one gamete with only one chromosome in the other gamete. If, for example, in Figure 3.20 chromosomes 11 (A), 22 (D) and the derivative 22 (C) segregate together to a gamete that is subsequently fertilized, this will result in the embryo being trisomic for the material present in the derivative 22 chromosome. This is sometimes referred to as **tertiary trisomy.** Experience has shown that, with this particular reciprocal translocation, tertiary trisomy for the derivative 22 chromosome is the only viable unbalanced product. All other patterns of malsegregation lead to early pregnancy loss. Unfortunately, tertiary trisomy for the derivative 22 chromosome is a serious condition in which affected children have multiple congenital abnormalities and severe learning difficulties.

Table 3.5 Patterns of Segregation of a Reciprocal Translocation (see Figures 3.20 and 3.21)

Pattern of Segregation	Segregating Chromosomes	Chromosome Constitution in Gamete
2:2		
Alternate	A + D	Normal
	B + C	Balanced translocation
Adjacent-1 (non-homologous centromeres segregate together)	A + C or B + D	Unbalanced, leading to a combination of partial monosomy and partial trisomy in the zygote
Adjacent-2 (homologous centromeres segregate together)	A + B or C + D	
3:1		
Three chromosomes	A + B + C	Unbalanced, leading to trisomy in the zygote
	A + B + D	
	A + C + D	
	B + C + D	
One chromosome	A	Unbalanced, leading to monosomy in the zygote
	B	
	C	
	D	

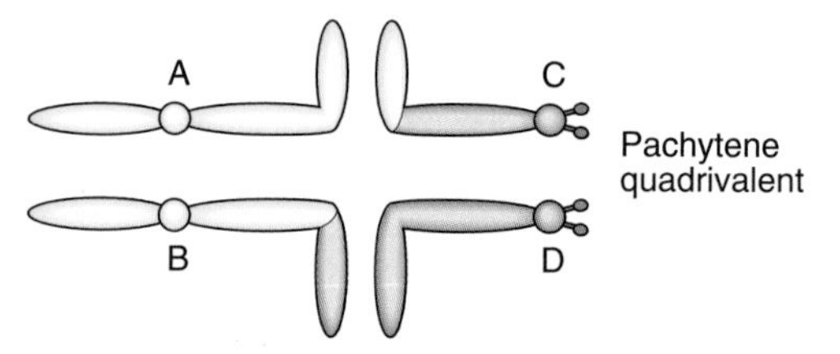

1 Alternate segregation yields normal or balanced haploid complement

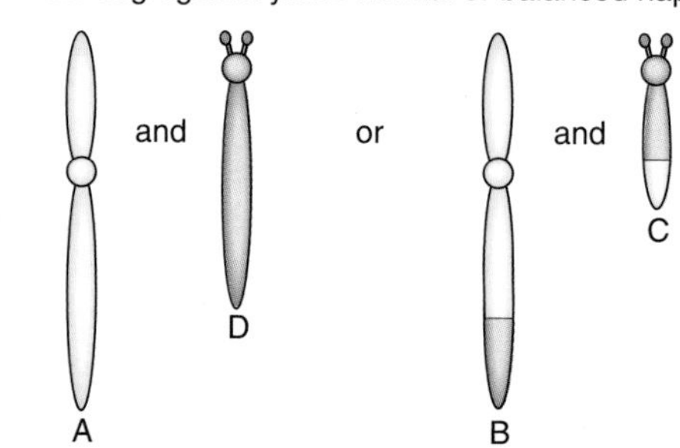

2 Adjacent–1 segregation yields unbalanced haploid complement

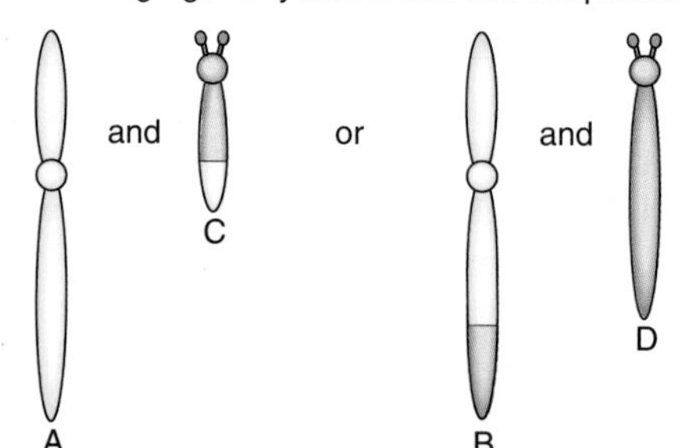

3 Adjacent–2 segregation yields unbalanced haploid complement

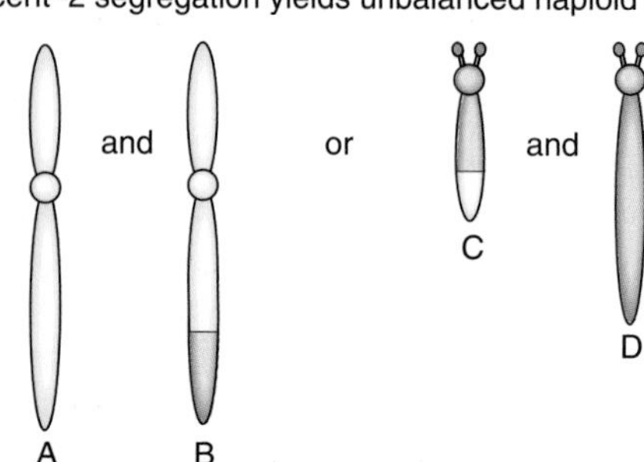

Figure 3.21 The different patterns of 2:2 segregation that can occur from the quadrivalent shown in Figure 3.20. (See Table 3.5.)

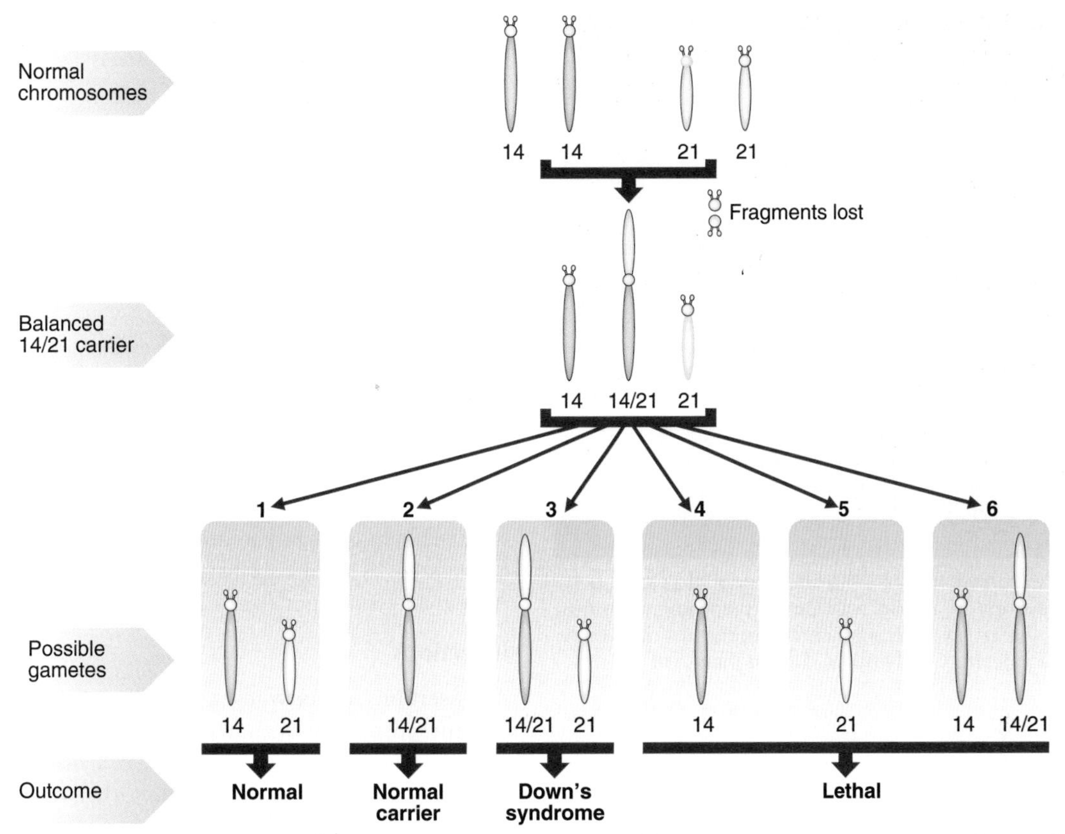

Figure 3.22 Formation of a 14q21q Robertsonian translocation and the possible gamete chromosome patterns that can be produced at meiosis.

Risks in Reciprocal Translocations. When counseling a carrier of a balanced translocation it is necessary to consider the particular rearrangement to determine whether it could result in the birth of an abnormal baby. This risk is usually somewhere between 1% and 10%. For carriers of the 11;22 translocation discussed, the risk has been shown to be 5%.

Robertsonian Translocations

A Robertsonian translocation results from the breakage of two acrocentric chromosomes (numbers 13, 14, 15, 21, and 22) at or close to their centromeres, with subsequent fusion of their long arms (see Figure 3.19). This is also referred to as **centric fusion**. The short arms of each chromosome are lost, this being of no clinical importance as they contain genes only for ribosomal RNA, for which there are multiple copies on the various other acrocentric chromosomes. The total chromosome number is reduced to 45. Because there is no loss or gain of important genetic material, this is a functionally balanced rearrangement. The overall incidence of Robertsonian translocations in the general population is approximately 1 in 1000, with by far the most common being fusion of the long arms of chromosomes 13 and 14 (13q14q).

Segregation at Meiosis. As with reciprocal translocations, the importance of Robertsonian translocations lies in their behavior at meiosis. For example, a carrier of a 14q21q translocation can produce gametes with (Figure 3.22):

1. A normal chromosome complement (i.e., a normal 14 and a normal 21).
2. A balanced chromosome complement (i.e., a 14q21q translocation chromosome).
3. An unbalanced chromosome complement possessing both the translocation chromosome and a normal 21. This will result in the fertilized embryo having Down syndrome.
4. An unbalanced chromosome complement with a normal 14 and a missing 21.
5. An unbalanced chromosome complement with a normal 21 and a missing 14.
6. An unbalanced chromosome complement with the translocation chromosome and a normal 14 chromosome.

The last three combinations will result in zygotes with monosomy 21, monosomy 14, and trisomy 14, respectively. All of these combinations are incompatible with survival beyond early pregnancy.

Translocation Down Syndrome. The major practical importance of Robertsonian translocations is that they can predispose to the birth of babies with Down syndrome as a result of the embryo inheriting two normal number 21 chromosomes (one from each parent) plus a translocation chromosome involving a number 21 chromosome

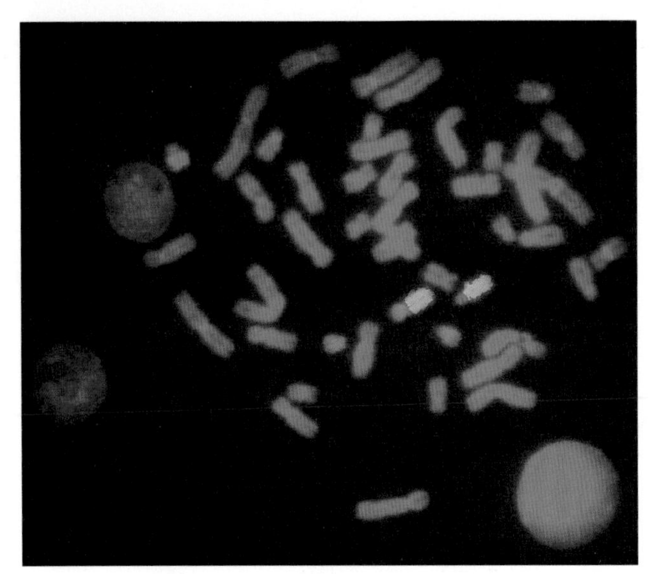

Figure 3.23 Chromosome painting showing a 14q21q Robertsonian translocation in a child with Down syndrome. Chromosome 21 is shown in *blue* and chromosome 14 in *yellow*. (Courtesy Meg Heath, City Hospital, Nottingham, UK.)

(Figure 3.23). The clinical consequences are exactly the same as those seen in pure trisomy 21. However, unlike trisomy 21, the parents of a child with translocation Down syndrome have a relatively high risk of having further affected children if one of them carries the rearrangement in a balanced form.

Consequently, the importance of performing a chromosome analysis in a child with Down syndrome lies not only in confirmation of the diagnosis, but also in identification of those children with a translocation. In roughly two-thirds of these latter children with Down syndrome, the translocation will have occurred as a new (de novo) event in the child, but in the remaining one-third one of the parents will be a carrier. Other relatives might also be carriers. Therefore it is regarded as essential that efforts are made to identify all adult translocation carriers in a family so that they can be alerted to possible risks to future offspring. This is sometimes referred to as translocation **tracing**, or 'chasing'.

Risks in Robertsonian translocations Studies have shown that the female carrier of either a 13q21q or a 14q21q Robertsonian translocation runs a risk of approximately 10% for having a baby with Down syndrome, whereas for male carriers the risk is 1% to 3%. It is worth sparing a thought for the unfortunate carrier of a 21q21q Robertsonian translocation. All gametes will be either nullisomic or disomic for chromosome 21. Consequently, all pregnancies will end either in spontaneous miscarriage or in the birth of a child with Down syndrome. This is one of the very rare situations in which offspring are at a risk of greater than 50% for having an abnormality. Other examples are parents who are both heterozygous for the same autosomal dominant disorder (p. 113), and parents who are both homozygous for the **same gene mutation** causing an autosomal recessive disorder, such as sensorineural deafness.

Deletions

A **deletion** involves loss of part of a chromosome and results in monosomy for that segment of the chromosome. A very large deletion is usually incompatible with survival to term, and as a general rule any deletion resulting in loss of more than 2% of the total haploid genome will have a lethal outcome.

Deletions are now recognized as existing at two levels. A 'large' chromosomal deletion can be visualized under the light microscope. Such deletion syndromes include Wolf-Hirschhorn and cri du chat, which involve loss of material from the short arms of chromosomes 4 and 5, respectively (p. 280). Submicroscopic microdeletions were identified with the help of high-resolution prometaphase cytogenetics augmented by FISH studies and include Prader-Willi and Angelman syndromes (pp. 122, 123).

Insertions

An **insertion** occurs when a segment of one chromosome becomes inserted into another chromosome. If the inserted material has moved from elsewhere in another chromosome then the karyotype is balanced. Otherwise an insertion causes an unbalanced chromosome complement. Carriers of a balanced deletion–insertion rearrangement are at a 50% risk of producing unbalanced gametes, as random chromosome segregation at meiosis will result in 50% of the gametes inheriting either the deletion or the insertion, but not both.

Inversions

An inversion is a two-break rearrangement involving a single chromosome in which a segment is reversed in position (i.e., inverted). If the inversion segment involves the centromere it is termed a **pericentric inversion** (Figure 3.24*A*). If it involves only one arm of the chromosome it is known as a **paracentric inversion** (Figure 3.24*B*).

Inversions are balanced rearrangements that rarely cause problems in carriers unless one of the breakpoints has disrupted an important gene. A pericentric inversion involving chromosome number 9 occurs as a common structural variant or polymorphism, also known as a **heteromorphism**, and is not thought to be of any functional importance. However, other inversions, although not causing any clinical problems in balanced carriers, can lead to significant chromosome imbalance in offspring, with important clinical consequences.

Segregation at Meiosis

Pericentric Inversions. An individual who carries a pericentric inversion can produce unbalanced gametes if a crossover occurs within the inversion segment during meiosis I, when an inversion loop forms as the chromosomes attempt to maintain homologous pairing at synapsis. For a pericentric inversion, a crossover within the loop will result in two complementary recombinant chromosomes, one with

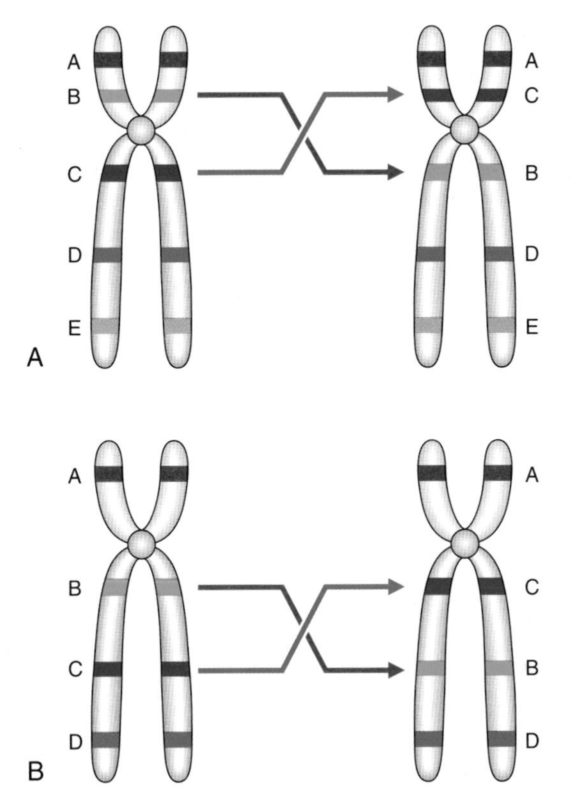

Figure 3.24 **A,** Pericentric and, **B,** paracentric inversions. (Courtesy Dr. J. Delhanty, Galton Laboratory, London.)

duplication of the distal non-inverted segment and deletion of the other end of the chromosome, and the other having the opposite arrangement (Figure 3.25*A*).

If a pericentric inversion involves only a small proportion of the total length of a chromosome then, in the event of crossing over within the loop, the duplicated and deleted segments will be relatively large. The larger these are, the more likely it is that their effects on the embryo will be so severe that miscarriage ensues. For a large pericentric inversion, the duplicated and deleted segments will be relatively small so that survival to term and beyond becomes more likely. Thus, in general, the larger the size of a pericentric inversion the more likely it becomes that it will result in the birth of an abnormal infant.

The pooled results of several studies have shown that a carrier of a balanced pericentric inversion runs a risk of approximately 5% to 10% for having a child with viable imbalance if that inversion has already resulted in the birth of an abnormal baby. The risk is nearer 1% if the inversion has been ascertained because of a history of recurrent miscarriage.

Paracentric Inversions. If a crossover occurs in the inverted segment of a paracentric inversion, this will result in recombinant chromosomes that are either acentric or dicentric (Figure 3.25*B*). Acentric chromosomes, which strictly speaking should be known as chromosomal **fragments**, cannot undergo mitotic division, so that survival of an embryo with such a rearrangement is extremely uncommon.

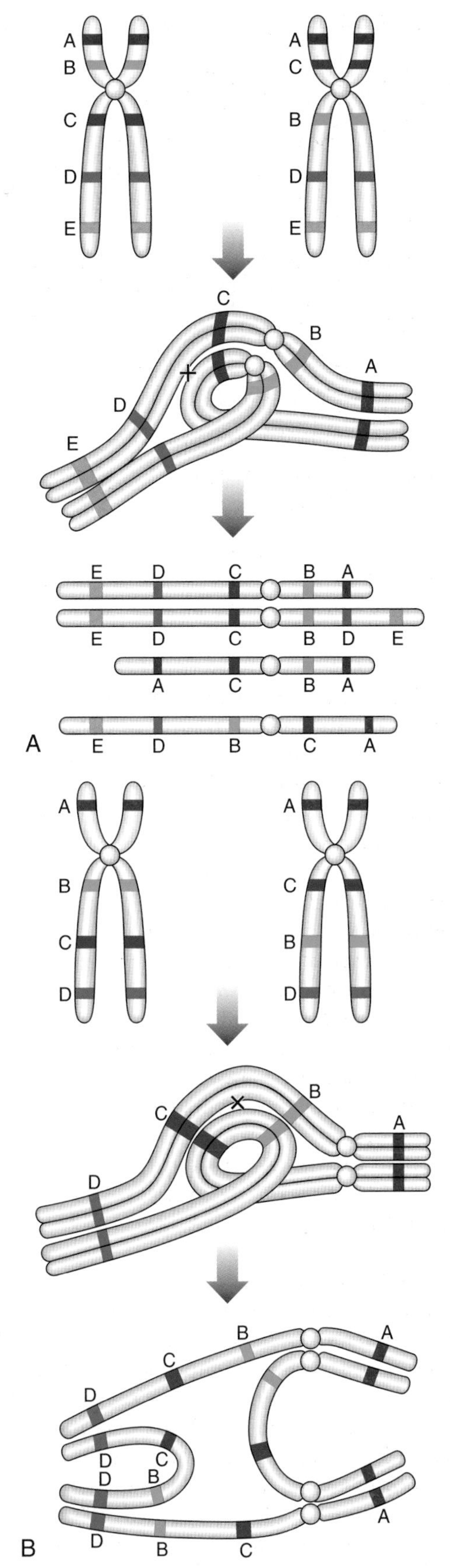

Figure 3.25 Mechanism of production of recombinant unbalanced chromosomes from, **A,** pericentric and, **B,** paracentric inversions by crossing over in an inversion loop. (Courtesy Dr. J. Delhanty, Galton Laboratory, London.)

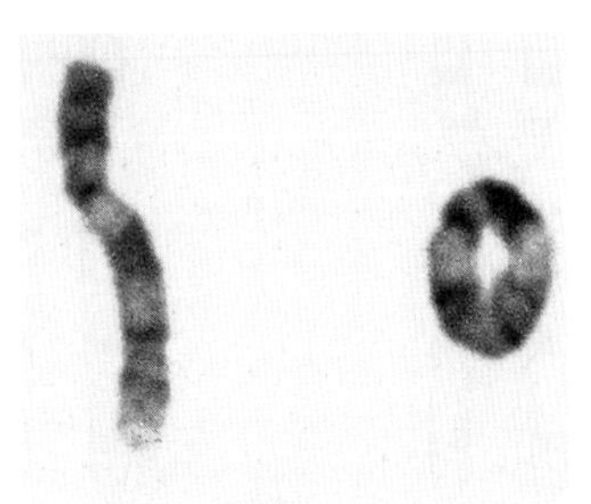

Figure 3.26 Partial karyotype showing a ring chromosome 9. (Courtesy Meg Heath, City Hospital, Nottingham.)

Dicentric chromosomes are inherently unstable during cell division and are, therefore, also unlikely to be compatible with survival of the embryo. Thus, overall, the likelihood that a balanced parental paracentric inversion will result in the birth of an abnormal baby is extremely low.

Ring Chromosomes

A **ring chromosome** is formed when a break occurs on each arm of a chromosome leaving two 'sticky' ends on the central portion that reunite as a ring (Figure 3.26). The two distal chromosomal fragments are lost so that, if the involved chromosome is an autosome, the effects are usually serious.

Ring chromosomes are often unstable in mitosis so that it is common to find a ring chromosome in only a proportion of cells. The other cells in the individual are usually monosomic because of the absence of the ring chromosome.

Isochromosomes

An isochromosome shows loss of one arm with duplication of the other. The most probable explanation for the formation of an isochromosome is that the centromere has divided transversely rather than longitudinally. The most commonly encountered isochromosome is that which consists of two long arms of the X chromosome. This accounts for up to 15% of all cases of Turner syndrome (p. 277).

Mosaicism and Chimerism (Mixoploidy)

Mosaicism

Mosaicism can be defined as the presence in an individual, or in a tissue, of two or more cell lines that differ in their genetic constitution but are derived from a single zygote, that is, they have the same genetic origin. Chromosome mosaicism usually results from non-disjunction in an early embryonic mitotic division with the persistence of more than one cell line. If, for example, the two chromatids of a number 21 chromosome failed to separate at the second mitotic division in a human zygote (Figure 3.27), this would result in the four-cell zygote having two cells with 46 chromosomes, one cell with 47 chromosomes (trisomy 21), and one cell with 45 chromosomes (monosomy 21). The ensuing cell line with 45 chromosomes would probably not survive, so that the resulting embryo would be expected to show approximately 33% mosaicism for trisomy 21. Mosaicism accounts for 1% to 2% of all clinically recognized cases of Down syndrome.

Mosaicism can also exist at a molecular level if a new mutation arises in a somatic or early germline cell division (p. 120). The possibility of germline or gonadal mosaicism is a particular concern when counseling the parents of a child in whom a condition such as Duchenne muscular dystrophy (p. 307) is an isolated case.

Chimerism

Chimerism can be defined as the presence in an individual of two or more genetically distinct cell lines derived from more than one zygote; that is, they have a different genetic origin. The word *chimera* is derived from the mythological Greek monster that had the head of a lion, the body of a goat and the tail of a dragon. Human chimeras are of two kinds: dispermic chimeras and blood chimeras.

Dispermic Chimeras

These are the result of double fertilization whereby two genetically different sperm fertilize two ova and the resulting two zygotes fuse to form one embryo. If the two zygotes are of different sex, the chimeric embryo can develop into an individual with true hermaphroditism (p. 287) and an XX/XY karyotype. Mouse chimeras of this type can now be produced experimentally in the laboratory to facilitate the study of gene transfer.

Blood Chimeras

Blood chimeras result from an exchange of cells, via the placenta, between non-identical twins in utero. For example, 90% of one twin's cells can have an XY karyotype with red blood cells showing predominantly blood group B, whereas 90% of the cells of the other twin can have an XX karyotype

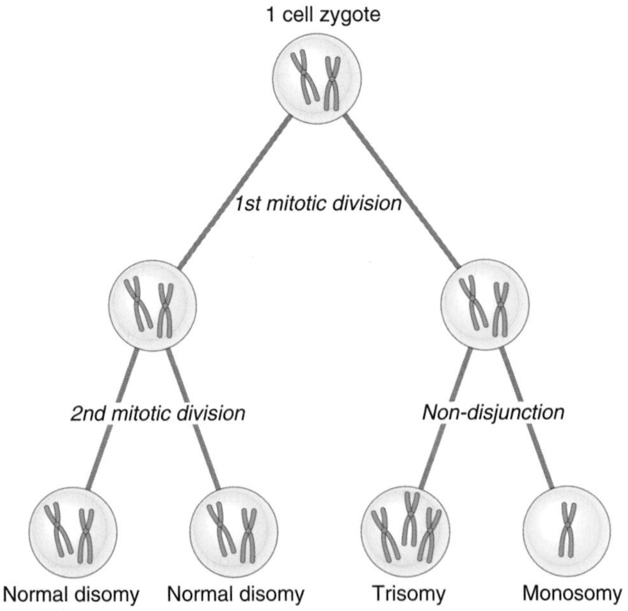

Figure 3.27 Generation of somatic mosaicism caused by mitotic non-disjunction.

with red blood cells showing predominantly blood group A. It has long been recognized that, when twin calves of opposite sex are born, the female can have ambiguous genitalia. It is now thought that this is because of gonadal chimerism in the female calves, which are known as **freemartins.**

FURTHER READING

Barch MJ, Knutsen T, Spurbeck JL, eds 1997 The AGT cytogenetics laboratory manual, 3rd ed. Philadelphia: Lippincott-Raven.

A large multiauthor laboratory handbook produced by the Association of Genetic Technologists.

Gersen SL, Keagle MB, eds 2011 The principles of clinical cytogenetics, 3rd ed. Totowa, NJ: Humana Press

A detailed multiauthor guide to all aspects of laboratory and clinical cytogenetics.

Shaffer LG, Slovak ML, Campbell LJ, eds 2009 An international system for human cytogenetic nomenclature. Basel: Karger

A report giving details of how chromosome abnormalities should be described.

Rooney DE, Czepulkowski BH 1997 Human chromosome preparation. Essential techniques. Chichester, UK: John Wiley

A laboratory handbook describing the different methods available for chromosome analysis.

Speicher MR, Carter NP 2006 The new cytogenetics: blurring the boundaries with molecular biology. Nat Rev Gen 6:782–792

A review of the exciting advances in FISH and array-based techniques.

Therman E, Susman M 1993 Human chromosomes. Structure, behavior and effects, 3rd ed. New York: Springer

A useful and comprehensive introduction to human cytogenetics.

Tjio JH, Levan A 1956 The chromosome number of man. Hereditas 42:1–6

A landmark paper that described a reliable method for studying human chromosomes and gave birth to the subject of clinical cytogenetics.

Website

National Center for Biotechnology Information. Microarrays: chipping away at the mysteries of science and medicine. Online. http://www.ncbi.nlm.nih.gov/About/primer/microarrays.html

ELEMENTS

1 The normal human karyotype is made up of 46 chromosomes consisting of 22 pairs of autosomes and a pair of sex chromosomes, XX in the female and XY in the male.

2 Each chromosome consists of a short (p) and long (q) arm joined at the centromere. Chromosomes are analyzed using cultured cells, and specific banding patterns can be identified by means of special staining techniques. Molecular cytogenetic techniques, such as fluorescence in-situ hybridization (FISH) and array CGH can be used to detect and characterize subtle chromosome abnormalities.

3 During mitosis in somatic cell division the two sister chromatids of each chromosome separate, with one chromatid passing to each daughter cell. During meiosis, which occurs during the final stage of gametogenesis, homologous chromosomes pair, exchange segments, and then segregate independently to the mature daughter gametes.

4 Chromosome abnormalities can be structural or numerical. Numerical abnormalities include trisomy and polyploidy. In trisomy a single extra chromosome is present, usually as a result of non-disjunction in the first or second meiotic division. In polyploidy, three or more complete haploid sets are present instead of the usual diploid complement.

5 Structural abnormalities include translocations, inversions, insertions, rings, and deletions. Translocations can be balanced or unbalanced. Carriers of balanced translocations are at risk of having children with unbalanced rearrangements; these children are usually physically and mentally handicapped.

CHAPTER 4

DNA Technology and Applications

In the history of medical genetics, the 'chromosome breakthrough' in the mid-1950s was revolutionary. In the past 4 decades, DNA technology has had a profound effect, not only in medical genetics (Figure 4.1), but also in many areas of biological science (Box 4.1).The seminal developments in the field are summarized in Table 4.1.

DNA technology can be split into two main areas: DNA cloning and methods of DNA analysis.

DNA Cloning

DNA cloning is the selective amplification of a specific DNA fragment or sequence to produce relatively large amounts of a homogeneous DNA fragment to enable its structure and function to be analyzed in detail.

DNA cloning falls into two main types: techniques that use natural in-vivo cell-based mechanisms of DNA replication and the more recently developed cell-free or in-vitro polymerase chain reaction.

In-vivo Cell-Based DNA Cloning

There are six basic steps in in-vivo cell-based DNA cloning.

Generation of DNA Fragments

Although fragments of DNA can be produced by mechanical shearing techniques, this is a haphazard process producing fragments that vary in size. In the early 1970s, it was recognized that certain microbes contain enzymes that cleave double-stranded DNA in or near a particular sequence of nucleotides. These enzymes restrict the entry of foreign DNA into bacterial cells and were therefore called **restriction enzymes**. They recognize a palindromic nucleotide sequence of DNA of between four and eight nucleotides in length (i.e., the same sequence of nucleotides occurring on the two complementary DNA strands when read in one direction of polarity, e.g., 5′ to 3′) (Table 4.2). The longer the nucleotide recognition sequence of the restriction enzyme, the less frequently that particular nucleotide sequence will occur by chance and therefore the larger the average size of the DNA fragments generated.

More than 300 different restriction enzymes have been isolated from various bacterial organisms. Restriction endonucleases are named according to the organism from which they are derived (e.g., *Eco*RI is from *Escherichia coli* and was the first restriction enzyme isolated from that organism).

The complementary pairing of bases in the DNA molecule means that cleavage of double-stranded DNA by a restriction endonuclease always creates double-stranded breaks, which, depending on the cleavage points of the particular restriction enzyme used, results in either a staggered or a blunt end (Figure 4.2).

Digestion of DNA from a specific source with a particular restriction enzyme will produce the same reproducible collection of DNA fragments each time the process is carried out.

Recombination of DNA Fragments

DNA from any source, when digested with the same restriction enzyme, will produce DNA fragments with identical complementary ends or termini. When DNA has been cleaved by a restriction enzyme that produces staggered termini, these are referred to as being 'sticky' or 'cohesive' because they will unite under appropriate conditions with complementary sequences produced by the same restriction enzyme on DNA from any source. Initially the cohesive termini are held together by hydrogen bonding but are covalently attached with the enzyme called **DNA ligase**. The union of two DNA fragments from different sources produces what is referred to as a **recombinant DNA molecule**.

Vectors

A **vector** is the term for the carrier DNA molecule used in the cloning process that, through its own independent replication within a host organism, will allow the production of multiple copies of itself. The incorporation of the **target** DNA into a vector allows the production of large amounts of that DNA fragment.

For naturally occurring vectors to be used for DNA cloning, they need to be modified to ensure that the target DNA is inserted at a specific location and that recombinant vectors containing target inserted DNA can be detected. Many of the early vectors were constructed so that insertion of the target DNA in a gene for antibiotic resistance resulted in loss of that function (Figure 4.3).

The five main types of vector commonly used include **plasmids**, **bacteriophages**, **cosmids**, and **bacterial** and **yeast artificial chromosomes** (**BACs** and **YACs**). The choice of vector used in cloning depends on a number of factors, such as the particular restriction enzyme being used and the size of the target DNA to be inserted. Some of the early vectors, such as plasmids and bacteriophages, were very limited in

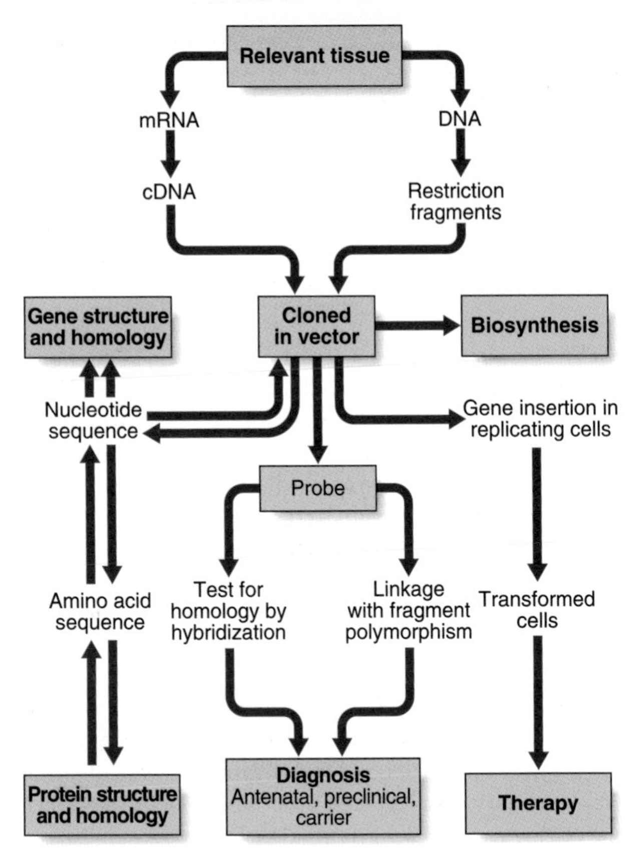

FIGURE 4.1 Some of the applications of DNA technology in medical genetics.

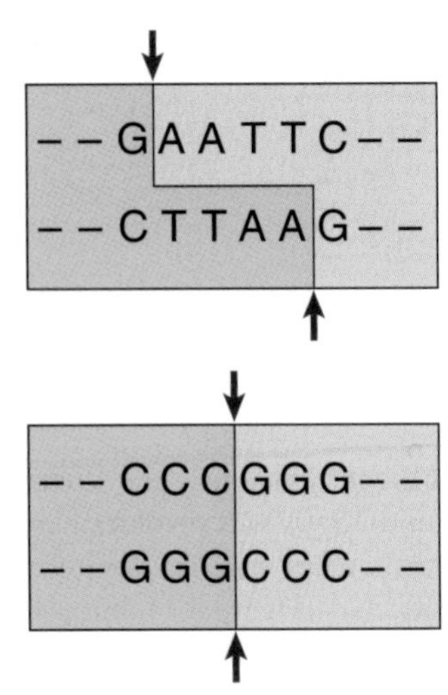

FIGURE 4.2 The staggered and blunt ends generated by restriction digest of double-stranded DNA by *Eco*RI and *Sma*I. Sites of cleavage of the DNA strands are indicated by arrows.

terms of the size of the target DNA fragment that could be inserted. Later generations of vectors, such as cosmids, can take inserts up to approximately 50 kb in size. A cosmid is essentially a plasmid that has had all but the minimum vector DNA necessary for propagation removed (i.e., the cos sequence), to enable insertion of the largest possible foreign DNA fragment and still allow replication.

The development of BACs and YACs allows the possibility of cloning DNA fragments of between 300 kb and

Table 4.1 Development of DNA Technology

Decade	Development	Examples of Application
1970s	Recombinant DNA technology, Southern blot, and Sanger sequencing	Recombinant erythropoietin (1987), DNA fingerprinting (1984), and DNA sequence of Epstein-Barr virus genome (1984)
1980s	Polymerase chain reaction (PCR)	Diagnosis of genetic disorders
1990s	Capillary sequencing and microarray technology	Draft human genome sequence (2001)
2000s	Next-generation 'clonal' sequencing	First acute myeloid leukaemia (AML) cancer genome sequenced (2008)

Box 4.1 Applications of DNA Technology

Gene structure/mapping/function
Population genetics
Clinical genetics
- Preimplantation genetic diagnosis
- Prenatal diagnosis
- Presymptomatic diagnosis
- Carrier detection

Diagnosis and pathogenesis of disease
- Genetic
- Acquired—infective, malignant

Biosynthesis
- (e.g., insulin, growth hormone, interferon, immunization)

Treatment of genetic disease
Gene therapy
Agriculture
- (e.g., nitrogen fixation)

Table 4.2 Some Examples of Restriction Endonucleases with Their Nucleotide Recognition Sequence and Cleavage Sites

Enzyme	Organism	Cleavage Site 5′ → 3′
BamHI	*Bacillus amyloliquefaciens* H	G·GATCC
EcoRI	*Escherichia coli* RY 13	G·AATTC
HaeIII	*Haemophilus aegyptius*	GG·CC
HindIII	*Haemophilus influenzae* Rd	A·AGCTT
HpaI	*Haemophilus parainfluenzae*	GTT·AAC
PstI	*Providencia stuartii*	CTGCA·G
SmaI	*Serratia marcescens*	CCC·GGG
SalI	*Streptomyces albus* G	G·TCGAC

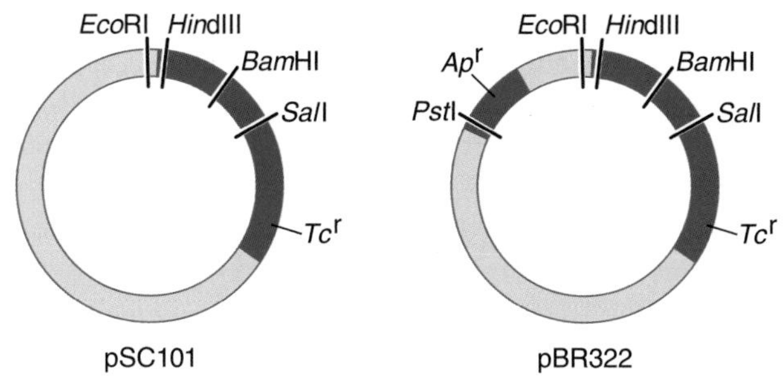

FIGURE 4.3 Two plasmids originally used in recombinant DNA technology showing drug resistance genes (Ap^r, ampicillin resistance; Tc^r, tetracycline resistance) and cleavage sites of restriction endonucleases that are present in the DNA only once for use as a cloning site.

1000 kb in size. YACs consist of a plasmid that contains within it the minimum DNA sequences necessary for centromere and telomere formation plus DNA sequences known as **autonomous replication sequences**, all of which are necessary for accurate replication within yeast. YACs have the advantage that they can incorporate DNA fragments of up to 1000 kb in size as well as allow replication of eukaryotic DNA with repetitive DNA sequences, which often cannot take place in bacterial cells. Many eukaryotic genes are very large, being up to 2 to 3 million base pairs (bp) in length (p. 388). YACs allow detailed mapping of genes of this size and their flanking regions, whereas the use of conventional vectors would require an inordinate number of overlapping clones.

Transformation of the Host Organism

After introducing the target DNA fragment into the vector, the recombinant vector is introduced into specially modified bacterial or yeast host cells. The bacterial cell membrane is not normally permeable to large molecules such as DNA fragments but can be made permeable by a variety of different methods, including exposure to certain salts or high voltage; this is known as becoming **competent**. Usually only a single DNA molecule is taken up by a host cell undergoing the process known as **transformation**. If the transformed cells are allowed to multiply, large quantities of identical copies of the original single target DNA or **clones** will be produced (Figure 4.4).

Screening for Recombinant Vectors

After the transformed cells have multiplied in culture medium, they are plated out on a master plate of nutrient agar in a Petri dish. Recombinant vectors can be screened for by a detection system; for example, loss of antibiotic resistance can be screened for by replica plating on agar containing the appropriate antibiotic (see Figure 4.3). Thus,

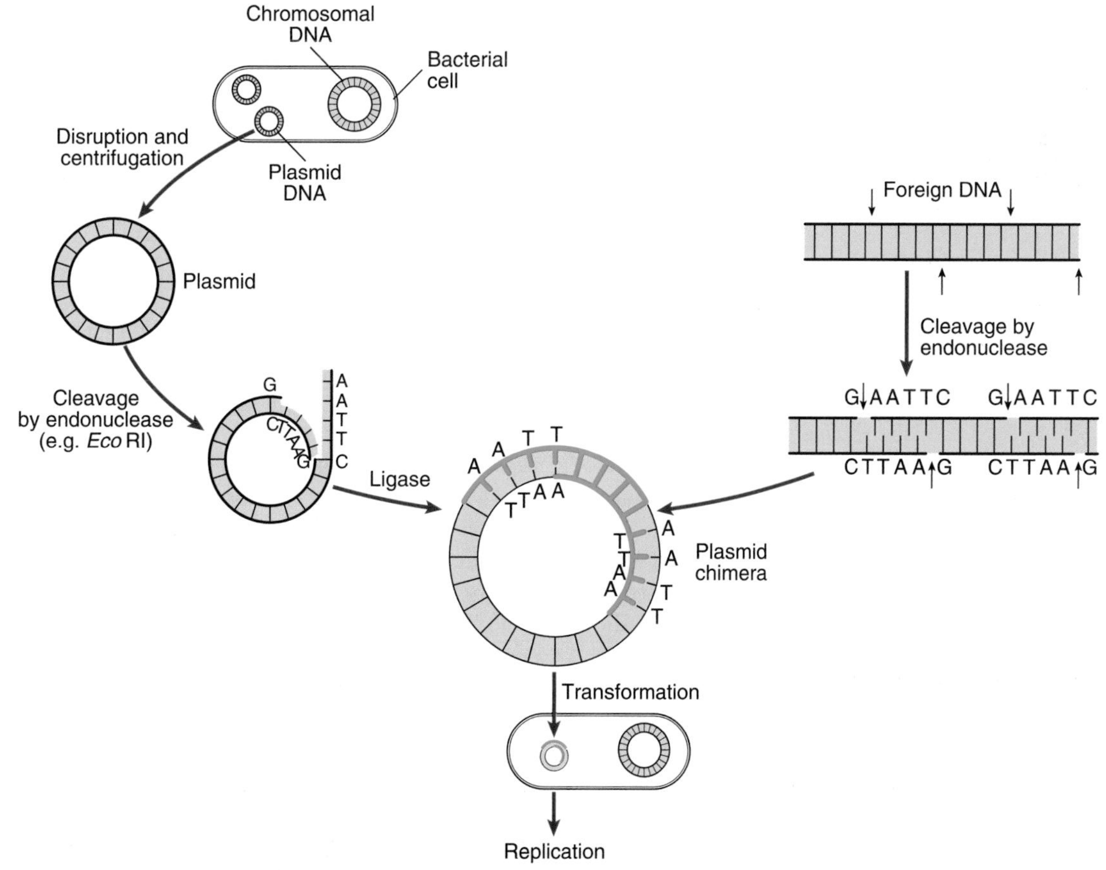

FIGURE 4.4 Generation of a recombinant plasmid using *Eco*RI and transformation of the host bacterial organism. (From Emery AE 1981 Recombinant DNA technology. Lancet ii:1406–1409, with permission.)

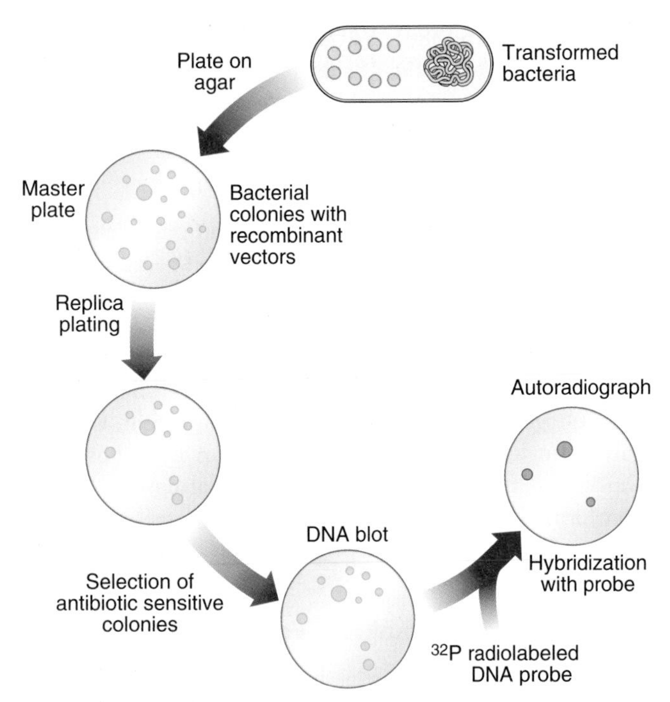

FIGURE 4.5 Identification of recombinant DNA clones with specific DNA inserts by loss of antibiotic resistance, nucleic acid hybridization, and autoradiography.

if the enzyme *Pst*I were used to generate DNA fragments and to cut the plasmid pBR322, any recombinant plasmids produced would make the bacterial host cells they transform sensitive to ampicillin, as this gene would no longer be functional, but they would remain resistant to tetracycline. Replica plating of the master plates from the cultures allows identification of individual specific recombinant clones.

Selection of Specific Clones

Several techniques have been developed to detect the presence of clones with specific DNA sequence inserts. The most widely used method is nucleic acid hybridization (p. 57). Colonies of transformed host bacteria with recombinant clones are used to make replica plates that are lyzed and then blotted on to a nitrocellulose filter to which nucleic acid binds. The DNA of the replica blot is then denatured to make the DNA single stranded, which will allow it to hybridize with single-stranded, radioactively labeled DNA or RNA probes (p. 58), which can then be detected by exposure to an x-ray film, or what is known as **autoradiography**. In this way, a transformed host bacterial colony containing a sequence complementary to the probe can be detected and, from its position on the replica plate, the colony containing that clone can be identified on the master plate, 'picked', and cultured separately (Figure 4.5).

DNA Libraries

Different sources of DNA can be used to make recombinant DNA molecules. DNA from nucleated cells is termed **total** or **genomic DNA**. DNA made by the action of the enzyme reverse transcriptase on messenger RNA (mRNA) is called **complementary DNA** or **cDNA**. It is possible to enrich for DNA sequences of particular interest by using a specific tissue or cell type as a source of mRNA; for instance, immature red blood cells (reticulocytes) containing predominantly globin mRNA resulted in cloning of the genes for the globin chains of hemoglobin (p. 156).

The collection of recombinant DNA molecules generated from a specific source is referred to as a **DNA library** (e.g., a genomic or cDNA library). A DNA library of the human genome using plasmids as a vector would need to consist of several hundred thousand clones to be likely to contain the whole of the human genome. The use of YACs as cloning vectors with DNA digested by infrequently cutting restriction enzymes means that the whole of the human genome can be contained in a library of 13,000 to 14,000 clones.

Cell-Free DNA Cloning

One of the most revolutionary developments in DNA technology is the technique first developed in the mid-1980s known as the **polymerase chain reaction** or **PCR**. PCR can be used to produce vast quantities of a target DNA fragment provided that the DNA sequence of that region is known.

The PCR

DNA sequence information is used to design two oligonucleotide primers (**amplimers**) of approximately 20 bp in length complementary to the DNA sequences flanking the target DNA fragment. The first step is to denature the double-stranded DNA by heating. The primers then bind to the complementary DNA sequences of the single-stranded DNA templates. DNA polymerase extends the primer DNA in the presence of the deoxynucleotide triphosphates (dATP, dCTP, dGTP, and dTTP) to synthesize the complementary DNA sequence. Subsequent heat denaturation of the double-stranded DNA, followed by annealing of the same primer sequences to the resulting single-stranded DNA, will result in the synthesis of further copies of the target DNA. Some 30 to 35 successive repeated cycles results in more than 1 million copies (**amplicons**) of the DNA target, sufficient for direct visualization by ultraviolet fluorescence after ethidium bromide staining, without the need to use indirect detection techniques (Figure 4.6).

PCR allows analysis of DNA from any cellular source containing nuclei; in addition to blood, this can include less invasive samples such as buccal scrapings or pathological archival material. It is also possible to start with quantities of DNA as small as that from a single cell, as is the case in preimplantation genetic diagnosis (p. 335). Great care has to be taken with PCR, however, because DNA from a contaminating extraneous source, such as desquamated skin from a laboratory worker, will also be amplified. This can lead to false-positive results unless the appropriate

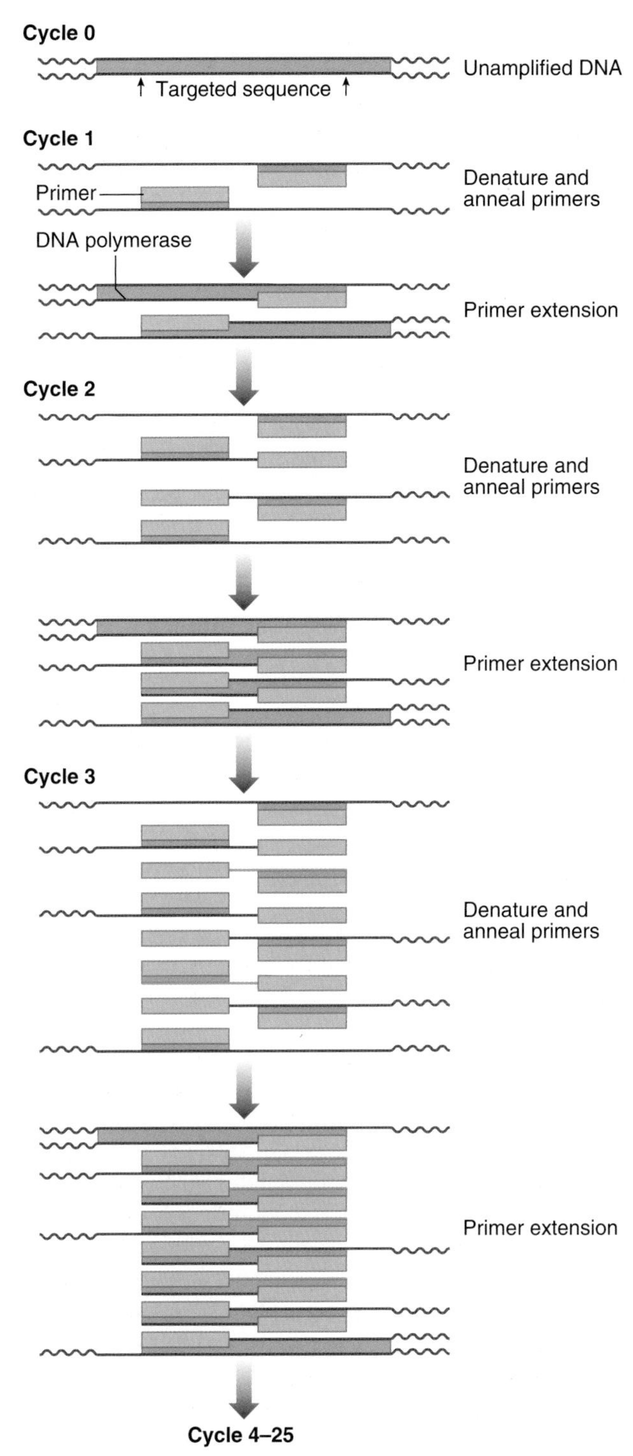

FIGURE 4.6 Diagram of the polymerase chain reaction showing serial denaturation of DNA, primer annealing, and extension with doubling of the target DNA fragment numbers in each cycle.

control studies are used to detect this possible source of error.

Another advantage of PCR is the rapid turnaround time of samples for analysis. Use of the heat-stable *Taq* DNA polymerase isolated from the bacterium *Thermophilus aquaticus*, which grows naturally in hot springs, generates PCR products in a matter of hours rather than the days or weeks required for cell-based in-vivo DNA cloning techniques.

Real-time PCR machines have reduced this time to less than 1 hour, and fluorescence technology is used to monitor the generation of PCR products during each cycle, thus eliminating the need for gel electrophoresis.

DNA cloning by PCR, in contrast to in-vivo cell-based techniques, has the disadvantage that it requires knowledge of the nucleotide sequence of the target DNA fragment and is best used to amplify DNA fragments of up to 1 kb, although long-range PCR allows the amplification of larger DNA fragments of up to 20 kb to 30 kb.

Techniques of DNA Analysis

Many methods of DNA analysis involve the use of nucleic acid probes and the process of nucleic acid hybridization.

Nucleic Acid Probes

Nucleic acid probes are usually single-stranded DNA sequences that have been radioactively or non-radioactively labeled and can be used to detect DNA or RNA fragments with sequence homology. DNA probes can come from a variety of sources, including random genomic DNA sequences, specific genes, cDNA sequences or oligonucleotide DNA sequences produced synthetically based on knowledge of the protein amino-acid sequence. A DNA probe can be labeled by a variety of processes, including isotopic labeling with ^{32}P and non-isotopic methods using modified nucleotides containing fluorophores (e.g., fluorescein or rhodamine). Hybridization of a radioactively labeled DNA probe with cDNA sequences on a nitrocellulose filter can be detected by autoradiography, whereas DNA fragments that are fluorescently labeled can be detected by exposure to the appropriate wavelength of light, for example fluorescent in-situ hybridization (p. 34).

Nucleic Acid Hybridization

Nucleic acid hybridization involves mixing DNA from two sources that have been denatured by heat or alkali to make them single stranded and then, under the appropriate conditions, allowing complementary base pairing of homologous sequences. If one of the DNA sources has been labeled in some way (i.e., is a DNA probe), this allows identification of specific DNA sequences in the other source. The two main methods of nucleic acid hybridization most commonly used are Southern and northern blotting.

Southern Blotting

Southern blotting, named after Edwin Southern (who developed the technique), involves digesting DNA by a restriction enzyme that is then subjected to electrophoresis on an agarose gel. This separates the DNA or restriction fragments by size, the smaller fragments migrating faster than the larger ones. The DNA fragments in the gel are then denaturated with alkali, making them single stranded. A

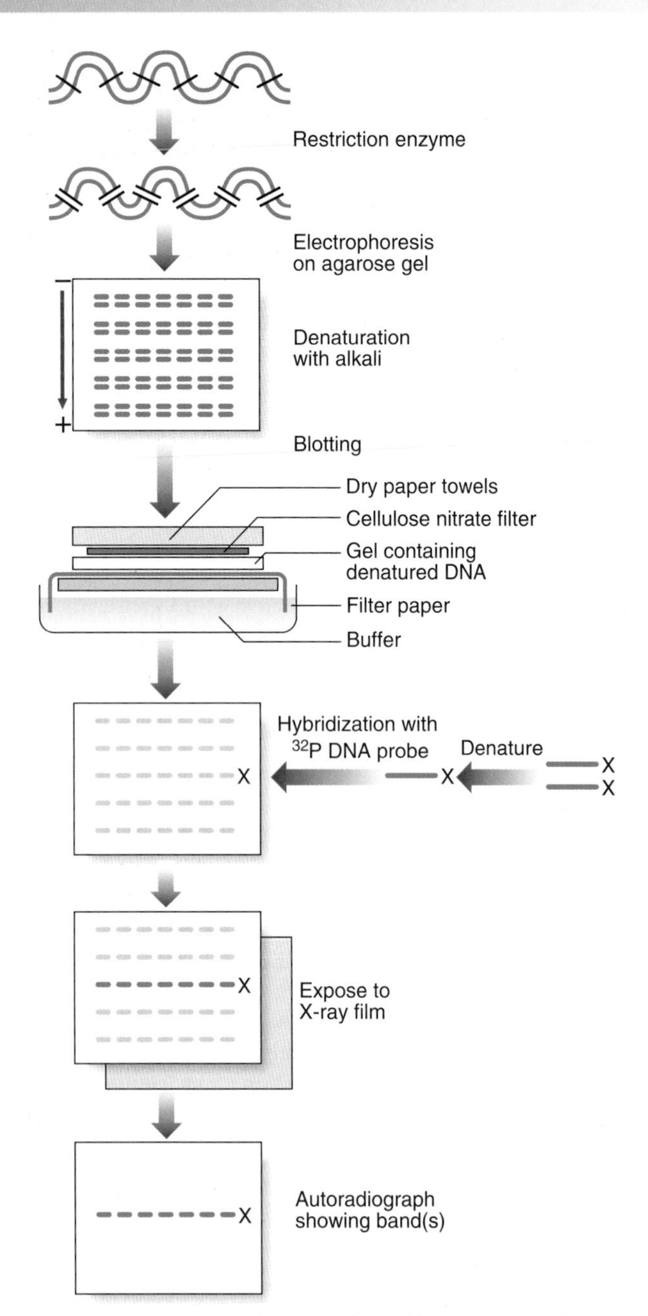

FIGURE 4.7 Diagram of the Southern blot technique showing size fractionation of the DNA fragments by gel electrophoresis, denaturation of the double-stranded DNA to become single stranded, and transfer to a nitrocellulose filter that is hybridized with a ^{32}P radioactively labeled DNA probe.

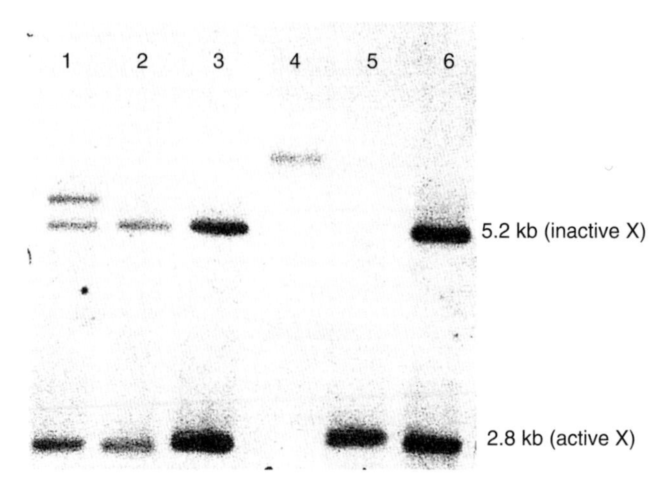

FIGURE 4.8 Southern blot to detect methylation of the *FMR1* promoter in patients with fragile X. DNA digested with *Eco*R1 and the methylation sensitive enzyme *Bst* Z1 was probed with Ox1.9, which hybridizes to a CpG island within the *FMR1* promoter. Patient 1 is a female with a methylated expansion, patients 2, 3, and 6 are normal females, patient 4 is an affected male and patient 5 is a normal male. (Courtesy A. Gardner, Department of Molecular Genetics, Southmead Hospital, Bristol, UK.)

'permanent' copy of these single-stranded fragments is made by transferring them on to a nitrocellulose filter that binds the single-stranded DNA, the so-called **Southern blot**. A particular target DNA fragment of interest from the collection on the filter can be visualized by adding a single-stranded ^{32}P radioactively labeled DNA probe that will hybridize with homologous DNA fragments in the Southern blot, which can then be detected by autoradiography (Figure 4.7). Non-radioactive Southern blotting techniques have been developed with the DNA probe labeled with digoxigenin and detected by chemiluminescence. This approach is safer and generates results more rapidly. An example of the use of Southern blotting for diagnostic fragile X testing in patients is shown in Figure 4.8.

Northern Blotting

Northern blotting differs from Southern blotting by the use of mRNA as the target nucleic acid in the same procedure; mRNA is very unstable because of intrinsic cellular ribonucleases. Use of ribonuclease inhibitors allows isolation of mRNA that, if run on an electrophoretic gel, can be transferred to a filter. Hybridizing the blot with a DNA probe allows determination of the size and quantity of the mRNA transcript, a so-called **Northern blot**. With the advent of real-time reverse transcriptase PCR, and microarray technology for gene expression studies, Northern blotting is used less often.

DNA Microarrays

DNA microarrays are based on the same principle of hybridization but on a miniaturized scale, which allows simultaneous analysis of several million targets. Short, fluorescently labeled oligonucleotides attached to a glass microscope slide can be used to detect hybridization of target DNA under appropriate conditions. The color pattern of the microarray is then analyzed automatically by computer. Four classes of application have been described: (1) expression studies to look at the differential expression of thousands of genes at the mRNA level; (2) analysis of DNA variation for mutation detection and single nucleotide polymorphism (SNP) typing

(p. 67); (3) testing for genomic gains and losses by array comparative genomic hybridization (CGH) (p. 36); and (4) a combination of the latter two, SNP–CGH, which allows the detection of copy-neutral genetic anomalies such as uniparental disomy (p. 121).

Mutation Detection

The choice of method depends primarily on whether the test is for a known sequence change or to identify the presence of any mutation within a particular gene. A number of techniques can be used to screen for mutations that differ in their ease of use and reliability. The choice of assay depends on many factors, including the sensitivity required, cost, equipment, and the size and structure (including number of polymorphisms) of the gene (Table 4.3). Identification of a possible sequence variant by one of the mutation screening methods requires confirmation by DNA sequencing. Some of the most common techniques in current use are described in the following section.

Size Analysis of PCR Products

Deletion or insertion mutations can sometimes be detected simply by determining the size of a PCR product. For example, the most common mutation that causes cystic fibrosis, p.Phe508del, is a 3-bp deletion that can be detected on a polyacrylamide gel. Some trinucleotide repeat expansion mutations can be amplified by PCR (Figure 4.9).

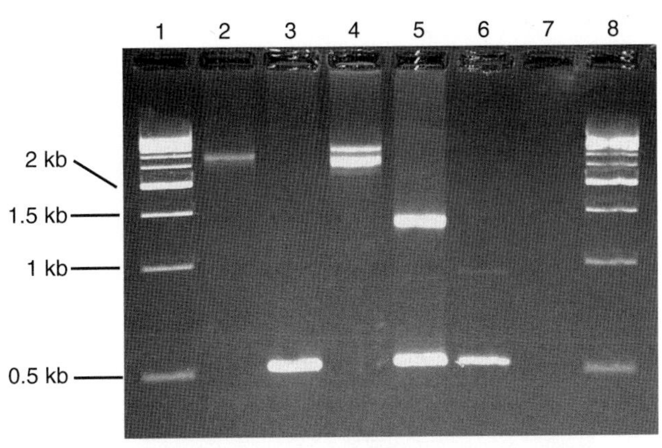

FIGURE 4.9 Amplification of the GAA repeat expansion mutation by polymerase chain reaction (PCR) to test for Friedreich ataxia. Products are stained with ethidium bromide and electrophoresed on a 1.5% agarose gel. Lanes 1 and 8 show 500-bp ladder-size standards, lanes 2 and 4 show patients with homozygous expansions, lanes 3 and 6 show unaffected controls, lane 5 shows a heterozygous expansion carrier, and lane 7 is the negative control. (Courtesy K. Thomson, Department of Molecular Genetics, Royal Devon and Exeter Hospital, Exeter, UK.)

Restriction Fragment Length Polymorphism

If a base substitution creates or abolishes the recognition site of a restriction enzyme, it is possible to test for the mutation by digesting a PCR product with the appropriate enzyme and separating the products by electrophoresis (Figure 4.10).

Table 4.3 Methods for Detecting Mutations

Method	Known/Unknown Mutations	Example	Advantages/Disadvantages
Southern blot	Known (or unknown rearrangement)	Trinucleotide expansions in fragile X and myotonic dystrophy	Laborious
Sizing of PCR products	Known	p.Phe508del *CFTR* mutation; trinucleotide expansions in *HD* and *SCA* genes	Simple, cheap
ARMS-PCR	Known	*CFTR* mutations	Multiplex possible
Oligonucleotide ligation	Known	*CFTR* mutations	Multiplex possible
Real-time PCR	Known	Factor V Leiden	Expensive equipment
Conformation-sensitive capillary electrophoresis	Unknown	Any gene	High-throughput method that can use capillary sequencer platform
High-resolution melt	Unknown	Any gene	High sensitivity; high-throughput method
Sanger sequencing	Known or unknown	Any gene	Gold standard
Pyrosequencing	Known or unknown	Any gene	Expensive equipment
DNA microarray	Known or unknown	Any gene	High throughput; expensive equipment. Sensitivity to detect some types of mutations is limited
Next-generation 'clonal' sequencing	Known or unknown	Any gene	Expensive equipment, enormous capacity but vast amount of data to analyse and interpretation of novel variants can be difficult

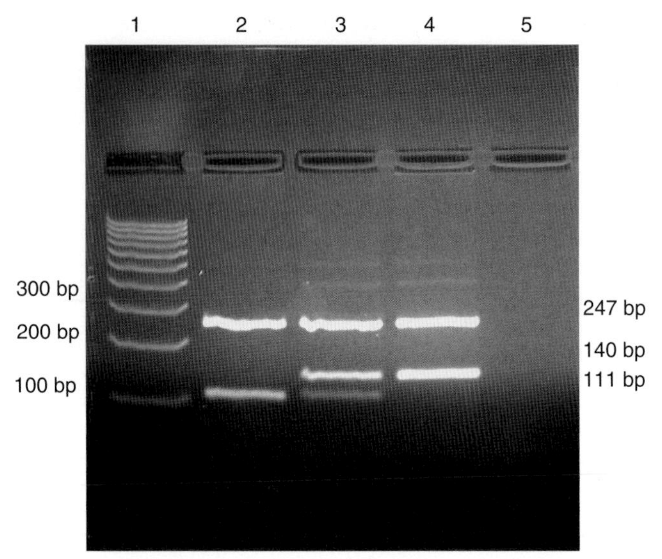

FIGURE 4.10 Detection of the *HFE* gene mutation C282Y by restriction fragment length polymorphisms (RFLP). The normal 387-bp polymerase chain reaction (PCR) product is digested with *RsaI* to give products of 247 and 140 bp. The C282Y mutation creates an additional recognition site for *RsaI*, giving products of 247, 111, and 29 bp. Lane 1 shows a 100-bp ladder-size standard. Lanes 2 though 4 show patients homozygous, heterozygous, and normal for the C282Y mutation, respectively. Lane 5 is the negative control. (Courtesy N. Goodman, Department of Molecular Genetics, Royal Devon and Exeter Hospital, Exeter, UK.)

Amplification-Refractory Mutation System (ARMS) PCR

Allele-specific PCR uses primers specific for the normal and mutant sequences. The most common design is a two-tube assay with normal and mutant primers in separate reactions together with control primers to ensure that the PCR reaction has worked. An example of a multiplex ARMS assay to detect 12 different cystic fibrosis mutations is shown in Figure 4.11.

Oligonucleotide Ligation Assay

A pair of oligonucleotides is designed to anneal to adjacent sequences within a PCR product. If the pair is perfectly hybridized, they can be joined by DNA ligase. Oligonucleotides complementary to the normal and mutant sequences are differentially labeled and the products identified by computer software (Figure 4.12).

Real-Time PCR

There are multiple hardware platforms for real-time PCR and 'fast' versions that can complete a PCR reaction in less than 30 minutes. TaqMan™ and LightCycler™ use fluorescence technology to detect mutations by allelic discrimination of PCR products. Figure 4.13 illustrates the factor V Leiden mutation detected by TaqMan™ methodology.

DNA Microarrays (DNA 'Chips')

DNA microarrays hold the promise of rapid mutation testing. They involve synthesizing custom-designed 20 bp to 25 bp oligonucleotide sequences for both the normal DNA sequence and known and/or possible single nucleotide substitutions of a gene. These are attached to a 'chip' in a structured arrangement in what is known as a **microarray**. The sample DNA being screened for a mutation is amplified by PCR, fluorescently labeled, and hybridized with the

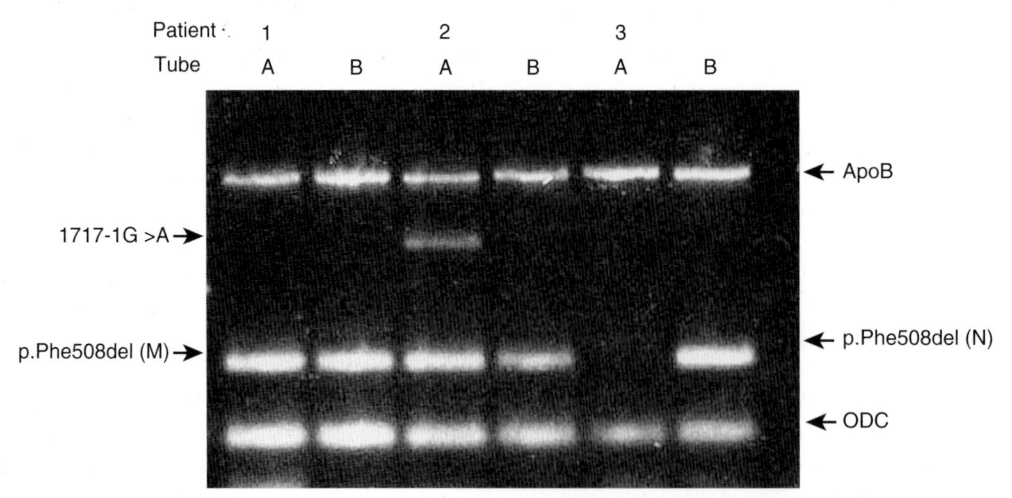

FIGURE 4.11 Detection of *CFTR* mutations by two-tube amplification-refractory mutation system (ARMS)-polymerase chain reaction (PCR). Patient 1 is heterozygous for ΔF508 (p.Phe508del). Patient 2 is a compound heterozygote for p.Phe508del and c.1717-1G>A. Patient 3 is homozygous normal for the 12 mutations tested. Primers for two internal controls (ApoB and ODC) are included in each tube.

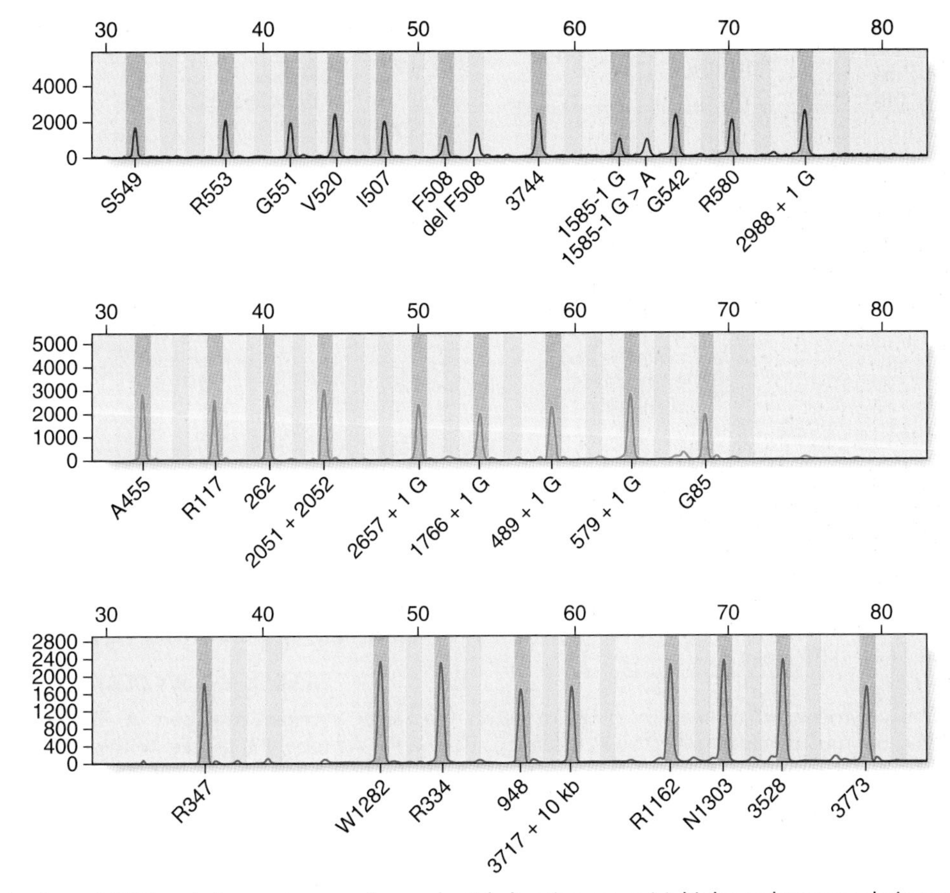

FIGURE 4.12 Detection of *CFTR* mutations using an oligonucleotide ligation assay. Multiplex polymerase chain reaction (PCR) amplifies 15 exons of the *CFTR* gene. Oligonucleotides are designed to anneal to the PCR products such that two oligonucleotides anneal to adjacent sequences for each mutation and are then joined by ligation. The 32 mutations are discriminated using a combination of size and differently colored fluorescent labels. This patient is a compound heterozygote for the ΔF508 (p.Phe508del) and c.1585-1G>A mutations. (Courtesy Karen Stals, Department of Molecular Genetics, Royal Devon and Exeter Hospital, Exeter, UK.)

oligonucleotides in the microarray (Figure 4.14). Computer analysis of the color pattern of the microarray generated after hybridization allows rapid automated mutation testing. The prospect of gene-specific DNA chip microarrays may lead to a revolution in the speed and reliability of mutation screening, provided the technology is affordable and the technique can be demonstrated to be robust. The detection of known base substitutions and SNPs has been very successful, but screening for insertion mutations is more limited.

Conformation-Sensitive Capillary Electrophoresis

Conformation-sensitive capillary electrophoresis is used to detect the presence of heteroduplexes using fluorescence technology. PCR products can be multiplexed by using multiple fluorescent dyes. An alteration in the DNA sequence can result in a different conformation, which has a different electrophoretic mobility, and an appropriate polymer can be used for identification.

High-Resolution Melt Curve Analysis

This technique employs a class of fluorescent dyes that intercalate with double-stranded, but not single-stranded, DNA. The intercalating dye is incorporated in the PCR reaction and the products are then heated to separate the two strands. Fluorescence levels decrease as the DNA strands dissociate and this 'melting' profile depends on the PCR product size and sequence (Figure 4.15). High-resolution melt curve analysis appears to be very sensitive and can be used for high-throughput mutation screening.

Sanger Sequencing

The 'gold standard' method of mutation screening is DNA sequencing using the dideoxy chain termination method developed in the 1970s by Fred Sanger. This method originally employed radioactive labeling with manual interpretation of data. The use of fluorescent labels detected by computerized laser systems has improved ease of use and increased throughput and accuracy. Today's capillary

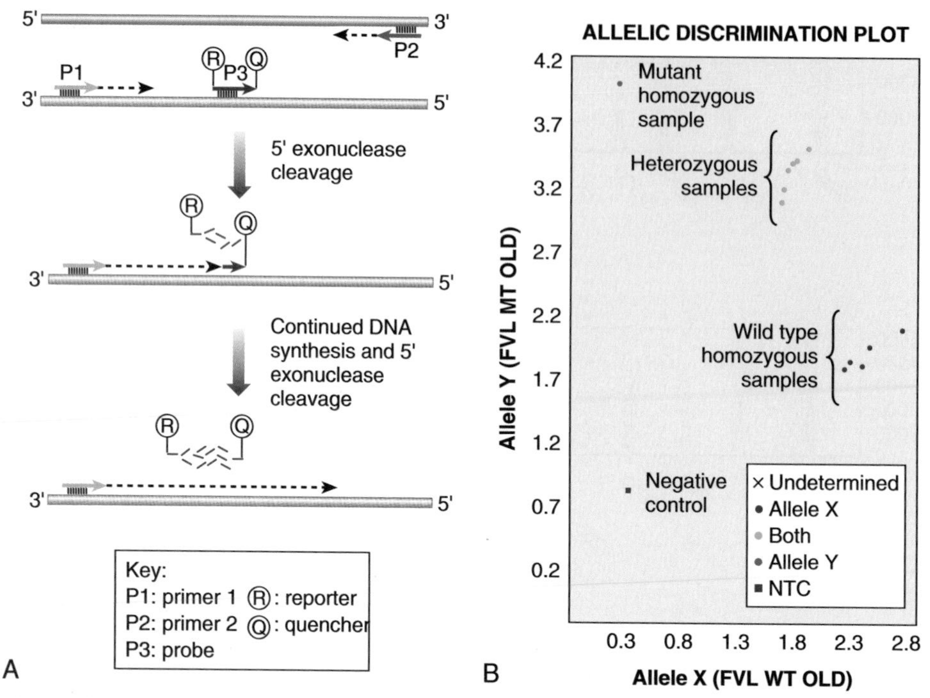

FIGURE 4.13 Real-time polymerase chain reaction (PCR) to detect the Factor V Leiden mutation. **A**, TaqMan™ technique. The sequence encompassing the mutation is amplified by PCR primers, P1 and P2. A probe, P3, specific to the mutation is labelled with two fluorophores. A reporter fluorophore, R, is attached to the 5′ end of the probe and a quencher fluorophore, Q, is attached to the 3′ end. During the PCR reaction, the 5′ exonuclease activity of the polymerase enzyme progressively degrades the probe, separating the reporter and quencher dyes, which results in fluorescent signal from the reporter fluorophore. **B**, TaqMan™ genotyping plot. Each sample is analysed with two probes, one specific for the wild-type and one for the mutation. The strength of fluorescence from each probe is plotted on a graph (wild-type on X-axis, mutant on Y-axis). Each sample is represented by a single point. The samples fall into 3 clusters representing the possible genotypes; homozygous wild-type, homozygous mutant or heterozygous. (Courtesy Dr. E. Young, Department of Molecular Genetics, Royal Devon and Exeter Hospital, Exeter, UK.)

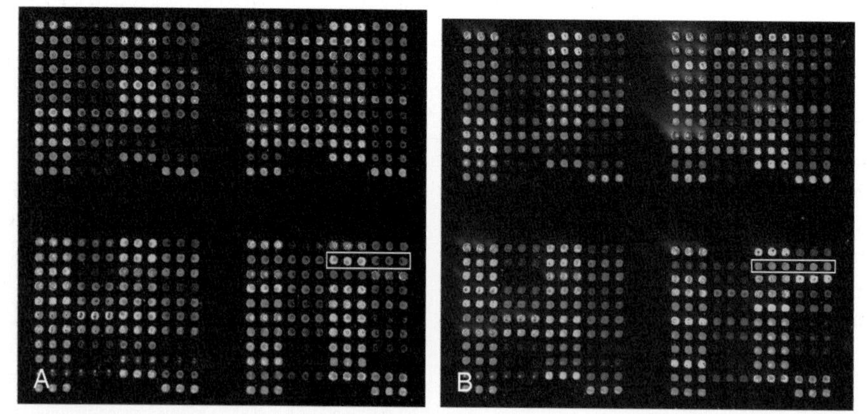

FIGURE 4.14 Detection of *HNF1A* mutations using a DNA microarray. The '*HNF1A* chip' contains normal and mutant probes for 75 different mutations spotted in triplicate. Patient DNA was amplified by multiplex polymerase chain reaction (PCR) to yield fluorescently labeled products that were hybridized to the chip. **A**, Control sample. **B**, Patient heterozygous for an *HNF1A* mutation. (Courtesy N. Huh, Samsung Advanced Institute of Technology, South Korea.)

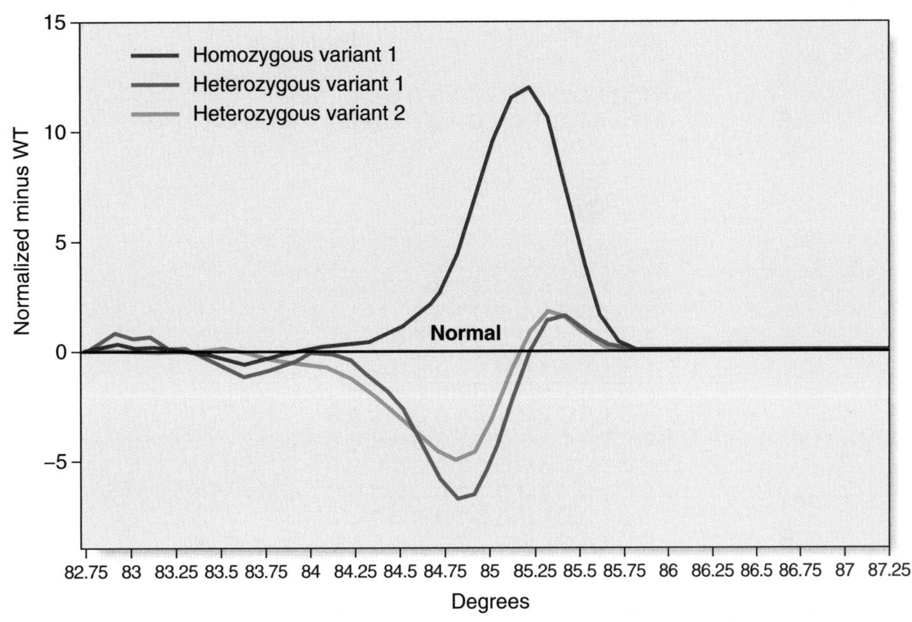

FIGURE 4.15 High-resolution melt curve analysis (HRM). Melting profiles for normal and mutant samples are shown after normalization to a control sample. Each variant has a different melting profile.

sequencers can sequence around 1 Mb (1 million bases) per day.

Dideoxy sequencing involves using a single-stranded DNA template (e.g., denatured PCR products) to synthesize new complementary strands using a DNA polymerase and an appropriate oligonucleotide primer. In addition to the four normal deoxynucleotides, a proportion of each of the four respective dideoxynucleotides is included, each labeled with a different fluorescent dye. The dideoxynucleotides lack a hydroxyl group at the 3′ carbon position; this prevents phosphodiester bonding, resulting in each reaction container consisting of a mixture of DNA fragments of different lengths that terminate in their respective dideoxynucleotide, owing to chain termination occurring at random in each reaction mixture at the respective nucleotide. When the reaction products are separated by capillary electrophoresis, a ladder of DNA sequences of differing lengths is produced. The DNA sequence complementary to the single-stranded DNA template is generated by the computer software and the position of a mutation may be highlighted with an appropriate software package (Figure 4.16).

Pyrosequencing

Pyrosequencing uses a sequencing by synthesis approach in which modified nucleotides are added and removed one at a time, with chemiluminescent signals produced after the addition of each nucleotide. This technology generates quantitative sequence data rapidly and an example of its application in the identification of *KRAS* mutations in patients with colorectal cancer is shown in Figure 4.17.

Next-Generation 'Clonal' Sequencing

The demand for low-cost sequencing has driven the development of high-throughput **sequencing** technologies that produce millions of sequences at once. Next (or second) generation 'clonal' sequencers use an in vitro cloning step to amplify individual DNA molecules by emulsion or bridge PCR (Figure 4.18). The cloned DNA molecules are then sequenced in parallel, either by pyrosequencing, by using reversible terminators or with a sequencing by ligation approach. A comparison with Sanger sequencing is shown in Table 4.5 and an example of a mutation identified by next generation sequencing is shown in Figure 4.19. So-called 'third generation' sequencers have recently been developed. They can generate massively parallel sequence data from single molecules due to their extremely sensitive lasers.

Dosage Analysis

Most of the methods described previously will detect point mutations, small insertions, and deletions. Deletions of one or more exons are common in boys with Duchenne muscular dystrophy and may be identified by a multiplex PCR that reveals the absence of one or more PCR products. However, these mutations are more difficult to detect in carrier females as the normal gene on the other X chromosome 'masks' the deletion.

Large deletion and duplication mutations have been reported in a number of disorders and may encompass a single exon, several exons, or an entire gene (e.g., HNPP [p. 297]; HMSN type 1 [p. 296]). Several techniques have

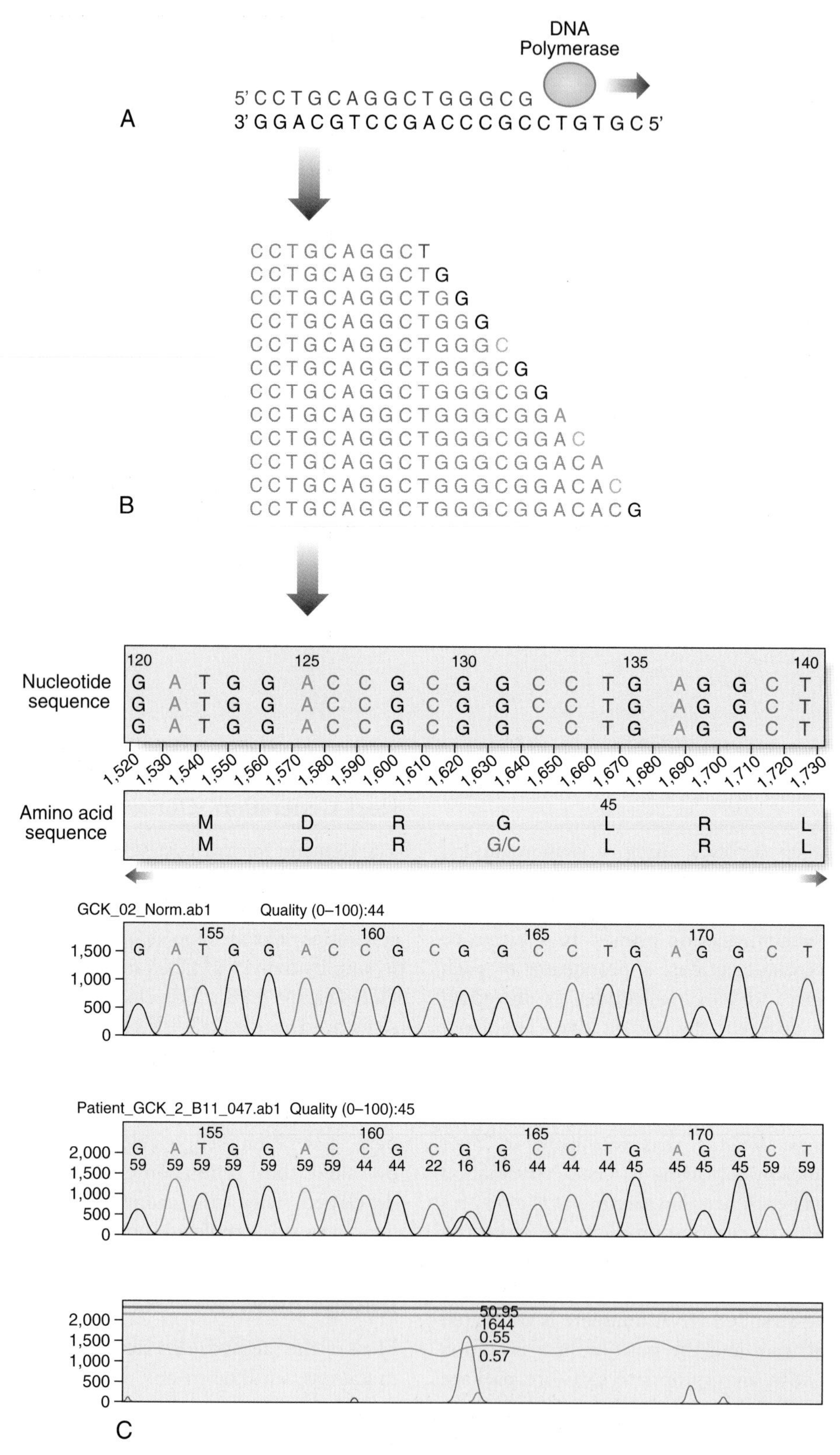

FIGURE 4.16 Fluorescent dideoxy DNA sequencing. The sequencing primer *(shown in red)* binds to the template and primes synthesis of a complementary DNA strand in the direction indicated (**A**). The sequencing reaction includes four dNTPs and four ddNTPs, each labeled with a different fluorescent dye. Competition between the dNTPs and ddNTPs results in the production of a collection of fragments (**B**), which are then separated by electrophoresis to generate an electropherogram (**C**). A heterozygous mutation, p.Gly44Cys (GGC>TGC; glycine>cysteine), is identified by the software.

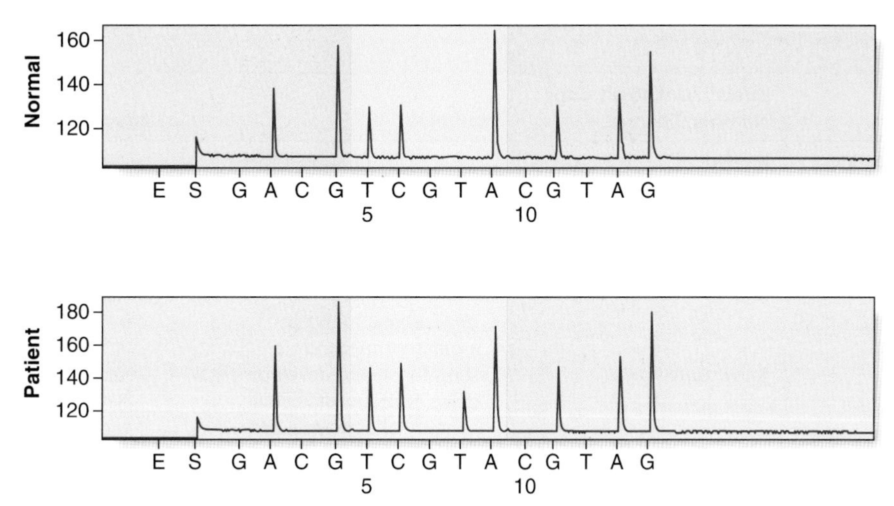

FIGURE 4.17 Detection of a *KRAS* mutation in a colorectal tumour by pyrosequencing. The upper panel shows a normal control, sequence A GGT CAA GAG G. In the lower panel is the tumour sample with the *KRAS* mutation p.Gln61Leu (c.182A>T). (Courtesy Dr. L. Meredith, Institute of Medical Genetics, University Hospital of Wales, Cardiff.)

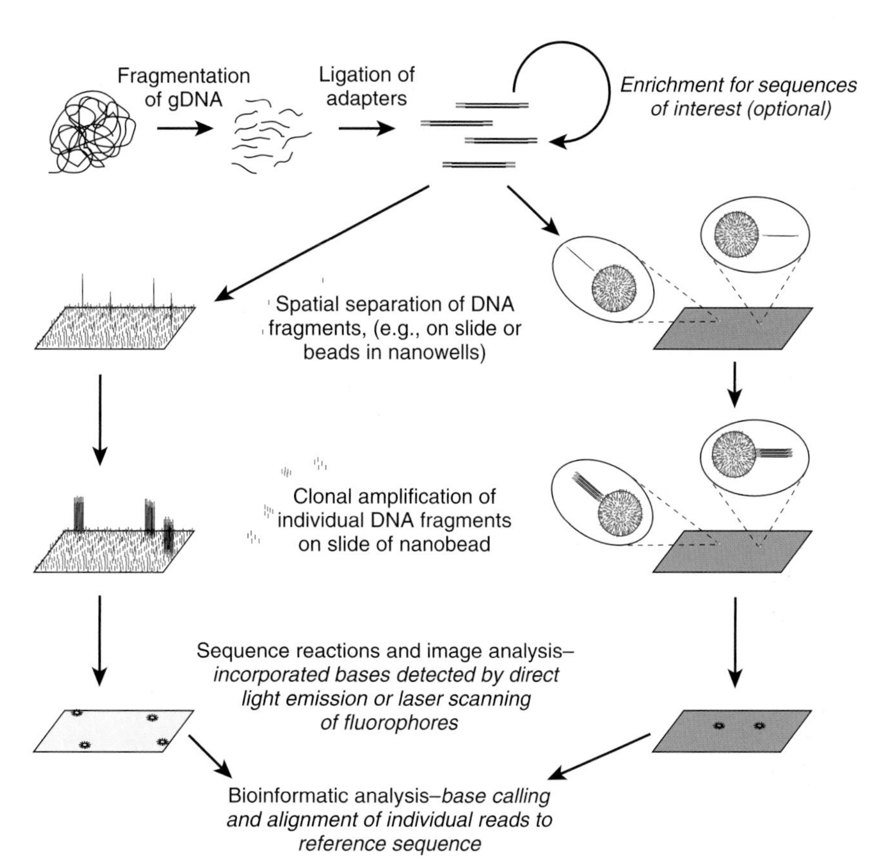

FIGURE 4.18 Next-generation 'clonal' sequencing. DNA is fragmented and adaptors ligated before clonal amplification on a bead or glass slide. Sequencing takes place in situ and incorporated bases are detected by direct light emission or scanning of fluorophores. Data analysis includes base calling and alignment to a reference sequence in order to identify mutations or polymorphisms. (Courtesy Dr. R. Caswell, Peninsula Medical School, Exeter.)

Table 4.4 Methods for Detecting Copy Number Changes

Method	Known/Unknown Copy Number Change	Example	Advantages/Disadvantages
Multiplex ligation-dependent probe amplification	Known	Gene-specific or subtelomere deletion analysis	Suited to the clinical diagnostic setting, but labor-intensive and requires good quality DNA
Quantitative fluorescent PCR	Known	Prenatal aneuploidy testing	Rapid but requires informative microsatellite markers
Real-time PCR	Known	Confirming deletions or duplications found by a different method	Flexible; uses standard PCR primers but gene-centric approach
Array CGH	Known/unknown	Testing for severe developmental delay, learning difficulties, congenital abnormalities	Detects any deletion or duplication but interpretation of novel variants can be difficult
Next-generation 'clonal' sequencing	Known/unknown		Expensive equipment, enormous capacity but vast amount of data to analyse and interpretation of novel variants can be difficult

ARMS, Amplification-refractory mutation system; *CGH*, comparative genomic hybridization; *PCR*, polymerase chain reaction.

Table 4.5 Sanger Sequencing Compared to Next-Generation 'Clonal' Sequencing

Sanger Sequencing	Next-Generation 'Clonal' Sequencing
One sequence read per sample	Massively parallel sequencing
500–1000 bases per read	100–400 bases per read
~1 million bases per day per machine	~2 billion bases per day per machine
~£1 per 1000 bases	~£0.02 per 1000 bases

been developed to identify such mutations (see Table 4.4). Multiplex ligation-dependent probe amplification (MLPA) is a high-resolution method used to detect deletions and duplications (Figure 4.20). Each MLPA probe consists of two fluorescently labeled oligonucleotides that can hybridize, adjacent to each other, to a target gene sequence. When hybridized, the two oligonucleotides are joined by a ligase and the probe is then amplified by PCR (each oligonucleotide includes a universal primer sequence at its terminus). The probes include a variable-length stuffer sequence that enables separation of the PCR products by capillary electrophoresis. Up to 40 probes can be amplified in a single reaction.

Dosage analysis by quantitative fluorescent PCR (QF-PCR) is routinely used for rapid aneuploidy screening; for example, in prenatal diagnosis (p. 325). Microsatellites (see the following section) located on chromosomes 13, 18, and 21 may be amplified within a multiplex and trisomies detected, either by the presence of three alleles or by a dosage effect where one allele is overrepresented (Figure 4.21).

Array CGH was introduced in Chapter 3 (p. 36) and provides a way to detect deletions and duplications on a

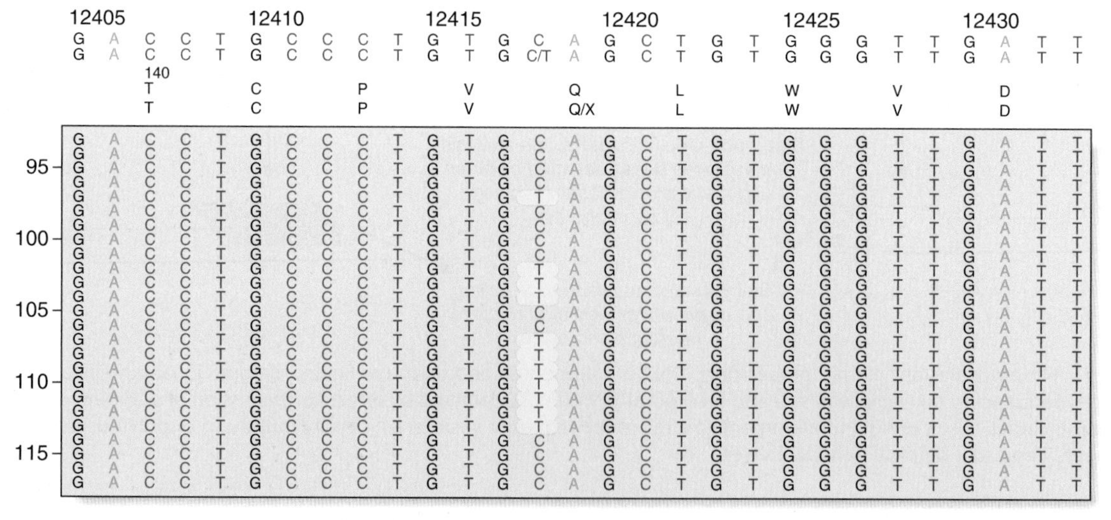

FIGURE 4.19 Detection of a *TP53* mutation in a patient with Li Fraumeni syndrome. The reference sequence is shown at the top and the patient sequence below. A heterozygous C>T mutation (c.430C>T; p.Gln144X) is visible in 11 of the 25 reads shown. (Courtesy Jo Morgan and Graham Taylor, Leeds Institute of Molecular Medicine, St. James's Hospital, Leeds, UK.)

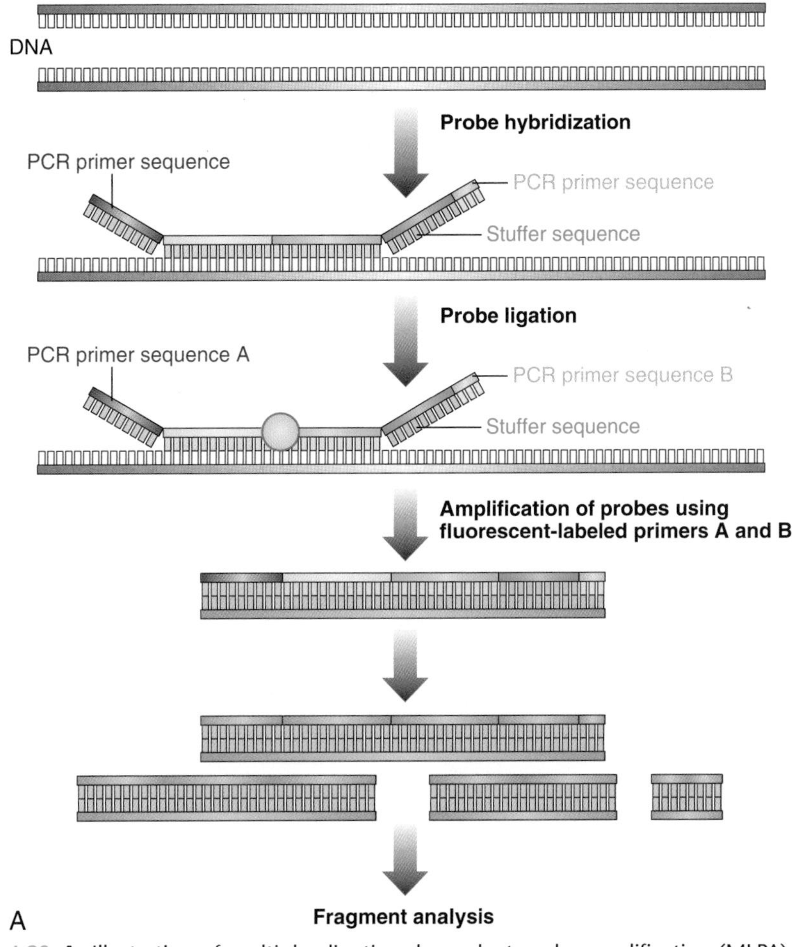

FIGURE 4.20 **A,** Illustration of multiplex ligation-dependent probe amplification (MLPA) method.

Continued

genome-wide scale (Figure 4.22). Arrays used in clinical diagnostic laboratories include both genome wide probes to detect novel mutations and probes targeted to known deletion/duplication syndromes. A comprehensive knowledge of normal copy number variation is essential for interpreting novel mutations.

It is also possible to obtain copy number data from next generation sequencing if genomic DNA, rather than PCR product, is used as the initial template for clonal amplification.

Application of DNA Sequence Polymorphisms

There is an enormous amount of DNA sequence variation in the human genome (p. 13). Two main types, SNPs and hypervariable tandem repeat DNA length polymorphisms, are predominantly used in genetic analysis.

Single Nucleotide Polymorphisms

Around 1 in 1000 bases within the human genome shows variation. SNPs are most frequently biallelic and occur in coding and non-coding regions. If an SNP lies within the recognition sequence of a restriction enzyme, the DNA fragments produced by that restriction enzyme will be of different lengths in different people. This can be recognized by the altered mobility of the restriction fragments on gel electrophoresis, so-called **restriction fragment length polymorphisms,** or **RFLPs.** Early genetic mapping studies used Southern blotting to detect RFLPs, but current technology enables the detection of any SNP. DNA microarrays have led to the creation of a dense SNP map of the human genome and assist genome searches for linkage studies in mapping single-gene disorders (p. 293) and association studies in common diseases.

Variable Number Tandem Repeats

Variable number tandem repeats (VNTRs) are highly polymorphic and are due to the presence of variable numbers of tandem repeats of a short DNA sequence that have been shown to be inherited in a mendelian co-dominant fashion (p. 113). The advantage of using VNTRs over SNPs is the large number of alleles for each VNTR compared with SNPs, which are mostly biallelic.

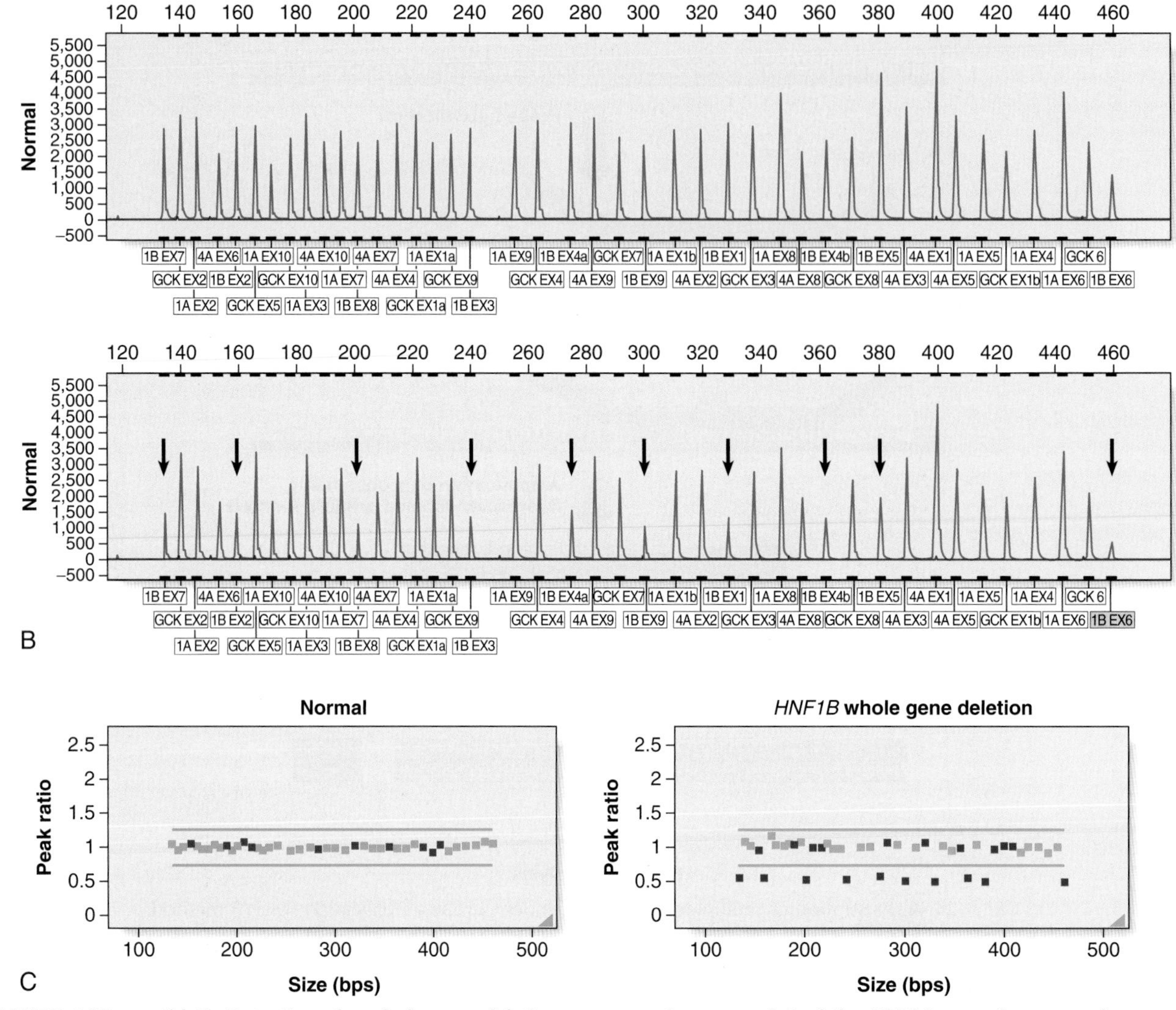

FIGURE 4.20, cont'd **B,** Detection of a whole gene deletion encompassing exons 1-9 of the *HNF1B* gene *(lower panel)* compared with a normal reference sample *(upper panel)*. This MLPA kit also includes probes for the *GCK, HNF1A,* and *HNF4A* genes. **C,** Peak ratio plots showing in graphical form the ratio of normalized peak intensities between the normal reference and patient sample. Each point represents one peak: green or blue=peak within the normal range (0.75–1.25), red=peak either deleted (ratio <0.75) or duplicated (>1.25). The data were analysed using GeneMarker®, SoftGenetics LLC. (Courtesy M. Owens, Department of Molecular Genetics, Royal Devon and Exeter Hospital, Exeter, UK.)

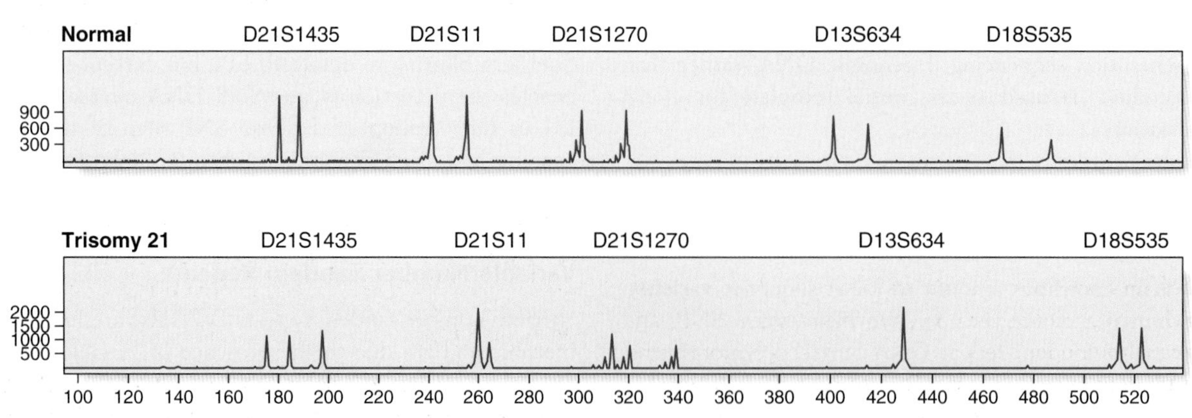

FIGURE 4.21 Quantitative fluorescent (QF)-polymerase chain reaction (PCR) for rapid prenatal aneuploidy testing. The upper panel shows a normal control, with two alleles for each microsatellite marker. The lower panel illustrates trisomy 21 with either three alleles (microsatellites D21S1435, D21S1270) or a dosage effect (D21S11). Microsatellite markers for chromosomes 13 and 18 show a normal profile. (Courtesy Chris Anderson, Institute of Medical Genetics, University Hospital of Wales, Cardiff, UK.)

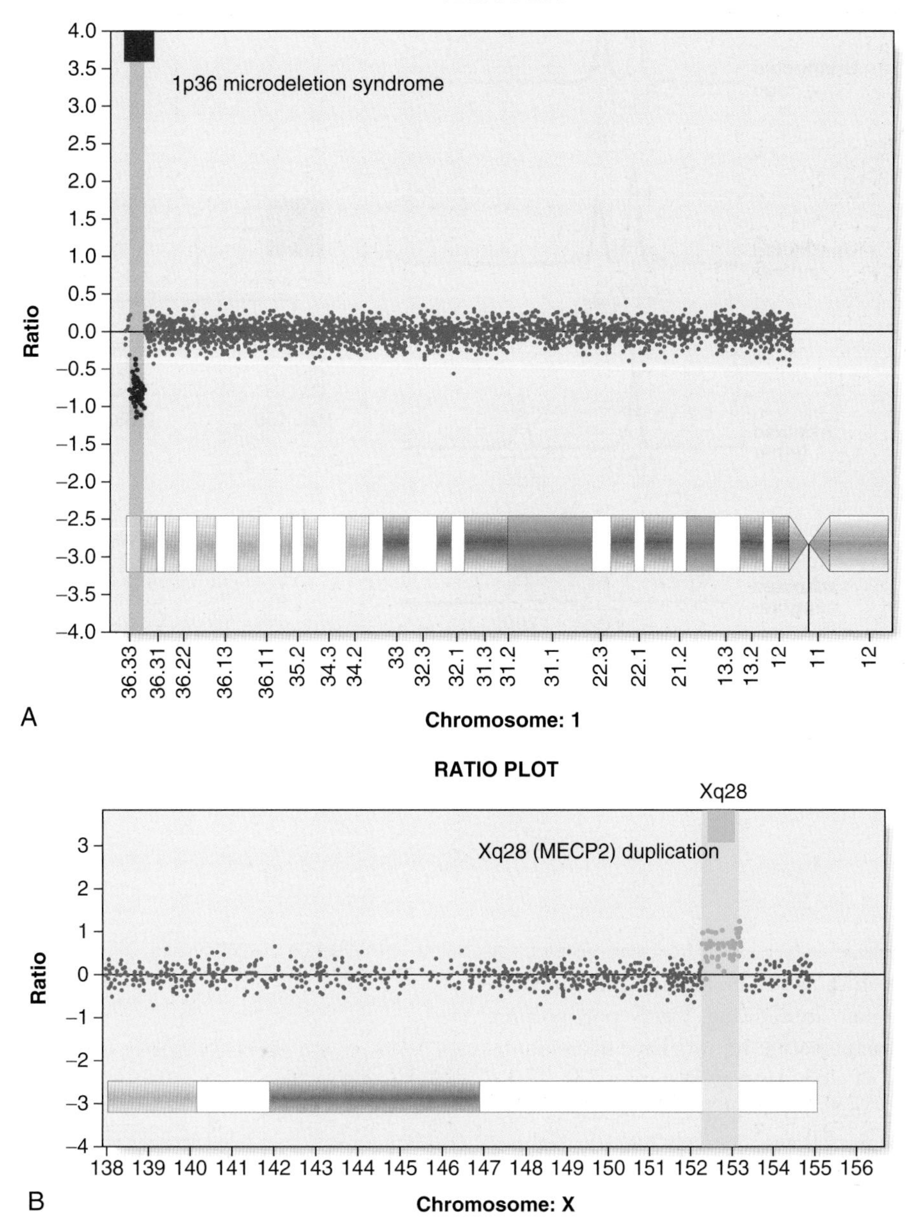

FIGURE 4.22 Identification of copy number changes by array comparative genomic hybridization (CGH) (this array includes 135,000 oligonucleotide probes). **A**, A patient with the 1p36 microdeletion syndrome. **B**, An *MECP2* duplication of chromosome Xq28. (Courtesy Rodger Palmer, North East Thames Regional Genetics Service Laboratories, Great Ormond Street Hospital for Children, London.)

Minisatellites

Alec Jeffreys identified a short 10-bp to 15-bp 'core' sequence with homology to many highly variable loci spread throughout the human genome (p. 17). Using a probe containing tandem repeats of this core sequence, a pattern of hypervariable DNA fragments could be identified. The multiple variable-size repeat sequences identified by the core sequence are known as **minisatellites**. These minisatellites are highly polymorphic, and a profile unique to an individual (unless they have an identical twin!) is described as a **DNA fingerprint**. The technique of DNA fingerprinting is used widely in paternity testing and for forensic purposes.

Microsatellites

The human genome contains some 50,000 to 100,000 blocks of a variable number of tandem repeats of the dinucleotide CA:GT, so-called CA repeats or **microsatellites** (p. 18). The difference in the number of CA repeats at any one site between individuals is highly polymorphic and

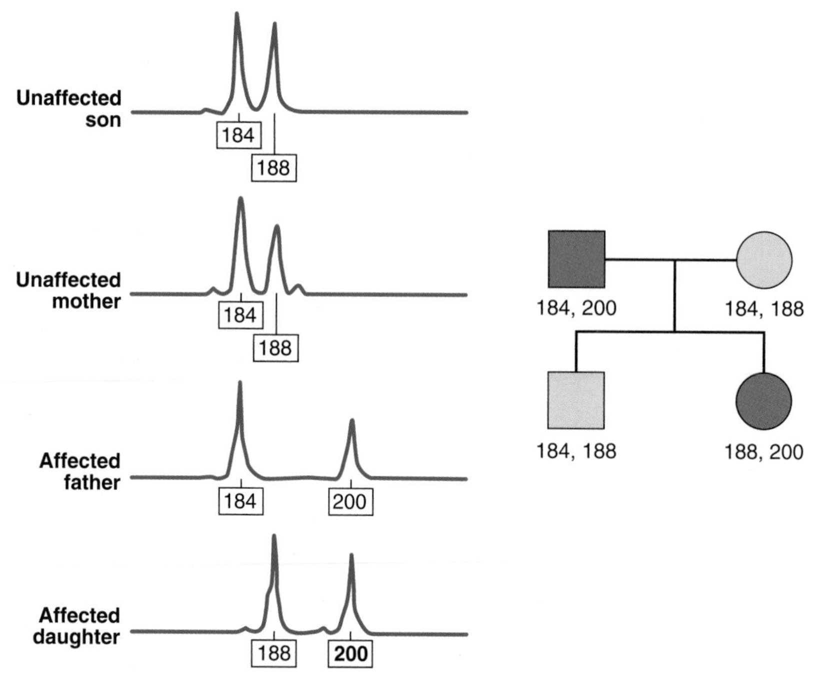

FIGURE 4.23 Analysis of a tetranucleotide microsatellite marker in a family with a dominant disorder. *Genotyper* software was used to label the peaks with the size of the polymerase chain reaction (PCR) products. The 200-bp allele is segregating with the disorder in the affected members of the family. (Courtesy M. Owens, Department of Molecular Genetics, Royal Devon and Exeter Hospital, Exeter, UK.)

these repeats have been shown to be inherited in a mendelian co-dominant manner. In addition, highly polymorphic trinucleotide and tetranucleotide repeats have been identified, and can be used in a similar way (Figure 4.23). These microsatellites can be analyzed by PCR and the use of fluorescent detection systems allows relatively high-throughput analysis. Consequently, microsatellite analysis has replaced DNA fingerprinting for paternity testing and establishing zygosity.

Clinical Applications of Gene Tracking

If a gene has been mapped by linkage studies but not identified, it is possible to use the linked markers to 'track' the mutant haplotype within a family. This approach may also be used for known genes where a familial mutation has not been found. Closely flanking or intragenic microsatellites are used most commonly, because of the lower likelihood of finding informative SNPs within families. Figure 4.24 illustrates a family in which gene tracking has been used to determine carrier risk in the absence of a known mutation. There are some pitfalls associated with this method: recombination between the microsatellite and the gene may give an incorrect risk estimate, and the possibility of genetic heterogeneity (where mutations in more than one gene cause a disease) should be borne in mind.

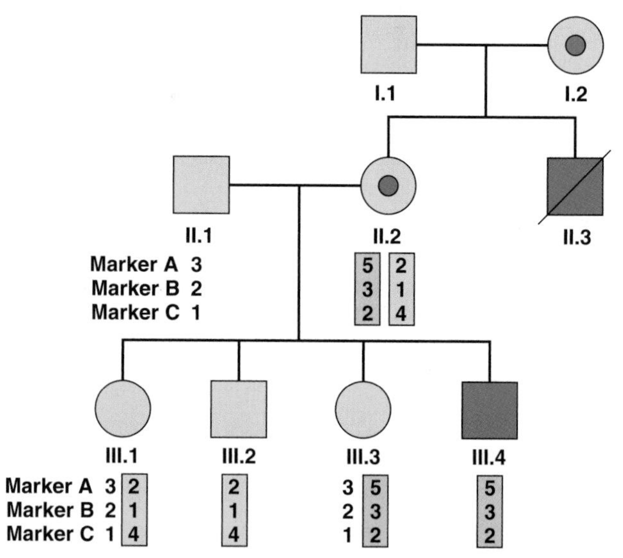

FIGURE 4.24 Gene tracking in a family with Duchenne muscular dystrophy where no mutation has been found in the affected proband, III.4. Analysis of markers A, B, and C has enabled the construction of haplotypes; the affected haplotype is shown by an orange box. Both of the proband's sisters were at 50% prior risk of being carriers. Gene tracking shows that III.1 has inherited the low-risk haplotype and is unlikely to be a carrier, but III.3 has inherited the high-risk haplotype and is therefore likely to be a carrier of Duchenne muscular dystrophy. The risk of recombination should not be forgotten.

Diagnosis in Non-Genetic Disease

DNA technology, especially PCR, has found application in the diagnosis and management of both infectious and malignant disease.

Infectious Disease

PCR can be used to detect the presence of DNA sequences specific to a particular infectious organism before conventional evidence such as an antibody response or the results of cultures is available. An example is the screening of blood products for the presence of DNA sequences from the human immunodeficiency virus (HIV) to ensure the safety of their use (e.g., screening pooled factor VIII concentrate for use in males with hemophilia A). Another example is the identification of DNA sequences specific to bacterial or viral organisms responsible for acute overwhelming infections, where early diagnosis allows prompt institution of the correct antibiotic or antiviral agent with the prospect of reducing morbidity and mortality. Real-time PCR techniques can generate rapid results, with some test results being available within 1 hour of a sample being taken. This methodology is particularly useful in the fight against methicillin-resistant *Staphylococcus aureus* (MRSA), as patients can be rapidly tested on admission to hospital. Anyone found to be MRSA-positive can be isolated to minimize the risk of infection to other patients.

Malignant Disease

PCR may assist in the diagnosis of lymphomas and leukemias by identifying translocations, for example t(9;22), which is characteristic of chronic myeloid leukemia (CML). The extreme sensitivity of PCR means that minimal residual disease may be detected after treatment for these disorders, and early indication of impending relapse will inform treatment options. For example, all patients with CML treated with the tyrosine kinase inhibitor Imatinib are regularly monitored as resistant clones may develop. After bone marrow transplantation, microsatellite markers may be used to monitor the success of engraftment by analysis of donor- and patient-specific alleles.

FURTHER READING

Elles R, Wallace A 2010 Molecular diagnosis of genetic disease, 3rd ed. Clifton, NJ: Humana Press

Key techniques used for genetic testing of common disorders in diagnostic laboratories.

Strachan T, Read AP 2011 Human molecular genetics, 4th ed. Garland Science, London

A comprehensive textbook of all aspects of molecular and cellular biology as related to inherited disease in humans.

Weatherall DJ 1991 The new genetics and clinical practice, 3rd ed. Oxford: Oxford Medical

One of the original texts that provided a lucid overview of the application of DNA techniques in clinical medicine.

ELEMENTS

1 Restriction enzymes allow DNA from any source to be cleaved into reproducible fragments based on the presence of specific nucleotide recognition sequences. These fragments can be made to recombine, enabling their incorporation into a suitable vector, with subsequent transformation of a host organism by the vector, leading to the production of clones containing a particular DNA sequence.

2 Polymerase chain reaction (PCR) has revolutionized medical genetics. Within hours, more than a million copies of a gene can be amplified from a patient's DNA sample. The PCR product may be analyzed for the presence of a pathogenic mutation, gene rearrangement, or infectious agent.

3 Techniques including Southern and Northern blotting, DNA sequencing, and mutation screening, real-time PCR, and microarray analysis can be used to identify or analyze specific DNA sequences of interest. These techniques can be used for analyzing normal gene structure and function as well as revealing the molecular pathology of inherited disease. This provides a means for presymptomatic diagnosis, carrier detection and prenatal diagnosis, either by direct mutational analysis or indirectly using polymorphic markers in family studies.

4 Single nucleotide polymorphism microarrays ('chips'), array comparative genomic hybridization, and next-generation sequencing techniques allow genome wide analysis of single nucleotide polymorphisms, copy number variants, and sequence variants. These methods have changed the scale of genetic analysis and provided novel insights into genetic disease.

Positional Cloning

Positional cloning describes the identification of a disease gene through its location in the human genome, without prior knowledge of its function. It is also described as **reverse genetics** as it involves an approach opposite to that of functional cloning, in which the protein is the starting point.

Linkage Analysis

Genetic mapping, or linkage analysis (p. 137), is based on genetic distances that are measured in centimorgans (cM). A genetic distance of 1 cM is the distance between two genes that show 1% recombination, that is, in 1% of meioses the genes will not be co-inherited and is equivalent to approximately 1 Mb (1 million bases). Linkage analysis is the first step in positional cloning that defines a genetic interval for further analysis.

Linkage analysis can be performed for a single, large family or for multiple families, although this assumes that there is no genetic heterogeneity (p. 378). The use of genetic markers located throughout the genome is described as a **genome-wide scan**. In the 1990s, genome-wide scans used microsatellite markers (a commercial set of 350 markers was popular), but microarrays with several million SNPs now provide greater statistical power.

Autozygosity mapping (also known as homozygosity mapping) is a powerful form of linkage analysis used to map autosomal recessive disorders in consanguineous pedigrees (p. 269). Autozygosity occurs when affected members of a family are homozygous at particular loci because they are identical by descent from a common ancestor.

Linkage of cystic fibrosis (CF) to chromosome 7 was found by testing nearly 50 white families with hundreds of DNA markers. The gene was mapped to a region of 500 kilobases (kb) between markers *MET* and *D7S8* at chromosome band 7q31-32, when it became evident that the majority of CF chromosomes had a particular set of alleles for these markers (shared haplotype) that was found in only 25% of non-CF chromosomes. This finding is described as **linkage disequilibrium** and suggests a common mutation from a founder effect (p. 378). Extensive physical mapping studies eventually led to the identification of four genes within the genetic interval identified by linkage analysis, and in 1989 a 3-bp deletion was found within the cystic fibrosis transmembrane receptor *(CFTR)* gene. This mutation (p.Phe508del) was present in approximately 70% of CF chromosomes and 2% to 3% of non-CF chromosomes, consistent with the carrier frequency of 1 in 25 in whites.

Contig Analysis

The aim of linkage analysis is to reduce the region of linkage as far as possible to identify a candidate region. Before publication of the human genome sequence, the next step was to construct a **contig**. This contig would contain a series of overlapping fragments of cloned DNA representing the entire candidate region. These cloned fragments were then used to screen cDNA libraries, to search for CpG islands (which are usually located close to genes), for zoo blotting (selection based on evolutionary conservation) and exon trapping (to identify coding regions via functional splice sites). The requirement for cloning the region of interest led to the phrase 'cloning the gene' for a particular disease.

Chromosome Abnormalities

Occasionally, individuals are recognized with single-gene disorders who are also found to have structural chromosomal abnormalities. The first clue that the gene responsible for Duchenne muscular dystrophy (DMD) (p. 307) was located on the short arm of the X chromosome was the identification of a number of females with DMD who were also found to have a chromosomal rearrangement between an autosome and a specific region of the short arm of one of their X chromosomes. Isolation of DNA clones spanning the region of the X chromosome involved in the rearrangement led in one such female to more detailed gene-mapping information as well as to the eventual cloning of the *DMD* or dystrophin gene (p. 307).

At the same time as these observations, a male was reported with three X-linked disorders: DMD, chronic granulomatous disease, and retinitis pigmentosa. He also had an unusual X-linked red cell group known as the McLeod phenotype. It was suggested that he could have a deletion of a number of genes on the short arm of his X chromosome, including the *DMD* gene, or what is now termed a **contiguous gene syndrome**. Detailed prometaphase chromosome analysis revealed this to be the case. DNA from this individual was used in vast excess to hybridize in competitive reassociation, under special conditions, with DNA from persons with multiple X chromosomes to enrich for DNA sequences that he lacked, the so-called ***p****henol* ***e****nhanced* ***r****eassociation* ***t****echnique*, or pERT, which allowed isolation of DNA clones containing portions of the *DMD* gene.

The occurrence of a chromosome abnormality and a single-gene disorder is rare, but identification of such individuals is important as it has led to the cloning of several other important disease genes in humans, such as tuberous sclerosis (p. 316) and familial adenomatous polyposis (p. 221).

Candidate Genes

Searching databases for genes with a function likely to be involved in the pathogenesis of the inherited disorder can also suggest what are known as **candidate genes**. If a disease has been mapped to a particular chromosomal region, any gene mapping to that region is a positional candidate gene. Data on the pattern of expression, the timing, and the distribution of tissue and cells types may suggest that a certain positional candidate gene or genes is more likely to be responsible for the phenotypic features seen in persons affected with a particular single-gene disorder. Several computer programs have been developed that can search genomic DNA sequence databases for sequence homology to known genes, as well as DNA sequences specific to all genes, such as the conserved intron–exon splice junctions, promoter sequences, polyadenylation sites and stretches of open reading frames (ORFs).

Identification of a gene with homology to a known gene causing a recognized inherited disorder can suggest it as a possible candidate gene for other inherited disorders with a similar phenotype. For example, the identification of mutations in the connexin 26 gene, which codes for one of the proteins that constitute the gap junctions between cells causing sensorineural hearing impairment or deafness, has led to the identification of other connexins responsible for inherited hearing impairment or deafness.

Confirmatory Testing that a Candidate Gene Is a Disease Gene

Mutations in candidate genes can be screened for by a variety of methods (p. 59) and confirmed by DNA sequencing (p. 61). Finding loss-of-function mutations or multiple different mutations that result in the same phenotype provides convincing evidence that a potential candidate gene is associated with a disorder. For example, in the absence of functional data to demonstrate the effect of the p.Phe508del mutation on the CFTR protein, confirmation that mutations in the *CFTR* gene caused cystic fibrosis was provided by the nonsense mutation p.Gly542X.

Further support is provided by the observation that the candidate gene is expressed in the appropriate tissues and at the relevant stages of development. The production of a transgenic animal model by the targeted introduction of the mutation into the homologous gene in another species that is shown to exhibit phenotypic features similar to those seen in persons affected with the disorder, or restoration of the normal phenotype by transfection of the normal gene into a cell line, provides final proof that the candidate gene and the disease gene are one and the same.

The Human Gene Map

The rate at which single-gene disorders and their genes are being mapped in humans is increasing exponentially (see Figure 1.6, p. 7). Many of the more common and clinically important monogenic disorders have been mapped to produce the 'morbid anatomy of the human genome' (Figure 5.3).

The Human Genome Project

Beginning and Organization of the Human Genome Project

The concept of a map of the human genome was proposed as long ago as 1969 by Victor McKusick (see Figure 1.5, p. 7), one of the founding fathers of medical genetics. Human gene mapping workshops were held regularly from 1973 to collate the mapping data. The idea of a dedicated human genome project came from a meeting organized by the US Department of Energy at Sante Fe, New Mexico, in 1986. The US Human Genome Project started in 1991 and is estimated to have cost around 2.7 billion US dollars. Other nations, notably France, the UK, and Japan, soon followed with their own major national human genome programs and were subsequently joined by a number of other countries. These individual national projects were all coordinated by the Human Genome Organization, which has three centers, one for the Americas based in Bethesda, Maryland, one for Europe located in London, and one for the Pacific in Tokyo.

Although the key objective of the Human Genome Project was to sequence all 3×10^9 base pairs of the human genome, this was just one of the six main objectives/areas of work of the Human Genome Project.

Human Gene Maps and Mapping of Human Inherited Diseases

Designated genome mapping centers with ear-marked funding were involved in the coordination and production of genetic or recombination and physical maps of the human genome. The genetic maps initially involved the production of fairly low-level resolution index, skeleton or framework maps, which were based on polymorphic variable-number di-, tri-, and tetranucleotide tandem repeats (p. 17) spaced at approximately 10-cM intervals throughout the genome.

The mapping information from these genetic maps was integrated with high-resolution physical maps (Figure 5.4). Access to the detailed information from these high-resolution genetic and physical maps allowed individual research groups, often interested in a specific or particular inherited disease or group of diseases, rapidly and precisely to localize or map a disease gene to a specific region of a chromosome.

Development of New DNA Technologies

A second major objective was the development of new DNA technologies for human genome research. For example, at the outset of the Human Genome Project, the technology involved in DNA sequencing was very time consuming, laborious and relatively expensive. The development of high-throughput automated capillary sequencers and robust fluorescent sequencing kits transformed the ease and cost of large-scale DNA sequencing projects.

Sequencing of the Human Genome

Although sequencing of the entire human genome would have been seen to be the obvious main focus of the Human Genome Project, initially it was not the straightforward proposal it seemed. The human genome contains large sections of repetitive DNA (p. 15) that were technically difficult to clone and sequence. In addition, it would seem a waste of time to collect sequence data on the entire genome when only a small proportion is made up of expressed sequences or genes, the latter being most likely to be the regions of greatest medical and biological importance. Furthermore, the sheer magnitude of the prospect of sequencing all 3×10^9 base pairs of the human genome seemed

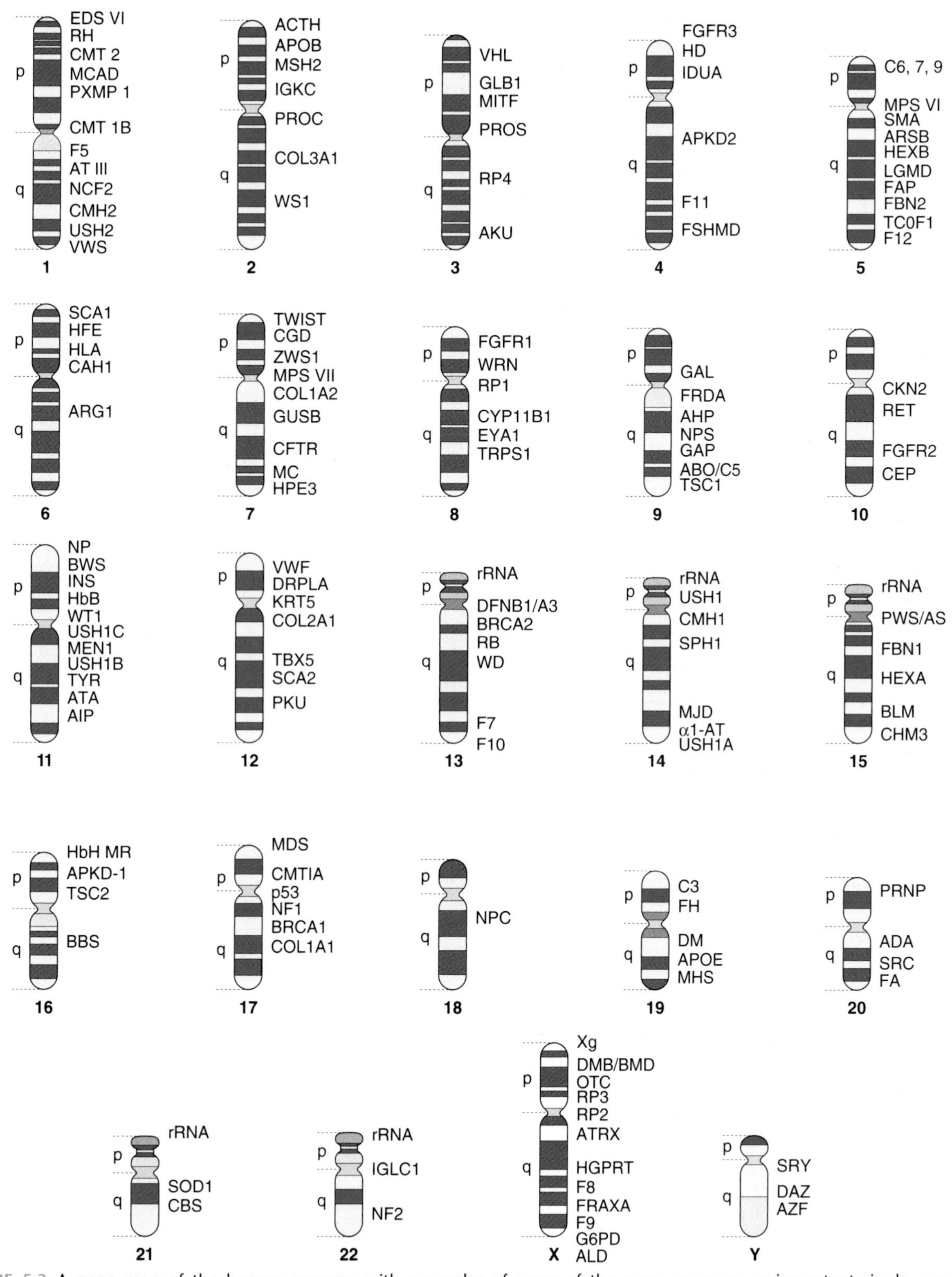

FIGURE 5.3 A gene map of the human genome with examples of some of the more common or important single genes and disorders.

α1-AT	14q32	α_1-Antitrypsin deficiency
ABO	9q34	ABO blood group
ACTH	2p25	Adrenocorticotrophic hormone deficiency
ADA	20q13.11	Severe combined immunodeficiency, ADA deficiency
AHP	9q34	Acute hepatic porphyria
AIP	11q23.3	Acute intermittent porphyria
AKU	3q2	Alkaptonuria

FIGURE 5.3, cont'd

ALD	Xq28	Adrenoleukodystrophy
APKD1	16p13	Adult polycystic kidney disease, locus 1
APKD2	4q21–23	Adult polycystic kidney disease, locus 2
APOB	2p24	Apolipoprotein B
APOE	19q.13.2	Apolipoprotein E
ARG1	6q23	Arginase deficiency, argininemia
ARSB	5q11–13	Mucopolysaccharidosis type VI, Maroteaux-Lamy syndrome
AS	15q11–13	Angelman syndrome
ATA	11q22.3	Ataxia telangiectasia
ATIII	1q23–25	Antithrombin III
ATRX	Xq13	α-Thalassemia mental retardation
AZF	Yq11	Azoospermia factor
BBS2	16q21	Bardet–Biedl syndrome
BLM	15q26.1	Bloom syndrome
BRCA1	17q21	Familial breast/ovarian cancer, locus 1
BRCA2	13q12.3	Familial breast/ovarian cancer, locus 2
BWS	11p15.4	Beckwith–Wiedemann syndrome
C3	19p13.2-13.3	Complement factor 3
C5	9q34.1	Complement factor 5
C6	5p13	Complement factor 6
C7	5p13	Complement factor 7
C9	5p13	Complement factor 9
CAH1	6p21.3	Congenital adrenal hyperplasia, 21-hydroxylase
CBS	21q22.3	Homocystinuria
CEP	10q25.2-26.3	Congenital erythropoietic porphyria
CFTR	7q31.2	Cystic fibrosis transmembrane conductance regulator
CKN2	10q11	Cockayne syndrome 2, late onset
CMH1	14q12	Hypertrophic obstructive cardiomyopathy type 1
CMH2	1q3	Hypertrophic obstructive cardiomyopathy type 2
CMH3	15q22	Hypertrophic obstructive cardiomyopathy type 3
CMT1A	17p11.2	Charcot–Marie–Tooth disease type 1A
CMT1B	1q22	Charcot–Marie–Tooth disease type 1B
CMT2	1p35–36	Charcot–Marie–Tooth disease type 2
COL1A1	17q21.31-22	Collagen type I, α_1 chain, osteogenesis imperfecta
COL1A2	7q22.1	Collagen type I, α_2 chain, osteogenesis imperfect
COL2A1	12q13.11-13.2	Collagen type II, Stickler syndrome
COL3A1	2q31	Collagen type III, α_1 chain, Ehlers-Danlos syndrome type IV
CYP11B1	8q21	Congenital adrenal hyperplasia, 11β-hydroxylase
DAZ	Yq11	Deleted in azoospermia
DFNB1/A3	13q12	Non-syndromic sensorineural deafness, first recessive, third dominant locus
DM	19q13.2-13.3	Myotonic dystrophy
DMD/BMD	Xp21.2	Dystrophin, Duchenne and Becker muscular dystrophy
DRPLA	12p13.1-12.3	Dentatorubropallidoluysian disease
EDSVI	1p36.2-36.3	Ehlers-Danlos syndrome type VI
EYA1	8q13.3	Brachio-otorenal syndrome
F5	1q23	Coagulation protein V
F7	13q34	Coagulation protein VII
F8	Xq28	Coagulation protein VIII, hemophilia A
F9	Xq27.1-27.2	Coagulation protein IX, Christmas disease, hemophilia B
F10	13q34	Coagulation protein X
F11	Xq27.1-27.2	Coagulation factor XI
F12	5q33-qter	Coagulation factor XII
FAP	5q21-22	Familial adenomatous polyposis, Gardner syndrome
FBN1	15q21.1	Fibrillin-1, Marfan syndrome
FBN2	5q23-31	Fibrillin-2, contractural arachnodactyly
FGFR1	8p11.1-11.2	Fibroblast growth factor receptor 1, Pfeiffer syndrome
FGFR2	10q26	Fibroblast growth factor receptor 2, Crouzon, Pfeiffer, Apert syndrome
FGFR3	4p16.3	Fibroblast growth factor receptor 3, achondroplasia, thanatophoric dysplasia
FH	19p13.1-13.2	Familial hypercholesterolemia
FRAXA (FMR1)	Xq27.3	Fragile X mental retardation
FRDA	9q13–21.1	Friedreich ataxia
FSHMD	4q35	Facioscapulohumeral muscular dystrophy
GAL	9p13	Galactosemia
GAP	9q31	Basal cell nevus syndrome, Gorlin syndrome
GLB1	3p21.33	GM1 gangliosidosis
G6PD	Xq28	Glucose-6-phosphate dehydrogenase
GUSB	7q21.11	Mucopolysaccharidosis type VII, Sly syndrome

FIGURE 5.3, cont'd

HbB	11p15.5	β-Globin gene
HD	4p16.3	Huntington disease
HEXA	15q23–24	Hexosaminidase A, Tay-Sachs disease
HEXB	5q13	Hexosaminidase B, Sandhoff disease
HFE	6p21.3	Hemochromatosis
HGPRT	Xq26-27.2	Hypoxanthine guanine phosphoribosyl transferase, Lesch-Nyhan syndrome
HLA	6p21.3	Major histocompatibility locus
HPE3	7q36	Holoprosencephaly
IDUA	4p16.3	Mucopolysaccharidosis type I, Hurler syndrome
IGKC	2p12	Immunoglobulin κ light chain
IGLC1	22q11	Immunoglobulin λ light chains
INS	11p15.5	Insulin-dependent diabetes mellitus type 2
KRT5	12q11-13	Epidermolysis bullosa simplex, Koebner type
LGMD7	5q31	Limb-girdle muscular dystrophy
MCAD	1p31	Acyl coenzyme-A dehydrogenase, medium chain
MDS	17p13.3	Miller-Dieker lissencephaly syndrome
MEN1	11q13	Multiple endocrine neoplasia syndrome type 1
MHS	19q13.1	Malignant hyperpyrexia susceptibility, locus 1
MITF	3p14.1	Waardenburg syndrome type 2
MJD	14q24.3-31	Machado-Joseph disease, spinocerebellar ataxia type 3
MPS VI	5q11-13	Maroteaux-Lamy syndrome
MSH2	2p15-16	Hereditary non-polyposis colorectal cancer type 1
NCF2	1q25	Chronic granulomatous disease, neutrophil cytosolic factor-2 deficiency
NF1	17q11.2	Neurofibromatosis type I, von Recklinghausen disease
NF2	22q12.2	Neurofibromatosis type II, bilateral acoustic neuroma
NP	11p15.1-15.4	Niemann-Pick disease type A and B
NPC	18q11-12	Niemann-Pick disease type C
NPS	9q43	Nail-patella syndrome
OTC	Xp21.1	Ornithine transcarbamylase
p53	17p13.1	p53 protein, Li-Fraumeni syndrome
PKU	12q24.1	Phenylketonuria
PROC	2q13-14	Protein C, coagulopathy disorder
PROS	3p11.1-q11.2	Protein S, coagulopathy disorder
PRNP	20p12-pter	Prion disease protein
PWS	15q11	Prader-Willi syndrome
PXMP1	1p21–22	Zellweger syndrome type 2
RB	13q14.1-14.2	Retinoblastoma
RET	10q11.2	Familial medullary thyroid carcinoma, MEN 2A and 2B, familial Hirschsprung disease
RH	1p34–36.2	Rhesus null disease, Rhesus blood group
RP1	8p11-q21	Retinitis pigmentosa, locus 1
RP2	Xp11.3	Retinitis pigmentosa, locus 2
RP3	Xp21.1	Retinitis pigmentosa, locus 3
rRNA		Ribosomal RNA
SCA1	6p23	Spinocerebellar ataxia, locus 1
SCA2	12q24	Spinocerebellar ataxia, locus 2
SPH1	14q22-23.2	Spherocytosis type I
SMA	5q12.2-13.3	Spinal muscular atrophy
SOD1	21q22.1	Superoxide dismutase, familial motor neuron disease
SRY	Yp11.3	Sex-determining region Y, testis-determining factor
TBX5	12q21.3-22	Holt-Oram syndrome
TCOF1	5q32-33.1	Treacher-Collins syndrome
TRPS1	8q24.12	Trichorhinophalangeal syndrome
TSC1	9q34	Tuberous sclerosis, locus 1
TSC2	16p13.3	Tuberous sclerosis, locus 2
TYR	11q14-21	Oculocutaneous albinism
USH1A	14q32	Usher syndrome type IA
USH1B	11q13.5	Usher syndrome type IB
USH1C	11p15.1	Usher syndrome type IC
USH2	1q41	Usher syndrome type II
VWS	1q32	van der Woude syndrome
VHL	3p25–26	von Hippel-Lindau syndrome
VWF	12p13.3	von Willebrand disease
WD	13q14.3-21.1	Wilson disease
WRN	8p11.2-12	Werner syndrome
WS1	2q35	Waardenburg syndrome type 1
WT1	11p13	Wilms tumor 1 gene
ZWS1	7q11.23	Zellweger syndrome type 1

overwhelming. With conventional sequencing technology, as was carried out in the early 1990s, it was estimated that a single laboratory worker could sequence up to approximately 2000 bp per day.

Projects involving sequencing of other organisms with smaller genomes showed how much work was involved as well as how the rate of producing sequence data increased with the development of new DNA technologies. For example, with initial efforts at producing genome sequence data for yeast, it took an international collaboration involving 35 laboratories in 17 countries from 1989 until 1995 to sequence just 315,000 bp of chromosome 3, one of the 16 chromosomes that make up the 14 million base pairs of the yeast genome. Advances in DNA technologies meant, however, that by the middle of 1995 more than half of the yeast genome had been sequenced, with the complete genomic sequence being reported the following year.

Further advances in DNA sequencing technology led to publication of the full sequence of the nematode *Caenorhabditis elegans* in 1998 and the 50 million base pairs of the DNA sequence of human chromosome 22 at the end of 1999. As a consequence of these technical developments, the 'working draft' sequence, covering 90% of the human genome, was published in February 2001. The finished sequence (more than 99% coverage) was announced more than 2 years ahead of schedule in April 2003, the 50th anniversary of the discovery of the DNA double helix. Researchers now have access to the full catalog of 25,000 to 30,000 genes, and the human genome sequence will underpin biomedical research for decades to come.

Although the Human Genome Sequencing Project is complete, a number of new projects have been initiated as a direct consequence, including the Cancer Genome, HapMap (p. 148), and 1000 Genomes (p. 150) projects.

Development of Bioinformatics

Bioinformatics was essential to the overall success of the Human Genome Project. This is the establishment of facilities for collecting, storing, organizing, interpreting, analyzing, and communicating the data from the project, which can be widely shared by the scientific community at large. It was vital for anyone involved in any aspect of the Human Genome Project to have rapid and easy access to the data/information arising from it. This dissemination of information was met by the establishment of a large number of electronic databases available on the World Wide Web on the Internet (see Appendix). These include protein and DNA sequence databases (e.g., GenBank, EMBL), databases of genetic maps for humans (such as the GDB, Genethon, CEPH, CHLC, and the Whitehead Institute sites) and other species (the Mouse Genome Database and the *C. elegans* database), linkage analysis programs (e.g., the Rockefeller University website), annotated genome data (Ensembl and UCSC Genome Bioinformatics) and the catalog of inherited diseases in humans (Online Mendelian Inheritance in Man, or OMIM).

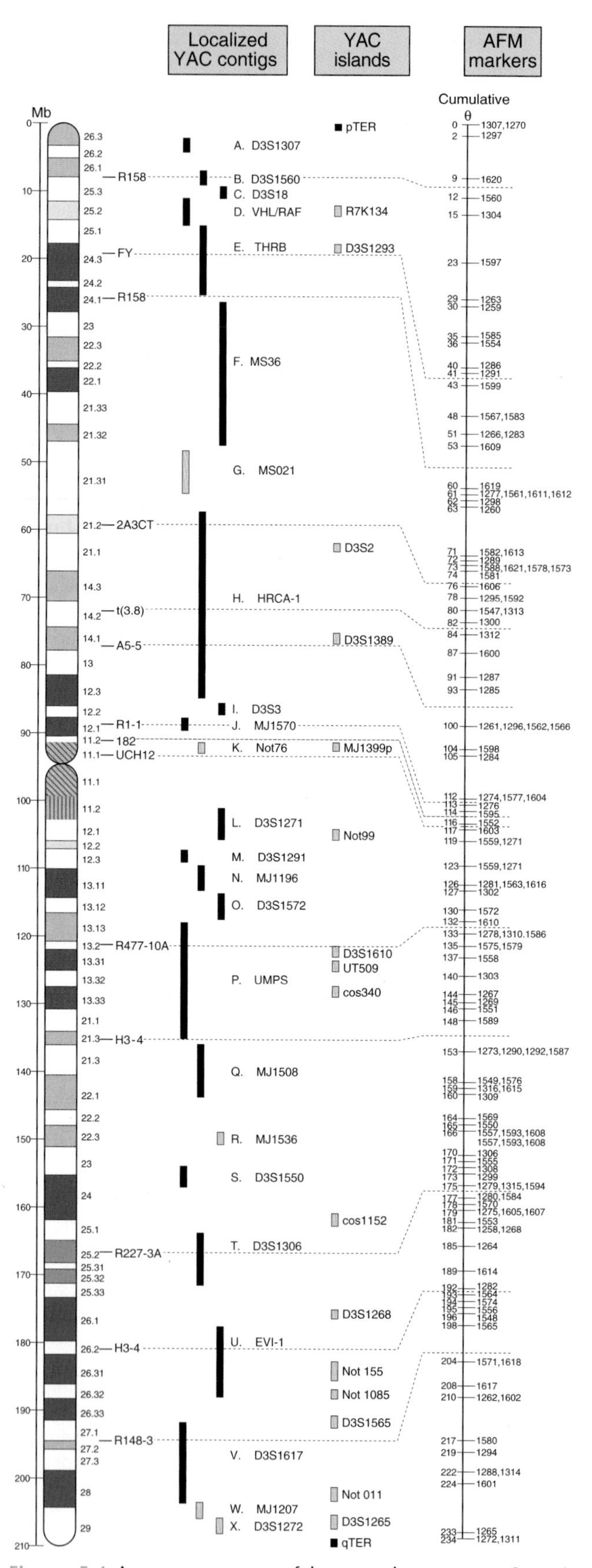

Figure 5.4 A summary map of human chromosome 3, estimated to be 210 Mb in size, which integrates physical mapping data covered by 24 YAC contigs and the Genethon genetic map with cumulative map distances. (From Gemmill RM, Chumakov I, Scott P, et al 1995 A second-generation YAC contig map of human chromosome 3. Nature 377:299–319; with permission.)

These developments in bioinformatics now allow the prospect of identifying coding sequences and determining their likely function(s) from homologies to known genes, leading to the prospect of identifying a new gene without the need for any laboratory experimental work, or what has been called 'cloning in silico'.

Comparative Genomics

In addition to the Human Genome Project, there were separate genome projects for a number of other species, for what are known as 'model organisms'. These included various prokaryotic organisms such as the bacteria *E. coli* and *Haemophilus influenzae*, as well as eukaryotic organisms such as *Saccharomyces cerevisiae* (yeast), C. *elegans* (flatworm), *Drosophila melanogaster* (fruit fly), *Mus musculus* (mouse), *Rattus norvegicus* (rat), *Fugu rubripes rupripes* (puffer fish), mosquito and zebrafish. These *comparative genomics* projects identified many novel genes and were of vital importance in the Human Genome Project because mapping the human homologs provided new 'candidate' genes for inherited diseases in humans.

Functional Genomics

The second major way in which model organisms proved to be invaluable in the Human Genome Project was by providing the means to follow the expression of genes and the function of their protein products in normal development as well as their dysfunction in inherited disorders. This is referred to as **functional genomics**.

The ability to introduce targeted mutations in specific genes, along with the production of transgenic animals (p. 102), for example in the mouse, allows the production of animal models to study the pathodevelopmental basis for inherited human disorders, as well as serve as a test system for the safety and efficacy of gene therapy and other treatment modalities (p. 350). Strategies using different model organisms in a complementary fashion, taking into account factors such as the ease or complexity of producing transgenic organisms and the generation times of different species, allow the possibility of relatively rapid analysis of gene expression, function and interactions in providing an understanding of the complex pathobiology of inherited diseases in humans.

Ethical, Legal, and Social Issues of the Human Genome Project

The rapid advances in the science and application of developments from the Human Genome Project have presented complex ethical issues for both the individual and society. These issues include ones of immediate practical relevance, such as who owns and should control genetic information with respect to privacy and confidentiality; who is entitled to access to it and how; whether it should be used by employers, schools, etc.; the psychological impact and potential stigmatization of persons positive for genetic testing; and the use of genetic testing in reproductive decision making. Other issues include the concept of disability/differences that have a genetic basis in relation to the treatment of genetic disorders or diseases by gene therapy and the possibility of genetic enhancement (i.e., using gene therapy to supply certain characteristics, such as height.) Last, issues need to be resolved with regard to the appropriateness and fairness of the use of the genetic technologies that come out of the Human Genome Project, with prioritization of the use of public resources and commercial involvement and property rights, especially with regard to patenting.

FURTHER READING

Botstein D, White RL, Skolnick M, Davis RW 1980 Construction of a genetic linkage map in man using restriction fragment length polymorphisms. Am J Hum Genet 32:314–331

One of the original papers describing the concept of linked restriction fragment length polymorphisms.

Kerem B, Rommens JM, Buchanan JA, et al 1989 Identification of the cystic fibrosis gene. Genetic analysis. Science 245:1073–1080

Original paper describing cloning of the cystic fibrosis gene.

McKusick VA 1998 Mendelian inheritance in man, 12th ed. Johns Hopkins University Press, London

A computerized catalog of the dominant, recessive, and X-linked mendelian traits and disorders in humans with a brief clinical commentary and details of the mutational basis, if known. Also available online, updated regularly.

Ng SB, Buckingham KJ, Lee C et al 2010 Exome sequencing identifies the cause of a mendelian disorder. Nat Genet 42:30–35

The first publication describing the use of next generation sequencing to elucidate the genetic aetiology of Miller syndrome.

Royer-Pokora B, Kunkel LM, Monaco AP, et al 1985 Cloning the gene for an inherited human disorder—chronic granulomatous disease—on the basis of its chromosomal location. Nature 322: 32–38

Original paper describing the identification of a disease gene through contiguous chromosome deletions.

Strachan T, Read AP 2011 Human molecular genetics, 4th ed. London: Garland Science

A comprehensive textbook of all aspects of molecular and cellular biology as related to inherited disease in humans.

Sulston J 2002 The common thread: a story of science, politics, ethics and the human genome. London: Joseph Henry Press

A personal account of the human genome sequencing project by the man who led the UK team of scientists.

ELEMENTS

1. Position-independent methods for the identification of monogenic disorders include functional cloning to identify genes from knowledge of the protein sequence and the use of animal models. A technique to identify novel trinucleotide repeat expansions led to the identification of the *SCA8* disease locus.

2. Positional cloning describes the identification of a gene on the basis of its location in the human genome. Chromosome abnormalities may assist this approach by highlighting particular chromosome regions of interest. Genetic databases with human genome sequence data now make the possibility of identifying genes 'in silico' a reality.

3. Confirmation that a specific gene is responsible for a particular inherited disorder can be obtained by tissue and developmental expression studies, in-vitro cell culture studies, or the introduction and analysis of mutations in a homologous gene in another species. As a consequence, the 'anatomy of the human genome' is continually being unraveled.

4. One of the goals of the Human Genome Project was to sequence the human genome. The sequencing was completed by an international consortium in 2003, and has greatly facilitated the identification of human disease genes.

5. The development of next generation 'clonal' sequencing methods will facilitate the identification of novel monogenic disease genes.

CHAPTER 6

Developmental Genetics

The history of man for the nine months preceding his birth would, probably, be far more interesting and contain events of greater moment than all the three score and ten years that follow it.
SAMUEL TAYLOR COLERIDGE

At fertilization the nucleus from a spermatozoon penetrates the cell membrane of an oocyte to form a zygote. This single cell divides to become two, then four, and when the number has doubled some 50 times the resulting organism comprises more than 200 distinct cell types and a total cell number of about 10,000 trillion. This is a fully formed human being with complex biochemistry and physiology, capable of exploring the cosmos and identifying subatomic particles. Not surprisingly, biologists and geneticists are intrigued by the mechanisms of early development and, whilst many mysteries remain, the rate of progress in understanding key events and signaling pathways is rapid.

A fetus is recognizably human after about 12 weeks of pregnancy—the first trimester. Normal development requires an optimum maternal environment but genetic integrity is fundamental; this has given rise to the field of developmental genetics. Most of what we know about the molecular processes inevitably comes from the study of animal models, with great emphasis on the mouse, whose genome closely resembles our own.

Prenatal life can be divided into three main stages: **pre-embryonic**, **embryonic**, and **fetal** (Table 6.1). During the pre-embryonic stage, a small collection of cells becomes distinguishable, first as a double-layered or **bilaminar disc**, and then as a triple-layered or **trilaminar disc** (Figure 6.1), which is destined to develop into the human infant. During the embryonic stage, craniocaudal, dorsoventral, and proximodistal axes are established, as cellular aggregation and differentiation lead to tissue and organ formation. The final fetal stage is characterized by rapid growth and development as the embryo, now known as a fetus, matures into a viable human infant.

On average, this extraordinary process takes approximately 38 weeks. By convention pregnancy is usually dated from the first day of the last menstrual period, which usually precedes conception by around 2 weeks, so that the normal period of gestation is often stated (incorrectly) as lasting 40 weeks.

Fertilization and Gastrulation

Fertilization, the process by which the male and female gametes fuse, occurs in the fallopian tube. Of the 100 to 200 million spermatozoa deposited in the female genital tract, only a few hundred reach the site of fertilization. Of these, usually only a single spermatozoon succeeds in penetrating first the corona radiata, then the zona pellucida, and finally the oocyte cell membrane, whereupon the oocyte completes its second meiotic division (see Figure 3.15, p. 40). After the sperm has penetrated the oocyte and the meiotic process has been completed, the two nuclei, known as pronuclei, fuse, thereby restoring the diploid number of 46 chromosomes. This is a potentially chaotic molecular encounter with a high chance of failure, as we know from observations of the early human embryo from in-vitro fertilization programs. It may be likened, somewhat flippantly, to 'speed dating', whereby couples test whether they might be compatible on the basis of only a few minutes conversation.

Germ cell and very early embryonic development are two periods characterized by widespread changes in DNA methylation patterns—epigenetic reprogramming (see p. 103). Primordial germ cells are globally *de*methylated as they mature and are subsequently methylated de novo during gametogenesis, the time when most DNA methylation **imprints** are established. After fertilization a second wave of change occurs. The oocyte rapidly removes the methyl imprints from the sperm's DNA, which has the effect of resetting the developmental stopwatch to zero. By contrast, the *maternal* genome is more passively *de*methylated in such a way that imprinting marks resist demethylation. A third wave of methylation, de novo, establishes the somatic cell pattern of DNA methylation after implantation. These alternating methylation states help to control which genes are active, or expressed, at a time when two genomes, initially alien to each other, collide.

The fertilized ovum or zygote undergoes a series of mitotic divisions to consist of two cells by 30 hours, four cells by 40 hours, and 12 to 16 cells by 3 days, when it is known as a morula. A key concept in development at all stages is the emergence of **polarity** within groups of cells—part of the process of differentiation that generates multiple cell types with unique identities. Although precise

Table 6.1 Main Events in the Development of a Human Infant

Stage	Time from Conception	Length of Embryo/Fetus
Pre-Embryonic		
First cell division	30 h	
Zygote reaches uterine cavity	4 d	
Implantation	5–6 d	
Formation of bilaminar disc	12 d	0.2 mm
Lyonization in female	16 d	
Formation of trilaminar disc and primitive streak	19 d	1 mm
Embryonic Stage		
Organogenesis	4–8 w	
Brain and spinal cord are forming, and first signs of heart and limb buds	4 w	4 mm
Brain, eyes, heart and limbs developing rapidly, and bowel and lungs beginning to develop	6 w	17 mm
Digits have appeared. Ears, kidneys, liver and muscle are developing	8 w	4 cm
Palate closes and joints form	10 w	6 cm
Sexual differentiation almost complete	12 w	9 cm
Fetal Stage		
Fetal movements felt	16–18 w	20 cm
Eyelids open. Fetus is now viable with specialized care	24–26 w	35 cm
Rapid weight gain due to growth and accumulation of fat as lungs mature	28–38 w	40–50 cm

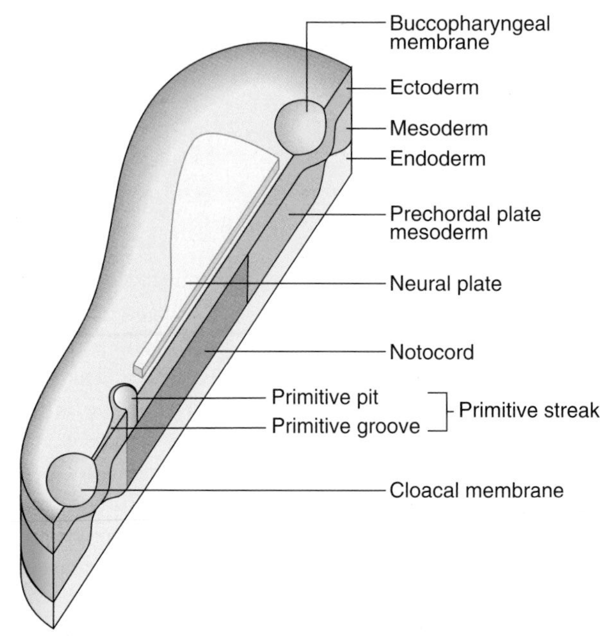

FIGURE 6.1 A schematic trilaminar disc, sectioned along the rostrocaudal axis. Cells from the future ectoderm *(top layer)* migrate through the primitive streak to form the endoderm *(bottom layer)* and mesoderm *(blue)*. Formation of the neutral plate in the overlying ectoderm, destined to be the central nervous system, involves sonic hedgehog signaling (p. 86) from the notochord and prechordial plate mesoderm. (Redrawn with permission from Larsen WJ 1998 Essentials of human embryology. New York: Churchill Livingstone.)

mechanisms remain elusive, observations suggest that this begins at the very outset; in the fertilized egg of the mouse, the point of entry of the sperm determines the plane through which the first cell cleavage division occurs. This seminal event is the first step in the development of the so-called dorso-ventral, or primary body, axis in the embryo.

Further cell division leads to formation of a **blastocyst**, which consists of an inner cell mass or **embryoblast**, destined to form the embryo, and an outer cell mass or **trophoblast**, which gives rise to the placenta. The process of converting the inner cell mass into first a bilaminar, and then a trilaminar, disc (see Figure 6.1) is known as **gastrulation**, and takes place between the beginning of the second and the end of the third weeks.

Between 4 and 8 weeks the body form is established, beginning with the formation of the primitive streak at the caudal end of the embryo. The germinal layers of the trilaminar disc give rise to **ectodermal**, **mesodermal,** and **endodermal** structures (Box 6.1). The neural tube is formed and neural crest cells migrate to form sensory ganglia, the sympathetic nervous system, pigment cells, and both bone and cartilage in parts of the face and branchial arches.

Disorders involving cells of neural crest origin, such as neurofibromatosis (p. 298), are sometimes referred to as **neurocristopathies**. This period between 4 and 8 weeks is described as the period of organogenesis, because during this interval all of the major organs are formed as regional specialization proceeds in a craniocaudal direction down the axis of the embryo.

Box 6.1 Organ and Tissue Origins

Ectodermal
Central nervous system
Peripheral nervous system
Epidermis, including hair and nails
Subcutaneous glands
Dental enamel

Mesodermal
Connective tissue
Cartilage and bone
Smooth and striated muscle
Cardiovascular system
Urogenital system

Endodermal
Thymus and thyroid
Gastrointestinal system
Liver and pancreas

Developmental Gene Families

Information about the genetic factors that initiate, maintain, and direct embryogenesis is incomplete. However, extensive genetic studies of the fruit fly, *Drosophila melanogaster*, and vertebrates such as mouse, chick, and zebrafish have identified several genes and gene families that play important roles in early developmental processes. It has also been possible through painstaking gene expression studies to identify several key developmental pathways, or cascades, to which more detail and complexity is continually being added. The gene families identified in vertebrates usually show strong sequence homology with developmental regulatory genes in *Drosophila*. Studies in humans have revealed that mutations in various members of these gene families can result in either isolated malformations or multiple congenital anomaly syndromes (see Table 16.5, p. 256). Many developmental genes produce proteins called **transcription factors** (p. 22), which control RNA transcription from the DNA template by binding to specific regulatory DNA sequences to form complexes that initiate transcription by RNA polymerase.

Transcription factors can switch genes on and off by activating or repressing gene expression. It is likely that important transcription factors control many other genes in coordinated sequential cascades and feedback loops involving the regulation of fundamental embryological processes such as **induction** (the process in which extracellular signals give rise to a change from one cell fate to another in a particular group of cells), **segmentation**, **migration**, **differentiation**, and **programmed cell death** (known as **apoptosis**). It is believed that these processes are mediated by growth factors, cell receptors, and chemicals known as **morphogens**. Across species the signaling molecules involved are very similar. The protein signals identified over and over again tend to be members of the *transforming growth factor-β (TGF-β)* family, the *wingless (Wnt)* family, and the *hedgehog (HH)* family (see the following section). In addition, it is clear within any given organism that the same molecular pathways are reused in different developmental domains. In addition, it has become clear that these pathways are closely interlinked with each other, with plenty of 'cross-talk'.

Early Patterning

The emergence of the mesoderm heralds the transition from the stage of **bilaminar** to **trilaminar** disc, or **gastrulation**. Induction of the mesoderm—the initiation, maintenance, and subsequent patterning of this layer—involves several key families of signaling factors. The *Nodal* family is involved in initiation, FGFs (fibroblast growth factors) and WNTs are involved in maintenance, and BMPs (bone morphogenetic proteins) are involved in patterning the mesoderm. Signaling pathways are activated when a key ligand binds specific membrane-bound protein receptors. This usually leads to the phosphorylation of a cytoplasmic factor, and this in turn leads to binding with other factor(s). These factors translocate to the nucleus where transcriptional activation of specific targets occurs.

In the case of Nodal and BMP pathways, ligand binding of a specific heterotetramer membrane-bound protein initiates the signaling, which is common to all members of the TGF-β family, the cytoplasmic mediators being SMAD factors (see the following section). The embryo appears to have gradients of Nodal activity along the dorsal-ventral axis, although the significance and role of these gradients in mesoderm induction are uncertain.

The WNT pathway has two main branches: one that is β-catenin–dependent (canonical) and the other independent of β-catenin. In the canonical pathway, Wnt ligand binds to a Frizzled/LRP heterodimer membrane-bound protein complex and the downstream intracellular signaling involves a G protein. The effect of this is to disrupt a large cytoplasmic protein complex that includes Axin, the adenomatous polyposis coli (APC; see p. 221) protein, and the glycogen synthase kinase-3β *(GSK-3β)* protein. This prevents the phosphorylation of β-catenin, but when β-catenin is not degraded, it accumulates and translocates to the nucleus where it activates the transcription of dorsal-specific regulatory genes. Binding of the ligand to the Fgf receptor results in dimerization of the receptor and transphosphorylation of the receptor's cytoplasmic domain, with activation of Ras and other kinases, one of which enters the nucleus and activates target transcription factors. Mutated *WNT10A* in man results in a form of ectodermal dysplasia (odonto-onychodermal dysplasia) but apart from the possibility of *WNT4* being implicated in a rare condition called Mayer-Rokitansky-Kuster syndrome, no other members of this gene family are yet implicated in human disease phenotypes.

The TGF-β Superfamily in Development and Disease

Thus far it recognized that there are 33 members of this ***cytokine*** family. Cytokines are a category of signaling molecules—polypeptide regulators—that enable cells to communicate. They differ from hormones in that they are not produced by discrete glands. These extracellular signaling polypeptides are transduced through a cascade to regulate gene expression within the cell nucleus. This is achieved through binding with cell surface receptors that, in a series of reactions, induces phosphorylation and activation of specific receptor kinases. This leads to the translocation of complexes into the nucleus, which execute transcriptional activation or repression of responsive target genes. The TGF-β family can be divided into two groups: (1) the BMPs and (2) the TGF-βs, activins, nodal, and myostatin, acting through various SMAD proteins. Ultimately, this superfamily is actively involved in a very broad range of cellular and developmental processes (Figure 6.2). This includes regulation of the cell cycle, cell migration, cell size, gastrulation and axis specification, and metabolic processes. In relation to health and disease, there are consequences for immunity, cancer, heart disease, diabetes, and Marfan syndrome

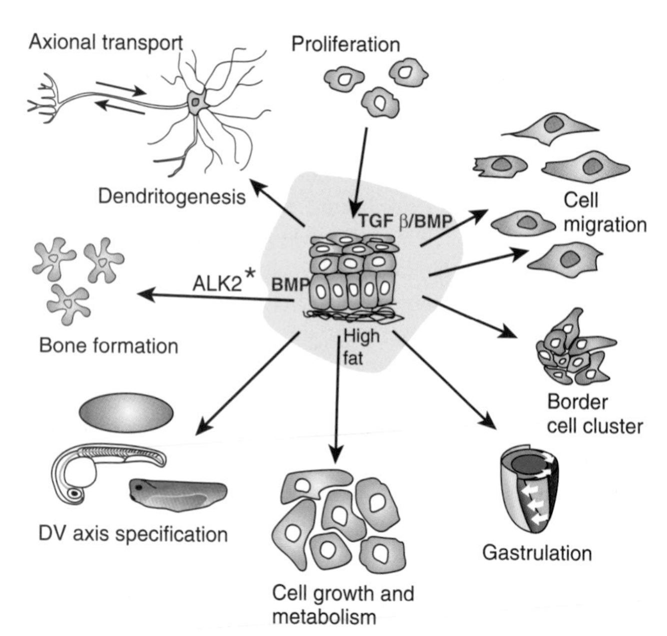

FIGURE 6.2 A summary of biological responses to TGF family signaling. The range of processes that come under the influence of this super family is very broad. (Modified from Wharton K, Derynck R 2009 TGFβ family signaling: novel insights in development and disease. Development 136[22]:3693.)

(p. 300). Hyperactive signalling (overexpression) of *BMP4* has been found in the rare bony condition fibrodysplasia ossificans progressiva, where disabling heterotopic bone deposition occurs, which is due to mutated *ACVR1*, encoding a BMP type 1 receptor. A mutated BMP receptor 2 has been shown to be a cause of familial primary pulmonary hypertension. BMP signalling is also involved in both dendritogenesis and axonal transport.

Somatogenesis and the Axial Skeleton

The vertebrate axis is closely linked to the development of the primary body axis during gastrulation, and during this process the presomitic mesoderm (PSM), where somites arise, is laid down in higher vertebrates. Wnt and FGF signals play vital roles in the specification of the PSM. The somites form as blocks of tissue from the PSM in a rostro-caudal direction (Figure 6.3), each being laid down with a precise periodicity that, in the 1970s, gave rise to the concept of the 'clock and wavefront' model. Since then, molecular techniques have given substance to this concept, and the key pathway here is **notch-delta signaling** and the 'oscillation clock'—a precise, temporally defined wave of **cycling gene** expression (*c-hairy* in the chick, *lunatic fringe* and *hes* genes in the mouse) that sweeps from the tail-bud region in a rostral direction and has a key role in the process leading to the defining of somite boundaries. Once again, not all of the components are fully understood, but the *notch receptor* and its ligands, *delta-like-1*, and *delta-like-3*, together with *presenilin-1* and *mesoderm posterior-2*, work in concert to establish rostro-caudal polarity within the PSM such that somite blocks are formed. Human phenotypes from mutated genes in this pathway are now well known and include presenile dementia *(presenilin-1)*, which is dominantly inherited, and spondylocostal dysostosis (*delta-like-3, mesoderm posterior-2, lunatic fringe*, and *hairy enhancer of split-7*), which is recessively inherited (Figure 6.4). Another component of the pathway is *JAGGED1*, which, when mutated, results in the dominantly inherited and very variable condition known as Alagille syndrome (arteriohepatic dysplasia) (Figure 6.5). Rarely, mutations in *NOTCH2* have been shown to cause some cases of Alagille syndrome, usually with renal malformations.

The Sonic Hedgehog–Patched GLI Pathway

The Sonic hedgehog gene *(SHH)* is as well known for its quirky name as for its function. *SHH* induces cell proliferation in a tissue-specific distribution and is expressed in the notochord, the brain, and the zone of polarizing activity of developing limbs. After cleavage and modification by the addition of a cholesterol moiety, the *SHH* protein binds with its receptor, Patched (Ptch), a transmembrane protein. The normal action of Ptch is to inhibit another transmembrane protein called Smoothened (Smo), but when bound by Shh this inhibition is released and a

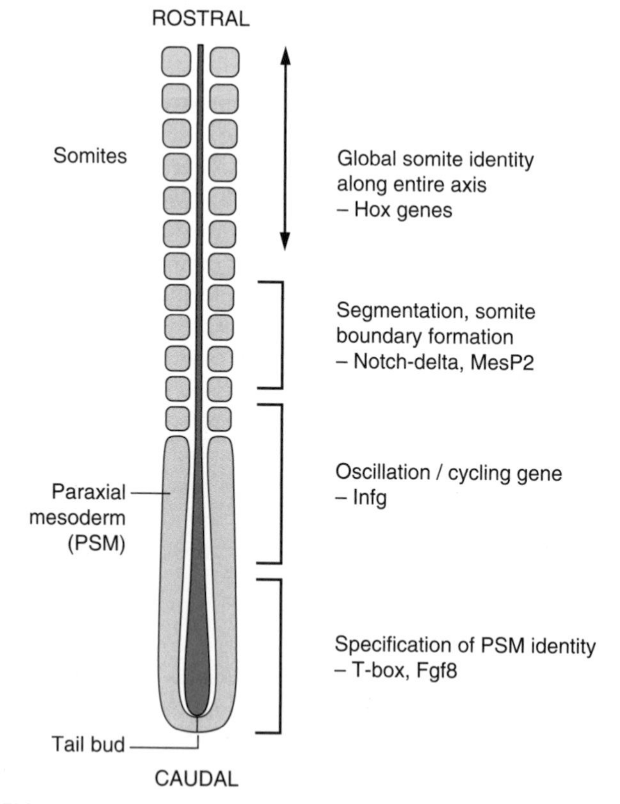

FIGURE 6.3 Somatogenesis and Notch-Delta pathway. T-box genes have a role in PSM specification, whereas the segmentation clock depends on oscillation, or cycling, genes that are important in somite boundary formation where genes of the Notch-Delta pathway establish rostro-caudal polarity. *HOX* genes have a global function in establishing somite identity along the entire rostro-caudal axis. (Adapted from Tickle C, ed 2003 Patterning in vertebrate development. Oxford: Oxford University Press.)

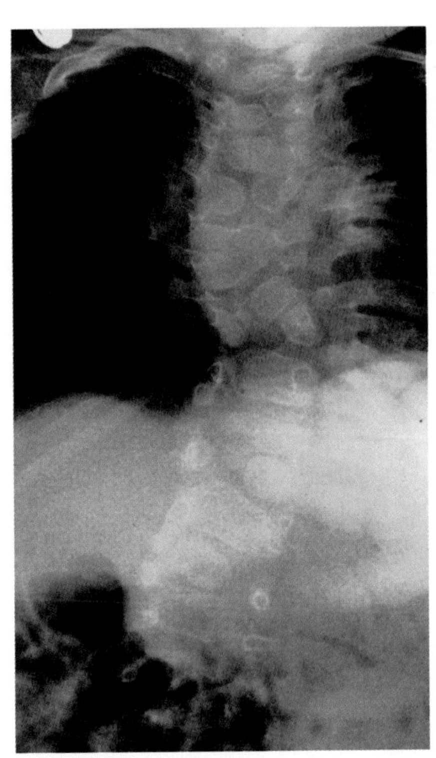

FIGURE 6.4 Disrupted development of the vertebrae in patient with spondylocostal dysostosis type 1 resulting from mutations in the *delta-like-3* gene, part of the notch signaling pathway. (Courtesy Dr. Meriel McEntagart, Kennedy-Galton Centre, London.)

signaling cascade within the cell is activated. The key intracellular targets are the *GLI* family of transcription factors (Figure 6.6).

Molecular defects in any part of this pathway lead to a number of apparently diverse malformation syndromes (see Figure 6.6). Mutations in, or deletions of, *SHH* (chromosome 7q36) cause holoprosencephaly (Figure 6.7), in which the primary defect is incomplete cleavage of the developing brain into separate hemispheres and ventricles. The most severe form of this malformation is cyclopia—the presence of a single central eye. (The complexity of early development can be appreciated by the fact that a dozen or so chromosomal regions have so far been implicated in the pathogenesis of holoprosencephaly [p. 257].) Mutations in *PTCH* (9q22) result in Gorlin syndrome (nevoid basal cell carcinoma syndrome; Figure 6.8), which comprises multiple basal cell carcinomas, odontogenic keratocysts, bifid ribs, calcification of the falx cerebri, and ovarian fibromata. Mutations in *SMO* (7q31) are found in some basal cell carcinomas and medulloblastomas. Mutations in *GLI3* (7p13) cause Pallister-Hall and Grieg syndromes, which are distinct entities with more or less the same body systems affected. However, there are also links to other conditions, in particular the very variable Smith-Lemli-Opitz syndrome (SLOS), which may include holoprosencephaly as well as some characteristic facial features, genital anomalies and syndactyly. This condition is due to a defect in the final step of cholesterol biosynthesis, which in turn may disrupt the binding of SHH with its receptor Ptch. Some, or all, of the features of SLOS may therefore be due to loss of integrity in this pathway (p. 288). Furthermore, a cofactor for the Gli proteins, *CREBBP* (16p13) is mutated in Rubenstein-Taybi syndrome (Figure 6.9). Disturbance to different components of the SHH is also clearly implicated in many types of tumor formation.

Homeobox *(HOX)* Genes

In *Drosophila* a class of genes known as the homeotic genes has been shown to determine segment identity. Incorrect expression of these genes results in major structural abnormalities; the *Antp* gene, for example, which is normally expressed in the second thoracic segment, will transform the adult fly's antennae into legs if incorrectly expressed in the head. Homeotic genes contain a conserved 180-base pair (bp) sequence known as the homeobox, which is

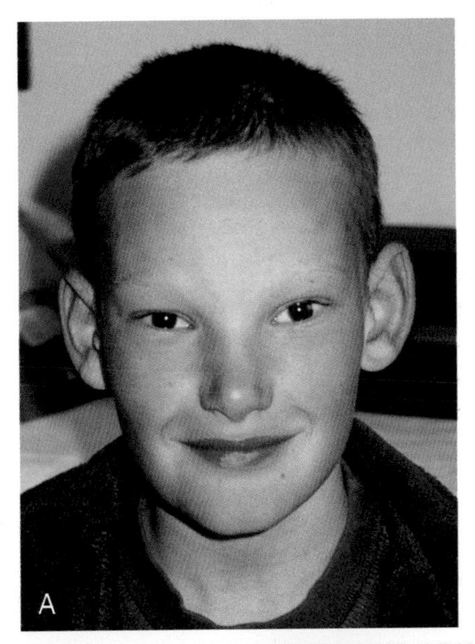

FIGURE 6.5 **A**, Boy with Alagille syndrome and confirmed mutation in *JAGGED1* who presented with congenital heart disease. **B**, The same boy a few years earlier with his parents. His mother has a pigmentary retinopathy and was positive for the same gene mutation.

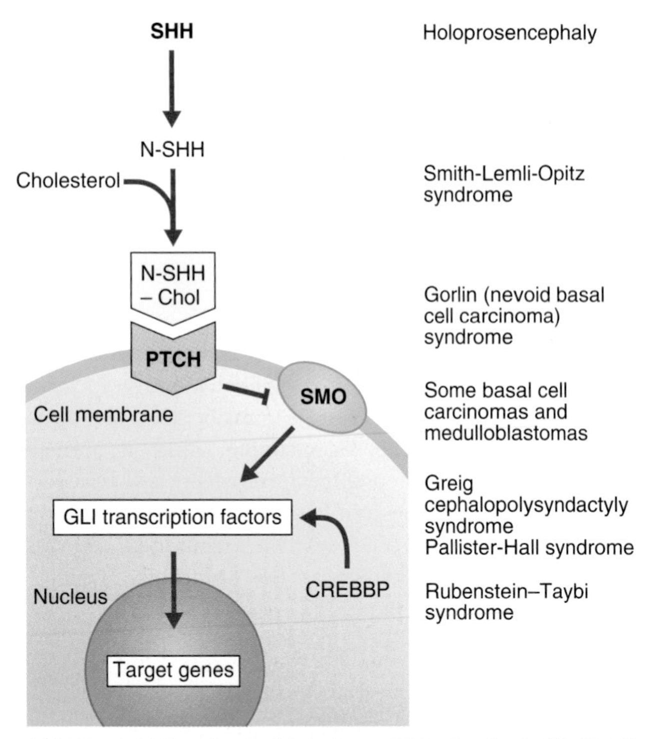

FIGURE 6.6 The Sonic hedgehog (Shh)-Patched (Ptch)-Gli pathway and connection with disease. Different elements in the pathway act as activators *(arrows)* or inhibitors *(bars)*. The Shh protein is initially cleaved to an active N-terminal form, which is then modified by the addition of cholesterol. The normal action of Ptch is to inhibit Smo, but when Ptch is bound by Shh this inhibition is removed and the downstream signaling proceeds. CREBBP, cAMP response element-binding binding protein.

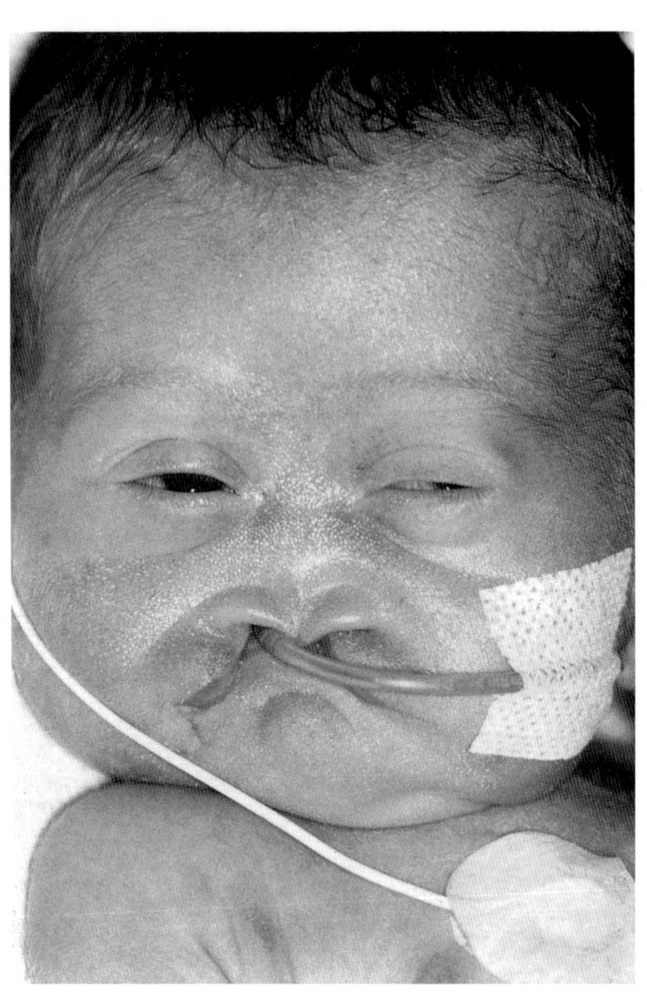

FIGURE 6.7 Facial features in holoprosencephaly. The eyes are close together and there is a midline cleft lip because of a failure of normal prolabia development.

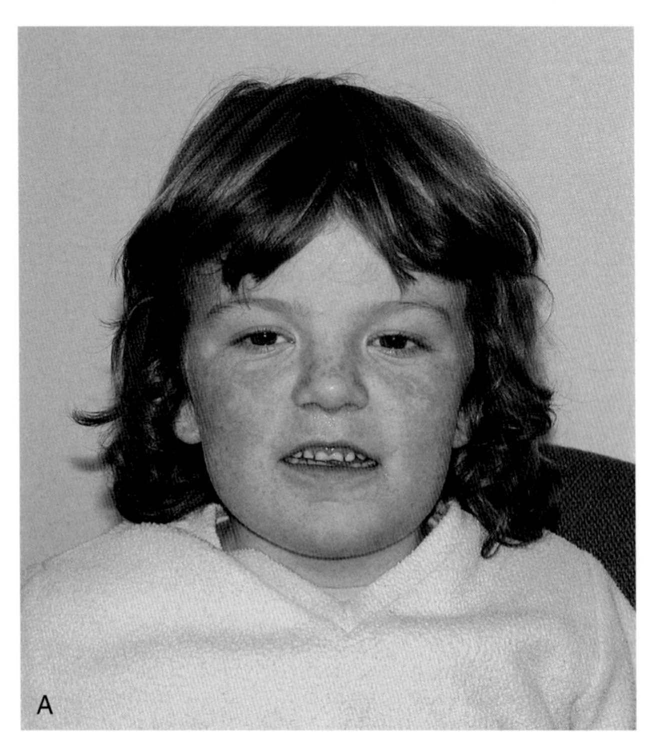

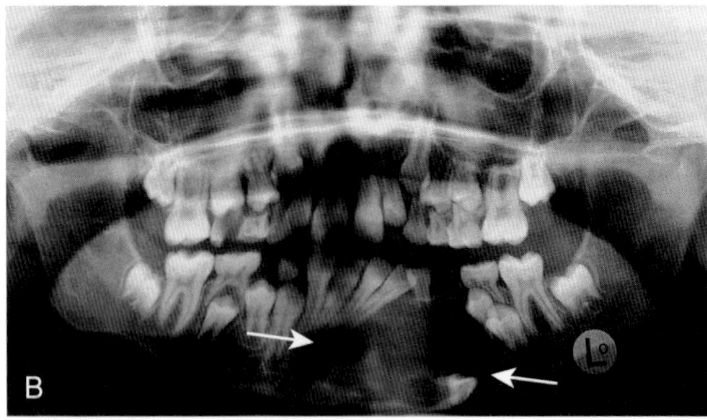

FIGURE 6.8 Gorlin (nevoid basal cell carcinoma) syndrome. **A**, This 6-year-old girl from a large family with Gorlin syndrome has macrocephaly and a cherubic appearance. **B**, Her affected sister developed a rapidly enlarging odontogenic keratocyst *(arrows)* in the mandible at the age of 9 years, displacing the roots of her teeth.

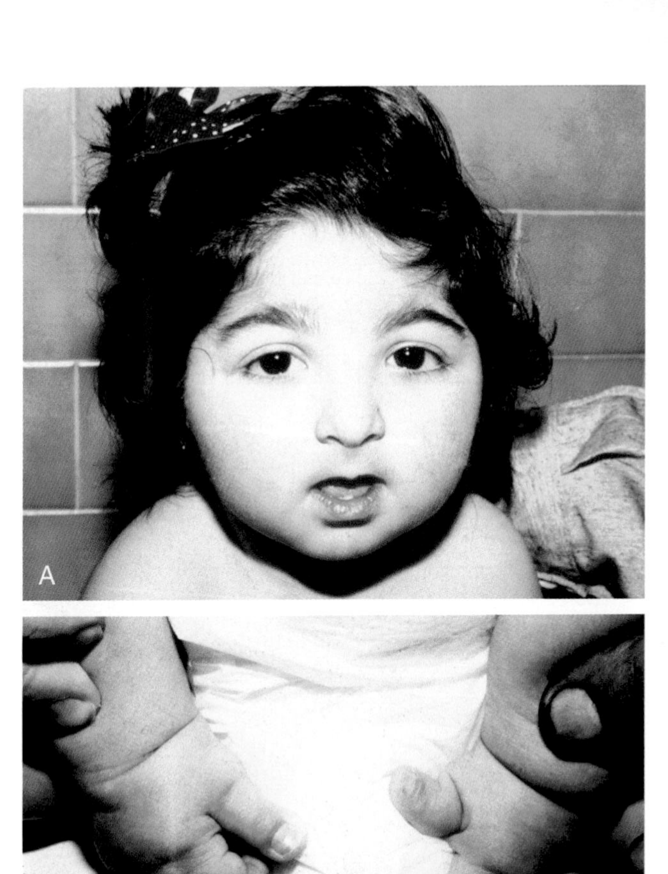

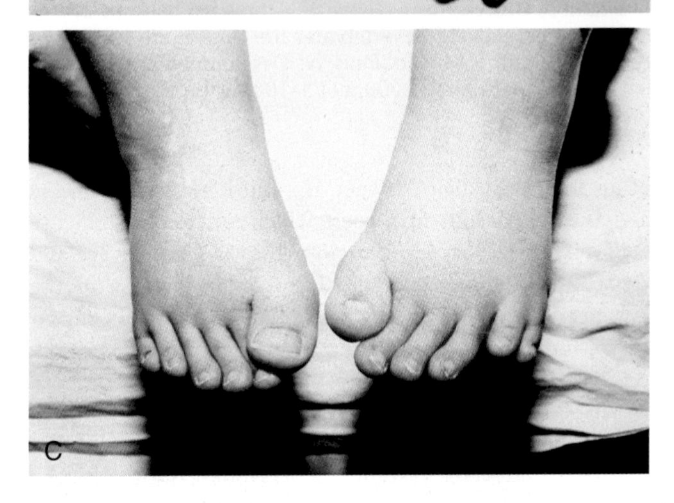

FIGURE 6.9 A baby with characteristic facial features (**A**) of Rubenstein-Taybi syndrome, angulated thumbs (**B**), and postaxial polydactyly of the feet (**C**). A young adult (**D**) with same condition, though more mildly affected.

believed to be characteristic of genes involved in spatial pattern control and development. This encodes a 60-amino-acid domain that binds to DNA in Hox-response enhancers. Proteins from homeobox-containing (or *HOX*) genes are therefore important transcription factors that activate and repress batteries of downstream genes. At least 35 downstream targets are known. The Hox proteins regulate other 'executive' genes that encode transcription factors or morphogen signals, as well as operating at many other levels, on genes that mediate cell adhesion, cell division rates, cell death, and cell movement. They specify cell fate and help to establish the embryonic pattern along the primary (rostro-caudal) axis as well as the secondary (genital and limb bud) axis. They therefore play a major part in the development of the central nervous system, axial skeleton and limbs, the gastrointestinal and urogenital tracts, and external genitalia.

Drosophila has eight *Hox* genes arranged in a single cluster, but in humans, as in most vertebrates, there are four homeobox gene clusters containing a total of 39 *HOX* genes

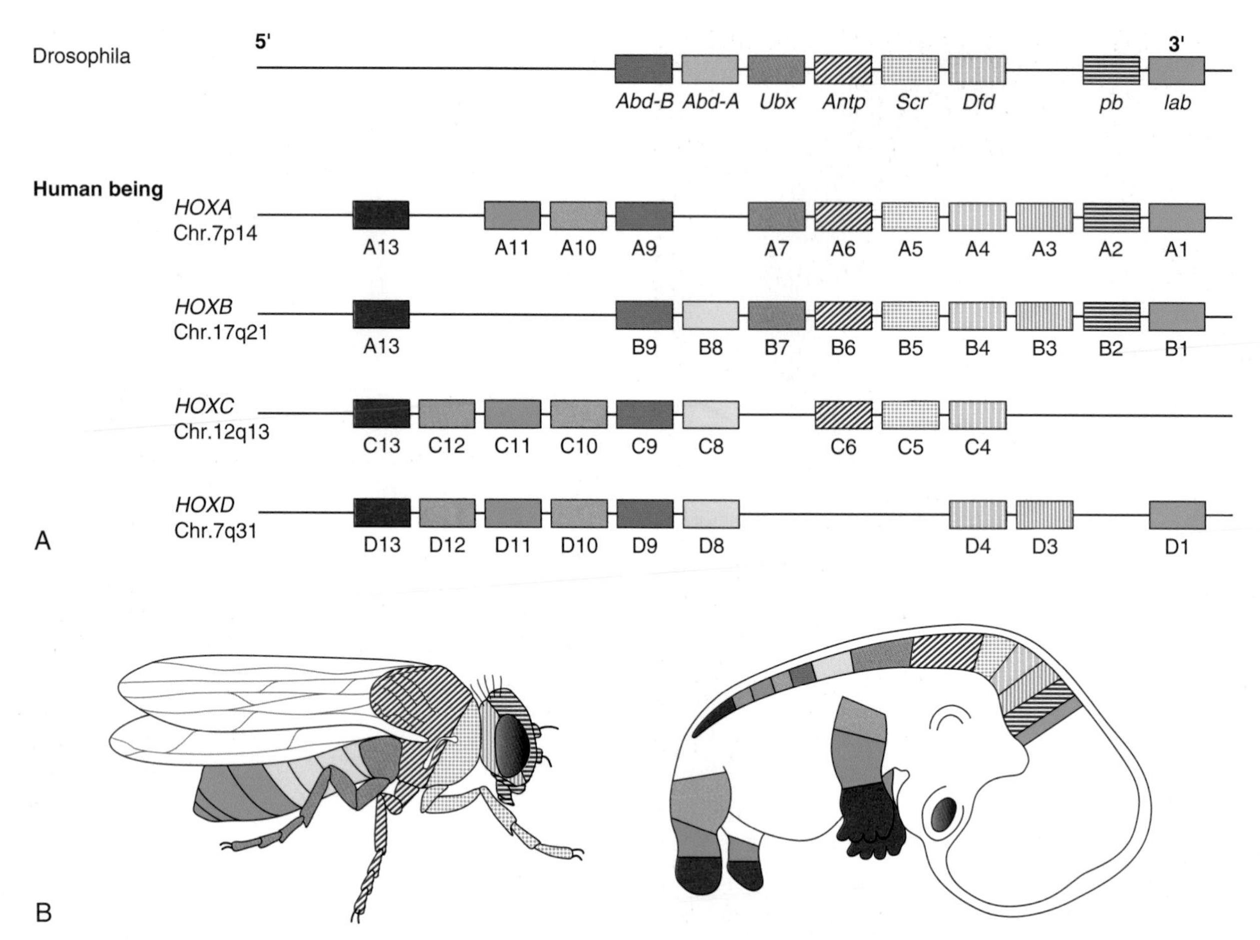

FIGURE 6.10 **A**, *Drosophila* has eight *Hox* genes in a single cluster whereas there are 39 *HOX* genes in humans, arranged in four clusters located on chromosomes 7p, 17q, 12q, and 2q for the A, B, C, and D clusters, respectively. **B**, Expression patterns of *Hox* and *HOX* genes along the rostro-caudal axis in invertebrates and vertebrates, respectively. In vertebrates the clusters are *paralogous* and appear to compensate for one another. (Redrawn from Veraksa A, Del Campo M, McGinnis W: Developmental patterning genes and their conserved functions: from model organisms to humans. Mol Genet Metab 2000;69:85–100, with permission.)

(Figure 6.10). Each cluster contains a series of closely linked genes. In vertebrates such as mice, it has been shown that these genes are expressed in segmental units in the hindbrain and in global patterning of the somites formed from axial presomitic mesoderm. In each *HOX* cluster, there is a direct linear correlation between the position of the gene and its temporal and spatial expression. These observations indicate that these genes play a crucial role in early morphogenesis. Thus, in the developing limb bud (p. 99) *HOXA9* is expressed both anterior to, and before, *HOX10*, and so on.

Mutations in *HOXA13* cause a rare condition known as the hand-foot-genital syndrome. This shows autosomal dominant inheritance and is characterized by shortening of the first and fifth digits, with hypospadias in males and bicornuate uterus in females. Experiments with mouse *Hoxa13* mutants have shown that expression of another gene, *EphA7*, is severely reduced. Therefore, if this gene is not activated by *Hoxa13*, there is failure to form the normal chondrogenic condensations in the distal limb primordial. Mutations in *HOXD13* result in an equally rare limb developmental abnormality known as synpolydactyly. This also shows autosomal dominant inheritance and is characterized by insertion of an additional digit between the third and fourth fingers and the fourth and fifth toes, which are webbed (Figure 6.11). The phenotype in homozygotes is more severe and reported mutations take the form of an increase in the number of residues in a polyalanine tract. This triplet-repeat expansion probably alters the structure and function of the protein, thereby constituting a gain-of-function mutation (p. 26). Mutated *HOXA1* has been found in the rare, recessively inherited Bosley-Saleh-Alorainy syndrome, consisting of central nervous system abnormalities, deafness, and cardiac and laryngotracheal anomalies. A mutation in *HOXD10* was found in isolated congenital vertical talus in a large family demonstrating autosomal dominant inheritance, and duplications of *HOXD* have recently been found in mesomelic limb abnomality syndromes.

Given that there are 39 *HOX* genes in mammals, it is surprising that so few syndromes or malformations have been attributed to *HOX* gene mutations. One possible explanation is that most *HOX* mutations are so devastating that the embryo cannot survive. Alternatively, the high degree of homology between *HOX* genes in the different clusters could lead to functional redundancy so that one *HOX* gene could compensate for a loss-of-function

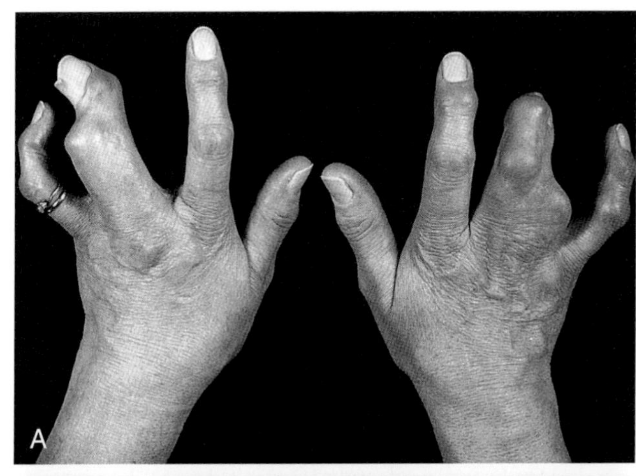

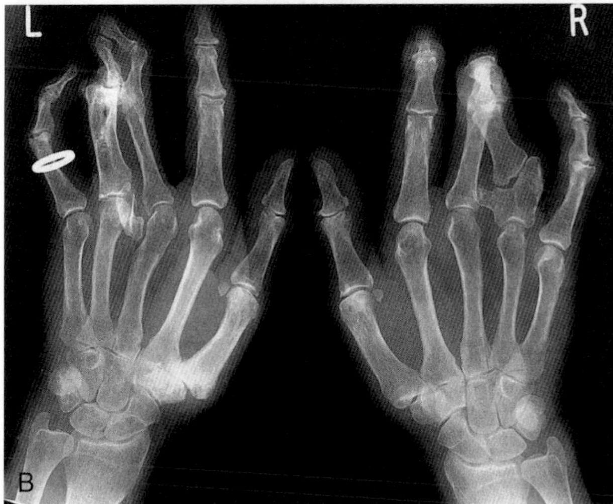

FIGURE 6.11 Clinical (**A**) and (**B**) radiographic views of the hands in synpolydactyly.

mutation in another. In this context *HOX* genes are said to be paralogous because family members from different clusters, such as *HOXA13* and *HOXD13*, are more similar than adjacent genes in the same cluster.

Several other developmental genes also contain a homeobox-like domain. These include *MSX2* and *EMX2*. Mutations in *MSX2* can cause craniosynostosis—premature fusion of the cranial sutures. Mutations in *EMX2* have been implicated in some cases of schizencephaly, in which there is a large full-thickness cleft in one or both cerebral hemispheres.

Paired-Box (*PAX*) Genes

The paired-box is a highly conserved DNA sequence that encodes a 130-amino-acid DNA-binding transcription regulator domain. Nine *PAX* genes have been identified in mice and humans. In mice these have been shown to play important roles in the developing nervous system and vertebral column. In humans, loss-of-function mutations in five *PAX* genes have been identified in association with developmental abnormalities (Table 6.2). Waardenburg syndrome type 1 is caused by mutations in *PAX3*. It shows autosomal dominant inheritance and is characterized by sensorineural hearing loss, areas of depigmentation in hair

Table 6.2 Developmental Abnormalities Associated with *PAX* Gene Mutations

Gene	Chromosome Location	Developmental Abnormality
PAX2	10q24	Renal-coloboma syndrome
PAX3	2q35	Waardenburg syndrome type 1
PAX6	11p13	Aniridia
PAX8	2q12	Absent or ectopic thyroid gland
PAX9	14q12	Oligodontia

and skin, abnormal patterns of pigmentation in the iris, and widely spaced inner canthi (Figure 6.12). Waardenburg syndrome shows genetic heterogeneity; the more common type 2 form, in which the inner canthi are not widely separated, is sometimes caused by mutations in the human microphthalmia *(MITF)* gene on chromosome 3.

The importance of expression of the *PAX* gene family in eye development is illustrated by the effects of mutations in *PAX2* and *PAX6*. Mutations in *PAX2* cause the renal-coloboma syndrome, in which renal malformations occur in association with structural defects in various parts of the eye, including the retina and optic nerve. Mutations in *PAX6* lead to absence of the iris, which is known as aniridia (Figure 6.13). This is a key feature of the WAGR syndrome

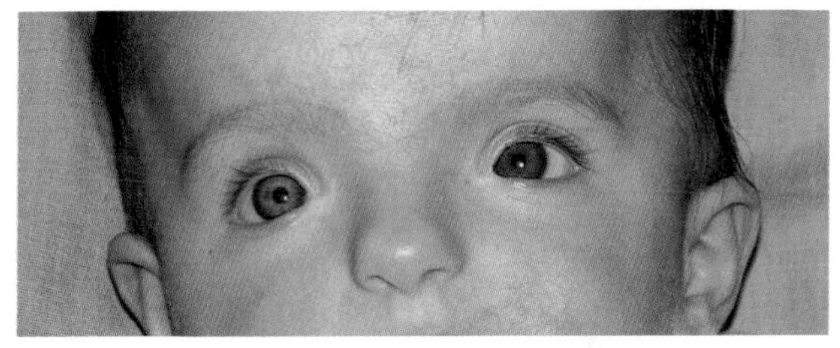

FIGURE 6.12 Iris heterochromia and marked dystopia canthorum in an infant with Waardenburg syndrome type 1, caused by a mutation in *PAX3*.

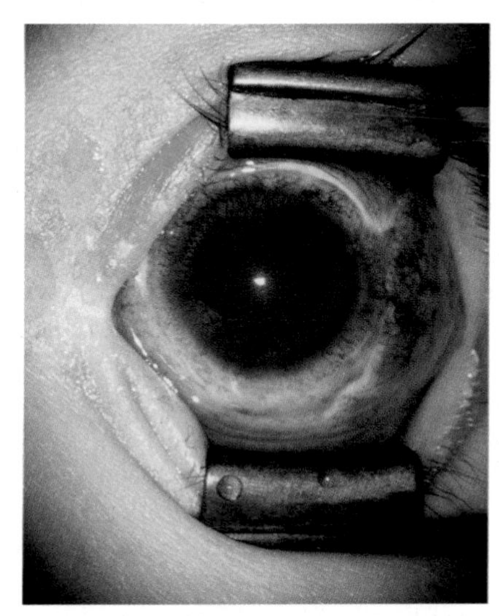

FIGURE 6.13 An eye showing absence of the iris (aniridia). The cornea shows abnormal vascularization. (Courtesy Mr. R. Gregson, Queen's Medical Centre, Nottingham, UK.)

(p. 282), which results from a contiguous gene deletion involving the *PAX6* locus on chromosome 11.

SRY-Type HMG Box (*SOX*) Genes

SRY is the Y-linked gene that plays a major role in male sex determination (p. 102). A family of genes known as the *SOX* genes shows homology with *SRY* by sharing a 79-amino-acid domain known as the HMG (high-mobility group) box. This HMG domain activates transcription by bending DNA in such a way that other regulatory factors can bind with the promoter regions of genes that encode for important structural proteins. These *SOX* genes are thus transcription regulators and are expressed in specific tissues during embryogenesis. For example, *SOX1*, *SOX2*, and *SOX3* are expressed in the developing mouse nervous system.

In humans it has been shown that loss-of-function mutations in *SOX9* on chromosome 17 cause campomelic dysplasia. This very rare disorder is characterized by bowing of the long bones, sex reversal in chromosomal males, and very poor long-term survival. In-situ hybridization studies in mice have shown that *SOX9* is expressed in the developing embryo in skeletal primordial tissue, where it regulates type II collagen expression, as well as in the genital ridges and early gonads. *SOX9* is now thought to be one of several genes that are expressed downstream of *SRY* in the process of male sex determination (p. 102). Mutations in *SOX10* on chromosome 22 cause a rare form of Waardenburg syndrome in which affected individuals have a high incidence of Hirschsprung disease. Mutations in *SOX2* (3q26) have been shown to cause anophthalmia or microphthalmia, but also a wider syndrome of esophageal atresia and genital hypoplasia in males: the anophthalmia-esophageal-genital syndrome.

T-Box (*TBX*) Genes

The T gene in mice plays an important role in specification of the paraxial mesoderm and notochord differentiation. Heterozygotes for loss-of-function mutations have a short tail and malformed sacral vertebrae. This gene, which is also known as *Brachyury*, encodes a transcription factor that contains both activator and repressor domains. It shows homology with a series of genes through the shared possession of the T domain, which is also referred to as the T-box. These T-box or *TBX* genes are dispersed throughout the human genome, with some family members existing in small clusters. One of these clusters on chromosome 12 contains *TBX3* and *TBX5*. Loss-of-function mutations in *TBX3* cause the ulnar-mammary syndrome in which ulnar ray developmental abnormalities in the upper limbs are associated with hypoplasia of the mammary glands. Loss-of-function mutations in *TBX5* cause the Holt-Oram syndrome. This autosomal dominant disorder is characterized by congenital heart abnormalities, most notably atrial septal defects, and upper limb radial ray reduction defects that can vary from mild hypoplasia (sometimes duplication) of the thumbs to almost complete absence of the forearms.

Zinc Finger Genes

The term **zinc finger** refers to a finger-like loop projection consisting of a series of four amino acids that form a complex with a zinc ion. Genes that contain a zinc finger motif act as transcription factors through binding of the zinc finger to DNA. Consequently they are good candidates for single-gene developmental disorders (Table 6.3).

For example, a zinc finger motif–containing gene known as *GLI3* on chromosome 7 (as mentioned, a component of the SHH pathway) has been implicated as the cause of two developmental disorders. Large deletions or translocations involving *GLI3* cause Greig cephalopolysyndactyly, which is characterized by head, hand and foot abnormalities such as polydactyly and syndactyly (Figure 6.14, *A*). In contrast, frameshift mutations in *GLI3* have been reported in the Pallister-Hall syndrome (Figure 6.14, *B*), in which the key features are polydactyly, hypothalamic hamartomata and imperforate anus.

Mutations in another zinc finger motif-containing gene known as *WT1* on chromosome 11 can cause both Wilms' tumour and a rare developmental disorder, the Denys-Drash syndrome, in which the external genitalia are ambiguous and there is progressive renal failure as a result of nephritis. Mutations in two other zinc finger motif–containing genes, *ZIC2* and *ZIC3*, have been shown to cause holoprosencephaly and laterality defects, respectively. Just as **polarity** is a key concept in development, so too is **laterality**, with implications for the establishment of a normal left-right body axis. In very early development, integrity of many of the same gene families previously mentioned—Nodal, SHH, and Notch—is essential to the establishment of this axis. Clinically, **situs solitus** is the term given to normal left-right asymmetry and **situs inversus** to reversal of the normal arrangement. Up to 25% of individuals with situs inversus have an autosomal recessive condition—Kartagener syndrome, or ciliary dyskinesia. Other terms used are **isomerism sequence**, **heterotaxy**, **asplenia/polyasplenia**, and **Ivemark syndrome**. Laterality defects are characterized by abnormal positioning of unpaired organs such as the heart, liver, and spleen, and more than 20 genes are now implicated from studies in vertebrates, with a number identified in humans by the

Table 6.3 Developmental Abnormalities Associated with Genes Containing a Zinc Finger Motif

Gene	Chromosome Location	Developmental Abnormality
GLI3	7p13	Greig syndrome and Pallister-Hall syndrome
WT1	11p13	Denys-Drash syndrome
ZIC2	13q32	Holoprosencephaly
ZIC3	Xq26	Laterality defects

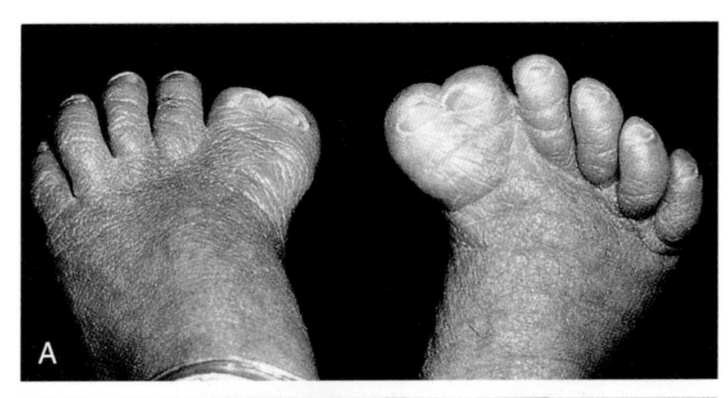

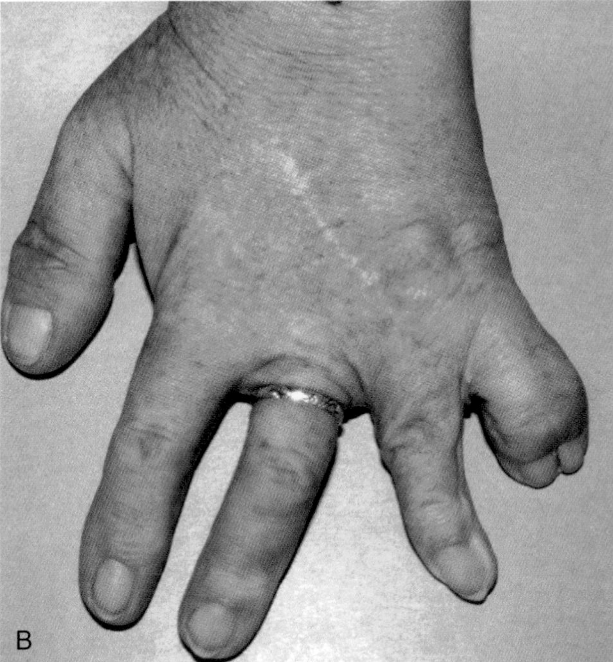

FIGURE 6.14 **A**, The feet of a child with Greig cephalopolysyndactyly. Note that they show both preaxial polydactyly (extra digits) and syndactyly (fused digits). **B**, The left hand of a woman with Pallister-Hall syndrome and a proven mutation in *GLI3*. Note the postaxial polydactyly and the surgical scar, where an extra digit arising from between the normal metacarpal rays (mesoaxial polydactyly) was removed.

study of affected families, with all of the main patterns of inheritance represented.

Signal Transduction ('Signaling') Genes

Signal transduction is the process whereby extracellular growth factors regulate cell division and differentiation by a complex pathway of genetically determined intermediate steps. Mutations in many of the genes involved in signal transduction play a role in causing cancer (p. 213). In some cases they can also cause developmental abnormalities.

The *RET* Proto-oncogene

The proto-oncogene *RET* on chromosome 10q11.2 encodes a cell-surface tyrosine kinase. Gain-of-function mutations, whether inherited or acquired, are found in a high proportion of thyroid cancers. Loss-of-function mutations in *RET* have been identified in approximately 50% of familial cases of Hirschsprung disease, in which there is failure of migration of ganglionic cells to the submucosal and myenteric plexuses of the large bowel. The clinical consequences are usually apparent shortly after birth when the child presents with abdominal distention and intestinal obstruction.

FGF Receptors

FGFs play key roles in embryogenesis, including cell division, migration, and differentiation. The transduction of extracellular FGF signals is mediated by a family of four transmembrane tyrosine kinase receptors. These are the fibroblast growth factor receptors (FGFRs), each of which contains three main components: an extracellular region with three immunoglobulin-like domains, a transmembrane segment, and two intracellular tyrosine kinase domains (Figure 6.15).

Mutations in the genes that code for FGFRs have been identified in two groups of developmental disorders (Table 6.4). These are the craniosynostosis syndromes and the achondroplasia family of skeletal dysplasias. The craniosynostosis syndromes, of which Apert syndrome (Figure 6.16) is the best known, are characterized by premature fusion of the cranial sutures, often in association with hand and foot abnormalities such as syndactyly (fusion of the digits). Apert syndrome is caused by a mutation in one of the adjacent *FGFR2* residues in the peptides that link the second and third immunoglobulin loops (see Figure 6.15). In contrast, mutations in the third immunoglobulin loop can cause either Crouzon syndrome, in which the limbs are normal, or Pfeiffer syndrome, in which the thumbs and big toes are broad. Achondroplasia is the most commonly encountered form of genetic short stature (Figure 6.17). The limbs show proximal ('rhizomelic') shortening and the head is enlarged with frontal bossing. Intelligence and life expectancy are entirely normal. Achondroplasia is almost always caused by a mutation in, or close to, the transmembrane domain of *FGFR3*. The common transmembrane domain mutation

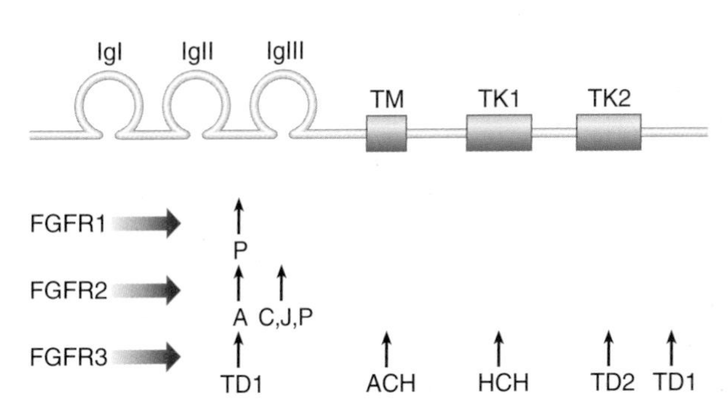

FIGURE 6.15 Structure of the fibroblast growth factor receptor (FGFR). Arrows indicate the location of mutations in the craniosynostosis syndromes and achondroplasia group of skeletal dysplasias. *Ig*, Immunoglobulin-like domain; *TM*, transmembrane domain; *TK*, tyrosine kinase domain; *P*, Pfeiffer syndrome; *A*, Apert syndrome; *C*, Crouzon syndrome; *J*, Jackson-Weiss syndrome; *TD*, thanatophoric dysplasia; *ACH*, achondroplasia; *HCH*, hypochondroplasia.

Table 6.4 Developmental Disorders Caused by Mutations in Fibroblast Growth Factor Receptors

Gene	Chromosome	Syndrome
Craniosynostosis Syndromes		
FGFR1	8p11	Pfeiffer
FGFR2	10q25	Apert
		Crouzon
		Jackson-Weiss
		Pfeiffer
FGFR3	4p16	Crouzon (with acanthosis nigricans)
Skeletal Dysplasias		
FGFR3	4p16	Achondroplasia
		Hypochondroplasia
		Thanatophoric dysplasia

leads to the replacement of a glycine amino-acid residue by an arginine—an amino acid that is never normally found in cell membranes. This in turn appears to enhance dimerization of the protein that catalyzes downstream signaling. Hypochondroplasia, a milder form of skeletal dysplasia with similar trunk and limb changes but normal head shape and size, is caused by mutations in the proximal tyrosine kinase domain (intracellular) of *FGFR3*. Finally, thanatophoric dysplasia, a much more severe and invariably lethal form of skeletal dysplasia, is caused by mutations in either the peptides linking the second and third immunoglobulin domains (extracellular) of *FGFR3*, or the distal *FGFR3* tyrosine kinase domain.

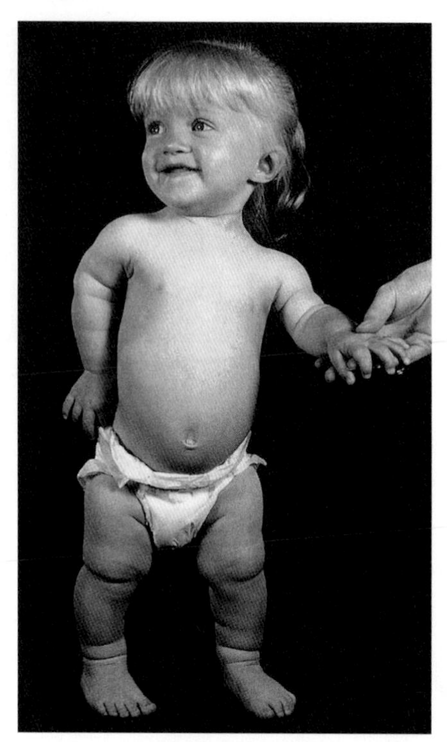

FIGURE 6.17 A young child with achondroplasia.

The mechanism by which these mutations cause skeletal shortening is not understood at present. The mutations cannot have loss-of-function effects as children with the Wolf-Hirschhorn syndrome (pp. 280–281), which is due to chromosome microdeletions that include *FGFR3*, do not

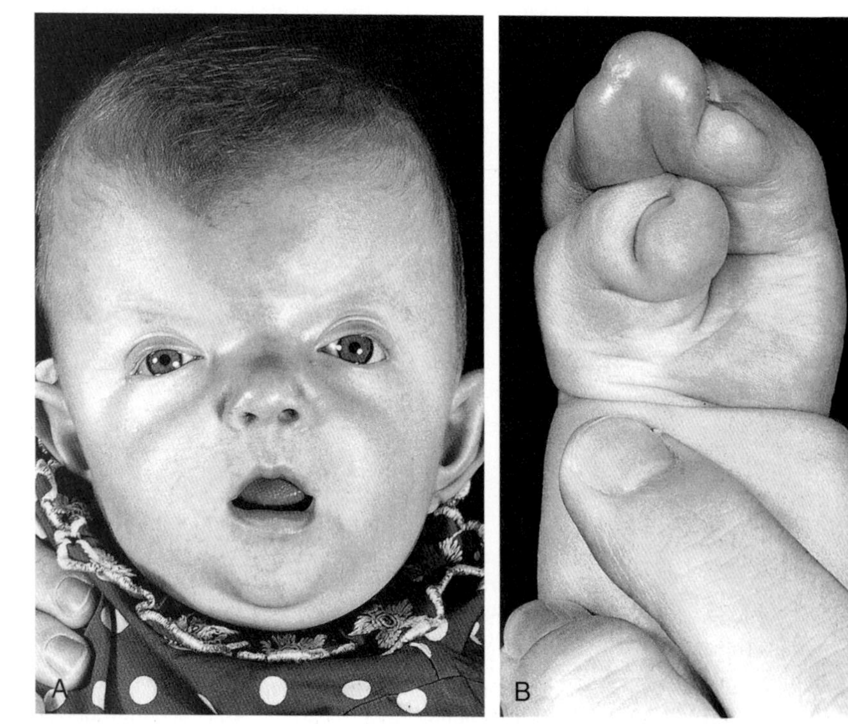

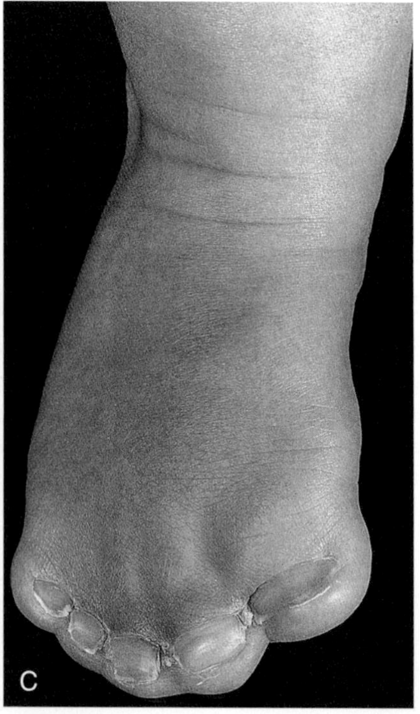

FIGURE 6.16 Views of the face (**A**), hand (**B**), and foot (**C**) of a child with Apert syndrome.

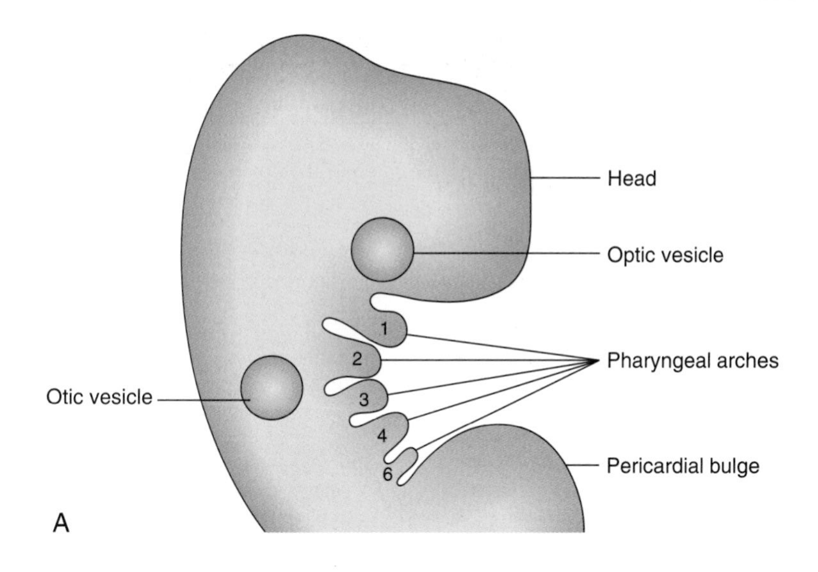

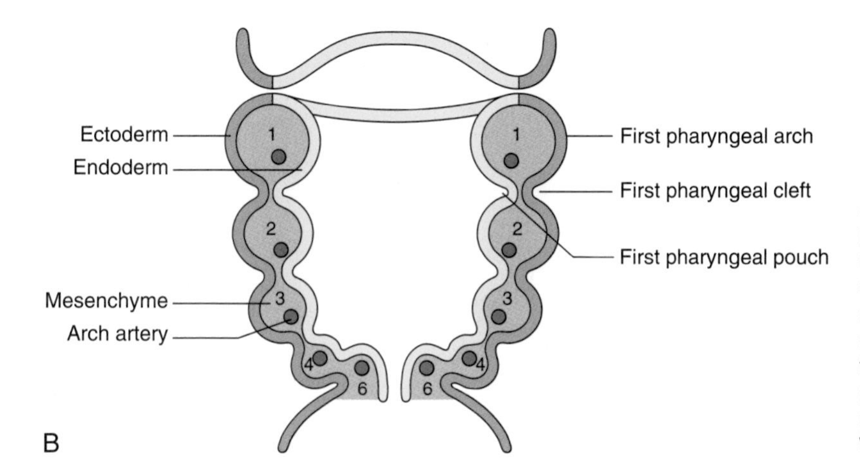

FIGURE 6.18 The pharyngeal (or branchial) apparatus. The lateral view (**A**) shows the five pharyngeal arches close to the embryonic head and the cross-section (**B**) shows the basic arrangement from which many head and neck structures, as well as the heart, develop. Humans and mice do not have arch no. 5. (Redrawn from Graham A, Smith A 2001 Patterning the pharyngeal arches. Bioessays 23:54–61, with permission of Wiley-Liss Inc., a subsidiary of John Wiley & Sons, Inc.)

show similar skeletal abnormalities. Instead, the mutations probably involve a gain of function mediated by increased ligand binding or receptor activation.

The Pharyngeal Arches

The pharyngeal (or branchial) arches correspond to the gill system of lower vertebrates and appear in the fourth and fifth weeks of development. Five (segmented) pharyngeal arches in humans arise lateral to the structures of the head (Figure 6.18) and each comprises cells from the three germ layers and the neural crest. The lining of the pharynx, thyroid, and parathyroids arises from the **endoderm**, and the outer epidermal layer arises from the **ectoderm**. The musculature arises from the **mesoderm**, and bony structures from **neural crest** cells. Separating the arches are the pharyngeal clefts externally and the pharyngeal pouches internally; these have important destinies. Numbered from the rostral end, the first arch forms the jaw and muscles of mastication, the first cleft is destined to be the external auditory meatus, and the first pouch the middle ear apparatus. The second arch forms the hyoid apparatus and muscles of facial expression, whereas the third pouch develops into the thymus, and the third and fourth pouches become the parathyroids. The arteries within the arches have important destinies too and, after remodeling, give rise to the aortic and pulmonary arterial systems. Some of the syndromes, malformations, inheritance patterns and genetic mechanisms associated with the first and second pharyngeal arches are listed in Table 6.5.

However, the most well known, and probably most common, condition due to disturbed development of pharyngeal structures—the third and fourth pouches—is DiGeorge syndrome (DGS), also known as velocardiofacial syndrome (VCFS), and well described even earlier by Sedláčková of Prague in 1955. This is described in more detail in Chapter 18 (pp. 282–283); it results from a 3Mb submicroscopic chromosome deletion of band 22q11 with the loss of some 30 genes. Studies in mice (the equivalent, or **syntenic**, region is on mouse chromosome 16) suggest that the most significant gene loss is that of *Tbx1*, strongly expressed throughout the pharyngeal apparatus. Heterozygous *Tbx1* knock-out mice show hypoplastic or absent

Table 6.5 Some Syndromes and Malformations Associated with the First and Second Pharyngeal Arches

Syndrome	Malformations	Inheritance	Mechanisms
Oculo-auriculo-vertebral spectrum (OAVS)	Hemifacial microsomia, ear malformations; epibulbar dermoids; occasional clefts; (cervical vertebral anomalies)	Usually sporadic. Occasional AD and AR families reported	Probable non-genetic factors Possible locus at 14q32.1
Treacher Collins syndrome	Hypoplasia of the maxilla and mandible; downslanting palpebral fissures with coloboma of the lower lid; cleft palate; hearing impairment	AD	Mutation in *TCOF1* gene
Branchio-oto-renal syndrome	Long, narrow face; aplasia or stenosis of lacrimal duct; ear anomalies—external and inner—and preauricular pits	AD	Mutation in *EYA1*, *SIX5*, possible locus at 1q31
Pierre-Robin sequence	Micrognathia, cleft palate, glossoptosis (posteriorly placed tongue); if syndromic, may be associated with limb anomalies and or congenital heart disease	Various: sporadic if occurring as an isolated malformation complex; syndromic forms—both AD and AR families reported	Sporadic cases may be a deformation sequence secondary to oligohydramnios. One AD form segregates with a 75kb deletion at 17q24, which may affect expression of *SOX9* (campomelic dysplasia)
Townes-Brock syndrome	Malformed ('satyr') ears, sensorineural hearing loss, preauricular skin tags; (imperforate anus, triphalangeal thumbs, cardiac/renal defects)	AD	Mutation in *SALL1*
Auriculo-condylar syndrome	Prominent, malformed ears; abnormal temporomandibular joint; microstomia	AD	One locus, mapped to 1p21-q23
Oro-facial-digital syndromes (types I to X)	Cleft or lobulated tongue; cleft palate; oral frenulae; (digital anomalies—brachydactyly, polydactyly, syndactyly, clinodactyly)	XLD (OFD1, OFD7) XLR (OFD8, OFD9) AR (OFD2, OFD3, OFD4, OFD5), OFD6, OFD9) AD (OFD7)	OFD1 due to mutation in *CXORF5* (Xp22)
Otopalatodigital syndrome	Prominent supraorbital ridge, wide nasal bridge, downslanting palpebral fissures, low set ears, microstomia, micrognathia; (skeletal abnormalities—restricted growth, narrow thorax, platyspondyly, bowed long bones)	XL semidominant	Mutation in *FLNA* (Xq28)

AD, Autosomal dominant; *AR*, autosomal recessive; *XLD*, X-linked dominant; *XLR*, X-linked recessive.

fourth pharyngeal arch arteries, suggesting that *TBX1* in humans is the key. Indeed, mutations in this gene have now been found in some congenital heart abnormalities and it is possible that *TBX1* is the key gene for other elements of the phenotype. However, there are still unanswered questions in the whole DGS/VCFS/Sedláčková story.

Role of Cilia in Developmental Abnormalities

In recent years the vital role of the humble cilium in driving movement or particle flow across epithelial surfaces has become increasingly apparent. As with other areas of cell and molecular biology, we learn most about motile cilia when they are dysfunctional—the result may be major developmental abnormalities.

Cilia are the equivalent of flagella in wider biology and they share structural identity. They are hairlike protrusions from the cell surface (Figure 6.19), up to 20 μm long and, present in large numbers on the apical cell surface, beat in coordinated waves. In cross-section they consist of a scaffold of nine microtubule doublets surrounding a central pair. The central and outer doublets are connected by radial spokes, which produce the force necessary to bend the cilium; dynein arms facilitate this movement. They clear mucous from the respiratory epithelium, drive sperm along the Fallopian tube, and move cerebrospinal fluid in the cavities of the central nervous system. In development the cilia at the organizational node of the vertebrate embryo conduct a circular motion, wafting molecules unidirectionally and helping to establish left-right asymmetry.

Apart from their obvious mechanical function, which conceptually is straightforward, it appears increasingly likely that cilia behave like molecular antennae that sense extracellular signaling molecules. The sonic hedgehog and Wnt signaling pathways depend to an extent on cilial

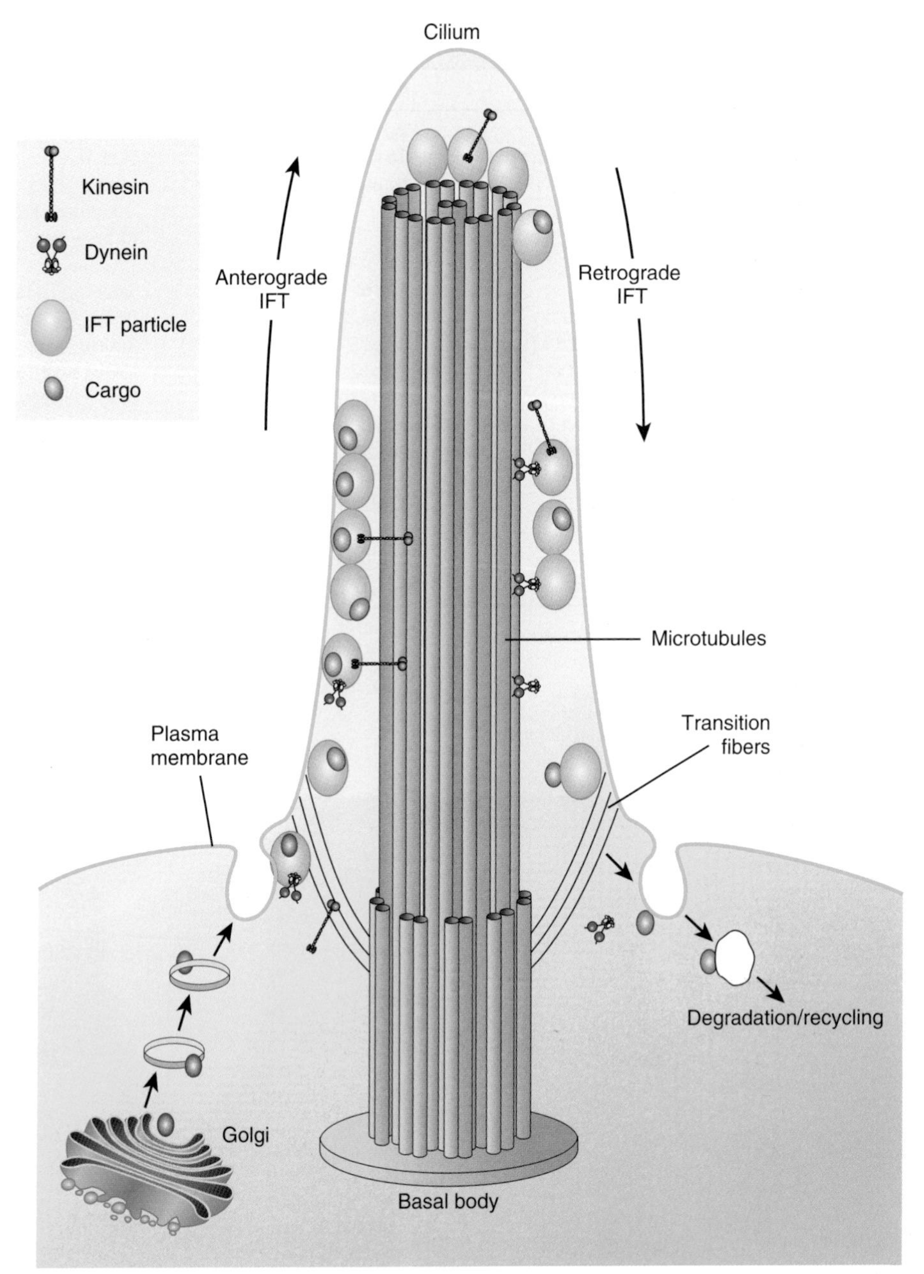

FIGURE 6.19 The structure of a cilium. Nine microtubule doublets provide the main scaffold and surround one microtubule doublet in the center.

functional integrity for optimal signaling. Defective cilial function can therefore impact on a broad range of developmental processes and pathways. Defects in the cilial proteins themselves lead to wide-ranging phenotypic effects that include retinal degeneration, anosmia, renal, hepatic and pancreatic cyst formation, postaxial polydactyly, and situs inversus. The growing list of recognizable syndromes is now referred to as 'ciliopathies' (Table 6.6). One of these, short-rib polydactyly syndrome, which follows autosomal recessive inheritance and is due to mutated *DYNC2H1*, is shown in Figure 6.20. The features of the listed syndromes in Table 6.6 overlap with many other multiple congenital abnormality syndromes, and we shall certainly see the relevance of cilia to developmental genetics increase in the years ahead.

The Limb as a Developmental Model

Four main phases are recognized in limb development: (1) initiation, (2) specification, (3) tissue differentiation, and (4) growth. Although none of these stages is fully understood, insight into the probable underlying mechanisms has been gleaned from the study of limb development in chicks and mice in particular.

Table 6.6 The Human Ciliopathies: Those Diseases of Development Known to Be Due to Defective Cilia

Disease/Syndrome	Gene	Chromosome Locus	Body System(s) Affected
Alstrom syndrome	*ALMS1*	2p13	Retina, adipose, endocrine, heart
Jeune asphyxiating thoracic dystrophy	*IFT80*	15q13	Skeleton
Bardet-Biedl syndrome	*BBS1-BBS14*	Multiple	Multisystem, including retina, kidney, skeleton
Cranioectodermal dysplasia (Sensenbrenner syndrome)			Kidney, liver
Ellis–van Crefeld syndrome	*EVC1, EVC2*	4p16	Skeleton, heart
Joubert syndrome	*JBTS1* (+ others)	9q34.3	Brain
Leber congenital amaurosis	*GUCY2D, RPE65* (+ others)	17p13, 11p31 (+ others)	Retina
McKusick-Kaufman syndrome	*BBS6*	20p12	Limb, heart, urogenital tract
Meckel-Gruber syndrome	*MKS1* (+ others)	17q23 (+ others)	Brain, kidney, liver
Nephronophthisis (types 1–4)	Nephrocyston (+ others)	Multiple	Kidney
Oro-facio-digital syndrome type 1	*OFD1* (+ others)	Xp.22 (+ others)	Skeleton (limb, face)
Polycystic kidney disease	Multiple	Multiple	Kidney
Primary ciliary dyskinesia (Kartegener syndrome)	Multiple	Multiple	Multi-system
Senior-Loken syndrome	Multiple	Multiple	Retina, kidney
Short-rib polydactyly syndrome	*DYNC2H1*	11q13	Skeleton, kidney, urogenital tract

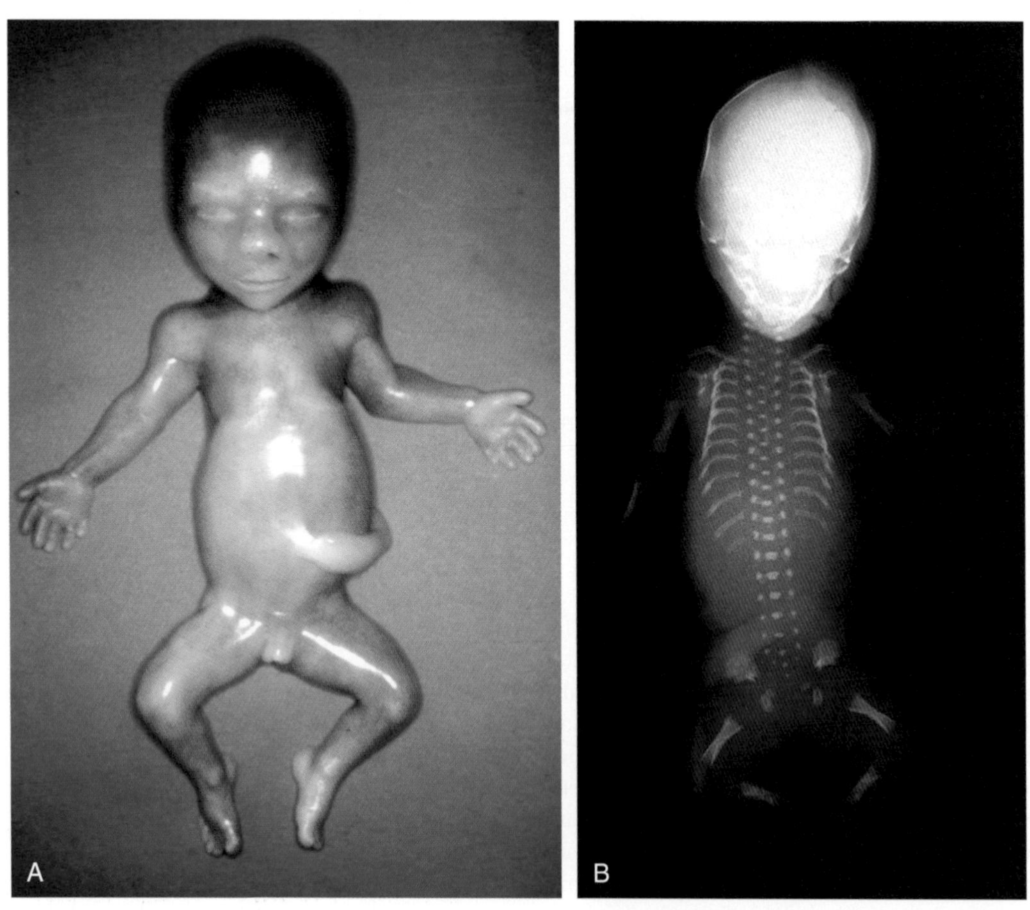

FIGURE 6.20 Short-rib polydactyly syndrome. **A,** The chest of the fetus is narrow and postaxial polydactyly affects all four limbs. **B,** As seen in this x-ray, the ribs of the fetus are very short.

Initiation and Specification

Limb bud formation is thought to be initiated at around 28 days by a member of the *FGF* family as illustrated by the development of an extra limb if *FGF1*, *FGF2*, or *FGF4* is applied to the side of a developing chick embryo. During normal limb initiation *FGF8* transcripts have been identified in mesenchyme near the initiation site. *FGF8* expression is probably controlled by *HOX* genes, which determine limb type (forelimb or hindlimb) and number.

Tissue Differentiation and Growth

Once limb formation has been initiated, a localized area of thickened ectoderm at the limb tip, known as the **apical ectodermal ridge** (AER), produces growth signals such as *FGF4* and *FGF8*, which maintain further growth and establish the proximo-distal axis (Figure 6.21). Expression of the gene *TP63* is crucial for sustaining the AER and, when this gene is mutated, split hand-foot (ectrodactyly) malformations result, often together with oral clefting and other anomalies—ectrodactyly-ectodermal dysplasia-clefting syndrome. Signals from another localized area on the posterior margin of the developing bud, known as the **zone of polarizing activity**, determine the anteroposterior axis. One of these signals is *SHH* (pp. 86–88), which acts in concert with other *FGF* genes, *GLI3*, and another gene family, which produces BMPs. Another morphogen, retinoic acid, is believed to play a major role at this stage in determining development at the anterior margin of the limb bud.

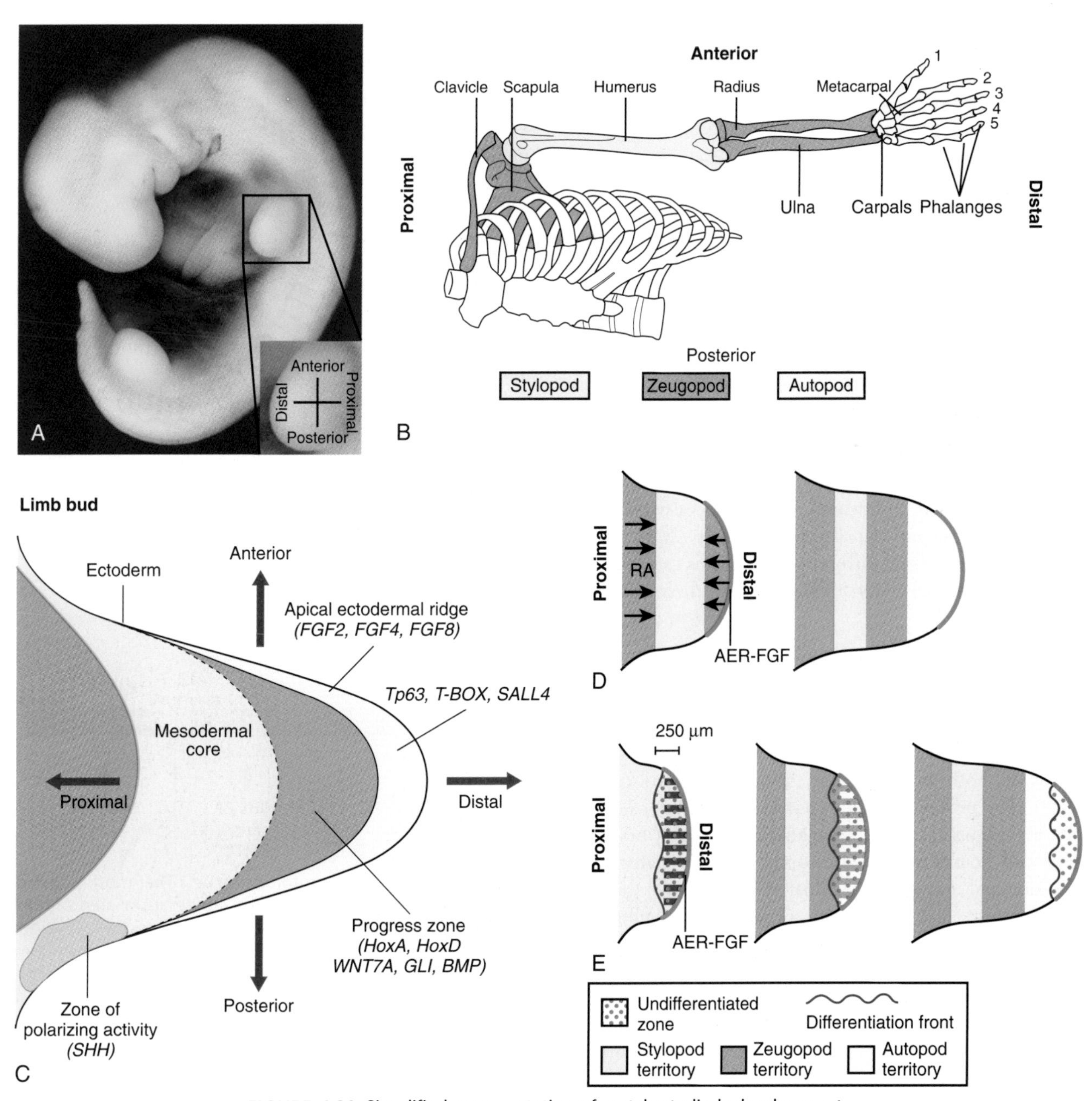

FIGURE 6.21 Simplified representation of vertebrate limb development.

Table 6.7 Genes that Can Cause Both Developmental Anomalies and Cancer

Gene	Chromosome	Developmental Anomaly	Cancer
PAX3	2q35	Waardenburg syndrome type 1	Alveolar rhabdomyosarcoma
KIT	4q12	Piebaldism	Mast cell leukemia
PTCH (Patched)	9q22	Gorlin syndrome	Basal cell carcinoma
RET	10p11	Hirschsprung disease	MEN 2A, MEN 2B, thyroid carcinoma
WT1	11p13	Denys-Drash syndrome	Wilms' tumor

Subsequent development involves the activation of genes from the *HOXA* and *HOXD* clusters in the undifferentiated proliferating mesenchymal cells beneath the AER. This area is known as the progress zone. Cells in different regions express different combinations of *HOX* genes that determine local cell proliferation, adhesion, and differentiation. Downstream targets of the *HOX* gene clusters remain to be identified. Other genes that clearly have a key role are those of the *T-box* family, already discussed, and *SALL4*, which is mutated in Okihiro syndrome (radial ray defects with abnormal eye movements resulting from congenital palsy affecting the sixth cranial nerve).

FGFs continue to be important during the later stages of limb development. In this context it becomes easy to understand why limb abnormalities are a feature of disorders such as Apert syndrome (see Figure 6.16), in which mutations have been identified in the extracellular domains of *FGFR2*.

Developmental Genes and Cancer

Several genes that play important roles in embryogenesis have also been shown to play a role in causing cancer (Table 6.7). This is not surprising, given that many developmental genes are expressed throughout life in processes such as signal transduction and signal transcription (p. 213). It has been shown that several different mechanisms can account for the phenotypic diversity demonstrated by these so-called teratogenes.

Gain-of-Function versus Loss-of-Function Mutations

Mention has already been made of the causal role of the *RET* proto-oncogene in familial Hirschsprung disease, as well as in both inherited and sporadic thyroid cancer (p. 93). The protein product encoded by *RET* consists of three main domains: an extracellular domain that binds to a glial cell line–derived neurotrophil factor, a transmembrane domain, and an intracellular tyrosine kinase domain that activates signal transduction (Figure 6.22). Mutations causing loss of function result in Hirschsprung disease. These include whole gene deletions, small intragenic deletions, nonsense mutations, and splicing mutations leading to synthesis of a truncated protein.

In contrast, mutations causing a gain-of-function effect result in either type 2A or type 2B multiple endocrine neoplasia (MEN). These disorders are characterized by a high incidence of medullary thyroid carcinoma and pheochromocytoma. The activating mutations that cause MEN-2A are clustered in five cysteine residues in the extracellular domain. MEN-2B, which differs from MEN-2A, in that affected individuals are tall and thin, is usually caused by a unique mutation in a methionine residue in the tyrosine kinase domain.

Somatic Rearrangements

Activation of the *RET* proto-oncogene can occur by a different mechanism whereby the genomic region encoding the intracellular domain is juxtaposed to one of several activating genes that are normally preferentially expressed in the thyroid gland. The newly formed hybrid *RET* gene produces a novel protein whose activity is not ligand dependent. These somatic rearrangements are found in a high proportion of papillary thyroid carcinomas, which show a particularly high incidence in children who were exposed to radiation following the Chernobyl accident in 1986.

PAX3 provides another example of a developmental gene that can cause cancer if it is fused to new DNA sequences. A specific translocation between chromosomes 2 and 13 that results in a new chimeric transcript leads to the development in children of a rare lung tumor called alveolar rhabdomyosarcoma.

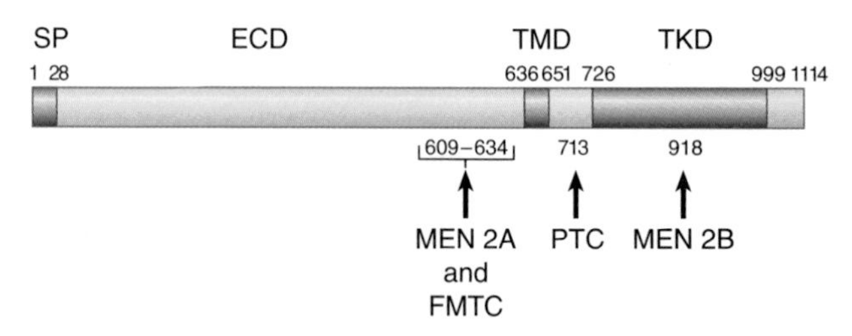

FIGURE 6.22 The *RET* proto-oncogene. The most common mutation sites in the different clinical entities associated with *RET* are indicated. Numbers refer to amino-acid residues. *SP*, signal peptide; *ECD*, extracellular domain; *TMD*, transmembrane domain; *TKD*, tyrosine kinase domain; *MEN*, multiple endocrine adenomatosis; *FMTC*, familial medullary thyroid carcinoma. The *arrow* above *PTC* (papillary thyroid carcinoma) indicates the somatic rearrangement site for the formation of new hybrid forms of *RET*. (Adapted from Pasini B, Ceccherini I, Romeo G 1996 RET mutations in human disease. Trends Genet 12:138–144.)

Table 6.8 Developmental Genes that Show a Position Effect

Gene	Chromosome	Developmental Anomaly
GLI3	7p13	Greig cephalopolysyndactyly
SHH	7q36	Holoprosencephaly
PAX6	11p13	Aniridia
SOX9	17q24	Campomelic dysplasia

Positional Effects and Developmental Genes

The discovery of a chromosomal abnormality, such as a translocation or inversion, in a person with a single-gene developmental syndrome provides a strong indication of the probable position of the disease locus, because it is likely that one of the breakpoints involved in the rearrangement will have disrupted the relevant gene. However, in a few instances, it has emerged that the chromosome breakpoint actually lies approximately 10 to 1000 kb upstream or downstream of the gene that is subsequently shown to be mutated in other affected individuals (Table 6.8). The probable explanation is that the breakpoint has separated the coding part of the gene from contiguous regulatory elements (p. 21). These observations have created obvious difficulties for those carrying out the original research when the putative disease gene in translocation families has been found not to contain an intragenic mutation.

Hydatidiform Moles

Occasionally conception results in an abnormal pregnancy in which the placenta consists of a proliferating disorganized mass known as a hydatidiform mole. These changes can be either partial or complete (Table 6.9).

Partial Hydatidiform Mole

Chromosome analysis of tissue from partial moles reveals the presence of 69 chromosomes—i.e., triploidy (p. 276). Using DNA polymorphisms, it has been shown that 46 of these chromosomes are always derived from the father, with the remaining 23 being maternal in origin. This doubling of the normal haploid paternal contribution of 23 chromosomes can be due to either fertilization by two sperm, which is known as **dispermy**, or to duplication of a haploid sperm chromosome set by a process known as **endoreduplication**.

Table 6.9 Characteristics of Partial and Complete Hydatidiform Moles

	Partial Mole	Complete Mole
No. of chromosomes	69	46
Parental origin of chromosomes	23 maternal 46 paternal	All 46 paternal
Fetus present	Yes, but not viable	No
Malignant potential	Very low	High

In these pregnancies the fetus rarely if ever survives to term. Triploid conceptions survive to term only when the additional chromosome complement is maternally derived, in which cases partial hydatidiform changes do not occur. Even in these situations, it is extremely uncommon for a triploid infant to survive for more than a few hours or days after birth.

Complete Hydatidiform Mole

Complete moles have only 46 chromosomes, but these are exclusively paternal in origin. A complete mole is caused by fertilization of an empty ovum either by two sperm or by a single sperm that undergoes endoreduplication. The opposite situation of an egg undergoing development without being fertilized by a sperm, a process known as parthenogenesis, occurs in lower animals such as arthropods but has been reported in a human on only one occasion, this being in the form of chimeric fusion with another cell line that had a normal male-derived complement.

The main importance of complete moles lies in their potential to undergo malignant change into invasive choriocarcinoma. This can usually be treated successfully by chemotherapy, but if untreated the outcome can be fatal. Malignant change is seen only very rarely with partial moles.

Different Parental Expression in Trophoblast and Embryoblast

Studies in mice have shown that when all nuclear genes in a zygote are derived from the father, the embryo fails to develop, whereas trophoblast development proceeds relatively unimpaired. In contrast, if all of the nuclear genes are maternal in origin, the embryo develops normally, but extra-embryonic development is poor. The observations outlined previously on partial and complete moles indicate that a comparable situation exists in humans, with paternally derived genes being essential for trophoblast development and maternally derived genes being necessary for early embryonic development. These phenomena are relevant to the concept of epigenetics (see the following section, p. 103) and genomic imprinting (p. 121).

Sexual Differentiation and Determination

The sex of an individual is determined by the X and Y chromosomes (p. 33). The presence of an intact Y chromosome leads to maleness regardless of the number of X chromosomes present. Absence of a Y chromosome results in female development.

Although the sex chromosomes are present from conception, differentiation into a phenotypic male or female does not commence until approximately 6 weeks. Up to this point both the müllerian and Wolffian duct systems are present and the embryonic gonads, although consisting of cortex and medulla, are still undifferentiated. From 6 weeks onwards, the embryo develops into a female unless the testis-determining factor initiates a sequence of events that prompt the undifferentiated gonads to develop into testes.

The Testis-Determining Factor—*SRY*

In 1990 it was shown that the testis-determining factor or gene is located on the short arm of the Y chromosome close to the pseudoautosomal region (p. 118). This gene is now referred to as being located in the sex-determining region of the Y chromosome *(SRY)*. It consists of a single exon that encodes a protein of 204 amino acids that include a 79-amino-acid HMG box (p. 92), indicating that it is likely to be a transcription regulator.

Evidence that the *SRY* gene is the primary factor that determines maleness comes from several observations:

1. *SRY* sequences are present in XX males. These are infertile phenotypic males who appear to have a normal 46,XX karyotype.
2. Mutations or deletions in the SRY sequences are found in many XY females. These are infertile phenotypic females who are found to have a 46,XY karyotype.
3. In mice the *SRY* gene is expressed only in the male gonadal ridge as the testes are developing in the embryo.
4. Transgenic XX mice that have a tiny portion of the Y chromosome containing the *SRY* region develop into males with testes.

From a biological viewpoint (i.e., the maintenance of the species), it would clearly be impossible for the *SRY* gene to be involved in crossing over with the X chromosome during meiosis I. Hence *SRY* has to lie outside the pseudoautosomal region. However, there has to be pairing of X and Y chromosomes, as otherwise they would segregate together into the same gamete during, on average, 50% of meioses. Nature's compromise has been to ensure that only a small portion of the X and Y chromosomes are homologous, and therefore pair during meiosis I. Unfortunately, the close proximity of *SRY* to the pseudoautosomal region means that, occasionally, it can get caught up in a recombination event. This almost certainly accounts for the majority of XX males, in whom molecular and fluorescent in-situ hybridization studies show evidence of Y-chromosome sequences at the distal end of one X-chromosome short arm (see Figure 18.22, p. 288).

Expression of *SRY* triggers off a series of events that involves other genes such as *SOX9* (17q24), leading to the medulla of the undifferentiated gonad developing into a testis, in which the Leydig cells begin to produce testosterone (Figure 6.23). This leads to stimulation of the Wolffian ducts, which form the male internal genitalia, and also to masculinization of the external genitalia. This latter step is mediated by dihydrotestosterone, which is produced from testosterone by the action of 5α-reductase (pp. 174, 287). The Sertoli cells in the testes produce a hormone known as

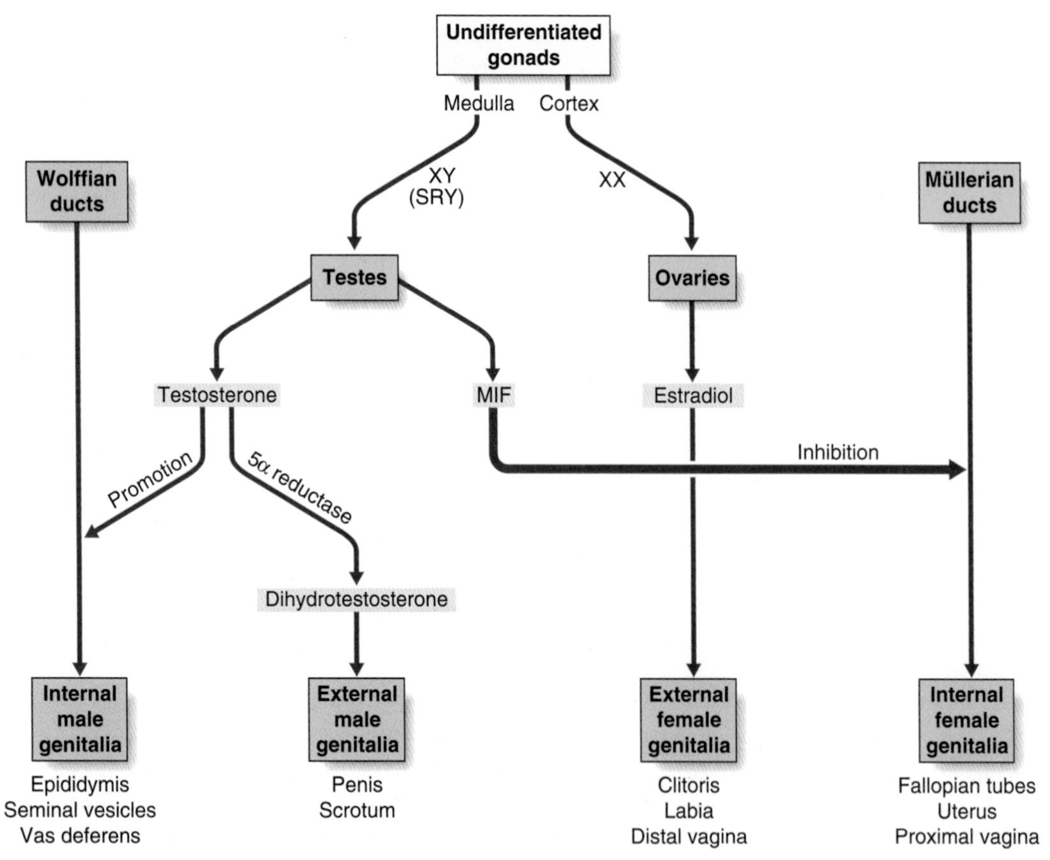

FIGURE 6.23 Summary of the main events involved in sex determination. *SRY*, Sex determining region of the Y chromosome; *MIF*, müllerian inhibitory factor.

müllerian inhibitory factor, which causes the müllerian duct system to regress. In campomelic dysplasia, resulting from mutated *SOX9*, ambiguous genitalia is common in cases with a 46,XY karyotype. Ambiguous genitalia, or sex reversal, is also frequent in cases of deletion 9p24.3 syndrome. This is probably due to haploinsufficiency for the *DMRT1* gene, a transcriptional regulator expressed in Sertoli cells, spermatogonia, and spermatocytes.

In the absence of normal *SRY* expression, the cortex of the undifferentiated gonad develops into an ovary. The müllerian duct forms the internal genitalia. The external genitalia fail to fuse and grow as in the male, and instead evolve into normal female external genitalia. This normal process of female development is sometimes referred to rather chauvinistically as the 'default' pathway. Without the stimulating effects of testosterone, the Wolffian duct system regresses.

Normally sexual differentiation is complete by 12 to 14 weeks' gestation, although the testes do not migrate into the scrotum until late pregnancy. Abnormalities of sexual differentiation are uncommon but they are important causes of infertility and sexual ambiguity. They are considered further in Chapter 18.

Epigenetics and Development

The concept of 'epigenetics' is not recent. Epigenesis was first mooted as a theme by Conrad Waddington in 1942 and referred, in essence, to the unfolding of developmental programs and processes from an undifferentiated zygote—the very heart of embryonic development. This roughly equates with our modern understanding of the control of developmental gene expression and signaling pathways. It incorporated the concept of epigenetic mechanisms being 'wiped clean' and 'reset' at one point in the life cycle. Although this is still valid, the term in current usage is extended to include heritable changes to gene expression that are *not* from differences in the genetic code. Such gene expression states may be transmitted stably through cell divisions—certainly mitosis but also meiosis (thereby not necessarily subject to a 'resetting' process). One genotype can therefore give rise to more than one phenotype, depending on the 'epigenetic state' of a locus, or loci.

The most common form of DNA modification—the biochemical mechanism for epigenesis—is direct covalent **methylation** of nucleotides. This appears to lead to a series of steps that alters local chromatin structure. In human genetics the best recognized epigenetic phenomena are X-chromosome inactivation, described in the following section, and parent-of-origin specific gene expression (parental imprinting), which is realized in Prader-Willi and Angelman syndromes (pp. 122–123), and Beckwith-Wiedemann and Russell-Silver syndromes (pp. 124–125)—i.e., when errors occur. There is much interest, however, in the possibility that epigenetic states can be influenced by environmental factors. In animal studies there is evidence that the nutritional and behavioral environment may lead

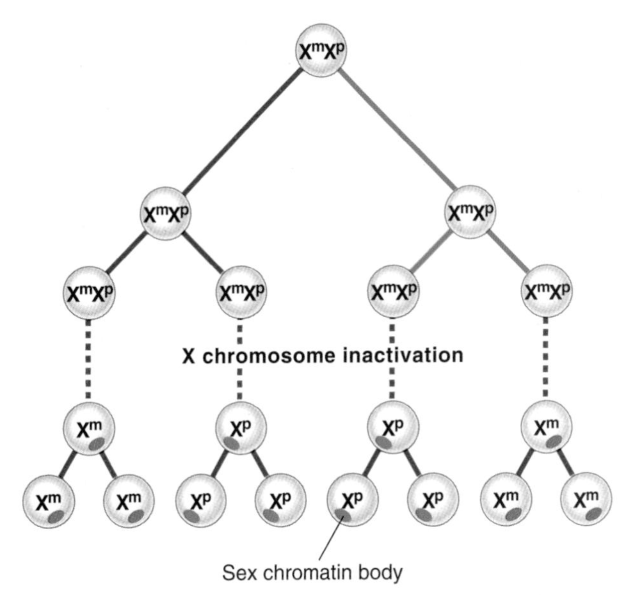

FIGURE 6.24 X-chromosome inactivation during development. The maternally and paternally derived X chromosomes are represented as X^m and X^p, respectively.

to different 'epialleles', and in human populations epidemiological studies have shown convincing correlations of maternal (and in some cases grandparental) nutritional status with late-onset cardiovascular and metabolic-endocrine disease.

X-Chromosome Inactivation

As techniques were developed for studying chromosomes, it was noted that in female mice one of the X chromosomes often differed from all other chromosomes in the extent to which it was condensed. In 1961 Dr Mary Lyon proposed that this heteropyknotic X chromosome was inactivated, citing as evidence her observations on the mosaic pattern of skin coloration seen in mice known to be heterozygous for X-linked genes that influence coat color. Subsequent events have confirmed the validity of Lyon's hypothesis, and in recognition of her foresight the process of X-chromosome inactivation (XCI) is often referred to as **lyonization**.

The process of XCI occurs early in development at around 15 to 16 days' gestation, when the embryo consists of approximately 5000 cells. Normally either of the two X chromosomes can be inactivated in any particular cell. Thereafter the same X chromosome is inactivated in all daughter cells (Figure 6.24). This differs from the case in marsupials, in which the paternally derived X chromosome is consistently inactivated.

The inactive X chromosome exists in a condensed form during interphase when it appears as a darkly staining mass of chromatin known as the sex chromatin, or Barr body. During mitosis the inactive X chromosome is late replicating. Laboratory techniques have been developed for distinguishing which of the X chromosomes is late replicating in each cell. This can be useful for confirming that one of the X chromosomes is structurally abnormal, as usually an abnormal X chromosome will be preferentially inactivated;

or, more correctly, only those hematopoietic stem cells in which the normal X chromosome is active will have survived. Apparent non-random inactivation also occurs when one of the X chromosomes is involved in a translocation with an autosome (p. 116).

The epigenetic process of XCI is achieved by differential methylation (a form of imprinting; see p. 121) and is initiated by a gene, *XIST* ('X inactivation specific transcript'), which maps within the X-inactivation centre at Xq13.3. *XIST* is expressed only from the inactive X chromosome and produces RNA that spreads an inactivation methylation signal up and down the X chromosome on which it is located. This differential methylation of the X chromosomes has been utilized in carrier detection studies for X-linked immunodeficiency diseases (e.g., Wiskott-Aldrich syndrome) using methylation-sensitive restriction enzymes (p. 204). Not all of the X chromosome is inactivated. Genes in the pseudoautosomal region at the tip of the short arm remain active, as do other loci elsewhere on the short and long arms, such as *XIST*. There are more genes that escape XCI in Xp compared with Xq. This probably explains why more severe phenotypic effects are seen in women with small Xp chromosome deletions compared with those in women with small deletions in Xq. If all loci on the X chromosome were inactivated, then all women would have the clinical features of Turner syndrome and the presence of more than one X chromosome in a male (e.g., 47,XXY) or two in a female (e.g., 47,XXX) would have no phenotypic effects. There are, in fact, quite characteristic clinical features in these disorders (p. 278).

XCI provides a satisfactory explanation for several observations, described below.

Barr Bodies

In men and women with more than one X chromosome, the number of Barr bodies (p. 39) visible at interphase is always one less than the total number of X chromosomes. For example, men with a 47,XXY karyotype have a single Barr body, whereas women with a 47,XXX karyotype have two Barr bodies.

Dosage Compensation

Women with two normal X chromosomes have the same blood levels of X-chromosome protein products, such as factor VIII, as normal men, who of course have only one X chromosome. An exception to this phenomenon of dosage compensation is the level of steroid sulfatase in blood, which is increased in women compared with men. Not surprisingly, perhaps, it has been shown that the locus for steroid sulfatase (deficiency of which causes a skin disorder known as ichthyosis) is in the pseudoautosomal region.

Mosaicism

Mice that are heterozygous for X-linked genes affecting coat color show mosaicism with alternating patches of different color rather than a homogeneous pattern. This is consistent with patches of skin being clonal in origin in that they are derived from a single stem cell in which one or other of the X chromosomes is expressed, but not both. Thus, each patch reflects which of the X chromosomes was active in the original stem cell. Similar effects are seen in tissues of clonal origin in women who are heterozygous for X-linked mutations such as ocular albinism (Figure 6.25).

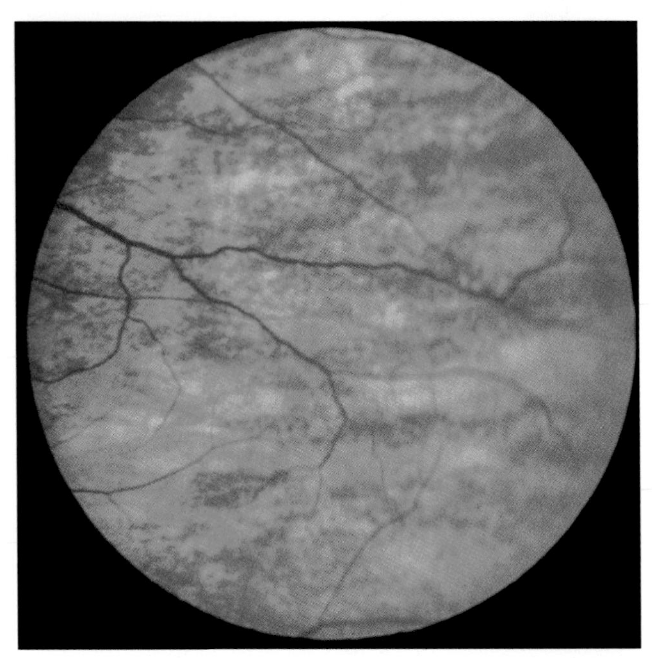

FIGURE 6.25 The fundus of a carrier of X-linked ocular albinism showing a mosaic pattern of retinal pigmentation. (Courtesy Mr. S.J. Charles, The Royal Eye Hospital, Manchester, UK.)

Other evidence confirming that X-inactivation leads to mosaicism in females comes from studies of the expression of the enzyme glucose-6-phosphate dehydrogenase (p. 187) in clones of cultured fibroblasts from women heterozygous for variants of this gene. Each clone is derived from a single cell and expresses one of the variants, but never both. The clonal origin of tumors can be confirmed in women who are heterozygous for such variants by demonstration of the expression of only one of the variants in the tumor.

Problems of Carrier Detection

Carrier detection for X-linked recessive disorders based only on examination of clinical features or on indirect assay of gene function is notoriously difficult and unreliable. Cells in which the X chromosome with the normal gene is active can have a selective advantage, or they can correct the defect in closely adjacent cells in which the X chromosome with the mutant gene is active. For example, only a proportion of carriers of Duchenne muscular dystrophy (DMD) show evidence of muscle damage as indicated by measurement of creatine kinase in serum (p. 314). Similarly, distorted ratios of very long chain fatty acids are seen in many, but not all, carriers of X-linked adrenoleukodystrophy.

Fortunately, the development of molecular methods for carrier detection in X-linked disorders can bypass these problems by use of PCR primers that distinguish the

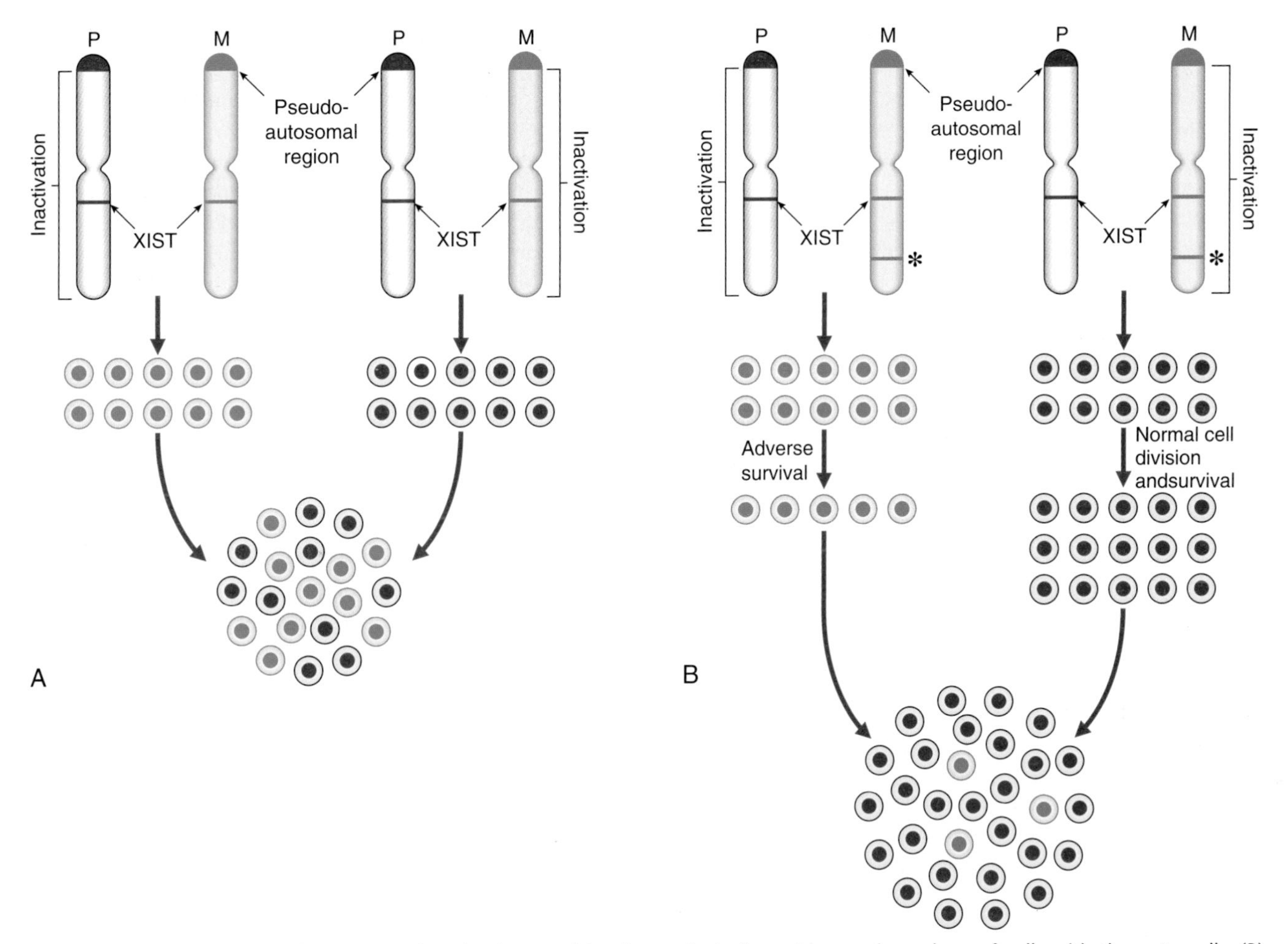

FIGURE 6.26 **A**, Normal X-chromosome inactivation resulting in survival of roughly equal numbers of cells with the paternally *(P)* and maternally (derived X chromosome active). **B**, In this situation the maternally derived X chromosome has a mutation (*) that results in selection against the cells in which it is active. Thus surviving cells show preferential expression of the paternally derived X chromosome.

products of methylated and unmethylated DNA—if there has been selection against the cell line in which the mutant-bearing X chromosome is active (Figure 6.26).

Manifesting Heterozygotes

Occasionally a woman is encountered who shows mild or even full expression of an X-linked recessive disorder, such as DMD or hemophilia A. One possible explanation is that she is a manifesting heterozygote in whom, by chance, the X chromosome bearing the normal gene has been inactivated in significantly more than 50% of relevant cells. This is referred to as skewed X-inactivation (p. 116). There is some evidence that X-chromosome inactivation can itself be under genetic control, because families with several manifesting carriers of disorders such as DMD and Fabry disease have been reported. In a few families, marked skewing of X-inactivation in several females has been shown to be associated with an underlying mutation in *XIST*.

The 46,Xr(X) Phenotype

A 46,Xr(X) karyotype is found in some women with typical features of Turner syndrome. This is consistent with the ring lacking X sequences, which are normally not inactivated and which are needed for a normal phenotype. Curiously, a few 46,Xr(X) women have congenital abnormalities and show intellectual impairment. In these women it has been shown that *XIST* is not expressed on the ring X, so their relatively severe phenotype is likely to be caused by functional disomy for those genes present on their ring X chromosome. Other studies in Turner syndrome, this time on pure 45,X cases, have shown some differences in social cognition and higher order executive function skills according to whether their X chromosome was paternal or maternal in origin. Those with a **paternal** X scored better, from which the existence of a locus for social cognition on the X chromosome can be postulated. If such a locus is not expressed from the maternal X, this could provide at least part of the explanation for the excess difficulty with language and social skills observed in 46,XY males, as their X is always maternal in origin.

Recent Research

Research has suggested that, just as XCI is not an all-or-none phenomenon for the whole chromosome, it is probably not all-or-none for every gene. In a study of skin fibroblasts, which express more than 600 of the 1098 genes identified on the X chromosome, about 20% were found to be inactivated in some but not all samples. About 15% escaped XCI completely, whereas only 65% were fully silenced and thus expressed in one dose. In addition to non-random XCI, the variable dosage of genes that escape XCI may account for variation among normal females as well as those who are heterozygous for X-linked disease genes.

Twinning

Twinning occurs frequently in humans, although the incidence in early pregnancy as diagnosed by ultrasonography is greater than at delivery, presumably as a result of death and subsequent resorption of one of the twins in a proportion of twin pregnancies. The overall incidence of twinning in the UK is approximately 1 in 80 of all pregnancies, so that approximately 1 in 40 (i.e., 2 of 80) of all individuals is a twin. However, the spontaneous twinning rate varies enormously, from approximately 1 in 125 pregnancies in Japan to 1 in 22 in Nigeria.

Twins can be identical or non-identical—i.e., **monozygotic** (MZ) (uniovular) or **dizygotic** (DZ) (biovular) —depending on whether they originate from a single conception or from two separate conceptions (Table 6.10). Comparison of the incidence of disease in MZ and DZ twins reared apart and together can provide information about the relative contributions of genetics and the environment to the cause of many of the common diseases of adult life (see Chapter 15), especially the study of mental health and behavior.

Monozygotic Twins

MZ twinning occurs in about 1 in 300 births in all populations that have been studied. MZ twins originate from a single egg that has been fertilized by a single sperm. A very early division, occurring in the zygote before separation of the cells that make the chorion, results in dichorionic twins. Division during the blastocyst stage from days 3 to 7 results in monochorionic diamniotic twins. Division after the first week leads to monoamniotic twins. However, the reason(s) why MZ twinning occurs at all in humans is not clear. As an event, the incidence is increased two- to five-fold in babies born by in-vitro fertilization. There are rare cases of familial MZ twinning that can be transmitted by the father or mother, suggesting a single-gene defect that predisposes to the phenomenon.

We tend to think of MZ twins as being genetically identical, and basically this is of course true. However, they can be discordant for structural birth defects that may be linked to the twinning process itself—especially those anomalies affecting midline structures. There is probably a two- to three-fold increased risk of congenital anomalies in MZ twins (i.e., 5% to 10% of MZ twins overall). Discordance for single-gene traits or chromosome abnormalities may occur because of a post-zygotic somatic mutation or non-disjunction, respectively. One example of the latter is the rare occurrence of MZ twins of different sex: one 46,XY and the other 45,X. Curiously, MZ female twins can show quite striking discrepancy in X-chromosome inactivation. There are several reports of female MZ twin pairs of which only one is affected by an X-linked recessive condition such as DMD or hemophilia A. In these rare examples, both twins have the mutation and both show non-random X-inactivation, but in opposite directions.

MZ twins have traditionally provided ideal research material for the study of genetic versus environmental influences. In a study of 40 pairs of MZ twins, geneticists measured levels of two epigenetic modifications, DNA methylation, and histone acetylation. Two-thirds of the twin pairs had essentially identical profiles, but significant differences were observed in the remaining third. These differences were broadly correlated with the age of the twins, with the amount of time spent apart and the differences in their medical histories, suggesting a cumulative effect on DNA modification over time. It also suggests a possible causal link between epigenetic modification and susceptibility to disease.

Very late division occurring more than 14 days after conception can result in conjoined twins. This occurs in about 1 in 100,000 pregnancies, or approximately 1 in 400 MZ twin births. Conjoined twins are sometimes referred to as Siamese, in memory of Chang and Eng, who were born in 1811 in Thailand, then known as Siam. They were joined at the upper abdomen and made a successful living as celebrities at traveling shows in the United States, where they settled and married. They both managed to have large numbers of children despite remaining conjoined until they died within a few hours of each other at age 61.

The sex ratio for conjoined twins is markedly distorted, with about 75% being female. The later the twinning event, the more distorted the sex ratio in favor of females, and X-inactivation studies suggest that MZ twinning occurs

Table 6.10 Summary of Differences between Monozygotic and Dizygotic Twins

	Monozygotic	Dizygotic
Origin	Single egg fertilized	Two eggs, each fertilized by a single sperm
Incidence	1 in 300 pregnancies	Varies from 1 in 100 to 1 in 500 pregnancies
Proportion of genes in common	100%	50% (on average)
Fetal membranes	70% monochorionic and diamniotic; 30% dichorionic and diamniotic; rarely monochorionic and monoamniotic	Always dichorionic and diamniotic

around the time of X-inactivation, a phenomenon limited to female zygotes, of course.

Dizygotic Twins

DZ twins result from the fertilization of two ova by two sperm and are no more closely related genetically than brothers and sisters, as they share, on average, 50% of the same genes from each parent. Hence they are sometimes referred to as **fraternal twins**. DZ twins are dichorionic and diamniotic, although they can have a single fused placenta if implantation occurs at closely adjacent sites. The incidence varies from approximately 1 in 100 deliveries in black Caribbean populations to 1 in 500 deliveries in Asia and Japan. In western European whites, the incidence is approximately 1 in 120 deliveries and has been observed to fall with both urbanization and starvation, but increases in relation to the amount of seasonal light (e.g., in northern Scandinavia during the summer). Factors that convey an increased risk for DZ twinning are: increased maternal age, a positive family history (from a familial increase in follicle-stimulating hormone levels), and the use of ovulation-inducing drugs such as clomiphene.

Determination of Zygosity

Zygosity used to be established by study of the placenta and membranes and also by analysis of polymorphic systems such as the blood groups, the human leukocyte antigens and other biochemical markers. Now it is determined most reliably by the use of highly polymorphic molecular (DNA) markers (pp. 69–70) and single nucleotide polymorphisms (SNPs).

FURTHER READING

Baker K, Beales PL 2009 Making sense of cilia in disease: the human ciliopathies. Am J Med Genet 151C:281–295.

Dreyer SD, Zhou G, Lee B 1998 The long and the short of it: developmental genetics of the skeletal dysplasias. Clin Genet 54:464–473

A short review of developmental genes known to cause abnormal skeletal development.

Hall JG 2003 Twinning. Lancet 2003; 362:735–743

Hammerschmidt M, Brook A, McMahon AP 1997 The world according to hedgehog. Trends Genet 13:14–20

A comprehensive account of the role of the hedgehog gene family in early vertebrate development.

Kleinjan DJ, van Heyningen V 1998 Position effect in human genetic disease. Hum Mol Genet 7:1611–1618

An outline of the various theories and mechanisms that have been proposed to account for observed positional effects in developmental gene expression.

Kornak U, Mundlos S 2003 Genetic disorders of the skeleton: a developmental approach. Am J Hum Genet 73:447–474

An up-to-date summary of current knowledge.

Lacombe D 1999 Transcription factors in dysmorphology. Clin Genet 55:137–143

As the title indicates, a description of the role of transcription regulatory genes in causing multiple congenital abnormality syndromes.

Lindor NM, Ney JA, Gaffey TA, et al 1992 A genetic review of complete and partial hydatidiform moles and nonmolar triploidy. Mayo Clin Proc 67:791–799

A detailed review of the mechanisms that can lead to the formation of hydatidiform moles.

Lyon MF 1961 Gene action in the X chromosome of the mouse (*Mus musculus* L). Nature 190:372–373

The original proposal of X-inactivation—very short and easily understood.

Manouvrier-Hanu S, Holder-Espinasse M, Lyonnet S 1999 Genetics of limb anomalies in humans. Trends Genet 15:409–417

A detailed and well-illustrated account of vertebrate limb development.

Muenke M, Schell U 1995 Fibroblast-growth-factor receptor mutations in human skeletal disorders. Trends Genet 11:308–313

A concise review of the functions of the fibroblast growth factors and their receptors.

Muragaki Y, Mundlos S, Upton J, Olsen BR 1996 Altered growth and branching patterns in synpolydactyly caused by mutations in *HOXD13*. Science 272:548–551.

The long-awaited first report of a human malformation caused by a mutation in a HOX gene.

Passos-Bueno MR, Ornelas CC, Fanganiello RD 2009 Syndromes of the first and second pharyngeal arches: a review. Am J Med Genet Part A 149A:1853–1859.

An excellent short overview of pharyngeal arches and associated syndromes.

Saga Y, Takeda H 2001 The making of the somite: molecular events in vertebrate segmentation. Nature Rev Genet 2:835–844

An excellent review of somite development.

Tickle C, ed 2003 Patterning in vertebrate development. Oxford University Press, Oxford

A detailed, multi-author collection handling very early development, from mainly molecular perspectives.

Villavicencio EH, Walterhouse DO, Iannaccone PM 2000 The Sonic hedgehog-patched-Gli pathway in human development and disease. Am J Hum Genet 67:1047–1054

An excellent short overview of the Sonic hedgehog pathway.

ELEMENTS

1 Several developmental gene families first identified in *Drosophila* and mice also play important roles in human morphogenesis. These include segment polarity genes, homeobox-containing genes *(HOX)* and paired-box containing genes *(PAX)*. Many of these genes act as transcription factors that regulate sequential developmental processes. Others are important in cell signaling. Several human malformations and multiple malformation syndromes are caused by mutations in these genes.

2 Several well-recognized syndromes are now known to be linked through their relationships to developmental signaling pathways (e.g., sonic-hedgehog).

3 For normal development a haploid chromosome set must be inherited from each parent. A paternal diploid complement results in a complete hydatidiform mole if there is no maternal contribution, and in triploidy with a partial hydatidiform mole if there is a haploid maternal contribution.

4 A testis-determining factor on the Y chromosome, known as *SRY*, stimulates the undifferentiated gonads to develop into testes. This, in turn, sets off a series of events leading to male development. In the absence of *SRY* expression the human embryo develops into a female.

5 In females one of the X chromosomes is inactivated in each cell in early embryogenesis. This can be either the maternally derived or the paternally derived X chromosome. Thereafter, in all daughter cells the same X chromosome is inactivated. This process, known as lyonization, explains the presence of the Barr body in female nuclei and achieves dosage compensation of X-chromosome gene products in males and females.

6 Twins can be monozygotic (identical) or dizygotic (fraternal). Monozygotic twins originate from a single zygote that divides into two during the first 2 weeks after conception. Monozygotic twins are genetically identical. Dizygotic twins originate from two separate zygotes and are no more genetically alike than brothers and sisters.

CHAPTER 7

Patterns of Inheritance

That the fundamental aspects of heredity should have turned out to be so extraordinarily simple supports us in the hope that nature may, after all, be entirely approachable.

THOMAS MORGAN (1919)

Family Studies

If we wish to investigate whether a particular trait or disorder in humans is genetic and hereditary, we usually have to rely either on observation of the way in which it is transmitted from one generation to another, or on study of its frequency among relatives.

An important reason for studying the pattern of inheritance of disorders within families is to enable advice to be given to members of a family regarding the likelihood of their developing it or passing it on to their children (i.e., **genetic counseling**; see Chapter 17). Taking a family history can, in itself, provide a diagnosis. For example, a child could come to the attention of a doctor with a fracture after a seemingly trivial injury. A family history of relatives with a similar tendency to fracture and blue sclerae would suggest the diagnosis of osteogenesis imperfecta. In the absence of a positive family history, other diagnoses would have to be considered.

Pedigree Drawing and Terminology

A family tree is a shorthand system of recording the pertinent information about a family. It usually begins with the person through whom the family came to the attention of the investigator. This person is referred to as the **index case**, **proband,** or **propositus**; or, if female, the **proposita**. The position of the proband in the family tree is indicated by an arrow. Information about the health of the rest of the family is obtained by asking direct questions about brothers, sisters, parents, and maternal and paternal relatives, with the relevant information about the sex of the individual, affection status, and relationship to other individuals being carefully recorded in the pedigree chart (Figure 7.1). Attention to detail can be crucial because patients do not always appreciate the important difference between siblings and *half*-siblings, or might overlook the fact, for example, that the child of a brother who is at risk of Huntington disease is actually a *step*-child and not a biological relative.

Mendelian Inheritance

More than 16,000 traits or disorders in humans exhibit single gene **unifactorial** or **mendelian inheritance**. However, characteristics such as height, and many common familial disorders, such as diabetes or hypertension, do not usually follow a simple pattern of mendelian inheritance (see Chapter 9).

A trait or disorder that is determined by a gene on an autosome is said to show **autosomal inheritance**, whereas a trait or disorder determined by a gene on one of the sex chromosomes is said to show **sex-linked inheritance**.

Autosomal Dominant Inheritance

An autosomal dominant trait is one that manifests in the heterozygous state, that is, in a person possessing both an abnormal or mutant allele and the normal allele. It is often possible to trace a dominantly inherited trait or disorder through many generations of a family (Figure 7.2). In South Africa the vast majority of cases of porphyria variegata can be traced back to one couple in the late seventeenth century. This is a metabolic disorder characterized by skin blistering as a result of increased sensitivity to sunlight (Figure 7.3), and the excretion of urine that becomes 'port wine' colored on standing as a result of the presence of porphyrins (p. 179). This pattern of inheritance is sometimes referred to as 'vertical' transmission and is confirmed when male–male (i.e., father to son) transmission is observed.

Genetic Risks

Each gamete from an individual with a dominant trait or disorder will contain either the normal allele or the mutant allele. If we represent the dominant mutant allele as 'D' and the normal allele as 'd', then the possible combinations of the gametes is seen in Figure 7.4. Any child born to a person affected with a dominant trait or disorder has a 1 in 2 (50%) chance of inheriting it and being similarly affected. These diagrams are often used in the genetic clinic to explain segregation to patients and are more user-friendly than a Punnett square (see Figs. 1.3 and 8.1).

Pleiotropy

Autosomal dominant traits may involve only one organ or part of the body, for example the eye in congenital cataracts. It is common, however, for autosomal dominant disorders to manifest in different systems of the body in a variety of

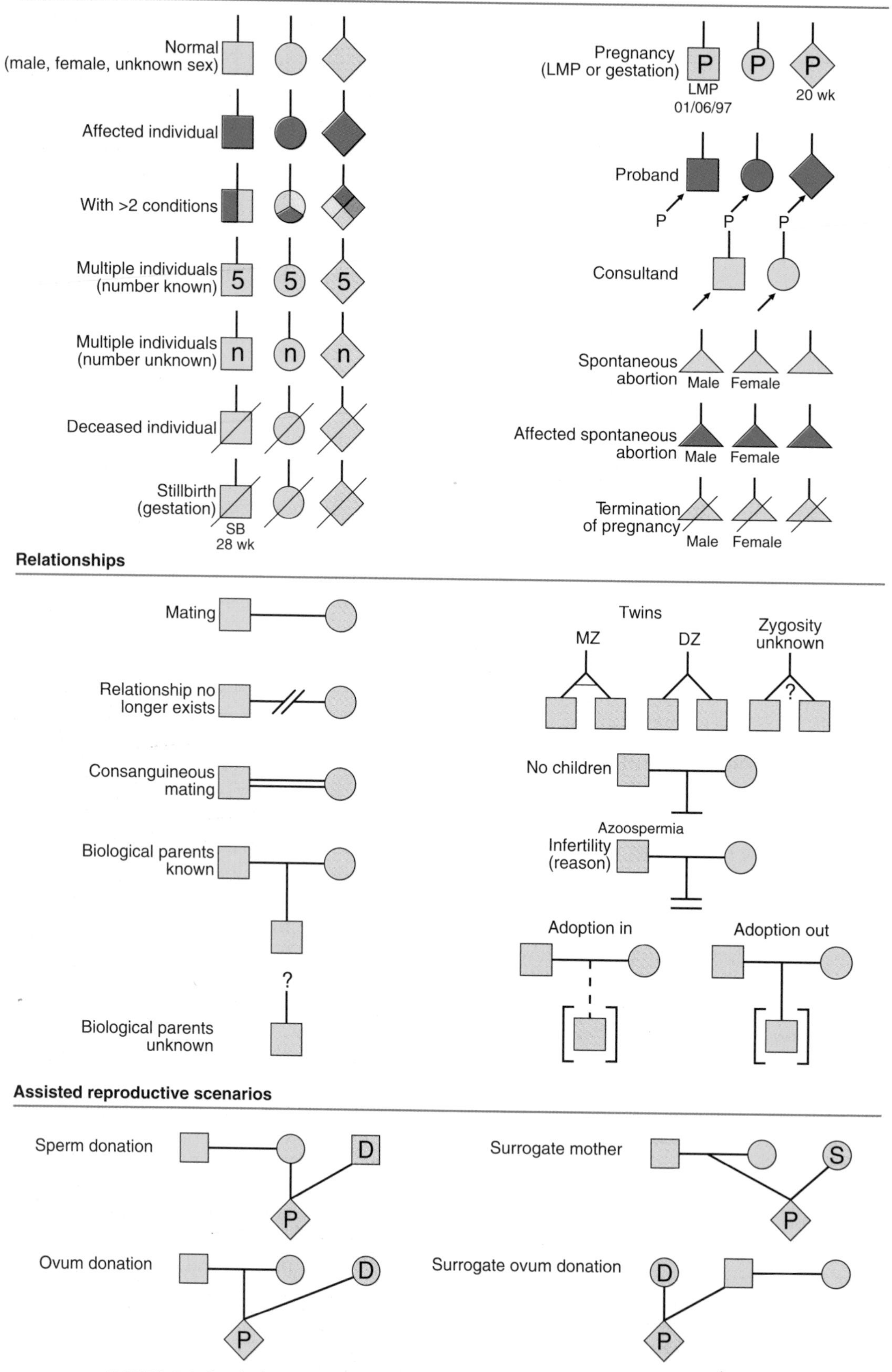

FIGURE 7.1 Symbols used to represent individuals and relationships in family trees.

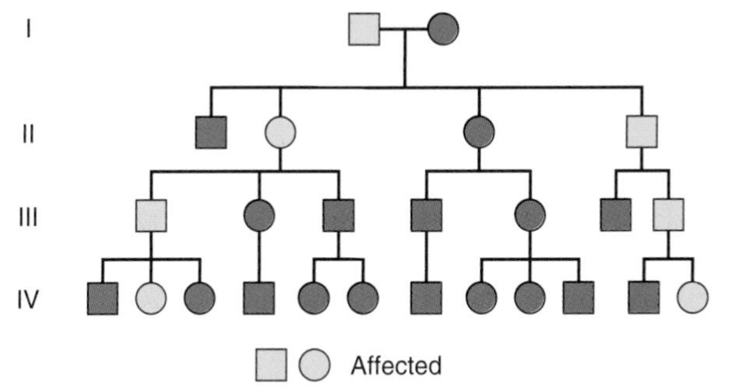

FIGURE 7.2 Family tree of an autosomal dominant trait. Note the presence of male-to-male transmission.

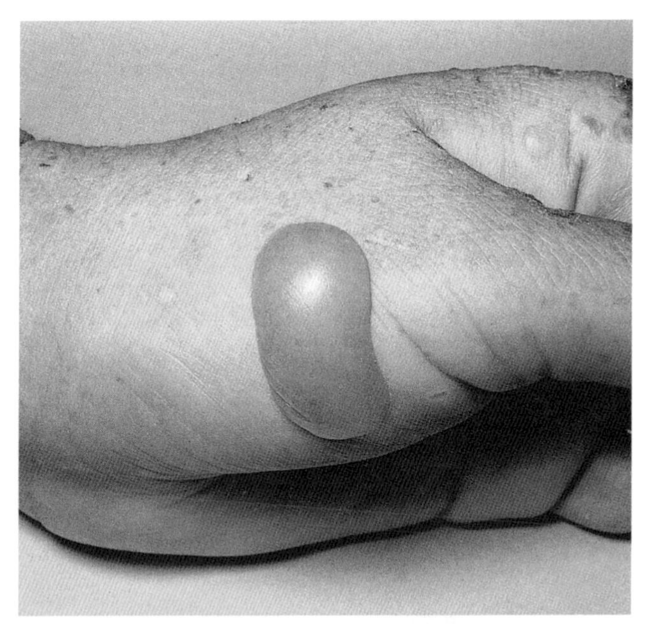

FIGURE 7.3 Blistering skin lesions on the hand in porphyria variegata.

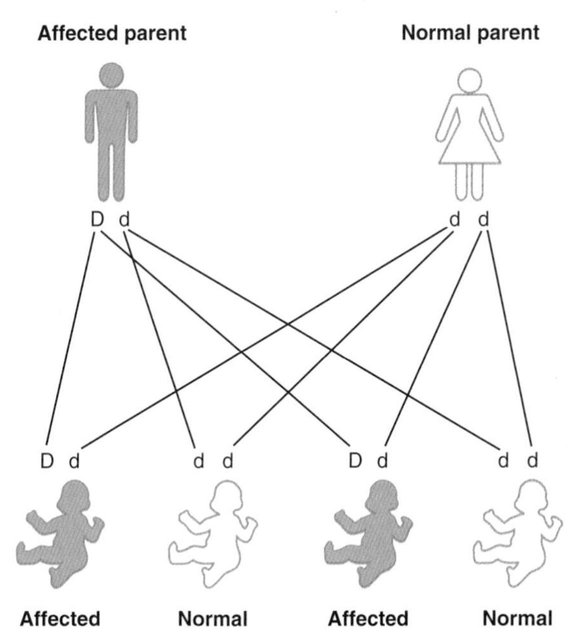

FIGURE 7.4 Segregation of alleles in autosomal dominant inheritance. *D* represents the mutated allele, whereas *d* represents the normal allele.

ways. This is pleiotropy—a single gene that may give rise to two or more apparently unrelated effects. In tuberous sclerosis affected individuals can present with a range of problems including learning difficulties, epilepsy, a facial rash known as adenoma sebaceum (histologically composed of blood vessels and fibrous tissue known as angiokeratoma) or subungual fibromas (Figure 7.5); some affected individuals have all features, whereas others may have almost none. Some discoveries are challenging our conceptual understanding of the term **pleiotropy** on account of the remarkably diverse syndromes that can result from different mutations in the same gene—for example, the *LMNA* gene (which encodes lamin A/C) and the X-linked filamin A

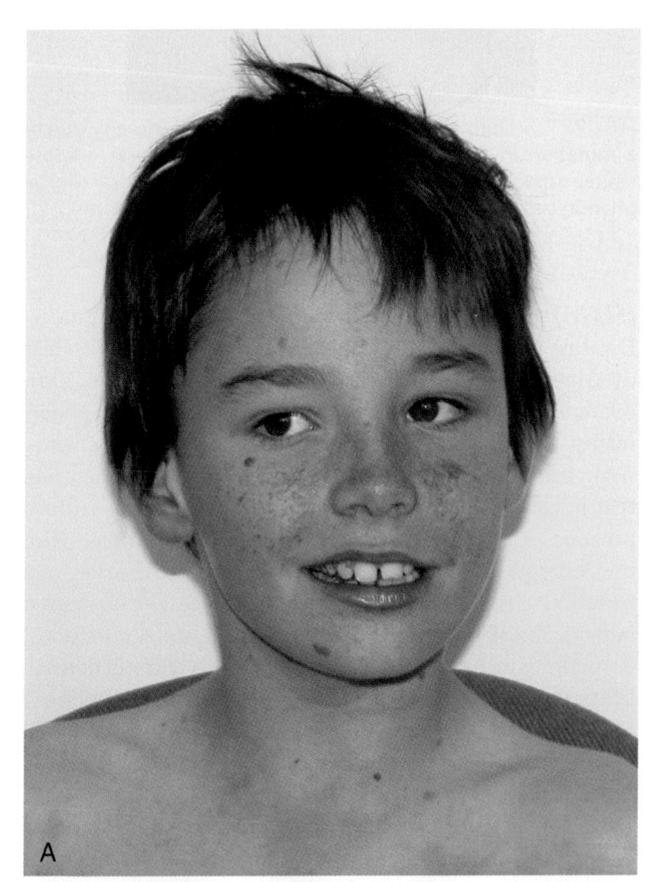

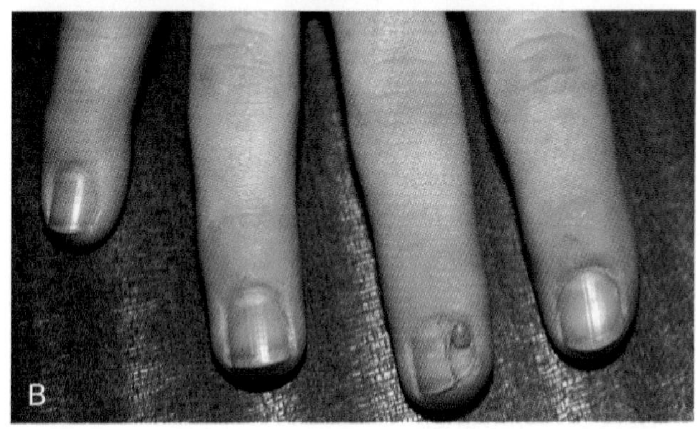

FIGURE 7.5 The facial rash (**A**) of angiokeratoma (adenoma sebaceum) in a male with tuberous sclerosis, and a typical subungual fibroma of the nail bed (**B**).

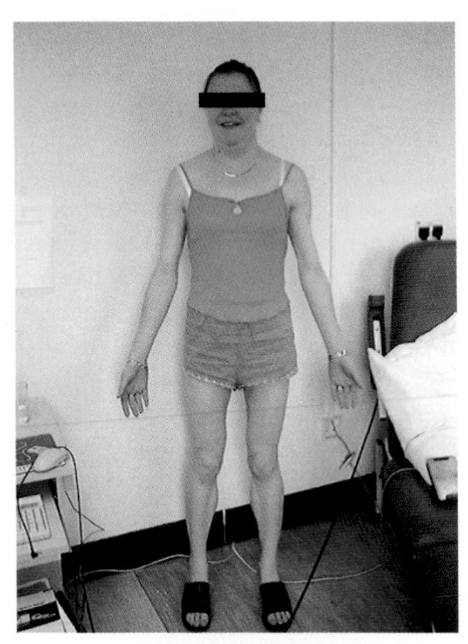

FIGURE 7.6 Dunnigan-type familial partial lipodystrophy due to a mutation in the lamin *A/C* gene. The patient lacks adipose tissue, especially in the distal limbs. A wide variety of clinical phenotypes is associated with mutations in this one gene.

(*FLNA*) gene. Mutations in *LMNA* may cause Emery-Dreifuss muscular dystrophy, a form of limb girdle muscular dystrophy, a form of Charcot-Marie-Tooth disease (p. 305), dilated cardiomyopathy (p. 296) with conduction abnormality, Dunnigan-type familial partial lipodystrophy (Figure 7.6), mandibuloacral dysplasia, and a very rare condition that has always been a great curiosity—Hutchinson-Gilford progeria. These are due to heterozygous mutations, with the exception of the Charcot-Marie-Tooth disease and mandibuloacral dysplasia, which are recessive—affected individuals are therefore homozygous for *LMNA* mutations. Sometimes an individual with a mutation is entirely normal. Mutations in the filamin A gene have been implicated in the distinct, though overlapping, X-linked dominant dysmorphic conditions oto-palato-digital syndrome, Melnick-Needles syndrome and frontometaphyseal dysplasia. However, it could not have been foreseen that a form of X-linked dominant epilepsy in women, called periventricular nodular heterotopia, is also due to mutations in this gene.

Variable Expressivity

The clinical features in autosomal dominant disorders can show striking variation from person to person, even in the same family. This difference between individuals is referred to as **variable expressivity**. In autosomal dominant polycystic kidney disease, for example, some affected individuals develop renal failure in early adulthood whereas others have just a few renal cysts that do not affect renal function significantly.

Reduced Penetrance

In some individuals heterozygous for gene mutations giving rise to certain autosomal dominant disorders, there may be no abnormal clinical features, representing so-called **reduced penetrance** or what is commonly referred to in lay terms as 'skipping a generation'. Reduced penetrance is thought to be the result of the modifying effects of other genes, as well as interaction of the gene with environmental factors. An individual who has no features of a disorder despite being heterozygous for a particular gene mutation is said to represent **non-penetrance**.

Reduced penetrance and variable expressivity, together with the pleiotropic effects of a mutant allele, all need to be taken into account when trying to interpret family history information for disorders that follow autosomal dominant inheritance. A good example of a very variable condition for which non-penetrance is frequently seen is Treacher-Collins syndrome. In its most obvious manifestation the facial features are unmistakable (Figure 7.7). However, the mother of the child illustrated is also known to harbor the gene (*TCOF1*) mutation as she has a number of close relatives with the same condition.

FIGURE 7.7 The baby in this picture has Treacher-Collins syndrome, resulting from a mutation in *TCOF1*. The mandible is small, the palpebral fissures slant downward, there is usually a defect (coloboma) of the lower eyelid, the ears may show microtia, and hearing impairment is common. The condition follows autosomal dominant inheritance but is very variable—the baby's mother also has the mutation but she shows no obvious signs of the condition.

New Mutations

In autosomal dominant disorders an affected person usually has an affected parent. However, this is not always the case and it is not unusual for a trait to appear in an individual when there is no family history of the disorder. A striking example is achondroplasia, a form of short-limbed dwarfism (pp. 93–94), in which the parents usually have normal stature. The sudden unexpected appearance of a condition arising as a result of a mistake occurring in the transmission of a gene is called a **new mutation**. The dominant mode of inheritance of achondroplasia could be confirmed only by the observation that the offspring of persons with achondroplasia had a 50% chance of having achondroplasia. In less dramatic conditions other explanations for the 'sudden' appearance of a disorder must be considered. This includes non-penetrance and variable expression, as mentioned in the previous section. However, the astute clinician also needs to be aware that the family relationships may not be as stated—i.e., there may be undisclosed **non-paternity** (p. 342) (or, occasionally, **non-maternity**).

New dominant mutations, in certain instances, have been associated with an increased age of the father. Traditionally, this is believed to be a consequence of the large number of mitotic divisions that male gamete stem cells undergo during a man's reproductive lifetime (p. 41). However, this may well be a simplistic view. In relation to mutations in *FGFR2* (craniosynostosis syndromes), ground-breaking work by Wilkie's group in Oxford demonstrated that causative gain-of-function mutations confer a selective advantage to spermatogonial stem cells, so that mutated cell lines accumulate in the testis.

Co-Dominance

Co-dominance is the term used for two allelic traits that are both expressed in the heterozygous state. In persons with blood group AB it is possible to demonstrate both A and B blood group substances on the red blood cells, so the A and B blood groups are therefore co-dominant (p. 205).

Homozygosity for Autosomal Dominant Traits

The rarity of most autosomal dominant disorders and diseases means that they usually occur only in the heterozygous state. There are, however, a few reports of children born to couples where both parents are heterozygous for a dominantly inherited disorder. Offspring of such couples are, therefore, at risk of being homozygous. In some instances, affected individuals appear either to be more severely affected, as has been reported with achondroplasia, or to have an earlier age of onset, as in familial hypercholesterolemia (p. 175). The heterozygote with a phenotype intermediate between the homozygotes for the normal and mutant alleles is consistent with a haploinsufficiency loss-of-function mutation (p. 26).

Conversely, with other dominantly inherited disorders, homozygous individuals are not more severely affected than heterozygotes—e.g., Huntington disease (p. 293) and myotonic dystrophy (p. 295).

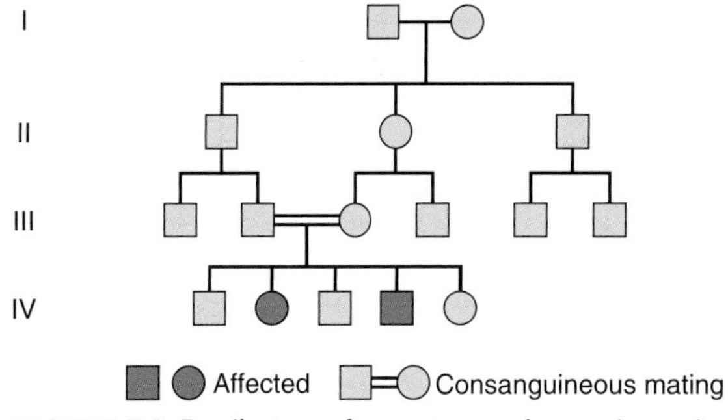

FIGURE 7.8 Family tree of an autosomal recessive trait.

Autosomal Recessive Inheritance

Recessive traits and disorders are manifest only when the mutant allele is present in a double dose (i.e., homozygosity). Individuals heterozygous for such mutant alleles show no features of the disorder and are perfectly healthy; they are described as **carriers**. The family tree for recessive traits (Figure 7.8) differs markedly from that seen in autosomal dominant traits. It is not possible to trace an autosomal recessive trait or disorder through the family, as all the affected individuals in a family are usually in a single **sibship** (i.e., brothers and sisters). This is sometimes referred to as 'horizontal' transmission, but this is an inappropriate and misleading term.

Consanguinity

Enquiry into the family history of individuals affected with rare recessive traits or disorders might reveal that their parents are related (i.e., **consanguineous**). The rarer a recessive trait or disorder, the greater the frequency of consanguinity among the parents of affected individuals. In cystic fibrosis, the most common 'serious' autosomal recessive disorder in western Europeans (p. 1), the frequency of parental consanguinity is only slightly greater than that seen in the general population. By contrast, in alkaptonuria, one of the original inborn errors of metabolism (p. 171), which is an exceedingly rare recessive disorder, Bateson and Garrod, in their original description of the disorder, observed that one-quarter or more of the parents were first cousins. They reasoned that rare alleles for disorders such as alkaptonuria are more likely to 'meet up' in the offspring of cousins than in the offspring of parents who are unrelated. In large inbred kindreds an autosomal recessive condition may be present in more than one branch of the family.

Genetic Risks

If we represent the normal dominant allele as 'R' and the recessive mutant allele as 'r', then each parental gamete carries either the mutant or the normal allele (Figure 7.9). The various possible combinations of gametes mean that the offspring of two heterozygotes have a 1 in 4 (25%) chance of being homozygous affected, a 1 in 2 (50%) chance of

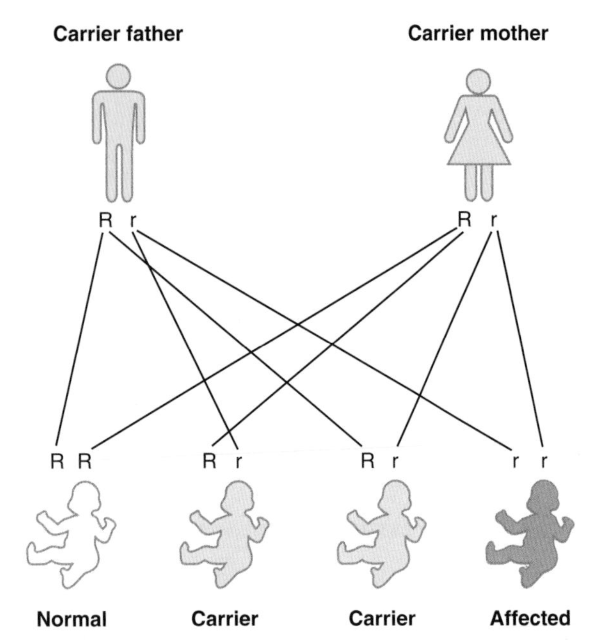

FIGURE 7.9 Segregation of alleles in autosomal recessive inheritance. *R* represents the normal allele, *r* the mutated allele.

being heterozygous unaffected, and a 1 in 4 (25%) chance of being homozygous unaffected.

Pseudodominance

If an individual who is homozygous for an autosomal recessive disorder has children with a carrier of the same disorder, their offspring have a 1 in 2 (50%) chance of being affected. Such a pedigree is said to exhibit **pseudodominance** (Figure 7.10).

Locus Heterogeneity

A disorder inherited in the same manner can be due to mutations in more than one gene, or what is known as **locus heterogeneity**. For example, it is recognized that sensorineural hearing impairment/deafness most commonly shows autosomal recessive inheritance. Deaf persons, by virtue of their schooling and involvement in the deaf community, often choose to have children with another deaf person. It would be expected that, if two deaf persons were homozygous for the same recessive gene, all of their children would be similarly affected. Families have been described in which all the children born to parents who are deaf due to autosomal recessive genes have had perfectly normal hearing because they are **double heterozygotes**. The explanation is that the parents were homozygous for mutant alleles at different loci (i.e., different genes can cause autosomal recessive sensorineural deafness). In fact, over the past 10 to 15 years, approximately 30 genes and a further 50 loci have been shown to be involved. A very similar story applies to autosomal recessive retinitis pigmentosa, and to a lesser extent primary autosomal recessive microcephaly.

Disorders with the same phenotype from different genetic loci are known as **genocopies**, whereas, when the same phenotype results from environmental causes it is known as a **phenocopy**.

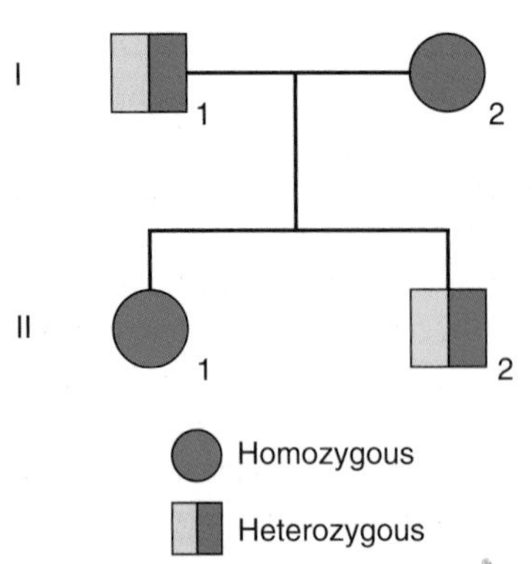

FIGURE 7.10 A pedigree with a woman (I_2) homozygous for an autosomal recessive disorder whose husband is heterozygous for the same disorder. They have a homozygous affected daughter so that the pedigree shows pseudodominant inheritance.

Mutational Heterogeneity

Heterogeneity can also occur at the allelic level. In the majority of single-gene disorders (e.g., β-thalassemia) a large number of different mutations have been identified as being responsible (p. 160). There are individuals who have two different mutations at the same locus and are known as **compound heterozygotes**, constituting what is known as allelic or **mutational heterogeneity**. Most individuals affected with an autosomal recessive disorder are probably compound heterozygotes rather than true homozygotes, unless their parents are related, when they are likely to be homozygous for the same mutation by descent, having inherited the same mutation from a common ancestor.

Sex-Linked Inheritance

Sex-linked inheritance refers to the pattern of inheritance shown by genes that are located on either of the sex chromosomes. Genes carried on the X chromosome are referred to as being X-linked, and those carried on the Y chromosome are referred to as exhibiting **Y-linked** or **holandric inheritance**.

X-Linked Recessive Inheritance

An X-linked recessive trait is one determined by a gene carried on the X chromosome and usually manifests only in males. A male with a mutant allele on his single X chromosome is said to be hemizygous for that allele. Diseases inherited in an X-linked manner are transmitted by healthy heterozygous female carriers to affected males, as well as by affected males to their obligate carrier daughters, with a consequent risk to male grandchildren through these daughters (Figure 7.11). This type of pedigree is sometimes said to show 'diagonal' or a 'knight's move' pattern of transmission.

The mode of inheritance whereby only males are affected by a disease that is transmitted by normal females was appreciated by the Jews nearly 2000 years ago. They excused from circumcision the sons of all the sisters of a

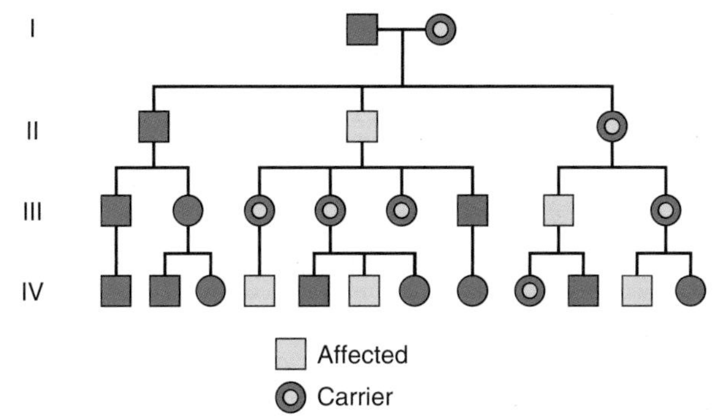

FIGURE 7.11 Family tree of an X-linked recessive trait in which affected males reproduce.

mother who had sons with the 'bleeding disease', in other words, hemophilia (p. 309). The sons of the father's siblings were not excused. Queen Victoria was a carrier of hemophilia, and her carrier daughters, who were perfectly healthy, introduced the gene into the Russian and Spanish royal families. Fortunately for the British royal family, Queen Victoria's son, Edward VII, did not inherit the gene and so could not transmit it to his descendants.

Genetic Risks

A male transmits his X chromosome to each of his daughters and his Y chromosome to each of his sons. If a male affected with hemophilia has children with a normal female, then all of his daughters will be **obligate carriers** but none of his sons will be affected (Figure 7.12). A male cannot transmit an X-linked trait to his son, with the very rare exception of uniparental heterodisomy (p. 121).

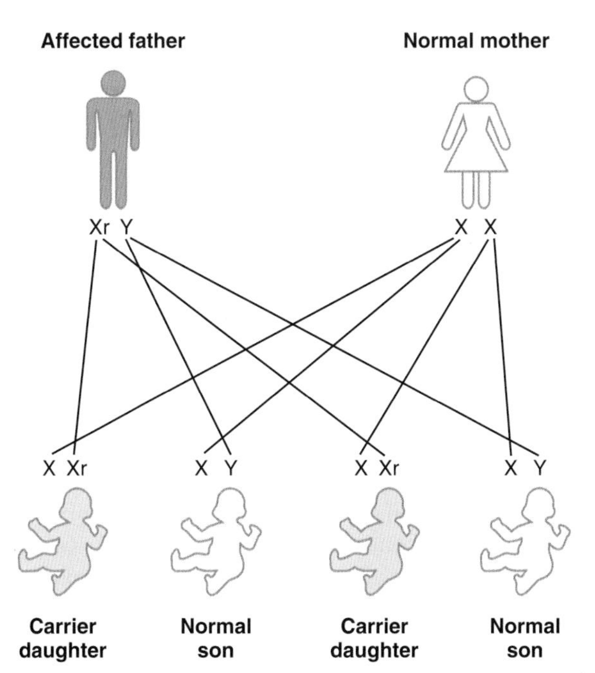

FIGURE 7.12 Segregation of alleles in X-linked recessive inheritance, relating to the offspring of an affected male. *r* represents the mutated allele.

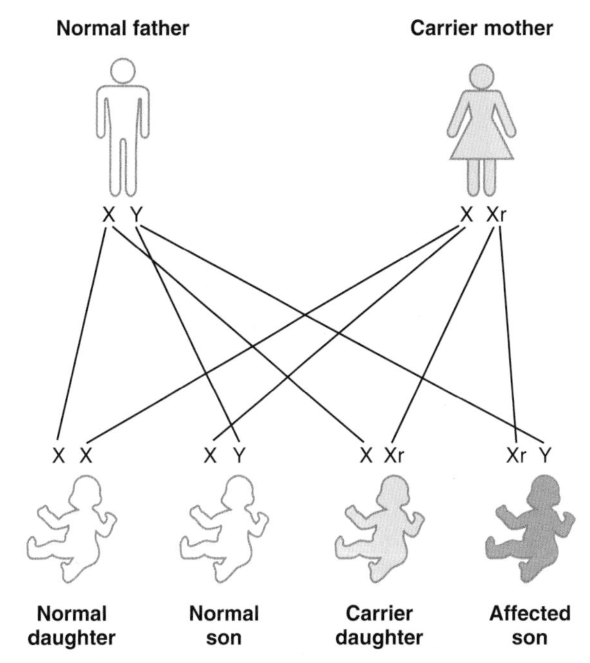

FIGURE 7.13 Segregation of alleles in X-linked recessive inheritance, relating to the offspring of a carrier female. *r* represents the mutated allele.

For a carrier female of an X-linked recessive disorder having children with a normal male, each son has a 1 in 2 (50%) chance of being affected and each daughter has a 1 in 2 (50%) chance of being a carrier (Figure 7.13).

Some X-linked disorders are not compatible with survival to reproductive age and are not, therefore, transmitted by affected males. Duchenne muscular dystrophy is the commonest muscular dystrophy and is a severe disease (p. 307). The first sign is delayed walking followed by a waddling gait, difficulty in climbing stairs unaided, and a tendency to fall easily. By about the age of 10 years affected boys usually need to use a wheelchair. The muscle weakness progresses gradually and affected males ultimately become confined to bed and often die in their late teenage years or early 20s (Figure 7.14). Because affected boys do not usually survive to reproduce, the disease is transmitted by healthy female carriers (Figure 7.15), or may arise as a new mutation.

Variable Expression in Heterozygous Females

In humans, several X-linked disorders are known in which heterozygous females have a mosaic phenotype with a mixture of features of the normal and mutant alleles. In X-linked ocular albinism, the iris and ocular fundus of affected males lack pigment. Careful examination of the ocular fundus in females heterozygous for ocular albinism reveals a mosaic pattern of pigmentation (see Figure 6.25, p. 104). This mosaic pattern of involvement can be explained by the random process of X-inactivation (p. 103). In the pigmented areas, the normal gene is on the active X chromosome, whereas in the depigmented areas the mutant allele is on the active X chromosome.

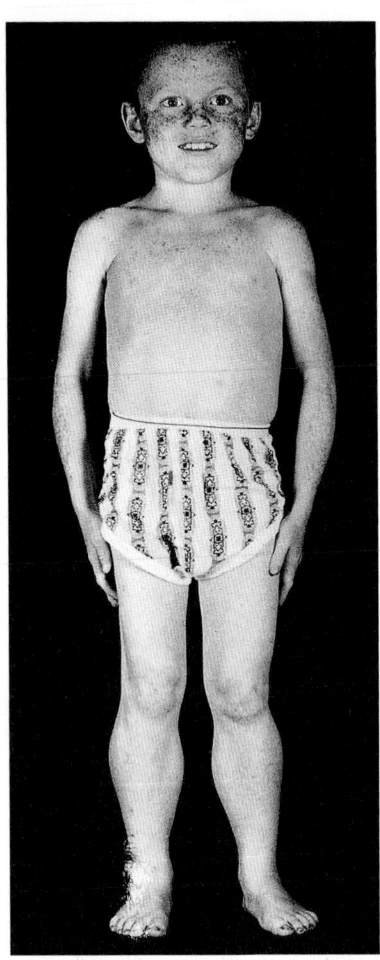

FIGURE 7.14 Boy with Duchenne muscular dystrophy; note the enlarged calves and wasting of the thigh muscles.

Females Affected with X-Linked Recessive Disorders

Occasionally a woman might manifest features of an X-linked recessive trait. There are several explanations for how this can happen.

Homozygosity for X-Linked Recessive Disorders. A common X-linked recessive trait is red–green color blindness—the inability to distinguish between the colors red and green. About 8% of males are red-green color blind and, although it is unusual, because of the high frequency of this allele in the population about 1 in 150 women are red-green color-blind by virtue of both parents having the allele on the X chromosome. Therefore, a female can be affected with an X-linked recessive disorder as a result of homozygosity for an X-linked allele, although the rarity of most X-linked conditions means that the phenomenon is uncommon. A female could also be homozygous if her father was affected and her mother was normal, but a new mutation occurred on the X chromosome transmitted to the daughter; alternatively, it could happen if her mother was a carrier and her father was normal, but a new mutation occurred on the X chromosome he transmitted to his daughter—but these scenarios are rare.

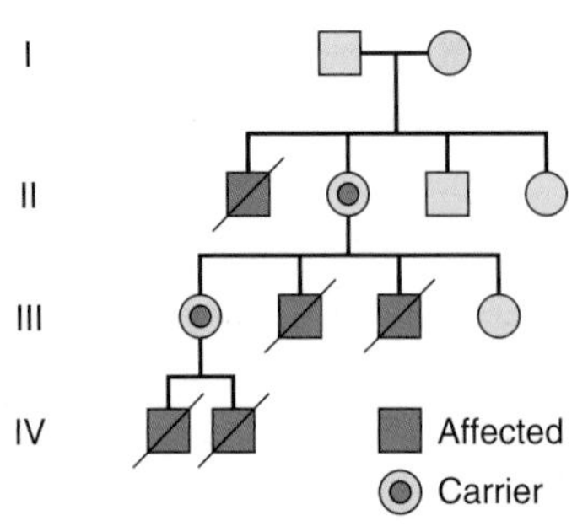

FIGURE 7.15 Family tree of Duchenne muscular dystrophy with the disorder being transmitted by carrier females and affecting males, who do not survive to transmit the disorder.

Skewed X-Inactivation. The process of X-inactivation (p. 103) usually occurs randomly, there being an equal chance of either of the two X chromosomes in a heterozygous female being inactivated in any one cell. After X-inactivation in embryogenesis, therefore, in roughly half the cells one of the X chromosomes is active, whereas in the other half it is the other X chromosome that is active. Sometimes this process is not random, allowing for the possibility that the active X chromosome in most of the cells of a heterozygous female carrier is the one bearing the mutant allele. If this happens, a carrier female would exhibit some of the symptoms and signs of the disease and be a so-called **manifesting heterozygote** or **carrier**. This has been reported in a number of X-linked disorders, including Duchenne muscular dystrophy and hemophilia A (pp. 307, 309). In addition, there are reports of several X-linked disorders in which there are a number of manifesting carriers in the same family, consistent with the coincidental inheritance of an abnormality of X-inactivation (p. 204).

Numerical X-Chromosome Abnormalities. A female could manifest an X-linked recessive disorder by being a carrier of an X-linked recessive mutation and having only a single X chromosome (i.e., Turner syndrome, see p. 207). Women with Turner syndrome and hemophilia A or Duchenne muscular dystrophy have been reported occasionally.

X-Autosome Translocations. Females with a translocation involving one of the X chromosomes and an autosome can be affected with an X-linked recessive disorder. If the breakpoint of the translocation disrupts a gene on the X chromosome, then a female can be affected. This is because the X chromosome involved in the translocation survives preferentially so as to maintain functional disomy of the autosomal genes (Figure 7.16). The observation of females affected with Duchenne muscular dystrophy with X-autosome translocations involving the same region of the short arm of the X chromosome helped to map the Duchenne muscular dystrophy gene (p. 307). This type of observation has been vital in the positional cloning of a number of genes in humans (p. 75).

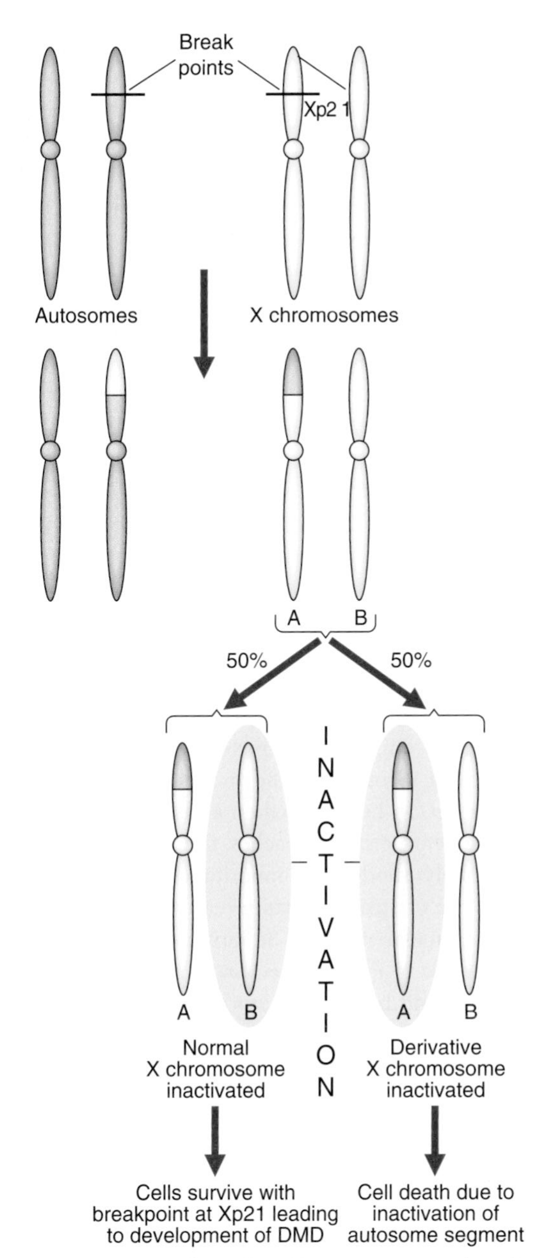

FIGURE 7.16 Generation of an X-autosome translocation with breakpoint in a female and how this results in the development of Duchenne muscular dystrophy.

X-Linked Dominant Inheritance

Although uncommon, there are disorders that are manifest in the heterozygous female as well as in the male who has the mutant allele on his single X chromosome. This is known as X-linked dominant inheritance (Figure 7.17). X-linked dominant inheritance superficially resembles that of an autosomal dominant trait because both the daughters and sons of an affected female have a 1 in 2 (50%) chance of being affected. There is, however, an important difference. With an X-linked dominant trait, an affected male transmits the trait to all his daughters but to none of his sons. Therefore, in families with an X-linked dominant disorder there is an excess of affected females and direct male-to-male transmission cannot occur.

An example of an X-linked dominant trait is X-linked hypophosphatemia, also known as vitamin D–resistant rickets. Rickets can be due to a dietary deficiency of vitamin D, but in vitamin D–resistant rickets the disorder occurs even when there is an adequate dietary intake of vitamin D. In the X-linked dominant form of vitamin D–resistant rickets, both males and females are affected with short stature due to short and often bowed long bones, although the females usually have less severe skeletal changes than the males. The X-linked form of Charcot-Marie-Tooth disease (hereditary motor and sensory neuropathy) is another example.

A mosaic pattern of involvement can be demonstrated in females heterozygous for some X-linked dominant disorders. An example is the mosaic pattern of abnormal pigmentation of the skin that follows developmental lines seen in females heterozygous for the X-linked dominant disorder incontinentia pigmenti (Figure 7.18). This is also an example of a disorder that is usually lethal for male embryos that inherit the mutated allele. Others include the neurological conditions Rett syndrome and periventricular nodular heterotopia.

Y-Linked Inheritance

Y-linked or **holandric inheritance** implies that only males are affected. An affected male transmits Y-linked traits to all of his sons but to none of his daughters. In the past it has been suggested that bizarre-sounding conditions such as porcupine skin, hairy ears and webbed toes are Y-linked traits. With the possible exception of hairy ears, these claims of holandric inheritance have not stood up to more careful study. Evidence clearly indicates, however, that the H-Y histocompatibility antigen (p. 200) and genes involved in spermatogenesis are carried on the Y chromosome and, therefore, manifest holandric inheritance. The latter, if deleted, leads to infertility from azoospermia (absence of the sperm in semen) in males. The recent advent of techniques of assisted reproduction, particularly the technique of intracytoplasmic sperm injection (ICSI), means that, if a pregnancy with a male conceptus results after the use of this technique, the child will also necessarily be infertile.

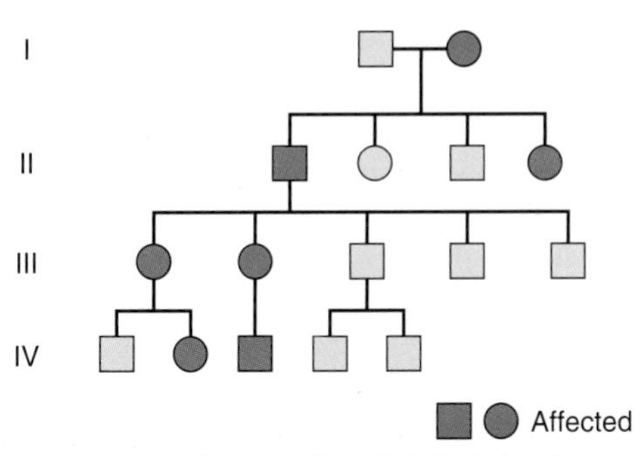

FIGURE 7.17 Family tree of an X-linked dominant trait.

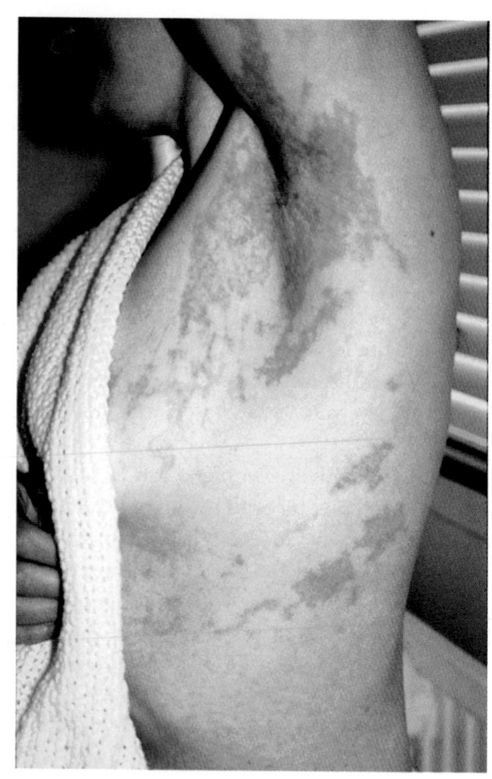

FIGURE 7.18 Mosaic pattern of skin pigmentation in a female with the X-linked dominant disorder, incontinentia pigmenti. The patient has a mutation in a gene on one of her X chromosomes; the pigmented areas indicate tissue in which the normal X chromosome has been inactivated. This developmental pattern follows Blaschko's lines (see Chapter 18, p. 276).

Partial Sex-Linkage

Partial sex-linkage has been used in the past to account for certain disorders that appear to exhibit autosomal dominant inheritance in some families and X-linked inheritance in others. This is now known to be likely to be because of genes carried on that portion of the X chromosome sharing homology with the Y chromosome, and which escapes X-inactivation. During meiosis, pairing occurs between the homologous distal parts of the short arms of the X and Y chromosomes, the so-called **pseudoautosomal region**. As a result of a cross-over, a gene could be transferred from the X to the Y chromosome, or vice versa, allowing the possibility of male-to-male transmission. The latter instances would be consistent with autosomal dominant inheritance. A rare skeletal dysplasia, Leri-Weil dyschondrosteosis, in which affected individuals have short stature and a characteristic wrist deformity (Madelung deformity), has been reported to show both autosomal dominant and X-linked inheritance. The disorder has been shown to be due to deletions of, or mutations in, the short stature homeobox *(SHOX)* gene, which is located in the pseudoautosomal region.

Sex Influence

Some autosomal traits are expressed more frequently in one sex than in another—so-called **sex influence**. Gout and presenile baldness are examples of sex-influenced autosomal dominant traits, males being predominantly affected in both cases. The influence of sex in these two examples is probably through the effect of male hormones. Gout, for example, is very rare in women before the menopause but the frequency increases in later life. Baldness does not occur in males who have been castrated. In hemochromatosis (p. 244), the most common autosomal recessive disorder in Western society, homozygous females are much less likely than homozygous males to develop iron overload and associated symptoms; the explanation usually given is that women have a form of natural blood loss through menstruation.

Sex Limitation

Sex limitation refers to the appearance of certain features only in individuals of a particular sex. Examples include virilization of female infants affected with the autosomal recessive endocrine disorder, congenital adrenal hyperplasia (p. 174).

Establishing the Mode of Inheritance of a Genetic Disorder

In experimental animals it is possible to arrange specific types of mating to establish the mode of inheritance of a trait or disorder. In humans, when a disorder is newly recognized, the geneticist approaches the problem indirectly by fitting likely models of inheritance to the observed outcome in the offspring. Certain features are necessary to support a particular mode of inheritance. Formally establishing the mode of inheritance is not usually possible with a single family and normally requires study of a number of families (Box 7.1).

Box 7.1 Features that Support the Single-Gene or Mendelian Patterns of Inheritance

Autosomal Dominant
Males and females affected in equal proportions
Affected individuals in multiple generations
Transmission by individuals of both sexes (i.e., male to male, female to female, male to female, and female to male)

Autosomal Recessives
Males and females affected in equal proportions
Affected individuals usually in only a single generation
Parents can be related (i.e., consanguineous)

X-Linked Recessive
Only males usually affected
Transmitted through unaffected females
Males cannot transmit the disorder to their sons (i.e., no male-to-male transmission)

X-Linked Dominant
Males and females affected but often an excess of females
Females less severely affected than males
Affected males can transmit the disorder to their daughters but not to sons

Y-Linked Inheritance
Affected males only
Affected males must transmit it to their sons

Autosomal Dominant Inheritance

To determine whether a trait or disorder is inherited in an autosomal dominant manner, there are three specific features that need to be observed. First, it should affect both males and females in equal proportions. Second, it is transmitted from one generation to the next. Third, all forms of transmission between the sexes are observed (i.e., male to male, female to female, male to female, and female to male). Male-to-male transmission excludes the possibility of the gene being on the X chromosome. In the case of sporadically occurring disorders, increased paternal age may suggest a new autosomal dominant mutation.

Autosomal Recessive Inheritance

There are three features that suggest the possibility of autosomal recessive inheritance. First, the disorder affects males and females in equal proportions. Second, it usually affects only individuals in one generation in a single sibship (i.e., brothers and sisters) and does not occur in previous and subsequent generations. Third, consanguinity in the parents provides further support for autosomal recessive inheritance.

X-Linked Recessive Inheritance

There are three main features necessary to establish X-linked recessive inheritance. First, the trait or disorder should affect males almost exclusively. Second, X-linked recessive disorders are transmitted through unaffected carrier females to their sons. Affected males, if they survive to reproduce, can have affected grandsons through their daughters who are obligate carriers. Thirdly, male-to-male transmission is not observed (i.e., affected males cannot transmit the disorder to their sons).

X-Linked Dominant Inheritance

There are three features necessary to establish X-linked dominant inheritance. First, males and females are affected but affected females are more frequent than affected males. Second, females are usually less severely affected than males. Third, although affected females can transmit the disorder to both male and female offspring, affected males can transmit the disorder only to their daughters (except in partial sex-linkage; see p. 118), all of whom will be affected. In the case of X-linked dominant disorders that are almost invariably lethal in male embryos (e.g., incontinentia pigmenti; see pp. 117–118), only females will be affected and families may show an excess of females over males as well as a number of miscarriages that are the affected male pregnancies.

Y-Linked Inheritance

There are two features necessary to establish a Y-linked pattern of inheritance. First, it affects only males. Second, affected males must transmit the disorder to their sons (e.g., male infertility by ICSI) (p. 117).

Multiple Alleles and Complex Traits

So far, each of the traits we have considered has involved only two alleles, the normal, and the mutant. However, some traits and diseases are neither **monogenic** nor **polygenic**. Some genes have more than two allelic forms (i.e., multiple alleles). Multiple alleles are the result of a normal gene having mutated to produce various different alleles, some of which can be dominant and others recessive to the normal allele. In the case of the ABO blood group system (p. 205), there are at least four alleles (A_1, A_2, B, and O). An individual can possess any two of these alleles, which may be the same or different (AO, A_2B, OO, and so on). Alleles are carried on homologous chromosomes and therefore a person transmits only one allele for a certain trait to any particular offspring. For example, a person with the genotype AB will transmit to any particular offspring either the A allele or the B allele, but never both or neither (Table 7.1). This relates only to genes located on the autosomes and does not apply to alleles on the X chromosome; in this instance a woman would have two alleles, either of which could be transmitted to offspring, whereas a man only has one allele to transmit.

The dramatic advances in genome wide scanning using multiple DNA probes has made it possible to begin investigating so-called **complex traits** (i.e., conditions that are usually much more common than mendelian disorders and likely to be due to the interaction of more than one gene). The effects may be additive, one may be rate limiting over the action of another, or one may enhance or multiply the effect of another; this is considered in more detail in Chapter 15. The possibility of a small number of gene loci being implicated in some disorders has given rise to the concept of **oligogenic** inheritance, examples of which include the following.

Digenic Inheritance

This refers to the situation where a disorder has been shown to be due to the additive effects of heterozygous mutations at two different gene loci, a concept referred to as **digenic inheritance**. This is seen in certain transgenic mice. Mice

Table 7.1 Possible Genotypes, Phenotypes, and Gametes Formed from the Four Alleles A_1, A_2, B, and O at the ABO Locus

Genotype	Phenotype	Gametes
A_1A_1	A_1	A_1
A_2A_2	A_2	A_2
BB	B	B
OO	O	O
A_1A_2	A_1	A_1 or A_2
A_1B	A_1B	A_1 or B
A_1O	A_1	A_1 or O
A_2B	A_2B	A_2 or B
A_2O	A_2	A_2 or O
BO	B	B or O

that are homozygotes for *rv* (rib-vertebrae) or *Dll1* *(Delta–like-1)* manifest abnormal phenotypes, whereas their respective heterozygotes are normal. However, mice that are **double heterozygotes** for *rv* and *Dll1* show vertebral defects. In humans, one form of retinitis pigmentosa, a disorder of progressive visual impairment, is caused by double heterozygosity for mutations in two unlinked genes, *ROM1* and *Peripherin*, which both encode proteins present in photoreceptors. Individuals with only one of these mutations are not affected. In the field of inherited cardiac arrhythmias and cardiomyopathies (p. 304), it is becoming clear that some cases of arrhythmogenic right ventricular dysplasia exhibit digenic inheritance.

Triallelic Inheritance

Bardet–Biedl syndrome is a rare dysmorphic condition (though relatively more common in some inbred communities) with obesity, polydactyly, renal abnormalities, retinal pigmentation, and learning disability. Seven different gene loci have been identified and, until recently, the syndrome was thought to follow straightforward autosomal recessive inheritance. However, it is now known that one form occurs only when an individual who is homozygous for mutations at one locus *is also* heterozygous for mutation at another Bardet-Biedl locus; this is referred to as **triallelic inheritance**.

Other patterns of inheritance that are not classically mendelian are also recognized and explain some unusual phenomena.

Anticipation

In some autosomal dominant traits or disorders, such as myotonic dystrophy, the onset of the disease occurs at an earlier age in the offspring than in the parents, or the disease occurs with increasing severity in subsequent generations. This phenomenon is called **anticipation**. It used to be believed that this effect was the result of a bias of ascertainment, because of the way in which the families were collected. It was argued that this arose because persons in whom the disease begins earlier, or is more severe, are more likely to be ascertained and only those individuals who are less severely affected tend to have children. In addition, it was thought that, because the observer is in the same generation as the affected presenting probands, many individuals who at present are unaffected will, by necessity, develop the disease later in life.

Recent studies, however, have shown that in a number of disorders, including Huntington disease and myotonic dystrophy, anticipation is, in fact, a real biological phenomenon occurring as a result of the expansion of unstable triplet repeat sequences (p. 24). An expansion of the CTG triplet repeat in the 3′ untranslated end of the myotonic dystrophy gene, occurring predominantly in **maternal** meiosis, appears to be the explanation for the severe neonatal form of myotonic dystrophy that usually only occurs when the gene is transmitted by the mother (Figure 7.19).

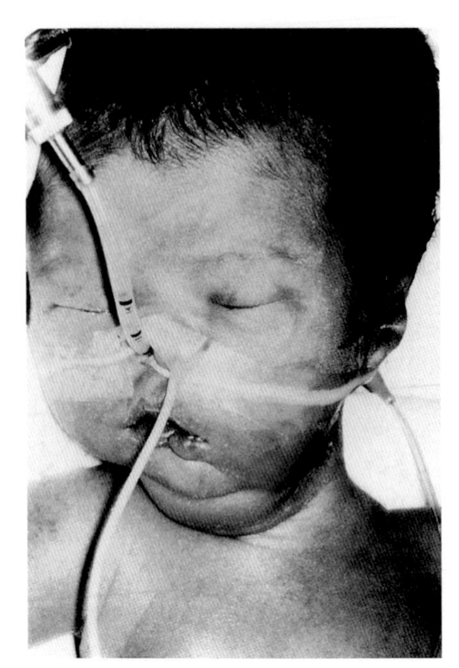

FIGURE 7.19 Newborn baby with severe hypotonia requiring ventilation as a result of having inherited myotonic dystrophy from his mother.

Fragile X syndrome (CGG repeats) (p. 278) behaves in a similar way, with major instability in the expansion occurring during **maternal** meiosis. A similar expansion—in this case CAG repeats—in the 5′ end of the Huntington disease gene (Figure 7.20) in **paternal** meiosis accounts for the increased risk of early onset Huntington disease, occasionally in childhood or adolescence, when the gene is transmitted by the father. The inherited spinocerebellar ataxia group of conditions is another example.

Mosaicism

An individual, or a particular tissue of the body, can consist of more than one cell type or line, through an error occurring during mitosis at any stage after conception. This is known as **mosaicism** (p. 50). Mosaicism of either somatic tissues or germ cells can account for some instances of unusual patterns of inheritance or phenotypic features in an affected individual.

Somatic Mosaicism

The possibility of somatic mosaicism is suggested by the features of a single-gene disorder being less severe in an individual than is usual, or by being confined to a particular part of the body in a segmental distribution; for example, as occurs occasionally in neurofibromatosis type I (p. 298). The timing of the mutation event in early development may determine whether it is transmitted to the next generation with full expression—this will depend on the mutation being present in all or some of the gonadal tissue, and hence germline cells.

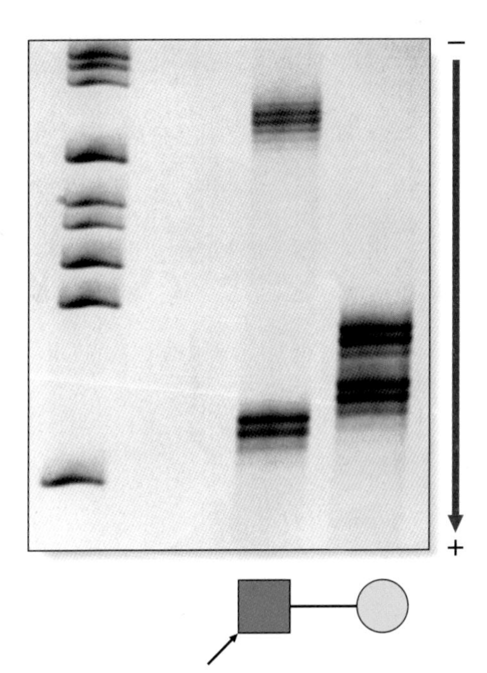

FIGURE 7.20 Silver staining of a 5% denaturing gel of the polymerase chain reaction products of the CAG triplet in the 5′ untranslated end of the Huntington disease gene from an affected male and his wife, showing her to have two similar-sized repeats in the normal range (20 and 24 copies) and him to have one normal-sized triplet repeat (18 copies) and an expanded triplet repeat (44 copies). The bands in the left lane are standard markers to allow sizing of the CAG repeat. (Courtesy Alan Dodge, Regional DNA Laboratory, St. Mary's Hospital, Manchester, UK.)

Gonadal Mosaicism

There have been many reports of families with autosomal dominant disorders, such as achondroplasia and osteogenesis imperfecta, and X-linked recessive disorders, such as Duchenne muscular dystrophy and hemophilia, in which the parents are phenotypically normal, and the results of investigations or genetic tests have also all been normal, but in which more than one of their children has been affected. The most favored explanation for these observations is gonadal, or germline, mosaicism in one of the parents; that is, the mutation is present in a proportion of the gonadal or germline cells. An elegant example of this was provided by the demonstration of a mutation in the collagen gene responsible for osteogenesis imperfecta in a proportion of individual sperm from a clinically normal father who had two affected infants with different partners. It is important to keep germline mosaicism in mind when providing recurrence risks in genetic counseling for apparently new autosomal dominant and X-linked recessive mutations (p. 343).

Uniparental Disomy

An individual normally inherits one of a pair of homologous chromosomes from each parent (p. 39). Over the past decade, with the advent of DNA technology, some individuals have been shown to have inherited both homologs of a chromosome pair from only one of their parents, so-called **uniparental disomy**. If an individual inherits two copies of the same homolog from one parent, through an error in meiosis II (p. 41), this is called **uniparental isodisomy** (Figure 7.21). If, however, the individual inherits the two different homologs from one parent through an error in meiosis I (p. 39), this is termed **uniparental heterodisomy**. In either instance, it is presumed that the conceptus would originally be trisomic, with early loss of a chromosome leading to the 'normal' disomic state. One-third of such chromosome losses, if they occurred with equal frequency, would result in uniparental disomy. Alternatively, it is postulated that uniparental disomy could arise as a result of a gamete from one parent that does not contain a particular chromosome homolog (i.e., a gamete that is **nullisomic**), being 'rescued' by fertilization with a gamete that, through a second separate chance error in meiosis, is disomic.

Using DNA techniques, uniparental disomy has been shown to be the cause of a father with hemophilia having an affected son and of a child with cystic fibrosis being born to a couple in which only the mother was a carrier (with proven paternity!). Uniparental paternal disomy for chromosome 15 may be linked to either Prader-Willi or Angelman syndrome, or for chromosome 11 with a proportion of cases of the overgrowth condition known as the Beckwith-Wiedemann syndrome (see the following section).

Genomic Imprinting

Genomic imprinting is an **epigenetic** phenomenon, referred to in Chapter 6 (p. 103). Epigenetics and genomic imprinting give the lie to Thomas Morgan's quotation at the start of this chapter! Although it was originally thought that genes on homologous chromosomes were expressed equally, it is now recognized that different clinical features can result, depending on whether a gene is inherited from the father or from the mother. This 'parent of origin' effect is referred to as **genomic imprinting**, and **methylation** of DNA is thought to be the main mechanism by which expression is modified. Methylation is the **imprint** applied to certain DNA sequences in their passage through gametogenesis, although only a small proportion of the human genome is in fact subject to this process. The differential allele expression (i.e., maternal or paternal) may occur in all somatic cells, or in specific tissues or stages of development. Thus far, at least 80 human genes are known to be imprinted and the regions involved are known as differentially methylated regions (DMRs). These DMRs include imprinting control regions (ICRs) that control gene expression across imprinted domains.

Evidence of genomic imprinting has been observed in two pairs of well known dysmorphic syndromes: Prader-Willi and Angelman syndromes (chromosome 15q), and Beckwith-Wiedemann and Russell-Silver syndromes

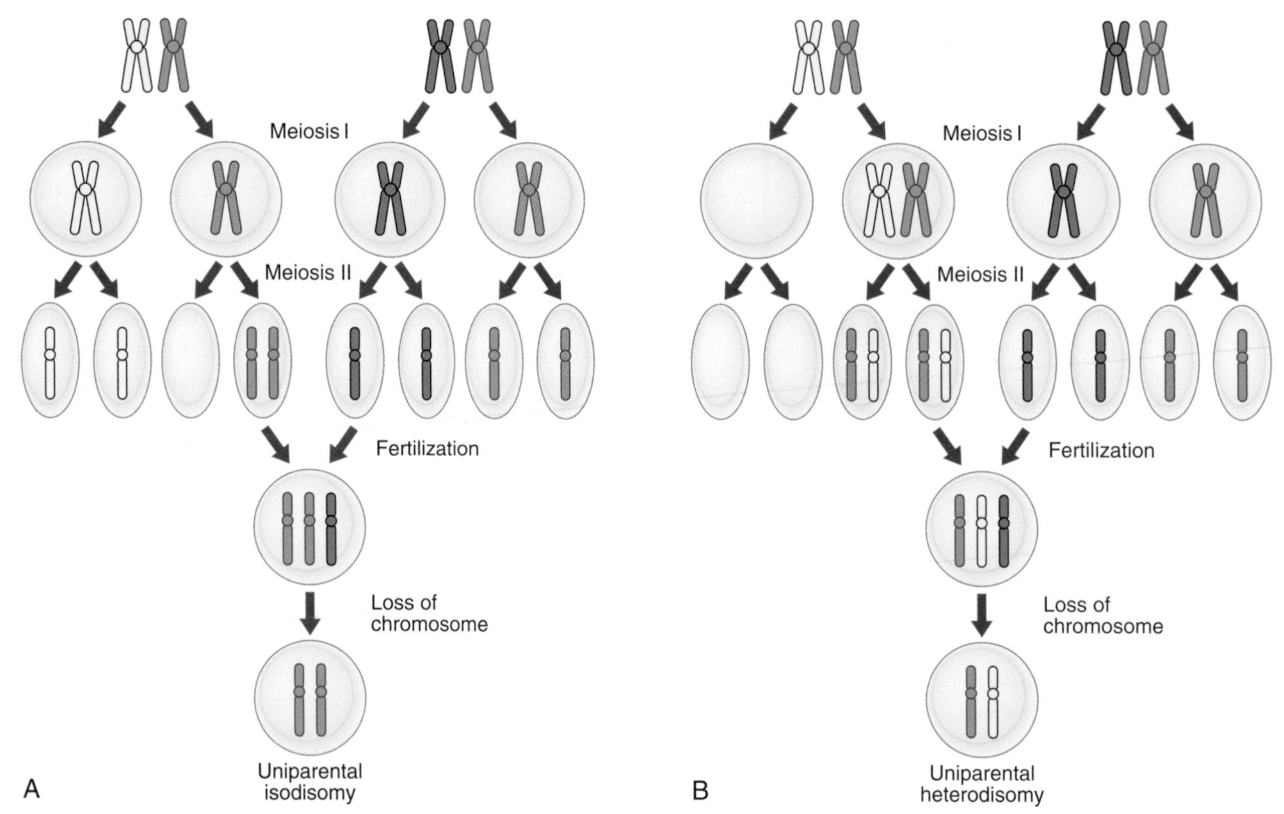

FIGURE 7.21 Mechanism of origin of uniparental disomy. **A**, Uniparental isodisomy occurring through a disomic gamete arising from non-disjunction in meiosis II fertilizing a monosomic gamete with loss of the chromosome from the parent contributing the single homolog. **B**, Uniparental heterodisomy occurring through a disomic gamete arising from non-disjunction in meiosis I fertilizing a monosomic gamete with loss of the chromosome from the parent contributing the single homolog.

(chromosome 11p). The mechanisms giving rise to these conditions, although complex, reveal much about imprinting and are therefore now considered in a little detail.

Prader-Willi Syndrome

Prader-Willi syndrome (PWS) (p. 282) occurs in approximately 1 in 20,000 births and is characterized by short stature, obesity, hypogonadism, and learning difficulty (Figure 7.22). Approximately 50% to 60% of individuals with PWS can be shown to have an interstitial deletion of the proximal portion of the long arm of chromosome 15, approximately 2 Mb at 15q11-q13, visible by conventional cytogenetic means, and in a further 15% a submicroscopic deletion can be demonstrated by fluorescent in-situ hybridization (see p. 34) or molecular means. DNA analysis has revealed that the chromosome deleted is almost always the paternally derived homolog. Most of the remaining 25% to 30% of individuals with PWS, without a chromosome deletion, have been shown to have maternal uniparental disomy. Functionally, this is equivalent to a deletion in the paternally derived chromosome 15.

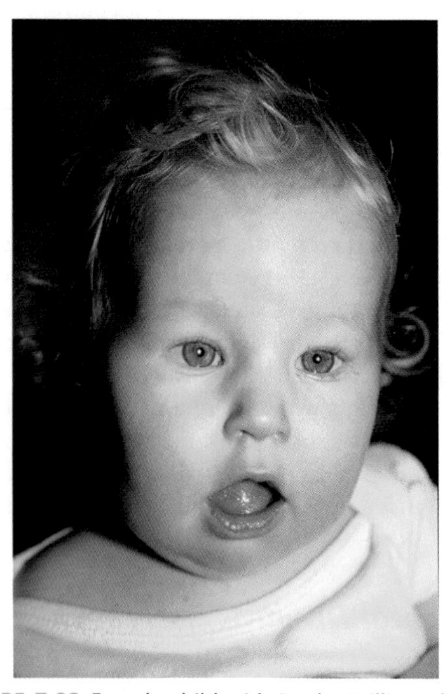

FIGURE 7.22 Female child with Prader-Willi syndrome.

Paternal allele

Centromere

Telomere

PWS ICR

UBE3A Antisense

5'

3'

MKRN3

MAGE-L2

NDN

AS ICR

SNURF/SNRPN

UBE3A

Maternal allele

FIGURE 7.23 Molecular organization (simplified) at 15q11-q13: Prader-Willi syndrome (PWS) and Angelman syndrome (AS). The imprinting control region (ICR) for this locus has two components. The more telomeric acts as the PWS ICR and contains the promoter of *SNURF/SNRPN. SNURF/SNRPN* produces several long and complex transcripts, one of which is believed to be an RNA antisense inhibitor of *UBE3A*. The more centromeric ICR acts as the AS ICR on *UBE3A*, which is the only gene whose maternal expression is lost in AS. The AS ICR also inhibits the PWS ICR on the maternal allele. The PWS ICR also acts on the upstream genes *MKRN3, MAGE-L2,* and *NDN,* which are unmethylated (○) on the paternal allele but methylated (•) on the maternal allele.

It is now known that only the paternally inherited allele of this critical region of 15q11-q13 is expressed. The molecular organization of the region is shown in Figure 7.23. PWS is a multigene disorder and in the normal situation the small nuclear ribonucleoprotein polypeptide N (*SNRPN*) and adjacent genes (*MKRN3*, etc.) are paternally expressed. Expression is under the control of a specific ICR. Analysis of DNA from patients with PWS and various submicroscopic deletions enabled the ICR to be mapped to a segment of about 4 kb, spanning the first exon and promoter of *SNRPN* and upstream reading frame *(SNURF)*. The 3′ end of the ICR is required for expression of the paternally expressed genes and also the origin of the long *SNURF/SNRPN* transcript. The maternally expressed genes are not differentially methylated but they are silenced on the paternal allele, probably by an antisense RNA generated from *SNURF/SNRPN*. In normal cells, the 5′ end of the ICR, needed for maternal expression and involved in Angelman syndrome (see below), is methylated on the maternal allele.

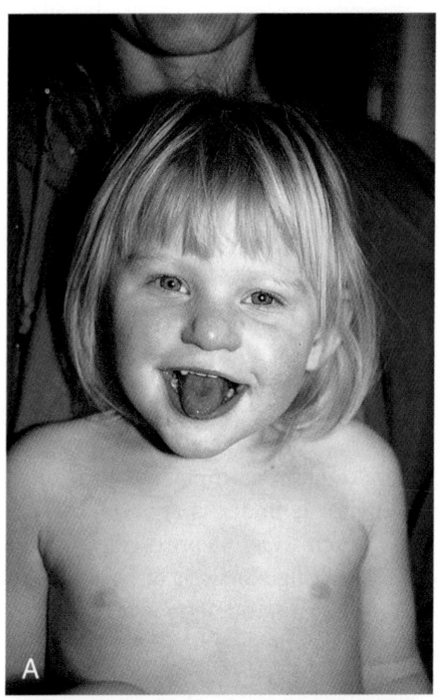

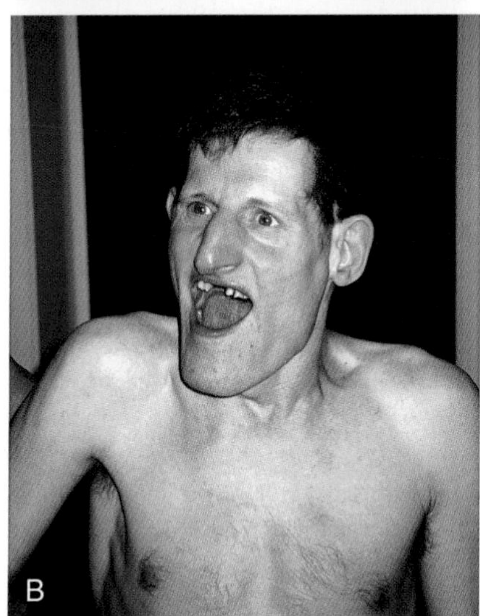

FIGURE 7.24 **A**, Female child with Angelman syndrome. **B**, Adult male with Angelman syndrome.

Angelman Syndrome (AS)

Angelman syndrome (p. 282) occurs in about 1 in 15,000 births and is characterized by epilepsy, severe learning difficulties, an unsteady or ataxic gait, and a happy affect (Figure 7.24). Approximately 70% of individuals with AS have been shown to have an interstitial deletion of the same 15q11-q13 region as is involved in PWS, but in this case on the maternally derived homolog. In a further 5% of individuals with AS, the syndrome can be shown to have arisen through paternal uniparental disomy. Unlike PWS, the features of AS arise through loss of a single gene, *UBE3A*. In up to 10% of individuals with AS, mutations have been identified in *UBE3A*, one of the ubiquitin genes, which appears to be preferentially or exclusively expressed from the maternally derived chromosome 15 in brain. How mutations in *UBE3A* lead to the features seen in persons with AS is not clear, but could involve ubiquitin-mediated destruction of proteins in the central nervous system in

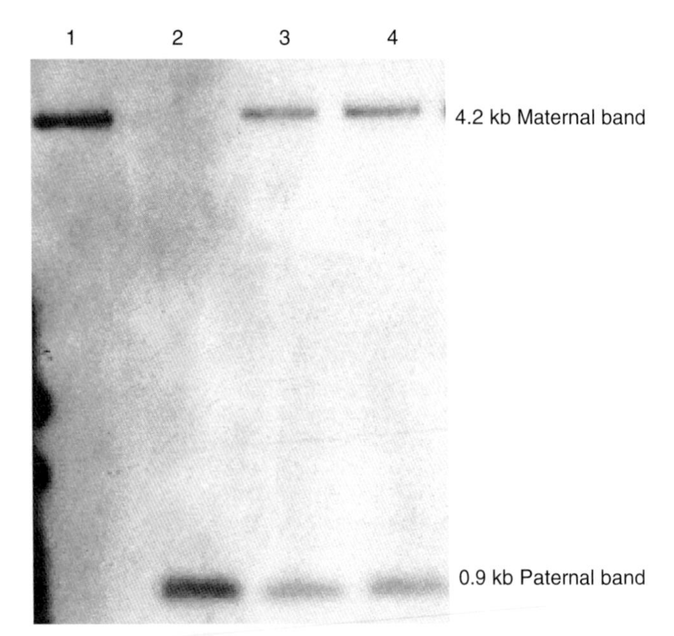

FIGURE 7.25 Southern blot to detect methylations of *SNRPN*. DNA digested with *Xba I* and *Not I* was probed with KB17, which hybridizes to a CpG island within exon a of *SNRPN*. Patient 1 has Prader-Willi syndrome, patient 2 has Angelman syndrome, and patients 3 and 4 are unaffected. (Courtesy A. Gardner, Department of Molecular Genetics, Southmead Hospital, Bristol.)

development, particularly where *UBE3A* is expressed most strongly, namely the hippocampus and Purkinje cells of the cerebellum. *UBE3A* is under control of the AS ICR (see Figure 7.23), which was mapped slightly upstream of *SNURF/SNRPN* through analysis of patients with AS who had various different microdeletions.

About 2% of individuals with PWS and approximately 5% of those with AS have abnormalities of the ICR itself; these patients tend to show the mildest phenotypes. Patients in this last group, unlike the other three, have a risk of recurrence. In the case of AS, if the mother carries the same mutation as the child, the recurrence risk is 50%, but even if she tests negative for the mutation, there is an appreciable recurrence risk from gonadal mosaicism.

Rare families have been reported in which a translocation of the proximal portion of the long arm of chromosome 15 is segregating. Depending on whether the translocation is transmitted by the father or mother, affected offspring within the family have had either PWS or AS. In approximately 10% of AS cases the molecular defect is unknown—but it may well be that some of these alleged cases have a different, albeit phenotypically similar, diagnosis.

In many genetics service laboratories a simple DNA test is used to diagnose both PWS and AS, exploiting the differential DNA methylation characteristics at the 15q11-q13 locus (Figure 7.25).

Beckwith-Wiedemann Syndrome

Beckwith–Wiedemann syndrome (BWS) is a clinically heterogeneous condition whose main underlying characteristic is overgrowth. First described in 1963 and 1964, the main features are macrosomia (prenatal and/or postnatal overgrowth), macroglossia (large tongue), abdominal wall defect (omphalocele, umbilical hernia, diastasis recti), and neonatal hypoglycemia (Figure 7.26). Hemihyperplasia may be present, as well as visceromegaly, renal abnormalities, ear anomalies (anterior earlobe creases, posterior helical pits) and cleft palate, and there may be embryonal tumors (particularly Wilms tumor).

BWS is, in a way, celebrated in medical genetics because of the multiple different (and complex) molecular mechanisms that underlie it. Genomic imprinting, somatic mosaicism, and multiple genes are involved, all within a 1 Mb region at chromosome 11p15 (Figure 7.27). Within this region lie two independently regulated imprinted domains. The more telomeric (differentially methylated region 1 [DMR1] under control of ICR1) contains paternally expressed *IGF2* (insulin growth factor 2) and maternally expressed *H19*. The more centromeric imprinted domain (DMR2, under control of ICR2) contains the maternally expressed *KCNQ1* (previously known as *KvLQT1*) and *CDKN1C* genes, and the paternally expressed antisense transcript *KCNQ1OT1*, the promoter for which is located within the *KCNQ1* gene.

Disruption to the normal regulation of methylation can give rise to altered gene expression dosage and,

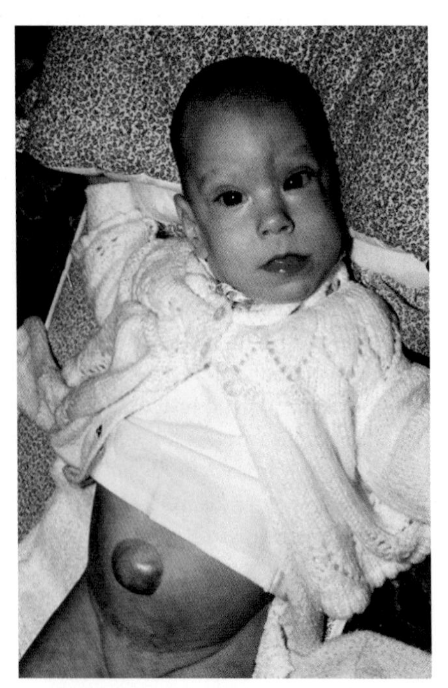

FIGURE 7.26 Baby girl with Beckwith-Wiedemann syndrome. Note the large tongue and umbilical hernia.

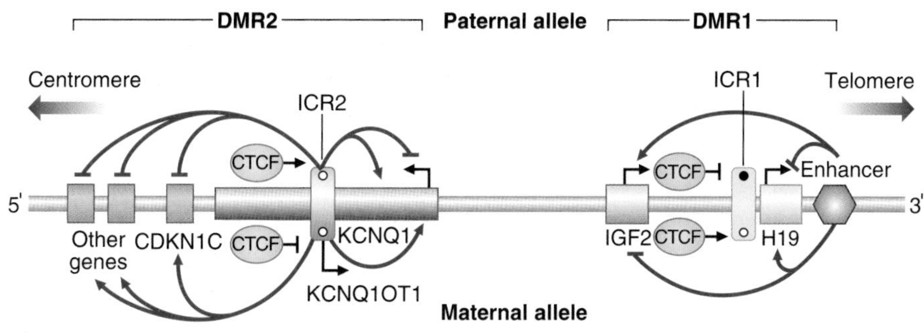

FIGURE 7.27 Molecular organization (simplified) at 11p15.5: Beckwith-Wiedemann and Russell-Silver syndromes. The region contains two imprinted domains (DMR1 and DMR2) that are regulated independently. The ICRs are differentially methylated (• methylated; ○ unmethylated). CCCTC-binding factor (CTCF) binds to the unmethylated alleles of both ICRs. In DMR1, coordinated regulation leads to expression of *IGF2* only on the paternal allele and *H19* expression only on the maternal allele. In DMR2, coordinated regulation leads to *maternal* expression of *KCNQ1* and *CDKN1C* (plus other genes), and *paternal* expression of *KCNQ1OT1* (a non-coding RNA with antisense transcription to *KCNQ1*). Angled black arrows show the direction of the transcripts.

consequentially, features of BWS. In DMR1, **gain of methylation** on the maternal allele leads to loss of *H19* expression and biallelic *IGF2* expression (i.e., effectively two copies of the paternal epigenotype). This occurs in up to 7% of BWS cases and is usually sporadic. In DMR2, **loss of methylation** results in two copies of the paternal epigenotype and a reduction in expression of *CDKN1C*; this mechanism is implicated in 50% to 60% of sporadic BWS cases. *CDKN1C* may be a growth inhibitory gene and mutations have been found in 5% to 10% of cases of BWS. About 15% of BWS cases are familial, and *CDKN1C* mutations are found in about half of these. In addition to imprinting errors in DMR1 and DMR2, other mechanisms may account for BWS: (1) paternally derived duplications of chromosome 11p5.5 (these cases were the first to identify the BWS locus); (2) paternal uniparental disomy for chromosome 11—invariably present in mosaic form—often associated with neonatal hypoglycemia and hemi-hypertrophy, and associated with the highest risk (about 25%) of embryonal tumors, particularly Wilms tumor; and (3) maternally inherited balanced translocations involving rearrangements of 11p15.

Russell–Silver Syndrome

This well-known condition has 'opposite' characteristics to BWS by virtue of marked prenatal and postnatal growth retardation. The head circumference is relatively normal, the face rather small and triangular, giving rise to a 'pseudohydrocephalic' appearance (Figure 7.28), and there may be body asymmetry. About 10% of cases appear to be due to maternal uniparental disomy, indicating that this chromosome is subject to imprinting. In contrast to paternally derived duplications of 11p15, which give rise to overgrowth and BWS, maternally derived duplications of this region are associated with growth retardation. Recently it has been shown that about a third of Russell–Silver syndrome (RSS) cases are due to abnormalities of imprinting at the 11p.15.5 locus. Whereas *hyper*methylation of DMR1 leads to upregulated *IGF2* and overgrowth, *hypo*methylation of *H19* leads to downregulated *IGF2*, the opposite molecular and biochemical consequence, and these patients have features of RSS. Interestingly, in contrast to BWS, there are no cases of RSS with altered methylation of the more centromeric DMR2 region.

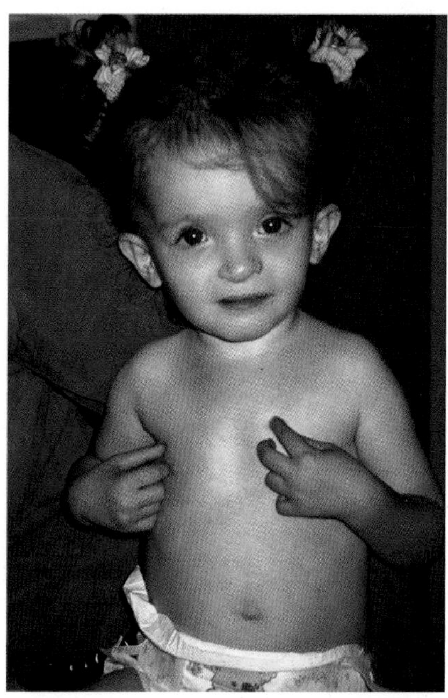

FIGURE 7.28 Girl with Russell-Silver syndrome. Note the bossed forehead, triangular face, and 'pseudohydrocephalic' appearance.

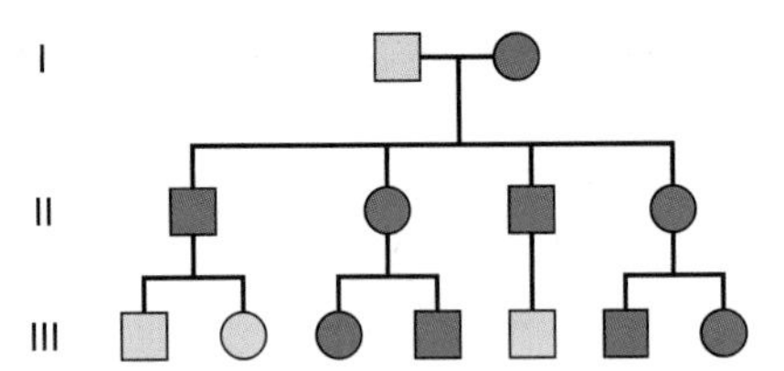

FIGURE 7.29 Family tree consistent with mitochondrial inheritance.

Mitochondrial Inheritance

Each cell contains thousands of copies of mitochondrial DNA with more being found in cells that have high energy requirements, such as brain and muscle. Mitochondria, and therefore their DNA, are inherited almost exclusively from the mother through the oocyte (p. 41). Mitochondrial DNA has a higher rate of spontaneous mutation than nuclear DNA, and the accumulation of mutations in mitochondrial DNA has been proposed as being responsible for some of the somatic effects seen with aging.

In humans, **cytoplasmic** or **mitochondrial inheritance** has been proposed as a possible explanation for the pattern of inheritance observed in some rare disorders that affect both males and females but are transmitted only through females, so-called maternal or matrilineal inheritance (Figure 7.29).

A number of rare disorders with unusual combinations of neurological and myopathic features, sometimes occurring in association with other conditions such as cardiomyopathy and conduction defects, diabetes, or deafness, have been characterized as being due to mutations in mitochondrial genes (p. 181). Because mitochondria have an important role in cellular metabolism through oxidative phosphorylation, it is not surprising that the organs most susceptible to mitochondrial mutations are the central nervous system, skeletal muscle and heart.

In most persons, the mitochondrial DNA from different mitochondria is identical, or shows what is termed **homoplasmy**. If a mutation occurs in the mitochondrial DNA of an individual, initially there will be two populations of mitochondrial DNA, so-called **heteroplasmy**. The proportion of mitochondria with a mutation in their DNA varies between cells and tissues, and this, together with mutational heterogeneity, is a possible explanation for the range of phenotypic severity seen in persons affected with mitochondrial disorders (Figure 7.30).

Whilst matrilineal inheritance applies to disorders that are directly because of mutations in mitochondrial DNA, it is also important to be aware that mitochondrial **proteins** are encoded mainly by nuclear genes. Mutations in these genes can have a devastating impact on respiratory chain functions within mitochondria. Examples include genes encoding proteins within the cytochrome *c* (COX) system, which follow autosomal recessive inheritance, and the *G4.5 (TAZ)* gene that is X-linked and causes Barth syndrome (endocardial fibroelastosis) in males (p. 182). There is even a mitochondrial myopathy following autosomal dominant inheritance in which multiple mitochondrial DNA deletions can be detected. Further space is devoted to mitochondrial disorders in Chapter 11 (p. 181).

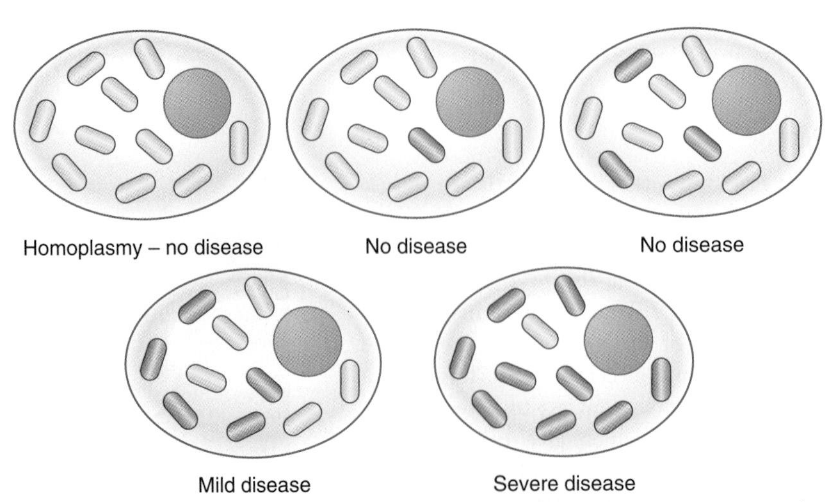

FIGURE 7.30 Progressive effects of heteroplasmy on the clinical severity of disease from mutations in the mitochondrial genome. Low proportions of mutant mitochondria are tolerated well, but as the proportion increases different thresholds for cellular, and hence tissue, dysfunction are breached (mauve circle represents the cell nucleus).

FURTHER READING

Bateson W, Saunders ER 1902 Experimental studies in the physiology of heredity, pp 132–134. Royal Society Reports to the Evolution Committee, 1902

Early observations on mendelian inheritance.

Bennet RL, Steinhaus KA, Uhrich SB, et al 1995 Recommendations for standardized human pedigree nomenclature. Am J Hum Genet 56:745–752

Goriely A, McVean GAT, Rojmyr M, et al 2003 Evidence for selective advantage of pathogenic *FGFR2* mutations in the male germ line. Science 301:643–646

Hall JG 1988 Somatic mosaicism: observations related to clinical genetics. Am J Hum Genet 43:355–363

Good review of findings arising from somatic mosaicism in clinical genetics.

Hall JG 1990 Genomic imprinting: review and relevance to human diseases. Am J Hum Genet 46:857–873

Extensive review of examples of imprinting in inherited diseases in humans.

Heinig RM 2000 The monk in the garden: the lost and found genius of Gregor Mendel. London: Houghton Mifflin

The life and work of Gregor Mendel as the history of the birth of genetics.

Kingston HM 1994 An ABC of clinical genetics, 2nd ed. London: British Medical Association

A simple outline primer of the basic principles of clinical genetics.

Reik W, Surami A, eds 1997 Genomic imprinting (frontiers in molecular biology). London: IRL Press

Detailed discussion of examples and mechanisms of genomic imprinting.

Vogel F, Motulsky AG 1996 Human genetics, 3rd ed. Berlin: Springer

This text has detailed explanations of many of the concepts in human genetics outlined in this chapter.

ELEMENTS

1 Family studies are often necessary to determine the mode of inheritance of a trait or disorder and to give appropriate genetic counseling. A standard shorthand convention exists for pedigree documentation of the family history.

2 Mendelian, or single-gene, disorders can be inherited in five ways: autosomal dominant, autosomal recessive, X-linked dominant, X-linked recessive, and, rarely, Y-linked inheritance.

3 Autosomal dominant alleles are manifest in the heterozygous state and are usually transmitted from one generation to the next but can occasionally arise as a new mutation. They usually affect both males and females equally. Each offspring of a parent with an autosomal dominant gene has a 1 in 2 chance of inheriting it from the affected parent. Autosomal dominant alleles can exhibit reduced penetrance, variable expressivity, and sex limitation.

4 Autosomal recessive disorders are manifest only in the homozygous state and normally only affect individuals in one generation, usually in one sibship in a family. They affect both males and females equally. Offspring of parents who are heterozygous for the same autosomal recessive allele have a 1 in 4 chance of being homozygous for that allele. The less common an autosomal recessive allele, the greater the likelihood that the parents of a homozygote are consanguineous.

5 X-linked recessive alleles are normally manifest only in males. Offspring of females heterozygous for an X-linked recessive allele have a 1 in 2 chance of inheriting the allele from their mother. Daughters of males with an X-linked recessive allele are obligate heterozygotes but sons cannot inherit the allele. Rarely, females manifest an X-linked recessive trait because they are homozygous for the allele, have a single X chromosome, have a structural rearrangement of one of their X chromosomes, or are heterozygous but show skewed or non-random X-inactivation.

6 There are only a few disorders known to be inherited in an X-linked dominant manner. In X-linked dominant disorders, hemizygous males are usually more severely affected than heterozygous females.

7 Unusual features in single-gene patterns of inheritance can be explained by phenomena such as genetic heterogeneity, mosaicism, anticipation, imprinting, uniparental disomy, and mitochondrial inheritance.

CHAPTER 8

Population and Mathematical Genetics

Do not worry about your difficulties in mathematics. I can assure you mine are still greater.

ALBERT EINSTEIN

In this chapter, some of the more mathematical aspects of gene inheritance are considered, together with how genes are distributed and maintained at particular frequencies in populations. This subject constitutes what is known as **population genetics**. Genetics lends itself to a numerical approach, with many of the most influential and pioneering figures in human genetics having come from a mathematical background. They were particularly attracted by the challenges of trying to determine the frequencies of genes in populations and the rates at which they mutate. Much of this early work impinges on the specialty of medical genetics, and in particular on genetic counseling, and by the end of this chapter it is hoped that the reader will have gained an understanding of the following.

1. Why a dominant trait does not increase in a population at the expense of a recessive one.
2. How the carrier frequency and mutation rate can be determined from the disease incidence.
3. Why a particular genetic disorder can be more common in one population or community than another.
4. How it can be confirmed that a genetic disorder shows a particular pattern of inheritance.
5. The concept of genetic linkage and how this differs from linkage disequilibrium.
6. The effects of medical intervention.

Allele Frequencies in Populations

On first reflection, it would be reasonable to predict that dominant genes and traits in a population would tend to increase at the expense of recessive ones. On average, three-quarters of the offspring of two heterozygotes will manifest the dominant trait, but only one-quarter will have the recessive trait. It might be thought, therefore, that eventually almost everyone in the population would have the dominant trait. However, it can be shown that in a large randomly mating population, in which there is no disturbance by outside influences, dominant traits do not increase at the expense of recessive ones. In fact, in such a population, the relative proportions of the different genotypes (and phenotypes) remain constant from one generation to another. This is known as the **Hardy-Weinberg principle**, as it was proposed, independently, by an English mathematician, G. H. Hardy, and a German physician, W. Weinberg, in 1908. This is a very important principle in human genetics.

The Hardy-Weinberg Principle

Consider an 'ideal' population in which there is an autosomal locus with two alleles, A and a, that have frequencies of p and q, respectively. These are the only alleles found at this locus, so that $p + q = 100\%$, or 1. The frequency of each genotype in the population can be determined by construction of a Punnett square, which shows how the different genes can combine (Figure 8.1).

From Figure 8.1, it can be seen that the frequencies of the different genotypes are:

Genotype	Phenotype	Frequency
AA	A	p^2
Aa	A	2pq
Aa	a	q^2

If there is random mating of sperm and ova, the frequencies of the different genotypes in the first generation will be as shown. If these individuals mate with one another to produce a second generation, Punnett square can again be used to show the different matings and their frequencies (Figure 8.2).

From Figure 8.2 the total frequency for each genotype in the second generation can be derived (Table 8.1). This shows that the relative frequency or proportion of each genotype is the same in the second generation as in the first. In fact, no matter how many generations are studied, the relative frequencies will remain constant. The actual numbers of individuals with each genotype will change as the population size increases or decreases, but their **relative** frequencies or proportions remain constant. This is the fundamental tenet of the Hardy-Weinberg principle. When studies confirm that the relative proportions of each genotype remain constant with frequencies of p^2, 2pq, and q^2, then that population is said to be in **Hardy-Weinberg equilibrium** for that particular genotype.

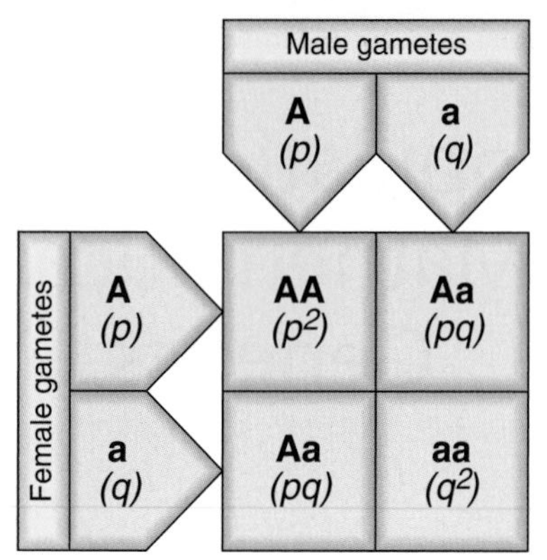

FIGURE 8.1 Punnett square showing allele frequencies and resulting genotype frequencies for a two-allele system in the first generation.

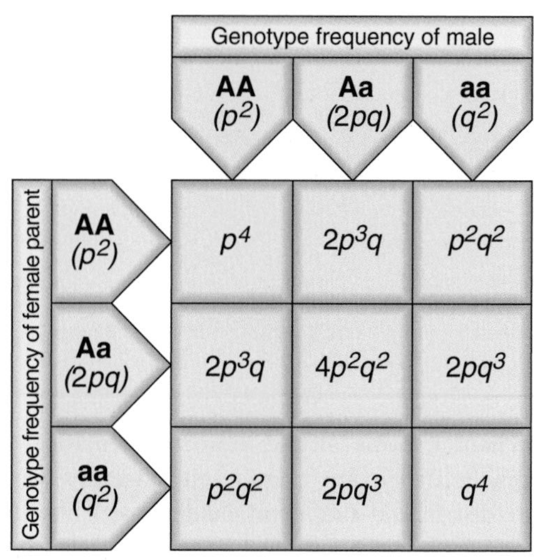

FIGURE 8.2 Punnett square showing frequencies of the different matings in the second generation.

Factors that Can Disturb Hardy-Weinberg Equilibrium

So far, this relates to an 'ideal' population. By definition such a population is large and shows random mating with no new mutations and no selection for or against any particular genotype. For some human characteristics, such as neutral genes for blood groups or enzyme variants, these criteria can be fulfilled. However, several factors can disturb Hardy-Weinberg equilibrium, either by influencing the distribution of genes in the population or by altering the gene frequencies. These factors include:

1. Non-random mating
2. Mutation
3. Selection
4. Small population size
5. Gene flow (migration).

Non-Random Mating

Random mating, or **panmixis**, refers to the selection of a partner regardless of that partner's genotype. Non-random mating can lead to an increase in the frequency of affected homozygotes by two mechanisms, either assortative mating or consanguinity.

Assortative Mating

This is the tendency for human beings to choose partners who share characteristics such as height, intelligence, and racial origin. If assortative mating extends to conditions such as autosomal recessive (AR) deafness, which accounts for a large proportion of all congenital hearing loss, this will lead to a small increase in the relative frequency of affected homozygotes.

Consanguinity

Consanguinity is the term used to describe marriage between blood relatives who have at least one common ancestor no more remote than a great-great-grandparent. Widespread consanguinity in a community will lead to a relative increase in the frequency of affected homozygotes but a relative decrease in the frequency of heterozygotes.

Mutation

The validity of the Hardy-Weinberg principle is based on the assumption that no new mutations occur. If a particular locus shows a high mutation rate, then there will be a steady increase in the proportion of mutant alleles in a population.

Table 8.1 Frequency of the Various Types of Offspring from the Matings Shown in Figure 8.2

		Frequency of Offspring		
Mating Type	Frequency	AA	Aa	aa
AA × AA	p4	p^4	—	—
AA × Aa	$4p^3q$	$2p^3q$	$2p^3q$	—
Aa × Aa	$4p^2q^2$	p^2q^2	$2p^2q^2$	p^2q^2
AA × aa	$2p^2q^2$	—	$2p^2q^2$	—
Aa × aa	$4pq^3$	—	$2pq^3$	$2pq^3$
aa × aa	q^4	—	—	q^4
Total		$p^2(p^2 + 2pq + q^2)$	$2pq(p^2 + 2pq + q^2)$	$q^2(p^2 + 2pq + q^2)$
Relative frequency		p^2	$2pq$	q^2

In practice, mutations do occur at almost all loci, albeit at different rates, but the effect of their introduction is usually balanced by the loss of mutant alleles due to reduced fitness of affected individuals. If a population is found to be in Hardy-Weinberg equilibrium, it is generally assumed that these two opposing factors have roughly equal effects. This is discussed further in the section that follows on the estimation of mutation rates.

Selection

In the 'ideal' population there is no selection for or against any particular genotype. In reality, for deleterious characteristics there is likely to be negative selection, with affected individuals having reduced reproductive (= biological = 'genetic') fitness. This implies that they do not have as many offspring as unaffected members of the population. In the absence of new mutations, this reduction in fitness will lead to a gradual reduction in the frequency of the mutant gene, and hence disturbance of Hardy-Weinberg equilibrium.

Selection can act in the opposite direction by increasing fitness. For some autosomal recessive disorders there is evidence that heterozygotes show a slight increase in biological fitness compared with unaffected homozygotes—referred to as **heterozygote advantage**. The best understood example is sickle-cell disease, in which affected homozygotes have severe anemia and often show persistent ill-health (p. 159). However, heterozygotes are relatively immune to infection with *Plasmodium falciparum* malaria because their red blood cells undergo sickling and are rapidly destroyed when invaded by the parasite. In areas where this form of malaria is endemic, carriers of sickle-cell anemia (sickle cell **trait**), have a biological advantage compared with unaffected homozygotes. Therefore, in these regions the proportion of heterozygotes tends to increase relative to the proportions of normal and affected homozygotes, and Hardy-Weinberg equilibrium is disturbed.

Small Population Size

In a large population, the numbers of children produced by individuals with different genotypes, assuming no alteration in fitness for any particular genotype, will tend to balance out, so that gene frequencies remain stable. However, in a small population it is possible that by random statistical fluctuation one allele could be transmitted to a high proportion of offspring by chance, resulting in marked changes in allele frequency from one generation to the next, so that Hardy-Weinberg equilibrium is disturbed. This is known as **random genetic drift**. If one allele is lost altogether, it is said to be **extinguished** and the other allele is described as having become **fixed** (Figure 8.3).

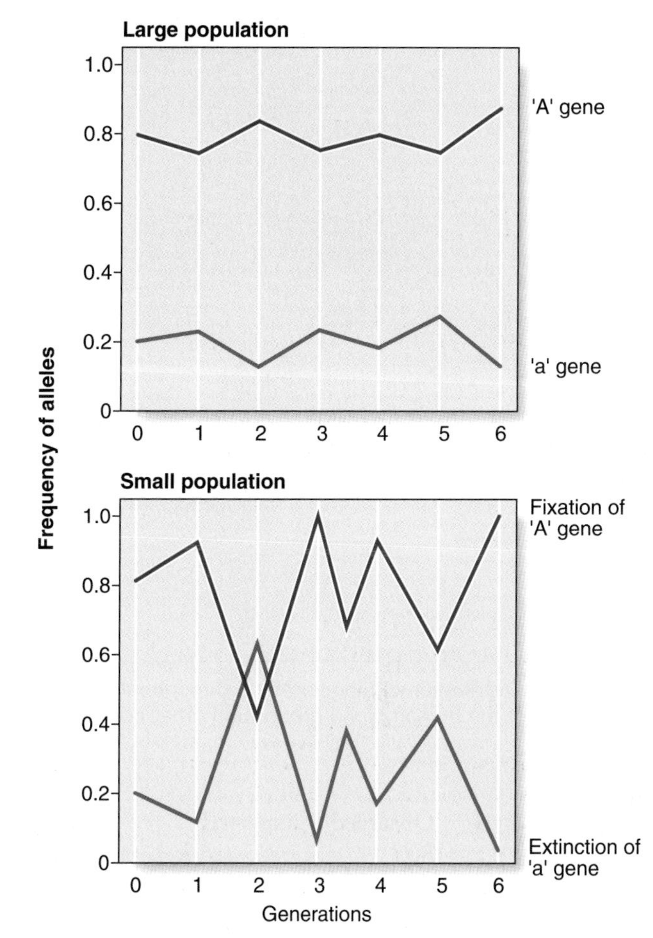

FIGURE 8.3 Possible effects of random genetic drift in large and small populations.

Gene Flow (Migration)

If new alleles are introduced into a population as a consequence of migration, with later intermarriage, a change in the relevant allele frequencies will result. This slow diffusion of alleles across racial or geographical boundaries is known as **gene flow**. The most widely quoted example is the gradient shown by the incidence of the B blood group allele throughout the world (Figure 8.4). This allele is thought to have originated in Asia and spread slowly westward as a result of admixture through invasion.

Validity of Hardy-Weinberg Equilibrium

It is relatively simple to establish whether a population is in Hardy-Weinberg equilibrium for a particular trait if all possible genotypes can be identified. Consider a system with two alleles, A and a, with three resulting genotypes, AA, Aa/aA, and aa. Among 1000 individuals selected at random, the following genotype distributions are observed:

AA	800
Aa/aA	185
aa	15

From these figures, the incidence of the A allele (p) equals $[(2 \times 800) + 185]/2000 = 0.8925$ and the incidence of the a allele (q) equals $[185 + (2 \times 15)]/2000 = 0.1075$.

Now consider what the expected genotype frequencies would be if the population were in Hardy-Weinberg equilibrium, and compare these with the observed values:

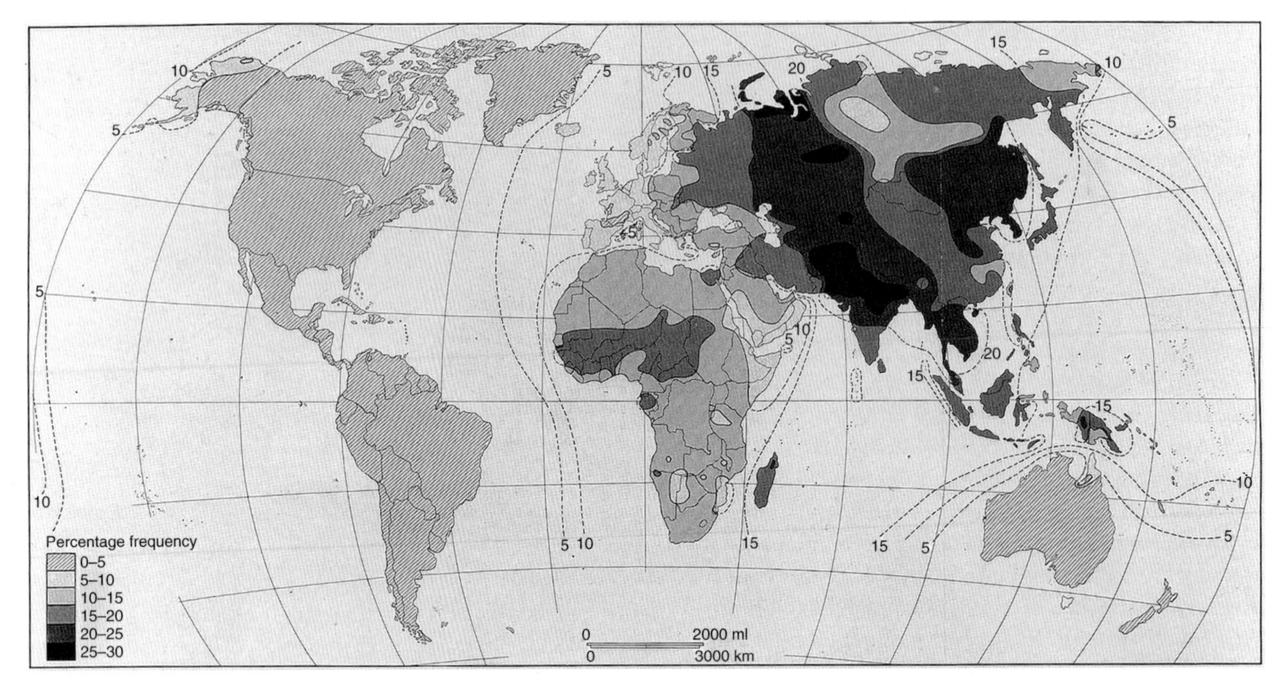

FIGURE 8.4 Distribution of blood group B throughout the world. (From Mourant AE, Kopéc AC, Domaniewska-Sobczak K 1976 The distribution of the human blood groups and other polymorphisms, 2nd ed. Oxford University Press, London, with permission.)

Genotype	Observed	Expected
AA	800	796.5 ($p^2 \times 1000$)
Aa/aA	185	192 ($2pq \times 1000$)
aa	15	11.5 ($q^2 \times 1000$)

These observed and expected values correspond closely and formal statistical analysis with a χ^2 test would confirm that the observed values do not differ significantly from those expected if the population is in equilibrium.

Next consider a different system with two alleles, B and b. Among 1000 randomly selected individuals the observed genotype distributions are:

BB	430
Bb/bB	540
bb	30

From these values, the incidence of the B allele (p) equals $[(2 \times 430) + 540]/2000 = 0.7$ and the incidence of the b allele (q) equals $[540 + (2 \times 30)]/2000 = 0.3$.

Using these values for p and q, the observed and expected genotype distributions can be compared:

Genotype	Observed	Expected
BB	430	490 ($p^2 \times 1000$)
Bb/bB	540	420 ($2pq \times 1000$)
bb	30	90 ($q^2 \times 1000$)

These values differ considerably, with an increased number of heterozygotes at the expense of homozygotes. Such deviation from Hardy-Weinberg equilibrium should prompt a search for factors that could result in increased numbers of heterozygotes, such as heterozygote advantage or negative assortative mating—i.e., the attraction of opposites!

Despite the number of factors that can disturb Hardy-Weinberg equilibrium, most populations are in equilibrium for most genetic traits, and significant deviations from expected genotype frequencies are unusual.

Applications of Hardy-Weinberg Equilibrium

Estimation of Carrier Frequencies

If the incidence of an AR disorder is known, it is possible to calculate the carrier frequency using some relatively simple algebra. For example, if the disease incidence is 1 in 10,000, then $q^2 = 1/10{,}000$ and $q = 1/100$. Because $p + q = 1$, therefore $p = 99/100$. The carrier frequency can then be calculated as $2 \times 99/100 \times 1/100$ (i.e., 2pq), which approximates to 1 in 50. Thus, a rough approximation of the carrier frequency can be obtained by doubling the square root of the disease incidence. Approximate values for gene frequency and carrier frequency derived from the disease incidence can be extremely useful in genetic risk counseling (p. 266) (Table 8.2). However, if the disease incidence includes cases resulting from consanguineous relationships, then it is not valid to use the Hardy-Weinberg principle to calculate heterozygote frequencies because a high incidence of consanguinity disturbs the equilibrium by leading to a relative increase in the proportion of affected homozygotes.

For an X-linked recessive (XLR) disorder, the frequency of affected males equals the frequency of the mutant allele, q. Thus, for a trait such as red-green color blindness, which affects approximately 1 in 12 male western European whites, $q = 1/12$ and $p = 11/12$. This means that the frequency of affected females (q^2) and carrier females (2pq) is $1/144$ and $22/144$, respectively.

Table 8.2 Approximate Values for Gene Frequency and Carrier Frequency Calculated from the Disease Incidence Assuming Hardy-Weinberg Equilibrium

Disease Incidence (q2)	Gene Frequency (q)	Carrier Frequency (2pq)
1/1000	1/32	1/16
1/2000	1/45	1/23
1/5000	1/71	1/36
1/10,000	1/100	1/50
1/50,000	1/224	1/112
1/100,000	1/316	1/158

Estimation of Mutation Rates

Direct Method

If an autosomal dominant (AD) disorder shows full penetrance, and is therefore always expressed in heterozygotes, an estimate of its mutation rate can be made relatively easily by counting the number of new cases in a defined number of births. Consider a sample of 100,000 children, 12 of whom have a particular AD disorder such as achondroplasia (p. 93). Only two of these children have an affected parent, so that the remaining 10 must have acquired their disorder as a result of new mutations. Therefore 10 new mutations have occurred among the 200,000 genes inherited by these children (because each child inherits two copies of each gene), giving a mutation rate of 1 per 20,000 gametes per generation. In fact, this example is unusual because all new mutations in achondroplasia occur on the paternally derived chromosome 4; therefore the mutation rate is 1 per 10,000 in spermatogenesis and, as far as we know, zero in oogenesis.

Indirect Method

For an AD disorder with reproductive fitness (f) equal to zero, all cases must result from new mutations. If the incidence of a disorder is denoted as I and the mutation rate as m, then as each child inherits two alleles, either of which can mutate to cause the disorder, the incidence equals twice the mutation rate (i.e., $I = 2\mu$).

If fitness is greater than zero, and the disorder is in Hardy-Weinberg equilibrium, then genes lost through reduced fitness must be counterbalanced by new mutations. Therefore, $2\mu = I(1 - f)$ or $\mu = [I(1 - f)]/2$.

Thus, if an estimate of genetic fitness can be made by comparing the average number of offspring born to affected parents, to the average number of offspring born to controls such as their unaffected siblings, it will be possible to calculate the mutation rate.

A similar approach can be used to estimate mutation rates for AR and XLR disorders. With an AR condition, two genes will be lost for each homozygote that fails to reproduce. These will be balanced by new mutations. Therefore, $2\mu = I(1 - f) \times 2$ or $\mu = I(1 - f)$.

For an XLR condition with an incidence in males equal to I^M, three X chromosomes are transmitted per couple per generation. Therefore, $3\mu = I^M(1 - f)$ or $\mu = [I^M(1 - f)]/3$.

Why Is It Helpful to Know Mutation Rates?

There is a tendency to either love or hate mathematical formulae but the link between mutation rates, disease incidence, and fitness does hold practical value.

Estimation of Gene Size

If a disorder has a high mutation rate the gene may be large. Alternatively, it may contain a high proportion of GC residues and be prone to copy error, or contain a high proportion of repeat sequences (p. 23), which could predispose to misalignment in meiosis resulting in deletion and duplication.

Determination of Mutagenic Potential

Accurate methods for determining mutation rates may be useful in relation to predicted and observed differences in disease incidence in the aftermath of events such as nuclear accidents, for example Chernobyl in 1986 (p. 26).

Consequences of Treatment of Genetic Disease

As discussed later, improved treatment for serious genetic disorders may increase biological fitness, which may result in an increase in disease incidence.

Why Are Some Genetic Disorders More Common than Others?

It follows that if a gene has a high mutation rate, the disease incidence may be relatively high. However, factors other than the mutation rate and biological fitness may be involved, as mentioned previously. These are now considered in the context of population size.

Small Populations

Several rare AR disorders show a relatively high incidence in certain population groups (Table 8.3). High allele frequencies are usually explained by the combination of a **founder effect** together with social, religious, or geographical isolation—hence the term **genetic isolates**. In some situations, genetic drift may have played a role.

For example, several very rare AR disorders occur at relatively high frequency in the Old Order Amish living in Pennsylvania—Christians originating from the Anabaptist movement who fled Europe during religious persecution in the eighteenth century. Original founders of the group must have carried abnormal alleles that became established at relatively high frequency due to the restricted number of partners available to members of the community.

Founder effects can also be observed in AD disorders. Variegate porphyria, which is characterized by photosensitivity and drug-induced neurovisceral disturbance, has a high incidence in the Afrikaner population of South Africa, believed to be due to one of the early Dutch settlers having

Table 8.3 Rare Recessive Disorders that Are Relatively Common in Certain Groups of People

Group	Disorder	Clinical Features
Finns	Congenital nephrotic syndrome	Edema, proteinuria, susceptibility to infection
	Aspartylglycosaminuria	Progressive mental and motor deterioration, coarse features
	Mulibrey nanism	*Mu*scle, *li*ver, *br*ain and *ey*e involvement
	Congenital chloride diarrhea	Reduced Cl^- absorption, diarrhea
	Diastrophic dysplasia	Progressive epiphyseal dysplasia with dwarfism and scoliosis
Amish	Cartilage–hair hypoplasia	Dwarfism, fine, light-colored and sparse hair
	Ellis–van Creveld syndrome	Dwarfism, polydactyly, congenital heart disease
	Glutaric aciduria type 1	Episodic encephalopathy and cerebral palsy-like dystonia
Hopi and San Blas Indians	Albinism	Lack of pigmentation
Ashkenazi Jews	Tay-Sachs disease	Progressive mental and motor deterioration, blindness
	Gaucher disease	Hepatosplenomegaly, bone lesions, skin pigmentation
	Dysautonomia	Indifference to pain, emotional lability, lack of tears, hyperhidrosis
Karaite Jews	Werdnig-Hoffmann disease	Infantile spinal muscular atrophy
Afrikaners	Sclerosteosis	Tall stature, overgrowth of craniofacial bones with cranial nerve palsies, syndactyly
	Lipoid proteinosis	Thickening of skin and mucous membranes
Ryukyan islands (off Japan)	'Ryukyan' spinal muscular atrophy	Muscle weakness, club foot, scoliosis

transmitted the condition to a large number of descendants (p. 109).

Interestingly, the Hopi Indians of Arizona show a high incidence of albinism. Affected males were excused from outdoor farming activities because of the health and visual problems of bright sunlight, thus providing more opportunity to reproduce relative to unaffected group members.

Large Populations

When a serious AR disorder, resulting in reduced fitness in affected homozygotes, has a high incidence in a large population, the explanation is presumed to lie in either a very high mutation rate and/or a heterozygote advantage. The latter explanation is the more probable for most AR disorders (Table 8.4).

Heterozygote Advantage

For sickle cell (SC) anemia (p. 159) and thalassemia (p. 161), there is very good evidence that heterozygote advantage results from reduced susceptibility to *Plasmodium falciparum* malaria, as explained in Chapter 10. Americans of Afro-Caribbean origin are no longer exposed to malaria, so it would be expected that the frequency of the SC allele in this group would gradually decline. However, the predicted rate of decline is so slow that it will be many generations before it is detectable.

For several AR disorders the mechanisms proposed for heterozygote advantage are largely speculative (see Table 8.4). The discovery of the cystic fibrosis (CF) gene, with the subsequent elucidation of the role of its protein product in membrane permeability (p. 301), supports the hypothesis of selective advantage through increased resistance to the effects of gastrointestinal infections, such as cholera and dysentery, in the heterozygote. This relative resistance could result from reduced loss of fluid and electrolytes. It is likely that this selective advantage was of greatest value several hundred years ago when these infections were endemic in Western Europe. If so, a gradual decline in the incidence of CF would be expected. However, if this theory is correct one has to ask why CF has not become relatively

Table 8.4 Presumed Increased Resistance in Heterozygotes that Could Account for the Maintenance of Various Genetic Disorders in Certain Populations

Disorder	Genetics	Region/Population	Resistance or Advantage
Sickle-cell disease	AR	Tropical Africa	Falciparum malaria
α- and β-thalassemia	AR	Southeast Asia and the Mediterranean	Falciparum malaria
G6PD deficiency	XLR	Mediterranean	Falciparum malaria
Cystic fibrosis	AR	Western Europe	Tuberculosis? The plague? Cholera?
Tay-Sachs disease	AR	Eastern European Jews	Tuberculosis?
Congenital adrenal hyperplasia	AR	Yupik Eskimos	Influenza B
Type 2 diabetes	AD	Pima Indians and others	Periodic starvation
Phenylketonuria	AR	Western Europe	Spontaneous abortion rate lower?

AR, Autosomal recessive; *XLR*, X-linked recessive; *AD*, autosomal dominant; *G6PD*, glucose 6-phosphate dehydrogenase.

common in other parts of the world where gastrointestinal infections are endemic, particularly the tropics; in fact, the opposite is the case, for CF is rare in these regions.

An alternative, but speculative, mechanism for the high incidence of a condition such as CF is that the mutant allele is preferentially transmitted at meiosis. This type of segregation distortion, whereby an allele at a particular locus is transmitted more often than would be expected by chance (i.e., in more than 50% of gametes), is referred to as **meiotic drive**. Firm evidence for this phenomenon in CF is lacking, although it has been demonstrated in the AD disorder myotonic dystrophy (p. 295).

A major practical problem when studying heterozygote advantage is that even a tiny increase in heterozygote fitness, compared with the fitness of unaffected homozygotes, can be sufficient to sustain a high allele frequency. For example, in CF, with an allele frequency of approximately 1 in 50, a heterozygote advantage of 2% to 3% would be sufficient to account for the high allele frequency.

Genetic Polymorphism

Polymorphism is the occurrence in a population of two or more genetically determined forms (alleles, sequence variants) in such frequencies that the rarest of them could not be maintained by mutation alone. By convention, a polymorphic locus is one at which there are at least two alleles, each with a frequency greater than 1%. Alleles with frequencies of less than 1% are referred to as rare variants.

In humans, at least 30% of structural gene loci are polymorphic, with each individual being heterozygous at between 10% and 20% of all loci. Known polymorphic protein systems include the ABO blood groups (p. 205) and many serum proteins, which may exhibit polymorphic electrophoretic differences—or **isozymes**.

DNA polymorphisms, including SNPs, have been crucial to positional cloning, gene mapping, and isolation of many disease genes (p. 75). They are also used in gene tracking (p. 70) in the clinical context of presymptomatic tests, prenatal diagnosis and carrier detection for many single-gene disorders where direct mutation analysis may not be possible. The value of a particular polymorphic system is assessed by determining its **polymorphic information content (PIC)**. The higher the PIC value, the more likely it is that a polymorphic marker will be of value in linkage analysis and gene tracking.

Segregation Analysis

Segregation analysis refers to the study of the way in which a disorder is transmitted in families so as to establish the underlying mode of inheritance. The mathematical aspects of segregation analysis are very complex and far beyond the scope of this book—and most doctors! However, it is important that those who encounter families with genetic disease have some understanding of the principles involved and some awareness of the pitfalls and problems.

Autosomal Dominant Inheritance

For an AD disorder, the simplest approach is to compare the observed numbers of affected offspring born to affected parents with what would be expected based on the disease penetrance (i.e., 50% if penetrance is complete). A χ^2 test can be used to see whether the observed and expected numbers differ significantly. Care must be taken to ensure that a bias is not introduced by excluding parents who were ascertained through an affected child.

Autosomal Recessive Inheritance

For disorders thought to follow AR inheritance, formal segregation analysis is much more difficult. This is because some couples who are both carriers will by chance not have affected children, therefore not feature in ascertainment. To illustrate this, consider 64 possible sibships of size 3 in which both parents are carriers, drawn from a large hypothetical population (Table 8.5). The sibship structure shown in Table 8.5 is that which would be expected, on average.

In this population, on average, 27 of the 64 sibships will not contain any affected individuals. This can be calculated simply by cubing $\frac{3}{4}$—i.e., $\frac{3}{4} \times \frac{3}{4} \times \frac{3}{4} = \frac{27}{64}$. Therefore, when the families are analyzed, these 27 sibships containing

Table 8.5 Expected Sibship Structure in a Hypothetical Population that Contains 64 Sibships Each of Size 3, in Which Both Parents are Carriers of an Autosomal Recessive Disorder. If No Allowance Is Made for Truncate Ascertainment, in that the 27 Sibships with No Affected Cases Will Not Be Ascertained, Then a Falsely High Segregation Ratio of 48/111 (= 0.43) Will Be Obtained

Number of Affected in Sibship	Structure of Sibship	Number of Sibships	Number of Affected	Total Number of Sibs
3	■■■	1	3	3
2	■■□	3	6	9
	□■■	3	6	9
	■□■	3	6	9
1	■□□	9	9	27
	□■□	9	9	27
	□□■	9	9	27
0	□□□	27	0	81
Total		64	48	192

only healthy individuals will not be ascertained—referred to as **incomplete ascertainment**. If this is not taken into account, a falsely *high* segregation ratio of 0.43 will be obtained instead of the correct value of 0.25.

Mathematical methods have been devised to cater for incomplete ascertainment, but analysis is usually further complicated by problems associated with achieving full or complete ascertainment. In practice 'proof' of AR inheritance requires accurate molecular or biochemical markers for carrier detection. Affected siblings (especially when at least one is female) born to unaffected parents usually suggests AR inheritance, but somatic and germline parental mosaicism (p. 121), non-paternity, and other possibilities need to be considered. There are some good examples of conditions originally reported to follow AR inheritance but subsequently shown to be dominant with germline or somatic mosaicism; for example, osteogenesis imperfecta and pseudoachondroplasia. However, a high incidence of parental consanguinity undoubtedly provides strong supportive evidence for AR inheritance, as first noted by Bateson and Garrod in 1902 (pp. 7, 113).

Genetic Linkage

Mendel's third law—the principle of independent assortment—states that members of different gene pairs assort to gametes independently of one another (p. 5). Stated more simply, the alleles of genes at different loci segregate independently. Although this is true for genes on different chromosomes, it is not always true for genes that are located on the same chromosome (i.e., close together, or **syntenic**).

Two loci positioned adjacent, or close, to each other on the same chromosome, will tend to be inherited together, and are said to be **linked**. The closer they are, the less likely they will be separated by a crossover, or recombination, during meiosis I (Figure 8.5).

Linked alleles on the same chromosome are said to be in **coupling**, whereas those on opposite homologous chromosomes are described as being in **repulsion**. This is known as the **linkage phase**. Thus in the parental chromosomes in Figure 8.5, C, A and B, as well as a and b, are in coupling, whereas A and b, as well as a and B, are in repulsion.

Recombination Fraction

The **recombination fraction**, usually designated as θ (Greek theta), is a measure of the distance separating two loci, or more precisely an indication of the likelihood that a crossover will occur between them. If two loci are not linked then θ equals 0.5 because, on average, genes at unlinked loci will segregate together during 50% of all meioses. If θ equals 0.05, this means that on average the syntenic alleles will segregate together 19 times out of 20 (i.e., a crossover will occur between them during, on average, only 1 in 20 meioses).

Centimorgans

The unit of measurement for genetic linkage is known as a **map unit** or **centimorgan** (cM). If two loci are 1 cM apart, a crossover occurs between them, on average, only once in every 100 meioses (i.e., θ = 0.01). Centimorgans are a measure of the **genetic**, or **linkage**, distance between two loci. This is *not* the same as **physical** distance, which is measured in base pairs (kb – kilobases: 1000 base pairs: Mb – megabases: 1,000,000 base pairs).

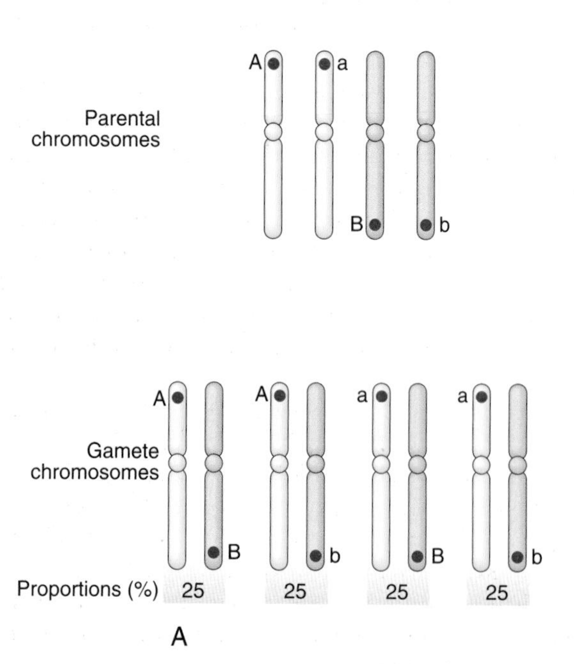

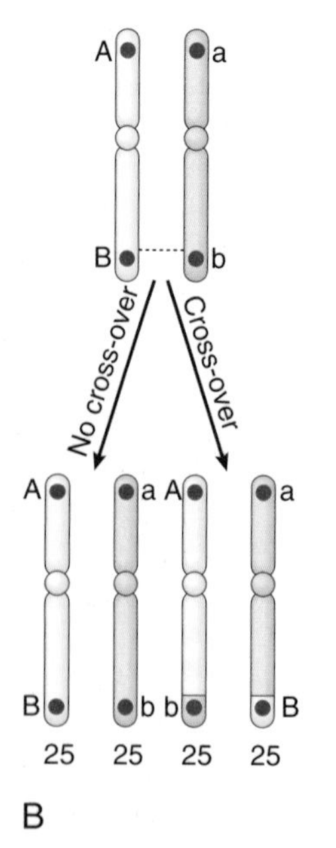

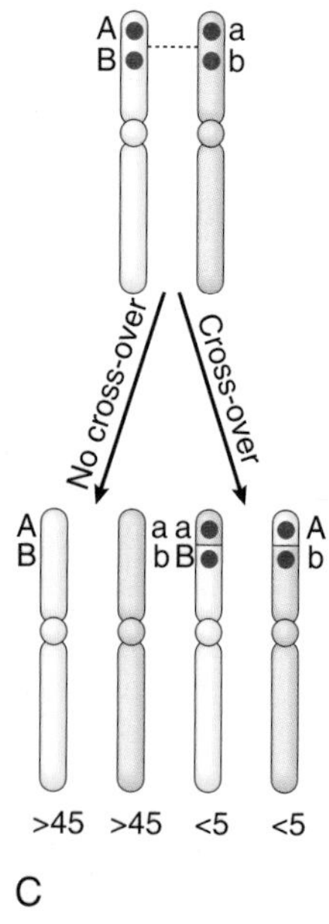

FIGURE 8.5 Segregation at meiosis of alleles at two loci. In **A** the loci are on different chromosomes and in **B** they are on the same chromosome but widely separated. Hence these loci are not linked and there is independent assortment. In **C** the loci are closely adjacent so that separation by a cross-over is unlikely (i.e., the loci are linked).

The human genome has been estimated by recombination studies to be about 3000 cM in length in males. Because the physical length of the haploid human genome is approximately 3×10^9 bp, 1 cM corresponds to approximately 10^6 bp (1 Mb or 1000 kb). However, the relationship between linkage map units and physical length is not linear. Some chromosome regions appear to be particularly prone to recombination—so-called 'hotspots'—and recombination occurs less often during meiosis in males than in females, in whom the genome 'linkage' length has been estimated to be 4200 cM. Generally, in humans one or two recombination events take place between each pair of homologous chromosomes in meiosis I, with a total of ~40 across the entire genome. Recombination events are rare close to the centromeres but relatively common in telomeric regions.

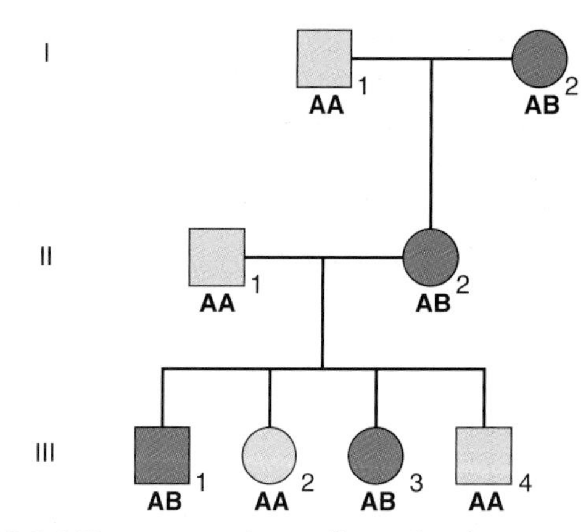

FIGURE 8.6 Three-generation pedigree showing segregation of an autosomal dominant disorder and alleles (A and B) at a locus that may or may not be linked to the disease locus.

Linkage Analysis

Linkage analysis has proved invaluable for mapping genes (see Chapter 5). It is based on studying the segregation of the disease with polymorphic markers from each chromosome—preferably in large families. Eventually a marker will be identified that co-segregates with the disease more often than would be expected by chance (i.e., the marker and disease locus are linked). The mathematical analysis tends to be very complex, particularly if many closely adjacent markers are being used, as in **multipoint linkage analysis**. However, the underlying principle is relatively straightforward and involves the use of likelihood ratios, the logarithms of which are known as LOD scores (*l*ogarithm of the *od*ds).

LOD Scores

When studying the segregation of alleles at two loci that could be linked, a series of likelihood ratios is calculated for different values of the recombination fraction (θ), ranging from θ = 0 to θ = 0.5. The likelihood ratio at a given value of θ equals the likelihood of the observed data, if the loci are linked at recombination value of θ, divided by the likelihood of the observed data if the loci are not linked (θ = 0.5). The logarithm to the base 10 of this ratio is known as the LOD score (Z)—i.e., LOD (θ) = $\log_{10}$ [Lθ/L(0.5)]. Logarithms are used because they allow results from different families to be added together.

For example, when a research paper reports that linkage of a disease with a DNA marker has been identified with a LOD score (Z) of 4 at recombination fraction (θ) 0.05, this means that the results, in the families studied, indicate that it is 10,000 (10^4) times more likely that the disease and marker loci are closely linked (i.e., 5 cM apart) than that they are not linked. It is generally agreed that a LOD score of +3 or more is confirmation of linkage. This would yield a ratio of 1000 to 1 in favor of linkage; however, because there is a prior probability of only 1 in 50 that any two given loci are linked, a LOD score of +3 means that the overall probability that the loci are linked is approximately 20 to 1—i.e., [1000 × $\frac{1}{50}$]:1. The importance of taking prior probabilities into account in probability theory is discussed in the section on Bayes' theorem (p. 339).

A 'Simple' Example

Consider a three-generation family in which several members have an AD disorder (Figure 8.6). A and B are alleles at a locus that is being tested for linkage to the disease locus.

To establish whether it is likely that these two loci are linked, the LOD score is calculated for various values of θ. The value of θ that gives the highest LOD score is taken as the best estimate of the recombination fraction. This is known as a **maximum likelihood** method.

To demonstrate the underlying principle, the LOD score is calculated for a value of θ equal to 0.05. If θ equals 0.05 then the loci are linked, in which case the disease gene and the B marker must be on the same chromosome in II2, as both of these characteristics have been inherited from the mother. Thus in II2 the linkage phase is known: the disease allele and the B allele are in coupling. Therefore the probability that III1 will be affected and will also inherit the B marker equals 0.95 (i.e., 1 – θ). A similar result is obtained for the remaining three members of the sibship in generation III, giving a value for the numerator of $(0.95)^4$. If the loci are not linked, the likelihood of observing both the disease and marker B in III1 equals 0.5. A similar result is obtained for his three siblings, giving a value for the denominator of $(0.5)^4$.

Therefore the LOD score for this family, given a value of θ = 0.05, equals $\log_{10} 0.95^4/0.5^4 = \log_{10} 13.032 = 1.12$. For a value of θ = 0, the LOD score equals $\log_{10} 1^4/0.5^4 = \log_{10} 16 = 1.20$. For a value of θ = 0.1, the LOD score equals $\log_{10} 0.9^4/0.5^4 = \log_{10} 10.498 = 1.02$. The highest LOD score is obtained for a value of θ equals 0, which is consistent with the fact that if the disease and marker loci are linked then no recombination has occurred between the two loci in members of generation III.

To confirm linkage other families would have to be studied by pooling all the results until a LOD score of +3

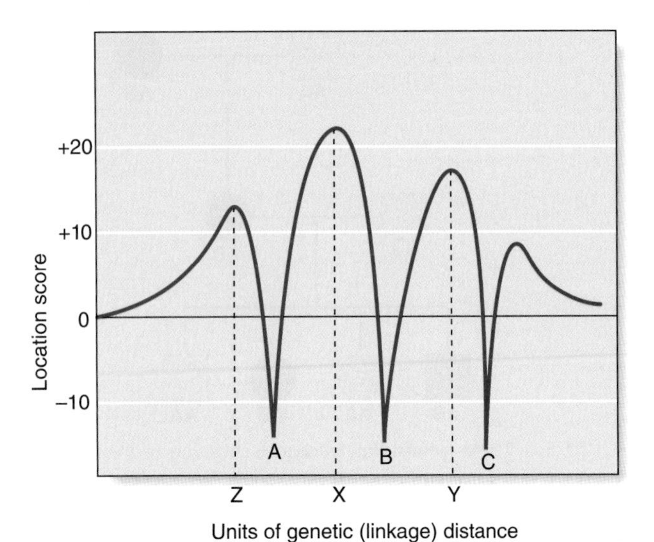

FIGURE 8.7 Multipoint linkage analysis. A, B, and C represent the known linkage relationships of three polymorphic marker loci. X, Y, and Z represent in descending order of likelihood the probable position of the disease locus.

or greater was obtained. A LOD score of −2 or less is taken as proof that the loci are not linked. This less stringent requirement for proof of non-linkage (i.e., a LOD score of −2 compared with +3 for proof of linkage) is due to the high prior probability of $^{49}/_{50}$ that any two loci are not linked.

Multipoint Linkage Analysis

Two-point linkage analysis is often used to map a disease locus to a specific chromosome region. This gives a rather rough or 'coarse' indication of the location of the disease locus. The next step often involves multipoint linkage analysis using a series of polymorphic markers that are known to map to the disease region. This process allows fine tuning of the probable position of the disease locus within the rough interval defined by the small number of polymorphic marker loci.

Using this approach the results of linkage studies with the various markers are analyzed by a computer program that calculates the overall likelihood of the position of the disease locus in relation to the marker loci. The results are presented in the form of a likelihood ratio known as a **location score**. This is calculated for different positions of the disease locus and a graph is drawn up of location score against map distance (Figure 8.7). On this graph the peaks represent possible positions of the disease locus, with the tallest peak being the most probable location. The troughs represent the positions of the polymorphic marker loci.

Multipoint linkage analysis is used to define the smallest possible interval in which a disease locus is located, so that physical mapping methods can then be applied to isolate the disease gene (see Chapter 5).

Autozygosity Mapping

This ingenious form of linkage analysis has been used to map many rare AR disorders. **Autozygosity** occurs when individuals are homozygous at particular loci by descent from a common ancestor. In an inbred pedigree containing two or more children with a rare AR disorder, it is very likely that the children will be homozygous not only at the disease locus but also at closely linked loci. In other words, all affected relatives in an inbred family will be homozygous for markers within the region surrounding the disease locus. Thus a search can be made for shared areas of homozygosity in affected relatives using highly polymorphic markers such as microsatellites (p. 69). In a pedigree with a relatively large number of affected individuals, only a small number of shared homozygous regions will be identified; one of these can be expected to harbor the relevant disease locus, which can then be isolated using physical mapping strategies.

Autozygosity mapping can be applied in both small inbred families (Figure 8.8) and in genetic isolates (p. 133) with a shared common genetic ancestry (e.g., the Old Order Amish). It is a particularly powerful technique in large inbred families in which more than one branch has affected individuals. Several of the genes that cause AR sensorineural hearing loss have been mapped in this way, as well as a number of skeletal dysplasias and primary microcephalies, for example.

Linkage Disequilibrium

Linkage disequilibrium is defined formally as the association of two alleles at linked loci more frequently than would be expected by chance, and is also referred to as **allelic association**. The concept and the term relate to the study of diseases in **populations** rather than **families**. In the latter, an association between specific alleles and the disease in question holds true only within an individual family; in a separate affected family a different pattern of alleles, or markers, at the same locus may show association with the disease—because the alleles themselves are **polymorphic**.

The rationale for studying allelic association in populations is based on the assumption that a mutation occurred in a **founder** case some generations previously and is still causative of the disease. If this is true, the pattern of markers in a small region close to the mutation will have been maintained and thus constitutes what is termed the **founder haplotype**. The underlying principles used in mapping are the same as those for linkage analysis in families, the difference being the degree of relatedness of the individuals under study. In the pedigree shown in Figure 8.6, support was obtained for linkage of the disease gene with the B marker allele. Assume that further studies confirm linkage of these loci and that the A and B alleles have an equal frequency of 0.5. It would be reasonable to expect that the disease gene would be in coupling with allele A in approximately 50% of families and with allele B in the remaining 50%. If, however, the disease allele was found to be in coupling exclusively with one particular marker allele, this would be an example of linkage disequilibrium.

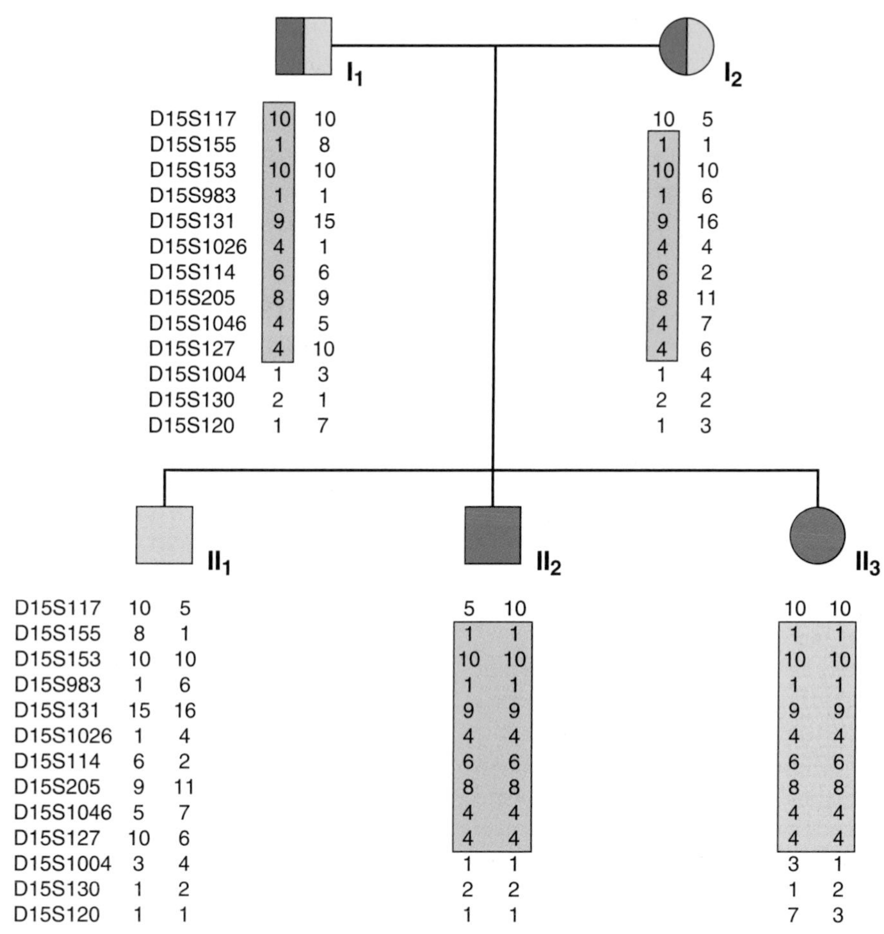

FIGURE 8.8 Autozygosity mapping in a family with spondylocostal dysostosis. The father of individual I_1 is the brother of I_2's grandfather. The region of homozygosity is defined by markers D15S155 and D15S127. A mutation in the *MESP2* gene was subsequently shown to be the cause of spondylocostal dysostosis in this pedigree.

The demonstration of linkage disequilibrium in a particular disease suggests that the mutation causing the disease has occurred relatively recently and that the marker locus studied is very closely linked to the disease locus. There may be pitfalls, however, in interpreting haplotype data that suggest linkage disequilibrium. Other possible reasons for linkage disequilibrium include: (1) the rapid growth of genetically isolated populations leading to large regions of allelic association throughout the genome; (2) selection, whereby particular alleles enhance or diminish reproductive fitness; and (3) population admixture, where population subgroups with different patterns of allele frequencies are combined into a single study. Allowance for the latter problem can be made by using family-based controls and analyzing the transmission of alleles using a method called the **transmission/disequilibrium test**. This uses the fact that transmitted and non-transmitted alleles from a given parent are paired observations, and examines the preferential transmission of one allele over the other in all heterozygous parents. The technique has been applied, amongst others, to studies based on sibling pairs that are discordant for the disease or condition under study.

Medical and Societal Intervention

Recent developments in molecular biology, such as the human genome project (p. 9) and pilot studies using gene therapy (p. 350), have reawakened concern that future generations could have to cope with an ever increasing burden of genetic disease. The term **eugenics** was first used by Charles Darwin's cousin, Francis Galton, to refer to the improvement of a population by selective breeding. The notion that this should be applied to human populations became popular during the early years of the twentieth century, culminating in the horrifying practices of Nazi Germany. Ensuing revulsion led to the abandonment of eugenic programs in humans, with universal condemnation and agreement that such programs have no place in modern medical practice. Sadly, however, these practices have continued by groups engaged in territorial conflicts—somewhat sanitized by the term 'ethnic cleansing.'

Doctors caring for patients and families with hereditary disease inevitably give priority to treatment and improving survival. By so doing biological fitness may be increased, leading to increased numbers of 'bad genes' in society, potentially adding adversely to humanity's future genetic load. Such long-term consequences generally carry no weight, but the approach has sometimes been interpreted as **dysgenic**.

The ethical debate is important but it is worth considering the possible long-term effects of artificial selection for or against genetic disorders, according to pattern of inheritance.

AD Disorders

If everyone with an AD disorder were successfully encouraged not to reproduce, the incidence of that disorder would decline rapidly, with all future cases being the result only of new mutations. This would have a particularly striking effect on the incidence of relatively mild conditions such as familial hypercholesterolemia, in which genetic fitness is close to 1.

Alternatively, if successful treatment became available for all patients with a serious AD disorder that at present is associated with a marked reduction in genetic fitness, there would be an immediate increase in the frequency of the disease gene followed by a more gradual leveling off at a new equilibrium level. If, at one time, all those with a serious AD disorder died in childhood (f = 0), then the incidence of affected individuals would be 2μ. If treatment raised the fitness from 0 to 0.9, the incidence of affected children in the next generation would rise to 2μ due to new mutations plus 1.8μ inherited, which equals 3.8μ. Eventually a new equilibrium would be reached, by which time the disease incidence would have risen tenfold to 20μ. This can be calculated relatively easily with the formula $\mu = [I(1 - f)]/2$ (p. 133), which can also be expressed as $I = 2\mu/(1 - f)$. The net result would be that the proportion of affected children who died would be lower (from 100% down to 10%), but the total number affected would be much greater, although the actual number who died from the disease would remain unchanged at 2μ.

Autosomal Recessive Disorders

In contrast to an AD disorder, artificial selection against an AR condition will have only a very slow effect.

The reason for this difference is that in AR conditions most of the genes in a population are present in healthy heterozygotes who would not be affected by eugenic measures. It can be shown that if there is complete selection against an AR disorder, so that no homozygotes reproduce, the number of generations (n) required for the allele frequency to change from q_0 to q_n equals $1/q_n - 1/q_0$. Therefore, for a condition with an incidence of approximately 1 in 2000 and an allele frequency of roughly 1 in 45, if all affected patients refrained from reproduction then it would take more than 500 years (18 generations) to reduce the disease incidence by half and more than 1200 years (45 generations) to reduce the gene frequency by half, assuming an average generation time of 27 years.

Now consider the opposite situation, where selection operating against a serious AR disorder is relaxed because of improvement in medical treatment. More affected individuals will reach adult life and transmit the mutant allele to their offspring. The result will be that the frequency of the mutant allele will increase until a new equilibrium is reached. Using the formula $\mu = I(1 - f)$, it can be shown that, when the new equilibrium is eventually reached, an increase in fitness from 0 to 0.9 will have resulted in a tenfold increase in the disease incidence.

X-Linked Recessive Disorders

When considering the effects of selection against these disorders, it is necessary to take into account the fact that a large proportion of the relevant genes are present in entirely healthy female carriers, who are often unaware of their carrier status. For a very serious condition, such as Duchenne muscular dystrophy (p. 307), with fitness equal to 0 in affected males, selection will have no effect unless female carriers choose to limit their families. If all female carriers opted not to have any children, the incidence would be reduced by two-thirds (i.e., from 3μ to μ).

A much more plausible possibility is that effective treatment will be forthcoming for these disorders. This will result in a steady increase in the disease incidence. For example, an increase in fitness from 0 to 0.5 will lead to a doubling of the disease incidence by the time a new equilibrium has been established. This can be calculated using the formula $\mu = [I^M(1 - f)]/3$ (p. 133).

Conclusion

In reality it is extremely difficult to predict the long-term impact of medical intervention on the incidence and burden of genetic disease. Although it is true that improvements in medical treatment could result in an increased genetic load in future generations, it is equally possible that successful gene therapy will ease the overall burden of these disorders in terms of human suffering. Some of these arguments could have been made many years ago for other major medical developments, such as the discovery of insulin and antibiotics, which have had immeasurable financial implications in terms of the pharmaceutical industry as well as contributing to an aging population. Ultimately, how society copes with these advances and challenges provides a measure of civilization.

FURTHER READING

Allison AC 1954 Protection afforded by sickle-cell trait against subtertian malarial infection. BMJ i:290–294

A landmark paper providing clear evidence that the sickle-cell trait provides protection against parasitemia by falciparum malaria.

Emery AEH 1986 Methodology in medical genetics, 2nd ed. Edinburgh: Churchill Livingstone

A useful handbook of basic population genetics and mathematical methods for analyzing the results of genetic studies.

Francomano CA, McKusick VA, Biesecker LG, eds 2003 Medical genetic studies in the Amish: historical perspective. Am J Med Genet C Semin Med Genet 121:1–4

Haldane JBS 1935 The rate of spontaneous mutation of a human gene. J Genet 31:317–326

The first estimate of the mutation rate for hemophilia using an indirect method.

Hardy GH 1908 Mendelian proportions in a mixed population. Science 28:49–50

A short letter in which Hardy pointed out that in a large randomly mating population dominant 'characters' would not increase at the expense of recessives.

Khoury MJ, Beaty TH, Cohen BH 1993 Fundamentals of genetic epidemiology. New York: Oxford University Press

A comprehensive textbook of population genetics and its areas of overlap with epidemiology.

Ott J 1991 Analysis of human genetic linkage. Baltimore: Johns Hopkins University Press

A detailed mathematical explanation of linkage analysis.

Vogel F, Motulsky AG 1997 Human genetics, problems and approaches, 3d ed. Berlin: Springer

The definitive textbook of human genetics with extensive coverage of mathematical aspects.

ELEMENTS

1. According to the Hardy-Weinberg principle, the relative proportions of the possible genotypes at a particular locus remain constant from one generation to the next.

2. Factors that may disturb Hardy-Weinberg equilibrium are non-random mating, mutation, selection for or against a particular genotype, small population size, and migration.

3. If an autosomal recessive disorder is in Hardy-Weinberg equilibrium, the carrier frequency can be estimated by doubling the square root of the disease incidence.

4. The mutation rate for an autosomal dominant disorder can be measured directly by estimating the proportion of new mutations among all members of one generation. Indirect estimates of mutation rates can be made using the formula:
   $= [I(1 - f)]/2$ for autosomal dominant inheritance
   $= I(1 - f)$ for autosomal recessive inheritance
   $= [I^M(1 - f)]/3$ for X-linked recessive inheritance.

5. Otherwise rare single-gene disorders can show a high incidence in a small population because of a founder effect coupled with genetic isolation.

6. When a serious autosomal recessive disorder has a relatively high incidence in a large population, it is likely due to heterozygote advantage.

7. Closely adjacent loci on the same chromosome are regarded as linked if genes at these loci segregate together during more than 50% of meioses. The recombination fraction (θ) indicates how often two such genes will be separated (recombine) at meiosis.

8. The logarithm of the odds (LOD) score is a mathematical indication of the relative likelihood that two loci are linked. A LOD score of +3 or greater is taken as confirmation of linkage. Two-point linkage analysis is used to map a disease locus to a chromosome region. Multipoint linkage analysis can then be used to determine the probable order of polymorphic loci within that region and to narrow down the size of the interval to be studied by physical mapping.

CHAPTER 9

Polygenic and Multifactorial Inheritance

Many disorders demonstrate familial clustering that does not conform to any recognized pattern of Mendelian inheritance. Examples include several of the most common congenital malformations and many common acquired diseases (Box 9.1). These conditions show a definite familial tendency, but the incidence in close relatives of affected individuals is much lower than would be seen if these conditions were caused by mutations in single genes.

Because it is likely that many factors, both genetic and environmental, are involved in causing these disorders, they are generally referred to as showing **multifactorial** inheritance. The prevailing view until recently has been that in multifactorial inheritance, environmental factors interact with many genes to generate a normally distributed susceptibility. According to this theory, individuals are affected if they lie at the wrong end of the distribution curve. This concept of a normal distribution generated by many genes, known as **polygenes**, each acting in an additive fashion, is plausible for physiological characteristics such as height and possibly blood pressure. However, for disease states such as type 1 diabetes mellitus (T1DM), the genetic contribution involves many loci, some of which play a much more important role than others.

Sequencing of the human genome has shown that the 3 billion base pairs are 99.9% identical in every person. This also means that individuals are, on average, 0.1% different genetically from every other person on the planet. And within that 0.1% lies the mystery of why some people are more susceptible to a particular illness, or more likely to be healthy, than another member of the population. Our increased knowledge of genetic variation at the level of single nucleotide polymorphisms (SNPs), together with high throughput SNP genotyping platforms, has recently revolutionized our ability to identify disease susceptibility loci for many common diseases.

Polygenic Inheritance and the Normal Distribution

Before considering the impact of recent research in detail, it is necessary to outline briefly the scientific basis of what is known as **polygenic** or **quantitative** inheritance. This involves the inheritance and expression of a phenotype being determined by many genes at different loci, with each gene exerting a small additive effect. **Additive** implies that the effects of the genes are cumulative, i.e. no one gene is dominant or recessive to another.

Several human characteristics (Box 9.2) show a continuous distribution in the general population, which closely resembles a normal distribution. This takes the form of a symmetrical bell-shaped curve distributed evenly about a mean (Figure 9.1). The spread of the distribution about the mean is determined by the standard deviation. Approximately 68%, 95%, and 99.7% of observations fall within the mean plus or minus one, two, or three standard deviations, respectively.

It is possible to show that a phenotype with a normal distribution in the general population can be generated by polygenic inheritance involving the action of many genes at different loci, each of which exerts an equal additive effect. This can be illustrated by considering a trait such as height. If height were to be determined by two equally frequent alleles, a (tall) and b (short), at a single locus, then this would result in a discontinuous phenotype with three groups in a ratio of 1 (tall-aa) to 2 (average-ab/ba) to 1 (short-bb). If the same trait were to be determined by two alleles at each of two loci interacting in a simple additive way, this would lead to a phenotypic distribution of five groups in a ratio of 1 (4 tall genes) to 4 (3 tall + 1 short) to 6 (2 tall + 2 short) to 4 (1 tall + 3 short) to 1 (4 short). For a system with three loci each with two alleles the phenotypic ratio would be 1-6-15-20-15-6-1 (Figure 9.2).

It can be seen that as the number of loci increases, the distribution increasingly comes to resemble a normal curve, thereby supporting the concept that characteristics such as height are determined by the additive effects of many genes at different loci. Further support for this concept comes from the study of familial correlations for characteristics such as height. **Correlation** is a statistical measure of the degree of resemblance or relationship between two parameters. First-degree relatives share, on average, 50% of their genes (Table 9.1). Therefore, if height is polygenic, the correlation between first-degree relatives should be 0.5. Several studies have shown that the sib–sib correlation for height is indeed close to 0.5.

In reality, human characteristics such as height and intelligence are also influenced by environment, and possibly also by genes that are not additive in that they exert a dominant effect. These factors probably account for the observed

Box 9.1 Disorders that Show Multifactorial Inheritance

Congenital Malformations
Cleft lip/palate
Congenital dislocation of the hip
Congenital heart defects
Neural tube defects
Pyloric stenosis
Talipes

Acquired Diseases of Childhood and Adult Life
Asthma
Autism
Diabetes mellitus
Epilepsy
Glaucoma
Hypertension
Inflammatory bowel disease (Crohn disease and ulcerative colitis)
Ischaemic heart disease
Ischaemic stroke
Manic depression
Multiple sclerosis
Parkinson disease
Psoriasis
Rheumatoid arthritis
Schizophrenia

Box 9.2 Human Characteristics that Show a Continuous Normal Distribution

Blood pressure
Dermatoglyphics (ridge count)
Head circumference
Height
Intelligence
Skin color

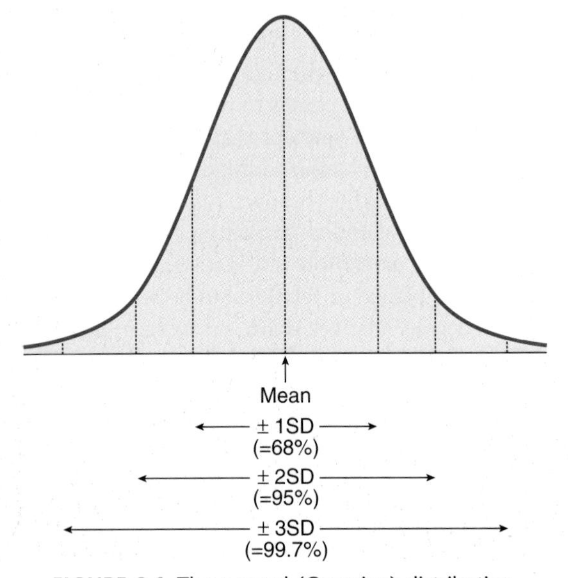

FIGURE 9.1 The normal (Gaussian) distribution.

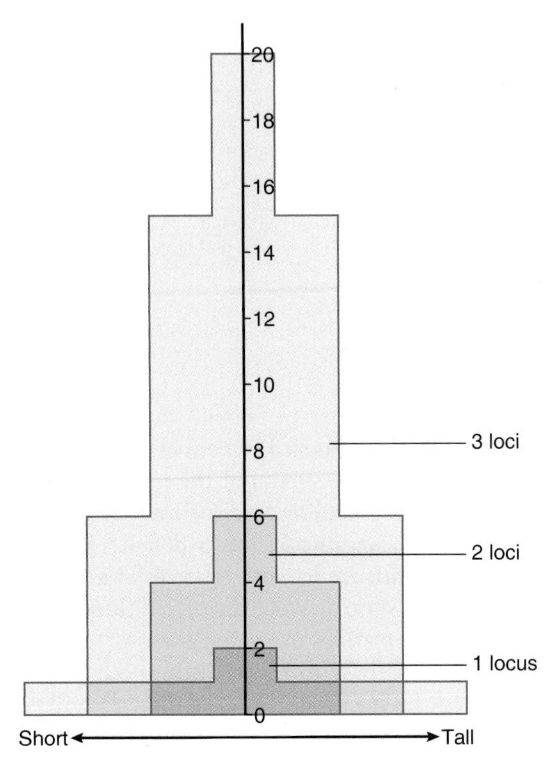

FIGURE 9.2 Distribution of genotypes for a characteristic such as height with 1, 2, and 3 loci each with two alleles of equal frequency. The values for each genotype can be obtained from the binomial expansion $(p+q)^{(2n)}$, where $p = q = 1/2$ and n equals the number of loci.

tendency of offspring to show what is known as **regression to the mean**. This is demonstrated by tall or intelligent parents (the two are not mutually exclusive!) having children whose average height or intelligence is slightly lower than the average or mid-parental value. Similarly, parents who are very short or of low intelligence tend to have children whose average height or intelligence is lower than the general population average, but higher than the average value of the parents. If a trait were to show true polygenic

Table 9.1 Degrees of Relationship

Relationship	Proportion of Genes Shared
First degree	½
Parents	
Siblings	
Children	
Second degree	¼
Uncles and aunts	
Nephews and nieces	
Grandparents	
Grandchildren	
Half-siblings	
Third degree	⅛
First cousins	
Great-grandparents	
Great-grandchildren	

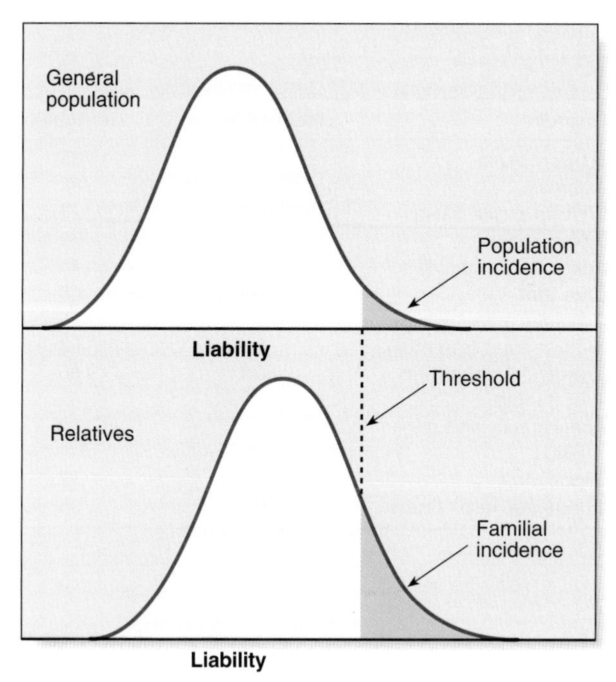

FIGURE 9.3 Hypothetical liability curves in the general population and in relatives for a hereditary disorder in which the genetic predisposition is multifactorial.

inheritance with no external influences, then the measurements in offspring would be distributed evenly around the mean of their parents' values.

Multifactorial Inheritance—The Liability/Threshold Model

Efforts have been made to extend the polygenic theory for the inheritance of quantitative or continuous traits to try to account for **discontinuous** multifactorial disorders. According to the **liability/threshold** model, all of the factors which influence the development of a multifactorial disorder, whether genetic or environmental, can be considered as a single entity known as liability. The liabilities of all individuals in a population form a continuous variable, which has a normal distribution in both the general population and in relatives of affected individuals. However, the curves for these relatives will be shifted to the right, with the extent to which they are shifted being directly related to the closeness of their relationship to the affected index case (Figure 9.3).

To account for a discontinuous phenotype (i.e., affected or not affected) with an underlying continuous distribution, it is proposed that a threshold exists above which the abnormal phenotype is expressed. In the general population, the proportion beyond the threshold is the population incidence, and among relatives the proportion beyond the threshold is the familial incidence.

It is important to emphasize again that liability includes all factors that contribute to the cause of the condition. Looked at very simply, a deleterious liability can be viewed as consisting of a combination of several 'bad' genes and adverse environmental factors. Liability cannot be measured but the mean liability of a group can be determined from the incidence of the disease in that group using statistics of the normal distribution. The units of measurement are standard deviations and these can be used to estimate the correlation between relatives.

Consequences of the Liability/Threshold Model

Part of the attraction of this model, and it should be emphasized again that this is a *hypothesis* rather than a proven fact, is that it provides a simple explanation for the observed patterns of familial risks in conditions such as cleft lip/palate, pyloric stenosis, and spina bifida.

1. The incidence of the condition is greatest among relatives of the most severely affected patients, presumably because they are the most extreme deviants along the liability curve. For example, in cleft lip/palate the proportion of affected first-degree relatives (parents, siblings, and offspring) is 6% if the index patient has bilateral cleft lip and palate, but only 2% if the index patient has a unilateral cleft lip (Figure 9.4).

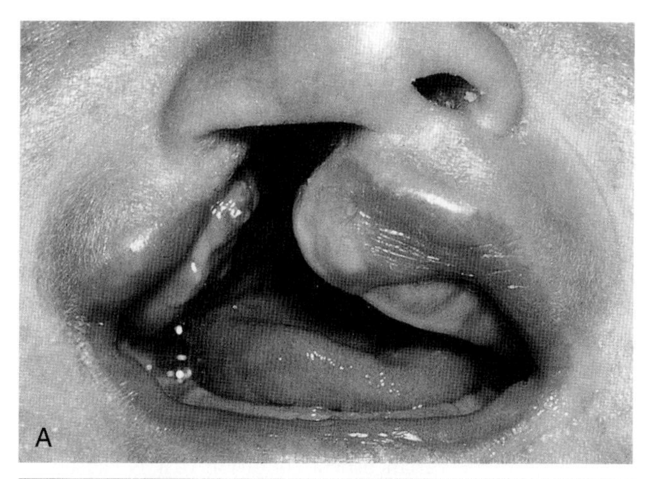

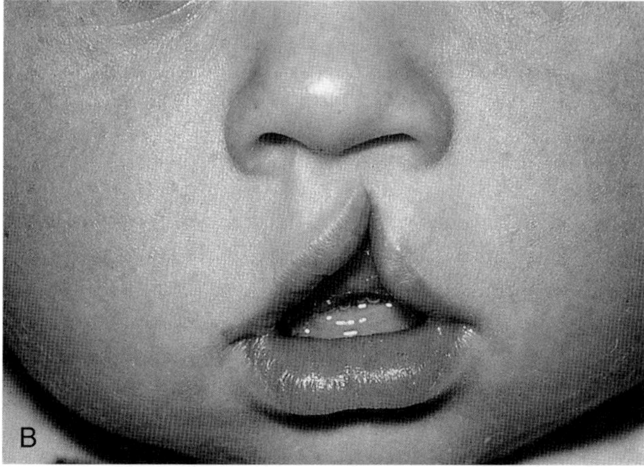

FIGURE 9.4 Severe (**A**) and mild (**B**) forms of cleft lip/palate.

2. The risk is greatest among close relatives of the index case and decreases rapidly in more distant relatives. For example, in spina bifida the risks to first-, second,- and third-degree relatives of the index case are approximately 4%, 1%, and less than 0.5%, respectively.
3. If there is more than one affected close relative, then the risks for other relatives are increased. In spina bifida, if one sibling is affected the risk to the next sibling (if folic acid is not taken by the mother periconceptionally) is approximately 4%; if two siblings are affected, the risk to a subsequent sibling is approximately 10%.
4. If the condition is more common in individuals of one sex, then relatives of an affected individual of the less frequently affected sex will be at higher risk than relatives of an affected individual of the more frequently affected sex. This is illustrated by the condition pyloric stenosis. Pyloric stenosis shows a male to female ratio of 5 to 1. The proportions of affected offspring of male index patients are 5.5% for sons and 2.4% for daughters, whereas the risks to the offspring of female index patients are 19.4% for sons and 7.3% for daughters. The probable explanation for these different risks is that for a female to be affected, she has to lie at the extreme of the liability curve, so that her close relatives will also have a very high liability for developing the condition. Because males are more susceptible to developing the disorder, risks in male offspring are higher than in female offspring regardless of the sex of the affected parent.
5. The risk of recurrence for first-degree relatives (i.e., siblings and offspring) approximates to the square root of the general population incidence. Thus if the incidence is 1 in 1000, the sibling and offspring risk will equal approximately 1 in 32, or 3%.

Heritability

Though it is not possible to assess an individual's liability for a particular disorder, it is possible to estimate what proportion of the etiology can be ascribed to genetic factors as opposed to environmental factors. This is referred to as **heritability**, which can be defined as the proportion of the total phenotypic variance of a condition that is caused by additive genetic variance. In statistical terms, variance equals the square of the standard deviation. Heritability is often depicted using the symbol h^2 and is expressed either as a proportion of 1 or as a percentage.

Estimates of the heritability of a condition or trait provide an indication of the relative importance of genetic factors in its causation, so that the greater the value for the heritability the greater the role of genetic factors.

Heritability is estimated from the degree of resemblance between relatives expressed in the form of a correlation coefficient, which is calculated using statistics of the normal distribution. Alternatively, heritability can be calculated using data on the concordance rates in monozygotic and dizygotic twins. In practice, it is desirable to try to derive heritability estimates using different types of relatives and to measure the disease incidence in relatives reared together and living apart so as to try to disentangle the possible effects of common environmental factors. Estimates of heritability for some common diseases are given in Table 9.2.

Table 9.2 Estimates of Heritability of Various Disorders

Disorder	Frequency (%)	Heritability
Schizophrenia	1	85
Asthma	4	80
Cleft lip ± cleft palate	0.1	76
Pyloric stenosis	0.3	75
Ankylosing spondylitis	0.2	70
Club foot	0.1	68
Coronary artery disease	3	65
Hypertension (essential)	5	62
Congenital dislocation of the hip	0.1	60
Anencephaly and spina bifida	0.3	60
Peptic ulcer	4	37
Congenital heart disease	0.5	35

The degree of familial clustering shown by a multifactorial disorder can be estimated by measuring the ratio of the risk to siblings of affected individuals compared to the general population incidence. This ratio of sib risk to population incidence is known as λ_s. For example, in type 1 diabetes, where the UK population incidence is 0.4% and the risk to siblings is 6%, λ_s is 15. For type 2 diabetes in Europe, λ_s is estimated at a more modest 3.5 (35% sibling risk; 10% population risk).

Identifying Genes that Cause Multifactorial Disorders

Multifactorial disorders are common and make a major contribution to human morbidity and mortality (p. 8). It is therefore not surprising that vigorous efforts are being made to try to identify genes that contribute to their etiology. Several strategies have been used to search for disease susceptibility genes.

Linkage Analysis

Linkage analysis has proved extremely valuable in mapping single gene disorders by studying the co-segregation of genetic markers with the disease (p. 137). However, this approach is much more difficult in multifactorial disorders. This is because it is extremely difficult mathematically to develop strategies for detecting linkage of additive **polygenes**, each of which makes only a small contribution to the phenotype. In addition, many multifactorial diseases show a variable age of onset so that the genetic status of unaffected family members cannot be known with certainty.

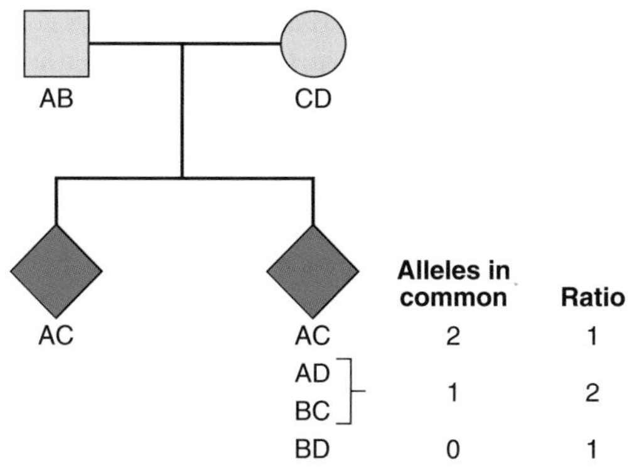

FIGURE 9.5 The probability that siblings will have 2, 1, or 0 parental alleles in common. Significant deviation from the 1 : 2 : 1 ratio indicates that the locus is causally related to the disease.

Despite the limitations, a small number of susceptibility loci have been identified using modifications of the approaches used for mapping single gene loci. Examples are given in Chapter 15.

Affected Sibling-Pair Analysis

Standard linkage analysis requires information regarding the mode of inheritance, gene frequencies, and penetrance. For multifactorial disorders, this information is not usually available. A possible solution is to look for regions of the genome that are **identical by descent** in affected sibling pairs. If affected siblings inherit a particular allele more or less often than would be expected by chance, this indicates that the allele or its locus is involved in some way in causing the disease.

Consider a set of parents with alleles AB (father) and CD (mother) at a particular locus. The probability that any two of their children will have both alleles in common is 1 in 4 (Figure 9.5). The probability that they will have one allele in common is 1 in 2 and the probability that they will have no alleles in common equals 1 in 4. If siblings who are affected with a particular disease show deviation from this 1 : 2 : 1 ratio for a particular variant, this implies that there is a causal relationship between the locus and the disease.

Linkage Disequilibrium Mapping

After a chromosome region that appears to confer susceptibility has been identified, the next step is to reduce the genetic interval by **fine mapping**. The most powerful method uses **linkage disequilibrium (LD)** (p. 138) mapping to construct haplotypes by genotyping SNPs within the region. Historical crossover points reduce the genetic interval by defining LD 'blocks' (Figure 9.6). Candidate genes within the region are then sequenced to find DNA variants that can be tested for association with the disease.

Many genome-wide linkage scans (p. 76) have been performed for various disorders and although a number of loci have been mapped, the number of disease susceptibility genes identified by this approach is disappointingly small. One probable reason is the complex nature of multifactorial disease, with numerous genetic variants of modest effect interacting with each other and the environment. Most linkage studies are simply underpowered to detect these effects, and it was shown by Risch and Merikangas that an alternative approach, the **association study**, would be a more powerful way of finding genetic variants underlying complex diseases.

Association Studies

Association studies are undertaken by comparing the frequency of a particular variant in affected patients with its frequency in a carefully matched control group. This approach is often described as a **case-control** study. If the frequencies in the two groups differ significantly, this provides evidence for an association.

The polymorphic HLA histocompatibility complex on chromosome 6 (p. 200) has been frequently studied. One

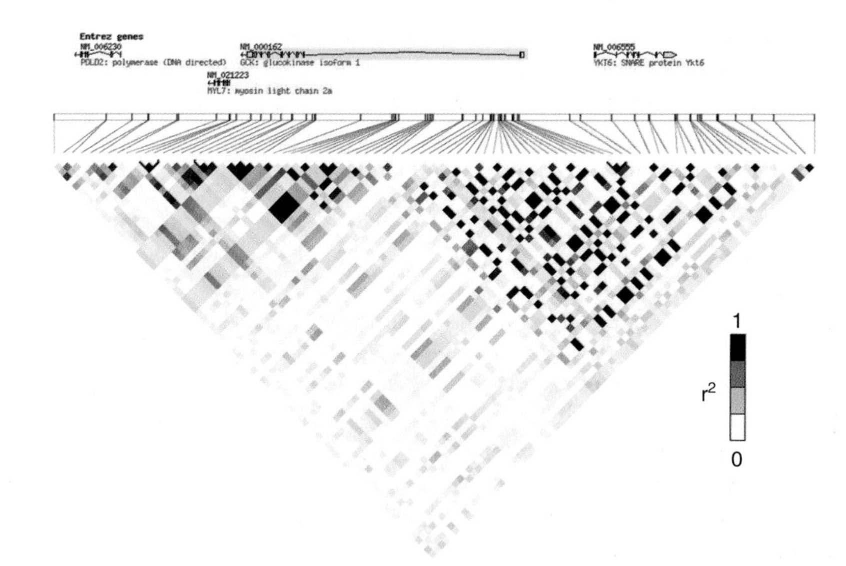

FIGURE 9.6 The linkage disequilibrium (LD) structure of glucokinase. r^2 values between the 84 single nucleotide polymorphisms (SNPs) across a 116-kb region are presented. An r^2 value of 1 indicates that two SNPs are linked. There are two blocks of LD within the glucokinase gene *(outlined in red)*.

Table 9.3 Calculation of Odds Ratio for a Disease Association

	Allele 1	Allele 2
Patients	A	b
Controls	C	d
Odds ratio	$= a/c \div b/d$	
	$= ad/bc$	

of the strongest known HLA associations is that between ankylosing spondylitis and the B27 allele. This is present in approximately 90% of all patients and in only 5% of controls. The strength of an association is indicated by the ratio of the odds of developing the disease in those with the antigen to the odds of developing the disease in those without the antigen (Table 9.3). This is known as the **odds ratio** and it gives an indication of how much more frequently the disease occurs in individuals with a specific marker than in those without that marker. For the HLA-ankylosing spondylitis association, the odds ratio is 171. However, for most markers associated with multifactorial disease, the frequency difference between cases and controls is small, giving rise to modest odds ratios (usually between 1.1 and 1.5).

If evidence for association is forthcoming, this suggests that the allele encoded by the marker is either directly involved in causing the disease (i.e., a susceptibility variant) or that the marker is in linkage disequilibrium with a closely linked susceptibility variant. When considering disease associations, it is important to remember that the identification of a susceptibility locus does not mean that the definitive disease gene has been identified. For example, although it is one of the strongest disease associations known, only 1% of all HLA B27 individuals develop ankylosing spondylitis, so that many other factors, genetic and/or environmental, must be involved in causing this condition.

Before 2006, association studies were carried out by first selecting a candidate gene or genomic region, which would either have plausible biological links to the disease of interest or be situated in a region of linkage. One or more genetic variants were selected from the gene or gene region and genotyped in cases and controls to test for association with the disease. Many studies showing evidence of association with candidate genes were published for a variety of diseases and traits. However, in numerous cases, these associations did not replicate in independent studies, leaving the validity of many of the initially reported associations unclear. The reasons for this inconsistency included (1) small sample sizes, (2) weak statistical support, and (3) the low prior probability of any of the few selected variants being genuinely associated with the disease. All of these features increased the chances of false-positive associations. An additional reason for false-positive associations is population stratification, in which the population contains subgroups of different ancestries and both the disease and the allele happen to be common within that subset. A famous example was reported in a study by Lander and Schork which showed, in a San Francisco population, that HLA-A1 is associated with the ability to eat with chopsticks. This association is simply explained by the fact that HLA-A1 is more common among Chinese than Europeans!

The candidate gene approach led to only a handful of widely replicated associations. Two important developments made it possible to move away from this approach, toward a genome-wide approach to association studies: the first was the development of microarray technology to genotype hundreds of thousands of SNPs in thousands of individuals quickly and at little cost; the second was the creation of a reference catalogue of SNPs and linkage disequilibrium, the International Haplotype Map (HapMap).

HapMap Project (www.hapmap.org)

Although it is estimated that there may be up to 10 million SNPs in the human genome, many SNPs are in linkage disequilibrium (p. 138) and therefore co-inherited. Regions of linked SNPs are known as haplotypes. The International HapMap project is identifying SNP frequencies and haplotypes in different populations (Table 9.4). By 2007, the project had genotyped more than 3 million SNPs in 270 samples from Europe, East Asia, and West Africa. It showed that most SNPs are strongly correlated to one or more others nearby. This means that by genotyping approximately 500,000 SNPs in most populations, we can capture information on the majority of common SNPs in the human genome (with minor allele frequency >5%). In African populations, the number needed is approximately 1 million SNPs because of lower overall linkage disequilibrium. Since 2007, genotype data have been added to the HapMap from seven other populations, whereas the original HapMap population samples have expanded. Together with

Table 9.4 Populations Studied in the International HapMap Project

Ancestry	Place of Residence	Number of Individuals Analysed
Yoruba	Ibadan, Nigeria	180*
Japanese	Tokyo, Japan	91
Han Chinese	Beijing, China	90
Northern and western European	Utah, USA	180*
Luhya	Webuye, Kenya	90
Maasai	Kinyawa, Kenya	180*
Tuscan	Italy	90
Gujarati Indian	Houston, Texas, USA	90
Metropolitan Chinese community	Denver, Colorado, USA	90
Mexican	Los Angeles, California, USA	90*
African	Southwestern USA	90*

*Sample contains DNA from family trios (mother, father, and child), whereas the others include only unrelated individuals.

high-throughput SNP genotyping, this valuable reference enabled a new generation of association studies, which could tackle the whole genome's common SNP variation in just one experiment.

Genome-Wide Association Studies

In **genome-wide association (GWA)** studies, researchers compare variants across the entire genome in a case control study, rather than looking at just one variant at a time. Since 2006, this powerful new method has produced an explosion in the number of widely replicated associations between SNPs and common diseases, which are catalogued at http://www.genome.gov/gwastudies/. By 2009, GWA studies had identified hundreds of reproducible associations with over 80 common diseases or traits. Examples of these associations are given in Chapter 15. The results of a GWA study of autism are shown in Figure 9.7. In a typical GWA study, 500,000 to 1 million SNPs are genotyped in each subject using a single microarray ('SNP chip').

A clear advantage of GWA studies over the candidate gene approach is that they are 'hypothesis-free'. No prior assumption is made about the genes likely to be involved in the disease, and as a result, associations have been uncovered which provide new insights into biological pathways, opening up new avenues for research. Examples are given in Chapter 15.

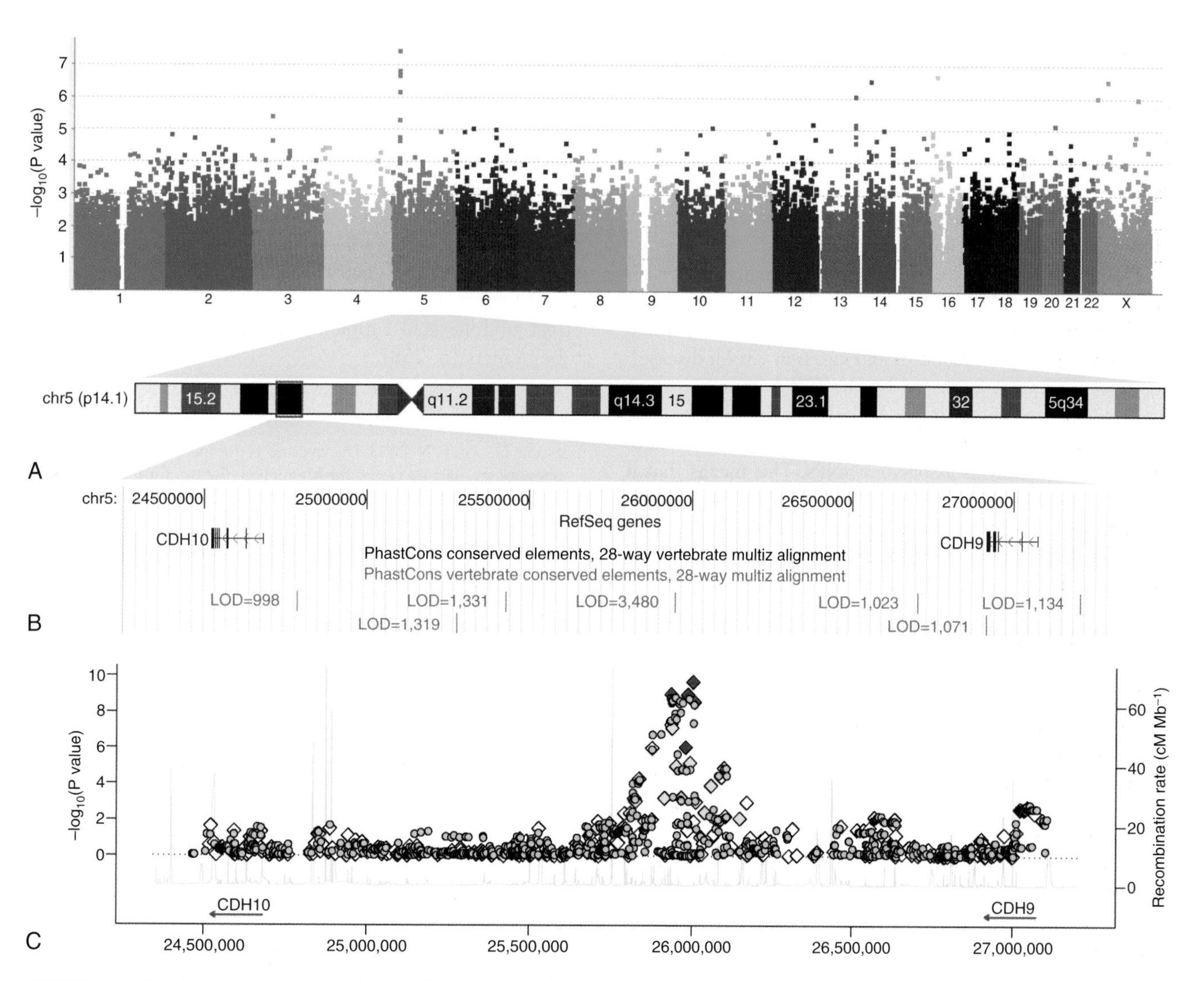

FIGURE 9.7 Results of a genome-wide association study of autism spectrum disorders. **A,** 'Manhattan plot' of –log10 (*P* value) against genomic position. Each data point represents the association between an individual single nucleotide polymorphism (SNP) and autism. SNPs are ordered according to their position in the genome and each chromosome is coloured differently. The higher the position on the y-axis, the stronger the evidence for association. SNPs on chromosome 5p14.1 show the strongest associations. **B,** The 5p14.1 genomic region as displayed in the UCSC genome browser (http://genome.ucsc.edu/). **C,** Zooming in on the 5p14.1 region: Both genotyped SNPs (diamonds) and imputed SNPs (inferred from linkage disequilibrium with genotyped SNPs; grey circles) are plotted with –log10 (*P* value) (y-axis) against genomic position (x-axis). Genotyped SNPs are colored on the basis of their correlation with the most strongly associated SNP (*red* = high, *yellow* = medium, *white* = low). Estimated recombination rates from HapMap data are plotted to reflect the local linkage disequilibrium structure. (From Wang K, et al 2009 Nature 459:528–533, with permission.)

It has been important to develop new statistical criteria for GWA studies. If we were to perform a statistical test of association comparing the frequency of one SNP between cases and controls, we might interpret a P value of <.05 as being unlikely to have occurred by chance. However, when testing associations with increasing numbers of SNPs, the P value threshold needs to change: 1 in 20 tests will have a P value <.05 just by chance. Based on HapMap European data, there are approximately 1 million common SNPs in the genome that are independent (i.e., in very low linkage disequilibrium with all others). Therefore, a comprehensive GWA study of common variants is equivalent to testing approximately 1 million hypotheses. Consequently, in GWA studies, $P = 5 \times 10^{-8}$ is the accepted threshold below which an association is unlikely to be a false positive. Large sample sizes are needed to achieve such low P values, and meta-analysis of two or more studies is a common approach to enlarge the sample size. Dense SNP data can be used to identify population stratification in GWA studies. For example, if an individual shows allele frequency differences from the rest of the study sample at thousands of SNPs, this may indicate that they are of different ancestry may lead to their exclusion from the study.

Despite the success of GWA studies, many challenges remain. To date, the associations identified only explain a small fraction of the heritability of each disease studied (e.g., <10% in type 2 diabetes and <20% in Crohn disease). Rarer variants, not captured by the GWA approach, may explain some of this missing heritability. In addition, the loci identified generally range from 10 to 100 kb in length and include numerous associated SNPs. This means that it has not been possible in most cases to identify the causal variants or even the causal genes. Further techniques, including resequencing of the associated regions, will be necessary to understand the associations fully.

Thousand Genomes Project (www.1000genomes.org)

The thousand genomes project is a new, large-scale initiative, launched in 2008, which aims to extend the publicly available catalogue of human variation by sequencing the genomes of 1000 people from around the world. This will provide an accurate map of alleles with frequencies as low as 1% and will capture not only SNPs but other types of variation including copy number polymorphisms (duplications and deletions). Together with ongoing GWA studies and improvements in high throughput sequencing technology, the thousand genomes project promises new insights into our understanding of the genetics of multifactorial disease in the next few years.

Conclusion

The term multifactorial has been coined to describe the pattern of inheritance displayed by a large number of common disorders that show familial clustering and that are probably caused by the interaction of genetic with environmental factors. The genetic mechanisms underlying these disorders are not well understood. The liability/threshold model should be viewed as an attractive hypothesis rather than as proven scientific fact.

Research in molecular biology is beginning to unravel some of the mysteries of multifactorial inheritance. Technological developments in SNP genotyping, together with an increased understanding of genetic variation, have enabled GWA studies to uncover many new susceptibility loci for polygenic diseases. Examples of progress to date are described in Chapter 15.

This emphasis on the underlying genetic contribution to multifactorial disorders should not in any way detract from the importance of trying to identify major environmental causal factors. This is amply demonstrated by the beneficial effect of folic acid supplementation in preventing neural tube defects (p. 258).

FURTHER READING

Botstein D, Risch N 2003 Discovering genotypes underlying human phenotypes: past successes for Mendelian disease, future approaches for complex disease. Nature Genet Suppl 33:228–237

Comprehensive review article that suggests future strategies for identifying genes underlying complex disease.

Falconer DS 1965 The inheritance of liability to certain diseases estimated from the incidence among relatives. Ann Hum Genet 29:51–76

The original exposition of the liability/threshold model and how correlations between relatives can be used to calculate heritability.

Fraser FC 1980 Evolution of a palatable multifactorial threshold model. Am J Hum Genet 32:796–813

An amusing and 'reader-friendly' account of models proposed to explain multifactorial inheritance.

McCarthy MI, Abecasis GR, Cardon LR, et al 2008 Genome-wide association studies for complex traits: consensus, uncertainty and challenges. Nat Rev Genet 9:356–369

Detailed review article on genome-wide association studies, which gives a comprehensive overview of the methods and highlights the various challenges which still need to be addressed in the search for complex disease genes.

ELEMENTS

1 The concept of multifactorial inheritance has been proposed to account for the common congenital malformations and acquired disorders that show non-Mendelian familial aggregation. These disorders are thought to result from the interaction of genetic and environmental factors.

2 Human characteristics such as height and intelligence, which show a normally distributed continuous distribution in the general population, are probably caused by the additive effects of many genes (i.e. polygenic inheritance).

3 According to the liability/threshold model for multifactorial inheritance the population's genetic and environmental susceptibility, which is known as liability, is normally distributed. Individuals are affected if their liability exceeds a threshold superimposed on the liability curve.

4 Recurrence risks to relatives for multifactorial disorders are influenced by the disease severity, the degree of relationship to the index case, the number of affected close relatives and, if there is a higher incidence in one particular sex, the sex of the index case.

5 Heritability is a measure of the proportion of the total variance of a character or disease that is due to the genetic variance.

6 Loci that contribute to susceptibility for multifactorial disorders can be identified by (a) a search for disease associations with variants in candidate genes; (b) linkage analysis, for example, looking for chromosomal regions that are identical by descent in affected sibling pairs; and (c) genome-wide association studies to compare genetic variation across the entire genome in large case-control studies. Approach (c) has been by far the most successful to date.

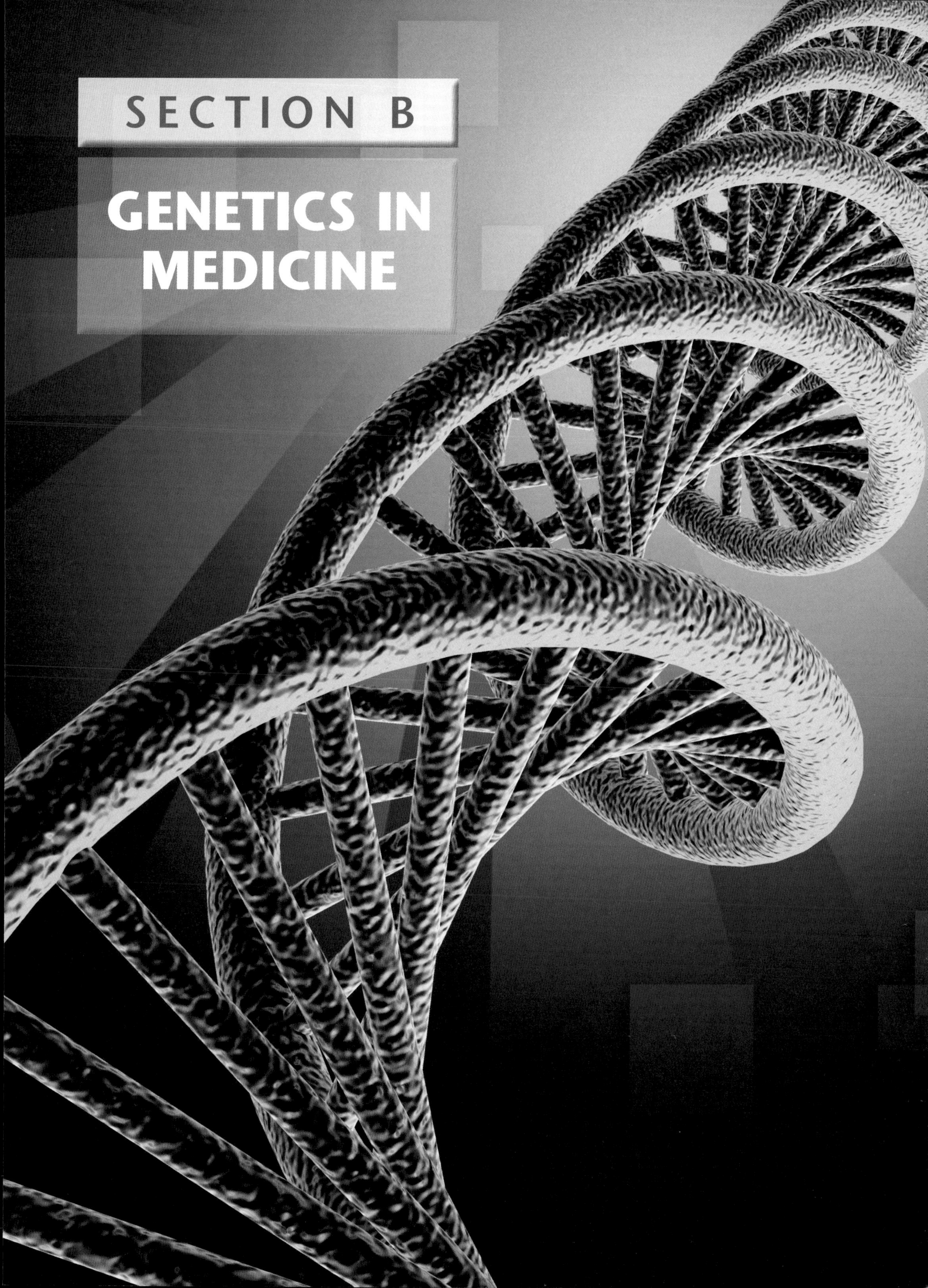
SECTION B
GENETICS IN MEDICINE

CHAPTER 10

Hemoglobin and the Hemoglobinopathies

Blood is a very special juice.

JOHANN WOLFGANG VON GOETHE, IN *FAUST I* (1808)

At least a quarter of a million people are born in the world each year with one of the disorders of the structure or synthesis of hemoglobin (Hb), the *hemoglobinopathies*. The hemoglobinopathies therefore have the greatest impact on morbidity and mortality of any single group of disorders following mendelian inheritance. The mobility of modern society means that new communities with a high frequency of hemoglobinopathies have become established in countries whose indigenous populations have a low frequency. Awareness of this group of disorders is therefore important and many countries have introduced screening programs. In England and Wales, there are an estimated 600,000 healthy carriers of Hb variants. It is noteworthy that the hemoglobinopathies have served as a paradigm for our understanding of the pathology of inherited disease at the clinical, protein, and DNA levels.

To understand the various hemoglobinopathies and their clinical consequences, it is first necessary to consider the structure, function, and synthesis of Hb.

Structure of Hb

Hb is the protein present in red blood cells that is responsible for oxygen transport. There are about 15 grams of Hb in every 100 ml of blood, making it amenable to analysis.

Protein Analysis

In 1956, by fractionating the peptide products of digestion of human Hb with the proteolytic enzyme, trypsin, Ingram found 30 discrete peptide fragments. Trypsin cuts polypeptide chains at the amino acids arginine and lysine. Analysis of the 580 amino acids of human Hb had previously shown there to be a total of 60 arginine and lysine residues, suggesting that Hb was made up of two identical peptide chains with 30 arginine and lysine residues on each chain.

At about the same time, a family was reported in which two hemoglobin variants, Hb S and Hb Hopkins II, were both present in some family members. Several members of the family who possessed both variants had children with normal Hb, offspring who were heterozygous for only one Hb variant, as well as offspring who, like their parents, were doubly heterozygous for the two Hb variants. These observations provided further support for the suggestion that at least two different genes were involved in the production of human Hb.

Shortly thereafter, the amino-terminal amino acid sequence of human Hb was determined and showed valine–leucine and valine–histidine sequences in equimolar proportions, with two moles of each of these sequences per mole of Hb. This was consistent with human Hb being made up of a tetramer consisting of two pairs of different polypeptides referred to as the α- and β-globin chains.

Analysis of the iron content of human Hb revealed that iron constituted 0.35% of its weight, from which it was calculated that human Hb should have a minimum molecular weight of 16,000 Da. In contrast, determination of the molecular weight of human Hb by physical methods gave values of the order of 64,000 Da, consistent with the suggested *tetrameric* structure, $\alpha_2\beta_2$, with each of the globin chains having its own iron-containing group—*heme* (Figure 10.1).

Subsequent investigators demonstrated that Hb from normal adults also contained a minor fraction, constituting 2% to 3% of the total Hb, with an electrophoretic mobility different from the majority of human Hb. The main component was called Hb A, whereas the minority component was called Hb A_2. Subsequent studies revealed Hb A_2 to be a tetramer of two normal α chains and two other polypeptide chains whose amino-acid sequence resembled most closely the β chain and was designated delta (δ).

Developmental Expression of Hemoglobin

Analysis of Hb from a human fetus revealed it to consist primarily of a Hb with a different electrophoretic mobility from normal Hb A, and was designated fetal Hb or Hb F. Subsequent analysis showed Hb F to be a tetramer of two α chains and two polypeptide chains whose sequence resembled the β chain and which were designated gamma (γ). Hb F makes up somewhere in the region of 0.5% of hemoglobin in the blood of normal adults.

Analysis of Hb from embryos earlier in gestation revealed a developmental, or ontological, succession of different embryonic Hbs: Hb Gower I and II, and Hb Portland, which are produced transiently in varying amounts at different gestational ages. These are in tetramers of various

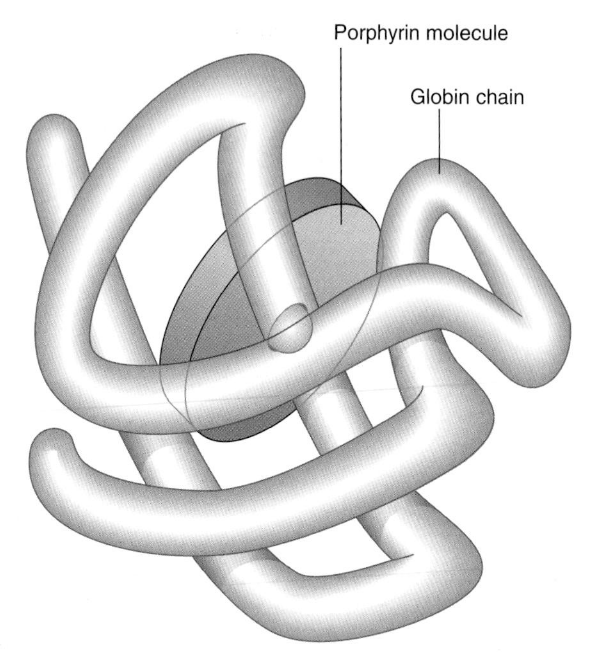

FIGURE 10.1 Diagrammatic representation of one of the globin chains and associated porphyrin molecule of human hemoglobin.

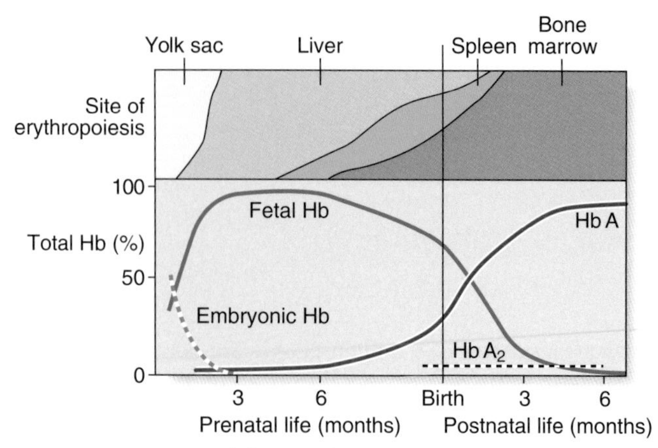

FIGURE 10.2 Hemoglobin synthesis during prenatal and postnatal development. There are several embryonic hemoglobins. (After Huehns ER, Shooter EM 1965 Human haemoglobins. J Med Genet 2:48–90, with permission.)

combinations of α, or α-like, zeta (ζ) chains with β, or β-like, γ- and epsilon (ε) chains (Table 10.1). Although both the ζ chain and ε chain are expressed transiently in early embryonic life, the α chain and γ chain are expressed throughout development, with increasing levels of expression of the β chain toward the end of fetal life (Figure 10.2).

Globin Chain Structure

Analysis of the structure of the individual globin chains was initially carried out at the protein level.

Protein Studies

Amino acid sequencing of the various globin polypeptides in the 1960s showed that the α chain was 141 amino acids long compared with the β chain's 146 amino acids. Their amino acid sequences were similar but by no means identical. The sequence of the δ chain differs from the β chain by 10 amino acids, and analysis of the γ chain showed that it also most closely resembles the β chain, differing by 39 amino acids. In addition, it was found that there were two types of HbF, in which the γ chain contain either the amino acid glycine or alanine at position 136, designated (G)γ and (A)γ, respectively. Partial sequence analyses of the ζ and ε chains of embryonic Hb suggest that ζ is similar in amino acid sequence to the α chain, whereas ε resembles the β chain.

Thus, there are two groups of globin chains, the α-like and β-like, possibly derived from an ancestral Hb gene that has undergone a number of gene duplications and diverged over evolutionary time.

Table 10.1 Human Hemoglobins

Stage in Development	Hemoglobin	Structure	Proportion in Normal Adult (%)
Embryonic	Gower I	$\zeta_2\varepsilon_2$	—
	Gower II	$\alpha_2\varepsilon_2$	—
	Portland I	$\zeta_2\gamma_2$	—
Fetal	F	$\alpha_2\gamma_2$	<1
Adult	A	$\alpha_2\beta_2$	97–98
	A_2	$\alpha_2\delta_2$	2–3

Globin Gene Mapping

The first evidence for the arrangement of the various globin structural genes on the human chromosomes was provided by analysis of the Hb electrophoretic variant, Hb Lepore. Comparison of trypsin digests of Hb Lepore with Hb from normal people revealed that the α chains were normal, whereas the non-α chains appeared to consist of an amino-terminal δ-like sequence and a carboxy-terminal β-like sequence. It was therefore proposed that Hb Lepore could represent a 'fusion' globin chain that had arisen as a result of a crossover coincidental with mispairing of the δ- and β-globin genes during meiosis as a result of the sequence similarity of the two genes and the close proximity of the δ- and β-globin genes on the same chromosome (Figure 10.3). If this hypothesis was correct, it was argued that there should also be an 'anti-Lepore' Hb (i.e., a β–δ-globin fusion product in which the non-α-globin chains contained β-chain residues at the amino-terminal end and δ-chain residues at the carboxy-terminal end). In the late 1960s, a new Hb electrophoretic variant, Hb Miyada, was identified in Japan, whereby analysis of trypsin digests showed it to contain β-globin sequence at the amino-terminal end and δ-globin sequence at the carboxy-terminal end, as predicted.

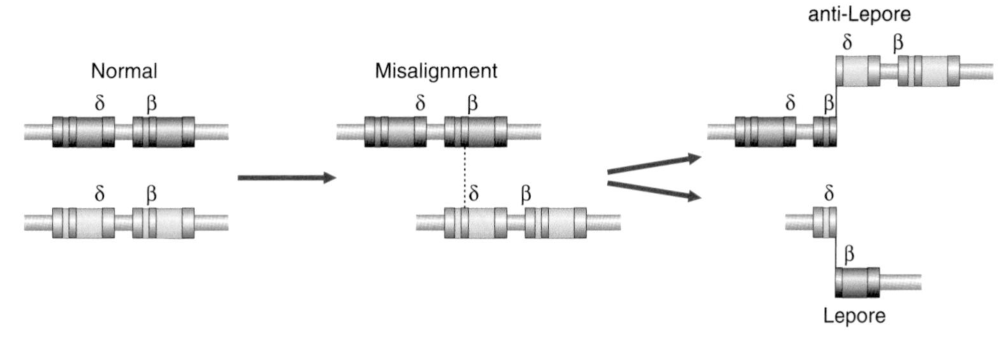

FIGURE 10.3 Mechanism of unequal crossing over which generates Hb Lepore and anti-Lepore. (Adapted from Weatherall DJ, Clegg JB 1981 The thalassaemia syndromes. Blackwell, Oxford.)

Further evidence at the protein level for the physical mapping of the human globin genes was provided by the report of another Hb electrophoretic variant, Hb Kenya. Amino acid sequence analysis suggested it was a γ–β fusion product with a crossover having occurred somewhere between amino acids 81 and 86 in the two globin chains. For this fusion polypeptide to have occurred, it was argued that the γ-globin structural gene must also be in close physical proximity to the β-globin gene.

Little evidence was forthcoming from protein studies about the mapping of the α-globin genes. The presence of normal Hb A in individuals who, from family studies, should have been homozygous for a particular α chain variant, or obligate compound (double) heterozygotes (p. 114), suggested there could be more than one α-globin gene. In addition, the proportion of the total Hb made up by the α chain variant, in those heterozygous for those variants, was consistently lower (less than 20%) than that seen with the β chain variants (usually more than 30%), suggesting there could be more than one α-globin structural gene.

Globin Gene Structure

The detailed structure of globin genes has been made possible by DNA analysis. Immature red blood cells, reticulocytes, provide a rich source of globin messenger RNA (mRNA) for the synthesis of complementary DNA (cDNA)—reticulocytes synthesize little else! Use of β-globin cDNA for restriction mapping studies of DNA from normal individuals revealed that the non-α, or β-like, globin genes are located in a 50-kilobase (kb) stretch on the short arm of chromosome 11 (Figure 10.4). The entire sequence of this 50-kb stretch containing the various globin structural genes is known. Of interest are non-functional regions with sequences similar to those of the globin structural genes—i.e., they produce no identifiable message or protein product and are pseudogenes.

Studies of the α-globin structural genes have shown that there are two α-globin structural genes—α_1 and α_2—located on chromosome 16p (see Figure 10.4). DNA sequencing has revealed nucleotide differences between these two structural genes even though the transcribed α-globin chains have an identical amino acid sequence—evidence for 'degeneracy' of the genetic code. In addition, there are pseudo-α, pseudo-ζ, and ζ genes to the 5′ side of the α-globin genes, as well as an additional theta (θ)-globin gene to the 3′ side of the α_1-globin gene. The θ-globin gene, whose function is unknown, is interesting because, unlike the globin pseudogenes, which are not expressed, its structure is compatible with expression. It has been suggested that it could be expressed in very early erythroid tissue such as the fetal liver and yolk sac.

Synthesis and Control of Hemoglobin Expression

Translation studies with reticulocyte mRNA have shown that α- and β-globin chains are synthesized in roughly equal proportions. In vitro studies have shown, however, that β-globin mRNA is slightly more efficient in protein synthesis than α-globin mRNA, and this difference is compensated for in red blood cell precursors by a relative excess of α-globin mRNA. The most important level of regulation of expression of the globin genes, as with other eukaryotic genes, appears to occur at the level of transcription (p. 18).

DNA studies of the β-globin genes and flanking regions have revealed that, in addition to promoter sequences in the 5′ flanking regions of the various globin genes, there are sequences 6 to 20 kb 5′ to the ε-globin gene necessary for the expression of the various β-like globin genes. This region has been called the **locus control region** (**lcr**), and is involved in the timing and tissue specificity of expression, or **switching**, of the β-like globin genes in development. There is a similar region 5′ to the α-globin genes involved in the control of their expression. Both are involved in the binding of proteins and transcription factors involved in the control of expression of the globin genes.

Disorders of Hemoglobin

The disorders of human Hb can be divided into two main groups: (1) structural globin chain variants, such as sickle-cell disease, and (2) disorders of synthesis of the globin chains, the thalassemias.

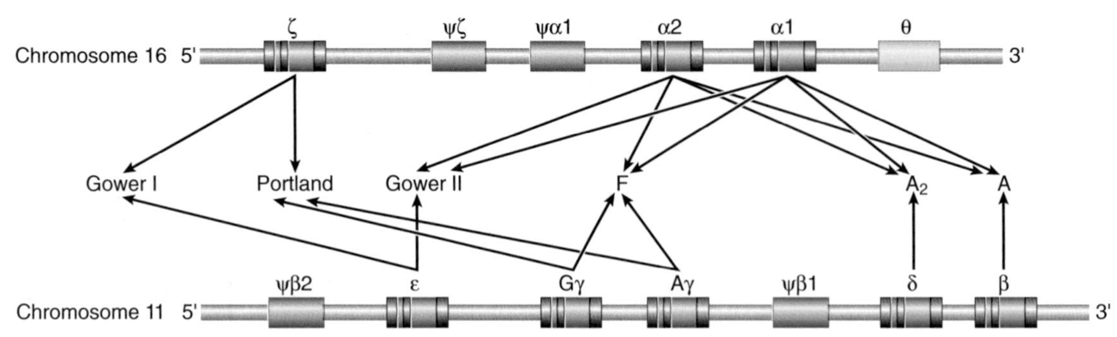

FIGURE 10.4 The α- and β-globin regions on chromosomes 16 and 11 showing the structural genes and pseudogenes (ψ) and the various hemoglobins produced. (Adapted from Carrell RW, Lehman H 1985 The haemoglobinopathies. In: Dawson AM, Besser G, Compston N eds. Recent advances in medicine 19, pp. 223–225. Churchill Livingstone, Edinburgh, UK.)

Structural Variants/Disorders

In 1975, Ingram demonstrated that the difference between Hb A and Hb S lay in the substitution of valine for glutamic acid in the β chain. Since then, more than 300 Hb electrophoretic variants have been described due to a variety of types of mutation (Table 10.2). Some 200 of these electrophoretic variants are single amino acid substitutions resulting from a point mutation. The majority are rare and not associated with clinical disease. A few are associated with disease and relatively prevalent in certain populations.

Types of Mutation

Point Mutation

A point mutation that results in substitution of one amino acid for another can lead to an altered hemoglobin, such as Hb S, C, or E, which are missense mutations (p. 23).

Deletion

There are a number of Hb variants in which one or more amino acids of one of the globin chains is missing or deleted (p. 24) (e.g., Hb Freiburg).

Insertion

Conversely, there are variants in which the globin chains are longer than normal because of insertions (p. 24), such as Hb Grady.

Frameshift Mutation

Frameshift mutations involve disruption of the normal triplet reading frame—i.e., the addition or removal of a number of bases that are not a multiple of three (p. 25). In this instance, translation of the mRNA continues until a termination codon is read 'in frame'. These variants can result in either an elongated or a shortened globin chain.

Table 10.2 Structural Variants of Hemoglobin

Type of Mutation	Examples	Chain/Residue(s)/Alteration
Point (>200 variants)	Hb S	β, 6 glu to val
	Hb C	β, 6 glu to lys
	Hb E	β, 26 glu to lys
Deletion (shortened chain)	Hb Freiburg	β, 23 to 0
	Hb Lyon	β, 17–18 to 0
	Hb Leiden	β, 6 or 7 to 0
	Hb Gun Hill	β, 92–96 or 93–97 to 0
Insertion (elongated chain)	Hb Grady	α, 116–118 (glu, phe, thr) duplicated
Frameshift (insertion or deletion of multiples other than 3 base pairs)	Hb Tak, Hb Cranston	β*, +11 residues, loss of termination codon, insertion of 2 base pairs in codon 146/147
	Hb Wayne	α*, +5 residues, due to loss of termination codon by single base-pair deletion in codon 138/139
	Hb McKees Rock	β*, –2 residues, point mutation in 145, generating premature termination codon
Chain termination	Hb Constant Spring	α*, +31 residues, point mutation in termination codon
Fusion chain (unequal crossing over)	Hb Lepore/anti-Lepore	Non-α, δ-like residues at N-terminal end and β-like residues at C-terminal end, and vice versa, respectively
	Hb Kenya/anti-Kenya	Non-α, γ-like residues at N-terminal end and β-like residues at C-terminal end, and vice versa, respectively

*Residues are either added (+) or lost (–).

Table 10.3 Functional Abnormalities of Structural Variants of Hemoglobin

Clinical Features	Examples
Hemolytic Anemia	
Sickling disorders	HbS/S, HbS/C disease, or HbS/O (Arab), HbS/D (Punjab), HBS/β-thalassemia, HbS/Lepore
	Other rare homozygous sickling mutations—HbS-Antilles, Hb S-Oman)
Unstable hemoglobin	Hb Köln
	Hb Gun Hill
	Hb Bristol
Cyanosis	
Hemoglobin M (methemoglobinemia)	Hb M (Boston)
	Hb M (Hyde Park)
Low oxygen affinity	Hb Kansas
Polycythemia	
High oxygen affinity	Hb Chesapeake
	Hb Heathrow

Chain Termination

A mutation in the termination codon itself can lead to an elongated globin chain (e.g., Hb Constant Spring).

Fusion Polypeptides

Unequal crossover events in meiosis can lead to structural variant called **fusion polypeptides**, of which Hbs Lepore and Kenya are examples (p. 157).

Clinical Aspects

Some Hb variants are associated with disease, but many are harmless and do not interfere with normal function, having been identified coincidentally in the course of population surveys. The more common that interfere with normal Hb function are shown in Table 10.3.

If the mutation is on the inside of the globin subunits, in close proximity to the heme pockets, or at the interchain contact areas, this can produce an unstable Hb molecule that precipitates in the red blood cell, damaging the membrane and resulting in hemolysis of the cell. Alternatively, mutations can interfere with the normal oxygen transport function of Hb, leading to either enhanced, or reduced, oxygen affinity, or an Hb that is stable in its reduced form, so-called **methemoglobin**.

The structural variants of Hb identified by electrophoretic techniques probably represent a minority of the total number of variants that exist, as it is predicted that only one-third of the possible Hb mutations that could occur will produce an altered charge in the Hb molecule, and thereby be detectable by electrophoresis (Figure 10.5).

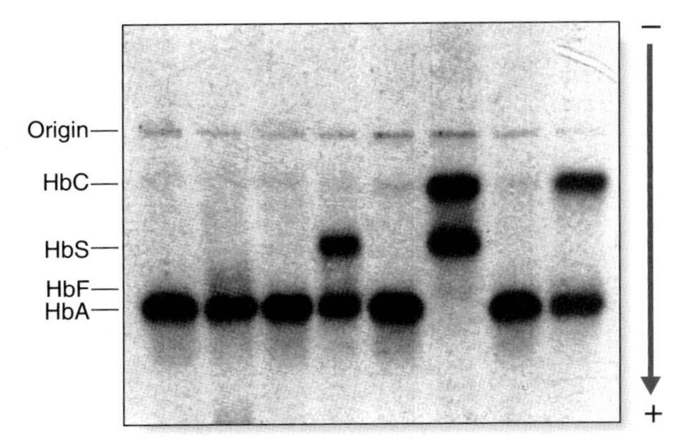

FIGURE 10.5 Hemoglobin electrophoresis showing hemoglobins A, C, and S. (Courtesy Dr. D. Norfolk, General Infirmary, Leeds, UK.)

Sickle Cell Disease

This severe hereditary hemolytic anemia was first recognized clinically early in the twentieth century, but in 1940 red blood cells from affected individuals with sickle cell (SC) disease were noted to appear birefringent when viewed in polarized light under the microscope, reflecting polymerization of the sickle hemoglobin. This distorts the shape of red blood corpuscles under deoxygenated conditions—so-called **sickling** (Figure 10.6). Pauling, in 1949, using electrophoresis, showed that it had different mobility to HbA and called it HbS, for sickle.

Clinical Aspects of SC Disease

SC disease, following autosomal recessive inheritance, is the most common hemoglobinopathy; in the United Kingdom, about 310,000 individuals are carriers, and approximately 400 pregnancies are affected annually. The disease is especially prevalent in those areas of the world where malaria is endemic. The parasite *Plasmodium falciparum* is disadvantaged because the red cells of SC heterozygotes are believed to express malarial or altered self-antigens more effectively,

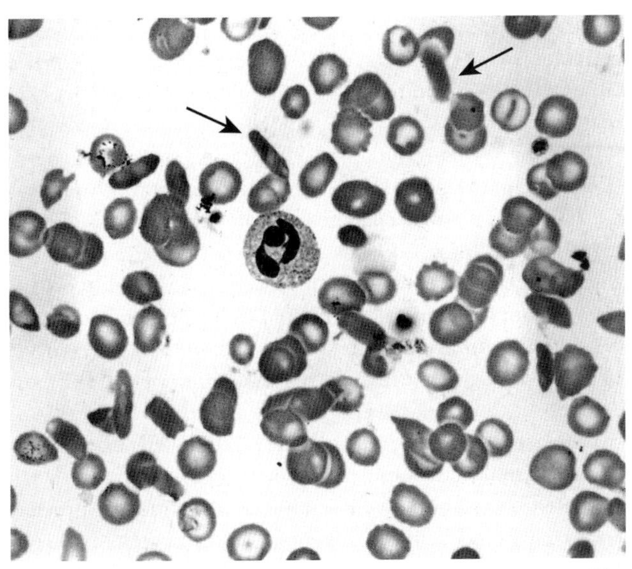

FIGURE 10.6 Blood film showing sickling of red cells in sickle-cell disease. Sickled cells are arrowed. (Courtesy Dr. D. Norfolk, General Infirmary, Leeds, UK.)

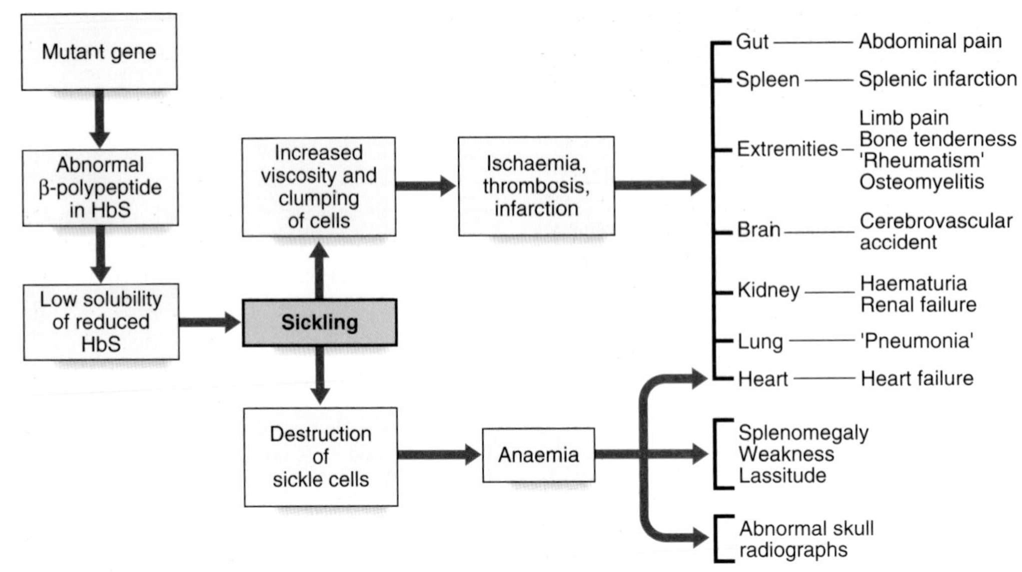

FIGURE 10.7 The pleiotropic effects of the gene for sickle-cell disease.

resulting in more rapid removal of parasitized cells from the circulation. SC heterozygotes are therefore relatively protected from malarial attacks. They are therefore biologically fitter, the SC gene can be passed on to the next generation, and over time this has resulted in relatively high gene frequency in malarial-infested regions (see Chapter 8). The clinical manifestations are manifold and include painful **sickle cell crisis**, chest crisis, aplastic crisis, splenic sequestration crisis, priapism, retinal disease, and cerebrovascular accident. Pulmonary hypertension may occur and heart failure can accompany severe anemia during aplastic or splenic sequestration crises. All of these are the result of deformed, sickle-shaped red cells, which are less able to change shape and tend to obstruct small arteries, thus reducing oxygen supply to the tissues (Figure 10.7). Sickled cells, with damaged cell membranes, are taken up by the reticuloendothelial system. The shorter red cell survival time leads to a more rapid red cell turnover and, consequently, anemia.

Sickling crises reduce life expectancy, so early recognition and treatment of the complications is vital. Prophylactic penicillin to prevent the risk of overwhelming sepsis from splenic infarction has been successful and increased survival. The other beneficial approach is the use of **hydroxyurea**, a simple chemical compound that can be taken orally. Once-daily administration has been shown to increase levels of HbF through pharmacological induction. The HbF percentage has been shown to predict the clinical severity of SC disease, preventing intracellular sickling, which decreases vasoocclusion and hemolysis. It has been suggested that a potential threshold of 20% HbF is required to prevent recurrent vasoocclusive events. Hydroxyurea is well-tolerated, safe, and has many features of an ideal drug. The US Food and Drug Administration approved hydroxyurea for adult patients with clinically severe some years ago, but it has been used only sparingly. There is ongoing debate about its wider use in less severe cases and children.

SC Trait

The heterozygous, or carrier, state for the SC allele is known as **sickle cell trait** and in general is not associated with any significant health risk. However, there may be a small increased risk of sudden death associated with strenuous exercise, possible risks from hypoxia on airplane flights, and anesthesia in pregnant women who are carriers.

Mutational Basis of SC Disease

The amino acid valine, at the sixth position of the β-globin chain, is substituted by glutamic acid. The mutation is therefore a single base-pair in the triplet code at this point, from G<u>A</u>G to G<u>T</u>G. Detecting this change in the laboratory used to be undertaken using a restriction enzyme, *Mst*II, for which the nucleotide recognition sequence is abolished by the point mutation in Hb S, resulting in an altered restriction fragment length polymorphism on a gel. This technique has been superseded by polymerase chain reaction.

Disorders of Hemoglobin Synthesis

The **thalassemias** are the commonest single group of inherited disorders in humans, occurring in persons from the Mediterranean region, Middle East, Indian subcontinent, and Southeast Asia. They are heterogeneous and classified according to the particular globin chain, or chains, synthesized in reduced amounts (e.g., α-, α-, δβ-thalassemia).

There are similarities in the pathophysiology of all forms of thalassemia, though excessive α chains are more haemolytic than excessive β chains. An imbalance of globin-chain production results in the accumulation of free globin chains in the red blood cell precursors which, being insoluble,

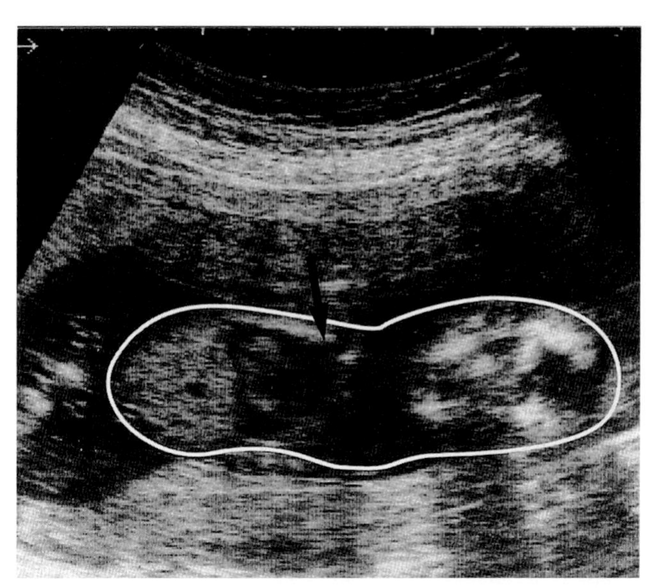

FIGURE 10.8 Longitudinal ultrasonographic scan of a coronal section of the head *(to the right)* and thorax of a fetus with hydrops fetalis from the severe form of α-thalassemia, Hb Barts, showing a large pleural effusion (*arrows*). (Courtesy Mr. J. Campbell, St. James's Hospital, Leeds, UK.)

precipitate, resulting in hemolysis of the red blood cells (i.e., a hemolytic anemia). The consequence is compensatory hyperplasia of the bone marrow.

α-Thalassemia

This results from underproduction of the α-globin chains and occurs most commonly in Southeast Asia but is also prevalent in the Mediterranean, Middle East, India, and sub-Saharan Africa, with carrier frequencies ranging from 15% to 30%. There are two main types of α-thalassemia, with different severity: the severe form, in which no α chains are produced, is associated with fetal death due to massive edema secondary to heart failure from severe anemia—*hydrops fetalis* (Figure 10.8). Analysis of Hb from such fetuses reveals a tetramer of γ chains, originally called Hb Barts.

In the milder forms of α-thalassemia compatible with survival, although some α chains are produced, there is still a relative excess of β chains, resulting in production of the β-globin tetramer Hb H—known as **Hb H disease**. Both Hb Barts and Hb H globin tetramers have an oxygen affinity similar to that of myoglobin and do not release oxygen as normal to peripheral tissues. Also, Hb H is unstable and precipitates, resulting in hemolysis of red blood cells.

Mutational Basis of α-Thalassemia

The absence of α chain synthesis in hydropic fetuses, and partial absence in Hb H disease, was confirmed using quantitative mRNA studies from reticulocytes. Studies comparing the quantitative hybridization of radioactively labeled α-globin cDNA to DNA from hydropic fetuses, and in Hb H disease, were consistent with the α-globin genes being deleted. Restriction mapping studies of the α-globin region revealed two α-globin structural genes on chromosome 16p. The various forms of α-thalassemia are to be mostly the result of deletions of one or more of these structural genes (Figure 10.9). These deletions are thought to have arisen as a result of unequal crossover events in meiosis, more likely to occur where genes with homologous sequences are in close proximity. Support for this hypothesis comes from the finding of the other product of such an event (i.e., individuals with three α-globin structural genes located on one chromosome).

These observations resulted in the recognition of two other milder forms of α-thalassemia that are not associated with anemia and can be detected only by the transient presence of Hb Barts in newborns. Mapping studies of the α-globin region showed that these milder forms of α-thalassemia are due to the deletion of one or two of the α-globin genes. Occasionally, non-deletion point mutations in the α-globin genes, as well as the 5′ transcriptional region, have been found to cause α-thalassemia.

An exception to this classification of α-thalassemias is the Hb variant Constant Spring, named after the town in the United States from which the original patient came. This was detected as an electrophoretic variant in a person with Hb H disease. Hb Constant Spring is due to an abnormally long α chain resulting from a mutation in the normal termination codon at position 142 in the α-globin gene. Translation of α-globin mRNA continues until another termination codon is reached, resulting in an abnormally long α-globin chain. The abnormal α-globin mRNA molecule is also unstable, leading to a relative deficiency of α chains and the presence of the β-globin tetramer, Hb H.

β-Thalassemia

By now the reader will deduce that this is caused by underproduction of the β-globin chain of Hb. Production of

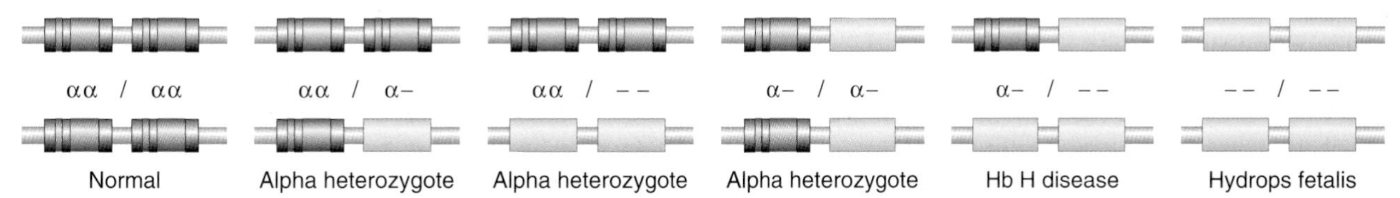

FIGURE 10.9 Structure of the normal and deleted α-globin structural genes in the various forms of α-thalassemia. (Adapted from Emery AEH 1984 An introduction to recombinant DNA. John Wiley, Chichester.)

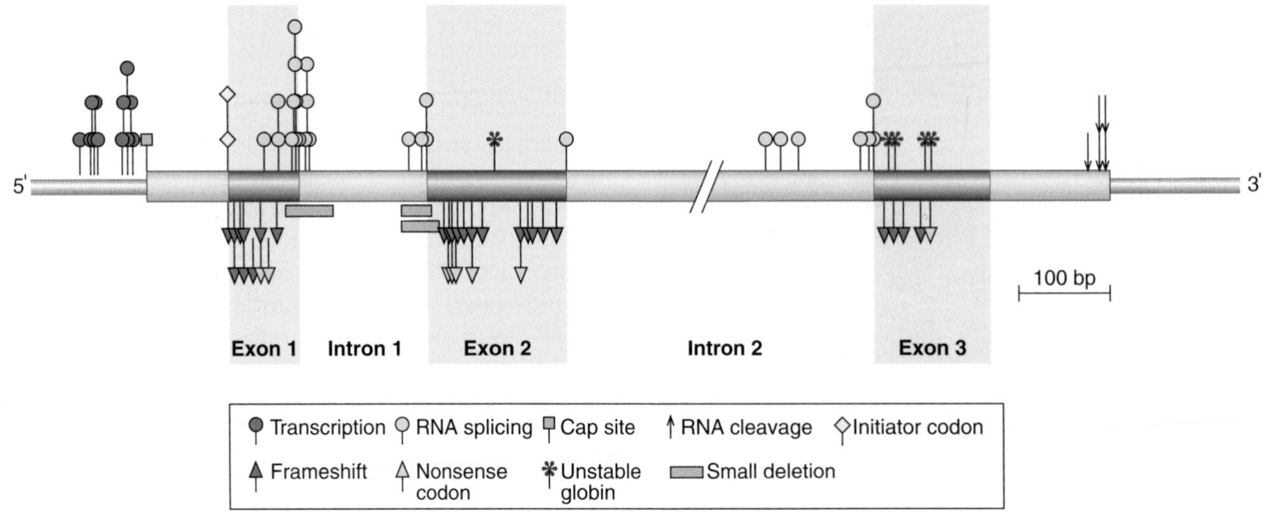

FIGURE 10.10 Location and some of the types of mutation in the β-globin gene and flanking region that result in β-thalassemia. (Adapted from Orkin SH, Kazazian HH 1984 The mutation and polymorphism of the human β-globin gene and its surrounding DNA. Annu Rev Genet 18:131–171.)

β-globin chains may be either reduced ($β^+$) or absent ($β^0$). Individuals homozygous for $β^0$-thalassemia mutations have severe, transfusion-dependent anemia. Approximately 1:1000 Northern Europeans are β-thalassemia carriers and in United Kingdom, on average, 22 babies with $β^0$ thalassemia are born annually and roughly 860 people live with the condition; there are an estimated 327,000 carriers.

Mutational Basis of β-Thalassemia

β-Thalassemia is rarely the result of gene deletion and DNA sequencing is often necessary to determine the molecular pathology. In excess of 100 different mutations have been shown to cause β-thalassemia, including point mutations, insertions, and base-pair deletions. These occur within both the coding and non-coding portions of the β-globin genes as well as the 5′ flanking promoter region, the 5′ capping sequences (p. 19) and the 3′ polyadenylation sequences (p. 19) (Figure 10.10). The various mutations are often unique to certain population groups and can be considered to fall into six main functional types.

Transcription mutations. Mutations in the 5′ flanking TATA box or the promoter region of the β-globin gene can result in reduced transcription levels of the β-globin mRNA.

mRNA splicing mutations. Mutations involving the invariant 5′ GT or 3′ AG dinucleotides of the introns in the β-globin gene or the consensus donor or acceptor sequences (p. 19) result in abnormal splicing with consequent reduced levels of β-globin mRNA. The most common Mediterranean β-thalassemia mutation leads to the creation of a new acceptor AG dinucleotide splice site sequence in the first intron of the β-globin gene, creating a 'cryptic' splice site (p. 25). The cryptic splice site competes with the normal splice site, leading to reduced levels of the normal β-globin mRNA. Mutations in the coding regions of the β-globin region can also lead to cryptic splice sites.

Polyadenylation signal mutations. Mutations in the 3′ end of the untranslated region of the β-globin gene can lead to loss of the signal for cleavage and polyadenylation of the β-globin gene transcript.

RNA modification mutations. Mutations in the 5′ and 3′ DNA sequences, involved respectively in the capping (p. 19) and polyadenylation (p. 19) of the mRNA, can result in abnormal processing and transportation of the β-globin mRNA to the cytoplasm, and therefore reduced levels of translation.

Chain termination mutations. Insertions, deletions, and point mutations can all generate a nonsense or chain termination codon, leading to premature termination of translation of the β-globin mRNA. Usually this results in a shortened β-globin mRNA that is unstable and more rapidly degraded, again leading to reduced levels of translation of an abnormal β-globin.

Missense mutations. Rarely, missense mutations lead to a highly unstable β-globin (e.g., Hb Indianapolis).

Clinical Aspects of β-Thalassemia

Children with thalassemia major, or Cooley's anemia as it was originally known, usually present in infancy with a severe transfusion-dependent anemia. Unless adequately transfused, compensatory expansion of the bone marrow results in an unusually shaped face and skull (Figure 10.11). Affected individuals used to die in their teens or early adulthood from complications resulting from iron overload from repeated transfusions. However, daily use of iron-chelating drugs, such as desferrioxamine, has greatly improved their long-term survival.

Individuals heterozygous for β-thalassemia—**thalassemia trait** or **thalassemia minor**—usually have no symptoms or signs but do have a mild hypochromic, microcytic anemia. This can easily be confused with iron deficiency anemia.

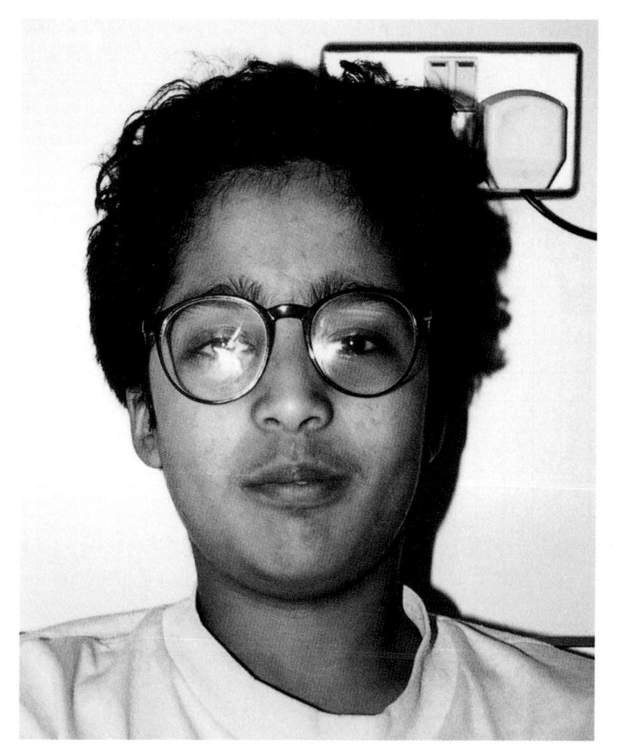

FIGURE 10.11 Facies of a child with β-thalassemia showing prominence of the forehead through changes in skull shape as a result of bone marrow hypertrophy. (Courtesy Dr. D. Norfolk, General Infirmary, Leeds, UK.)

δβ-Thalassemia

In this hemoglobinopathy, there is underproduction of both the δ and β chains. Homozygous individuals produce no δ- or β-globin chains, which one might expect to cause a profound illness. However, they have only mildly anemia because of increased production of γ chains, with Hb F levels being much higher than the mild compensatory increase seen in β^0 thalassemia.

Mutational Basis of δβ-thalassemia

The cause is extensive deletions in the β-globin region involving the δ- and β-globin structural genes (Figure 10.12). Some large deletions include the Aγ-globin gene so that only the Gγ-globin chain is synthesized.

Hereditary Persistence of Fetal Hemoglobin

Hereditary persistence of fetal Hb (HPFH), in which HbF production persists into childhood and beyond, is included in the thalassemias. It is usually a form of δβ-thalassemia in which continued γ-chain synthesis compensates for the lack of δ and β chains. HbF may account for 20% to 30% of total Hb in heterozygotes and 100% in homozygotes. Individuals are usually symptom free.

Mutational Basis of HPFH

Some forms of HPFH are due to deletions of the δ- and β-globin genes, whereas non-deletion forms may have point mutations in the 5′ flanking promoter region of either the Gγ or Aγ globin genes near the CAAT box sequences (pp. 21–22), which are involved in the control of Hb gene expression.

Clinical Variation of the Hemoglobinopathies

The marked mutational heterogeneity of β-thalassemia means that affected individuals are often compound heterozygotes (p. 114), i.e., they have different mutations in their β-globin genes, leading to a broad spectrum of severity, including intermediate forms—**thalassemia intermedia**—which require less frequent transfusions.

Certain areas of the world show a high prevalence of all the hemoglobinopathies and, not unexpectedly, individuals may have two different disorders of Hb. In the past, precise

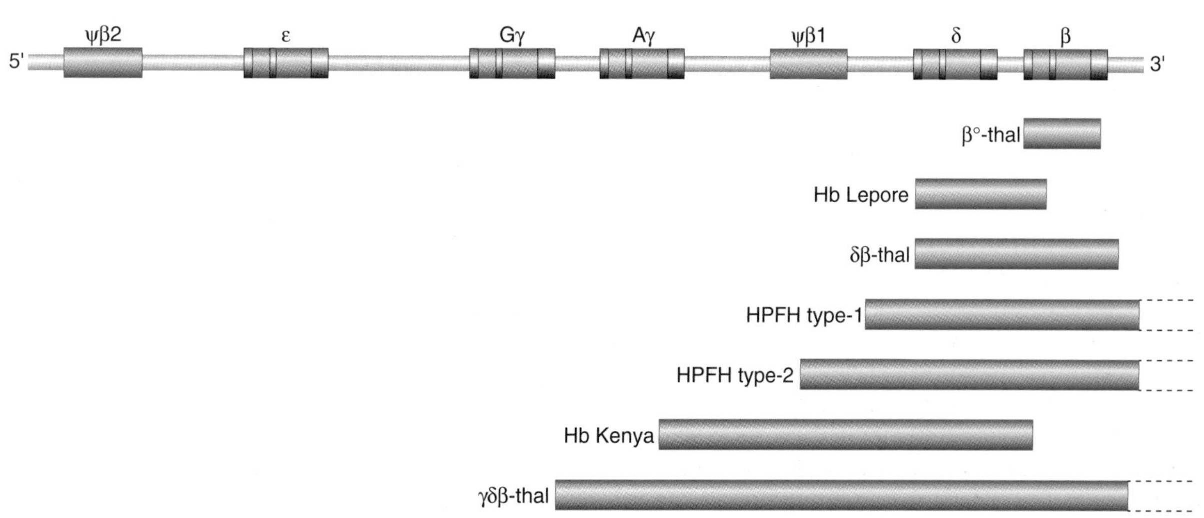

FIGURE 10.12 Some of the deletions in the β-globin region that result in some forms of thalassemia and hereditary persistence of fetal hemoglobin.

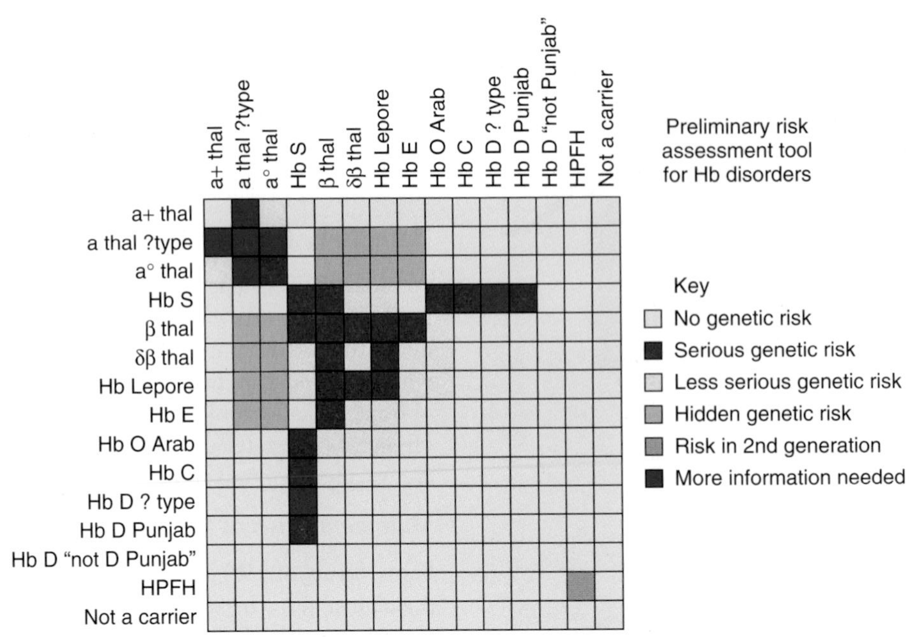

FIGURE 10.13 A hemoglobinopathy tool depicting the anticipated clinical severity associated with the occurrence of different homozygous or compound heterozygous states.

diagnoses were difficult but the arrival of DNA sequencing has greatly helped to solve conundrums—e.g., individuals heterozygous for both Hb S and β-thalassemia (compound heterozygotes; see p. 114). Certain combinations can result in a previously unexplained mild form of what would reasonably be anticipated to be a severe hemoglobinopathy. For example, deletion of one or two of the α-globin genes in a person homozygous for β-thalassemia results in a milder illness because there is less of an imbalance in globin chain production. Similarly, the presence of one form of HPFH in a person homozygous for β-thalassemia or sickle cell can contribute to amelioration of the disease as the increased production of γ-globin chains compensates for the deficient β-globin chain production. The relative severity of different homozygous or compound heterozygous hemoglobinopathy states is helpfully summarized in a risk assessment tool produced by the NHS Sickle Cell and Thalassaemia Screening Programme (Figure 10.13).

Antenatal and Newborn Hemoglobinopathy Screening

In the United Kingdom in 2005, SC and thalassemia screening was introduced in newborns, and in some areas antenatal screening is under way. The purpose is to identify infants with serious hemoglobinopathies at an early stage so that early treatment can be instituted and long-term complications minimized. The genetic risk to future pregnancies is also identified early. The programs also mean that parental testing, cascade screening through the wider family, and genetic counseling, can offered when a carrier infant is identified.

The decision to perform antenatal screening is in some regions guided by the findings of an ethnicity and family history questionnaire administered to the pregnant woman. Initial screening is undertaken on a simple full blood count, looking for anemia (Hb <11 g/dl) and microcytosis MCH (mean corpuscular hemoglobin/0 <27 pg). These findings prompt electrophoresis by high performance liquid chromatography, summarized in Figure 10.14. As with all screening, this program aims to reduce the burden of health care in the long term, which is entirely appropriate for the most common single-gene conditions in humans.

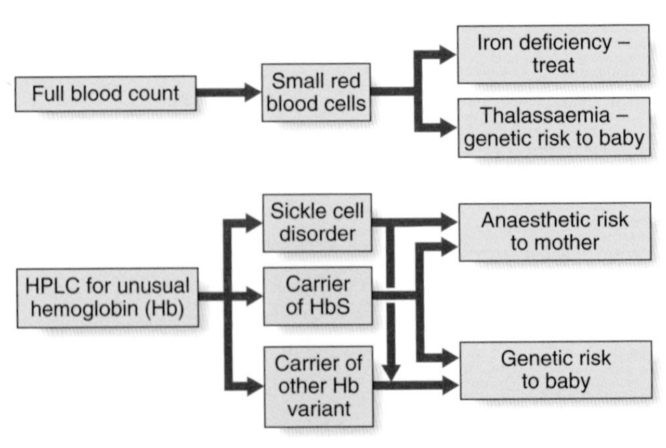

FIGURE 10.14 Antenatal screening for hemoglobinopathies.

FURTHER READING

Cay JC, Phillips JA, Kazazian HH 1996 Haemoglobinopathies and thalassemias. In: Rimoin DL, Connor JM, Pyeritz RE eds. Principles and practice of medical genetics, 3 ed, pp 1599–1626. Edinburgh: Churchill Livingstone

A useful, up-to-date, concise summary of the hemoglobinopathies.

Cooley TB, Lee P 1925 A series of cases of splenomegaly in children with anemia and peculiar bone changes. Trans Am Pediatr Soc 37:29–40

The original description of β-thalassemia.

Pauling L, Itano HA, Singer SJ, Wells IC 1949 Sickle-cell anaemia, a molecular disease. Science 110:543–548

The first genetic disease in which a molecular basis was described, leading to a Nobel Prize.

Serjeant GR 1992 Sickle cell disease, 2 ed. Oxford: Oxford University Press

Excellent, comprehensive text covering all aspects of this important disorder.

Weatherall DJ, Clegg JB, Higgs DR, Woods WG 1995 The hemoglobinopathies. In: Scriver CR, Beaudet AL, Sly WS, Valle D eds. The metabolic and molecular basis of inherited disease, 7 ed, pp. 3417–3484. New York: McGraw Hill

A very comprehensive, detailed account of hemoglobin and the hemoglobinopathies.

ELEMENTS

1. Hemoglobin (Hb), the protein present in red blood cells responsible for oxygen transport, is a tetramer made up of two dissimilar pairs of polypeptide chains and the iron-containing molecule heme.

2. Human Hb is heterogeneous. During development, it comprises a succession of different globin chains that are expressed differentially during embryonic, fetal, and adult life (e.g., $\alpha_2\varepsilon_2$, $\alpha_2\gamma_2$, $\alpha_2\delta_2$, $\alpha_2\beta_2$).

3. The disorders of Hb—the hemoglobinopathies—can be divided into two main groups: the structural disorders, such as sickle-cell Hb or Hb S, and disorders of synthesis, the thalassemias. The former can be subdivided by the way in which they interfere with the normal function of Hb and/or the red blood cell (e.g., abnormal oxygen affinity, hemolytic anemia). The latter can be subdivided according to which globin chain is synthesized abnormally (i.e., α-, β-, or δβ-thalassemia).

4. Family studies of the various disorders of human Hb and analysis of the mutations responsible for these at the protein and DNA levels have led to an understanding of the normal structure, function, and synthesis of Hb. This has allowed demonstration of the molecular pathology of these disorders, and prenatal diagnosis of a number of the inherited disorders of human Hb is possible.

CHAPTER 11

Biochemical Genetics

Life ... is a relationship between molecules.
LINUS PAULING

The existence of chemical individuality follows of necessity from that of chemical specificity, but we should expect the differences between individuals to be still more subtle and difficult of detection.
ARCHIBALD GARROD (1908)

In this chapter, we consider single-gene biochemical or metabolic diseases, including mitochondrial disorders. The range of known disorders is vast, so only an overview is possible, but it is hoped that the reader will gain a flavor of this fascinating area of medicine. At the beginning of the twentieth century, Garrod introduced the concept of 'chemical individuality', leading in turn to the concept of the **inborn error of metabolism** (IEM). Beadle and Tatum later developed the idea that metabolic processes, whether in humans or any other organism, proceed by steps. They proposed that each step was controlled by a particular enzyme and that this, in turn, was the product of a particular gene. This was referred to as the **one gene–one enzyme (or protein)** concept.

Inborn Errors of Metabolism

In excess of 200 IEMs are known that can be grouped by either the metabolite, metabolic pathway, function of the enzyme, or cellular organelle involved (Table 11.1). Most follow autosomal recessive or X-linked recessive inheritance, with only a few being autosomal dominant. This is because the defective protein in most cases is a diffusible enzyme, and there is usually sufficient residual activity in the heterozygous state (loss-of-function, see p. 26) for the enzyme to function normally in most situations. If, however, the reaction catalysed by an enzyme is rate limiting (haploinsufficiency, see p. 26) or the gene product is part of a multimeric complex (dominant-negative, see p. 26), the disorder can manifest in the heterozygous state and follow dominant inheritance (p. 109).

Disorders of Amino Acid Metabolism

There are a number of disorders of amino acid metabolism, the best known of which is phenylketonuria.

Phenylketonuria

Children with phenylketonuria (PKU), if untreated, are severely intellectually impaired and often develop seizures. There is a deficiency of the enzyme required for the conversion of phenylalanine to tyrosine, phenylalanine hydroxylase (PAH)—causing a 'genetic block' in the metabolic pathway (Figure 11.1).

PKU was the first genetic disorder in humans shown to be caused by a specific enzyme deficiency, by Jervis in 1953. As a result of the enzyme defect, phenylalanine accumulates and is converted into phenylpyruvic acid and other metabolites that are excreted in the urine. The enzyme block leads to a deficiency of tyrosine, with a consequent reduction in melanin formation, and children therefore often have blond hair and blue eyes (Figure 11.2). In addition, areas of the brain that are usually pigmented, such as the substantia nigra, may also lack pigment.

Treatment of PKU

An obvious method of treating children with PKU would be to replace the missing enzyme, but this is not simply achieved (p. 349). Bickel, just 1 year after the enzyme deficiency had been identified, suggested that PKU could be treated by removal of phenylalanine from the diet and this has proved effective. If PKU is detected early enough in infancy, intellectual impairment can be prevented by giving a phenylalanine restricted diet. Phenylalanine is an essential amino acid and therefore cannot be removed entirely from the diet. By monitoring the level of phenylalanine in the blood, it is possible to supply sufficient amounts to meet normal requirements but avoiding toxic levels, resulting in mental retardation. After brain development is complete, dietary restriction can be relaxed—from adolescence onward.

The intellectual impairment seen in children with phenylketonuria is likely due to toxic levels of phenylalanine, and/or its metabolites, rather than a deficiency of tyrosine, of which adequate amounts are present in a normal diet. Both prenatal and postnatal factors may be responsible for developmental delay in untreated PKU.

Diagnosis of PKU

PKU affects approximately 1 in 10,000 people in western Europe and was the first IEM routinely screened for in

Table 11.1 Characteristics of Some Inborn Errors of Metabolism

Type of Defect	Genetics	Deficiency	Main Clinical Features
Amino-Acid Metabolism			
Phenylketonuria	AR	Phenylalanine hydroxylase	Mental retardation, fair skin, eczema, epilepsy
Alkaptonuria	AR	Homogentisic acid oxidase	Arthritis
Oculocutaneous albinism	AR	Tyrosinase	Lack of skin and hair pigment, eye defects
Homocystinuria	AR	Cystathione α-synthase	Mental retardation, dislocation of lens, thrombosis, skeletal abnormalities
Maple syrup urine disease	AR	Branched-chain β-ketoacid decarboxylase	Mental retardation
Urea Cycle Disorders			
Carbamyl synthase deficiency	AR	Carbamyl synthase	Hyperammonemia, coma, death
Ornithine carbamyl transferase	XD	Ornithine carbamyl transferase deficiency	Hyperammonemia, death in early infancy
Citrullinemia	AR	Arginosuccinic acid synthase	Variable clinical course
Argininosuccinic aciduria	AR	Arginosuccinic acid lyase	Hyperammonemia, mild mental retardation, protein intolerance
Hyperargininemia	AR	Arginase	Hyperammonemia, progressive spasticity, intellectual deterioration
Carbohydrate Metabolism			
MONOSACCHARIDE METABOLISM			
Galactosemia	AR	Galactose 1-phosphate uridyl transferase	Cataracts, mental retardation, cirrhosis
Hereditary fructose intolerance	AR	Fructose 1-phosphate aldolase	Failure to thrive, vomiting, jaundice, convulsions
Glycogen Storage Diseases			
PRIMARILY AFFECTING LIVER			
von Gierke disease (GSD-I)	AR	Glucose-6-phosphatase	Hepatomegaly, hypoglycemia
Cori disease (GSD-III)	AR	Amylo-1,6-glucosidase	Hepatomegaly, hypoglycemia
Anderson disease (GSD-IV)	AR	Glycogen debrancher enzyme	Abnormal liver function/failure
Hepatic phosphorylase deficiency (GSD-VI)	AR/X-linked	Hepatic phosphorylase	Hepatomegaly, hypoglycemia, failure to thrive
PRIMARILY AFFECTING MUSCLE			
McArdle disease (GSD-V)	AR	Muscle phosphorylase	Muscle cramps
Pompe disease (GSD-II)	AR	Lysosomal α-1,4-glucosidase	Heart failure, muscle weakness
Steroid Metabolism			
Congenital adrenal hyperplasia	AR	21-Hydroxylase, 11β-hydroxylase, 3β-dehydrogenase	Virilization, salt losing
Androgen insensitivity	XR	Androgen receptor	Female external genitalia, testes, male chromosomes
Lipid Metabolism			
Familial hypercholesterolemia	AD	Low-density lipoprotein receptor	Early coronary artery disease
Lysosomal Storage Diseases			
MUCOPOLYSACCHARIDOSES			
Hurler syndrome (MPS-I)	AR	α-L-Iduronidase	Mental retardation, skeletal abnormalities, hepatosplenomegaly, corneal clouding
Hunter syndrome (MPS-II)	XR	Iduronate sulfate sulfatase	Mental retardation, skeletal abnormalities, hepatosplenomegaly
Sanfilippo syndrome (MPS-III)	AR	Heparan-*S*-sulfaminidase (MPS-III A), *N*-ac-α-D-glucosaminidase (MPS-III B), Ac-CoA-α-glucosaminidase, *N*-acetyltransferase (MPS-III C) *N*-ac-glucosaminine-6-sulfate sulfatase (MPS-III D)	Behavioral problems, dementia, fits
Morquio syndrome (MPS-IV)	AR	Galactosamine-6-sulfatase (MPS-IV A), β-galactosidase (MPS-IV B)	Corneal opacities, short stature, skeletal abnormalities
MPS-V (formerly Scheie disease, now known to be a mild allelic form of MPS-I) Maroteaux-Lamy syndrome	AR	Arylsulfatase B, *N*-acetyl-galactosamine, α-4-sulfate sulfatase	Corneal clouding, skeletal abnormalities, cardiac (MPS-VI) abnormalities

Table 11.1 Characteristics of Some Inborn Errors of Metabolism—cont'd

Type of Defect	Genetics	Deficiency	Main Clinical Features
Sly syndrome (MPS-VII)	AR	β-Glucuronidase	Variable presentation, skeletal and cardiac abnormalities, hepatosplenomegaly, corneal clouding, mental retardation
SPHINGOLIPIDOSES			
Tay-Sachs disease	AR	Hexosaminidase-A	Developmental regression, blindness, cherry-red spot, deafness
Gaucher disease	AR	Glucosylceramide β-Glucosidase	Type I—joint and limb pains, splenomegaly Type II—spasticity, fits, death
Niemann-Pick disease	AR	Sphingomyelinase	Failure to thrive, hepatomegaly, cherry-red spot, developmental regression
Purine/Pyrimidine Metabolism			
Lesch-Nyhan disease	XR	Hypoxanthine guanine phosphoribosyltransferase	Mental retardation, uncontrolled movements, self-mutilation
Adenosine deaminase deficiency	AR	Adenosine deaminase	Severe combined immunodeficiency
Purine nucleoside phosphorylase	AR	Purine nucleoside phosphorylase	Severe viral infections due to impaired T-cell function
Hereditary orotic aciduria	AR	Orotate phosphoribosyltransferase, orotidine 5′-phosphate decarboxylase	Megaloblastic anemia, failure to thrive, developmental delay
Porphyrin Metabolism			
HEPATIC PORPHYRIAS			
Acute intermittent porphyria (AIP)	AD	Uroporphyrinogen I synthetase	Abdominal pain, CNS effects
Hereditary coproporphyria	AD	Coproporphyrinogen oxidase	As for AIP, photosensitivity
Porphyria variegata	AD	Protoporphyrinogen oxidase	Photosensitivity, as for AIP
ERYTHROPOIETIC PORPHYRIAS			
Congenital erythropoietic porphyria	AR	Uroporphyrinogen III synthase	Hemolytic anemia, photosensitivity
Erythropoietic protoporphyria	AD	Ferrochelatase	Photosensitivity, liver disease
Organic-Acid Disorders			
Methylmalonic acidemia	AR	Methylmalonyl-CoA mutase	Hypotonia, poor feeding, acidosis, developmental delay
Propionic acidemia	AR	Propionyl-CoA carboxylase	Poor feeding, failure to thrive, vomiting, acidosis, hypoglycemia
Copper Metabolism			
Wilson disease	AR	ATPase membrane copper transport protein	Spasticity, rigidity, dysphagia, cirrhosis
Menkes disease	XR	ATPase membrane copper transport protein	Failure to thrive, neurological deterioration
Peroxisomal Disorders			
PEROXISOMAL BIOGENESIS DISORDERS			
Zellweger syndrome	AR	All peroxisomal enzymes	Dysmorphic features, hypotonia, large liver, renal cysts
ISOLATED PEROXISOMAL ENZYME DEFICIENCY			
Adrenoleukodystrophy	XR	Very long-chain fatty acid-CoA synthase	Mental deterioration, fits, behavioral changes, adrenal failure
Disorders Involving Mitochondria			
MERFF	Mt	Mutation in lysine tRNA (m.8344G>A substitution, m.8356T>C substitution)	Myoclonus, seizures, optic atrophy, hearing impairment, dementia, myopathy
MELAS	Mt	Mutation in leucine (UUR) tRNA (m.3243A>G mutation)	Encephalomyopathy, stroke-like episodes, seizures, dementia, migraine, lactic acidosis
Leigh disease	Mt	Mutation in subunit 6 of ATPase (usually m.8993T>G substitution—'NARP' mutation)	Hypotonia, psychomotor regression, ataxia, spastic quadriparesis
Leber hereditary optic neuropathy	Mt	Mutations in *ND*, *ND4*, *ND6* (m.11778A mutations)	Retinal degeneration, occasional cardiac conduction defects

Table 11.1 Characteristics of Some Inborn Errors of Metabolism—cont'd

Type of Defect	Genetics	Deficiency	Main Clinical Features
Barth syndrome	XR	Uncertain; deficient mitochondrial cardiolipins and raised urinary 3-methylglutaconic acid levels	Cardioskeletal myopathy, growth retardation, neutropenia
Fatty-Acid Oxidation Disorders			
MCAD	AR	Medium-chain acyl-CoA dehydrogenase	Episodic hypoketotic hypoglycemia
Glutaric aciduria type I	AR	Glutaryl-CoA dehydrogenase	Episodic encephalopathy, cerebral palsy-like dystonia
Glutaric aciduria type II	AR	Multiple acyl-CoA dehydrogenase	Hypotonia, hepatomegaly, acidosis, hypoglycemia

AR, Autosomal recessive; *AD,* autosomal dominant; *CNS,* central nervous system; *MCAD,* medium-chain acyl-CoA dehydrogenase; *MELAS,* mitochondrial encephalomyopathy, lactic acidosis, and stroke-like episodes; MERRF, myoclonic epilepsy associated with ragged red fibres; *XR,* X-linked recessive; *XD,* X-linked dominant; *Mt,* mitochondrial.

newborns. The test detects the presence of the metabolite of phenylalanine, phenylpyruvic acid, in the urine by its reaction with ferric chloride, or through increased levels of phenylalanine in the blood. The latter, known as the Guthrie test, involved analysing blood from newborns and comparing the amount of growth induced by the sample, against standards, in a strain of the bacterium *Bacillus subtilis,* which requires phenylalanine for growth. This has been replaced by the use of a variety of biochemical assays of phenylalanine levels.

Heterogeneity of Hyperphenylalaninemia

Raised phenylalanine levels in the newborn period may be due to causes other than PKU. Rarely, newborns have a condition called benign hyperphenylalaninemia, caused by a transient immaturity of liver cells to metabolize phenylalanine. Treatment is not necessary because they are not at risk of developing mental retardation. However, there are two other rare but serious causes of hyperphenylalanemia, in which levels of the enzyme PAH are normal, but there is a deficiency of either dihydropteridine reductase or dihydrobiopterin synthase. These two enzymes help synthesize tetrahydrobiopterin, a cofactor necessary for normal activity of PAH. Both disorders are more serious than classic PKU because there is a high risk of mental retardation despite satisfactory management of phenylalanine levels.

Mutational Basis of PKU

More than 450 different mutations in the *PAH* gene have been identified. Certain mutations are more prevalent in specific population groups and, in western Europeans with PKU, they are found on the background of a limited number

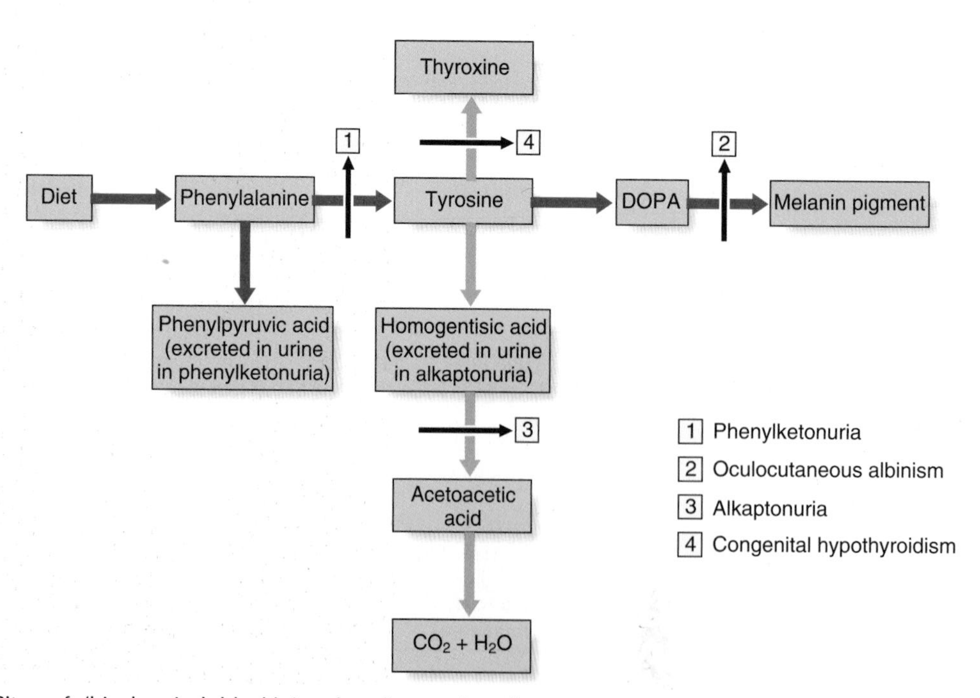

FIGURE 11.1 Sites of 'biochemical block' in phenylketonuria, alkaptonuria, congenital hypothyroidism, and oculocutaneous albinism.

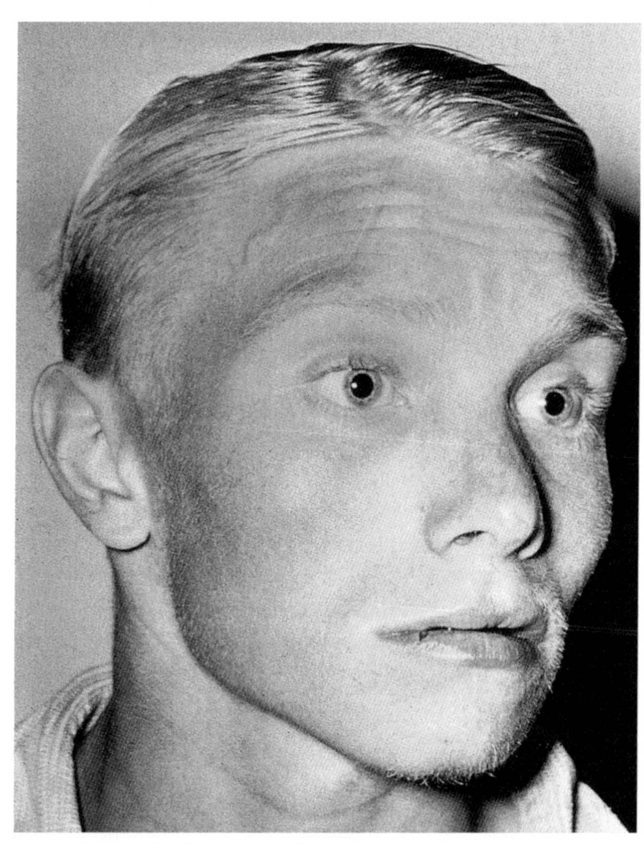

FIGURE 11.2 Facies of a male with phenylketonuria; note the fair complexion.

of DNA haplotypes. Interestingly though, a variety of different individual mutations has been found in association with some of these haplotypes.

Maternal Phenylketonuria

Children born to mothers with phenylketonuria have an increased risk of mental retardation even when their mothers are on closely controlled dietary restriction. It has been suggested that the reduced ability of the mother with PKU to deliver an appropriate amount of tyrosine to her fetus in utero may cause reduced fetal brain growth.

Alkaptonuria

Alkaptonuria was the original autosomal recessive IEM described by Garrod. Here there is a block in the breakdown of homogentisic acid, a metabolite of tyrosine, because of a deficiency of the enzyme homogentisic acid oxidase (see Figure 11.1). As a consequence, homogentisic acid accumulates and is excreted in the urine, which then darkens on exposure to air. Dark pigment is also deposited in certain tissues, such as the ear wax, cartilage, and joints, where it is known as ochronosis, which in joints can lead to arthritis later in life.

Oculocutaneous Albinism

Oculocutaneous albinism (OCA) is an autosomal recessive disorder resulting from a deficiency of the enzyme tyrosinase, which is necessary for the formation of melanin from tyrosine (see Figure 11.1). In OCA there is a lack of pigment in the skin, hair, iris, and ocular fundus (Figure 11.3), and the lack of eye pigment results in poor visual acuity and uncontrolled pendular eye movements—nystagmus. Reduced fundal pigmentation leads to underdevelopment of part of retina for fine vision—the fovea—and abnormal projection of the visual pathways to the optic cortex.

Heterogeneity of OCA

OCA is genetically and biochemically heterogeneous. Cells from those with classic albinism have no measurable tyrosinase activity, the so-called **tyrosinase-negative** form. However, cells from some persons with albinism show reduced but residual tyrosinase activity and are termed **tyrosinase positive**. This is usually reflected clinically by variable development of pigmentation of their hair and skin with age. Both types are known as tyrosinase gene-related OCA type 1.

OCA type 1 is due to mutations in the tyrosinase gene on chromosome 11q. However, linkage studies in some families with tyrosinase-positive OCA have excluded the tyrosinase gene. Some of these have a mutation in the *P* gene, the human homolog of a mouse gene called *pink-eyed dilution*, or 'pink-eye', located on chromosome 15q. This has been termed OCA type 2. In addition, in a proportion of families with OCA, linkage to both of these two loci has been excluded, consistent with the existence of a third locus responsible for OCA.

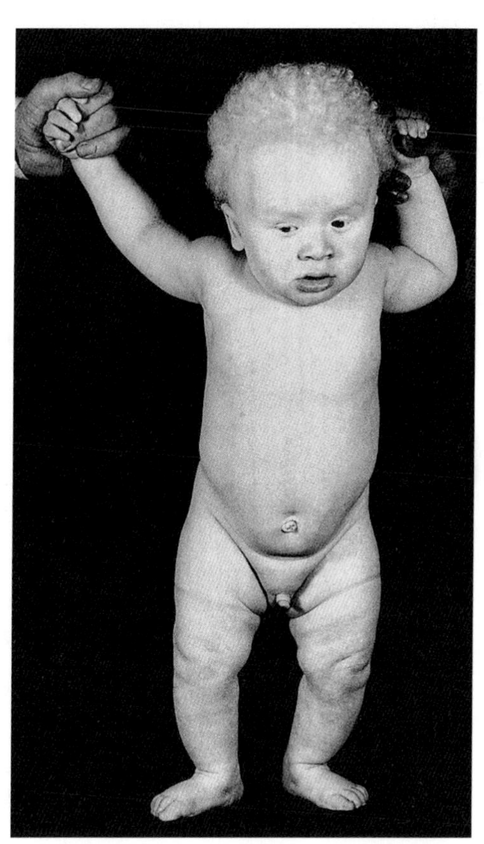

FIGURE 11.3 Oculocutaneous albinism in a child of Afro-Caribbean origin. (Courtesy of Dr V. A. McKusick.)

Homocystinuria

Homocystinuria is a recessively inherited sulfur amino-acid IEM characterized by learning disability, seizures, thrombophilia, osteoporosis, scoliosis, pectus excavatum, long fingers and toes (arachnodactyly), and a tendency to dislocation of the lenses. The somatic features therefore resemble the autosomal dominant disorder Marfan syndrome (p. 300).

Homocystinuria results from deficiency of the enzyme cystathionine β-synthase and can be screened for by means of a positive cyanide nitroprusside test, which detects the presence of increased levels of homocystine in the urine. The diagnosis is confirmed by raised plasma homocystine levels. Treatment involves a low-methionine diet with cystine supplementation. A proportion of individuals with homocystinuria are responsive to the enzyme cofactor pyridoxine (i.e., the pyridoxine-responsive form). A small proportion of affected individuals have mutations in genes leading to deficiencies of enzymes involved in the synthesis of cofactors for cystathionine β-synthase.

Disorders of Branched-Chain Amino Acid Metabolism

The essential branched-chain amino acids leucine, isoleucine and valine have a part of their metabolic pathways in common. Deficiency of the enzyme involved results in maple syrup urine disease.

Maple Syrup Urine Disease

Newborn infants with this autosomal recessive disorder present in the first week of life with vomiting, then alternating tone, leading to death within a few weeks if untreated. The name derives from the odor of the urine—likened to that of maple syrup. The disorder is caused by a deficiency of the branched-chain ketoacid decarboxylase, producing increased urinary excretion of the branched-chain amino acids valine, leucine, and isoleucine, the presence of which suggests the diagnosis, confirmed by demonstration of the three essential branched-chain amino acids in blood. Treatment involves a diet restricting the intake of these three amino acids to the amounts necessary for growth. Affected individuals are particularly susceptible to deterioration, particularly in association with intercurrent illnesses leading to catabolic protein degradation.

Urea Cycle Disorders

The urea cycle is a five-step metabolic pathway that takes place primarily in liver cells for the removal of waste nitrogen from the amino groups of amino acids arising from the normal turnover of protein. It converts two molecules of ammonia and one of bicarbonate into urea (Figure 11.4). Deficiencies of enzymes in the urea cycle result in intolerance to protein from the accumulation of ammonia in the body—hyperammonemia. Increased ammonia levels are toxic to the central nervous system and can lead to coma and, with some untreated urea cycle disorders, death. They are collectively and individually rare and, with the exception of X-linked ornithine transcarbamylase deficiency, inherited as autosomal recessive traits.

Disorders of Carbohydrate Metabolism

The inborn errors of carbohydrate metabolism can be considered in two main groups: disorders of monosaccharide metabolism and the glycogen storage disorders.

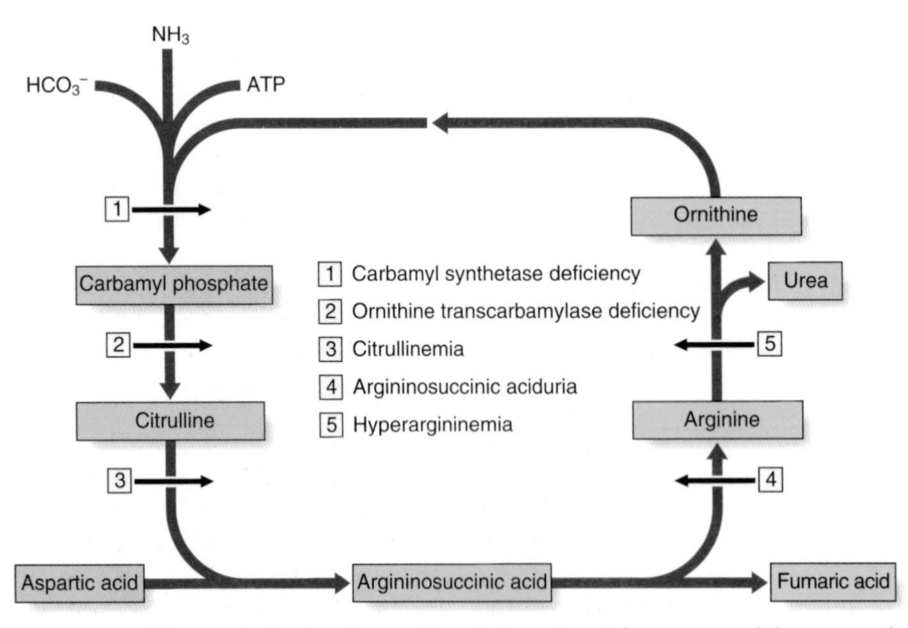

FIGURE 11.4 Diagram indicating the position of the various inborn errors of the urea cycle.

Disorders of Monosaccharide Metabolism

Two examples of disorders of monosaccharide metabolism are galactosemia and hereditary fructose intolerance.

Galactosemia

Galactosemia is an autosomal recessive disorder resulting from a deficiency of the enzyme galactose 1-phosphate uridyl transferase, necessary for the metabolism of the dietary sugar galactose. Newborns with galactosemia present with vomiting, lethargy, failure to thrive, and jaundice in the second week of life. If untreated, they develop complications that include mental retardation, cataracts, and liver cirrhosis. Complications can be prevented by early diagnosis and feeding infants with milk substitutes that do not contain galactose or lactose—the sugar found in milk that is broken down into galactose. Early diagnosis is essential and galactosemia can be screened for by the presence of reducing substances in the urine with specific testing for galactose.

Hereditary Fructose Intolerance

Hereditary fructose intolerance is an autosomal recessive disorder resulting from a deficiency of the enzyme fructose 1-phosphate aldolase. Dietary fructose is present in honey, fruit, and certain vegetables, and in combination with glucose in the disaccharide sucrose in cane sugar. Individuals with hereditary fructose intolerance present at different ages, depending on when fructose is introduced into the diet. Symptoms can be minimal but might also be as severe as those seen in galactosemia, which include failure to thrive, vomiting, jaundice, and seizures. The diagnosis is confirmed by the presence of fructose in the urine and enzyme assay on an intestinal mucosal or a liver biopsy sample. Dietary restriction of fructose is associated with a good long-term prognosis.

Glycogen Storage Diseases

Glycogen is the form in which the sugar glucose is stored in muscle and liver as a polymer, acting as a reserve energy source. In the glycogen storage diseases (GSDs) glycogen accumulates in excessive amounts in skeletal muscle, cardiac muscle, and/or liver because of a variety of inborn errors of the enzymes involved in synthesis and degradation of glycogen. In addition, because of the metabolic block, glycogen is unavailable as a normal glucose source. This can result in hypoglycemia, impairment of liver function and neurological abnormalities.

In each of the six major types of GSD, there is a specific enzyme defect involving one of the steps in the metabolic pathways of glycogen synthesis or degradation. The various types can be grouped according to whether they affect primarily the liver or muscle. All six types are inherited as autosomal recessive disorders, although there are variants of the hepatic phosphorylase that are X-linked.

Glycogen Storage Diseases that Primarily Affect Liver

von Gierke Disease (GSD-I)

von Gierke disease was the first described disorder of glycogen metabolism and results from a deficiency of the enzyme glucose-6-phosphatase, which is responsible for degradation of liver glycogen to release glucose. Infants present with an enlarged liver (hepatomegaly) and/or sweating and a fast heart rate due to hypoglycemia, which can occur after fasting of only 3 to 4 hours duration. Treatment is simple—frequent feeding and avoidance of fasting to maintain the blood sugar concentration.

Cori Disease (GSD-III)

Cori disease is caused by deficiency of the enzyme amylo-1,6-glucosidase, which is also known as the debrancher enzyme. Deficiency results in glycogen accumulation in the liver and other tissues because of the inability to cleave the 'branching' links of the glycogen polymer. Infants may present with hepatomegaly because of glycogen accumulation and/or muscle weakness. Treatment involves avoiding hypoglycemia by frequent feeding and avoiding prolonged periods of fasting.

Anderson Disease (GSD-IV)

Anderson disease results from deficiency of glycogen brancher enzyme leading to the formation of abnormal glycogen consisting of long chains with few branches that cannot be broken down by the enzymes normally responsible for glycogen degradation. Infants present with hypotonia and abnormal liver function in their first year, progressing rapidly to liver failure. No effective treatment is available apart from the possibility of a liver transplant.

Hepatic Phosphorylase Deficiency (GSD-VI)

Hepatic phosphorylase is a multimeric enzyme complex with subunits coded for by both autosomal and X-linked genes. Deficiency of hepatic phosphorylase obstructs glycogen degradation, which results in children presenting in the first 2 years of life with hepatomegaly, hypoglycemia, and failure to thrive. Treatment is with carbohydrate supplementation.

Glycogen Storage Diseases that Primarily Affect Muscle

Pompe Disease (GSD-II)

Infants with Pompe disease usually present in the first few months of life with floppiness (hypotonia) and delay in the gross motor milestones because of muscle weakness. They then develop an enlarged heart and die from cardiac failure in the first or second year. Voluntary and cardiac muscle accumulates glycogen because of a deficiency of the lysosomal enzyme α-1,4-glucosidase, which is needed to break

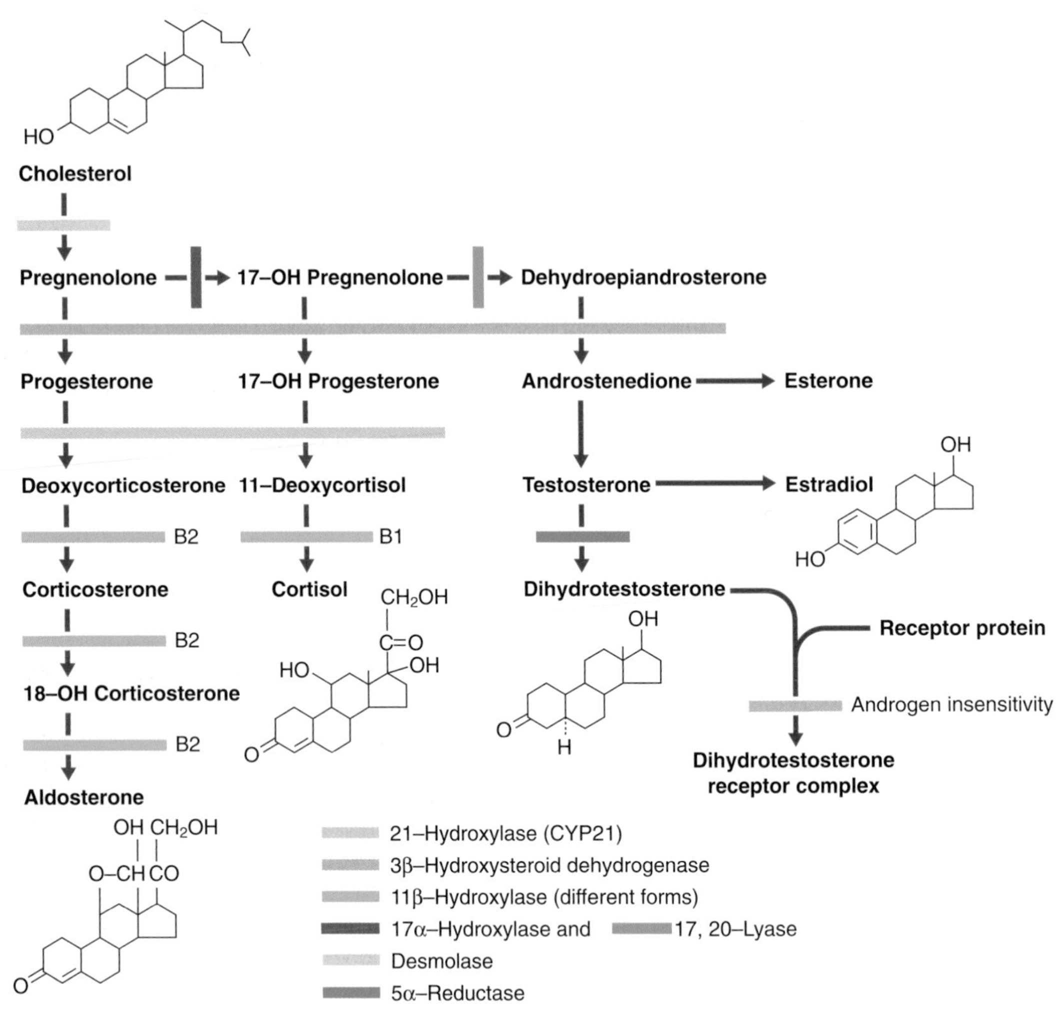

FIGURE 11.5 Steroid biosynthesis indicating the site of the common inborn errors of steroid biosynthesis.

down glycogen. The diagnosis can be confirmed by enzyme assay of white blood cells or fibroblasts. Early reports of enzyme replacement therapy appear promising.

McArdle Disease (GSD-V)

People with McArdle disease present with muscle cramps on exercise in the teenage years. The condition is caused by a deficiency of muscle phosphorylase, which is necessary for degradation of muscle glycogen. There is no effective treatment, although in some the muscle cramps tend to decline if exercise is continued, probably as a result of other energy sources becoming available from alternative metabolic pathways.

Disorders of Steroid Metabolism

The disorders of steroid metabolism include a number of autosomal recessive inborn errors of the biosynthetic pathways of cortisol. Virilization of a female fetus may occur together with salt loss in infants of either sex from a deficiency of the hormone aldosterone. In addition, defects of the androgen receptor result in lack of virilization of chromosomally male individuals (Figure 11.5).

Congenital Adrenal Hyperplasia (Adrenogenital Syndrome)

The diagnosis of congenital adrenal hyperplasia (CAH) should be considered in any newborn female infant presenting with virilization of the external genitalia, because this is the most common cause of ambiguous genitalia in female newborns (p. 288) (Figure 11.6). 21-Hydroxylase deficiency accounts for more than 90% of cases. Approximately 25% have the salt-losing form, presenting in the second or third week of life with circulatory collapse, hyponatremia, and hyperkalemia. Less commonly, CAH is a result of deficiency of the enzymes 11β-hydroxylase or 3β-dehydrogenase, and very rarely occurs as a result of deficiencies of enzymes 17α-hydroxylase and 17,20-lyase. Desmolase deficiency is very rare, with all pathways blocked, causing a reversed phenotype of ambiguous genitalia in males, and severe addisonian crises. Males with the rare 5α-reductase deficiency are significantly under-masculinized but do not suffer other metabolic problems and are likely to be raised as females. At puberty, however, the surge in androgen production is sufficient to stimulate growth of the phallus, making gender identity and assignment problematic.

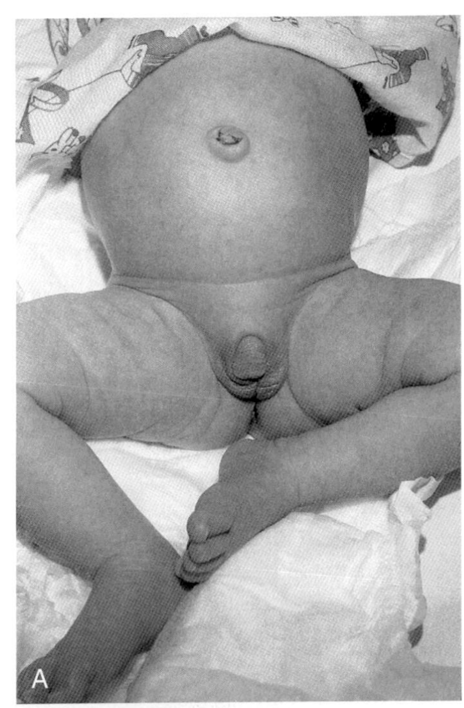

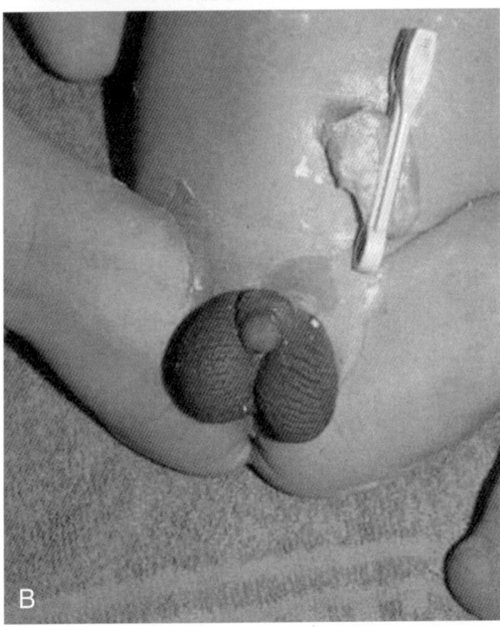

FIGURE 11.6 **A,** Virilized external genitalia in a female with congenital adrenal hyperplasia. **B,** A male baby with hypospadias who clearly has testes in the scrotal sacs.

Affected females with classic CAH are virilized from accumulation of the adrenocortical steroids proximal to the enzyme block in the steroid biosynthetic pathway, many of which have testosterone-like activity (see Figure 11.5). However, they have normal müllerian-derived internal organs. The possibility of CAH should not be forgotten, of course, in male infants presenting with circulatory collapse in the first few weeks of life.

Affected infants, in addition to requiring urgent correct assignment of gender, are treated with replacement cortisol, along with fludrocortisone if they have the salt-losing form. Virilized females may require plastic surgery later. Steroid replacement is lifelong and should be increased during intercurrent illness or stress, such as surgery. Menarche in girls with salt-losing CAH is late, menstruation irregular, and they are subfertile.

Androgen Insensitivity Syndrome

Individuals with the androgen insensitivity syndrome have female external genitalia and undergo breast development in puberty (p. 288). They classically present either with primary amenorrhea or with an inguinal hernia containing a gonad that turns out to be a testis. Inguinal hernia is uncommon in girls and if present, especially if bilateral, androgen insensitivity syndrome should be considered. There is often scanty secondary sexual hair and investigation of the internal genitalia reveals an absent uterus and fallopian tubes with a blind-ending vagina. Chromosome analysis reveals a normal male karyotype, 46,XY.

Androgen production by the testes is normal but androgen does not bind normally because of an abnormal androgen receptor (see Figure 11.5)—the androgen receptor gene on the X chromosome is mutated. This can be functionally assayed in skin fibroblasts. Some individuals have incomplete or partial androgen insensitivity, and under-virilization is variable. Affected subjects may have a female sexual orientation and are sterile. Testes must be removed because of an increased malignancy risk, and estrogen given for secondary sexual development and prevention of long-term osteoporosis.

Disorders of Lipid Metabolism

Familial hypercholesterolemia is the most common autosomal dominant single-gene disorder in Western society and is associated with high morbidity and mortality rates through premature coronary artery disease (p. 241).

Familial Hypercholesterolemia

Persons with familial hypercholesterolemia (FH) have raised cholesterol levels with a significant risk of developing early coronary artery disease (p. 241). They can present in childhood or adolescence with subcutaneous deposition of lipid, known as xanthomata (Figure 11.7). Starting with families who presented with early coronary artery disease, Brown and Goldstein unraveled the biology of the low-density lipoprotein (LDL) receptor (p. 241) and the pathological basis of FH.

Cells normally derive cholesterol from either endogenous synthesis or dietary uptake from LDL receptors on the cell surface. Intracellular cholesterol levels are maintained by a feedback system, with free cholesterol inhibiting LDL receptor synthesis as well as reducing the level of de novo endogenous synthesis.

High cholesterol levels in FH are due to deficient or defective function of the LDL receptors leading to increased levels of endogenous cholesterol synthesis. Four main classes

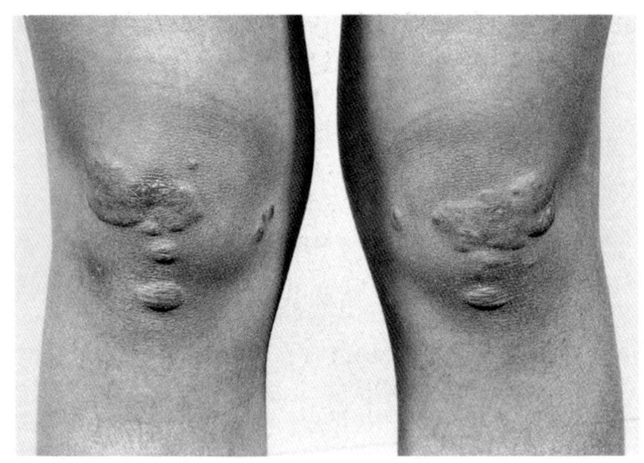

FIGURE 11.7 Legs of a person homozygous for familial hypercholesterolemia, showing multiple xanthomata. (Courtesy Dr. E. Wraith, Royal Manchester Children's Hospital, Manchester, UK.)

of mutation in the LDL receptor have been identified: (1) reduced or defective biosynthesis of the receptor; (2) reduced or defective transport of the receptor from the endoplasmic reticulum to the Golgi apparatus; (3) abnormal binding of LDL by the receptor; and (4) abnormal internalization of LDL by the receptor (Figure 11.8). Specific mutations are more prevalent in certain ethnic groups because of founder effects (p. 133).

The mainstay of management is dietary restriction of cholesterol intake and drug treatment with 'statins' that reduce the endogenous synthesis of cholesterol by inhibiting the enzyme 3-hydroxy-3-methylglutaryl coenzyme A (CoA) reductase. Cholesterol levels in affected families are variable and lipid assays do not necessarily identify those with mutations. There is therefore interest in the introduction of widespread genetic testing, though most mutations are missense, which may pose problems of interpretation.

Lysosomal Storage Disorders

In addition to the IEMs in which an enzyme defect leads to deficiency of an essential metabolite and accumulation of intermediate metabolic precursors, there are a number of disorders in which deficiency of a lysosomal enzyme involved in the degradation of complex macromolecules leads to their accumulation. This accumulation occurs because macromolecules are normally in a constant state of flux, with a delicate balance between their rates of synthesis and breakdown. Children born with lysosomal storage diseases are usually normal initially but with the passage of time commence a downhill course of variable duration

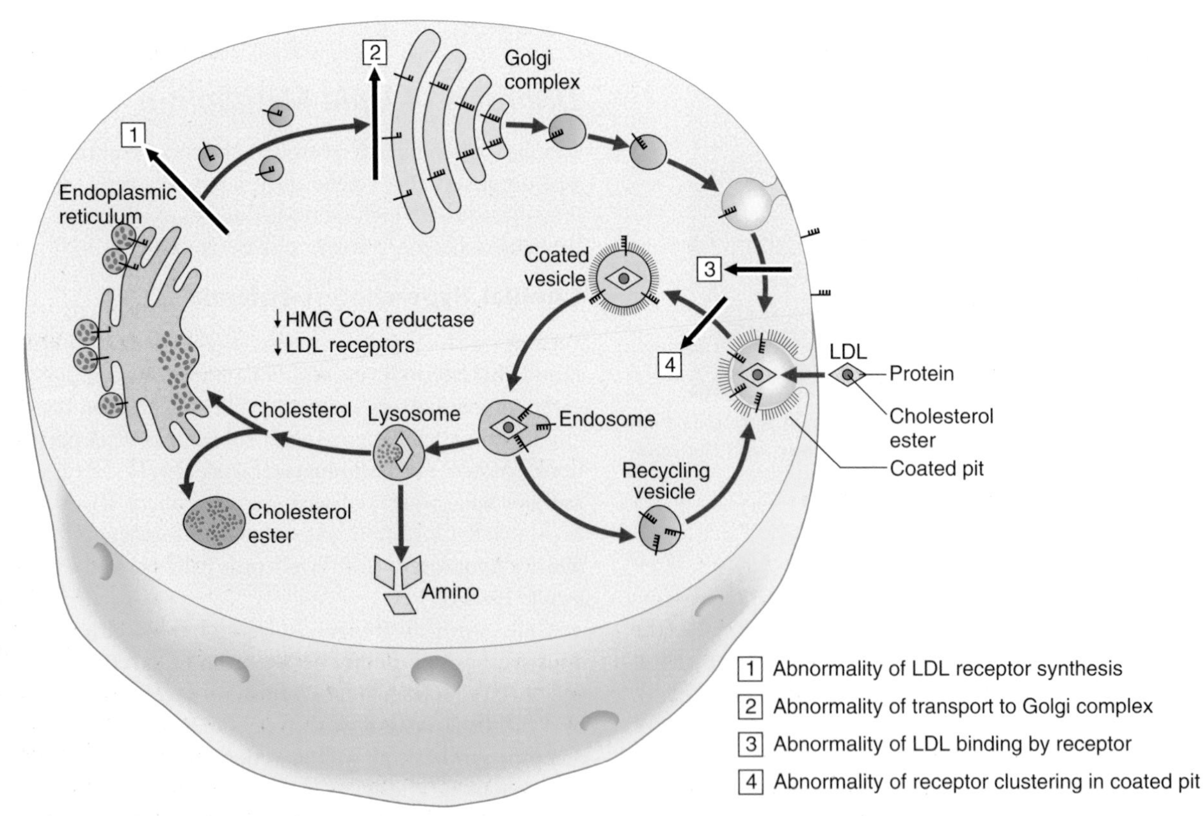

FIGURE 11.8 Stages in cholesterol biosynthesis and in the metabolism of low-density lipoprotein (LDL) receptors, indicating the types of mutation in familial hypercholesterolemia. (Adapted from Brown MS, Goldstein JL 1986 A receptor-mediated pathway for cholesterol homeostasis. Science 232:34–47.)

owing to the accumulation of one or more of a variety or type of macromolecules.

Mucopolysaccharidoses

Children with one of the mucopolysaccharidoses (MPSs) present with skeletal, vascular, or central nervous system findings along with coarsening of the facial features. These features are due to progressive accumulation of sulfated polysaccharides (also known as glycosaminoglycans) caused by defective degradation of the carbohydrate side-chain of acid mucopolysaccharide.

Six different MPSs are recognized, based on clinical and genetic differences. Each specific MPS type has a characteristic pattern of excretion in the urine of the glycosaminoglycans, dermatan, heparan, keratan, and chondroitin sulfate. Subsequent biochemical investigation has revealed the various types to be due to deficiency of different individual enzymes. All but Hunter syndrome, which is X-linked, are autosomal recessive disorders.

Hurler Syndrome (MPS-I)

Hurler syndrome is the most severe MPS. Infants present in the first year with corneal clouding, a characteristic curvature of the lower spine and subsequent poor growth. They develop hearing loss, coarse facial features, an enlarged liver and spleen, joint stiffness, and vertebral changes in the second year. These features progress together with mental deterioration and eventually death by mid-adolescence from a combination of cardiac failure and respiratory infections.

The diagnosis of Hurler syndrome was initially made by demonstrating the presence of metachromatic granules in the cells (i.e., lysosomes distended by the storage material that is primarily dermatan sulfate). Increased urinary excretion of dermatan and heparan sulfate is commonly used as a screening test, but confirmation of the diagnosis involves demonstration of reduced activity of the lysosomal hydrolase, α-L-iduronidase, and direct gene analysis. Less severe allelic forms of Hurler syndrome, caused by varying levels of residual α-L-iduronidase activity, were previously classified separately as Scheie disease (MPS-I S) and Hurler/Scheie disease (MPS-I H/S).

Hunter Syndrome (MPS-II)

Males with Hunter syndrome usually present between 2 and 5 years with hearing loss, recurrent infections, diarrhea, and poor growth. Facial features are characteristic with coarsening (Figure 11.9), the liver and spleen are enlarged, and joint stiffness occurs. Spinal radiographs show abnormal shape of the vertebrae. Progressive physical and mental deterioration occurs with death usually in adolescence.

The diagnosis is confirmed by the presence of excess amounts of dermatan and heparan sulfate in the urine, deficient or decreased activity of the enzyme iduronate sulfate sulfatase in serum or white blood cells, and direct gene analysis.

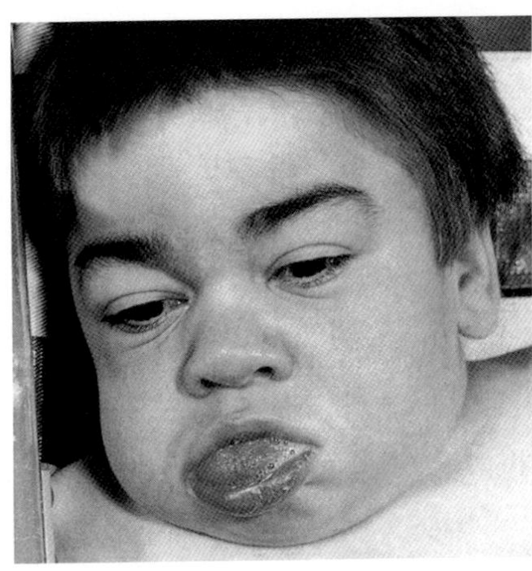

FIGURE 11.9 Facies of a male with the mucopolysaccharidosis, Hunter syndrome. (Courtesy Dr. E. Wraith, Royal Manchester Children's Hospital, Manchester, UK.)

Sanfilippo Syndrome (MPS-III)

Sanfilippo syndrome is the most common MPS. Affected individuals present in their second year with mild coarsening of features, skeletal changes, and progressive intellectual loss with behavioral problems, seizures, and death in early adult life. The diagnosis is confirmed by the presence of increased urinary heparan and chondroitin sulfate excretion, and deficiency of one of four enzymes involved in the degradation of heparan sulfate: sulfaminidase (MPS-III A), *N*-acetyl-α-D-glucosaminidase (MPS-III B), acetyl-CoA-α-glucosaminidase-*N*-acetyltransferase (MPS-III C), or *N*-acetyl-glucosamine-6-sulfate sulfatase (MPS-III D). Individuals with these different enzyme deficiencies cannot be distinguished clinically. Direct gene testing is not so reliable.

Morquio Syndrome (MPS-IV)

Children with Morquio syndrome present age 2 to 3 years with short stature, thoracic deformity, and curvature of the spine (kyphoscoliosis). Intelligence is normal and survival is long term, but there is a risk of spinal-cord compression from progression of the skeletal involvement. The diagnosis is confirmed by the presence of keratan sulfate in the urine and deficiency of either galactosamine-6-sulphatase (MPS-IV A) or β-galactosidase (MPS-IV B).

Maroteaux–Lamy Syndrome (MPS-VI)

This MPS presents with Hurler-like features in early childhood, including coarse facial features, short stature with thoracic deformity, kyphosis, and restriction of joint mobility. In addition, corneal clouding and cardiac valve abnormalities develop; intelligence is normal. A milder form presents later with survival into late adulthood, in contrast to the severe form in which survival is usually only to the third decade. The diagnosis is confirmed by the presence of increased urinary dermatan sulfate excretion and arylsulfatase B deficiency in white blood cells or fibroblasts.

Sly Syndrome (MPS-VII)

This is an extremely variable MPS. Presentation ranges from skeletal features that include mild kyphoscoliosis and hip dysplasia to coarse facial features, hepatosplenomegaly, corneal clouding, cardiac abnormalities, and mental retardation, with death in childhood or adolescence. Increased urinary glycosaminoglycans excretion and β-glucuronidase deficiency in serum, white blood cells, or fibroblasts confirm the diagnosis.

Treatment of the MPSs

Treatment of these disorders by enzyme replacement has proved difficult in practice (p. 349). However, bone marrow transplantation has been attempted with varying success, biochemically, and clinically in relation to the skeletal and cerebral features.

Sphingolipidoses (Lipid Storage Diseases)

In the sphingolipidoses, there is an inability to degrade sphingolipid, resulting in the progressive deposition of lipid or glycolipid, primarily in the brain, liver, and spleen. Central nervous system involvement results in progressive mental deterioration, often with seizures, leading to death in childhood. There are at least 10 different types, with specific enzyme deficiencies, Tay-Sachs, Gaucher, and Niemann-Pick diseases being the most common.

Tay-Sachs Disease

This well-known sphingolipidosis has an incidence of ~1:3600 in Ashkenazi Jews (pp. 313–314). Infants usually present by 6 months of age with poor feeding, lethargy, and floppiness. Developmental regression usually becomes apparent in late infancy, feeding becomes increasingly difficult, and the infant progressively deteriorates, with deafness, visual impairment, and spasticity, which progresses to rigidity. Death usually occurs by the age of 3 years from respiratory infection. Less severe juvenile, adult, and chronic forms are reported.

The diagnosis is supported clinically by the presence of a 'cherry-red' spot in the center of the macula of the fundus. Biochemical confirmation of Tay-Sachs disease is by demonstration of reduced hexosaminidase A levels in serum, white blood cells or cultured fibroblasts, and direct gene analysis is available. Reduced hexosaminidase A activity is due to deficiency of the a subunit of the enzyme β-hexosaminidase that leads to accumulation of the sphingolipid GM_2 ganglioside. This deficiency leads to reduced activity of the isozyme, hexosaminidase B, causing the other GM_2 gangliosidosis, Sandhoff disease, which presents with similar clinical features.

Gaucher Disease

This is the most common sphingolipidosis and, as with Tay-Sachs, is relatively more frequent among Ashkenazi Jews. There are two main types based on the age of onset.

Type I, with adult onset, is the more common form and presents with febrile episodes, pain in limbs, joints, or trunk, and a tendency to pathological fractures. Clinical examination usually reveals hepatosplenomegaly and investigations show mild anemia and radiological changes in the vertebrae and proximal femora. The central nervous system is spared.

In type II, infantile Gaucher disease, central nervous system involvement is a major feature and presents age 3 to 6 months with failure to thrive and hepatosplenomegaly. By 6 months, developmental regression and neurological deterioration occur with spasticity and seizures. Recurrent pulmonary infections cause death in the second year.

The diagnosis is confirmed by reduced activity of the enzyme glucosylceramide β-glucosidase in white blood cells or cultured fibroblasts.

Treatment in type 1 involves symptomatic analgesia, and sometimes splenectomy to prevent premature sequestration of red blood cells (hypersplenism). Initial attempts to treat adults by enzyme replacement therapy met with little success because of difficulty in obtaining sufficient quantities of enzyme and in targeting the appropriate sites. However, modification of β-glucosidase by the addition of mannose 6-phosphate, which targets the enzyme to macrophage lysosomes, has led to dramatic alleviation of symptoms and regression of organomegaly. The treatment is expensive, and regimens using lower doses and alternative methods to target the enzyme may be more rational.

Niemann-Pick Disease

Infants with Niemann-Pick disease present with failure to thrive and hepatomegaly, and a cherry-red spot may be found on their macula. Developmental regression progresses rapidly by the end of the first year, with death by 4 years of age. A characteristic finding is the presence of what are called foam cells in the bone marrow from sphingomyelin accumulation. Confirmation of the diagnosis is by demonstration of deficiency of the enzyme sphingomyelinase. A less severe form without neurological involvement has been reported. As with Tay-Sachs and Gaucher diseases, it is more common in Ashkenazi Jews from eastern Europe.

Disorders of Purine/Pyrimidine Metabolism

Gout is the classic disorder of abnormal purine metabolism. Joint pain, swelling, and tenderness are a result of the inflammatory response of the body to deposits of crystals of a salt of uric acid. In fact, only a minority of persons with gout have an IEM. The cause in most instances results from a combination of genetic and environmental factors; however, it is always important to consider disorders that can result in an increased turnover of purines (e.g., a malignancy such as leukemia) or reduced secretion of the metabolites (e.g., renal impairment) as a possible underlying precipitating cause.

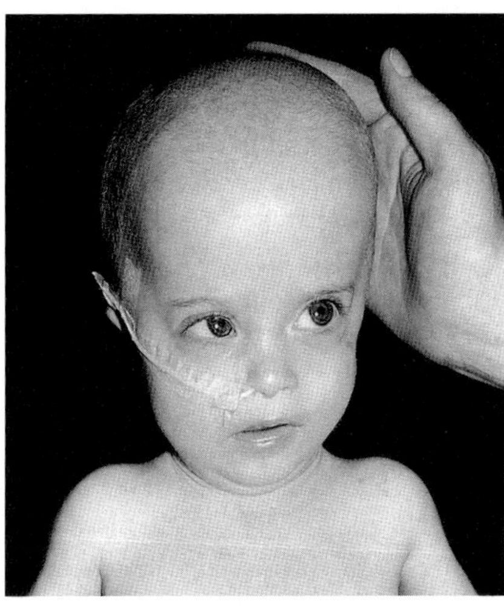

FIGURE 11.10 Facies of an infant with Zellweger syndrome showing a prominent forehead.

It is unusual for IEMs to give rise to a dysmorphic syndrome (p. 249), but another is Smith-Lemli-Opitz syndrome, an inborn error of cholesterol biosynthesis from a mutation in the sterol delta-7-reductase (*DHCR7*) gene.

Adrenoleukodystrophy

Males with the X-linked disorder adrenoleukodystrophy (ALD) classically present in late childhood with deteriorating school performance, though presentation may occur at any age; occasionally, there are no symptoms. Some males present in adult life with less severe neurological features and adrenal insufficiency, so-called adrenomyeloneuropathy. ALD has been shown to be associated with a deficiency of the enzyme very long-chain fatty acid CoA synthase, but is secondary to deficiency of a peroxisomal membrane protein, due to mutation in the *ABCD1* gene.

Treatment of ALD with a diet that uses an oil with low levels of very long-chain fatty acids—'Lorenzo's' oil—has proved disappointing.

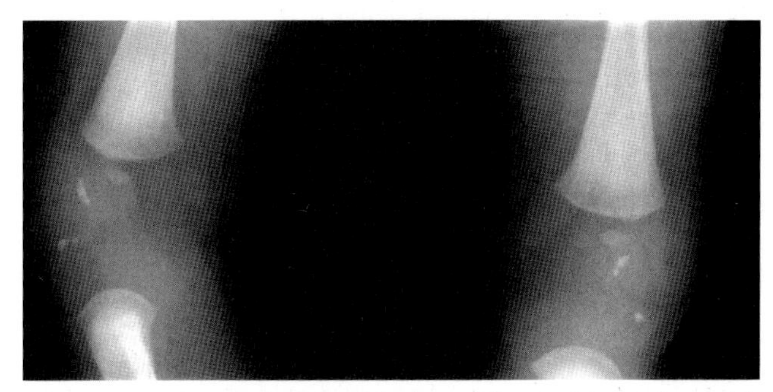

FIGURE 11.11 Radiograph of the knee of a newborn infant with Zellweger syndrome showing abnormal punctate calcification of the distal femoral epiphyses.

Disorders Affecting Mitochondrial Function

Mitochondrial disease was first identified in 1962 in a patient whose mitochondria showed structural abnormalities and loss of coupling between oxidation and phosphorylation, although it was not until 20 years later that the relevance of mutated mitochondrial DNA (mtDNA) to human disease began to be appreciated. The small circular double-stranded mtDNA (see Figure 2.7, p. 18) contains genes coding for ribosomal RNA (rRNA) production and various transfer RNAs (tRNA) required for mitochondrial protein biosynthesis, as well as some of the proteins involved in electron transport. There are 5523 codons and a total of 37 gene products. Guanine and cytosine nucleotides are asymmetrically distributed between the two mtDNA strands—the guanine-rich strand being called the heavy (H) strand and the cytosine-rich the light (L) strand. Replication and transcription is controlled by a 1122-bp sequence of mtDNA known as the displacement loop (D-loop). Oxidative phosphorylation (OXPHOS) is the biochemical process responsible for generating much of the ATP required for cellular energy. The process is mediated by five intramitochondrial enzyme complexes, referred to as complexes I–V, and the mtDNA encodes 13 OXPHOS subunits, 22 tRNAs, and 2 rRNAs.

The 'complexes' are aptly named. Analysis of complex I, for example, has revealed approximately 41 different subunits, of which 7 are polypeptides encoded by mtDNA genes known as *ND1*, *ND2*, *ND3*, *NDL4*, *ND4*, *ND4L*, *ND5*, and *ND6*, with the remaining 34 subunits encoded by nuclear DNA genes. Complex V comprises 12 or 13 subunits, of which two, ATPase 6 and 8, are encoded by mtDNA. Maximal activity of complex V appears to require tight linking with cardiolipin (see Barth syndrome, p. 182), encoded by nuclear DNA.

Because most mitochondrial proteins, including subunits involved in electron transport, are encoded by nuclear genes, these most often follow autosomal recessive inheritance. As with other metabolic autosomal recessive diseases, disorders resulting from mutations in these genes tend to breed true. However, the disorders resulting from mutations in mtDNA are extremely variable owing to the phenomenon of heteroplasmy (see Figure 7.30, p. 126). The clinical features are mainly a combination of neurological signs—encephalopathy, dementia, ataxia, dystonia, neuropathy, and seizures—and myopathic signs—hypotonia, weakness, and cardiomyopathy with conduction defects. Other symptoms and signs may include deafness, diabetes mellitus, retinal pigmentation, and acidosis may occur. The clinical manifestations are so variable that a mitochondrial cytopathy should be considered as a possibility at any age when the presenting illness has a neurological or myopathic component. Several distinct clinical entities have been determined and, although some of them overlap considerably, there is a degree of genotype–phenotype correlation.

Myoclonic Epilepsy and Ragged Red Fiber Disease (MERRF)

Myoclonic epilepsy and ragged red fiber (MERRF) disease was first described in 1973 and so called because Gomori's trichrome staining of muscle revealed abnormal deposits of mitochondria as 'ragged red'. In 1988 it was determined that the condition was maternally inherited. The classic picture is of progressive myoclonic epilepsy, myopathy, and slowly progressive dementia. Optic atrophy is frequently present and the electroencephalogram is characteristically abnormal. Post-mortem brain examination reveals widespread neurodegeneration. In 1990 it was reported that MERRF results from a point mutation in the gene for lysine tRNA.

Mitochondrial Encephalomyopathy, Lactic Acidosis, and Stroke-Like Episodes (MELAS)

First delineated in 1984, this extremely variable condition is now recognized as one of the commonest mitochondrial disorders. Short stature may be a feature, but it is stroke-like episodes that mark out this particular disorder, although these episodes do not necessarily occur in all affected family members. When they do occur, they may manifest as vomiting, headache, or visual disturbance, and sometimes lead to transient hemiplegia or hemianopia. A common presenting feature of MELAS is type 2 diabetes mellitus, and a sensorineural hearing loss may also occur (described as maternally inherited diabetes and deafness). These latter clinical features are associated with the most common mutation, an A>G substitution at nucleotide m.3243, which affects tRNA leucine^{UUR}. This is found in about 80% of patients, followed by a T>C transition at nucleotide m.3271, also affecting tRNA leucine^{UUR}.

Neurodegeneration, Ataxia, and Retinitis Pigmentosa (NARP)

The early presenting feature is night blindness, which may be followed years later by neurological symptoms. Dementia may occur in older patients, but seizures can present at almost any age and younger patients show developmental delay. The majority of cases are due to a single mutation—the T>G substitution at nucleotide m.8993, which occurs in the coding region of subunit 6 of ATPase. This change is often referred to as the NARP mutation.

Leigh Disease

This condition is characterized by its neuropathology, consisting of typical spongiform lesions of the basal ganglia, thalamus, substantia nigra, and tegmental brainstem. In its severe form, death occurs in infancy or early childhood, and it was in such a patient that the m.8993T>G NARP mutation was first identified. In effect, therefore, one form of Leigh disease is simply a severe form of NARP, and higher proportions of mutant mtDNA have been reported in these cases. However, variability is again sometimes marked and the author knows one family in which a mother, whose daughter died in early childhood, was found to have low levels of the 8993 mutation and her only symptom was slow recovery from a general anesthetic.

The same or very similar pathology, and a similar clinical course, has now been described in patients with different molecular defects. Cytochrome *c* deficiency has been reported in a number of patients and some of these have been shown to have mutations in *SURF1*, a nuclear gene. These cases follow autosomal recessive inheritance. Leigh disease is therefore genetically heterogeneous, and there is even an X-linked form.

Leber Hereditary Optic Neuropathy

Leber hereditary optic neuropathy was the first human disease to be shown to result from an mtDNA point mutation; about a dozen different mutations have now been described. The most common mutation occurs at nucleotide m.11,778 (*ND4* gene), accounting for up to 70% of cases in Europe and more than 90% of cases in Japan. It presents with acute, or subacute, loss of central visual acuity without pain, which typically occurs between 12 and 30 years of age. Males in affected pedigrees are much more likely to develop visual loss than females. In some Leber hereditary optic neuropathy pedigrees, additional neurological problems occur.

Barth Syndrome

Also known as X-linked cardioskeletal myopathy, this is characterized by congenital dilated cardiomyopathy (including endocardial fibroelastosis), a generalized myopathy, and growth retardation. Abnormal mitochondria are found in many tissues, deficient in cardiolipin, and skeletal muscle shows increased lipid levels. A variable and sometimes fluctuating increase in urinary levels of 3-methylglutaconic acid may be useful in achieving a diagnosis, and mutations have been identified in the *G4.5((TAZ)* gene, but the enzyme defect leading to 3-methylglutaconic aciduria is unknown.

Disorders of Mitochondrial Fatty-Acid Oxidation

In the 1970s, the first reports appeared of patients with skeletal muscle weakness and abnormal muscle fatty-acid metabolism associated with decreased muscle carnitine. The carnitine cycle is a biochemical pathway required for the transport of long-chain fatty acids into the mitochondrial matrix, and those less than 10 carbons in length are then activated to form acyl-CoA esters. The carnitine cycle is one part of the pathway of mitochondrial b-oxidation that plays a major role in energy production, especially during periods of fasting. Carnitine deficiency is a secondary feature of the β-oxidation disorders, with the exception of the carnitine transport defect where it is primary, and this rare condition responds dramatically to carnitine replacement. The more common fatty-acid oxidation disorders are outlined.

Medium-Chain acyl-CoA Dehydrogenase Deficiency

Medium-chain acyl-CoA dehydrogenase deficiency is the most common of this group of disorders, presenting most frequently as episodic hypoketotic hypoglycemia provoked by fasting. The onset is often in the first 2 years of life and, tragically, is occasionally fatal, resembling sudden infant death syndrome. Management rests on maintaining adequate caloric intake and avoidance of fasting, which can be challenging in young children with intercurrent illnesses. Inherited as an autosomal recessive disorder, 90% of alleles result from a single point mutation, and neonatal population screening is now undertaken in many countries.

Long-Chain and Short-Chain (SCAD) acyl-CoA, and Long-Chain 3-Hydroxyacyl-CoA Dehydrogenase Deficiencies

These rare conditions all show autosomal recessive inheritance and present early in life with a variable combination of skeletal features and cardiomyopathy, hepatocellular dysfunction with hepatomegaly, and encephalopathy. Treatment revolves around nutritional maintenance and avoidance of fasting, but is not very rewarding in short-chain acyl-CoA deficiencies.

Glutaric Acidurias

Glutaric acidurias types I (glutaryl-CoA dehydrogenase deficiency) and II (multiple acyl-CoA dehydrogenase deficiency) are included as examples of organic acidurias that are intermediate in fatty-acid oxidation; both show autosomal recessive inheritance. In type I, macrocephaly is present at birth and infants suffer episodes of encephalopathy with spasticity, dystonia, seizures, and developmental delay. Treatment is by dietary restriction of glutarigenic amino acids—lysine, tryptophan, and hydroxylysine. Common among the Old Order Amish of Pennsylvania, neonatal screening has been introduced in the area.

Type II glutaric aciduria is variable, with two severe forms having neonatal onset, one of these including urogenital anomalies. In both of these severe types, hypotonia, hepatomegaly, metabolic acidosis, and hypoketotic hypoglycemia occur. The late-onset form may present in early childhood, rather than the neonatal period, with failure to thrive, metabolic acidosis, hypoglycemia, and encephalopathy. Treatment of the severe forms is supportive only, but in the milder form riboflavin, carnitine, and diets low in protein and fat have been more successful.

Prenatal Diagnosis of Inborn Errors of Metabolism

For the majority of inborn errors of metabolism in which an abnormal or deficient gene product can be identified, prenatal diagnosis is possible. Biochemical analysis of cultured amniocytes obtained at mid-trimester amniocentesis is possible but has largely given way to earlier testing using direct or cultured chorionic villi (CV), which allows a diagnosis to be made by 12 to 14 weeks' gestation (p. 325). For many conditions a biochemical analysis on cultured CV tissue is the appropriate test but, increasingly, direct mutation analysis is possible. This avoids the inherent delay of culturing CV tissue and is of particular value for inborn errors for which the biochemical basis is not clearly identified, or where the enzyme is not expressed in amniocytes or CV.

Prenatal diagnosis of mitochondrial disorders from mtDNA mutations presents particular difficulties because of the problem of heteroplasmy and the inability to predict the outcome for any result obtained, whether positive or negative for the mutation in question. This presents challenging counseling issues and also raises consideration of other reproductive options, such as ovum donation and perhaps, in the future, nuclear transfer technology to circumvent maternal mtDNA.

FURTHER READING

Benson PF, Fensom AH 1985 Genetic biochemical disorders. Oxford: Oxford University Press

A good reference source for detailed basic further information on the inborn errors of metabolism.

Clarke JTR 1996 A clinical guide to inherited metabolic diseases. Cambridge: Cambridge University Press

A good basic text, problem based and clinically oriented.

Cohn RM, Roth KS 1983 Metabolic disease: a guide to early recognition. Philadelphia: WB Saunders

A useful text as it considers the inborn errors from their mode of presentation rather than starting from the diagnosis.

Garrod AE 1908 Inborn errors of metabolism. Lancet ii:1–7, 73–79, 142–148, 214–220

Reports of the first inborn errors of metabolism.

Nyhan WL, Ozand PT 1998 Atlas of metabolic diseases. London: Chapman & Hall

A detailed text but very readable and full of excellent illustrations and clinical images.

Rimoin DL, Connor JM, Pyeritz RE, Korf BR eds 2001 Principles and practice of medical genetics, 4th edn. Edinburgh: Churchill Livingstone

The section on metabolic disorders includes 13 chapters covering in succinct detail the various groups of metabolic disorders.

Scriver CR, Beaudet AL, Sly WS, Valle D eds 2000: The metabolic basis of inherited disease, 8 edn. New York: McGraw Hill

A huge multi-author three-volume comprehensive detailed text on biochemical genetics with an exhaustive reference list and, with this edition, a CD-ROM.

ELEMENTS

1 Metabolic processes in all species occur in steps, each being controlled by a particular enzyme which is the product of a specific gene, leading to the one gene–one enzyme concept.

2 A block in a metabolic pathway results in the accumulation of metabolic intermediates and/or a deficiency of the end-product of the particular metabolic pathway concerned, a so-called inborn error of metabolism.

3 The majority of the inborn errors of metabolism are inherited as autosomal recessive or X-linked recessive traits. A few are inherited as autosomal dominant disorders involving rate-limiting enzymes, cell-surface receptors, or multimeric enzymes through haploinsufficiency or dominant negative mutations.

4 A number of the inborn errors of metabolism can be screened for in the newborn period and treated successfully by dietary restriction or supplementation.

5 Prenatal diagnosis of many of the inborn errors of metabolism is possible by either conventional biochemical methods, the use of linked DNA markers or direct mutation detection.

CHAPTER 12

Pharmacogenetics

If it were not for the great variability among individuals medicine might as well be a science and not an art.

SIR WILLIAM OSLER (1892)

Definition

Some individuals can be especially sensitive to the effects of a particular drug, whereas others can be quite resistant. Such individual variation can be the result of factors that are not genetic. For example, both the young and the elderly are very sensitive to morphine and its derivatives, as are people with liver disease. Individual differences in response to drugs in humans are, however, often genetically determined.

The term **pharmacogenetics** was introduced by Vogel in 1959 for the study of genetically determined variations that are revealed solely by the effects of drugs. Pharmacogenetics is now used to describe the influence of genes on the efficacy and side effects of drugs. **Pharmacogenomics** describes the interaction between drugs and the genome (i.e., multiple genes), but the two terms are often used interchangeably. Pharmacogenetics/pharmacogenomics is important because adverse drug reactions are a major cause of morbidity and mortality. It is also likely to be of increasing importance in the future, particularly as a result of the development of new drugs from information that has become available from the Human Genome Project (see Chapter 5). The human genome influences the effects of drugs in at least three ways. **Pharmacokinetics** describes the metabolism of drugs, including the uptake of drugs, their conversion to active metabolites, and detoxification or breakdown. **Pharmacodynamics** refers to the interaction between drugs and their molecular targets. An example would be the binding of a drug to its receptor. The third way relates to palliative drugs that do not act directly on the cause of a disease, but rather on its symptoms. Analgesics, for example, do not influence the cause of pain but merely the perception of pain in the brain.

Drug Metabolism

The metabolism of a drug usually follows a common sequence of events (Figure 12.1). A drug is first absorbed from the gut, passes into the bloodstream, and becomes distributed and partitioned in the various tissues and tissue fluids. Only a small proportion of the total dose of a drug will be responsible for producing a specific pharmacological effect, most of it being broken down or excreted unchanged.

Biochemical Modification

The actual breakdown process, which usually takes place in the liver, varies with different drugs. Some are oxidized completely to carbon dioxide, which is exhaled through the lungs. Others are excreted in modified forms either via the kidneys into the urine, or by the liver into the bile and thence the feces. Many drugs undergo biochemical modifications that increase their solubility, resulting in their being more readily excreted.

One important biochemical modification of many drugs is conjugation, which involves union with the carbohydrate glucuronic acid. Glucuronide conjugation occurs primarily in the liver. The elimination of morphine and its derivatives, such as codeine, is dependent almost entirely on this process. Isoniazid, used in the treatment of tuberculosis, and a number of other drugs, including the sulfonamides, are modified by the introduction of an acetyl group into the molecule, a process known as acetylation (Figure 12.2).

Kinetics of Drug Metabolism

The study of the metabolism and effects of a particular drug usually involves giving a standard dose of the drug and then, after a suitable time interval, determining the response, measuring the amount of the drug circulating in the blood or determining the rate at which it is metabolized. Such studies show that there is considerable variation in the way different individuals respond to certain drugs. This variability in response can be continuous or discontinuous.

If a dose–response test is carried out on a large number of subjects, their results can be plotted. A number of different possible responses can be seen (Figure 12.3). In continuous variation, the results form a bell-shaped or unimodal distribution. With discontinuous variation the curve is bimodal or sometimes even trimodal. A discontinuous response suggests that the metabolism of the drug is under monogenic control. For example, if the normal metabolism of a drug is controlled by a dominant gene, *R*, and if some people are unable to metabolize the drug because they are homozygous for a recessive gene, *r*, there will be three classes of individual: RR, Rr, and rr. If the responses of RR and Rr are indistinguishable, a bimodal distribution will result. If RR and Rr are distinguishable, a trimodal distribution will result, each peak or mode representing a different genotype. A unimodal distribution implies that the

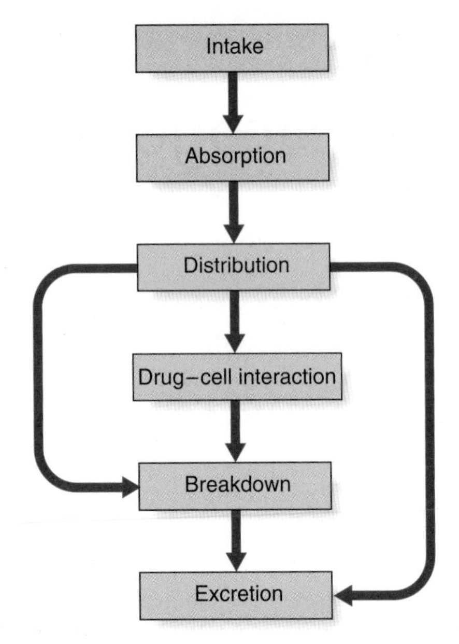

FIGURE 12.1 Stages of metabolism of a drug.

metabolism of the drug in question is under the control of many genes—i.e., is polygenic (p. 143).

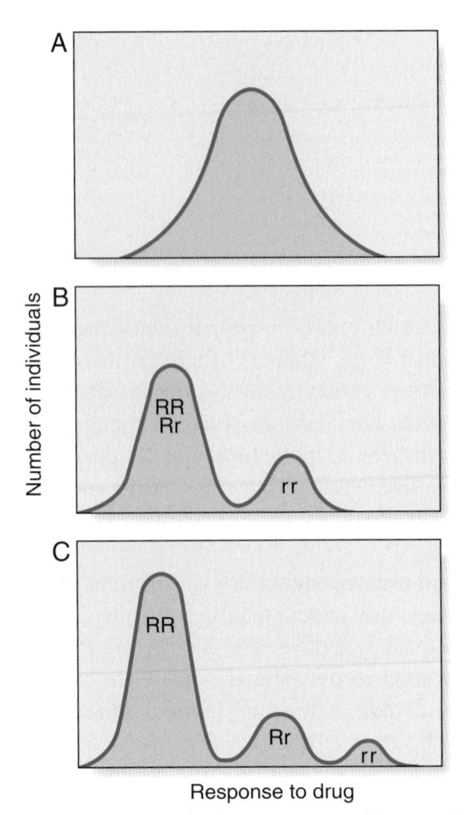

FIGURE 12.3 Various types of response to different drugs consistent with polygenic and monogenic control of drug metabolism. **A**, Continuous variation, multifactorial control of drug metabolism. **B**, Discontinuous bimodal variation. **C**, Discontinuous trimodal variation.

Genetic Variations Revealed by the Effects of Drugs

Among the best known examples of drugs that have been responsible for revealing genetic variation in response are isoniazid, succinylcholine, primaquine, coumarin anticoagulants, certain anesthetic agents, the thiopurines, and debrisoquine.

N-Acetyltransferase Activity

Isoniazid is one of the drugs used in the treatment of tuberculosis. It is rapidly absorbed from the gut, resulting in an initial high blood level that is slowly reduced as the drug is inactivated and excreted. The metabolism of isoniazid allows two groups to be distinguished: rapid and slow inactivators. In the former, blood levels of the drug fall rapidly after an oral dose; in the latter, blood levels remain high for some time. Family studies have shown that slow inactivators of isoniazid are homozygous for an autosomal recessive allele of the liver enzyme *N*-acetyltransferase, with lower activity levels. *N*-acetyltransferase activity varies in different populations. In the United States and Western Europe, about 50% of the population are slow inactivators, in contrast to the Japanese, who are predominately rapid inactivators.

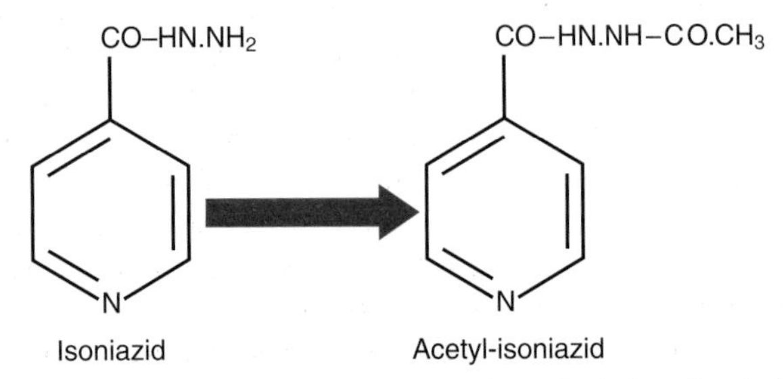

FIGURE 12.2 Acetylation of the antituberculosis drug isoniazid.

In some individuals, isoniazid can cause side effects such as polyneuritis, a systemic lupus erythematosus–like disorder, or liver damage. Blood levels of isoniazid remain higher for longer periods in slow inactivators than in rapid inactivators on equivalent doses. Slow inactivators have a significantly greater risk of developing side effects on doses that rapid inactivators require to ensure adequate blood levels for successful treatment of tuberculosis. Conversely, rapid inactivators have an increased risk of liver damage from isoniazid. Several other drugs are also metabolized by *N*-acetyltransferase, and therefore slow inactivators of isoniazid are also more likely to exhibit side effects. These drugs include hydralazine, which is an antihypertensive, and sulfasalazine, which is a sulfonamide derivative used to treat Crohn disease.

Studies in other animal species led to the cloning of the genes responsible for *N*-acetyltransferase activity in humans. This has revealed that there are three genes, one of which is not expressed and represents a pseudogene *(NATP)*, one that does not exhibit differences in activity between individuals *(NAT1)*, and a third *(NAT2)*, mutations in which are responsible for the inherited polymorphic variation. These inherited variations in *NAT2* have been reported to modify the risk of developing a number of cancers, including bladder, colorectal, breast, and lung cancer. This is thought to be through differences in acetylation of aromatic and heterocyclic amine carcinogens.

Succinylcholine Sensitivity

Curare is a plant extract used in hunting by certain South American Indian tribes that produces profound muscular paralysis. Medically, curare is used in surgical operations because of the muscular relaxation it produces. Succinylcholine, also known as suxamethonium, is another drug that produces muscular relaxation, though by a different mechanism from curare. Suxamethonium has the advantage over curare that the relaxation of skeletal and respiratory muscles and the consequent apnea (cessation of breathing) it induces is only short-lived. Therefore it is used most often in the induction phase of anesthesia for intubation. The anesthetist, therefore, needs to maintain respiration by artificial means for only 2 to 3 minutes before it returns spontaneously. However, about one patient in every 2000 has a period of apnea that can last 1 hour or more after the use of suxamethonium. It was found that the apnea in such instances could be corrected by transfusion of blood or plasma from a normal person. When a suxamethonium-induced apnea occurs the anesthetist has to maintain respiration until the effects of the drug have worn off.

Succinylcholine is normally destroyed in the body by the plasma enzyme pseudocholinesterase. In patients who are highly sensitive to succinylcholine, the plasma pseudocholinesterase in their blood destroys the drug at a markedly slower rate than normal, or in some very rare cases is entirely deficient. Succinylcholine sensitivity is inherited as an autosomal recessive trait due to mutations of the *CHE1* gene, and genetic testing may be offered to the relatives of a patient in whom a genetic predisposition has been identified.

Glucose 6-Phosphate Dehydrogenase Variants

For many years, quinine was the drug of choice in the treatment of malaria. Although it has been very effective in acute attacks, it is not effective in preventing relapses. In 1926 primaquine was introduced and proved to be much better than quinine in preventing relapses. However, it was not long after primaquine was introduced that some people were found to be sensitive to the drug. The drug could be taken for a few days with no apparent ill effects, and then suddenly some individuals would begin to pass very dark, often black, urine. Jaundice developed and the red cell count and hemoglobin concentration gradually fell as a consequence of hemolysis of the red blood cells. Affected individuals usually recovered from such a hemolytic episode, but occasionally the destruction of the red cells was extensive enough to be fatal. The cause of such cases of primaquine sensitivity was subsequently shown to be a deficiency in the red cell enzyme glucose 6-phosphate dehydrogenase (G6PD).

G6PD deficiency is inherited as an X-linked recessive trait (p. 114), rare in Caucasians but affecting about 10% of Afro-Caribbean males and relatively common in the Mediterranean. It is thought to be relatively common in these populations as a result of conferring increased resistance to the malarial parasite. These individuals are sensitive not only to primaquine, but also to many other compounds, including phenacetin, nitrofurantoin, and certain sulfonamides. G6PD deficiency is thought to be the first recognized pharmacogenetic disorder, having been described by Pythagoras around 500 BC.

Coumarin Metabolism

Coumarin anticoagulant drugs, such as warfarin, are used in the treatment of a number of different disorders to prevent the blood from clotting (e.g., after a deep venous thrombosis). Warfarin is metabolized by the cytochrome P450 enzyme encoded by the *CYP2C9* gene, and two variants (CYP2C9*2 and CYP2C9*3) result in decreased metabolism. Consequently, these patients require a lower warfarin dose to maintain their target international normalized ratio range and may be at increased risk of bleeding.

Debrisoquine Metabolism

Debrisoquine is a drug that was used frequently in the past for the treatment of hypertension. There is a bimodal distribution in the response to the drug in the general population. Approximately 5% to 10% of persons of European origin are poor metabolizers, being homozygotes for an autosomal recessive gene with reduced hydroxylation activity.

Molecular studies revealed that the gene involved in debrisoquine metabolism is one of the P450 family of genes on chromosome 22, known as *CYP2D6*. The mutations responsible for the poor metabolizer phenotype are heterogeneous; 18 different variants have been described.

CYP2D6 variation is important because this enzyme is involved in the metabolism of more than 20% of prescribed drugs, including the β-blockers metoprolol and carvedilol, the antidepressants fluoxetine and imipramine, the antipsychotics thioridazine and haloperidol, the painkiller codeine, and the anti-cancer drug tamoxifen.

Malignant Hyperthermia

Malignant hyperthermia (MH) is a rare complication of anesthesia. Susceptible individuals develop muscle rigidity as well as an increased temperature (hyperthermia), often as high as 42.3°C (108°F) during anesthesia. This usually occurs when halothane is used as the anesthetic agent, particularly when succinylcholine is used as the muscle relaxant for intubation. If it is not recognized rapidly and treated with vigorous cooling, the affected individual will die.

MH susceptibility is inherited as an autosomal dominant trait affecting approximately 1 in 10,000 people. The most reliable prediction of an individual's susceptibility status requires a muscle biopsy with in vitro muscle contracture testing in response to exposure to halothane and caffeine.

MH is genetically heterogeneous, but the most common cause is a mutation in the ryanodine receptor *(RYR1)* gene. Seven other candidate genes have been identified and variants in these genes may influence susceptibility within individual families. This observation may explain the discordant results of the in-vitro contracture test and genotype in members of some families that segregate *RYR1* mutations.

Thiopurine Methyltransferase

A group of potentially toxic substances known as the thiopurines, which include 6-mercaptopurine, 6-thioguanine, and azathioprine, are used extensively in the treatment of leukemia to suppress the immune response in patients with autoimmune disorders such as systemic lupus erythematosus and to prevent rejection of organ transplants. They are effective drugs clinically but have serious side effects, such as leukopenia and severe liver damage. Azathioprine is reported to cause toxicity in 10% to 15% of patients and it may be possible to predict those patients susceptible to side effects by analyzing genetic variation within the thiopurine methyltransferase (*TPMT*) gene. This gene encodes an enzyme responsible for methylation of thiopurines, and approximately two-thirds of patients who experience toxicity have one or more variant alleles.

Dihydropyrimidine Dehydrogenase

Dihydropyrimidine dehydrogenase (DPYD) is the initial and rate-limiting enzyme in the catabolism of the chemotherapeutic drug 5-fluorouracil (5FU). Deficiency of DPYD is recognized as an important pharmacogenetic factor in the etiology of severe 5FU-associated toxicity. Measurement of DPYD activity in peripheral blood mononuclear cells or genetic testing for the most common *DPYD* gene mutation (a splice site mutation, IVS14+1G>A, which results in the deletion of exon 14) may be warranted in cancer patients before the administration of 5FU.

Pharmacogenetics

Increased understanding of the influence of genes on the efficacy and side effects of drugs has led to the promise of **personalized** or **individualized medicine**, where the treatment for a particular disease is dependent on the individual's genotype.

Maturity-Onset Diabetes of the Young

Maturity-onset diabetes of the young is a monogenic form of diabetes characterized by young age of onset (often before the age of 25 years), dominant inheritance and β-cell dysfunction (p. 236). Patients with mutations in the *HNF1A* or *HNF4A* genes are sensitive to sulfonylureas (Figure 12.4) and may experience episodes of hypoglycemia on standard doses. However, this sensitivity is advantageous at lower

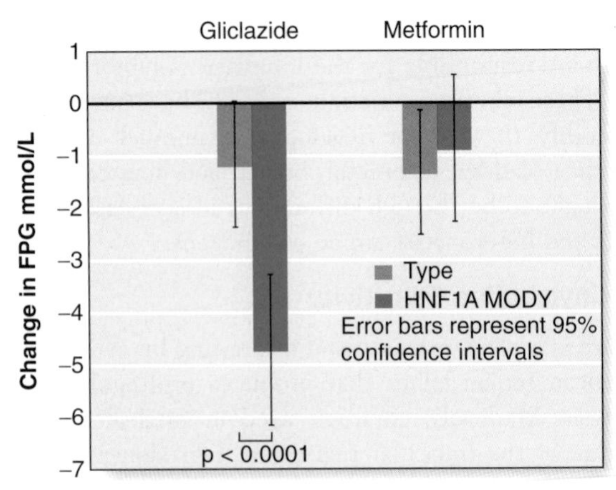

FIGURE 12.4 Response to the sulphonylurea gliclazide and the type 2 diabetes drug metformin in patients with HNF1A Maturity-onset diabetes of the young (MODY) and type 2 diabetes. FPG is fasting plasma glucose. Patients (*n* = 18 in each group) were treated with each drug for 6 weeks in a randomized trial. (Modified from Pearson ER, Starkey BJ, Powell RJ, et al 2003 Genetic cause of hyperglycaemia and response to treatment in diabetes. Lancet 362[9392]:1275–1281.)

doses, and sulfonylureas are the recommended oral treatment in this genetic subgroup.

Neonatal Diabetes

The most frequent cause of permanent neonatal diabetes is an activating mutation in the *KCNJ11* or *ABCC8* genes, which encode the Kir6.2 and SUR1 subunits of the ATP-sensitive potassium (K-ATP) channel in the pancreatic β cell (p. 236). The effect of such mutations is to prevent K-ATP channel closure by reducing the response to ATP. Because channel closure is the trigger for insulin secretion, these mutations result in diabetes. Defining the genetic etiology for this rare subtype of diabetes has led to improved treatment, because the majority of patients can be treated successfully with sulfonylurea tablets instead of insulin. These drugs bind to the sulfonylurea receptor subunits of the K-ATP channel to cause closure independently of ATP, thereby triggering insulin secretion (Figure 12.5). High-dose sulfonylurea therapy results in improved glycemic control with fewer hypoglycemic episodes and, for some patients, an Hb A1c level (this is a measure of glycemic control) within the normal range.

Pharmacogenomics

Pharmacogenomics is defined as the study of the interaction of an individual's genetic makeup and response to a drug. The key distinction between pharmacogenetics and pharmacogenomics is that the former describes the study of variability in drug responses attributed to individual genes and the latter describes the study of the entire genome related to drug response. The expectation is that inherited variation at the DNA level results in functional variation in the gene products that play an essential role in determining the variability in responses, both therapeutic and adverse,

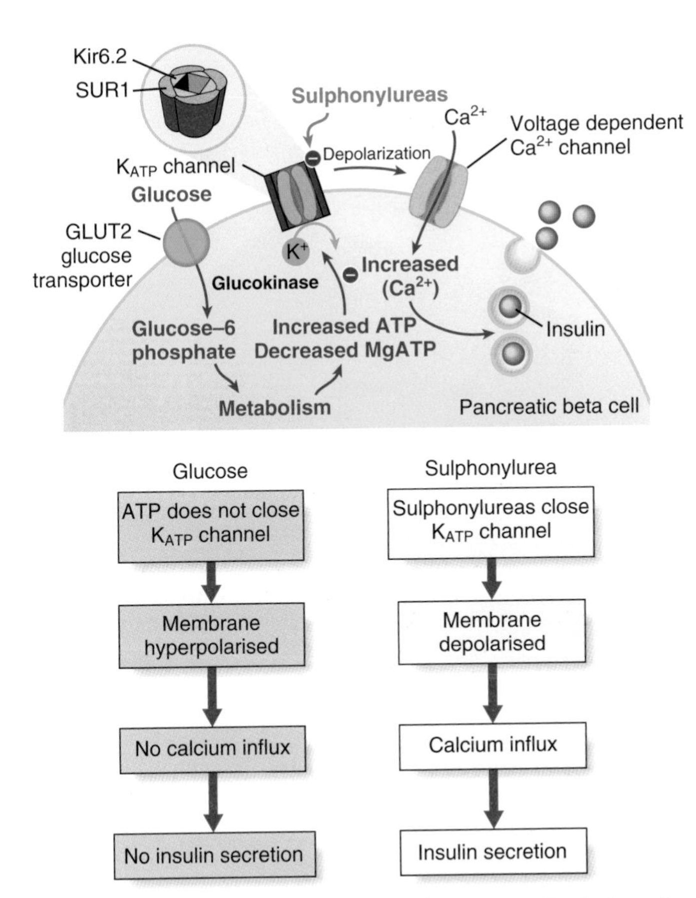

FIGURE 12.5 Insulin secretion in the pancreatic beta cell. Activating mutations in the genes encoding the K_{ATP} channel subunits Kir6.2 and SUR1 prevent closure of the channel in the presence of glucose. Sulphonylureas bind to the SUR1 subunit to close the channel and restore insulin secretion. (Courtesy Professor A.T. Hattersley, Peninsula Medical School, Exeter, UK.)

to a drug. If polymorphic DNA sequence variation occurs in the coding portion or regulatory regions of genes, it is likely to result in variation in the gene product through alteration of function, activity, or level of expression. Automated analysis of genome-wide single nucleotide polymorphisms (SNPs) (p. 67) allows the possibility of identifying genes involved in drug metabolism, transport and receptors that are likely to play a role in determining the variability in efficacy, side effects and toxicity of a drug.

The availability of whole-genome SNP maps will enable an SNP profile to be created for patients who experience adverse events or who respond clinically to the drug (efficacy). An individual's whole-genome SNP type has been described as an 'SNP print'. However, this raises issues pertaining to the disclosure of information of uncertain significance that is later shown to be associated with an adverse outcome unrelated to the reason for the original test. An example is apolipoprotein E (ApoE) genotyping, where ApoE ε4 was first reported to be associated with variation in cholesterol levels but later with age of onset of Alzheimer disease.

Adverse Events

It is estimated that around 15% of hospital inpatients will be affected by an adverse drug reaction. The objective of adverse-event pharmacogenetics is to identify a genetic profile that characterizes patients who are more likely to suffer such an adverse event. The best known example is abacavir, a reverse transcriptase inhibitor used to treat human immunodeficiency virus (HIV) infection. Approximately 5% of patients show potentially fatal hypersensitivity to abacavir and this limited its use. A strong association with the human leukocyte antigen allele B*5701 was proven in 2002. Today testing for B*5701 is routine practice before abacavir is prescribed.

At least 10% of Africans, North Americans, and Europeans are homozygous for a variant in the promoter of the *UGT1A1* gene (*UGT1A1**28). This results in reduced glucuronidation of irinotecan, a drug used to treat colorectal cancer, and increases the risk of severe neutropenia if exposed to the standard dose. A simple polymerase chain reaction–based test for *UGT1A1**28 can be used to determine the appropriate treatment dose.

Efficacy

There is no doubt that the cost-effectiveness of drugs is improved if they are prescribed only to those patients likely to respond to them. Several drugs developed for the treatment of various cancers have different efficacy depending on the molecular biology of the tumour (see Table 12.1). For example, herceptin (trastuzumab) is an antibody that targets overexpression of *HER2/neu* protein observed in approximately one-third of patients with breast cancer. Consequently, patients are prescribed herceptin only if their tumor has been shown to overexpress *HER2/neu*.

Gleevec (imatinib) is a protein tyrosine kinase inhibitor that has been used to treat chronic myeloid leukemia since 2001. It is a very effective treatment that works by binding the BCR-ABL fusion protein resulting from the t(9;22) translocation. This is an example of effective drug design resulting from knowledge of the molecular etiology. More recently it has been also shown to be effective in the treatment of gastrointestinal stromal tumours that harbour *KIT* mutations.

Table 12.1 Examples of Drugs Effective for the Treatment of Specific Cancers

Type of Cancer	Characteristic	Drug
Breast	*HER2* overexpression	Herceptin (trastuzumab)
Chronic myeloid leukemia	t(9;22) BCR-*ABL* fusion	Gleevec (imatinib)
Non–small-cell lung cancer	*EGFR* activating mutation	Iressa (gefitinib) or Tarceva (erlotinib)
Gastrointestinal stromal tumour	*KIT* or *PDGFRA* activating mutation	Gleevec (imatinib)

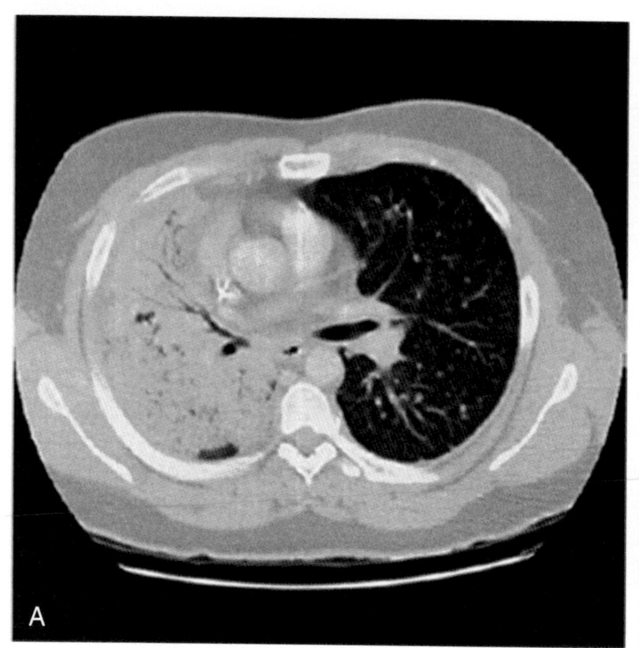

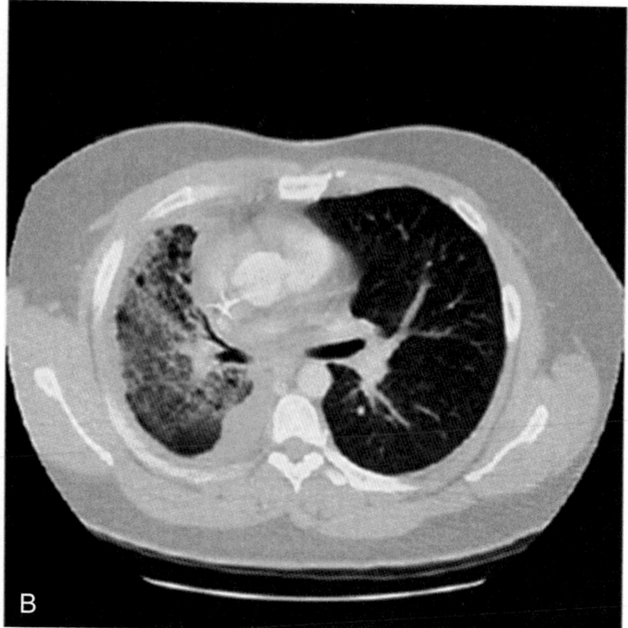

FIGURE 12.6 Example of the response to gefitinib in a patient with non–small-cell lung cancer and an activating *EGFR* mutation. A computed tomographic scan of the chest shows a large mass in the right lung before treatment (**A**) and marked improvement 6 weeks after gefitinib was initiated (**B**). (Reproduced with permission from Lynch TJ, Bell DW, Sordella R, et al 2004 Activating mutations in the epidermal growth factor receptor underlying responsiveness of non-small-cell lung cancer to gefitinib. N Engl J Med 350(21):2129–2139.)

Approximately 13% of patients with non–small-cell lung cancer have an activating *EGFR* mutation. These mutations increase the activity of the epidermal growth factor receptor tyrosine kinase domain so that the receptor is constitutionally active in the absence of epidermal growth factor. This leads to increased proliferation, angiogenesis and metastasis. Drugs designed to block the EGFR tyrosine kinase domain and inhibit these effects have been developed. Patients with lung tumours harbouring an activating *EGFR* mutation can show a dramatic response to treatment with these drugs (gefitinib and erlotinib) as shown in Figure 12.6.

Genetic profiling is a step toward personalized medicine. This information can be used to select the appropriate treatment at the correct dosage and to avoid adverse drug reactions.

FURTHER READING

Beutler E 1991 Glucose-6-phosphate dehydrogenase deficiency. N Engl J Med 324:169–174

Review of an important ethnic pharmacogenetic polymorphism.

Goldstein DB, Tate SK, Sisodiya SM 2003 Pharmacogenetics goes genomic. Nature Genet Rev 4:937–947

Review of pharmacogenetics/genomics.

Neumann DA, Kimmel CA 1998 Human variability in response in chemical exposures: measures, modelling, and risk assessment. London: CRC Press

A detailed discussion of the inherited human variability to exposure to the toxic effects of environmental chemicals.

Newman W, Payne K 2008 Removing barriers to a clinical pharmacogenetics service. Personalized Med 5:471–480

A review article describing the application of pharmacogenetics to current clinical practice.

Pearson ER, Flechtner I, Njolstad PR, et al 2006 Switching from insulin to oral sulfonylureas in patients with diabetes due to Kir6.2 mutations. Neonatal Diabetes International Collaborative Group. N Engl J Med 355:467–477

Pharmacogenetic treatment of monogenic diabetes.

Roses AD 2008 Pharmacogenetics in drug discovery and development: a translational perspective. Nat Rev Drug Discovery 7:807–817

A recent article describing the role of pharmacogenetics in drug development.

Vogel F, Buselmaier W, Reichert W, Kellerman G, Berg P eds 1978 Human genetic variation in response to medical and environmental agents: pharmacogenetics and ecogenetics. Berlin: Springer

One of the early definitive outlines of the field of pharmacogenetics.

ELEMENTS

1 Pharmacogenomics is defined as the study of the interaction of an individual's genetic makeup and response to a drug. The key distinction between pharmacogenetics and pharmacogenomics is that the former describes the study of variability in drug responses attributed to individual genes and the latter describes the study of the entire genome related to drug response.

2 The metabolism of many drugs involves biochemical modification, often by conjugation with another molecule, which usually takes place in the liver. This biochemical transformation facilitates excretion of the drug.

3 The ways in which many drugs are metabolized vary from person to person and can be genetically determined. In some instances, the biochemical basis is understood. For example, persons differ in the rate at which they inactivate the antituberculosis drug isoniazid by acetylation in the liver, being either rapid or slow inactivators. Slow inactivators have an increased risk of toxic side effects associated with isoniazid therapy.

4 In some instances, genetic variation can be revealed exclusively by exposure to drugs. One such example is malignant hyperthermia. This rare disorder is associated with the use of certain anesthetic agents and muscle relaxants in general anesthesia.

5 Knowledge regarding the genetic etiology of disease can lead to tailored treatments. Examples include sulfonylurea therapy for certain monogenic subtypes of diabetes, Herceptin for breast cancers showing *HER2* overexpression, and Gleevec (imatinib) for chronic myeloid leukemia. Testing for B*5701 status before prescribing abacavir is now routine for patients with HIV infection to reduce the risk of potentially fatal hypersensitivity.

CHAPTER 13

Immunogenetics

Medicinal discovery,
It moves in mighty leaps,
It leapt straight past the common cold
And gave it us for keeps.
PAM AYERS

Immunity

The immune system in all its forms is our defense mechanism against the armies of microorganisms, insects, and other infectious agents that, numerically, dwarf the human population. Effective defense mechanisms are absolutely essential to mankind's survival; in order to understand the inherited disorders of immunity, we must first understand the fundamentals of the genetic basis of immunity.

Immune defense mechanisms can be divided into two main types: **innate immunity**, which includes a number of non-specific systems that do not require or involve prior contact with the infectious agent, and **specific acquired** or **adaptive immunity**, which involves a tailor-made immune response that occurs after exposure to an infectious agent. Both types can involve either **humoral immunity**, which combats extracellular infections, or **cell-mediated immunity**, which fights intracellular infections.

Innate Immunity

The first simple defense against infection is a mechanical barrier. The skin functions most of the time as an impermeable barrier, but in addition the acidic pH of sweat is inhibitory to bacterial growth. The membranes lining the respiratory and gastrointestinal tracts are protected by mucus. In the respiratory tract, further protection is provided by ciliary movement, whereas other bodily fluids contain a variety of bactericidal agents, such as lysozymes in tears. If an organism succeeds in invading the body, a healthy immune system reacts immediately by recognizing the alien intruder and a chain of response is triggered.

Cell-Mediated Innate Immunity

Phagocytosis

Two major cell types go on the offensive when a foreign microorganism invades—macrophages and neutrophils. Macrophages are the mature form of circulating monocytes that migrate into tissues and occur primarily around the basement membrane of blood vessels in connective tissue, lung, liver, and the lining of the sinusoids of the spleen and the medullary sinuses of the lymph nodes. They are believed to play a key role in the orchestration of both the innate and adaptive responses, and can recognise invading microorganisms through surface receptors able to distinguishing between self and pathogen. Recognition of the foreign material leads to phagocytosis by the macrophage, followed rapidly by neutrophils recruited from the circulation during the inflammatory process. The activation of the macrophage triggers the inflammatory process through the release of inflammatory mediators. The invading organism is destroyed by fusion with intracellular granules of the phagocyte and exposure to the action of hydrogen peroxide, hydroxyl radicals, and nitrous oxide (Figure 13.1).

The Toll-like Receptor Pathway

A key component of cell-mediated immunity is the **Toll-like receptor (TLR)** pathway. TLRs are conserved transmembrane receptors which in fruit fly embryos play a critical role in dorsal-ventral development. However, their mammalian homologs function in innate immune responses and microbial recognition (in adult *Drosophila*, the pathway is responsible for the formation of antimicrobial peptides) and belong to the interleukin-1/TLR superfamily. The superfamily has two subgroups based on the extracellular characteristics of the receptor—i.e., whether they possess an immunoglobulin-like domain or leucine-rich repeats. TLRs typically have extracellular leucine-rich repeats.

There are 10 TLRs in man, each receptor being responsible for recognition of a specific set of pathogen-associated molecular patterns. TLR2 has been well characterised and has an essential role in the detection of invading pathogens, recognizing peptidoglycans and lipoproteins associated with gram-positive bacteria, as well as a host of other microbial and endogenous ligands. TLR2's primary function is therefore lipoprotein-mediated signaling, and activation of the pathway by recognition of its ligand results in activation of the transcription factor NF-κB, which in turn results in the increased expression of co-stimulatory molecules and inflammatory cytokines (Figure 13.3). These cytokines help mediate migration of dendritic cells from infected tissue to lymph nodes, where they may encounter and activate leukocytes involved in the adaptive immune response. The signaling pathways used by TLRs share many of the same proteins as the interleukin-1 receptor (IL-1R) pathway (Figure 13.2). Activation of TLR leads to recruitment of the MyD88 (this is sometimes known as the MyD88-dependent pathway) which mediates the interaction between IL-1R associated kinases 1 and 4 (IRAK1 and IRAK4).

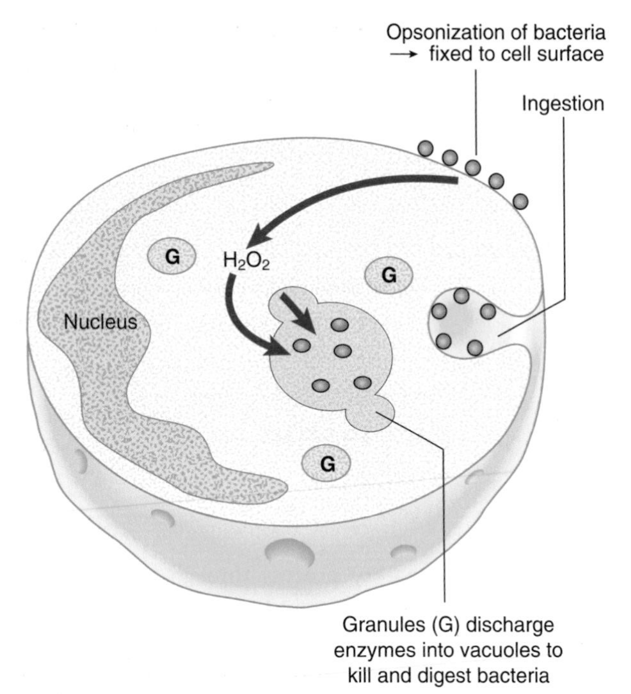

FIGURE 13.1 Phagocytosis and the pathways involved in intracellular killing of microorganisms.

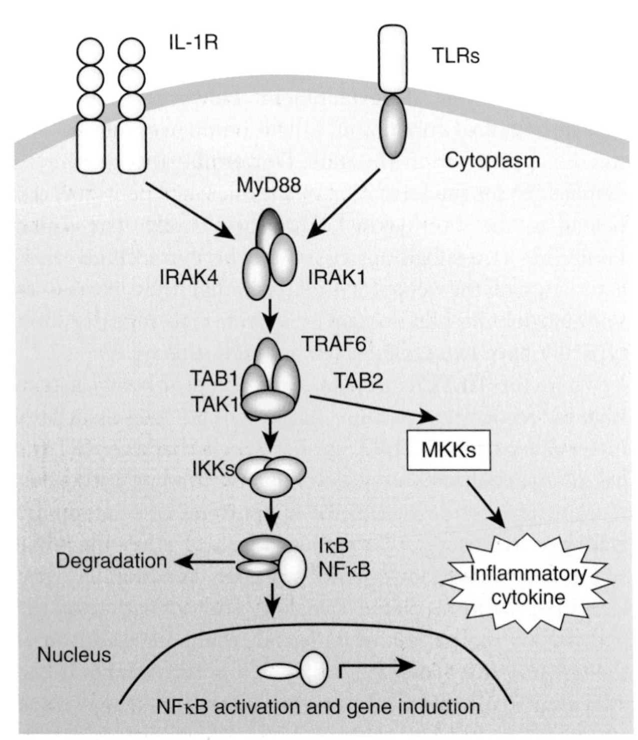

FIGURE 13.2 The Toll-like receptor (TLR) and interleukin-1 receptor (IL-1R) pathways, which share many of the same proteins. Activation of TLR2 and other TLRs, via NFκB activation and gene induction, leads to dendritic cell maturation, upregulation of expression of the major histocompatibility complex and co-stimulatory molecules, and production of immuno-stimulatory cytokines. *IKK,* I kappa kinase; *IkB,* NFκB inhibitor; *IRAK,* IL-1R–associated protein kinases; *MKK,* MAP kinases; *MyD88,* adapter molecule; *TAB1,* TAK1-binding protein 1; *TAB2,* TAK1-binding protein 2; *TAK1,* transforming growth factor-β–activated kinase; *TRAF6,* tumor necrosis factor receptor-associated factor 6.

The activation of the Toll pathway has several important effects in inducing innate immunity. These effects include the production of cytokines and chemokines, including IL-1, IL-6, and TNF-α (tumor necrosis factor-alpha), which have local effects in containing infection and systemic effects with the generation of fever and induction of acute phase responses, including production of C-reactive protein. One important medical condition related to the Toll pathway is septic shock, as activation of the Toll pathway by certain ligands induces systemic release of TNF-α. There are also important health-related consequences that result from *TLR2* deficiency or mutation. *TLR2* deficient mice are susceptible to infection by Gram-positive bacteria as well as meningitis from *Streptococcus pneumoniae*.

Extracellular Killing

Virally infected cells can be killed by large granular lymphocytes, known as **natural killer (NK) cells**. These have carbohydrate-binding receptors on their cell surface that recognize high molecular weight glycoproteins expressed on the surface of the infected cell as a result of the virus taking over the cellular replicative functions. NK cells play an early role in viral infections and are activated by cytokines from macrophages. They recognise virally infected cells through either changes in glycoproteins or in the expression of major histocompatibility complex (MHC) class 1 on virally infected host cells. Attachment to the infected cells results in the release of a number of agents, which in turn results in damage to the membrane of the infected cell, leading to cell death.

Humoral Innate Immunity

Several soluble factors are involved in innate immunity; they help to minimize tissue injury by limiting the spread of infectious microorganisms. These are called the **acute-phase proteins** and include C-reactive protein, mannose-binding protein, and serum amyloid P component. The first two act by facilitating the attachment of one of the components of complement, C3b, to the surface of the microorganism, which becomes opsonized (made ready) for adherence to phagocytes, whereas the latter binds lysosomal enzymes to connective tissues. In addition, cells infected by virus synthesize and secrete interferon-α and interferon-β, which have a role in promoting the cellular response to viral infection by NK cell activation and upregulation of the MHC class I. In addition, **interferon** interferes with viral replication by reducing messenger RNA (mRNA) stability and interfering with translation.

Complement

The complement system is a complex of 20 or so plasma proteins that cooperate to attack extracellular pathogens. Although the critical role of the system is to opsonize pathogens, it also recruits inflammatory cells and kills pathogens directly through membrane attack complexes. The complement system can be activated through three pathways: the classical pathway, the alternative pathway,

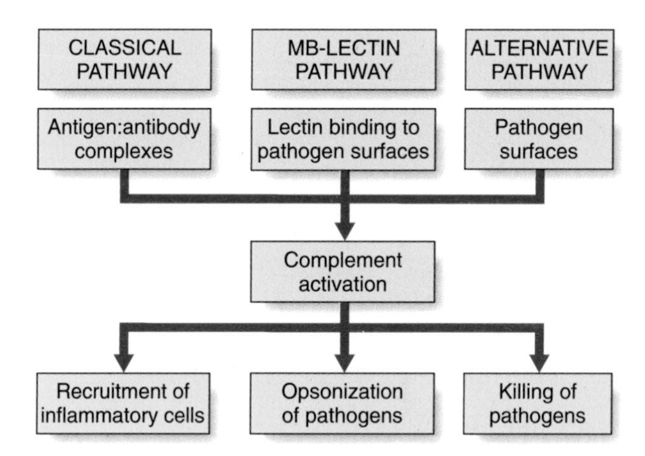

FIGURE 13.3 The classic and alternative pathways of complement activation. The main functions of complement are recruitment of inflammatory cells, opsonization of pathogens, and killing of pathogens.
MB, mannose-binding

and the mannose-binding **lectin** (MBL) pathway (see Figure 13.3).

Complement nomenclature, like much else in immunology, can be confusing. Each component is designated by the letter C, followed by a number. But they were numbered in order of their discovery rather than the sequence of reactions. The reaction sequence is C1, C4, C3, C5, C6, C7, C8, and C9. The product of each cleavage reaction is designated by letters, the larger fragment being 'b' (b = big), and the smaller fragment 'a'. In the lectin pathway, MBL in the blood binds another protein, a serine protease called MASP (MBL-associated serine protease). When MBL binds to its target (for example, mannose on the surface of a bacterium), the MASP protein functions like a convertase to clip C3 into C3a and C3b. C3 is abundant in the blood, so this happens very efficiently. The other two complement pathways also converge toward C3 convertase, which cleaves C3. C3a mediates inflammation while C3b binds to the pathogen surface, coating it and acting as an opsonin. The effector roles of the major complement proteins can be summarized according to function as follows (Figure 13.4):

1. Opsonisation: C3b and C4b are opsonins that coat foreign organisms, greatly enhancing their phagocytosis—phagocytes have receptors that recognize complement proteins bound to pathogen.
2. Inflammation: C5a, as well as C4a and C3a, are inflammatory activators that induce vascular permeability, and recruit and activate phagocytes.
3. Lysis: C5b binds and recruits C6 and C7, eventually forming a complex with C8 Maturity-onset diabetes of the young C5b678—which catalyses the polymerisation of the final component C9, forming a transmembrane pore of ~10 nm diameter, and cell lysis. This assembly is known as the membrane attack complex (MAC).
4. Immune complex clearance: Complement has a critical role in removing immune complexes from the circulation. The immune complex binds C4b and C3b, which then binds to receptors on red blood cells and the complexes are transported to the liver and spleen, where the complexes are given up to phagocytes for destruction.

There are clinical consequences relating to mutations in the genes of these pathways. The frequency of mutations of the *MBL2* gene in the general population may be 5% to 10%. Although most individuals with MBL deficiency from mutations and promoter polymorphisms in *MBL2* are healthy, there is an increased risk, severity, and frequency of infections and autoimmunity. The deficiency has been reported to be particularly common in infants with recurrent respiratory tract infection, otitis media, and chronic diarrhea.

Specific Acquired Immunity

Many infective microorganisms have, through mutation and selective pressures, developed strategies to overcome or evade the mechanisms associated with innate immunity. There is a need, therefore, to be able to generate specific acquired or adaptive immunity. This can, as with innate immunity, be separated into both humoral and cell-mediated processes.

Humoral Specific Acquired Immunity

The main mediators of humoral specific acquired immunity are immunoglobulins or antibodies. Antibodies are able to recognize and bind to surface antigens of infecting microorganisms, leading to the activation of phagocytes and the initiation of the **classic pathway** of complement, resulting in the generation of the MAC (see Figure 13.4) and availability of other complement effector functions. Exposure to a specific antigen results in the clonal proliferation of a small lymphocyte derived from the bone marrow (hence 'B' lymphocytes), resulting in mature antibody-producing cells or **plasma cells**.

Lymphocytes capable of producing antibodies express on their surface copies of the immunoglobulin (Ig) for which they code, which acts as a surface receptor for antigen. Binding of the antigen, in conjunction with other MASPs, results in signal transduction leading to the clonal expansion and production of antibody. In the first instance this results in the **primary response** with production of IgM and subsequently IgG. Re-exposure to the same antigen results in enhanced antibody levels in a shorter period of time, known as the **secondary response**, reflecting what is known as antigen-specific **immunological memory**.

Immunoglobulins

The immunoglobulins, or antibodies, are one of the major classes of serum protein. Their function, both in the recognition of antigenic variability and in effector activities, was initially revealed by protein studies of their structure, and later by DNA studies.

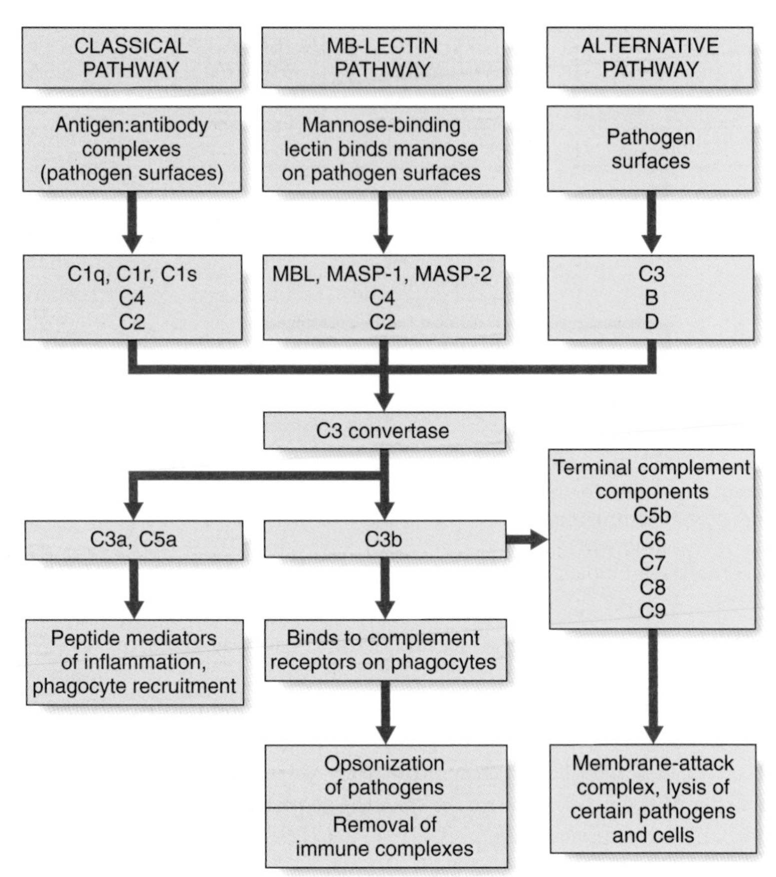

FIGURE 13.4 Overview of the main components and effector actions of complement. Note that the MBL pathway involves the MBL protein, MASP-1, MASP-2, C4, and C2. MASP acts as a C3 convertase, creating a C3b fragment from C3. C3b attaches to the pathogen surface and binds to receptors on phagocytes, leading to opsonization. C3b can also combine with other proteins on the pathogen surface and form a membrane attack complex.

Immunoglobulin Structure

Papaine, a proteolytic enzyme, splits the immunoglobulin molecule into three fragments. Two of the fragments are similar, each containing an antibody site capable of combining with a specific antigen and therefore referred to as the **antigen-binding fragment** or **Fab**. The third fragment can be crystalized and was therefore called **Fc**. The Fc fragment determines the secondary biological functions of antibody molecules, binding complement and Fc receptors on a number of different cell types involved in the immune response.

The immunoglobulin molecule is made up of four polypeptide chains—two 'light' (L) and two 'heavy' (H)—of approximately 220 and 440 amino acids in length, respectively. They are held together in a Y-shape by disulfide bonds and non-covalent interactions. Each Fab fragment is composed of L chains linked to the amino-terminal portion of the H chains, whereas each Fc fragment is composed only of the carboxy-terminal portion of the H chains (Figure 13.5).

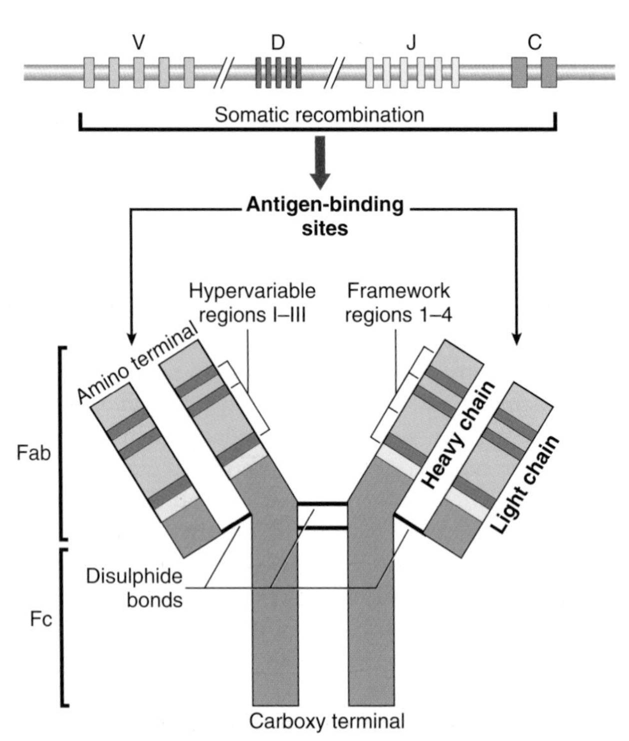

FIGURE 13.5 Model of antibody molecule structure.

Immunoglobulin Isotypes, Subclasses, and Idiotypes

There are five different types of heavy chain, designated respectively as γ, μ, α, δ, and ε, one each for the five major

Table 13.1 Classes of Human Immunoglobulin (Ig)

Class	Mol. wt (Da)	Serum Concentration (mg/mL)	Antibody Activity	Complement Fixation	Placental Transfer
IgG	150,000	8–16	Binds to microorganisms and neutralizes bacterial toxins	+	+
IgM	900,000	0.5–2	Produced in early immune response, especially in bacteremia	+	–
IgA	160,000	1.4–4	Guards mucosal surfaces	+	–
IgD	185,000	0–0.4	On lymphocyte cell surface, involved in control of activation and suppression	–	–
IgE	200,000	Trace	In parasitic and allergic reactions	–	–

antibody classes—the **isotypes**: IgG, IgM, IgA, IgD, and IgE, respectively. The L chains are of two types—kappa (κ) or lambda (λ), and these occur in all five classes of antibody, but only one type occurs in each individual antibody. Thus, the molecular formula for IgG is $\lambda_2\gamma_2$ or $\kappa_2\gamma_2$. The characteristics of the various classes of antibody are outlined in Table 13.1. In addition, there are four IgG subclasses—IgG1, IgG2, IgG3, and IgG4—and two IgA subclasses—IgA1 and IgA2—that differ in their amino acid sequence and interchain disulfide bonds. Individual antibody molecules that recognize specific antigens are known as idiotypes.

Immunoglobulin Allotypes

The five immunoglobulin classes occur in all normal individuals, but allelic variants, or what are known as antibody **allotypes** of these classes, have also been identified. These are the *Gm* system associated with the heavy chain of IgG, the *Am* system associated with the IgA heavy chain, the *Km* and *Inv* systems associated with the κ light chain, the Oz system for the λ light chain and the *Em* allotype for the IgE heavy chain. The Gm and Km systems are independent of each other and are polymorphic (p. 135), the frequencies of the different alleles varying in different ethnic groups.

Generation of Antibody Diversity

It could seem paradoxical for a single protein molecule to exhibit sufficient structural heterogeneity to have specificity for a large number of different antigens. Different combinations of H and L chains could, to some extent, account for this diversity. It would, however, require thousands of structural genes for each chain type to provide sufficient variability for the large number of antibodies produced in response to the equally large number of antigens to which individuals can be exposed. Our initial understanding of how this could occur came from persons with a malignancy of antibody-producing cells—**multiple myeloma**.

Multiple Myeloma

People with multiple myeloma make a single or monoclonal antibody species in large abundance, which in a proportion of patients is detected in their urine. This is known as **Bence Jones protein** and consists of antibody L chains. The amino-terminal ends of this protein molecule in different patients are quite variable in sequence, whereas the carboxy-terminal ends are relatively constant. These are called the **variable**, or **V**, and **constant**, or **C**, regions, respectively. However, the V regions of different myeloma proteins show four regions that vary little from one antibody to another, known as **framework regions** (FR 1–4), and three markedly variable regions interspersed between these, known as **hypervariable regions** (HV I–III) (see Figure 13.5).

DNA Studies of Antibody Diversity

In 1965 Dreyer and Bennett proposed that an antibody could be encoded by separate 'genes' in germline cells that undergo rearrangement or, as they termed it, 'scrambling', in lymphocyte development. Comparison of the restriction maps of the DNA segments coding for the C and V regions of the immunoglobulin λ light chains in embryonic and antibody-producing cells revealed that they were far apart in the former but close together in the latter. Detailed analysis revealed that the DNA segments coding for the V and C regions of the light chain are separated by some 1500 base-pairs (bp) in antibody-producing cells. The intervening DNA segment was found to code for a **joining**, or **J**, region immediately adjacent to the V region of the light chain. The κ L-chain was shown to have the same structure. Cloning and DNA sequencing of H-chain genes in germline cells revealed that they have a fourth region, called **diversity**, or **D**, between the V and J regions.

There are estimated to be some 60 different DNA segments coding for the V region of the H-chain, 40 for the V region of the κ L-chain, and 30 for the λ L-chain V region. Six functional DNA segments code for the J region of the H-chain, five for the J region of the κ L-chain, and four for the J region of the λ L-chain. A single DNA segment codes for the C region of the κ L-chain, seven for the C region of the λ L-chain and 11 functional DNA segments code for the C region of the different classes of H-chain. There are also 27 functional DNA segments coding for the D region of the H-chain (Figure 13.6).

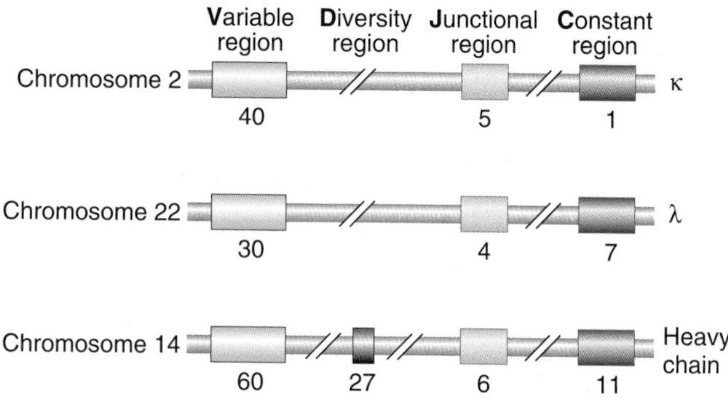

FIGURE 13.6 Estimated number of the various DNA segments coding for the κ, λ, and various heavy chains.

The genomic regions in question also contain a large number of unexpressed DNA sequences or pseudogenes (p. 17). Although the coding DNA segments for the various regions of the antibody molecule can be referred to as 'genes', use of this term in regard to antibodies has deliberately been avoided because they could be considered an exception to the general rule of 'one gene–one enzyme (or protein)' (p. 167).

Antibody Gene Rearrangement

The genes for the κ and λ L-chains and the H-chains are located on chromosomes 2, 22, and 14, respectively. Only one of each of the relevant types of DNA segment is expressed in any single antibody molecule. The DNA coding segments for the various portions of the antibody chains on these chromosomes are separated by DNA that is non-coding. Somatic recombinational events involved in antibody production involve short conserved recombination signal sequences that flank each germline DNA segment (Figure 13.7). Further diversity occurs by variable mRNA splicing at the V–J junction in RNA processing and by somatic mutation of the antibody genes. These mechanisms readily account for the antibody diversity seen in nature, even though it is still not entirely clear how particular DNA segments are selected to produce an antibody to a specific antigen.

Class Switching of Antibodies

There is a normal switch of antibody class produced by B cells on continued, or further, exposure to antigen—from IgM, the initial class of antibody produced in response to exposure to an antigen, to IgA or IgG. This **class switching** involves retention of the specificity of the antibody to the same antigen. Analysis of class switching in a population of cells derived from a single B cell has shown that both classes of antibody have the same antigen-binding sites, having the same V region but differing only in their C region. Class switching occurs by a somatic recombination event that involves DNA segments designated S (for switching) that lead to looping out and deletion of the intervening DNA. The result is to eliminate the DNA segment coding for the C region of the H-chain of the IgM molecule, and to bring the gene segment encoding the C region of the new class of H-chain adjacent to the segment encoding the V region (see Figure 13.7).

The Immunoglobulin Gene Superfamily

Several other molecules involved in the immune response have been shown to have structural and DNA sequence homology to the immunoglobulins. This involves a 110–amino acid sequence characterized by a centrally placed disulfide bridge that stabilizes a series of antiparallel β strands into an 'antibody fold'. This group of structurally similar molecules has been called the **immunoglobulin superfamily** (p. 16). It consists of eight multigene families that, in addition to the κ and λ L-chains and different classes of H-chain, include the chains of the T-cell receptor (p. 16), the class I and II MHC, or human leukocyte antigens (HLA) (p. 200). Other molecules in this group include the T-cell CD4 and CD8 cell surface receptors, which cooperate with T-cell receptors in antigen recognition, and the intercellular adhesion molecules-1, -2, and -3, which are involved in leukocyte-endothelial adhesion and extravasation, T-cell activation, and T-cell homing.

Antibody Engineering

At the beginning of the 20th century, Paul Ehrlich proposed the idea of the 'magic bullet'—the hope that one day there might be a compound that would selectively target a disease-causing organism. Today we have monoclonal antibodies (mAb) and, for almost any substance, it is possible to create a specific antibody that binds that substance. Monoclonal antibodies are the same because they are made by one type of immune cell which are all clones of a unique parent cell.

In the 1970s it was understood that the B-cell cancer multiple myeloma produced a single type of antibody—

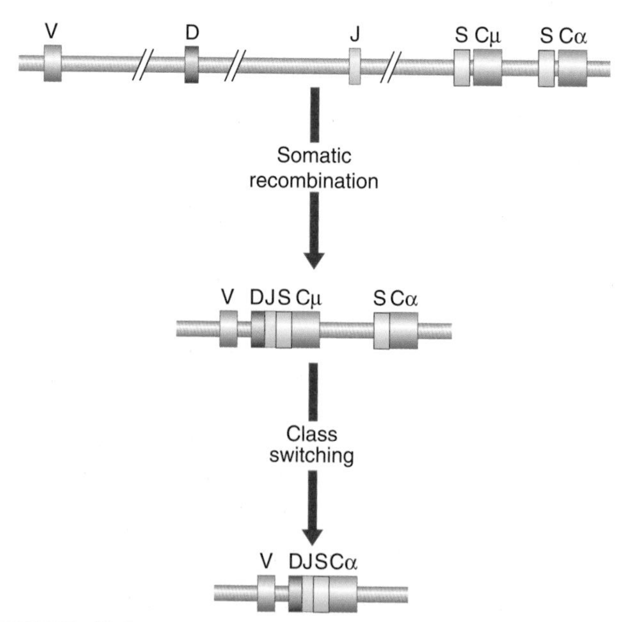

FIGURE 13.7 Immunoglobulin heavy-chain gene rearrangement and class switching.

a paraprotein. The structure of antibodies was studied from this but it was not possible to produce identical antibodies specific to a given antigen. Myeloma cells cannot grow because they lack hypoxanthine-guanine-phosphoribosyl transferase, which is necessary for DNA replication. Typically, mAb are made by fusing myeloma cells with spleen cells from a mouse (or rabbit) that has been immunized with the desired antigen. They are then grown in medium which is selective for these hybrids—the spleen cell partner supplies hypoxanthine-guanine-phosphoribosyl transferase and the myeloma has immortal properties because it is a cancer cell. The cell mixture is diluted and clones grown from single parent cells. The antibodies secreted by different clones are assayed for their ability to bind to the antigen in question, with the healthiest clone selected for future use. The hybrids can also be injected into the peritoneal cavity of mice to produce tumors containing antibody-rich ascitic fluid, and the mAb then has to be extracted and purified.

To overcome the problem of purification, recombinant DNA technologies have been used since the 1980s. DNA that encodes the binding portion of mouse mAb is merged with human antibody-producing DNA. Mammalian cell culture is then used to express this DNA, producing chimeric antibodies. The goal, of course, is to creat of 'fully human' mAb, which has met with success in 'phage display-generated' antibodies and mice that have been genetically modified to produce more human-like antibodies.

Specific mAb have now been developed and approved for the treatment of cancer, cardiovascular disease, inflammatory diseases, macular degeneration, and transplant rejection, among others. A mAb that inhibits TNF-α has applications in rheumatoid arthritis, Crohn disease, and ulcerative colitis; one that inhibits IL-2 on activated T cells is used in preventing rejection of transplanted kidneys; and one that inhibits vascular endothelial growth factor (VEGF) has a role in antiangiogenic cancer therapy.

Cell-Mediated Specific Acquired Immunity

Certain microorganisms, viruses, and parasites live inside host cells. As a result, a separate form of specific acquired immunity has developed to combat intracellular infections involving lymphocytes differentiated and mature in the thymus—hence **T cells**. T lymphocytes have specialized receptors on the cell surface, known as **T-cell surface antigen receptors**, which in conjunction with the MHC on the cell surface of the infected cell result in the involvement of different subsets of T cells, each with a distinct function—**T helper cells** and **cytotoxic T cells**. The battle against intracellular infections is a cooperative, coordinated response from these separate components of the immune system, leading to death of the infected cell (Figure 13.8).

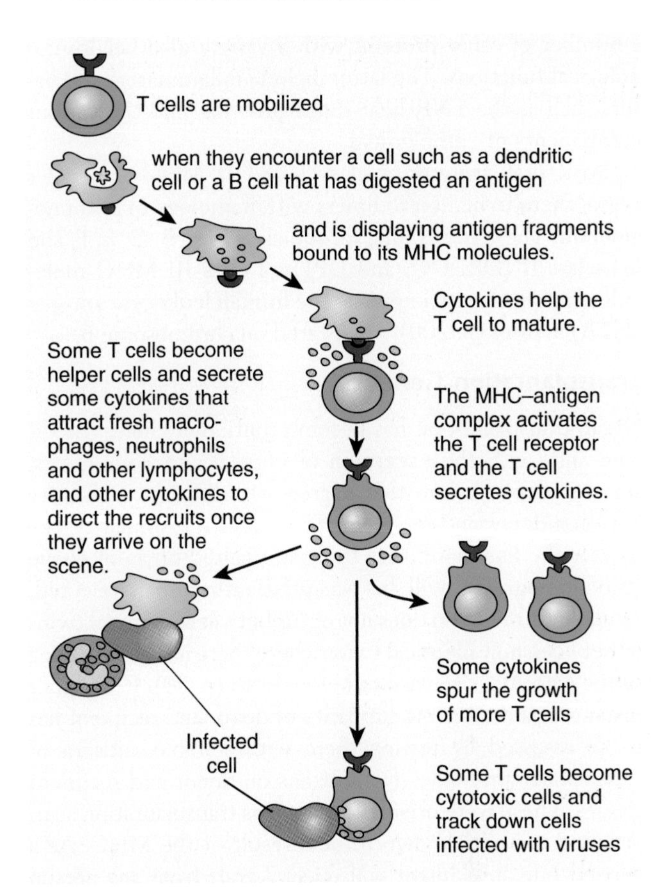

FIGURE 13.8 T cells and the cooperative response resulting in death of an infected cell.
MHC, Major histocompatibility complex.

T-Cell Surface Antigen Receptor

T cells express on their surface an antigen receptor, which distinguishes them from other lymphocyte types, such as B cells and NK cells. The antigen consists of two different polypeptide chains, linked by a disulfide bridge, that both contain two immunoglobulin-like domains, one that is relatively invariant in structure, the other highly variable like the Fab portion of an immunoglobulin. The diversity in T-cell receptors required for recognition of the range of antigenic variation that can occur is generated by a process similar to that seen with immunoglobulins. Rearrangement of variable (V), diversity (D), junctional (J), and constant (C) DNA segments during T-cell maturation, through a similar recombination mechanism as occurs in B cells, results in a contiguous VDJ sequence. Binding of antigen to the T-cell receptor, in conjunction with an associated complex of transmembrane peptides, results in signaling the cell to differentiate and divide.

The Major Histocompatibility Complex

The MHC plays a central role in the immune system. Its role is to bind antigen peptides processed intracellularly and present this material on the cell surface, with co-stimulatory molecules, where it can be recognised by T cells. MHC molecules occur in three classes: class I occur on virtually all cells and are responsible for presenting cytotoxic T cells; class II occur on B cells and macrophages and are involved in signaling T-helper cells to present further B cells and macrophages; and the non-classic class III molecules include a number of other proteins with a variety of other immunological functions. The latter include inflammatory mediators such as the TNF, heat-shock proteins, and the various components of complement.

Structural analysis of class I and II MHC molecules reveal them to be heterodimers with homology to immunoglobulin. The genes coding for the class I (A, B, C, E, F, and G), class II (DR, DQ, and DP) and class III MHC molecules, or what is also known as the **human leukocyte antigen** (HLA) system (p. 200), are located on chromosome 6.

Transplantation Genetics

Organ transplantation has become routine in clinical medicine and, with the exception of corneal and bone grafts, success depends on the degree of antigenic similarity between donor and recipient. The closer the similarity, the greater the likelihood that the transplanted organ or tissue (the **homograft**), will be accepted rather than rejected. Homograft rejection does not occur between identical twins or between non-identical twins where there has been mixing of the placental circulations before birth (p. 50). In all other instances, the antigenic similarity of donor and recipient has to be assessed by testing them with suitable antisera or monoclonal antibodies for antigens on donor and recipient tissues. These were originally known as transplantation antigens but are now known to be a result of the MHC. As a general rule, a recipient will reject a graft from any person who has antigens that the recipient lacks. HLA typing of an individual is carried out using PCR-based techniques (p. 56).

The HLA system is highly polymorphic (Table 13.2). A virtually infinite number of phenotypes resulting from different combinations of the various alleles at these loci is theoretically possible. Two unrelated individuals are therefore very unlikely to have identical HLA phenotypes. The close linkage of the HLA loci means that they tend to be inherited en bloc, the term **haplotype** being used to indicate the particular HLA alleles that an individual carries on each of the two copies of chromosome 6. Thus, any individual will have a 25% chance of having identical HLA antigens with a sibling, as there are only four possible combinations of the two paternal haplotypes (say P and Q) and the two maternal haplotypes (say R and S), i.e., PR, PS, QR and QS. The siblings of a particular recipient are more likely to be antigenically similar than either of his or her parents, and the latter more than a non-relative. Therefore, a sibling is frequently selected as a potential donor.

Table 13.2 Alleles at the HLA Loci

HLA Locus	Number of Alleles
A	57
B	111
C	34
D	228

HLA, Human leukocyte antigen.

Although recombination occurs within the HLA region, certain alleles tend to occur together more frequently than would be expected by chance, i.e. they tend to exhibit linkage disequilibrium (p. 138). An example is the association of the HLA antigens A1 and B8 in populations of western European origin.

H-Y Antigen

In a number of different animal species, it was noted that tissue grafts from males were rejected by females of the same inbred strain. These incompatibilities were found to be due to a histocompatibility antigen known as H-Y. However, H-Y seems to play little part in transplantation in humans. The H-Y antigen (not the same as the *SRY* gene, see p. 92) is important for testicular differentiation and function but its expression does not correlate with the presence or absence of testicular tissue.

HLA Polymorphisms and Disease Associations

The association of certain diseases with certain HLA types (Table 13.3) should shed light on the pathogenesis of the disease, but in reality this not well understood. The best documented is between ankylosing spondylitis and HLA-B27. Narcolepsy, a condition of unknown etiology characterized by a periodic uncontrollable tendency to fall asleep, is almost invariably associated with HLA-DR2. The possession of a particular HLA antigen does not mean that an individual will necessarily develop the associated disease, only that the *relative* risk of being affected is greater than the general population (p. 384). In a family, the risks to first-degree relatives of those affected are low, usually no more than 5%.

Explanations for the various HLA-associated disease susceptibilities include close linkage to a susceptibility gene near the HLA complex, cross-reactivity of antibodies to environmental antigens or pathogens with specific HLA antigens, and abnormal recognition of 'self' antigens through defects in T-cell receptors or antigen processing. These conditions are known as **autoimmune diseases**. An example of close linkage is congenital adrenal hyperplasia from a 21-hydroxylase deficiency (p. 174) from mutated *CYP21*,

Table 13.3 Some HLA-Associated Diseases

Disease	HLA
Ankylosing spondylitis	B27
Celiac disease	DR4
21-Hydroxylase deficiency	A3/Bw47/DR7
Hemochromatosis	A3
Insulin-dependent diabetes (type 1)	DR3/4
Myasthenia gravis	B8
Narcolepsy	DR2
Rheumatoid arthritis	DR4
Systemic lupus erythematosus	DR2/DR3
Thyrotoxicosis (Graves disease)	DR3

HLA, Human leukocyte antigen.

which lies within the HLA major histocompatibility locus. This form of congenital adrenal hyperplasia is strongly associated HLA-A3/Bw47/DR7 in northern European populations. Non-classical 21-hydroxylase deficiency is associated with HLA-B14/DR1, and HLA-A1/B8/DR3 is *negatively* associated with 21-hydroxylase deficiency.

Inherited Immunodeficiency Disorders

Inherited immunodeficiency disorders are uncommon and sometimes severe but, with early diagnosis and optimum management many patients with primary immune deficiency (PID) can remain very well. Prompt diagnosis is very important in order that treatment, for example antimicrobials, immunoglobulin, or bone marrow transplant, be instituted before significant irreversible end-organ damage takes place. Presentation is variable but often in childhood for more severe immune defects, especially after the benefits of maternal transplacental immunity have declined. New diagnoses of PID are sometimes made in adults. Investigation of immune function should be considered in all patients with recurrent infections and in children with failure to thrive. Failure to thrive, diarrhea, and hepatosplenomegaly may also be features.

Primary Inherited Disorders of Immunity

The manifestations of at least some of the PID diseases in humans can be understood by considering whether they are disorders of innate immunity or of specific acquired immunity. Abnormalities of humoral immunity are associated with reduced resistance to bacterial infections and may be lethal in infancy. Abnormalities of cell-mediated specific acquired immunity are associated with increased susceptibility to viral infections and are manifest experimentally in animals by prolonged survival of skin homografts.

Disorders of Innate Immunity

Primary disorders of innate immunity are considered under humoral and cell-mediated immunity categories.

Disorders of Innate Humoral Immunity

A variety of defects of complement can lead to disordered innate immunity.

Disorders of complement. If a complement defect is suspected, investigation of the integrity of the classical and alternative pathways should begin with functional assays looking at the entire pathway. If functional abnormalities are found, measurement of the individual components of that pathway can be undertaken.

The clinical effects of MBL deficiency have been described previously. Defects of the third component of complement, C3, lead to abnormalities of opsonization of bacteria, resulting in difficulties in combating pyogenic infections. Defects in the later components of complement—those involved in the formation of the MAC (p. 195)—also result in susceptibility to bacterial infection, though in particular *Neisseria* (meningococcal infections). This includes deficiency of properdin (factor P), a plasma protein active in the alternative complement pathway.

C1 inhibitor deficiency follows autosomal dominant inheritance and there are two forms—type 1 due to low levels, and type 2 resulting from non-functioning protein. Inappropriate activation and poor control of the complement pathway occurs with breakdown of C2 and C4, and production of inflammatory mediators. C1 inhibitor also controls the kinin-bradykinin pathway and when deficient an accumulation of bradykinin in the tissues occurs, and is believed to be the main cause of oedema, triggered by episodes of surgery, dental work, trauma, and some drugs. Attacks vary in severity from mild cutaneous to abdominal pain and swelling, which can be severe—laryngeal oedema is potentially fatal. This is known as **hereditary angio-edema.** Acute attacks are treated with C1 inhibitor concentrate, a blood product, which has superseded fresh frozen plasma when available. In due course, a recombinant C1 inhibitor may become the treatment of choice. The drug Danazol, an androgen, is the mainstay of long-term prevention.

Other associations with disease include homozygous C2 deficiency. There are various case reports of individuals who developed cutaneous vasculitis, Henoch-Schonlein purpura, seropositive rheumatoid arthritis, polyarteritis, membrano-proliferative glomerulonephritis, and an association with systemic lupus erythematosus (SLE). Similarly, C4 is associated with SLE. The copy number of C4 genes in a diploid human genome varies from two to six in the white population. Each of these genes encodes either a C4A or C4B protein. Subjects with only two copies of total C4 are at significantly increased risk of SLE, whereas those with five copies or more are at decreased risk.

Defects in NFκB Signaling

Inappropriate **activation** of nuclear factor kappa-B (NFκB) has been linked to inflammation associated with autoimmune arthritis, asthma, septic shock, lung fibrosis, glomerulonephritis, atherosclerosis, and AIDS. Conversely, persistent **inhibition** of NFκB has been linked directly to apoptosis, abnormal immune cell development, and delayed cell growth.

Since 2000, mutations have occasionally been found in the X-linked *IKK-gamma* gene, part of the TLR pathway (p. 193), in children demonstrating failure to thrive, recurrent digestive tract infections, often with intractable diarrhea, and recurrent ulcerations, respiratory tract infections with bronchiectasis, and recurrent skin infections, presenting in infancy, suggesting susceptibility to various gram-positive and gram-negative bacteria. Sparse scalp hair is sometimes a feature and in older children oligodontia and conical-shaped maxillary lateral incisors have been noted. Survival ranged from 9 months to 17 years in one study. IgG is low and IgM usually high. Interestingly, *IKKg* is the same as *NEMO*, the gene that causes X-linked dominant incontinentia pigmenti (p. 117–118). However, in this condition of the immune system mutations occur in exon 10 of the gene.

IRAK4 is another component of the TLR pathway and deficiency leads to recurrent infections, mainly from gram-positive microorganisms, though also fungi. There is a reduced inflammatory response. Infections begin early in life but become less frequent with age, some patients requiring no treatment by late childhood. It follows autosomal recessive inheritance.

Disorders of Innate Cell-Mediated Immunity

An important mechanism in innate cell-mediated immunity is phagocytosis, as previously discussed, which results in subsequent cell-mediated killing of microorganisms.

Chronic granulomatous disease. Chronic granulomatous disease (CGD) is the best known example of a disorder of phagocytic function, and follows either an X-linked or an autosomal recessive inheritance. It results from an inability of phagocytes to kill ingested microbes, because of any of several defects in the NADPH oxidase enzyme complex which generates the so-called microbicidal 'respiratory burst' (see Figure 13.1). Hypergammaglobulinemia may be present. CGD is therefore associated with recurrent bacterial or fungal infections, and may present as suppurative lymphadenitis, hepatosplenomegaly, pulmonary infiltrates, and/or eczematoid dermatitis. Childhood mortality was high until the advent of supportive treatment and prophylactic antibiotics. Bone marrow transplant has been successful, as well as transplantation of peripheral blood stem cells from an HLA-identical sibling. The X-linked gene mutated in CGD, *CYBB*, was the first human disease gene cloned by positional cloning (p. 75).

The neutropenias. The neutropenias are a heterogeneous group of disorders of varying severity, following different patterns of inheritance, and characterised by very low neutrophil counts. Autosomal dominant or sporadic congenital neutropenia (SCN1) is caused by mutation in the neutrophil elastase gene (*ELA2*), and mutation in the protooncogene *GFI1*, which targets *ELA2*, also causes dominantly inherited neutropenia (SCN2). Mutation in the *HAX1* gene causes autosomal recessive SCN3 ('classical' SCN—Kostmann disease), whereas autosomal recessive SCN4 is caused by mutation in the *G6PC3* gene. SCN patients with **acquired** mutations in the granulocyte colony-stimulating factor receptor *(CSF3R)* gene in hematopoietic cells are at high risk for developing acute myeloid leukemia.

In SCN, hematopoiesis is characterized by a maturation arrest of granulopoiesis at the promyelocyte level; peripheral absolute neutrophil counts are below 0.5×10^9/L and there is early onset of severe bacterial infections. As well as dominantly inherited SCN1, there is an X-linked form caused by a constitutively activating mutation in the *WAS* gene, mutated in Wiskott-Aldrich syndrome (see the following section).

Cyclic neutropenia rare, characterized by regular 21-day fluctuations in the numbers of blood neutrophils, monocytes, eosinophils, lymphocytes, platelets, and reticulocytes. This results in patients experiencing periodic symptoms of fever, malaise, mucosal ulcers, and occasionally life-threatening infections. As with SCN1, it is due to mutated *ELA2*.

Leukocyte adhesion deficiency. Individuals affected with **leukocyte adhesion deficiency (LAD)** present with life-threatening bacterial infections of the skin and mucous membranes and impaired pus formation. The increased susceptibility to infections occurs because of defective migration of phagocytes from abnormal adhesion-related functions of chemotaxis and phagocytosis. This disorder is fatal unless antibiotics are given, both for infection and prophylactically, until bone marrow transplantation can be offered. Three different forms of LAD are recognised, each with unique clinical features, though leukocytosis is a constant feature. LAD I and LAD II follow autosomal recessive inheritance while the mode of inheritance of LAD III is unclear; LAD II and LAD III are very rare.

LAD I is characterized by delayed separation of the umbilical cord, omphalitis, and severe recurrent infections with no pus formation. It is due to mutated *ITGB2*, located on chromosome 21, and encodes the β_2 subunit of the integrin molecule.

LAD II patients have the rare Bombay blood group and suffer from psychomotor retardation and growth delay. It is caused by mutations in the gene encoding the Golgi-specific GDP-fucose transporter.

LAD III is similar to LAD I but includes severe neonatal bleeding tendency. Various defects in leukocyte chemotaxis and adhesion to endothelial cells have been found and the definitive diagnosis is reached by the showing defects in the integrin activation process, whereas the CD18 molecule is structurally intact. The precise genetic defect in LAD III is not known.

Autoimmune-Poly Endocrinopathy-Candidosis-Ectodermal Dysplasia Syndrome

Autoimmune polyendocrinopathy syndrome type I is characterized by the presence of two of three major clinical symptoms: Addison disease, hypoparathyroidism, chronic mucocutaneous candidiasis, and is caused by mutations in the autoimmune regulator gene *(AIRE)*. Malabsorption and diarrhea can be striking and dominate the clinical picture, and immune disorders may be present, though diabetes mellitus and thyroid disease are infrequent. The onset of Addison disease is mostly in childhood or early adulthood, and frequently accompanied by chronic active hepatitis, malabsorption, juvenile-onset pernicious anemia, alopecia, and primary hypogonadism.

Disorders of Specific Acquired Immunity

Again, these can be considered under the categories of disorders of humoral and cell-mediated specific acquired immunity.

Disorders of Humoral Acquired Immunity

Abnormalities of immunoglobulin function lead to an increased tendency to develop bacterial infections.

Bruton-type agammaglobulinemia. Boys with this X-linked immunodeficiency usually develop multiple recurrent bacterial infections of the respiratory tract and skin after the first few months of life, having been protected initially by placentally transferred maternal IgG. Features similar to rheumatoid arthritis develop in many and they are not prone to viral infection. Treatment of life-threatening infections with antibiotics and the use of prophylactic intravenous immunoglobulins have improved survival prospects, but children with this disorder can still die from respiratory failure through complications of repeated lung infections. The diagnosis of this type of immunodeficiency is confirmed by demonstration of immunoglobulin deficiency and absence of B lymphocytes. The disorder has been shown to result from mutations in a tyrosine kinase specific to B cells (Btk) that result in loss of the signal for B cells to differentiate to mature antibody-producing plasma cells. A rarer, autosomal recessive, form of agammaglobulinemia shows marked depression of the circulating lymphocytes, and lymphocytes are absent from the lymphoid tissue.

Hyper-IgM syndrome (HIGM). HIGM is another genetically heterogeneous condition that includes increased levels of IgM, and also usually of IgD, with levels of the other immunoglobulins being decreased or virtually absent. Patients are susceptible to recurrent pyogenic infections, as well as opportunistic infections such as *Pneumocystis* and *Cryptosporidium*, because of primary T-cell abnormality. In the X-linked form (HIGM1) the mutated gene encodes a cell surface molecule on activated T cells called CD40 ligand (renamed *TNFSF5*). When the gene is not functioning, immunoglobulin class switches are inefficient, so that IgM production cannot be readily switched to IgA or IgG. IgM levels are therefore high, and IgG levels reduced. At least four other types are recognised, including autosomal recessive forms HIGM2 (CD40 deficiency) and HIGM3 (activation-induced cytidine deaminase, AICDA) deficiency.

Hyper-IgE syndrome. Again heterogeneous, this condition is sometimes known as Job syndrome and is a PID characterized by chronic eczema, recurrent staphylococcal infections, increased serum IgE, and eosinophilia. Abscesses may be 'cold', i.e. they lack of surrounding warmth, erythema, or tenderness. Patients have a distinctive coarse facial appearance, abnormal dentition, hyperextensibility of the joints, and bone fractures. Autosomal dominant HIES is caused by mutation in the *STAT3* gene and autosomal recessive by mutation in *DOCK8*.

Common variable immunodeficiency (CVID). CVID constitutes the most common group of B-cell deficiencies but is very heterogeneous and the causes are basically unknown. The presentation is similar to that for other forms of immune deficiency, including nodular lymphoid hyperplasia. The sexes are equally affected and presentation can begin at any age. Affecting approximately 1:800 Caucasians, selective IgA deficiency is the most frequently recognized PID. Many affected people have no obvious health problems, but others may have recurrent infections, gastrointestinal disorders, autoimmune diseases, allergies, or malignancies. The pathogenesis is arrest of B-cell differentiation, giving rise to a normal number of IgA-bearing B-cell precursors but a profound deficit in IgA-producing plasma cells. The response to immunization with protein and polysaccharide antigens is abnormal.

CVID is regarded as a 'wastebasket' category that includes a number of immune disorders; however, most individuals with CVID show a distinctive phenotype characterized by normal numbers of immunoglobulin-bearing B-cell precursors and a broad deficiency of immunoglobulin isotypes. CD40 ligand deficiency has been found in some patients in this group.

Disorders of Cell-Mediated Specific Acquired Immunity

The most common inherited disorder of cell-mediated specific acquired immunity is severe combined immunodeficiency (SCID).

Severe combined immunodeficiency. SCID, as the name indicates, is associated with an increased susceptibility to both viral and bacterial infections because of profoundly abnormal humoral and cell-mediated immunity. Common to all forms of SCID is the absence of T cell–mediated cellular immunity from defective T-cell development. Presentation is in its infancy with recurrent, persistent, opportunistic infections by many organisms, including *Candida albicans*, *Pneumocystis carinii*, and cytomegalovirus. The incidence of all types of SCID is approximately 1:75,000. Death usually occurs in infancy because of overwhelming infection, unless a bone marrow transplant is performed. SCID is genetically heterogeneous and can be inherited as either an X-linked or autosomal recessive disorder. The X-linked form (SCIDX1) is the most common form of SCID in males, accounting for 50% to 60% overall, and has been shown to be due to mutations in the γ chain of the cytokine receptor for IL-2 *(IL2RG)*. In approximately one-third to one-half of children with SCID that is not X-linked, inheritance is autosomal recessive (SCID1) and the different forms are classified according to whether they are B-cell negative (T-B–) or B-cell positive (T-B+). The presence or absence of NK cells is variable.

T-B+ SCID, apart from SCIDX1, includes deficiency of the protein tyrosine phosphatase receptor type C (or CD45) deficiency. CD45 suppresses Janus kinases (JAK), and there is a specific B-cell–positive SCID due to JAK3 deficiency, which can be very variable—from subclinical to life threatening in early childhood. Other rare autosomal recessive forms of SCID include mutation in the *IL7R* gene—IL2RG is dependent on a functional interleukin-7 receptor.

T-B-SCID includes adenosine deaminase deficiency, which accounts for approximately 15% of all SCID and one-third of autosomal recessive SCID. The phenotypic spectrum is variable, the most severe being SCID presenting in infancy and usually resulting in early death. Ten to 15% of patients have a 'delayed' clinical onset by age 6 to 24 months, and a smaller percentage of patients have 'later' onset, diagnosed from ages 4 years to adulthood, showing less severe infections and gradual immunologic deterioration. The immune system is affected through the accumulation of purine degradation products that are selectively toxic to T cells. Rare forms of B-cell negative SCID include mutated *RAG1/RAG2* (recombination activating genes), which are normally responsible for VDJ recombinations (p. 199) that lead to mature immunoglobulin chains and T-cell receptors. In addition, cases occur due to mutation in the Artemis gene (DNA cross-link repair protein 1c—*DCLRE1C*). The latter forms are both sensitive to ionizing radiation. Lastly, reticular dysgenesis is a rare and very severe form of SCID characterized by congenital agranulocytosis, lymphopenia, and lymphoid and thymic hypoplasia with absent cellular and humoral immunity functions. It is due to mutation in the mitochondrial *Adenylate kinase-2* gene *(AK2)*.

Secondary or Associated Immunodeficiency

There are a number of hereditary disorders in which immunological abnormalities occur as one of a number of associated features as part of a syndrome.

DiGeorge/Sedláčková Syndrome

Children with the **DiGeorge syndrome** (also well described by Sedláčková, 10 years earlier than DiGeorge) present with recurrent viral illnesses and are found to have abnormal cellular immunity as characterized by reduced numbers of T lymphocytes, as well as abnormal antibody production. This has been found to be associated with partial absence of the thymus gland, leading to defects in cell-mediated immunity and T cell–dependent antibody production. Usually these defects are relatively mild and improve with age, as the immune system matures, but occasionally the immune deficiency is very severe because no T cells are produced and bone marrow transplantation is indicated. It is important for all patients diagnosed to be investigated by taking a full blood count with differential CD3, CD4, and CD8 counts, and immunoglobulins. The levels of diphtheria and tetanus antibodies can indicate the ability of the immune system to respond. These patients usually also have a number of characteristic congenital abnormalities, which can include heart disease and absent parathyroid glands. The latter finding can result in affected individuals presenting in the newborn period with tetany due to low serum calcium levels secondary to low parathyroid hormone levels. This syndrome has been recognized to be part of the spectrum of phenotypes caused by abnormalities of the third and fourth pharyngeal pouches (p. 95) as a consequence of a microdeletion of chromosome band 22q11.2 (p. 282).

Ataxia Telangiectasia

Ataxia telangiectasia is an autosomal recessive disorder in which children present in early childhood with signs of cerebellar ataxia, dilated blood vessels on the sclerae of the eyes, ears, and face (oculocutaneous telangiectasia), and a susceptibility to sinus and pulmonary infections. Low serum IgA levels occur and a hypoplastic thymus as a result of a defect in the cellular response to DNA damage. The diagnosis is made by the demonstration of low or absent serum IgA and IgG as well as characteristic chromosome abnormalities on culture of peripheral blood lymphocytes—a form of chromosome instability (p. 288). Patients have an increased risk of developing leukemia or lymphoid malignancies.

Wiskott-Aldrich Syndrome

Wiskott-Aldrich syndrome is an X-linked recessive disorder in which affected boys have eczema, diarrhea, recurrent infections, thrombocytopenia, and, usually, low serum IgM levels and impaired T-cell function and numbers. Mutations in the gene responsible have been shown to result in loss of cytotoxic T-cell responses and T-cell help for B-cell response, leading to an impaired response to bacterial infections. Until the advent of bone marrow transplantation, the majority of affected boys died by mid-adolescence from hemorrhage or B-cell malignancy.

Carrier Tests for X-Linked Immunodeficiencies

Before it was possible to sequence the genes responsible for Wiskott-Aldrich syndrome, Bruton-type hypogammaglobulinemia, and X-linked SCID, the availability of closely linked DNA markers allowed female carrier testing by studies of the pattern of X-inactivation (p. 103) in the lymphocytes of females at risk. A female relative of a sporadically affected male with an X-linked immunodeficiency would be confirmed as a carrier by the demonstration of a non-random pattern of X-inactivation in the T-lymphocyte population, indicating that all her peripheral blood T lymphocytes had the same chromosome inactivated (Figure 13.9).

The carrier (C) and non-carrier (NC) are both heterozygous for an *Hpa*II/*Msp*I restriction site polymorphism. *Hpa*II and *Msp*I recognize the same nucleotide recognition sequence, but *Msp*I cuts double-stranded DNA whether it is methylated or not, whereas *Hpa*II cuts only unmethylated DNA (i.e., only the active X chromosome). In the carrier female, the mutation in the SCID gene is on the X chromosome on which the *Hpa*II/*Msp*I restriction site is present. *Eco*RI/*Msp*I double digests of T lymphocytes result in 6, 4, and 2-kilobase (kb) DNA fragments on gel analysis of the restriction fragments for both the carrier and non-carrier females. *Eco*RI/*Hpa*II double digests of T-lymphocyte DNA result, however, in a single 6-kb fragment in the carrier female. This is because in a carrier the only T cells to survive will be those in which the normal gene is on the active unmethylated X chromosome. Thus, inactivation appears to be non-random in a carrier, although, strictly speaking, it is cell population survival that is non-random.

Blood Groups

Blood groups reflect the antigenic determinants on red cells and were one of the first areas in which an understanding

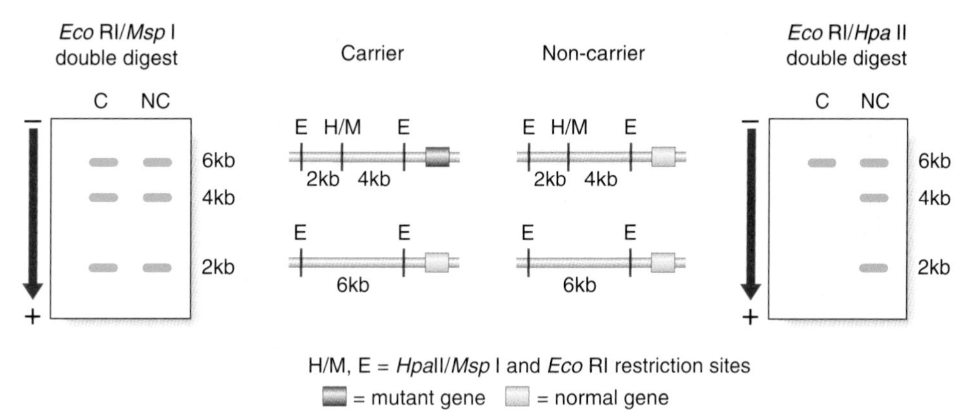

FIGURE 13.9 Non-random inactivation in T lymphocytes for carrier testing in X-linked SCID.

of basic biology led to significant advances in clinical medicine. Our knowledge of the ABO and Rhesus blood groups has resulted in safe blood transfusion and the prevention of Rhesus hemolytic disease of the newborn.

The ABO Blood Groups

The ABO blood groups were discovered by Landsteiner early in the twentieth century. In some cases blood transfusion resulted in rapid hemolysis because of incompatibility. Four major ABO blood groups were discovered: A, B, AB, and O. Those with blood group A possess the antigen A on the surface of their red blood cells, blood group B has antigen B, AB has both antigens, and those with blood group O have neither. People of blood group A have naturally occurring anti-B antibodies, and blood group B have anti-A, whereas blood group O have both. The alleles at the ABO blood group locus are inherited in a co-dominant manner but are both dominant to the gene for the O antigen. There are, therefore, six possible genotypes (Table 13.4).

Blood group AB individuals do not produce A or B antibodies, so they can receive a blood transfusion from people of all other ABO blood groups, and are therefore referred to as **universal recipients**. On the other hand, because individuals of group O do not express either A or B antigens on their red cells, they are referred to as **universal donors**. Antisera can differentiate two subgroups of blood group A, A1, and A2, but this is of little practical importance as far as blood transfusions are concerned.

Individuals with blood groups A, B, and AB possess enzymes with glycosyltransferase activity that convert the basic blood group, which is known as the 'H' antigen, into the oligosaccharide antigens 'A' or 'B'. The alleles for blood groups A and B differ in seven single base substitutions that result in different A and B transferase activities, the A allele being associated with the addition of *N*-acetylgalactosaminyl groups and the B allele with the addition of D-galactosyl groups. The O allele results from a critical single base-pair deletion that results in an inactive protein incapable of modifying the H antigen.

Table 13.4 ABO Blood Group Phenotypes and Genotypes

Red Blood Cells		React with Antiserum		
Phenotype	Genotype	Antibodies	Anti-A	Anti-B
O	OO	Anti-A,B	–	–
A	AA, AO	Anti-B	+	–
B	BB, BO	Anti-A	–	+
AB	AB	–	+	+

Rhesus Blood Group

The Rhesus (Rh) blood group system involves three sets of closely linked antigens, Cc, Dd, and Ee. D is very strongly antigenic and persons are, for practical purposes, either Rh positive (possessing the D antigen) or Rh negative (lacking the D antigen).

Rhesus Hemolytic Disease of the Newborn

A proportion of women who are Rh-negative have an increased chance of having a child who will either die in utero or be born severely anemic because of hemolysis, unless transfused in utero. This occurs because if Rh-positive blood is given to persons who are Rh-negative, the majority will develop anti-Rh antibodies. Such sensitization occurs with exposure to very small quantities of blood and, once a person is sensitized, further exposure results in the production of very high antibody titers.

In the case of an Rh-negative mother carrying an Rh-positive fetus, fetal red cells that cross to the mother's circulation can induce the formation of maternal Rh antibodies. In a subsequent pregnancy, these antibodies can cross the placenta from the mother to the fetus, leading to hemolysis and severe anemia. In its most severe form, this is known as **erythroblastosis fetalis**, or **hemolytic disease of the newborn**. After a woman has been sensitized. there is a significantly greater risk that a child in a subsequent pregnancy, if Rh-positive, will be more severely affected.

To avoid sensitizing an Rh-negative woman, Rh-compatible blood must always be used in any blood transfusion. Furthermore, the development of sensitization, and therefore

Rh incompatibility after delivery, can be prevented by giving the mother an injection of Rh antibodies—anti-D—so that fetal cells in the maternal circulation are destroyed before the mother can become sensitized.

It is routine to screen all Rh-negative women during pregnancy for the development of Rh antibodies. Despite these measures, a small proportion of women do become sensitized. If Rh antibodies appear, tests are carried out to see whether the fetus is affected. If so, there is a delicate balance between the choice of early delivery, with the risks of prematurity and exchange transfusion, and treating the fetus in utero with blood transfusions.

Molecular Basis of the Rh Blood Group

There are two types of Rh red cell membrane polypeptide. One corresponds to the D antigen and the other to the C and E series of antigens. We now know that two genes code for the Rh system: one for *D* and *d*, and a second for both C and *c* and *E* and *e*. The D locus is present in most persons and codes for the major D antigen present in those who are Rh-positive. Rh-negative individuals are homozygous for a deletion of the *D* gene. Therefore, an antibody has never been raised to d!

Analysis of complementary DNA from reticulocytes in Rh-negative persons who were homozygous for dCe, dcE, and dce allowed identification of the genomic DNA sequences responsible for the different antigenic variants at the second locus, revealing that they are produced by alternative splicing of the mRNA transcript. The Ee polypeptide is a full-length product of the *CcEe* gene, very similar in sequence to the D polypeptide. The E and e antigens differ by a point mutation in exon 5. The Cc polypeptides are, in contrast, products of a shorter transcript of the same gene through splicing. The difference between C and c is four amino-acid substitutions in exons 1 and 2.

Other Blood Groups

There are approximately a further 12 'common' blood group systems of clinical importance in humans, including Duffy, Lewis, MN, and S. These are usually of concern only when cross-matching blood for persons who, because of repeated transfusions, have developed antibodies to one of these other blood group antigens. Until the advent of DNA fingerprinting (p. 69), they were used in linkage studies (p. 137) and paternity testing (p. 270).

FURTHER READING

Bell JI, Todd JA, McDevitt HO 1989 The molecular basis of HLA–disease association. Adv Hum Genet 18:1–41

Good review of the HLA–disease associations.

Dreyer WJ, Bennet JC 1965 The molecular basis of antibody formation: a paradox. Proc Natl Acad Sci 54:864–869

The proposal of the generation of antibody diversity.

Hunkapiller T, Hood L 1989 Diversity of the immunoglobulin gene superfamily. Adv Immunol 44:1–63

Good review of the structure of the immunoglobulin gene superfamily.

Lachmann PJ, Peters K, Rosen FS, Walport MJ 1993 Clinical aspects of immunology, 5 edn. Oxford: Blackwell

A comprehensive three-volume multiauthor text covering both basic and clinical immunology.

Murphy KM, Travers P, Walport M 2007 Janeway's immunobiology, 7 edn. Oxford: Garland Science

Good, well-illustrated, textbook of the biology of immunology.

Roitt I 1997 Essential immunology, 9th edn. Oxford: Blackwell

Excellent basic immunology textbook.

ELEMENTS

1 The immune response can be divided into two main types, innate and specific acquired, or adaptive, immunity. Both can be further subdivided into humoral and cell-mediated immunity.

2 Innate humoral immunity involves acute-phase proteins that act to minimize tissue injury by limiting the spread of infective organisms that, through the alternative pathway of complement activation, results in a localized inflammatory response and the attraction of phagocytes and opsonization of microorganisms. Complement, which consists of a series of inactive blood proteins that are activated sequentially in a cascade, can also be activated through the classic pathway by antibody binding to antigen.

3 Innate cell-mediated immunity involves phagocytosis of microorganisms by macrophages and their intracellular destruction.

4 Specific acquired humoral immunity involves production of antibodies by mature B cells or plasma cells in response to antigen. Antibodies are Y-shaped molecules composed of two identical heavy (H) chains and two identical light (L) chains. The antibody molecule has two parts that differ in their function: two identical antigen-binding sites (Fab) and a single binding site for complement (Fc). There are five classes of antibody, immunoglobulin (Ig)A, IgD, IgE, IgG, and IgM, each with a specific heavy chain. The L chain of any class of antibody can be made up of either kappa (κ) or lambda (λ) chains.

5 Each Ig L or H chain has a variable (V) region of approximately 110 amino acids at the amino-terminal end. The carboxy-terminal end consists of a constant (C) region of approximately 110 amino acids in the κ and λ L chains and three to four times that length in the H chain. Most of the amino-acid sequence variation in both the L and H chains occurs within several small hypervariable regions, which are thought to be the sites of antigen binding. The Ig chains are produced from combinations of separate groups of DNA segments. These consist of one from a variable number of DNA segments coding for the constant (C), variable (V), and joining (J) regions between the V and C regions for the κ and λ L chains and the various types of H chains. The H chains also contain a diversity (D) region located between the V and J regions. The total number of possible antibodies that could be produced by various combinations of these DNA segments accounts for the antibody diversity seen in humans.

6 Cell-mediated specific acquired immunity primarily involves T cells that, through the T-cell surface antigen receptor, in conjunction with the major histocompatibility complex (MHC) molecules on the surface of infected cells, engage T helper cells and cytotoxic T cells to combat intracellular infections.

7 The MHC or human leukocyte antigen (HLA) system consists of a series closely linked loci on chromosome 6. The many different alleles that can occur at each locus mean that a very large number of different combinations can result. The HLA loci are inherited 'en bloc' as a haplotype. The closer the match of HLA antigens between the donor and recipient in organ transplantation, the greater the likelihood of long-term survival of the homograft. Possession of certain HLA antigens is associated with an increased relative risk of developing specific diseases.

8 An understanding of the ABO and Rhesus blood groups has resulted in safe blood transfusions and the prevention of Rhesus hemolytic disease of the newborn.

CHAPTER 14

Cancer Genetics

All cancer is genetic, but some cancers are more genetic than others.

PARAPHRASED FROM *ANIMAL FARM*, BY GEORGE ORWELL

Cell biology and molecular genetics have revolutionized our understanding of cancer in recent years; all cancer is a genetic disease of **somatic** cells because of aberrant cell division or loss of normal programmed cell death, but a small proportion is strongly predisposed by inherited **germline** mutations behaving as mendelian traits. However, this does not contradict our traditional understanding that, for many cancers, environmental factors are etiologically important, whereas heredity plays a lesser role. The latter is certainly true of the 'industrial cancers', which result from prolonged exposure to carcinogenic chemicals. Examples include cancer of the skin in tar workers, cancer of the bladder in aniline dye workers, angiosarcoma of the liver in process workers making polyvinyl chloride, and cancer of the lung (mesothelioma) in asbestos workers. Even so, for those who have been exposed to these substances and are unfortunate enough to suffer, it is possible that a significant proportion may have a genetic predisposition to the activity of the carcinogen. The link between cigarette smoking and lung cancer (as well as some other cancers) has been recognized for nearly half a century, but not all smokers develop a tobacco-related malignancy. Studies have shown that smokers with short chromosome telomeres (p. 31) appear to be at substantially greater risk for tobacco-related cancers than people with short telomeres who have never smoked, or smokers who have long telomeres, and another gene variant has been found to be more frequent in non-smokers who developed lung cancer.

The recognition that a number of rare cancer-predisposing syndromes, as well as a small but significant proportion of common cancers having a hereditary basis, has led over the past 25 years to an explosion in our understanding of the genetic basis and cellular biology of cancer in humans. As a general principle, it is now clear that cancers arise as the end result of an accumulation of both inherited and somatic mutations in proto-oncogenes and tumor suppressor genes. A third class of genes—the DNA mismatch repair genes—are also important because their inactivation is thought to contribute to the genesis of mutations in other genes directly affecting the survival and proliferation of cells. Germline mutations in at least 70 genes, and somatic mutations in at least 350 genes are known to contribute to the total burden of human cancer.

Differentiation between Genetic and Environmental Factors in Cancer

In many cancers, the differentiation between genetic and environmental etiological factors is not always obvious. In the majority of cancers in humans, there is no clear-cut mode of inheritance, nor is there any clearly defined environmental cause. In certain of the common cancers, such as breast and bowel, genetic factors play an important, but not exclusive, role in the etiology. Evidence to help differentiate environmental and genetic factors can come from a combination of epidemiological, family and twin studies, disease associations, biochemical factors, and animal studies.

Epidemiological Studies

Breast cancer is the most common cancer in women. Reproductive and menstrual histories are well-recognized risk factors. Women who have borne children have a lower risk of developing breast cancer than nulliparous women. In addition, the younger the age at which a woman has her first pregnancy, the lower her risk of developing breast cancer; the later the age at menarche, the lower the breast cancer risk.

The incidence of breast cancer varies greatly between different populations, being highest in women in North America and Western Europe, and up to eight times lower in women of Japanese and Chinese origin. Although these differences could be attributed to genetic differences between these population groups, study of immigrant populations moving from an area with a low incidence to one with a high incidence has shown that the risk of developing breast cancer rises with time to that of the native population, supporting the view that non-genetic factors are highly significant. Some of this changing risk may be accounted for by **epigenetic** factors (see the following section).

It has long been recognized that people from lower socio-economic groups have an increased risk of developing gastric cancer. Specific dietary irritants, such as salts and preservatives, or potential environmental agents, such as nitrates, have been suggested as possible carcinogens. Gastric cancer also shows variations in incidence in different populations, being up to eight times more common in Japanese and Chinese populations than in those of western European origin. Migration studies have shown that the risk of gastric cancer for immigrants from high-risk populations does not fall to that of the native low-risk population until two to three generations later. It has been suggested previously that this could be due to exposure to environmental factors at an early critical age. This may include early infection with *Helicobacter pylori*, which causes chronic gastric inflammation, and is associated with a five- to sixfold increased gastric cancer risk.

Family Studies

The frequency with which other family members develop the same cancer can provide evidence supporting a genetic contribution. The lifetime risk of developing breast cancer for a woman who lives until her mid-70s in Western Europe is at least 1 in 10. Family studies have shown that, for a woman who has a first-degree relative with breast cancer, the risk that she will also develop breast cancer is between 1.5 and 3 times the risk for the general population. The risk varies according to the age of onset in the affected family member: the earlier the age at diagnosis, the greater the risk to close relatives (p. 224).

Similar studies in gastric cancer have shown that first-degree relatives of those with cancer of the stomach have a twofold to threefold increased risk compared with the general population. The increased risk of developing gastric cancer in close relatives is, however, relatively small, suggesting that environmental factors are likely to be more important.

Twin Studies

Concordance rates for breast cancer in both monozygotic (MZ) and dizygotic (DZ) twins are low, being only slightly greater in MZ female twins, at 17%, than the 13% found in DZ female twins. This suggests, overall, that environmental factors are more important than genetic factors. Twin studies in gastric cancer have not shown an increased concordance rate in either MZ or DZ twins.

Disease Associations

Blood groups are genetically determined, and therefore association of a particular blood group with a disease suggests a possible genetic contribution to the etiology. A large number of studies from a variety of countries has shown an association between blood group A and gastric cancer. It is estimated that those with blood group A have a 20% increased risk of developing gastric cancer. Blood group A is associated with an increased risk of developing pernicious anemia, which is also closely associated with chronic gastritis. It appears, however, that pernicious anemia has a separate association with gastric cancer, as affected individuals have a three- to sixfold increased risk of developing gastric cancer.

Biochemical Factors

Biochemical factors can determine the susceptibility to environmental carcinogens. Examples include the association between **slow-acetylator** status and **debrisoquine metabolizer** status (p. 187) and a predisposition to bladder cancer, as well as glutathione *S*-transferase activity, which influences the risk of developing lung cancer in smokers.

Animal Studies

Certain inbred strains of mice have been developed that have a high chance of developing a particular type of tumor. The A (albino) Bittner strain is especially prone to develop tumors of the lung and breast; the C3H strain is particularly prone to develop breast tumors as well as tumors of the liver; the C58 strain is prone to develop leukemia. Breeding experiments with mice have shown that the tendency to develop cancer is influenced by environmental factors. In strains with a high incidence of breast tumors, the frequency of these tumors is reduced by dietary restrictions and increased by high temperature.

Viral Factors

Animal studies undertaken by Peyton Rous early in the twentieth century, among others, showed that transmission of a tumor was possible in the absence of body cells. Bittner later showed that susceptibility to breast tumors in certain strains of mice depended on a combination of genetic factors as well as a transmissible factor present in the milk, known as the 'milk agent'. In high-incidence strains, both genetic susceptibility and the milk agent are involved, but in low-incidence strains there is no milk agent. By using foster mothers from cancer-free strains to suckle newborn mice from strains with a high cancer susceptibility, it was possible to reduce the incidence of breast cancer from 100% to less than 50%. Conversely, an increased incidence was observed in cancer-free strains by suckling the newborn mice with foster mothers from high cancer-prone strains. The milk agent was shown to be a virus that was usually transmitted by the mother's milk, but could also be transmitted by the father's sperm.

Subsequent studies have shown that certain viruses are tumor-forming or **oncogenic** in humans. A limited number of DNA viruses are associated with certain types of human tumors (Table 14.1), whereas a variety of RNA viruses, or **retroviruses**, cause neoplasia in animals. The study of the genetics and replicative processes of oncogenic retroviruses has revealed some of the cellular biological processes involved in carcinogenesis.

Table 14.1 Human DNA Viruses Implicated in Carcinogenesis

Virus Family	Type	Tumor
Papova	Papilloma (HPV)	Warts (plantar and genital), urogenital cancers (cervical, vulval, and penile), skin cancer
Herpes	Epstein-Barr (EBV)	Burkitt lymphoma,* nasopharyngeal carcinoma, lymphomas in immunocompromised hosts
Hepadna	Hepatitis B (HBV)	Hepatocellular carcinoma[a]

*For full oncogenicity, 'co-carcinogens' are necessary (e.g., aflatoxin B_1 in hepatitis B–associated hepatocellular carcinoma).

Retroviruses

Retroviruses have their genetic information encoded in RNA and replicate through DNA by coding for an enzyme known as reverse transcriptase (p. 17), which makes a double-stranded DNA copy of the viral RNA. This DNA intermediate integrates into the host cell genome, allowing the appropriate proteins to be manufactured, resulting in repackaging of new progeny virions.

Naturally occurring retroviruses have only the three genes necessary to ensure replication: *gag*, encoding the structural proteins for the core antigens; *pol*, coding for reverse transcriptase; and *env*, the gene for the glycoprotein envelope proteins (Figure 14.1). Study of the virus responsible for the transmissible tumor in chickens, the so-called Rous sarcoma virus, identified a fourth gene that results in *transformation* of cells in culture, a model for malignancy in vivo. This viral gene, which *transforms* the host cell, is known as an **oncogene**.

Oncogenes

Oncogenes are the altered forms of normal genes—**proto-oncogenes**—that have key roles in cell growth and differentiation pathways. In normal mammalian cells are sequences of DNA that are homologous to viral oncogenes; these that are named **proto-oncogenes** or **cellular oncogenes**. Although the terms proto-oncogene and cellular oncogene are often used interchangeably, strictly speaking proto-oncogene is reserved for the normal gene and cellular oncogene, or c-*onc*, refers to a mutated proto-oncogene, which has oncogenic properties such as the viral oncogenes, or v-*onc*. At least 50 oncogenes have been identified.

Relationship Between C-*ONC* and V-*ONC*

Cellular oncogenes are highly conserved in evolution, suggesting that they have important roles as regulators of cell growth, maintaining the ordered progression through the cell cycle, cell division, and differentiation. Retroviral oncogenes are thought to acquire their dominant transforming activity during viral transduction through errors in the replication of the retrovirus genome following their random integration into the host DNA. The end result is a viral gene that is structurally similar to its cellular counterpart but is persistently different in its function.

Identification of Oncogenes

Oncogenes have been identified by two types of cytogenetic finding in association with certain types of leukemia and tumor in humans. These include the location of oncogenes at chromosomal translocation breakpoints, or their amplification in double-minute chromosomes or homogeneously staining regions of chromosomes (p. 212). In addition, a number of oncogenes have also been identified by the ability of tumor DNA to induce tumors in vitro by DNA transfection.

Identification of Oncogenes at Chromosomal Translocation Breakpoints

Chromosome aberrations are common in malignant cells, which often show marked variation in chromosome number

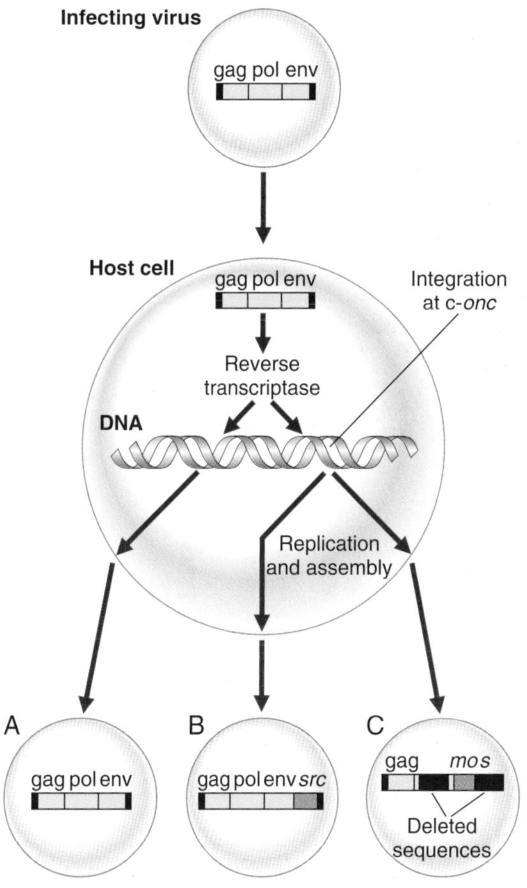

FIGURE 14.1 Model for acquisition of transforming ability in retroviruses. **A**, Normal retroviral replication. **B**, The Rous sarcoma virus has integrated near a cellular oncogene. The transforming ability of this virus is due to the acquired homolog of the cellular oncogene, v-*src*. **C**, A defective transforming virus carries an oncogene similar to *src* but is defective in the structural genes, (e.g., Moloney murine sarcoma virus, which carries *mos*).

and structure. Certain chromosomes seemed to be more commonly involved and it was initially thought that these changes were secondary to the transformed state rather than causal. This attitude changed when evidence suggested that chromosomal structural changes, often translocations (p. 44), resulted in rearrangements within or adjacent to proto-oncogenes. It has been found that chromosomal translocations can lead to novel chimeric genes with altered biochemical function or level of proto-oncogene activity. There are numerous examples of both types, of which chronic myeloid leukemia is an example of the former and Burkitt lymphoma an example of the latter.

Chronic Myeloid Leukemia

In 1960, investigators in Philadelphia were the first to describe an abnormal chromosome in white blood cells from patients with chronic myeloid leukemia (CML). The abnormal chromosome, referred to as the **Philadelphia**, or **Ph**[1], **chromosome**, is an acquired abnormality found in blood or bone marrow cells but not in other tissues from these patients. The Ph[1] is a tiny chromosome that is now known to be a chromosome 22 from which long arm material has been reciprocally translocated to and from the long arm of chromosome 9 (Figure 14.2), i.e., t(9;22)(q34;q11). This chromosomal rearrangement is seen in 90% of those with CML. This translocation has been found to transfer the cellular *ABL (Abelson)* oncogene from chromosome 9 into a region of chromosome 22 known as the **breakpoint cluster**, or *BCR*, region, resulting in a chimeric transcript derived from both the c-*ABL* (70%) and the *BCR* genes. This results in a chimeric gene expressing a fusion protein consisting of the BCR protein at the amino end and ABL protein at the carboxy end, which is associated with transforming activity.

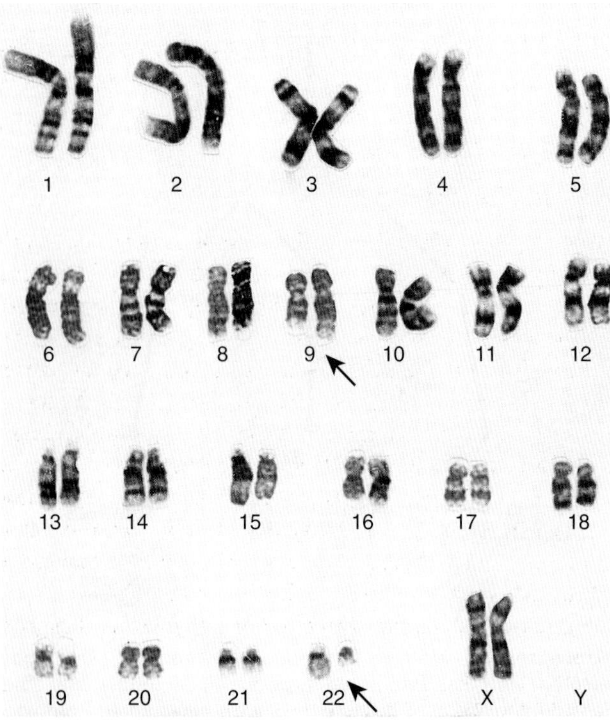

FIGURE 14.2 Karyotype from a patient with chronic myeloid leukemia showing the chromosome 22 (*arrow*) or Philadelphia chromosome, which has material translocated to the long arm of one of the number 9 chromosomes (*arrow*).

Burkitt Lymphoma

An unusual form of neoplasia seen in children in Africa is a lymphoma that involves the jaw, known as Burkitt lymphoma, named after Dennis Burkitt, a medical missionary who first described the condition in the late 1950s. Chromosomal analysis has revealed the majority (90%) of affected children to have a translocation of the c-*MYC* oncogene from the long arm of chromosome 8 on to heavy (H) chain immunoglobulin locus on chromosome 14. Less commonly the *MYC* oncogene is translocated to regions of chromosome 2 or 22, which encode genes for the kappa (κ) and lambda (λ) light chains, respectively (pp. 196–197). As a consequence of these translocations, *MYC* comes under the influence of the regulatory sequences of the respective immunoglobulin gene and is overexpressed 10-fold or more.

Oncogene Amplification

Proto-oncogenes can also be activated by the production of multiple copies of the gene or what is known as **gene amplification**, a mechanism known to have survival value when cells encounter environmental stress. For example, when leukemic cells are exposed to the chemotherapeutic agent methotrexate, the cells acquire resistance to the drug by making multiple copies of the gene for dihydrofolate reductase, the target enzyme for methotrexate.

Gene amplification can increase the number of copies of the oncogene per cell up to several hundred times, leading to greater amounts of the corresponding oncoprotein. In mammals the amplified sequence of DNA in tumor cells can be recognized by the presence of small extra chromosomes known as **double-minute chromosomes** or **homogeneously staining regions** of the chromosomes. These changes are seen in approximately 10% of tumors and are often present more commonly in the later rather than the early stages of the malignant process.

Amplification of specific proto-oncogenes appears to be a feature of certain tumors and is frequently seen with the *MYC* family of genes. For example, N-*MYC* is amplified in approximately 30% of neuroblastomas, but in advanced cases the proportion rises to 50%, where gene amplification can be up to 1000-fold. Human small cell carcinomas of the lung also show amplification of *MYC*, N-*MYC*, and L-*MYC*.

Amplification of *ERB*-B2, *MYC*, and cyclin D1 is a feature in 20% of breast carcinomas, where it has been suggested that it correlates with a number of well-established prognostic factors such as lymph node status, estrogen and progestogen receptor status, tumor size, and histological grade.

Detection of Oncogenes by DNA Transfection Studies

The ability of DNA from a human bladder carcinoma cell line to transform a well established mouse fibroblast cell line called NIH3T3, as demonstrated by the loss of contact inhibition of the cells in culture, or what is known as DNA **transfection**, led to the discovery of the human sequence homologous to the *ras* gene of the Harvey murine sarcoma virus. The human *RAS* gene family consists of three closely related members, H-*RAS*, K-*RAS*, and N-*RAS*. The RAS proteins are closely homologous to their viral counterparts and differ from one another only near the carboxy termini. Oncogenicity of the *ras* proto-oncogenes has been shown to arise by acquisition of point mutations in the nucleotide sequence. In approximately 50% of colorectal cancers and 95% of pancreatic cancers, as well as in a proportion of thyroid and lung cancers, a mutation in a *ras* gene can be demonstrated. The *RAS* gene family has been shown to be the key pathway (*RAS*-MAPK) in neurofibromatosis type 1 (p. 298) and the Noonan/cardio-facio-cutaneous/Costello syndromes, all of which demonstrate some increased risk of tumor formation.

DNA transfection studies have also led to identification of other oncogenes that have not been demonstrated through retroviral studies. These include *MET* (hereditary papillary renal cell carcinoma), *TRK*, *MAS*, and *RET* (multiple endocrine neoplasia type 2, see Tables 14.5, 14.9).

Function of Oncogenes

Cancers have characteristics that indicate, at the cellular level, loss of the normal function of oncogene products consistent with a role in the control of cellular proliferation and differentiation in the process known as **signal transduction**. Signal transduction is a complex multistep pathway from the cell membrane, through the cytoplasm to the nucleus, involving a variety of types of proto-oncogene product involved in positive and negative feedback loops necessary for accurate cell proliferation and differentiation (Figure 14.3).

Proto-oncogenes have been highly conserved during evolution, being present in a variety of different species, indicating that the protein products they encode are likely to have essential biological functions. Proto-oncogenes act in three main ways in the process of signal transduction: (1) through phosphorylation of serine, threonine, and tyrosine residues of proteins by the transfer of phosphate groups from ATP; this leads to alteration of the configuration activating the kinase activity of proteins and generating docking sites for target proteins, resulting in signal transduction; (2) through guanosine triphosphatase (GTPase) that function as molecular switches through the guanosine diphosphate–guanosine triphosphate (GDP–GTP) cycle as intermediates relaying the transduction signal from membrane-associated tyrosine kinases to serine threonine kinases; this includes the *RAS* family of proto-oncogenes; or (3) through proteins located in the nucleus that control

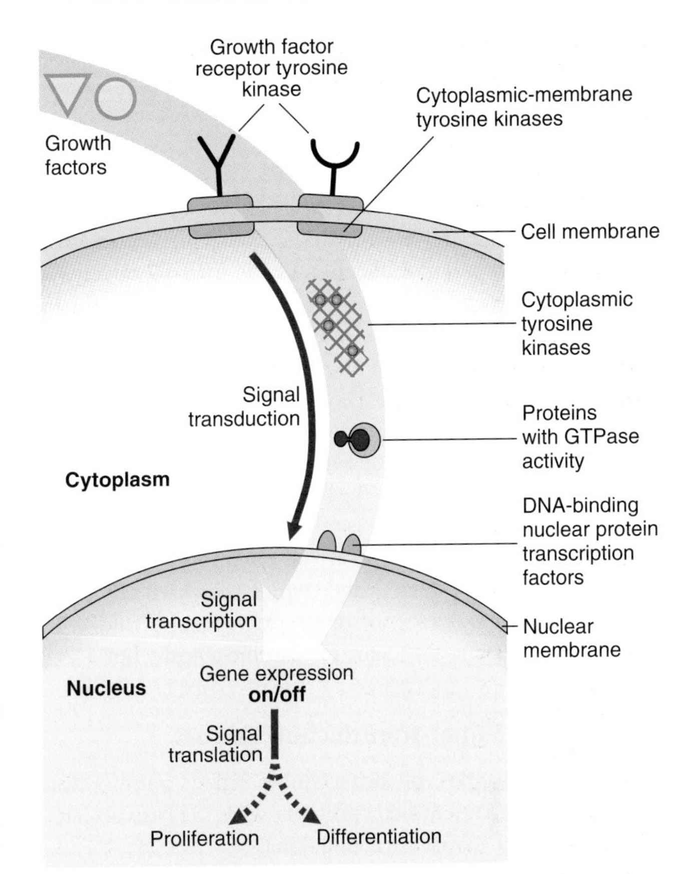

FIGURE 14.3 Simplified schema of the steps in signal transduction and transcription from cell surface to nucleus. The intracellular pathway amplifies the signal by a cascade that involves one or more of the steps.

progress through the cell cycle, DNA replication, and the expression of genes.

Types of Oncogene

Growth Factors

The transition of a cell from G_0 to the start of the cell cycle (p. 39) is governed by substances called growth factors. Growth factors stimulate cells to grow by binding to growth factor receptors. The best known oncogene that acts as a growth factor is the v-*SIS* oncogene, which encodes part of the biologically active platelet-derived growth factor B subunit. When v-SIS oncoprotein is added to the NIH 3T3 cultures, the cells are transformed, behaving like neoplastic cells; that is, their growth rate increases and they lose contact inhibition. In vivo they form tumors when injected into nude mice. Oncogene products showing homology to fibroblast growth factors include *HST* and *INT*-2, which are amplified in stomach cancers and in malignant melanomas, respectively.

Growth Factor Receptors

Many oncogenes encode proteins that form growth factor receptors, with tyrosine kinase activity possessing tyrosine kinase domains that allow cells to bypass the normal control mechanisms. More than 40 different tyrosine kinases have been identified and can be divided into two main types: those that span the cell membrane (growth factor receptor tyrosine kinases) and those located in the cytoplasm (non-receptor tyrosine kinases). Examples of tyrosine kinases include *ERB*-B, which encodes the epidermal growth factor receptor, and the related *ERB*-B2 oncogene. Mutations, rearrangements, and amplification of the *ERB*-B2 oncogene result in ligand-independent activation, which has been associated with cancer of the stomach, pancreas, and ovary. Mutations in *KIT* occur in the hereditary gastrointestinal stromal tumor syndrome. These oncogenes are not activated by **translocation** (as in Burkitt lymphoma) but rather by **point mutations**. When germline or inherited, the mutations are not lethal, nor are they sufficient by themselves to cause carcinogenesis. In the case of *MET* (located on chromosome 7), the papillary renal cell carcinoma tumors are trisomic for chromosome 7 and two of the three copies of *MET* are mutant. A ratio of one mutant to one wild-type copy of *MET* is not sufficient for carcinogenesis, but a 2:1 ratio is.

Intracellular Signal Transduction Factors

Two different types of intracellular signal transduction factor have been identified, proteins with GTPase activity and cytoplasmic serine threonine kinases.

Proteins with GTPase Activity

Proteins with GTPase activity are intracellular membrane proteins that bind GTP to become active and through their intrinsic GTPase activity generate GDP, which inactivates the protein. Mutations in the *ras* genes result in increased or sustained GTPase activity, leading to unrestrained growth.

Cytoplasmic Serine Threonine Kinases

Several soluble cytoplasmic gene products are recognized to be part of the signal transduction pathway. The *RAF* oncogene product modulates the normal signaling transduction cascade. Mutations in the gene can result in sustained or increased transmission of a growth-promoting signal to the nucleus.

DNA-Binding Nuclear Proteins

The *FOS*, *JUN*, and *ERB*-A oncogenes encode proteins that are specific transcription factors that regulate gene expression by activating or suppressing nearby DNA sequences. The function of *MYC* and related genes remains uncertain but appears to be related to alterations in control of the cell cycle. The *MYC* and *MYB* oncoproteins stimulate cells to progress from the G_1 into the S phase of the cell cycle (p. 39). Their overproduction prevents cells from entering a prolonged resting phase, resulting in persistent cellular proliferation.

Cell-Cycle Factors

Cancer cells can increase in number by increased growth and division, or accumulate through decreased cell death. In vivo, most cells are in a non-dividing state. Progress through the cell cycle (p. 39) is regulated at two points: one in G_1 when a cell becomes committed to DNA synthesis in the S phase, and another in G_2 for cell division in the M (mitosis) phase, through factors known as cyclin-dependent kinases. Abnormalities in regulation of the cell cycle through growth factors, growth factor receptors, GTPases or nuclear proteins, or loss of inhibitory factors lead to activation of the cyclin-dependent kinases, such as cyclin D1, resulting in cellular transformation with uncontrolled cell division. Alternatively, loss of the factors that lead to normal programmed cell death, a process known as **apoptosis** (p. 85), can result in the accumulation of cells through prolonged cell survival as a mechanism of development of some tumors. Activation of the *bcl*-2 oncogene through chromosomal rearrangements is associated with inhibition of apoptosis, leading to certain types of lymphoma.

Signal Transduction and Phakomatoses

Phakomatosis derives from the Greek *phakos*, meaning 'lentil' (in this context 'lentil-shaped object'), and originally referred to three diseases that included scattered benign lesions: neurofibromatosis, tuberous sclerosis, and von Hippel-Lindau disease. To this list has now been added nevoid basal cell carcinoma (Gorlin) syndrome, Cowden disease, familial adenomatous polyposis, Peutz-Jegher syndrome, and juvenile polyposis. The genes for all of these conditions are now known and are normally active within intracellular signal transduction, and their protein products are tumor suppressors.

Tumor Suppressor Genes

Although the study of oncogenes has revealed much about the cellular biology of the somatic genetic events in the malignant process, the study of hereditary cancer in humans has revealed the existence of what are known as **tumor suppressor genes**, which constitute the largest group of cloned hereditary cancer genes.

Studies carried out by Harris and colleagues in the late 1960s, which involved fusion of malignant cells with non-malignant cells in culture, resulted in the suppression of the malignant phenotype in the hybrid cells. The recurrence of the malignant phenotype with loss of certain chromosomes from the hybrid cells suggested that normal cells contain a gene(s) with tumor suppressor activity that, if lost or inactive, can lead to malignancy and was acting like a recessive trait. Such genes were initially referred to as **anti-oncogenes**. This term was considered inappropriate because anti-oncogenes do not oppose the action of the oncogenes and are more correctly known as tumor suppressor genes. The paradigm for our understanding of the biology of tumor suppressor genes is the eye tumor retinoblastoma. It is important to appreciate, however, that a germline mutation in a tumor suppressor gene (as with an oncogene) does not by itself provoke carcinogenesis: further somatic mutation at one or more loci is necessary and environmental factors, such as ionizing radiation, may be significant in the process. At least 20 tumor suppressor genes have been identified.

Retinoblastoma

Retinoblastoma (Rb) is a relatively rare, highly malignant, childhood cancer of the developing retinal cells of the eye that usually occurs before the age of 5 years (Figure 14.4). If diagnosed and treated at an early stage, it is associated with a good long-term outcome.

Rb can occur either sporadically, the so-called non-hereditary form, or be familial, the so-called hereditary form, which is inherited in an autosomal dominant manner. Non-hereditary cases usually involve only one eye, whereas hereditary cases can be unilateral but are more commonly bilateral or occur in more than one site in one eye (i.e., are multifocal). The familial form also tends to present at an earlier age than the sporadic form.

'Two-Hit' Hypothesis

In 1971, Knudson carried out an epidemiological study of a large number of cases of both types of Rb and advanced a 'two-hit' hypothesis to explain the occurrence of this rare tumor in patients with and without a positive family history. He proposed that affected individuals with a positive family history had inherited one non-functional gene that was present in all cells of the individual, known as a **germline mutation**, with the second gene at the same locus becoming inactivated somatically in a developing retinal cell (Figure 14.5). The occurrence of a second mutation was likely given the large number of retinal cells, explaining the autosomal dominant pattern of inheritance. This would also explain the observation that in hereditary Rb the tumors were often bilateral and multifocal. In contrast, in the non-heritable or sporadic form, *two* inactivating **somatic mutations** would need to occur independently in the same retinoblast cell (see Figure 14.5), which was much less likely to occur, explaining the fact that tumors in these patients were often unilateral and unifocal, and usually occurred at a later age than in the hereditary form. Hence, although the hereditary form of Rb follows an autosomal dominant pattern of inheritance, at the molecular level it is *recessive* because a tumor occurs only after the loss of both alleles.

It was also recognized, however, that approximately 5% of children presenting with Rb had other physical abnormalities along with developmental concerns. Detailed cytogenetic analysis of blood samples from these children revealed some of them to have an interstitial deletion involving the long arm of one of their number 13 chromosome pair. Comparison of the regions deleted revealed a common 'smallest region of overlap' involving the sub-band 13q14 (Figure 14.6). The detection of a specific chromosomal region involved in the etiology of these cases of Rb suggested that it could also be the locus involved in the autosomal dominant familial form of Rb. Family studies using a polymorphic enzyme, esterase D, which had previously been mapped to that region, rapidly confirmed linkage of the hereditary form of Rb to that locus.

Loss of Heterozygosity

Analyses of the DNA sequences in this region of chromosome 13 in the peripheral blood and in Rb tumor material from children who had inherited the gene for Rb showed them to have loss of an allele at the Rb locus in the tumor material, known as **loss of heterozygosity** (LOH), or sometimes as loss of **constitutional** heterozygosity. An example

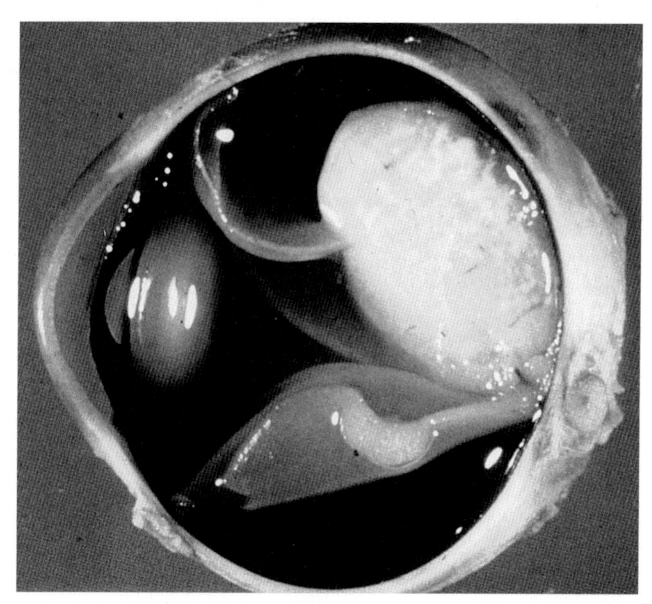

FIGURE 14.4 Section of an eye showing a retinoblastoma in situ.

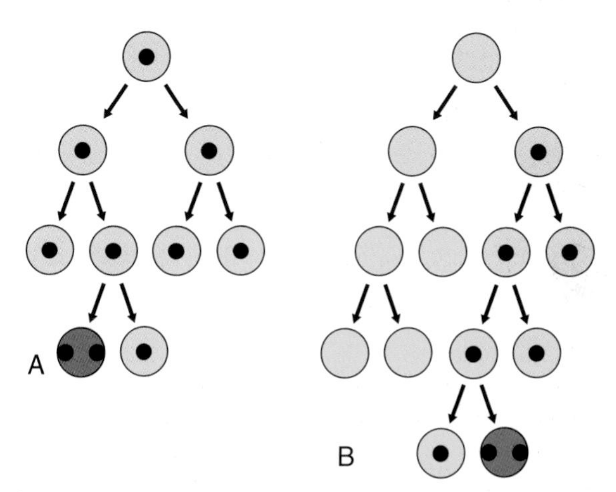

FIGURE 14.5 Retinoblastoma and Knudson's 'two-hit' hypothesis. All cells in the hereditary form (**A**) have one mutated copy of the gene, *RB1* (i.e., the mutation is in the *germline*). In the non-hereditary form (**B**) a mutation in *RB1* arises as a post-zygotic *(somatic)* event some time early in development. The retinoblastoma tumor occurs only when both *RB1* genes are mutated—i.e., after a(nother) *somatic* event, which is more likely to be earlier in life in the hereditary form compared with the non-hereditary form; it is also more likely to give rise to bilateral and multifocal tumors.

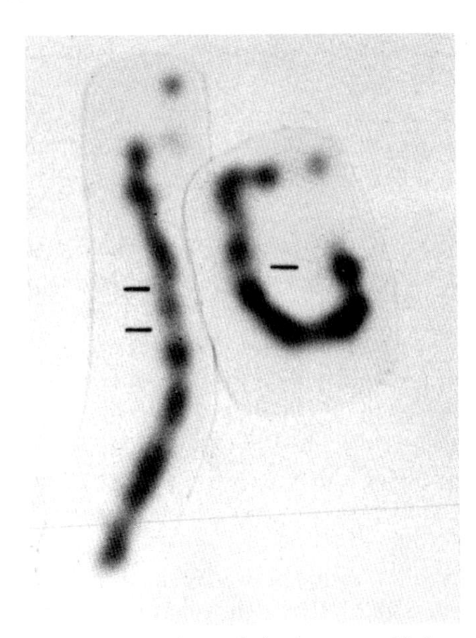

FIGURE 14.6 Two homologs of chromosome 13 from a patient with retinoblastoma showing an interstitial deletion of 13q14 in the right-hand homolog, as indicated.

of this is shown in Figure 14.7, *A*, in which the mother transmits the Rb gene along with allele 2 at a closely linked marker locus. The father is homozygous for allele 1 at this same locus, with the result that the child is constitutionally an obligate heterozygote at this locus. Analysis of the tumor tissue reveals apparent homozygosity for allele 2. In fact, there has been loss of the paternally derived allele 1 (i.e., LOH in the tumor material). This LOH is consistent with the 'two-hit' hypothesis leading to development of the malignancy as proposed by Knudson.

LOH can occur through several mechanisms, which include loss of a chromosome through mitotic nondisjunction (p. 43), a deletion on the chromosome carrying the corresponding allele, or a crossover between the two homologous genes leading to homozygosity for the mutant allele (Figure 14.7, *B*). Observation of consistent cytogenetic rearrangements in other malignancies has led to demonstration of LOH in a number of other cancers (Table 14.2). Subsequent to the observation of LOH, linkage studies of familial cases can be carried out to determine whether the familial cases of a specific type of malignancy are due to mutations at the same locus and thus lead to the identification of the gene responsible, as occurred with the isolation of the *RB1* gene.

Function of Tumor Suppressor Genes

Although familial Rb was classically considered to be an autosomal dominant trait, demonstration of the action of the Rb gene as a tumor suppressor gene is consistent with it being a recessive trait, as originally suggested in the somatic cell hybridization studies carried out by Harris and colleagues. In other words, absence of the gene product in the homozygous state leads to the development of this particular tumor. In contrast to oncogenes, tumor suppressor genes are a class of cellular genes whose normal function is to suppress inappropriate cell proliferation (i.e., the development of a malignancy is due to a loss of function mutation, see p. 26).

The tumor suppressor activity of the Rb gene has been demonstrated in vitro in cancer cells. In addition, further support for the *RB1* gene acting as a tumor suppressor gene comes from the recognition that individuals with the hereditary form of Rb have an increased risk of developing second new malignancies later in life, including osteosarcoma, fibrosarcoma, and chondrosarcoma.

The *RB1* Gene/p110^{RB} Protein

The *RB1* gene specifies a 4.7-kilobase (kb) transcript that encodes a nuclear protein called p110^{RB}, which associates with DNA and is involved in the regulation of the cell cycle. Fortuitously, research on the mechanism of action of the *E1A* oncogene of human adenovirus demonstrated that p110^{RB} forms a complex with *E2F*-1, which is an *E1A* oncogene-regulated inhibitor of the transcription factor *E2F*. The complex so formed interferes with the ability of *E2F* to activate transcription of some key proteins required for DNA synthesis. When p110^{RB} is in a hyperphosphorylated state, it does not interact readily with *E2F*-1, so permitting the cell cycle to proceed into the S phase (p. 39). Retinoblasts fail to differentiate normally in the presence of mutant p110^{RB}.

Table 14.2 Syndromes and Cancers that Show Loss of Heterozygosity and Their Chromosomal Location

Syndrome or Cancer	Chromosomal Location
Retinoblastoma	13q14
Osteosarcoma	13q, 17p
Wilms tumor	11p13, 11p15, 16q
Renal carcinoma	3p25, 17p13
von Hippel-Lindau disease	3p25
Bladder carcinoma	9q21, 11p15, 17p13
Lung carcinoma	3p, 13q14, 17p
Breast carcinoma	11p15, 11q, 13q12, 13q14, 17p13, 17q21
Rhabdomyosarcoma	11p15, 17p13
Hepatoblastoma	5q, 11p15
Gastric cancer	1p, 5q, 7q, 11p, 13q, 17p, 18p
Familial adenomatous polyposis	5q21
Colorectal carcinoma	1p, 5q21, 8p, 17p13, 18q21
Neurofibromatosis I (NF1, von Recklinghausen disease)	17q
Neurofibromatosis II (NF2)	22q
Meningioma	22q
Multiple endocrine neoplasia type I (MEN1)	11q
Melanoma	9p21, 17q
Ovarian	11q25, 16q, 17q
Pancreatic	9p21, 13q14, 17p13
Prostate cancer	1p36, 7q, 8p, 10q, 13q, 16q

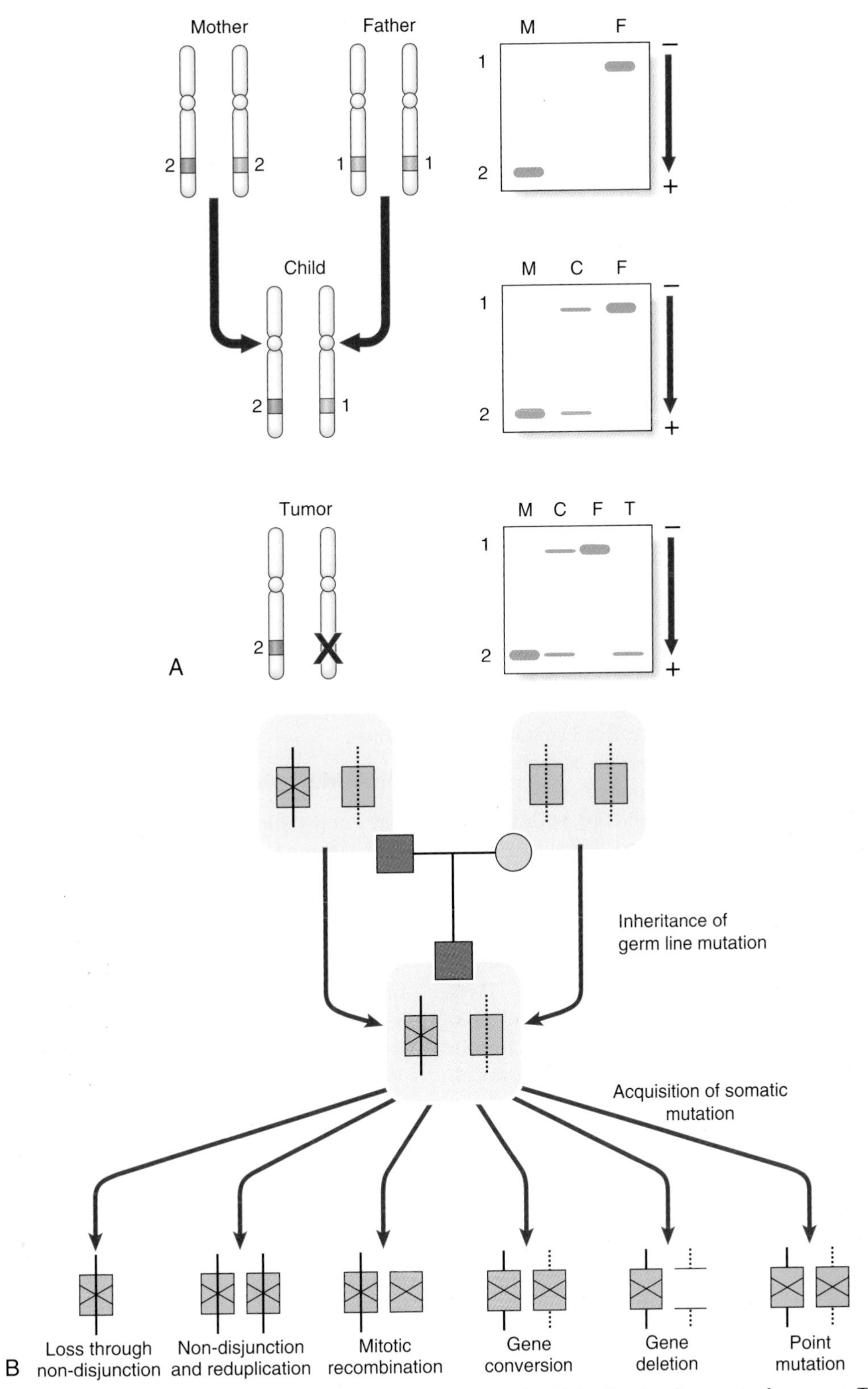

FIGURE 14.7 **A,** Diagrammatic representation of the loss of heterozygosity (LOH) in the development of a tumor. The mother (M) and father (F) are both homozygous for different alleles at the same locus, 2–2 and 1–1, respectively. The child (C) will therefore be constitutionally heterozygous, 1–2. If an analysis of DNA from a tumor at that locus reveals only a single allele, 2, this is consistent with LOH. **B,** Diagrammatic representations of the mechanisms by causing the 'second hit' leading to the development of retinoblastoma.

p110[RB] interacts with several viral oncoproteins, such as the transforming proteins of simian virus (SV) 40 (large T antigen) and papilloma virus (E7 protein), and is inactivated, thereby liberating cells from normal growth constraints. These findings yield insight into the mechanisms of interaction between oncogenes, tumor suppressor genes, and the cell cycle.

TP53

The p53 protein was first identified as a host cell protein bound to T antigen, the dominant transforming oncogene of the DNA tumor virus SV40. After the murine *p53* gene was cloned it was shown to be able to cooperate with activated *RAS* and act as an oncogene transforming primary rodent cells in vitro, even though the rodent cells expressed the wild-type or normal *p53*. Subsequently, inactivation of *p53* was frequently found in murine Friend virus–induced erythroleukemia cells, which led to the proposal that the *TP53* gene was, in fact, a tumor suppressor gene.

The *TP53* gene is the most frequently mutated of all the known cancer genes. Some 20% to 25% of breast and more than 50% of bladder, colon, and lung cancers have been found to have *TP53* mutations that, although occurring in different codons, are clustered in highly conserved regions in exons 5 to 10. This is in contrast to *TP53* mutations in hepatocellular carcinoma, which occur in a 'hotspot' in codon 249. The base change in this mutated codon, usually G to T, could be the result of an interaction with the carcinogen aflatoxin B_1, which is associated with liver cancer in China and South Africa, or with the hepatitis B virus that is also implicated as a risk factor in hepatomas. Interestingly, aflatoxin B_1, a ubiquitous food-contaminating aflatoxin in these areas, is a mutagen in many animal species and induces G to T substitutions in mutagenesis experiments. If an interaction between hepatitis B viral proteins and non-mutated *TP53* can be demonstrated, this will further support the role of this virus in the etiology of hepatocellular carcinoma.

Cancers frequently have a decreased cell death rate through **apoptosis**, and a major factor in the activation of apoptosis is *TP53*—p53 has been coined the 'guardian of the genome'. The p53 protein is a multimeric complex and it functions as a checkpoint control site in the cell cycle at G_1 before the S phase, interacting with other factors, including cyclins and p21, preventing DNA damaged through normal 'wear and tear' from being replicated. Mutant p53 protein monomers are more stable than the normal p53 proteins and can form complexes with the normal wild-type *TP53*, acting in a dominant-negative manner to inactivate it.

Li-Fraumeni Syndrome

Because mutations in *TP53* appear to be a common event in the genesis of many cancers, an inherited or germline mutation of *TP53* would be expected to have serious consequences. This hypothesis was substantiated with the discovery of such a defect in persons with **Li-Fraumeni syndrome**. Members of families with this rare syndrome (p. 225), which is inherited as an autosomal dominant trait, are highly susceptible to developing a variety of malignancies at an early age, including sarcomas, adrenal carcinomas, and breast cancer. Point mutations in highly conserved regions of the *TP53* gene (codons 245 to 258) have been identified in the germline of family members, with analysis of the tumor revealing loss of the normal allele.

Epigenetics and Cancer

Much of this chapter discusses familial cancer syndromes that follow mendelian inheritance, characterized by mutations in disease-specific genes. However, no discussion about cancer genetics is complete without considering epigenetic mechanisms. As discussed in Chapter 6 (p. 103), **epigenetics** refers to heritable changes to gene expression that are *not* due to differences in the genetic code. Such gene expression can be transmitted stably through cell divisions, both mitosis and meiosis. In cancer, much is now known about alterations to methylation status of the genome, both *hypo*methylation and *hyper*methylation, and in this section we also discuss telomere length and cancer.

DNA Methylation and Genomic Imprinting

The methylation of DNA is an **epigenetic** phenomenon (p. 103), and is the mechanism responsible for X-inactivation (p. 103) and genomic imprinting (p. 121). Methylation of DNA has the effect of silencing gene expression and maintaining stability of the genome, especially in areas where there is a vast quantity of repetitive DNA (heterochromatin), which might otherwise become erroneously involved in recombination events leading to altered regulation of adjacent genes. The relevance of this for cancer emerged in 1983 when studies showed that the genomes of cancer cells were *hypo*methylated compared with those of normal cells, primarily within repetitive DNA. This **loss of imprinting (LOI)** may lead to activation of an allele that is normally silent, and hence the high expression of a product that confers advantageous cellular growth. This appears to be an early event in many cancers and may correlate with disease severity. Chromosomal instability is strongly associated with increased tumor frequency, which has been clearly observed in mouse models, and all the 'chromosome breakage' syndromes (p. 288), which in humans are associated with a significant increased risk of cancer, particularly leukemia and lymphoma. LOI and removal of normal gene silencing may lead to oncogene activation, and hence cancer risk. LOI has been studied extensively at the *IGF2/H19* locus on chromosome 11p15.5, previously discussed in Chapter 7 (p. 124). Insulin-like growth factor 2 *(IGF2)* and *H19* are normally expressed from the paternal and maternal alleles, respectively (see Figure 7.27), but relaxed silencing of the

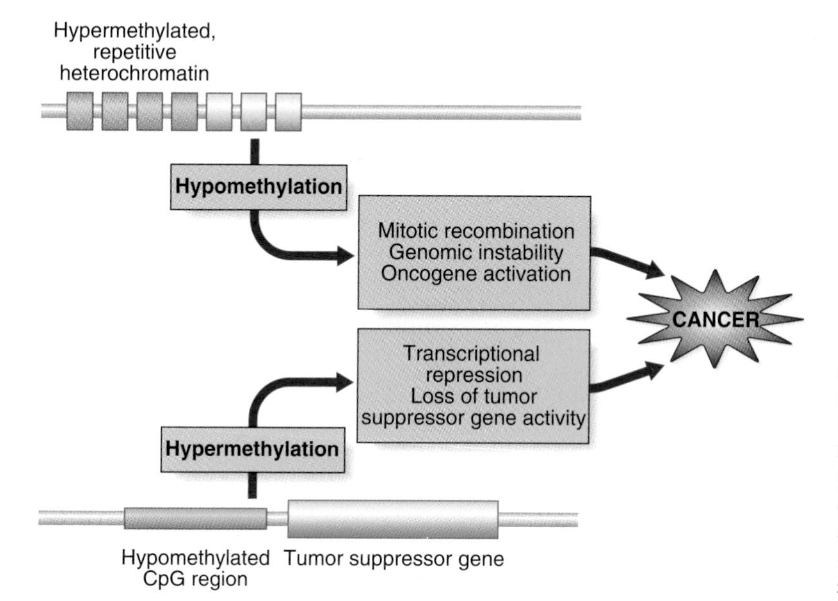

FIGURE 14.8 Methylation of DNA and cancer. The top schema shows a region of hypermethylated repetitive DNA sequence (heterochromatin). When this loses its methylation imprint, chromosome instability may result, which may lead to activation of oncogene(s). In the lower panel, hypomethylated stretches of CpG sequence (p. 219) become methylated, resulting in transcriptional suppression of tumor suppressor and cell regulatory genes.

maternal allele (i.e., hypomethylation, results in increased *IGF2* expression). This has been shown to be the most common LOI event across a wide range of common tumor types (e.g., lung, liver, colon, ovary) as well as Wilms tumor, in which it was first identified.

Just as *hypo*methylation may lead to activation of oncogenes, the opposite effect of *hyper*methylation may also give rise to an increased cancer risk, in this case through silencing of tumor suppressor genes whose normal functions include inhibition of cell growth. The aberrant hypermethylation usually affects CpG nucleotide islands (C and G adjacent to each, p-phosphodiester bond), which are mostly unmethylated in somatic cells. This results in changes in chromatin structure (hypoacetylation of histone) that effectively silence transcription. When the genes involved in all sorts of cell regulatory activity are silenced, cells have a growth advantage. Early hypermethylation has been detected in colonic cancer. The effects of altered methylation leading to cancer are summarized in Figure 14.8, although the mechanism(s) that initiate the processes are poorly understood.

Telomere Length and Cancer

The ends of the chromosomes are known as telomeres (p. 31) and they are specialized chromatin structures that have a protective function. The sequence of DNA is specific and consists of multiple double-stranded tandem repeats as follows: TTAGGG. This sequence is typically about 10 to 15 kb long in human cells and is bound by specific proteins. It is also the substrate for telomerase, an enzyme that can lengthen the telomeres in those cells in which it is expressed. The final length of DNA at the very tip of the telomere is a single-stranded overhang of 150 to 200 nucleotides. Telomerase recognizes the 3′ end of the overhang, allowing lengthening to proceed.

Every cell division appears to result in the loss of TTAGGG repeats because conventional DNA polymerases cannot replicate a linear chromosome in its entirety, known as the 'end-replication problem'. This progressive loss of telomere length is a form of cellular clock believed to be linked to both aging and human disease. When telomeres reach a critically short length, there is loss of protection and a consequence is chromosomal, and therefore genomic, instability, which means the cell is no longer viable. Short telomeres are now known to be a feature of the premature aging syndromes, such as ataxia telangiectasia, and other chromosome breakage disorders (p. 288), all of which are associated with premature onset of various cancers. It appears that the *rate* of telomere shortening is markedly increased in these conditions, so that cells and tissues literally 'age' more quickly. It is of great interest that some cancer cells express high levels of telomerase, so that cell viability is maintained. Most metastases have been shown to contain telomerase-positive cells, suggesting that telomerase is required to sustain such growth. However, cancer cells generally have shorter telomeres than the normal cells surrounding them, so telomerase activation in cancer rescues short telomeres and perpetuates genomically unstable cells.

Telomere length is therefore almost certainly a key concept in many cancers, as well as aging processes, even though the exact mechanisms remain to be elucidated. The relationship of telomere length to age and disease is displayed graphically in Figure 14.9.

Genetics of Common Cancers

It is estimated that about 5% of colorectal and breast cancers arise as a result of an inherited cancer susceptibility gene. A similar proportion of many other cancers are due to inherited predisposing genetic factors, but there are some

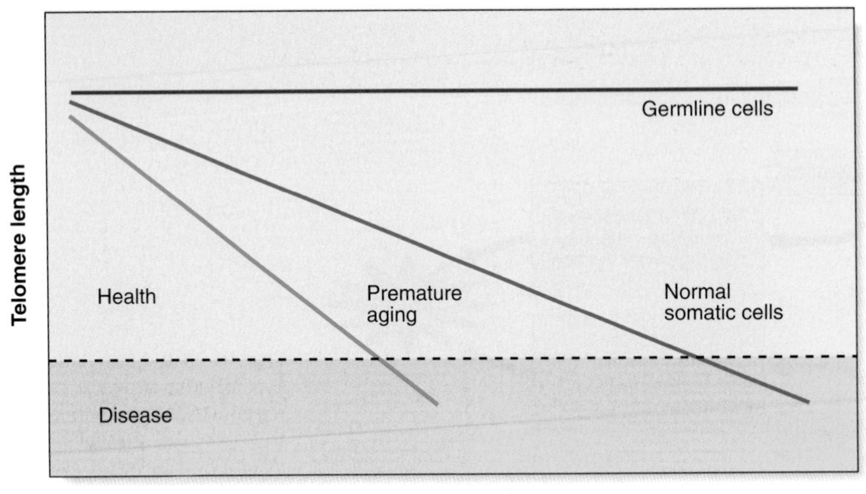

FIGURE 14.9 Telomere length over age, in normal life and premature aging syndromes. The only cells in the body that maintain telomere length throughout life, and have high levels of telomerase, are those of the germline. Somatic cells, in the absence of disease, undergo a slowly progressive decrease in telomere length throughout life, so that disease and cancer become an increasing risk in the elderly. In premature aging syndromes, the process of telomere shortening is accelerated, and the risk of cancer becomes high from early adult life onward.

notable exceptions in which only very low incidences of dominantly inherited carcinomas are recorded. These include the lung and cervix, as well as leukemias, lymphomas, and sarcomas. Here external agents or stimuli, and/or stochastic genetic events, are presumed to be the main factors. Nevertheless, studies of the common cancers—bowel or colorectal and breast cancer—have provided further insights into the genetics of cancer.

Colorectal Cancer

Approximately 1 in 40 people in the developed countries of Western Europe and North America will develop cancer of the bowel or colon. An understanding of the development of colorectal cancer has shed light on the process of carcinogenesis.

Multistage Process of Carcinogenesis

The majority of colorectal cancers are thought to develop from 'benign' adenomas. Conversely, only a small proportion of adenomas proceed to invasive cancer. Histologically, adenomatous polyps smaller than 1 cm in diameter rarely contain areas of carcinomatous change, whereas the risk of carcinomatous change increases to 5% to 10% when an adenoma reaches 2 cm in diameter. The transition from a small adenomatous polyp to an invasive cancer is thought to take between 5 and 10 years. Adenomatous polyps less than 1 cm in diameter have mutations in the *ras* gene in less than 10% of cases. As the size of the polyp increases to between 1 and 2 cm, the prevalence of *ras* gene mutations is in the region of 40%, rising to approximately 50% in full-blown colorectal cancers.

Similarly, allele loss of chromosome 5 markers occurs in approximately 40% of adenomatous polyps and 70% of carcinomas. Deletions on chromosome 17p in the region containing the *TP53* gene occur in more than 75% of carcinomas, but this is an uncommon finding in small or intermediate-sized polyps. A region on 18q is deleted in about 10% of small adenomas, rising to almost 50% when the adenoma shows foci of invasive carcinoma, and in more than 70% of carcinomas (Figure 14.10). Genes at this locus include *DCC* (deleted in colorectal cancer), *SMAD2*, and *SMAD4*, the latter being part of the transforming growth factor-β (TGF-β) pathway. In some colorectal cancers mutations in the *TGF-β* receptor gene have been identified.

It appears, therefore, that mutations of the *RAS* and *TP53* genes and LOH on 5q and 18q accumulate during the transition from a small 'benign' adenoma to carcinoma. The accumulation of alterations, rather than the order, appears to be more important in the development of carcinoma. More than one of these four alterations is seen in only 7% of small, early adenomas. Two or more such alterations are seen with increasing frequency when adenomas progress in size and show histological features of malignancy. More than 90% of carcinomas show two or more such alterations, and approximately 40% show three.

The multistage process of the development of cancer is likely, of course, to be an oversimplification. The distinction between oncogenes and tumor suppressor genes (Table 14.3) has not always been clear-cut—e.g., the *RET* oncogene and *MEN2* (p. 93, Table 14.5). In addition, the same mutation in some of the inherited cancer syndromes (p. 225) can result in cancers at various sites in different individuals, perhaps as a consequence of the effect of interactions with inherited polymorphic variation in a number of other genes or a variety of environmental agents.

Further insight into the processes involved in the development of colorectal cancer came from a rare cause of

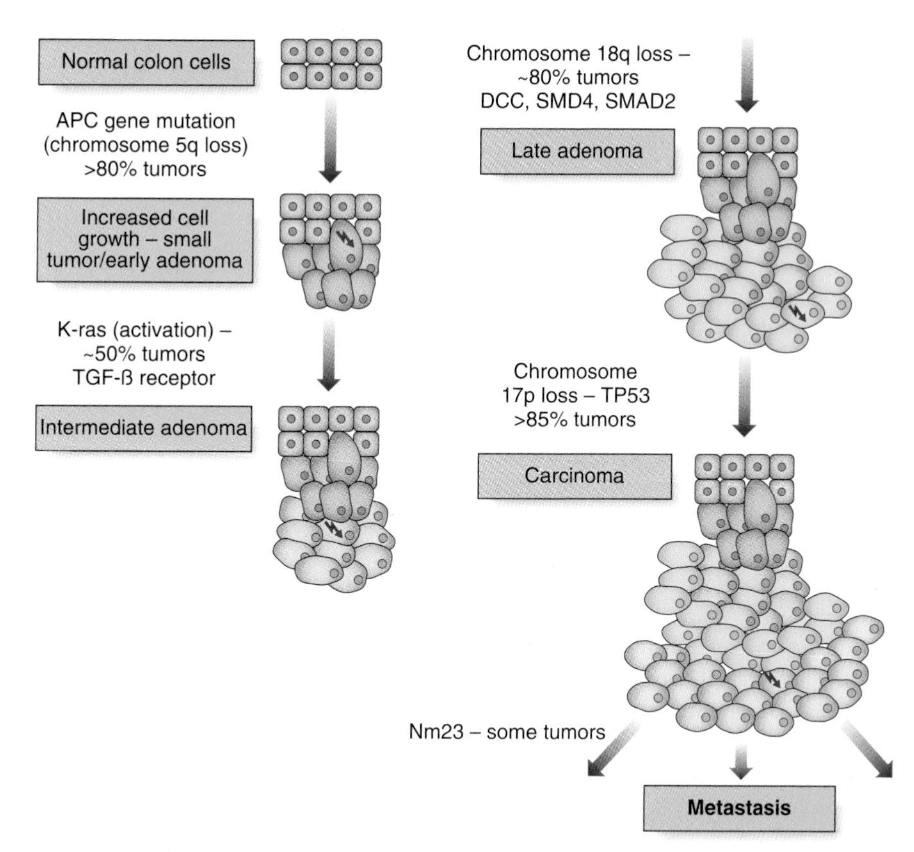

FIGURE 14.10 The development of colorectal cancer is a multistage process of accumulating genetic errors in cells. The *red arrows* represent a new critical mutation event, followed by clonal expansion. At the stage of carcinoma, the proliferating cells contain all the genetic errors that have accumulated.

familial colonic cancer known as familial adenomatous polyposis.

Familial Adenomatous Polyposis

Approximately 1% of persons who develop colorectal cancer do so through inheritance of an autosomal dominant disorder known as familial adenomatous polyposis (FAP). Affected persons develop numerous polyps of the large bowel, which can involve its entirety (Figure 14.11). There is a high risk of carcinomatous change taking place in these polyps, with more than 90% of persons with FAP eventually developing bowel cancer.

The identification of an individual with FAP and an interstitial deletion of a particular region of the long arm of chromosome 5 (5q21) led to the demonstration of linkage of FAP to DNA markers in that region. Subsequent studies

Table 14.3 Some Familial Cancers or Cancer Syndromes Due to Tumor Suppressor Mutations

Disorder	Gene	Locus
Retinoblastoma	*RB1*	13q14
Familial adenomatous polyposis	*APC*	5q31
Li-Fraumeni syndrome	*Tp53*	17p13
von Hippel–Lindau syndrome	*VHL*	3p25-26
Multiple endocrine neoplasia type II	*RET*	10q11.2
Breast–ovarian cancer	*BRCA1*	17q21
Breast cancer	*BRCA2*	13q12-13
Gastric cancer	*CDH1*	16q22.1
Wilms tumor	*WT1*	11p13
Neurofibromatosis I	*NF1*	17q12-22

FIGURE 14.11 Large bowel from a person with polyposis coli opened up to show multiple polyps throughout the colon. (Courtesy Mr. P. Finan, Department of Surgery, General Infirmary, Leeds.)

led to the isolation of the adenomatous polyposis coli *(APC)* gene. Analyses of the markers linked to the *APC* gene in cancers from persons who have inherited the gene for this disorder have shown LOH, suggesting a similar mechanism of gene action in the development of this type of bowel cancer.

Studies in the common, non-hereditary form of bowel cancer have shown similar LOH at 5q in the tumor material, with the FAP gene being deleted in 40% and 70% of sporadically occurring adenomas and carcinomas of the colon. LOH has also been reported at a number of different sites in colonic cancer tumors that include the regions 18q21-qter and 17p12-p13, the latter region including the *TP53* gene, as well as another gene at 5q21 known as the 'mutated in colorectal cancer' *(MCC)* gene, consistent with the development of the common form of colonic cancer being a multistage process.

'Deleted in Colorectal Cancer'

Allele loss on chromosome 18q is seen in more than 70% of colorectal carcinomas. The original candidate gene for this region, called *DCC*, has been identified and cloned; it has a high degree of homology with the family of genes encoding cell adhesion molecules. The *DCC* gene is expressed in normal colonic mucosa but is either reduced or absent in colorectal carcinomas. As with *TP53*, somatic mutations in the *DCC* gene of the remaining allele occur in some cancers where gene expression is absent. The known homology suggests that loss of *DCC* plays a role in cell–cell and cell–basement membrane interactions, features that are lost in overt malignancy. However, mutations in the *DCC* gene have been found in only a small proportion of colonic cancers. Other genes deleted in this region in colorectal tumors include *SMAD4* and *SMAD2*.

Hereditary Non-Polyposis Colorectal Cancer—Lynch Syndrome

A proportion of individuals with familial colonic cancer may have a small number of polyps, and the cancers occur more frequently in the proximal, or right side, of the colon, which is sometimes called 'site-specific' colonic cancer. The average age of onset for colonic cancer in this condition is the mid-forties. This familial cancer-predisposing syndrome is inherited as an autosomal dominant disorder and has been known as hereditary non-polyposis colorectal cancer (HNPCC)—even though polyps may be present (the name helps to distinguish the condition from FAP). There is now a preference to return to the original eponymous designation of Lynch syndrome, specifically Lynch syndrome type I (site-specific, e.g., colorectal). There is also a risk of small intestinal cancers, including stomach, endometrial cancer, and a variety of others (see Table 14.5).

DNA Mismatch Repair Genes

When looking for LOH, comparison of polymorphic microsatellite markers in tumor tissue and constitutional cells in persons with HNPCC somewhat surprisingly revealed the presence of new, rather than fewer, alleles in the DNA from tumor tissue. In contrast to the site-specific chromosome rearrangements seen with certain malignancies (pp. 211–212), this phenomenon, known as **microsatellite instability (MSI)**, is generalized, occurring with all microsatellite markers analyzed, irrespective of their chromosomal location.

Table 14.4 Mismatch Repair Genes Associated with Hereditary Non-Polyposis Colorectal Cancer

Human Gene	Chromosomal Locus	*E. coli* Homolog	HNPCC (%)
hMSH2	2p22-21	MutS	31
hMSH6	2p16	MutS	Rare
hMLH1	3p21	MutL	33
hPMS1	2q31	MutL	Rare
hPMS2	7p22	MutL	4
TACTSTD1 (deletions affect hMSH2, which is immediately downstream)	2p21		Rare
Undetermined loci			~30

HNPCC, Hereditary non-polyposis colorectal cancer.

This phenomenon was recognized to be similar to that seen in association with mutations in genes known as mutator genes, such as the *MutHLS* genes in yeast and *Escherichia coli*. In addition, the human homolog of the mutator genes were located in regions of the human chromosomes to which HNPCC had previously been mapped, leading to rapid cloning of the genes responsible for HNPCC in humans (Table 14.4). The mutator genes code for a system of 'proof-reading' enzymes and are known as **mismatch repair genes**, which detect mismatched base pairs arising through errors in DNA replication or acquired causes (e.g., mutagens). The place of the *TACSTD1* gene is unusual. It lies directly upstream of *MSH2* and, when the last exons of the gene are deleted, transcription of *TACSTD1* extends into *MSH2*, causing epigenetic inactivation of the *MSH2* allele. However, deletions in this gene appear to be a rare cause of HNPCC.

Individuals who inherit a mutation in one of the mismatch repair genes responsible for HNPCC are constitutionally heterozygous for a loss-of-function mutation (p. 26). Loss of function of the second copy through any of the mechanisms discussed in relation to LOH (p. 215) results in defective mismatch repair leading to an increased mutation rate associated with an increased risk of developing malignancy. Certain germline mutations, however, seem to have dominant-negative effects (p. 26). Although HNPCC accounts for a small proportion of colonic cancers, estimated as 2% to 4% overall, approximately 15% of *all* colorectal cancers exhibit MSI, the proportion being greater in tumors from persons who developed colorectal cancer at a younger age. Some of these individuals will have inherited constitutional mutations in one of the mismatch repair genes in the absence of a family history of colonic cancer. In addition, for women with a constitutional mismatch repair gene mutation, the lifetime risk of endometrial cancer is up to 50%.

Analysis of tumor DNA for evidence of MSI has become a routine first test in cases where a diagnosis of HNPCC is a possibility. High levels of MSI are suggestive of the presence of HNPCC-related mutations in the tumor, some of which will be somatic in origin whereas in others there will be a germline mutation plus a 'second hit' in the normal allele. An additional technique, **immunohistochemistry (IHC)**, is also proving useful as an investigation to discriminate those cases suitable for direct mutation analysis. Taking paraffin-embedded tumor tissue, loss of expression of specific mismatch repair genes can be tested using antibodies against the proteins hMSH2, hMLH1, hMSH6, and hPMS2. Where tumor cells fail to stain (in contrast to surrounding normal cells), a loss of expression of that protein has occurred and direct gene mutation analysis can be justified.

Other Polyposis Syndromes

Although isolated intestinal polyps are common, occurring in about 1% of children, there are familial forms of multiple polyposis that are distinct from FAP but showing heterogeneity.

MYH Polyposis

In a large study, nearly 20% of familial polyposis cases showed neither dominant inheritance nor evidence of an *APC* gene mutation. Of these families, more than 20% were found to have mutations in the *MYH* gene, and affected individuals were compound heterozygotes. In contrast to the other polyposis conditions described in the following section, MYH polyposis is an autosomal recessive trait, thus significantly affecting genetic counseling as well as the need for screening in the wider family. The gene, located on chromosome band 1p33, is the human homolog of *mutY* in *E. coli*. This bacterial mismatch repair operates in conjunction with *mutM* to correct A/G and A/C base-pair mismatches. In tumors studied, an excess of G:C to T:A transversions was observed in the *APC* gene. Mutations that effectively knock out the *MYH* gene, therefore, lead to defects in the base excision-repair pathway; this is a form of DNA mismatch repair that, unusually, follows autosomal recessive inheritance.

Juvenile Polyposis Syndrome

Autosomal dominant transmission is well described for a rare form of juvenile polyposis that may present in variety of ways, including bleeding with anemia, pain, intussusception, and failure to thrive. The polyps carry an approximate 13-fold increased cancer risk and, once diagnosed, regular surveillance and polypectomy should be undertaken. The average age at diagnosis of cancer is in the third decade, so that colectomy in adult life may be advisable. Two genes have been identified as causative: *SMAD4* (18q) and *BMPR1A* (10q22). Both are components of the TGF-β signaling pathway (p. 85) and *SMAD4* mutations, which account for about 60% of cases, appear to carry a higher malignancy potential and the possibility of large numbers of gastric polyps.

Cowden Disease

Also known as **multiple hamartoma syndrome**, Cowden disease is autosomal dominant but very variable. Gastrointestinal polyps are found in about half of the cases and are generally benign hamartomas or adenomas. Multiple lipomas occur with similar frequency and the oral mucosa may have a 'cobblestone' appearance. Significant macrocephaly is very common in this condition. Importantly, however, there is a high incidence (50%) of breast cancer in females, usually occurring at a young age, and papillary thyroid carcinoma affects about 7% of the patients. Testicular seminoma can occur in males. Mutations in the tumor suppressor *PTEN* gene on chromosome 10q23, encoding a tyrosine phosphatase, cause Cowden disease. A related phenotype with many overlapping features, which glories in the eponymous name Bannayan-Riley-Ruvalcaba syndrome, has also been shown to be due to mutations in *PTEN* in a large proportion of cases.

Peutz-Jegher Syndrome

Also autosomal dominant, this condition is characterized by the presence of dark melanin spots on the lips, around the mouth (Figure 14.12), on the palms and plantar areas, and other extremities. These are usually present in childhood and can fade in adult life. Patients often present with colicky abdominal pain from childhood from the development of multiple polyps that occur throughout the gastrointestinal tract, although they are most common in the small

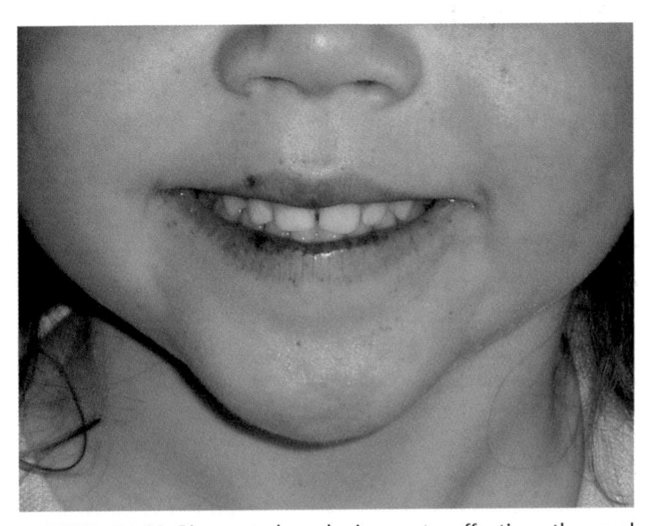

FIGURE 14.12 Pigmented melanin spots affecting the oral mucosa of a child with Peutz-Jegher syndrome, which are usually more prominent in childhood compared with adult life. Affected individuals are at risk of multiple polypoid hamartomas throughout the gastrointestinal tract, which may undergo malignant change.

intestine. These are hamartomas but there is a significant risk of malignant transformation. There is an increased risk of cancers at other sites, particularly breast, uterus, ovary, and testis, and these tend to occur in early adult life. Regular screening for these cancers throughout life, from early adulthood, is warranted. Mutations in a serine threonine kinase gene, *STK11*, located on chromosome 19p, cause Peutz-Jegher syndrome.

Breast Cancer

Approximately 1 in 12 women in Western societies will develop breast cancer, this being the most common cancer in women between 40 and 55 years of age, with approximately 1 in 3 affected women going on to develop metastatic disease. Some 15% to 20% of women who develop breast cancer have a family history of the disorder. Family studies have shown that the risk of a woman developing breast cancer is greater when one or more of the following factors is present in the family history: (1) a clustering of cases in close female relatives; (2) early age (<50 years) of presentation; (3) the occurrence of bilateral disease; (4) and the additional occurrence of ovarian cancer.

Molecular studies of breast cancer tumors have revealed a variety of different findings that included amplification of *erb*-B1, *erb*-B2, *myc*, and *int*-2 oncogenes as well as LOH at a number of chromosomal sites, including (in descending order of frequency) 7q, 16q, 13q, 17p, 8p, 21q, 3p, 18q, 2q, and 19p, as well as several other regions with known candidate genes or fragile sites. In many breast tumors showing LOH, allele loss occurs at two to four of the sites, again suggesting that the accumulation of alterations, rather than their order, is important in the evolution of breast cancer. One potentially key element in the development of sporadic breast cancer, and sporadic ovarian cancer, is the gene named *EMSY*. This was found to be amplified in 13% of breast cancers and 17% of ovarian cancers, and was ascertained when looking for DNA sequences that interact with *BRCA2*. The normal function of *EMSY* may be to switch off *BRCA2*; this may point to an important pathway of control of cell growth in these tissues.

BRCA1 and *BRCA2* Genes

Family studies of early-onset or premenopausal breast cancer showed that it behaved like a dominant trait in many families. Linkage analysis in these families showed that the tendency to develop breast cancer mapped to the long arm of chromosome 17, eventually leading to identification of the *BRCA1* gene. A proportion of families with early-onset breast cancer that did not show linkage to this region showed linkage to the long arm of chromosome 13, resulting in the identification of the *BRCA2* gene.

Approximately 40% to 50% of families with early-onset autosomal dominant breast cancer have a mutation in the *BRCA1* gene and have been shown to have a 60% to 85% lifetime risk of developing breast cancer. Females with a *BRCA1* mutation have an increased risk of developing ovarian cancer, and males an increased risk of developing prostate cancer. Mutations in the *BRCA2* gene account for 30% to 40% of families with early-onset autosomal dominant breast cancer, and the lifetime risk of developing breast cancer is similar. Although initially mutations in the *BRCA2* gene were not thought to be associated with an increased risk of other cancers, women heterozygous for a mutation also have an increased risk of developing ovarian cancer, and males an increased risk of prostate cancer. In some of the original familial breast cancer families recruited for linkage studies, a number had males who developed breast cancer. Although breast cancer in males is very rare, males with mutations in the *BRCA2* gene have a 6% lifetime risk of developing breast cancer, approximately a 100-fold increase in the population risk of breast cancer in males.

Ovarian Cancer

Approximately 1 in 70 women develops ovarian cancer, the incidence increasing with age. The majority arise as a result of genetic alterations within the ovarian surface epithelium and are therefore referred to as *epithelial* ovarian cancer. In general it is a poorly understood disease, although studies have shown a high frequency of LOH at 11q25 in tumor tissue. A possible tumor suppressor gene at this locus is *OPCML*, which encodes a cell adhesion molecule that includes an immunoglobulin domain. Approximately 5% of women with ovarian cancer have a family history of the disorder and it is estimated that 1% of all ovarian cancer follows dominant inheritance because it is strongly predisposed by single-gene mutations. In families with multiple women affected with ovarian cancer, the age of presentation is 10 to 15 years earlier than with non-hereditary ovarian cancer in the general population. Mutations in *BRCA1*, *BRCA2*, and less commonly the genes responsible for HNPCC/Lynch syndrome, are responsible in a proportion of these families, but a susceptibility gene locus for site-specific ovarian cancer has not been identified.

Prostate Cancer

Prostate cancer is the most common cancer overall after breast cancer, and is the most common cancer affecting men, who have a lifetime risk of 10% of developing the disease and a 3% chance of dying from it. Enquiries into the family history of males presenting with prostate cancer have revealed a significant proportion (about 15%) to have a first-degree male relative with prostate cancer. Family studies have shown that first-degree male relatives of a man presenting with prostate cancer have between two and five times the population risk of developing prostate cancer.

Analysis of prostate cancer tumor material has revealed LOH at several chromosomal locations. Segregation analysis of family studies of prostate cancer suggested that a single dominant susceptibility locus could be responsible, accounting for 9% of all prostate cancers and up to 40% of early-onset prostate cancers (diagnosed before age 55 years). Linkage analysis studies identified two major susceptibility loci, hereditary prostate cancer-1 and -2 (HPC1 and HPC2), and genome wide association studies have highlighted a number of other susceptibility loci of variable significance. It is possible in due course that testing of multiple susceptibility loci will enable identification of high risk individuals who can be offered surveillance. Mutations in the ribonuclease L gene *(RNASEL)* were identified in two families showing linkage to the HPC1 locus at 1q25. Mutations have been found in the *ELAC2* gene at 17p11, the HPC2 locus, and, rarely, mutations in three genes—*PTEN*, *MXI1*, and *KAI1*—have been identified in a minority of families with familial prostate cancer. A small proportion of familial prostate cancer is associated with *BRCA1* or *BRCA2*. Men who carry mutations in either *BRCA1* or *BRCA2* have an increased risk, and in one study, conducted in Ashkenazi Jews, men with such mutations had a 16% risk of prostate cancer by age 70 years, compared with 3.8% for the general population.

Although the majority of prostate cancers occur in men older than age 65, individuals with a family history of prostate cancer, consistent with the possibility of a dominant gene being responsible, are at increased risk of developing the disease at a relatively younger age (younger than 55 years). Screening by measuring prostate-specific antigen levels and performing digital rectal examination is often offered, but problems with specificity and sensitivity mean that interpretation of results is difficult.

Genetic Counseling in Familial Cancer

Recognition of individuals with an inherited susceptibility to cancer usually relies on taking a careful family history to document the presence or absence of other family members with similar or related cancers. The malignancies that develop in susceptible individuals are often the same as those that occur in the population in general. There are a number of other features that can suggest an inherited cancer susceptibility syndrome in a family (Box 14.1).

Inherited Cancer-Predisposing Syndromes

Although most cancers from an inherited cancer syndrome occur at a specific site, families have been described in which cancers occur at more than one site in an individual or at different sites in various members of the family more commonly than would be expected. These families are referred to as having a **familial cancer-predisposing** syndrome. The majority of the rare inherited familial cancer-predisposing syndromes currently recognized are dominantly inherited, with offspring of affected individuals having a 50% chance of inheriting the gene and therefore of being at increased risk of developing cancer (Table 14.5). For the clinician, it is important to be aware of the physical signs that may point to as diagnosis, for example melanin spots around the mouth and lips (Peutz-Jegher syndrome), macrocephaly (Cowden disease), and dome-shaped skin papules (trichodiscomas; Figure 14.13) over the face and neck (Birt-Hogg-Dubé syndrome). In the latter condition, pneumothorax may be a presenting feature. There are also a number of syndromes, usually inherited as autosomal recessive disorders, with an increased risk of developing cancer associated with an increased number of abnormalities in the chromosomes when cultured, or what are known as the **chromosomal breakage** syndromes (p. 288).

Those with an inherited familial cancer-predisposing syndrome are at risk of developing a second tumor (multifocal or bilateral in the case of breast cancer), have an increased risk of developing a cancer at a relatively younger age than those with the sporadic form, and can develop tumors at different sites in the body, although one type of cancer is usually predominant.

A number of different familial cancer-predisposing syndromes have been described, depending on the patterns of cancer occurring in a family. For example, persons with the Li-Fraumeni syndrome (p. 218) are at risk of developing adrenocortical tumors, soft-tissue sarcomas, breast cancer, brain tumors, and leukemia—sometimes at a strikingly young age. The cancer-predisposing syndrome HNPCC, also known as Lynch type I (site-specific colorectal cancer) and Lynch type II (confusingly, once known as the **cancer family syndrome**), in which family members are also at risk for a number of other cancers, including stomach, endometrial, breast, and renal transitional cell carcinomas. Progress at the molecular level (see Table 14.4) has highlighted the difficulty of classifying two types of Lynch syndrome—but despite this there is now a preference to return to this name. Further confusion can arise in consideration of Turcot syndrome, which is due to mutations in the *APC* gene and two of the mismatch repair genes, whereas Muir-Torré syndrome results from mutations in the *hMSH2* mismatch repair gene. Individuals at risk in such families should, however, be screened for the appropriate cancers.

Box 14.1 Features Suggestive of an Inherited Cancer Susceptibility Syndrome in a Family

- Several close (first- or second-degree) relatives with a common cancer
- Several close relatives with related cancers (e.g., breast and ovary or bowel and endometrial)
- Two family members with the same rare cancer
- An unusually early age of onset
- Bilateral tumors in paired organs
- Synchronous or successive tumors
- Tumors in two different organ systems in one individual

Inherited Susceptibility for the Common Cancers

The majority of persons at an increased risk of developing cancer because of their family history do not have one of the cancer-predisposing syndromes. The level of risk for persons with a family history of one of the common cancers such as bowel or breast cancer depends on several factors. These include the number of persons with cancer

Table 14.5 Inherited Family Cancer Syndromes, Mode of Inheritance, Gene Responsible and Chromosomal Site

Syndrome	Mode of Inheritance	Gene	Chromosomal Site	Main Cancer(s)
Breast/ovary families	AD	*BRCA1*	17q21	Breast, ovary, colon, prostate
Breast (+ ovary) families	AD	*BRCA2*	13q12	Breast, ovary
Familial adenomatous polyposis	AD	*APC*	5q21	Colorectal, duodenal, thyroid
Turcot syndrome	AD	*APC*	5q21	Colorectal, brain
		hMLH1	3p21	
		hMSH2	2p22-21	
Hereditary non-polyposis colorectal cancer (HNPCC)				
Lynch I	AD	*hMSH2*	2p22-21	Colorectal
		hMSH6	2p16	
		hMLH1	3p21	
		hPMS1	2q31	
		hPMS2	7p22	
		TACSTD1	2p21	
Lynch II	AD	*hMSH2*	2p22-21	Colorectal, endometrial, urinary tract, ovarian, gastric, small bowel, hepatobiliary
		hMLH1	3p21	
		hPMS1	2q31	
		hPMS2	7p22	
MYH polyposis	AR	*MYH*	1p33	
Muir-Torré syndrome	AD	*hMSH2*	2p22-21	As Lynch II plus sebaceous tumors, laryngeal
Juvenile polypois	AD	*SMAD4/DPC4*	18q21.1	Colorectal
		BMPR1A	10q22	
Peutz-Jegher syndrome	AD	*STK11*	19p13.3	Gastrointestinal, breast, uterus, ovary, testis
Cowden disease	AD	*PTEN*	10q23	Breast (females), thyroid (papillary), testicular (seminoma)
Familial retinoblastoma	AD	*RB1*	13q14	Retinoblastoma
Li-Fraumeni syndrome	AD	*TP53*	17p13	Sarcoma, breast, brain, leukemia, adrenal cortex
Multiple endocrine neoplasia (MEN)				
Type I (MEN1)	AD	*MEN1*	11q13	Parathyroid, thyroid, anterior pituitary, pancreatic islet cells, adrenal
Type II (MEN2)	AD	*RET*	10q11.2	Thyroid (medullary), pheochromocytoma
von Hippel–Lindau disease	AD	*VHL*	3p25-26	CNS hemangioblastoma, renal, pancreatic, pheochromocytoma
Gorlin (nevoid basal cell carcinoma) syndrome	AD	*PTCH*	9q22	Basal cell carcinomas, syndrome medulloblastoma, ovarian fibromas, (odontogenic keratocysts)
Birt-Hogg-Dubé syndrome	AD	*FLCN*	17p11.2	Renal
Dysplastic nevus syndrome	AD	*CMM1*	1p	Melanoma (familial atypical mole melanoma, FAMM)

AD, Autosomal dominant; *CNS*, central nervous system.

in the family, how closely related the person at risk is to the affected individuals, and the age at which the affected family member(s) developed cancer. A few families with a large number of members affected with one of the common cancers are consistent with a dominantly inherited cancer susceptibility gene. In most instances, there are only a few individuals with cancer in a family, and there is doubt about whether a cancer susceptibility gene is responsible or not. In such an instance, one relies on empirical data gained from epidemiological studies to provide risk estimates (Tables 14.6 and 14.7). With respect to mainly breast and ovarian cancers, in recent years the Manchester Scoring System (Table 14.8) has gained acceptance as a method of determining the likelihood of identifying a *BRCA1* or *BRCA2* mutation based on family history information. The derived score discriminates the likelihood of finding a mutation in one of these genes, and this provides a very useful clinical guide to genetic testing, which in many centers is set at a threshold of approximately 20%.

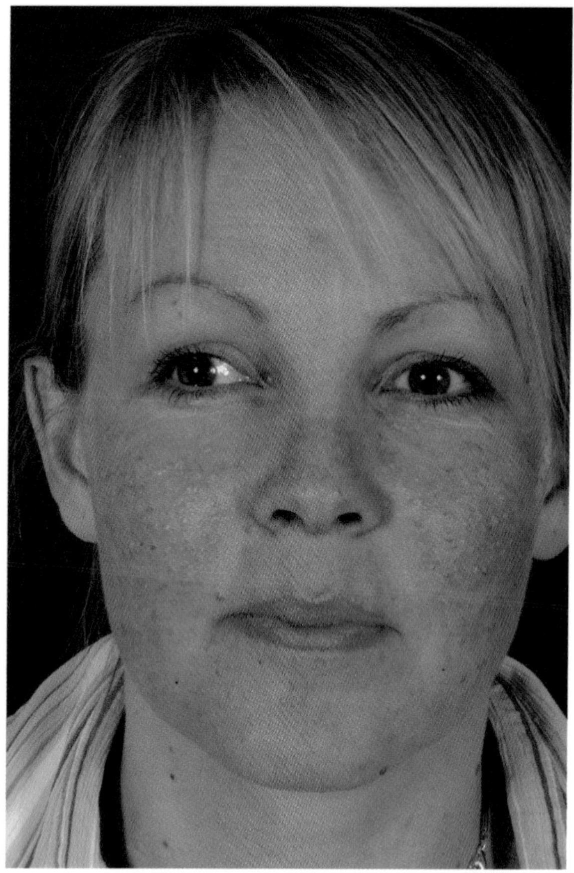

FIGURE 14.13 Facial trichodiscomas—the pale, dome-shaped papules found on the head and neck of patients with Birt-Hogg-Dubé syndrome. Affected individuals are at risk of renal cell carcinoma.

Table 14.6 Lifetime Risk of Colorectal Cancer for an Individual According to the Family History of Colorectal Cancer

Population risk	1 in 50
One first-degree relative affected	1 in 17
One first-degree relative and one second-degree relative affected	1 in 12
One relative younger than age 45 years affected	1 in 10
Two first-degree relatives affected	1 in 6
Three or more first-degree relatives affected	1 in 2

Data from Houlston RS, Murday V, Harocopos C, Williams CB, Slack J 1990 Screening and genetic counselling for relatives of patients with colorectal cancer in a family screening clinic. Br Med J 301:366–368.

Table 14.7 Lifetime Risk of Breast Cancer in Females According to the Family History of Breast Cancer

Population risk	1 in 10
Sister diagnosed at 65–70 years of age	1 in 8
Sister diagnosed younger than age 40 years	1 in 4
Two first-degree relatives affected younger than age 40 years	1 in 3

Screening for Familial Cancer

Prevention or early detection of cancer is the ultimate goal of screening individuals at risk of familial cancer. The means of prevention for certain cancers can include a change in lifestyle or diet, drug therapy, prophylactic surgery or screening.

Screening of those at risk of familial cancer is usually directed at detecting the phenotypic expression of the genotype (i.e., surveillance for a particular cancer or its precursor). Screening can also include diagnostic tests that indirectly reveal the genotype, looking for other clinical features that are evidence of the presence or absence of the gene. For example, individuals at risk for FAP can be screened for evidence of the *APC* gene by retinal examination looking for areas of *c*ongenital *h*ypertrophy of the *r*etinal *p*igment *e*pithelium, or what is known as *CHRPEs*. The finding of CHRPEs increases the likelihood of an individual at risk being heterozygous for the *APC* gene and therefore developing polyposis and malignancy. We now know that CHRPEs are seen in persons with FAP when mutations occur in the first part of the *APC* gene, an example of a genotype–phenotype correlation (p. 26).

More recently, identification of the gene responsible for a number of the cancer-predisposing syndromes, and determination of the genotypic status (i.e., presymptomatic testing, see p. 316), of an individual at risk allows more efficient delivery of surveillance screening for the phenotypic expression—e.g., renal cancer, central nervous system tumors and pheochromocytomas in von Hippel–Lindau disease (Table 14.9). For those who test negative for the

Table 14.8 The Manchester Scoring System for Predicting the Likelihood that either a *BRCA1* or *BRCA2* Mutation Will Be Identified, Based on Family History Information

Cancer, and Age at Diagnosis			
Female	**Male**	***BRCA1***	***BRCA2***
Breast <30		6	5
Breast 30–39		4	4
Breast 40–49		3	3
Breast 50–59		2	2
Breast >59		1	1
	Breast <60	5	8
	Breast >59	5	5
Ovarian <60		8	5
Ovarian >59		5	5
	Prostate <60	0	2
	Prostate >59	0	1
Pancreatic		0	1

In bilateral breast cancer each tumor is counted separately and DCIS (ductal carcinoma in situ) is included.

Example: in the family the proband is a female diagnosed with breast cancer at age 28 (*BRCA1*, 6; *BRCA2*, 5); her mother had breast cancer at age 46 (*BRCA1*, 3; *BRCA2*, 3); a maternal aunt had breast cancer at age 54 (*BRCA1*, 2; *BRCA2*, 2); in addition, a paternal aunt had breast cancer at age 57 (*BRCA1*, 2; *BRCA2*, 2), but this is discounted because this does not provide the highest score in a direct lineage. The total score is therefore 21, which reaches the threshold for testing the *BRCA* genes in most centers.

Table 14.9 Suggested Screening Guidelines for Persons at Significant Risk of Cancer: Familial Cancer-Predisposing Syndromes and Common Cancers

Condition/Cancer	Screening Test	Frequency	Starting Age (Years)
Familial Susceptibility for the Common Cancers			
BREAST CANCER			
Breast	Mammography	Annual	40–50 (3 yearly from 50 unless very high risk, e.g., *BRCA1* or *BRCA2* gene mutation carrier)
BREAST/OVARY			
Breast	Mammography	Annual	40–50 (as above)
Ovary	US/Doppler, CA125	Annual	35 (under review)
HNPCC—LYNCH I			
Colorectal—high risk families	Colonoscopy	2–3 yearly	25 or 5 years before the earliest diagnosis in the family
Colorectal—intermediate risk families	Colonoscopy	At first consultation or age 35–40 years	Repeat at age 55
HNPCC MATURITY-ONSET DIABETES OF THE YOUNG LYNCH II			
Colorectal	Colonoscopy	As above, for Lynch I	As above, for Lynch I
Endometrial	US (under evaluation)	Annual	35–65
Ovary	US	Annual	35
Renal tract	US	Annual	35
Gastric	Gastroscopy	2 yearly	25, if definite Lynch II syndrome
Small bowel	None		
Hepatobiliary	None		
Breast	Mammography	Annual	40–50
Familial Cancer-Predisposing Syndromes			
Familial adenomatous polyposis	Retinal examination (CHRPE)*		Childhood
Colorectal	Sigmoid/colonoscopy*,†	Annual	12
Duodenal	Gastroscopy	3 yearly	20
Thyroid (women)	None/US?	Annual	20
LI–FRAUMENI			
Breast	Mammography	Annual	40?
Sarcoma	None		
Brain	None		
Leukemia	None		
Adrenal cortex	None		
Retinoblastoma	Retinal examination	Frequently	From birth
MULTIPLE ENDOCRINE NEOPLASIA			
Type 1	Ca^{2+}, PTH, pituitary hormones, pancreatic hormones	Annual	8 years, up to age 50 years
Type 2	Calcitonin provocation test*		10
Medullary thyroid	US	?	10
Pheochromocytoma	Urinary VMA	Annual	10
Parathyroid adenoma	Ca^{2+}, PO_4, PTH	Annual	10
VON HIPPEL–LINDAU			
Retinal angioma	Retinal examination*	Annual	5
Hemangioblastoma	CNS CT/MRI	3 yearly	15 (5 yearly from age 40 years)
Pheochromocytoma	Urinary VMA	Annual	10
Renal	Abdominal CT	3 yearly	20
	Abdominal US	Annual	20
GORLIN (NEVOID BASAL CELL CARCINOMA) SYNDROME			
BCCs	Clinical surveillance	Annual	10
Medulloblastoma	Clinical surveillance	Annual	Infancy
Odontogenic keratocysts	Orthopantomography	6 monthly	10

*Test to detect heterozygous state.

†In individuals found to be affected, annual colonoscopy prior to colectomy and lifelong 4–6 monthly surveillance of the rectal stump, after subtotal colectomy.

US, Ultrasonography; *CHRPE*, congenital hypertrophy of the retinal pigment epithelium; *CT*, computed tomography; *MRI*, magnetic resonance imaging; *PTH*, parathyroid hormone; *VMA*, vanillyl mandelic acid; *BCC*, basal cell carcinoma.

family mutation, expensive and time-consuming screening is unnecessary. As more genes for cancer susceptibility are discovered, there will be an increasing number of conditions for which DNA testing will enable presymptomatic determination of genotypic status.

Although the potential for prevention of cancer through screening persons at high risk is considerable, it is important to remember that the impact on the overall rate of cancer in the population in general will be small as only a minority of all common cancers are due to gene mutations demonstrating mendelian inheritance. For many familial cancers, there has been a strong move toward nationally agreed screening protocols, especially in countries such as in the United Kingdom, where the bulk of health care is provided by the state. The provision of screening must increasingly be evidence based with demonstrable cost-benefits. In the United Kingdom, screening guidelines produced by the National Institute for Health and Clinical Excellence are seen as broadly determining what is available within the UK National Health Service, although it is important to appreciate that this is an evolving area and screening recommendations are subject to change. Furthermore, individualized screening strategies are often devised for women from families with *BRCA1*, *BRCA2*, and *TP53*, as well as families with a high risk of colorectal cancer.

Table 14.10 Conditions in Which Prophylactic Surgery is an Accepted Treatment, and Treatments that Are Under Evaluation, as an Option for the Familial Cancer-Predisposing Syndromes or Individuals at Increased Risk for the Common Cancers

Disorder	Treatment
Accepted Treatment	
Familial adenomatous polyposis	Total colectomy
Ovarian cancer families	Oophorectomy
Breast cancer families	Bilateral mastectomy
MEN2	Total thyroidectomy
Under Evaluation	
Familial adenomatous polyposis	Non-digestible starch—to delay onset of polyposis
	Sulindac—to reduce rectal and duodenal adenomas
Breast cancer families	Tamoxifen—to prevent development of breast cancer
	Avoidance of oral contraceptives and hormone replacement therapy

Familial Cancer-Predisposing Syndromes

Many familial cancer-predisposing syndromes are inherited as autosomal dominant traits that are fully penetrant, with the consequent risk for heterozygotes of developing cancer approaching 100%. This level of risk means that more invasive means of screening with more frequent and earlier initiation of screening protocols are justified than would be acceptable for the population in general (Table 14.10).

Inherited Susceptibility for the Common Cancers

Screening for the common cancers arising from inherited susceptibility has only relatively recently been established and by its very nature involves a long-term undertaking for the individual at risk as well as his or her physician or surgeon. It is important to emphasize that the natural enthusiasm for screening needs to be balanced with the paucity of hard data in many instances on the relative benefits and risks. However, recommended screening protocols are increasingly evidence based as more data become available (Box 14.2).

Who Should Be Screened?

In the case of the rare, dominantly inherited, single-gene familial cancer-predisposing syndromes such as FAP, von Hippel-Lindau, and multiple endocrine neoplasia (MEN), those who should be screened can be identified on a simple mendelian basis. However, for Rb, for example, the situation is more complex. If no *RB1* mutation has been identified, presymptomatic genetic testing cannot be offered. Some individuals with the non-hereditary form have bilateral tumors, whereas some with the hereditary form have no tumor (i.e., the condition is non-penetrant) or a unilateral tumor. It may be impossible to distinguish which form is present, and screening of second-degree, as well as first-degree, relatives may be appropriate given that early detection can successfully prevent blindness. For those with a family history of the common cancers, such as bowel or breast cancer, the risk levels at which screening is recommended, and below which screening is not likely to be of benefit, will vary. At each extreme of risk the decision is usually straightforward, but with intermediate-level risks there can be doubt as to relative benefits and risks of screening.

Box 14.2 Requirements of a Screening Test for Persons at Risk for a Familial Cancer-Predisposing Syndrome or at Increased Risk for the Common Cancers

The test should detect a malignant or premalignant condition at a stage before its producing symptoms, with high sensitivity and specificity

The treatment of persons detected by screening should improve the prognosis

The benefit of early detection should outweigh potential harm from the screening test

The test should preferably be non-invasive as most at-risk individuals require long-term surveillance

Adequate provision for prescreening counseling and follow-up should be available

What Age and How Often?

Cancer in persons with a familial cancer-predisposing syndrome tends to occur at a relatively earlier age than in the general population and screening programs must reflect this. With the exception of FAP, in which it is recommended that sigmoidoscopy to detect rectal polyps should start in the teenage years, most cancer screening programs do not start until 25 years of age or later. The highest-risk age band for most inherited susceptibilities is 35 to 50 years, but because cancer can still develop in those at risk at a later age, screening is usually continued thereafter. In some families the age of onset of cancer can be especially early and it is recommended that screening of at-risk individuals in these families commences 5 years before the age of onset in the earliest affected member. Again, Rb is an exception to the usual rule because, as it is a cancer of early childhood, screening starts in the postnatal period with frequent ophthalmic examination.

The recommended interval between repeated screening procedures should be determined from the natural history of the particular cancer. The development of colorectal cancer from an adenoma is believed to take place over a number of years, and as a result it is thought that 5-year screening intervals will suffice. If, however, a polyp is found, the interval between screening procedures is usually brought down to 3 years. Breast cancer is not detectable in a premalignant stage and early diagnosis is critical if there is to be a good prognosis. Annual mammography for females at high risk is therefore recommended from the age of 35 years.

What Sites Should Be Screened?

Having decided who is at risk within a family, one has to judge which types of cancer are most likely to occur and which systems of the body should be screened.

Familial Cancer-Predisposing Syndromes

This can be a very difficult problem with some of the family cancer syndromes, such as the Lynch type II form of HNPCC, in which a person at risk can develop cancer at a number of different sites. Screening for every possible cancer that can occur would mean frequent investigation by a variety of different specialists and/or investigations. This would result in an unwieldy and unpleasant protocol. People at risk for HNPCC should have regular colonoscopy, and females may be offered pelvic screening for gynecological malignancies, although the efficacy is very debatable. Some of the other cancers that can occur in those at risk for Lynch type II, such as stomach cancer, are not seen in every family, and so screening is usually restricted to people from those families in which these cancers have affected at least one family member. In those at risk for Li-Fraumeni syndrome, a wide spectrum of cancers can occur. However, apart from regular mammography, no satisfactory screening is available for the other malignancies (see Table 14.9).

Inherited Susceptibility for the Common Cancers

Colorectal Cancer

Colorectal carcinoma holds the greatest promise for prevention by screening. Endoscopy provides a sensitive and specific means of examination of the colorectal mucosa and polypectomy can be carried out with relative ease so that screening, diagnosis, and treatment can take place concurrently. Although colonoscopy is the preferred screening method, it requires a skilled operator and, because it is an invasive procedure, it has a small but consequent morbidity. Because of this, and to target screening on those most likely to benefit, most genetic centers have adopted the so-called **Amsterdam criteria** to select high-risk individuals. These minimal criteria suggest a familial form of colonic cancer:

1. At least three affected relatives (related to one another by first-degree relationships, FAP excluded).
2. At least two successive generations affected.
3. Cancer diagnosed before age 50 years in at least one relative.

Failure to visualize the right side of the colon with colonoscopy necessitates a barium enema to view this region, particularly in persons at risk for HNPCC, in which proximal right-sided involvement commonly occurs. For persons with a moderately increased risk of developing colorectal cancer, the majority of cancers occur in the distal (left-sided) colon and at a relatively later age. Flexible sigmoidoscopy, which is much less invasive than colonoscopy, provides an adequate screening tool for persons in this risk group and can be employed from the age of 50 years.

Breast Cancer

In the UK screening of women age 50 years and older for breast cancer by regular mammography has become established as a national program as a result of studies demonstrating improved survival of women detected as having early breast cancer. For women with an increased risk of developing breast cancer because of their family history, there is conflicting evidence of the relative benefit of screening with respect to the frequency of mammography and the chance of developing breast cancer in the interval between the screening procedures (i.e., 'interval' cancer). One reason is that cancer detection rates are lower in premenopausal than in postmenopausal breast tissue.

It is also argued that the radiation exposure associated with annual mammography could be detrimental if started at an early age, leading to an increased risk of breast cancer through screening when carried out over a long period. This is of particular concern in families with Li-Fraumeni syndrome, because mutations in the *TP53* gene have been shown experimentally in vitro to impair the repair of DNA damaged by X-irradiation. However, most experts believe that there is a greater relative benefit than risk in identifying and treating breast cancer in women from this high-risk group, although formal evaluation of such screening programs continues.

Mammography is usually offered only to women at increased risk of breast cancer after age 35 years, because interpretation of mammograms is difficult before this age because of the density of the breasts. As a consequence, women at increased risk of developing breast cancer should be taught breast self-examination and undergo regular clinical examination.

Ovarian Cancer

Ovarian cancer in the early stages is frequently asymptomatic and often incurable by the time a woman presents with symptoms. Early diagnosis of ovarian cancer in individuals at high risk is vital, with prophylactic oophorectomy being the only logical, if radical, alternative. The position of the ovaries within the pelvis makes screening difficult. Ultrasonography provides the most sensitive means of screening. Transvaginal scanning is more sensitive than conventional transabdominal scanning, and the use of color Doppler blood flow imaging further enhances screening of women at increased risk. If a suspicious feature is seen on scanning and confirmed on further investigation, laparoscopy or a laparotomy is usually required to confirm the diagnosis. Screening should be carried out annually as interval cancers can develop if screening is carried out less frequently.

Measuring the levels of CA125, an antigenic determinant of a glycoprotein that is present in increased levels in the blood of women with ovarian cancer, can be also be used as a screening test for women at increased risk of developing ovarian cancer. CA125 levels are not specific to ovarian cancer, as they are also increased in women with a number of other disorders, such as endometriosis. In addition, there are problems with sensitivity (p. 319), because CA125 levels are not necessarily increased in all women with ovarian cancer. Because of the problems outlined with these various screening modalities, many women with an increased risk of developing ovarian cancer choose to have their ovaries removed prophylactically after their family is complete. However, this in turn raises the issue of the benefits and risks associated with taking hormone replacement therapy.

What Treatment Is Appropriate?

Surgical intervention is the treatment of choice for persons at risk for some of the familial cancer-predisposing syndromes—e.g., prophylactic thyroidectomy in MEN type 2 (especially MEN2B) or colectomy in FAP. For those with a high risk from an inherited susceptibility for one of the common cancers (e.g., colon or breast/ovary), prophylactic surgery is also an accepted option, but the decision is more complex and dependent on the individual patient's choice. The option of prophylactic mastectomy in women at high risk of developing breast cancer is very appealing to some patients but totally abhorrent to others, and alternative management in the form of frequent surveillance, and possibly drugs such as the anti-estrogen tamoxifen, can be offered. For patients at high risk of colonic cancer, dietary modification such as the use of non-digestible starch, or the use of drugs such as the aspirin-like non-steroidal anti-inflammatory sulindac, may have value (see Table 14.10).

Those at an increased risk of developing cancer, especially if it is one of the single-gene dominantly inherited cancer-predisposing syndromes, or one of the single-gene causes of the common cancers, find themselves in an unenviable situation concerning both their health and the possibility of transmitting the condition to their children. Unfortunately, they are also likely in future to experience increasing difficulties in other areas of life, such as insurance and employment (p. 366).

FURTHER READING

Cowell JK (ed) 1995 Molecular genetics of cancer. Oxford: Bios Scientific

A multiauthor text covering the cancer family syndromes and the common cancers.

Eeles RA, Ponder BAJ, Easton DF, Horwich A eds 1996 Genetic predisposition to cancer. London: Chapman & Hall

A good multiauthor text reviewing the various cancer-predisposing syndromes and the common familial cancers, as well as the accepted and controversial areas of their management.

Harris H, Miller OJ, Klein G, Worst P, Tachibam T 1969 Suppression of malignancy by cell fusion. Nature 350:377–378.

Studies that eventually led to the concept of tumor suppressor genes.

Hodgson SV, Maher ER 2007 A practical guide to human cancer genetics, 3rd ed. Cambridge: Cambridge University Press

An up-to-date second edition of this text covering the developing field of human cancer genetics.

King RA, Rotter JI, Motulsky AG eds 1992 The genetic basis of common diseases. Oxford: Oxford University Press

Six chapters of this text cover the basic biology, epidemiology, and familial aspects of cancer.

Knudson AG 1971 Mutation and cancer: statistical study of retinoblastoma. Proc Natl Acad Sci 68:820–823

Proposal of the 'two-hit' hypothesis for the development of retinoblastoma.

Lalloo F, Kerr B, Friedman J, Evans G 2005 Risk assessment and management in cancer genetics. Oxford: Oxford University Press

Li FP, Fraumeni JF 1969 Soft tissue sarcomas, breast cancer, and other neoplasms: a familial syndrome? Ann Intern Med 71:747–752

The original description of the Li-Fraumeni syndrome.

Lynch HT 1967 'Cancer families': adenocarcinomas (endometrial and colon carcinoma) and multiple primary malignant neoplasms. Recent Results Cancer Res 12:125–142

The description of the cancer family syndrome now known as Lynch II.

Offit K 1998 Clinical cancer genetics. Chichester: Wiley-Liss

Text covering the clinical aspects of the various familial cancer-predisposing syndromes as well as the common cancers, along with the basic cellular biology and ethical and legal aspects.

Volgelstein B, Kinzler KW 2002 The genetic basis of human cancer. London: McGraw-Hill

Very comprehensive book covering in detail the cellular biology of cancer and the clinical aspects of the familial cancer-predisposing syndromes and the familial common cancers.

ELEMENTS

1 Cancer has both genetic and environmental causes.

2 Genetic and environmental factors in the etiology of cancer can be differentiated by epidemiological, family and twin studies, and by analysis of disease, biochemical, and viral associations.

3 Studies of tumor viruses have revealed genes present in humans known as oncogenes that are involved in carcinogenesis by altering cellular control mechanisms.

4 Study of rare, dominantly inherited tumors in humans, such as retinoblastoma, has led to the identification of tumor suppressor genes, consistent with the hypothesis that the development of cancer involves a minimum of two 'hits'. Persons at risk of familial cancer inherit the first 'hit' in the germ cell, the second 'hit' occurring in somatic cells in mitosis. In persons with sporadically occurring cancer, both 'hits' occur in somatic cells.

5 Some 5% of the common cancers, such as breast and bowel cancer, arise as a result of an inherited cancer susceptibility. Familial susceptibility for cancer can occur as an inherited susceptibility for a single type of cancer or for a number of different types of cancer as part of a familial cancer-predisposing syndrome.

6 Persons at risk of an inherited cancer susceptibility can be screened for associated features of a familial cancer predisposing syndrome or for particular cancers.

CHAPTER 15

Genetic Factors in Common Diseases

Medical genetics usually concentrates on the study of rare unifactorial chromosomal and single-gene disorders. Diseases such as diabetes, cancer, cardiovascular and coronary artery disease, mental health, and neurodegenerative disorders are responsible, however, for the majority of the morbidity and mortality in developed countries. These so-called common diseases are likely to be of even greater importance in the future, with the elderly accounting for an increasing proportion of the population.

The common diseases do not usually show a simple pattern of inheritance. Instead, the contributing genetic factors are often multiple, interacting with each other and environmental factors in a complex manner. In fact, it is uncommon for either genetic or environmental factors to be entirely responsible for a particular common disorder or disease in a single individual. In most instances, both genetic and environmental factors are contributory, although sometimes one can appear more important than the other (Figure 15.1).

At one extreme are diseases such as Duchenne muscular dystrophy; these are exclusively genetic in origin, and the environment plays little or no direct part in the aetiology. At the other extreme are infectious diseases that are almost entirely the result of environmental factors. Between these two extremes are the common diseases and disorders such as diabetes mellitus, hypertension, cerebrovascular and coronary artery disease, schizophrenia, the common cancers, and certain congenital abnormalities in which both genetic and environmental factors are involved.

Genetic Susceptibility to Common Disease

For many of the common diseases, a small but significant proportion of cases have single-gene causes, but the major proportion of the genetic basis of common diseases can be considered to be the result of an inherited predisposition or genetic susceptibility. Common diseases result from a complex interaction of the effects of multiple different genes, or what is known as **polygenic inheritance**, with environmental factors and influences due to what is known as **multifactorial inheritance** (see Chapter 9).

If the underlying genetic factors were understood, it would be possible to offer genetic testing to identify those persons genetically susceptible to a particular disorder. However, the utility of such a test would depend upon the action taken subsequently to reduce other risk factors—lifestyle changes, for example.

Types and Mechanisms of Genetic Susceptibility

Genetic susceptibility for a particular disease can occur through single-gene inheritance of an abnormal gene product involved in a particular metabolic pathway, such as occurs in early coronary artery disease arising from familial hypercholesterolemia (FH) (p. 175). In an individual with a mutation in the *FH* gene, the genetic susceptibility is the main determinant of the development of coronary artery disease, but this can be modified by environmental alteration like reduction in dietary cholesterol and avoidance of other risk factors such as obesity, lack of exercise, and smoking.

Inheritance of single-gene susceptibility does not, however, necessarily lead to development of a disease. For some diseases, exposure to specific environmental factors will be the main determinant in the development of the disease (e.g., smoking or occupational dust exposure in the development of pulmonary emphysema in persons with α_1-antitrypsin deficiency [p. 320, Table 23.1]).

In other instances, the mechanism of the genetic susceptibility is less clear-cut. This can involve inheritance of a single gene polymorphism (p. 67) that leads to differences in susceptibility to a disease (e.g., acetaldehyde dehydrogenase activity and alcoholism). In addition, inherited single-gene polymorphisms appear to determine the response to as yet undefined environmental factors—for example, the antigens of the major histocompatibility (HLA) complex and specific disease associations (p. 200) such as type 1 diabetes, rheumatoid arthritis, and coeliac disease. Lastly, genetic susceptibility can determine differences in responses to medical treatment; isoniazid inactivation status in the treatment of tuberculosis (p. 186) is a good example.

Approaches to Demonstrating Genetic Susceptibility to Common Diseases

In attempting to understand the genetics of a particular condition, the investigator can approach the problem in a

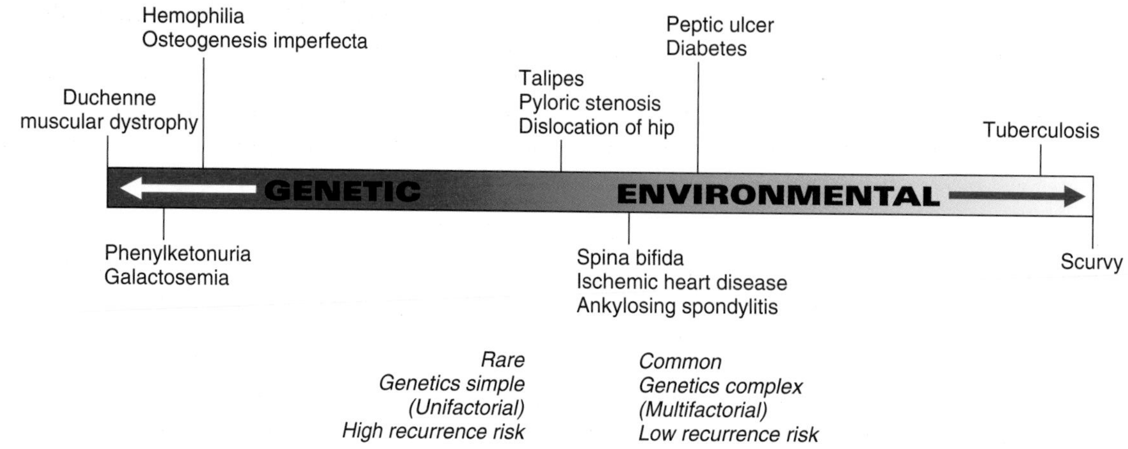

FIGURE 15.1 Human diseases represented as being on a spectrum ranging from those that are largely environmental in causation to those that are entirely genetic.

number of ways (Box 15.1). These can include comparing the prevalence and incidence in various different population groups, the effects of migration, studying the incidence of the disease among relatives in family studies, comparing the incidence in identical and nonidentical twins, determining the effect of environmental changes by adoption studies, and studying the association of the disease with DNA polymorphisms. In addition, study can be made of the pathological components or biochemical factors of the disease in relatives (e.g., serum lipids among the relatives of patients with coronary artery disease). Study of diseases in animals that are homologous to diseases that occur in humans can also be helpful (p. 73). Before considering the use of these different approaches in a number of the common diseases in humans, specific aspects of some of these approaches will be discussed in more detail.

Population/Migration Studies

Differences in the incidence of a particular disease in different population groups suggest the possibility of genetic factors being important. They could, however, also be explained by differences in environmental factors. Studies of migrant groups moving from a population group with a low incidence of a disease to one with a high incidence, in which the incidence of the disease in the migrant group rises to that of its new population group, would suggest that environmental factors are more important. Conversely, maintenance of a low incidence of the disease in the migrant group would suggest that genetic factors are more important.

Box 15.1 Types of Genetic Approach to the Common Diseases

Population/migration studies
Family studies
Twin studies
Adoption studies
Polymorphism associations
Biochemical studies
Animal models

Family Studies

Genetic susceptibility to a disease can be suggested by the finding of a higher frequency of the disease in relatives than in the general population. The proportion of affected relatives of a specific relationship—first degree, second degree, and so forth—can provide information for empirical recurrence risks in genetic counseling (p. 346), as well as evidence supporting a genetic contribution (see Chapter 9). Familial aggregation does not, however, prove a genetic susceptibility, since families share a common environment. The frequency of the disease in spouses who share the same environment but who will usually have a different genetic background can be used as a control, particularly for possible environmental factors in adult life.

Twin Studies

If both members of a pair of identical twins have the same trait, this could be thought to prove that the trait is hereditary. This is not necessarily so. Since twins tend to share the same environment, it is possible they will be exposed to the same environmental factors. For example, if one of a twin pair contracts a contagious disease such as impetigo, it is likely the other twin will also become affected. This problem can be partly resolved by comparing differences in the frequency of a disease or disorder between nonidentical or dizygotic (DZ) and identical or monozygotic (MZ) twin pairs.

Both members of a pair of twins are said to be **concordant** when either both are affected or neither is affected. The term **discordant** is used when only one member of a pair of twins is affected. Both types of twins will have a tendency to share the same environment but, whereas identical twins basically have identical genotypes (p. 106), nonidentical twins are no more similar genetically than brothers and sisters. If a disease is entirely genetically determined, then apart from rare events such as chromosome nondisjunction or a new mutation occurring in one of a twin pair, both members of a pair of identical twins will be similarly affected, but nonidentical twins are more likely to differ. If a disease is entirely caused by environmental factors, then identical and nonidentical twins will have similar concordance rates.

Although all twins tend to share the same environment, it is probable this is more likely in identical twins than in nonidentical twins. Similarities between identical twins can therefore reflect their shared environment as much as their identical genotypes. One way of getting round this difficulty is to study differences between identical twin pairs who, through unusual family circumstances, have been reared apart from an early age. If a particular disease is entirely genetically determined, then if one identical twin is affected, the other will also be affected, even if they have been brought up in different environments. It is rare, however, for identical twins to be separated from early childhood, so only a limited number of studies for any one disorder exist. In one study of identical twins reared separately, the data clearly showed that each pair of twins differed little in height but differed considerably in body weight. These observations suggest that heredity could play a bigger part in determining stature than it does in determining body weight.

Adoption Studies

Another approach that helps differentiate between genetic and environmental factors is to compare the frequency of a disease in individuals who remain with their biological parents with those who are adopted out of their biological family. Adopted individuals take their genes with them to a new environment. If the frequency of a disease in the individuals adopted out of a family is similar to that seen in those who remain with their biological parents, then genetic factors are likely to be more important. If, conversely, the frequency of the disease in the adopted individuals is similar to that of their adoptive parents, then environmental factors are likely to be more important.

Polymorphism Association Studies

The widespread existence of inherited biochemical, protein, enzyme, and DNA variants (p. 147) allows the possibility of determining whether particular variants occur more commonly in individuals affected with a particular disease than in the population in general, or what is known as **association**. Although demonstration of a polymorphic association can suggest that the inherited variation is involved in the aetiology of the disorder, such as the demonstration of HLA associations in the immune response in the causation of the autoimmune disorders (p. 200), it may only reflect that a gene nearby in linkage disequilibrium (p. 138) is involved in causation of the disorder.

The human genome contains approximately 10 million single nucleotide polymorphisms (SNPs). Developments in high-throughput microarray SNP genotyping, together with information about SNP haplotypes (from the HapMap project [p. 148]) and the availability of large collections of DNA samples from patients with common diseases, have collectively enabled genome-wide association (GWA) studies (p. 149) to reliably identify numerous loci harbouring such variants.

Biochemical Studies

Analysis of metabolite or enzyme activity levels in biochemical or metabolic pathways likely to be involved in the causation of a particular disorder can provide evidence of genetic contribution to some of the common diseases—the hormones involved in the control of blood pressure in hypertension or the regulation of lipid levels in atherosclerosis, for example. However, in disorders where we have a limited understanding of the biological basis (e.g., schizophrenia), this approach has been of very little use.

Animal Models

Recognition of the same disease/disorder that occurs in humans and in another species such as the mouse allows the possibility of experimental studies that are often not possible in humans. In many instances, however, the disorder in the animal model will have a single-gene basis that has been identified or bred for. Nevertheless, spontaneously occurring or transgenic animal models (p. 73), experimentally induced for mutations in single genes involved in the metabolic processes or disease pathways of common diseases, will provide vital insights into the genetic contribution to these disorders.

Disease Models for Multifactorial Inheritance

The search for susceptibility loci and polygenes, sometimes also referred to as **quantitative trait loci**, in human multifactorial disorders has met with increasing success in recent years. This is largely due to the success of GWA studies (p. 149). Examples of recent research in some common conditions will be considered to illustrate the progress to date and the extent of the challenges that lie ahead.

Diabetes Mellitus

There are two main forms of diabetes mellitus (DM) that are clinically distinct. Type 1 (T1DM) is the rarer juvenile-onset, insulin-dependent form (previous abbreviation IDDM) which affects 0.4% of the population and shows a high incidence of potentially serious renal, retinal, and vascular complications. T1DM has a peak age of onset in adolescence and can only be controlled by regular injection of insulin. Type 2 is the more common later-onset, non–insulin-dependent form that affects up to 10% of the population. It usually affects older persons and may respond to simple dietary restriction of carbohydrate intake, although many persons with T2DM require oral hypoglycemic medication, and some require insulin. An additional 1% to 2% of

Table 15.1 Subtypes of Diabetes

	Type 1 Diabetes	Type 2 Diabetes	MODY	Neonatal Diabetes
Prevalence	<1%	<10%	<0.01%	<0.001%
Age of onset	Childhood/adolescence	Middle/old age	Adolescence/early adulthood	Before 6 months
Inheritance	Polygenic	Polygenic	Monogenic	Monogenic
Number of genes	Numerous (>40 confirmed loci)	Numerous (>25 confirmed loci)	At least 8 genes	At least 14 genes
Pathophysiology	Autoimmune	Insulin secretion/resistance	β-Cell dysfunction	β-Cell dysfunction

MODY, Maturity-onset diabetes of the young.

persons with diabetes have **monogenic** (single gene) forms of diabetes (Table 15.1).

Around 1% to 3% of women develop glucose intolerance during pregnancy. This is known as **gestational** diabetes. Their abnormal glucose tolerance usually reverts to normal after the pregnancy, although approximately half to three-quarters of these women go on to develop T2DM later in life.

Diabetes can also occur secondary to a variety of other rare genetic syndromes and nongenetic disorders. Examples include Prader-Willi syndrome (p. 122), Bardet-Biedl syndrome, Wolfram syndrome, and Friedreich ataxia (Table 2.5, p. 24). Diabetes mellitus is therefore aetiologically heterogeneous.

Monogenic Forms of Diabetes

Rare forms of diabetes that show high penetrance within families are usually due to mutations in single genes. More than 20 monogenic forms of diabetes have been identified (see Table 15.1).

Maturity-Onset Diabetes of the Young

Maturity-onset diabetes of the young (MODY) is an autosomal dominant form of diabetes characterized by pancreatic β-cell dysfunction. It shows clinical heterogeneity that can now be explained by genetic heterogeneity. Mutations in the glucokinase gene cause mild hyperglycemia (blood glucose levels are usually between 5.5 and 8 mmol/L), which is stable throughout life and often treated by diet alone. Glucokinase is described as the pancreatic glucose sensor because it catalyses the rate-limiting step of glucose metabolism in the pancreatic β cell. It was therefore an obvious candidate gene. Many patients with glucokinase mutations are asymptomatic, and their hyperglycemia is detected during routine screening—for example, during pregnancy or employment medicals. The mild phenotype means that finding a glucokinase mutation is 'good news.'

Mutations in five additional genes which encode transcription factors required for development of the β cell have been reported. The hepatocyte nuclear factor 1α (*HNF1A*) and hepatocyte nuclear factor 4α (*HNF4A*) genes were identified through positional cloning efforts and are associated with a more severe, progressive form of diabetes usually diagnosed during adolescence or early adulthood. These patients are sensitive to treatment with sulphonylurea tablets; this is an example of pharmacogenetics (see Chapter 12). Good glycemic control is important, as patients have a long duration of diabetes and may suffer from diabetic complications. Mutations in the *HNF1A* gene are the most common cause of MODY in most populations (65% of UK MODY), and *HNF4A* mutations are less frequent.

Hepatocyte nuclear factor 1β (HNF-1β) plays a key role in development of the kidney. Mutations cause renal cysts and diabetes (RCAD), and some female patients also have genital tract malformations. Insulin promoter factor 1 (*IPF-1*), *NEUROD1*, *INS* and *CEL* mutations are rare causes of MODY but highlight the possibility that further genes encoding β-cell transcription factors may be mutated in MODY (so-called MODYX genes).

Neonatal Diabetes

Analysis of HLA genotypes in children diagnosed with diabetes before the age of 6 months has shown that these patients have a similar frequency of high-risk alleles for type 1 diabetes as that found in the general population. This suggests that type 1 diabetes is rare before 6 months of age and implies a genetic cause.

Although definitions of the neonatal period vary, we know that diabetes is rare before 6 months of age, and the incidence is estimated at around 1 in 100,000 live births. There has been great progress over the past 10 years in defining the genetics of this rare condition.

Neonatal diabetes can be transient or permanent. More than 70% of cases of transient neonatal diabetes result from the overexpression of paternally expressed genes on chromosome 6q24. The inheritance and extent of this abnormality of imprinting (p. 121) are variable. However, in a small percentage of cases, patients have imprinting abnormalities besides 6q24 at multiple loci in the genome; these are associated with mutations in the zinc-finger transcription factor gene, *ZFP57*. Patients with a 6q24 abnormality are usually diagnosed in the first week of life and treated with insulin. Apparent remission occurs by 3 months, but there is a tendency for children to develop diabetes in later life.

Permanent neonatal diabetes does not remit, and until recently, patients were treated with insulin for life. The most common causes (>50%) are mutations in the *KCNJ11* or *ABCC8* genes which encode the Kir6.2 and SUR1 subunits of the adenosine triphosphate (ATP)-sensitive potassium (K-ATP) channel in the pancreatic β cell. Closure of these channels in response to ATP generated from glucose metabolism is the key signal for insulin release. The effect of activating mutations in these genes is to prevent channel closure by reducing the response to ATP, and hence insulin secretion. The most exciting aspect of this recent discovery is that most patients with this genetic aetiology can be treated with sulphonylurea drugs that bind to the channel and cause closure independently of ATP. Not only have patients been able to stop insulin injections and take sulphonylurea tablets instead, but they achieve better glycaemic control, which improves both their quality of life and later risk of diabetic complications.

The second most common cause of permanent neonatal diabetes is a mutation in the insulin gene (*INS*). Heterozygous *INS* mutations result in misfolding of proinsulin, which leads to death of the β cell secondary to endoplasmic reticulum stress and apoptosis. Homozygous *INS* mutations cause decreased biosynthesis of insulin through a variety of mechanisms, including gene deletion, mRNA instability, and abnormal transcription. They result in a more severe phenotype (e.g., earlier age of diagnosis) than the heterozygous mutations.

Neonatal diabetes is genetically heterogeneous, and mutations in the following genes have also been reported in a small number of patients: homozygous or compound heterozygous *GCK* or *PDX1* (also known as *IPF1*) mutations (heterozygous mutations cause MODY); heterozygous *HNF1B* mutations (also cause renal cysts) or *SLC2A2* mutations (cause Fanconi-Bickel syndrome); homozygous *PTF1A* or *GLIS3* mutations result in diabetes with cerebellar aplasia or hypothyroidism respectively. Neonatal diabetes is also a feature of Wolcott-Rallison Syndrome (homozygous or compound heterozygous *EIF2AK3* mutations) and the X-linked IPEX syndrome (*FOXP3* mutations). There are still many patients without a genetic diagnosis, which suggests that there are likely to be other monogenic aetiologies.

Type 1 Diabetes

Initial research into the genetics of diabetes tended to focus on type 1 diabetes, where there is greater evidence for familial clustering (λ_s is 15 for T1DM versus 3.5 for T2DM [p. 146]). The concordance rates in monozygotic and dizygotic twins are around 50% and 12%, respectively. These observations point to a multifactorial aetiology with both environmental and genetic contributions. Known environmental factors include diet, viral exposure in early childhood, and certain drugs. The disease process involves irreversible destruction of insulin-producing islet β cells in the pancreas by the body's own immune system, perhaps as a result of an interaction between infection and an abnormal genetically programmed immune response.

The first major breakthrough came with the recognition of strong associations with the HLA region on chromosome 6p21. The original associations were with the HLA B8 and B15 antigens that are in linkage disequilibrium with the DR3 and DR4 alleles (pp. 147, 200). It is with these that the T1DM association is strongest, with 95% of affected individuals having DR3 and/or DR4 compared with 50% of the general population. Following the development of PCR analysis for the HLA region, it was shown that the HLA contribution to T1DM susceptibility is determined by the 57th amino acid residue at the DQ locus, where aspartic acid conveys protection, in contrast to other alleles that increase susceptibility.

The next locus to be identified was the insulin gene on chromosome 11p15, where it was shown that variation in the number of tandem repeats of a 14-bp sequence upstream to the gene (known as the **INS VNTR**) influences disease susceptibility. It is hypothesized that long repeats convey protection by increasing expression of the insulin gene in the fetal thymus gland, thereby reducing the likelihood that insulin-producing β cells will be viewed as foreign by the mature immune system.

These two loci contribute λ_s values of approximately 3 and 1.3, respectively. However, the total risk ratio for T1DM is around 15. Confirmation that other loci are involved came first of all from linkage analysis using breeding experiments with the nonobese diabetic (NOD) strain of mice. These mice show a very high incidence of T1DM, with immunopathological features similar to those seen in humans. These linkage data pointed to the existence of 9 or 10 different susceptibility loci in mice. Following this, the results of numerous genome-wide linkage scans in humans provided evidence for the existence of between 15 and 20 susceptibility loci. However, apart from HLA, many of the other regions were not consistently replicated in independent studies. The largest and most recent linkage study in 2009 (2496 families and 2658 affected sibling pairs, p. 147) provided evidence of linkage to only one previously identified region of linkage besides HLA and *INS*: a region near to *CTLA4*. This locus was one of the only three (others encompassing the *PTPN22* and *IL2RA* [*CD25*] genes) identified and confirmed in candidate gene association studies conducted between 1996 and 2007.

Since 2006, GWA studies p. 149) of increasing size have led to an explosion in the number of T1DM susceptibility loci supported by robust statistical evidence, bringing the total to over 40 distinct genomic locations. It is likely that many more remain to be identified through future, even larger, efforts. Most of the identified loci confer a modest increase in the risk of T1DM, with odds ratios (p. 148) ranging from 1.1 to 1.3 for each inherited allele, in contrast with the much larger role of the HLA locus. In most cases, the causal genes and variants underlying the associations have yet to be identified. However, the regions of association often encompass strong biological candidates—for example, the interleukin genes, *IL10*, *IL19*, *IL20*, and *IL27*. In two notable cases, follow-up studies have already enabled the causal gene to be confirmed, deepening our understanding of the biological pathways behind the associations.

The first example was a study of the *IL2RA* (*CD25*) locus by Dendrou et al. (2009). It used the UK-based Cambridge BioResource, a collection of approximately 5000 volunteers who can be recalled to participate in research on the basis of their genotype. Using fewer than 200 of these individuals, and by means of flow cytometry to assay the levels of CD25 protein expressed on the surface of T-regulatory cells, the study showed that people with the T1DM-protective haplotype expressed higher CD25 levels. This confirmed that *IL2RA* is indeed the causal gene and that the genotype-phenotype association is mediated via differences in expression of the gene product.

In the second study by Nejentsev et al. (2009), the exons and splice sites of 10 candidate genes situated in regions of genome-wide association were resequenced in 480 T1DM patients and 480 controls. Variants identified were then tested for association with the disease in 30,000 further subjects. Four rare variants (minor allele frequency ≈ 1% to 2%) in the *IFIH1* gene were identified, each of which independently reduced the odds of T1DM by about 50%. This finding demonstrated that the *IFIH1* gene is important in the aetiology of T1DM. Since its function is to mediate the induction of an interferon response to viral RNA, it adds to the evidence implicating viral infection in the development of the disease. These results also demonstrate that there may be both high- and low-frequency susceptibility variants at the same locus, with varying effect sizes. Future follow-up by resequencing of other loci, both in T1DM and in other diseases, should lead to the identification of even more of these variants and a better understanding of the loci.

Type 2 Diabetes

The prevalence of T2DM is increasing and is predicted to reach 300 million affected worldwide by 2025. Although commonly believed to be more benign than the earlier-onset, insulin-dependent type 1 diabetes, patients with T2DM are also prone to both macrovascular and microvascular diabetic complications, with corresponding excess morbidity and mortality.

Table 15.2 lists the known susceptibility loci for T2DM. There is no overlap with the T1DM loci, illustrating that these two diseases have very different aetiologies. Unlike the HLA and *INS* VNTR loci in T1DM, there are no major predisposing loci associated with T2DM. Most odds ratios are modest (between 1.05 and 1.3 per allele). As a result, large candidate gene and GWA studies (p. 146) have enabled progress in identifying the loci, whereas linkage studies (p. 146) have been underpowered and therefore unsuccessful.

Human models have proven useful for identifying candidate genes in T2DM. Mutations in all five of the genes identified in this way (*PPARG, KCNJ11, WFS1, HNF1B, HNF1A*) cause rare monogenic forms of diabetes. One of the variants is population specific: the G319S variant in the *HNF1A* gene has only been found in the Oji-Cree population in Ontario, Canada.

The *TCF7L2* locus has the largest odds ratio of all T2DM loci found in multiple populations. Individuals who inherit two risk alleles (approximately 9% of Europeans) are at nearly twice the risk of T2DM as those who inherit none. The locus was discovered in large-scale association studies of a region on chromosome 10, which was originally identified in linkage studies. However, the *TCF7L2* variant does not account for the linkage in this region, suggesting that other rarer but more penetrant variants may be close by. As with many of the other loci, the *TCF7L2* risk allele is associated with impaired β-cell function, highlighting the importance of the β cell in T2DM aetiology.

By far the greatest progress in identifying T2DM susceptibility loci has come from GWA studies. To date, 16 loci have been confirmed in T2DM case-control GWA studies, and 6 loci that were originally associated genome-wide with fasting glucose levels were later shown to predispose to T2DM. Many of the genes situated in the identified loci were never thought to be biological candidates, so their discovery has opened up new avenues for research. For example, the *FTO* gene, which harbours variants associated with body fat mass, was of previously unknown function. Subsequent research has shown, using bioinformatics and animal models, that it has a potential role in nucleic acid demethylation and is expressed in the hypothalamic nuclei of the brain, which govern energy balance and appetite.

In 2008, analysis of the combined effects of 18 of these loci by Lango, Weedon, and colleagues suggested that they increase the likelihood of disease in an additive way. The 1.2% of Europeans in the study who inherited more than 24 risk alleles were over 4 times more likely to develop T2DM than the 2% who inherited only 10 to 12 risk alleles. It is likely that many more loci will be identified through future meta-analyses of GWA studies and that detailed follow-up of the associated regions will lead to identification of the causal variants. The large number of predisposing loci highlights multiple targets for intervention, but there is much work to be done to translate these findings into useful clinical applications.

Crohn Disease

Inflammatory bowel disease (IBD) includes two clinical subtypes: Crohn disease and ulcerative colitis. Its prevalence in Western countries is 1% to 2%, and the estimated λ_s is 25. Positional cloning for IBD identified a striking linkage peak at chromosome 16p12, which was linked to Crohn disease but not ulcerative colitis in the majority of studies. Crohn disease is characterized by perturbed control of inflammation in the gut and with its interaction with bacteria.

In 2001, two groups working independently and using different approaches identified disease-predisposing variants in the *CARD15* gene (previously known as *NOD2*). One of the groups, Ogura et al., had previously identified a Toll-like receptor (p. 193), NOD2, which activates nuclear factor Kappa-B (NFκB) (p. 201), making it responsive to

Table 15.2 Known Susceptibility Loci for Type 2 Diabetes

Locus (Candidate Gene or Genes)	OR per Copy of Risk Allele	Biological Function Underlying the Association (Hypothesized)	Strategy Used to Identify the Locus
PPARG	1.23	Adipocyte differentiation and function	Candidate gene association study
KCNJ11	1.15	β-Cell potassium ATP channel function	Candidate gene association study
WFS1	1.11	Unknown	Candidate gene association study
HNF1B	1.08	β-Cell development and function	Candidate gene association study
HNF1A	1.97	β-Cell development and function	Candidate from human model. Private missense *G319S* variant found in Oji Cree
TCF7L2	1.37	Incretin signaling; β-cell function	Large-scale association study in a region of linkage
HHEX/IDE	1.13	β-Cell development and function	GWA study of T2DM
SLC30A8	1.12	β-Cell function (Zn transport)	GWA study of T2DM
FTO	1.23	Primary effect on body mass index and fat mass	GWA study of T2DM
CDKAL1	1.16	β-Cell function	GWA study of T2DM
CDKN2A/CDKN2B	1.19	Cell cycle regulation in the β cell	GWA study of T2DM
IRS1	1.19	Insulin resistance	GWA study of T2DM
IGF2BP2	1.11	β-Cell function	GWA study of T2DM
NOTCH2	1.11	Unknown	GWA study of T2DM
JAZF1	1.10	β-Cell function	GWA study of T2DM
CDC123/CAMK1D	1.09	Unknown	GWA study of T2DM
TSPAN8/LGR5	1.09	Unknown	GWA study of T2DM
THADA	1.12	Unknown	GWA study of T2DM
ADAMTS9	1.06	Unknown	GWA study of T2DM
KCNQ1 (2 independent loci)	1.29	β-Cell function; risk of diabetes associated with maternally inherited allele.	GWA study of T2DM and parent-of-origin analysis
KLF14	1.06	Unknown; risk of diabetes associated with maternally inherited allele.	GWA study of T2DM and parent-of-origin analysis
11p15	1.11	Unknown; risk of diabetes associated with paternally inherited allele.	GWA study of T2DM and parent-of-origin analysis
MTNR1B	1.09	β-Cell function via melatonin signaling	GWA study of FPG
ADCY5	1.12	Unknown	GWA study of FPG
GCK	1.07	Glucose sensing in the β cell	GWA study of FPG
GCKR	1.06	Unknown	GWA study of FPG
PROX1	1.07	Unknown	GWA study of FPG
DGKB-TMEM195	1.06	Unknown	GWA study of FPG

The gene name in the first column is either the most likely candidate at a locus (based on known biology) or is the nearest gene (or genes) to the associated variants. In most cases, the causal gene has not been proven. The odds ratios presented here are from studies of Europeans (apart from *HNF1A*, which is private to the Oji-Cree population).

ATP, Adenosine triphosphate; *FPG*, fasting plasma glucose; *GWA*, genome-wide association; *OR*, odds ratio; *T2DM*, type 2 diabetes mellitus.

bacterial lipopolysaccharides. The *CARD15* gene is located within the 16p12 region and was therefore a good positional and functional candidate. Sequence analysis revealed three variants (R702W, G908R and 3020insC) that were shown by case-control and transmission disequilibrium tests to be associated with Crohn disease. The second group, Hugot et al., fine-mapped the 16p12 region by genotyping SNPs within the 20Mb interval and also arrived at the same variants within the *CARD15* gene. These variants are found in up to 15% of patients with Crohn disease but only 5% of controls. The relative risk conferred by heterozygous and homozygous genotypes was approximately 2.5 and 40, respectively. For therapy, drugs which target the NFκB complex (p. 201) are already the most effective drugs currently available.

Since 2006, GWA studies have identified over 30 susceptibility loci for Crohn disease, all of which confer more modest risks of disease than the *CARD15* variants (odds ratios per allele between 1.1 and 2.5). Discoveries of loci containing the *IRGM* and *ATG16L1* genes were particularly exciting findings, as these genes are essential for autophagy, a biological pathway whose relevance to the disease was previously unsuspected. Further studies of the *IRGM* locus by McCarroll and colleagues (2008) identified that the causal variant is a 20-kb deletion immediately upstream of *IRGM*, which is in linkage disequilibrium (p. 138) with the associated SNPs. The deletion results in altered patterns of gene expression, which in turn were shown to modulate the autophagy of bacteria inside cells. Efforts are already underway to translate this finding into therapeutic applications.

Hypertension

Hypertension (chronically elevated blood pressure) leads to increased morbidity and mortality through a greater risk of

Table 15.3 Recurrence Risks for Hypertension

Group	%
Population	5
2 Normotensive parents	4
1 Hypertensive parent	8–28
2 Hypertensive parents	25–45

From Burke W, Motulsky AG 1992 Hypertension. Chapter 10 in King RA, Rotter JI, Motulsky AG eds The genetic basis of common diseases. New York: Oxford University Press

Table 15.4 Coefficient of Correlation for Blood Pressure in Various Relatives

Group	Correlation Coefficient
Siblings	0.12–0.34
Parent/child	0.12–0.37
Dizygotic twins	0.25–0.27
Monozygotic twins	0.55–0.72

From Burke W, Motulsky AG 1992 Hypertension. Chapter 10 in King RA, Rotter JI, Motulsky AG eds The genetic basis of common diseases. New York: Oxford University Press

stroke and coronary artery and renal disease. Various studies have shown that between 10% and 25% of the population is hypertensive, but the prevalence is age dependent, with up to 40% of 75- to 79-year-olds being hypertensive. Elevated blood pressure may contribute up to 50% of the global cardiovascular disease epidemic. There is substantial evidence that treatment of hypertension prevents development of these complications.

Persons with hypertension fall into two groups. In one, the onset is a consequence of another disorder such as kidney disease. In the other more common group, hypertension usually begins in middle age and has no recognized cause. This is known as **essential hypertension**. The following discussion is concerned with only essential hypertension.

Environmental Factors in Hypertension

Environmental factors such as high dietary sodium levels, obesity, alcohol intake, and reduced exercise are recognized as being associated with an increased risk of hypertension. Hypertension is also more prevalent in persons from poorer socioeconomic groups. Studies of adopted children have shown lower correlation of their blood pressure with their biological parents than with children remaining with their biological parents. In addition, migration studies involving persons moving from a population with a low prevalence to one with a high prevalence of hypertension have shown that the immigrant group acquires the frequency of hypertension of their new population group during the course of one to two generations. This suggests that environmental factors are of major importance in the aetiology of hypertension.

Genetic Factors in Hypertension

Family and twin studies have shown that hypertension is familial (Table 15.3) and that blood pressure correlates with the degree of relationship (Table 15.4). These findings suggest the importance of genetic factors in the aetiology of hypertension. In addition, there are differences in the prevalence of hypertension between populations, hypertension being more common in persons of Afro-Caribbean origin and less common in Eskimos, Australian Aborigines, and Central and South American Indians.

Susceptibility Genes

Rare homozygous mutations in the renal salt-handling genes, *SLC12A3*, *SLC12A1*, and *KCNJ1*, cause recessive diseases characterized by severe reductions in blood pressure. However, resequencing of these genes by Ji et al. (2008) showed that heterozygous rare variants (minor allele frequency < 0.1%) are present in healthy individuals and contribute to blood pressure variation in the general population. These variants cause clinically relevant reductions in blood pressure and protect against hypertension.

Common genetic variants also influence normal blood pressure variation. In 2009, very large meta-analyses and replication of GWA studies (p. 149) were published by the Global Blood Pressure Genetics (Global BPgen) consortium (N > 100,000 Europeans and > 12,000 Indian Asians) and the Cohorts for Heart and Aging Research in Genome Epidemiology (CHARGE) consortium (N > 29,136 Europeans). Fourteen genetic loci were robustly associated with either systolic or diastolic blood pressure, and all showed evidence of association with hypertension risk. Although the causal genes have not yet been confirmed, the loci highlighted likely candidates, including *CYP17A1*, rare mutations of which cause a form of adrenal hyperplasia (p. 174) characterized by hypertension.

A particularly interesting association was found on chromosome 12q24. The region of association is large, with linkage disequilibrium extending over 1.6 Mb and encompassing 15 genes. The locus is intriguing because other GWA studies have reported association between the same haplotype that raises blood pressure and type 1 diabetes, coeliac disease, myocardial infarction, eosinophil count, and platelet count. These pleiotropic effects are thought to be the result of selection which rapidly increased the frequency of the haplotype in Europeans approximately 3400 years ago at a time when human settlements were enlarging. Further studies will clarify whether the haplotype contains multiple functional variants that give rise to the pleiotropy.

Coronary Artery Disease

Coronary artery disease is the most common cause of death in industrialized countries and is rapidly increasing in prevalence in developing countries. It results from atherosclerosis, a process taking place over many years which involves the deposition of fibrous plaques in the subendothelial space (intima) of arteries, with a consequent narrowing of their lumina. Narrowing of the coronary arteries compromises the metabolic needs of the heart muscle, leading to myocardial ischemia, which if severe, results in myocardial infarction.

For the majority of persons, their risk of coronary artery disease is multifactorial or polygenic in origin. A variety of different genetic and environmental risk factors have been identified that predispose to early onset of the atherosclerotic process, including lack of exercise, dietary saturated fat, and smoking.

Lipid Metabolism

The metabolic pathways by which the body absorbs, synthesizes, transports, and catabolizes dietary and endogenous lipids are complex. Lipids are packaged in intestinal cells as a complex with various proteins known as **apolipoproteins** to form triglyceride-rich chylomicrons. These are secreted into the lymph and transported to the liver, where, in association with endogenous synthesis of triglyceride and cholesterol, they are packaged and secreted into the circulation as triglyceride-rich very low-density lipoproteins (VLDLs). VLDL is degraded to intermediate-density lipoprotein (IDL), which is further broken down into cholesterol-rich low-density lipoprotein (LDL). High-density lipoproteins (HDLs) are formed from lipoproteins secreted by the liver, chylomicrons, and VLDL remnants.

High levels of LDLs are associated with an increased risk of coronary artery disease. Conversely, high levels of HDLs are inversely correlated with a risk of coronary artery disease. Consequently, the LDL:HDL ratio has been used as a risk predictor for coronary artery disease and as an indicator for therapeutic intervention. Statins are effective drugs for lowering LDL cholesterol levels.

Family and Twin Studies

The risk to a first-degree relative of a person with premature coronary artery disease, defined as occurring before age 55 in males and age 65 in females, varies between 2 and 7 times that for the general population (Table 15.5). Twin studies of concordance for coronary artery disease vary from 15% to 25% for dizygotic twins and from 39% to 48% for monozygotic twins. Although these figures support the involvement of genetic factors, the low concordance rate for monozygotic twins clearly supports the importance of environmental factors.

Table 15.5 Recurrence Risks for Premature Coronary Artery Disease

Proband	Relative Risk
Male (<55 Years)	
Brother	5
Sister	2.5
Female (<65 Years)	
Siblings	7

Data from Slack J, Evans KA 1966 The increased risk of death from ischaemic heart disease in first degree relatives of 121 men and 96 women with ischaemic heart disease. J Med Genet 3:239–257

Single-Gene Disorders of Lipid Metabolism Leading to Coronary Artery Disease

Although there are a number of individually rare inherited disorders of specific lipoproteins, levels of the various lipoproteins and the hyperlipidemias are determined by a complex interaction of genetic and environmental factors. Family studies of some of the hyperlipidemias are, however, consistent with a single gene being a major factor determining genetic susceptibility.

Familial Hypercholesterolemia

The best-known disorder of lipid metabolism is familial hypercholesterolemia (FH) (p. 175). FH is associated with a significantly increased risk of early coronary artery disease and is inherited as an autosomal dominant disorder. It has been estimated that about 1 person in 500 in the general population, and about 1 in 20 persons presenting with early coronary artery disease, are heterozygous for a mutation in the *LDLR* (low-density lipoprotein receptor) gene. Molecular studies in FH have revealed that it is due to a variety of defects in the number, function, or processing of the LDL receptors on the cell surface (p. 176).

Susceptibility Genes

Since 2007, numerous large-scale GWA and follow-up replication studies have identified 12 susceptibility loci for coronary artery disease and myocardial infarction. The strongest association identified is on chromosome 9p21 (odds ratio per allele ≈ 1.3). The nearest genes, *CDKN2A* and *CDKN2B*, are over 100 kb away. Interestingly, the SNPs most strongly associated with coronary artery disease are only 10 kb away from those associated with type 2 diabetes (see Table 15.2). However, the two disease associations are independent and not in linkage disequilibrium with one another. Much work is already being done to investigate the role of *ANRIL*, a large, noncoding RNA which overlaps with the coronary artery disease–associated haplotype. It is expressed in tissues associated with atherosclerosis, and initial studies have shown correlations between expression of *ANRIL* transcripts and severity of atherosclerosis. However, additional evidence from large-scale association studies has shown that the same haplotype on 9p21 is associated with abdominal aortic aneurysm and intracranial aneurysm, suggesting that its role is not limited to atherosclerotic disease. Together with the other 11 loci, the locus at 9p21 only explains a small fraction of the heritability of coronary artery disease, and it is likely that many more loci will be identified.

Progress in uncovering susceptibility loci has also come from large GWA studies of lipid levels. Common variants at at least 30 loci are now robustly associated with circulating levels of lipids, with over one-third of these associated with LDL levels. Kathiresan and colleagues (2008) showed that individuals who inherit a higher number of LDL-raising alleles at these loci are more likely to have clinically high LDL cholesterol (>160 mg/dL) than those who inherit few

Table 15.6 Proportions (%) of First-Degree Relatives of Individuals with Schizophrenia Who Are Similarly Affected or Have a Schizoid Disorder

	Proportion (%) of Relatives		
Relatives	**Schizophrenia***	**Schizoid**	**Total Schizophrenia + Schizoid**
Identical twins	46	41	87
Offspring (of 1 schizophrenic)	16	33	49
Siblings	14	32	46
Parents	9	35	44
Offspring (of 2 schizophrenics)	34	32	66
General population	1	3	4

*Age corrected.
From Heston LL 1970 The genetics of schizophrenia and schizoid disease. Science 167:249–256

alleles. The frequency of these LDL-raising alleles is higher in patients with coronary artery disease than in controls, indicating that they predispose to the disease via their primary effect on LDL levels. In many cases, the genes implicated by the loci are already associated with single-gene disorders. For example, *PCSK9* harbours a full spectrum of LDL-altering alleles, from rare mutations which cause large differences in LDL (>100 mg/dL), through low-frequency variants with more modest effects (e.g., *PCSK9* R46L has a 1% minor allele frequency and a 16 mg/dL effect size), to common variants at 20% minor allele frequency which change LDL levels by less than 5 mg/dL. The resequencing of further loci is likely to uncover rarer variants and mutations at lipid trait loci, which may further explain genetic susceptibility to coronary artery disease.

Schizophrenia

Schizophrenia is a serious psychotic illness with an onset usually in late adolescence or early adult life. It is characterized by grossly disorganized thought processes and behavior, together with a marked deterioration of social and occupational functioning, and can be accompanied by hallucinations and delusions.

Epidemiology

Schizophrenia is a principal cause of chronic mental illness. There is a 1% lifetime risk for a person to develop schizophrenia, and at any one time, approximately 0.2% of the population is affected. Schizophrenia occurs more commonly in individuals of poorer socioeconomic status and has an earlier age of onset and worse prognosis in males. There is an excess of winter births in schizophrenic individuals, which has suggested that environmental factors such as certain viral infections or nutritional factors could be contributory.

Evidence for Genetic Factors

The nature and extent of the genetic contribution to schizophrenia is unclear. This is partly because of past and continuing controversy concerning the definition of schizophrenia and the term *schizoid*. The latter term refers to the schizophrenia-like traits often seen in relatives of schizophrenics. The problem arises because clinical criteria to distinguish schizoid from normal personality are lacking. For the sake of simplicity, we can regard the term *schizoid* as referring to a person with the fundamental symptoms of schizophrenia but in a milder form. It has been estimated that roughly 4% of the general population have schizophrenia or a schizoid personality disorder.

Family and Twin Studies

The results of several studies of the prevalence of schizophrenia and schizoid disorder among the relatives of schizophrenics are summarized in Table 15.6. If only schizophrenia is considered, the concordance rate for identical twins is only 46%, suggesting the importance of environmental factors. If, however, schizophrenia and schizoid personality disorder are considered together, then almost 90% of identical co-twins are concordant.

Susceptibility Genes

Genome-wide association studies of copy number variations (CNVs) have identified large (>500 kb) deletions associated with the condition—for example on chromosomes 1q21.1, 15q13.3, and 22q11.2 (p. 282). These deletions are rare but penetrant: the odds ratio for the 15q13.3 deletion has been estimated at between 16 and 18 in two independent studies. A key observation is that these deletions are not only associated with schizophrenia. The 1q21.1 deletion (pp. 284, 287) has also been associated with autism, learning disability, and epilepsy. Thus, current clinically defined disease boundaries are not mirrored by the underlying genetics. While these deletions explain some of the genetic susceptibility to schizophrenia, they also explain susceptibility to other conditions. It is likely that a better understanding of the genetics will lead to better definition of clinical phenotypes.

Common genetic variants are also implicated in the aetiology of schizophrenia. Recent meta-analyses of GWA studies have identified associations with the HLA region on chromosome 6p21.3-6p22.1, suggesting an immune system component to the risk of disease. Robust associations have also been observed with variants near the *NRGN* gene and in the *TCF4* gene, which implicate biological pathways involved in brain development, cognition, and memory. Analyses of existing GWA data conducted by the International Schizophrenia Consortium in 2009 suggested that there are likely to be thousands more common variants of small effect that collectively explain much of the heritability of schizophrenia.

Alzheimer Disease

Dementia is characterized by an irreversible and progressive global impairment of intellect, memory, social skills, and control of emotional reactions in the presence of normal consciousness. Dementia is aetiologically heterogeneous, occurring secondarily to both a variety of nongenetic causes such as vascular disease and infections such as AIDS, as well as genetic causes. Alzheimer disease (AD) is the most common cause of dementia in persons with either early-onset dementia (less than the age of 60 years, or presenile) or late onset (greater than age 60 years, or senile). The classic neuropathological finding in persons with AD is the presence at postmortem examination of amyloid deposits in neurofibrillary tangles and neuronal or senile plaques. In addition, individuals with Down syndrome have an increased risk of developing dementia (p. 273), which at postmortem has identical CNS findings to those seen in persons with typical AD.

Epidemiology

Limited numbers of studies of the incidence and prevalence of AD are available, owing to problems of ascertainment. However, the risk of developing AD clearly increases dramatically with age (Table 15.7).

Twin and Family Studies

Differences in the age of onset of AD in identical twins are consistent with the importance of environmental factors, but there are difficulties with family studies in AD. Many studies are based on a clinical diagnosis. However, a significant proportion of persons with a clinical diagnosis of AD are found to have other causes at postmortem, such as cerebrovascular atherosclerotic disease. Attempts to confirm diagnoses in relatives who have died previously are often unsuccessful. Obviously, given the age of onset, it is generally neither practical nor possible to obtain funding for prospective studies of the risk to offspring. Therefore, family studies of the risk to siblings are the only practical type of family study to provide reliable data. Although there are numerous retrospective reports of families with AD that are consistent with autosomal dominant inheritance, recurrence risks in a number of studies for first-degree relatives are less than 10%. The risks are age related and greater the younger the age at diagnosis in the affected individual.

Table 15.7 Estimates of Age-Specific Cumulative Prevalence of Dementia

Age Interval (Years)	Prevalence (%)
<70	1.3
70–74	2.3
75–79	6.4
80–84	15.3
85–89	23.7
90–94	42.9
>95	50.9

From Heston LL 1992 Alzheimer's disease. Chapter 39 in King RA, Rotter JI, Motulsky AG eds The genetic basis of common diseases. New York: Oxford University Press

Biochemical Studies

The amyloid deposits in the neurofibrillary tangles and neuronal plaques have been shown to consist of the amyloid-β A4 precursor protein (APP). The major protein component of the neurofibrillary tangles has been shown to be derived from a microtubule-associated protein (MAP) called *tau* (τ). Along with other MAPs, it interacts with β-tubulin to stabilize microtubules.

Single-Gene Disorders

The identification of APP in the amyloid deposits of the neuronal plaques, its mapping in or near to the critical region of the distal part of chromosome 21q associated with the phenotypic features of Down syndrome (p. 273), and the increased risk of AD in persons with Down syndrome led to the suggestion that duplication of the *APP* gene could be a cause of AD. Evidence of linkage to the *APP* locus was found in studies of families with early-onset AD, and it is now known that mutations in the *APP* gene account for a small proportion of cases.

Evidence of linkage to early-onset AD was found for another locus on chromosome 14q. Mutations were identified in a proportion of affected individuals in one of a novel class of genes known as *presenilin-1* (*PSEN1*), now known to be a component of the notch signaling pathway (p. 86). A large number of mutations in *PSEN1* have now been identified and account for up to 70% of familial early-onset AD. A second gene, presenilin-2 (*PSEN2*), with homology to *PSEN1*, was mapped to chromosome 1q and has been shown to have mutations in a limited number of families with AD. *PSEN1* and *PSEN2* are integral membrane proteins containing multiple transmembrane domains that localize to the endoplasmic reticulum and the Golgi complex. All of the presenile dementias following autosomal dominant inheritance demonstrate high penetrance.

Susceptibility Genes

Polymorphisms in the apolipoprotein E (*APOE*) gene are the most important genetic risk factor identified for late-onset AD. The locus was initially identified in the early 1990s through linkage studies. The *APOE* gene has three major protein isoforms, ε2, ε3, and ε4. Numerous studies in various populations and ethnic groups have shown an increased frequency of the ε4 allele in persons with both sporadic and late-onset familial AD. In addition, the ε2 allele is associated with a decreased risk of the disease. The finding of apolipoprotein E in senile plaques and neurofibrillary tangles, along with its role in lipid transport, possibly in relation to the nerve injury and regeneration seen in AD, provides further evidence for a possible role in the acceleration of the neurodegenerative process in AD.

Although the *APOE* ε4 allele, found in up to 40% of cases, is a clearly important risk factor, the strongest association is with the age of onset rather than absolute risk of developing AD. The *APOE* ε4 allele is therefore neither necessary nor sufficient for the development of AD, emphasizing the importance of other genetic and environmental aetiological factors.

In 2009, large meta-analyses of GWA studies, involving up to 6000 cases and 10,000 controls, began to extend our knowledge of AD susceptibility loci. Robust evidence of association was found for common variants in the clusterin (*CLU*) gene (previously known as *apolipoprotein J*), which is found in amyloid plaques and is an excellent functional candidate. Associations were also confirmed at loci marked by the *CR1* and *PICALM* genes. Further work will be needed to uncover the mechanisms underlying these associations, and it is likely that many more loci with more modest effects remain to be discovered.

Hemochromatosis

Hemochromatosis is a common disorder of iron metabolism that results in accumulation of iron. The liver is the most commonly damaged tissue, with iron deposition leading to cirrhosis and liver failure. Patients are at increased risk of hepatocellular carcinoma. Other organs that may be affected include the pancreas, heart, pituitary gland, skin, and joints. The iron overload is easily treated by venesection, and this is very effective at reducing morbidity and mortality. The ratio of affected males to females is 5:1, and the disease is underdiagnosed in the general population but overdiagnosed in patients with secondary iron overload.

Linkage and Gene Identification

In 1996, the *HFE* gene was discovered close to the HLA region on 6p21. Two variants were described, *C282Y* and *H63D*. Between 85% and 100% (depending on population) of affected individuals were found to be homozygous for *C282Y*, and the carrier frequency in Northern Europe is approximately 1 in 10. *H63D* is more common in the general population, and homozygosity for this variant is associated with only a modest increased (about fourfold) risk of hemochromatosis. Compound heterozygosity for *C282Y* and *H63D* is associated with reduced penetrance; only 1% are thought likely to develop symptoms. Homozygosity for *C282Y* was thought to confer a high risk of hemochromatosis, and it was suggested that population screening would be useful, since the iron overload is easily treated. However, population-based studies have suggested that the penetrance of hereditary hemochromatosis due to homozygosity for *C282Y* may be as low as 1%.

Genetic Testing

Genetic testing is now commonplace in families where an index case is found to be homozygous for *C282Y* or compoundly heterozygous for *C282Y* and *H63D*. Current strategy involves first testing the spouse for carrier status, as there is a 1 in 10 chance they will be heterozygous for *C282Y*. If the offspring have inherited a predisposing genotype, annual monitoring of serum ferritin is recommended to detect iron overload at an early stage and initiate prompt treatment.

Genetic Heterogeneity

Hemochromatosis is a genetically heterogeneous disorder (Table 15.8), with mutations also reported in the transferrin receptor 2 (*TFR2*) gene and the *SLC40A1* gene which encodes ferroportin. In addition to the common recessive adult-onset form, there is a rare juvenile form with iron overload and organ failure before the age of 30 years, which is lethal if untreated. Neonatal hemochromatosis is a severe form of unknown aetiology.

Venous Thrombosis

Venous thrombosis represents a major health problem worldwide, with increasing incidence from 1 in 100,000 during childhood to 1 in 100 in old age. Venous thromboembolism, including deep vein thrombosis and pulmonary embolism, is a complex disease that results from multiple interactions between inherited and acquired risk factors (Box 15.2). Inherited thrombophilias also increase the risk of fetal loss, both stillbirths and early miscarriages.

Identification of Genes

The identification of two low-frequency variants prevalent in Europeans has increased our understanding of venous thrombosis. The factor V Leiden variant (R506Q) renders

Table 15.8 Genetic Heterogeneity in Hemochromatosis

Type	Inheritance	Gene	Chromosome Location	Age of Onset
HFE	Autosomal recessive	*HFE*	6p21.3	40–60 years
HFE2A	Autosomal recessive	*HFE2*	1q21	<30 years (Juvenile)
HFE2B	Autosomal recessive	*HAMP*	19q13	<30 years (Juvenile)
HFE3	Autosomal recessive	*TFR2*	7q22	40–60 years
HFE4	Autosomal dominant	*SLC40A1*	2q32	up to 60/70* years

*Up to 60 years in males and 70 years in females.

Box 15.2 Inherited and Acquired Causes of Venous Thrombosis

Inherited

Common

- Factor V Leiden (R506Q)
- Prothrombin variant G20210A
- Homozygous C677T mutation in the methylenetetrahydrofolate reductase gene (*MTHFR*)

Rare

- Antithrombin deficiency
- Protein C deficiency
- Protein S deficiency

Acquired

- Surgery and trauma
- Prolonged immobilization
- Previous thrombosis
- Pregnancy
- Oral contraceptives or hormone replacement therapy
- Older age

the factor V protein resistant to cleavage by activated protein C, thus increasing generation of thrombin. The prothrombin variant G20210A is located in the 3' UTR and is associated with increased prothrombin levels. These variants confer a four- to fivefold increased risk of thrombosis in the heterozygous state (Table 15.9), but individuals homozygous for one or heterozygous for both are at significantly elevated risk (up to 80-fold).

Recently, GWA studies have identified associations between common variants near the *HIVEP1* gene and venous thrombosis. This locus has a much more modest effect than the factor V Leiden and prothrombin variants, with only a 1.2-fold increased risk in the heterozygous state, but the variants are more common (20% frequency in Europeans). The biological mechanism underlying this association is unknown, but the finding suggests that common genetic variants contribute to the heritability of the condition.

Genetic Testing

More than 50% of venous thromboembolism cases can be explained by factor V Leiden and the prothrombin variant. Testing is now commonplace, but there is no evidence that the detection of a heritable thrombophilic defect alters management of the index case. However, if an inherited predisposition is identified, testing can be offered to first-degree relatives. Those who test positive may be able to reduce their risk of developing thrombosis by, for example, short-term thromboprophylaxis in periods of increased thrombotic risk such as surgery. Knowledge of a genetic predisposition to thrombosis will also influence a woman's choice of oral contraceptives.

Age-Related Macular Degeneration

Age-related macular degeneration (AMD) is a leading cause of vision loss and blindness, affecting around 50 million elderly people throughout the world. AMD is characterized by a progressive loss of central vision attributable to degenerative and neovascular changes that occur at the interface between the neural retina and the underlying choroid.

Aetiological research suggests that AMD is a complex multifactorial disease. Familial studies have provided strong evidence for the heritability of AMD, with a higher risk in first-degree relatives of AMD patients and a higher concordance among monozygotic than dizygotic twins.

Known genetic susceptibility variants highlight a key role for the complement system in AMD aetiology. Since 2005, numerous studies have shown that common variants within the *CFH* gene are associated with AMD. This gene encodes factor H, the major inhibitor of the alternative complement pathway, which accumulates within drusen, the characteristic lesions of AMD. By analyzing 1536 SNPs across the *CFH* locus in over 1200 patients and 900 controls, Maller and colleagues (2006) demonstrated two independent genetic associations: the first is marked by the common coding variant, Y402H, and the second by the common SNP, rs1410996, in an intron of *CFH*. These variants have substantial effect sizes. Individuals homozygous for the high-risk haplotype formed by both variants are at a 15-fold higher risk of disease than those homozygous for the low-risk haplotype.

Associations have subsequently been confirmed with variants in the locus containing the complement component 2 (*C2*) and complement factor B (*CFB*) genes and in the complement component 3 (*C3*) gene. In addition, there is strong and well-replicated evidence for association at *ARMS2* on chromosome 10q26. The known variants collectively account for a substantial proportion of the heritability of AMD, explaining at least half of the risk to siblings.

Table 15.9 Frequency of Factor V Leiden and Prothrombin 20210A Variant in the UK Population and Corresponding Risk of Venous Thrombosis

Risk Factor	Status	Population Frequency	Approximate Increased Risk*
Factor V Leiden	Heterozygous	4% (1 in 25)	4.9
	Homozygous	0.15% (1 in 625)	10–80
Prothrombin	Heterozygous	2% (1 in 50)	3.8
	Homozygous	0.04% (1 in 2500)	Not known
FVL and prothrombin	Heterozygous for both	0.08% (1 in 1250)	20

*These figures will be increased in the presence of antithrombin, protein C, or protein S deficiency.

Their cumulative effects on risk are additive, with no evidence of epistasis. Molecules involved in complement activation and its regulation are now prime targets for therapeutic intervention in AMD.

FURTHER READING

Adams PC 2002 Hemochromatosis: clinical implications. Medscape Gastroenterology eJournal 4

A summary of clinical aspects of hemochromatosis and the role of genetic testing.

Barrett JC, Hansoul S, Nicolae DL et al 2008 Genome-wide association defines more than 30 distinct susceptibility loci for Crohn's disease. Nat Genet 40:955-962

The latest and largest meta-analysis of genome-wide association studies for Crohn disease, which is an excellent example of state-of-the-art methodology. Useful information is included on noteworthy genes found within the associated loci.

Maller J, George S, Purcell S et al 2006 Common variation in three genes, including a noncoding variant in *CFH*, strongly influences risk of age-related macular degeneration. Nat Genet 38:1055–1059

A publication describing the discovery or confirmation of five important susceptibility variants for age-related macular degeneration, which demonstrates that there is no epistasis between them.

Heston LL 1966 Psychiatric disorders in foster home reared children of schizophrenic mothers. Br J Psychiatry 112:819–825

A classic paper demonstrating genetic factors in the etiology of schizophrenia.

Hugot JP, Chamaillard M, Zouali H et al 2001 Association of NOD2 leucine-rich repeat variants with susceptibility to Crohn disease. Nature 411:599–603

Description of the positional cloning strategy which led to the identification of the NOD2/CARD15 *susceptibility gene for Crohn disease.*

Kathiresan S, Willer CJ, Peloso G et al 2008 Common variants at 30 loci contribute to polygenic dyslipidemia. Nat Genet 41:56–65

Meta-analysis of genome-wide association studies which identified and confirmed 30 distinct loci associated with circulating lipid levels. The paper also considers how the allelic dosage at these loci influences the proportion of individuals with clinically high lipid concentrations and discusses the spectrum of alleles that exist at the loci.

Lango H; UK Type 2 Diabetes Genetics Consortium, Palmer CN, Morris AD, Zeggini E et al 2008 Assessing the combined impact of 18 common genetic variants of modest effect sizes on type 2 diabetes risk. Diabetes 57:2911–2914

Recent publication describing the effects of multiple variants that predispose to type 2 diabetes.

Nejentsev S, Walker N, Riches D et al 2009 Rare variants of *IFIH1*, a gene implicated in antiviral responses, protect against type 1 diabetes. Science 324:387–389

A resequencing study demonstrating that rare variants in the IFIH1 *gene predispose to type 1 diabetes. This is an excellent example of GWA study follow-up to identify a causal gene.*

Ogura Y, Bonen DK, Inohara N et al 2001 A frameshift mutation in NOD2 associated with Crohn disease. Nature 411:603–606

Description of the positional cloning strategy that led to identification of the NOD2/CARD15 *susceptibility gene for Crohn disease.*

Prokopenko I, McCarthy MI, Lindgren C 2008 Type 2 diabetes: new genes, new understanding. Trends Genet 24:613–621

A review of the first genome-wide association studies for type 2 diabetes: loci identified, insights gained, and challenges remaining.

Seligsohn MD, Lubetsky A 2001 Genetic susceptibility to venous thrombosis. N Engl J Med 344:1222–1231

A comprehensive review of hereditary thrombophilia.

Wicker LS, Clark J, Fraser HI et al 2005 Type 1 diabetes genes and pathways shared by humans and NOD mice. J Autoimmun 25 Suppl:29–33

A review of the value of the NOD mouse model in identifying susceptibility genes for type 1 diabetes.

ELEMENTS

1 Both genetic and environmental factors are involved in the aetiology of many of the common diseases affecting humans. The genetic factors can be considered to be due to an inherited predisposition or genetic susceptibility.

2 The genetic contribution in a particular condition can be assessed by studying the incidence of disease in relatives, comparing concordance rates in identical and nonidentical twins, studying differences between populations and the effects of migration, studying the effects of adoption, biochemical studies, evaluating any possible association with other inherited factors, and studying animal models.

3 For common disorders such as diabetes mellitus, Crohn disease, hypertension, coronary artery disease, schizophrenia, Alzheimer disease, hemochromatosis, venous thrombosis, and age-related macular degeneration, a multifactorial mode of inheritance is most likely. It is also becoming clear that many of these conditions are heterogeneous, with different subtypes caused by different combinations of genetic and environmental factors.

4 Prevention of the common diseases involves determining causative environmental agents as well as identifying those individuals who are genetically susceptible to such agents.

5 Genetic susceptibility loci for multiple common diseases have been identified. Major progress has been enabled in very recent years by genome-wide association studies. Follow-up studies are beginning to identify the causal variants at these loci. For example, resequencing of type 1 diabetes loci led to the identification of low-frequency variants in *IFIH1*.

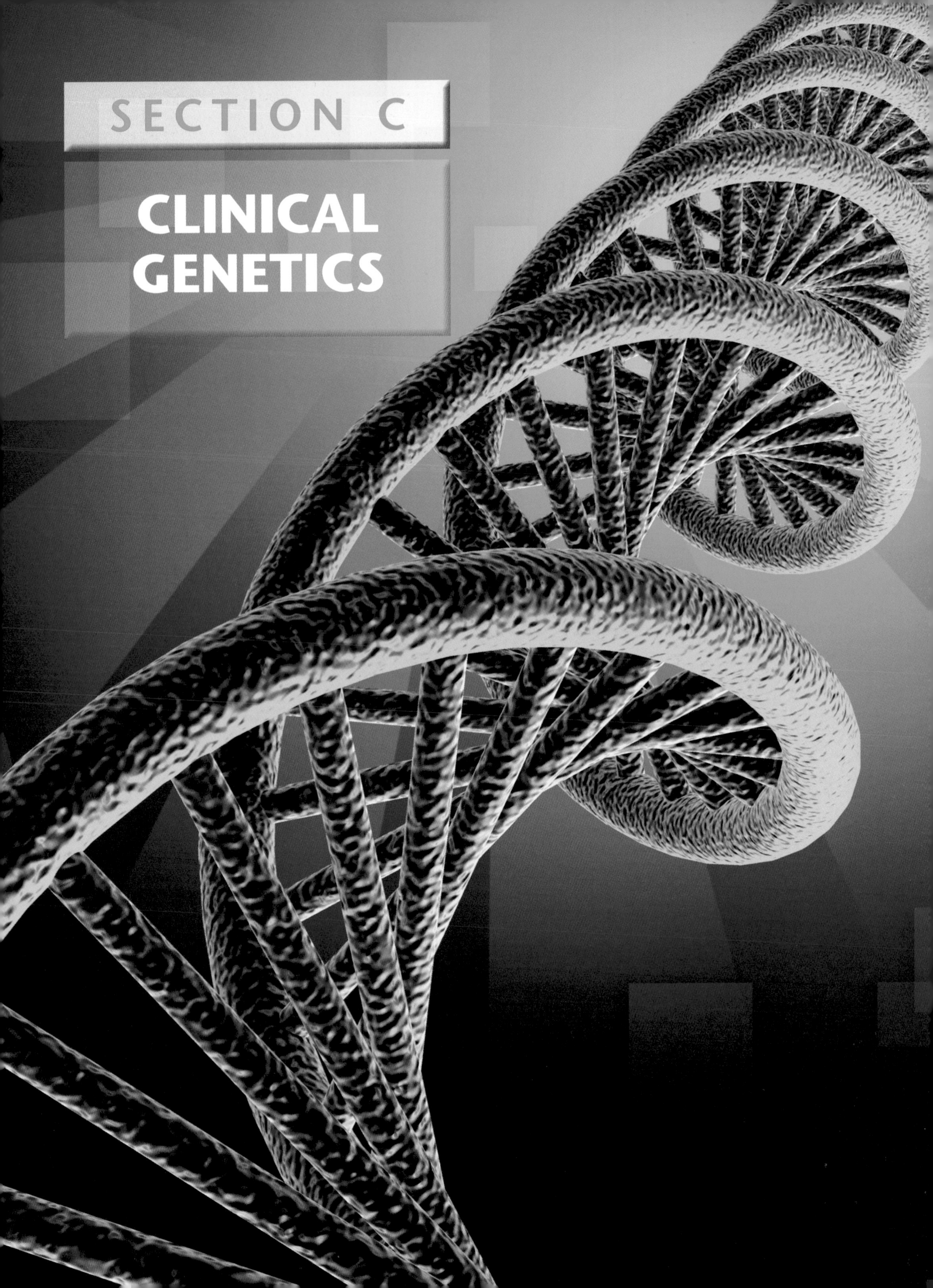

SECTION C

CLINICAL GENETICS

CHAPTER 16

Congenital Abnormalities and Dysmorphic Syndromes

They certainly give very strange names to diseases.
PLATO

The formation of a human being, a process sometimes known as **morphogenesis**, involves extremely complicated cell biology that, though only partially understood, is beginning to yield its mysteries (see Chapter 6). Given the complexity, it is not surprising that on occasion it goes wrong. Nor is it surprising that in many congenital abnormalities genetic factors can clearly be implicated. Approximately 2400 dysmorphic syndromes are described that are thought to be due to molecular pathology in single genes, and for at least 500 the genes have been identified and more than 200 mapped. A further 500 or so sporadically occurring syndromes are recognized, for which the precise cause remains elusive. In this chapter, we shall consider the overall impact of abnormalities in morphogenesis by reviewing the following.

1. The incidence of abnormalities at various stages from conception onwards.
2. Their nature and the ways in which they can be classified.
3. Their causes, when known, with particular emphasis on the role of genetics.

Incidence

Spontaneous First-Trimester Pregnancy Loss

It has been estimated that around 50% of all human conceptions are lost either before implantation at 5 to 6 days postconception or shortly afterwards (i.e., before the woman realizes she is pregnant). Among recognized pregnancies, at least 15% end in spontaneous miscarriage before 12 weeks' gestation. Even when material from the abortus can be obtained, it is often very difficult to establish why a pregnancy loss has occurred. However, careful study of large numbers of spontaneously aborted embryos has shown that gross structural abnormalities are present in 80% to 85%. These abnormalities vary from complete absence of an embryo in the developing pregnancy sac—a **blighted ovum**—to a very distorted body shape, or a specific abnormality in a single body system.

Chromosome abnormalities such as trisomy, monosomy, or triploidy are found in about 50% of all spontaneous abortions. This incidence rises to 60% when a gross structural abnormality is present and it is very likely that submicroscopic or de novo single-gene abnormalities account for a proportion of the remainder.

Congenital Abnormalities and Perinatal Mortality

Perinatal mortality figures include all infants who are stillborn after 28 weeks' gestation plus deaths during the first week of life. Of all perinatal deaths, 25% to 30% occur as a result of a serious structural abnormality and in 80% of these cases genetic factors can be implicated. The *relative* contribution of structural abnormalities to perinatal mortality is lower in developing countries, where environmental factors and health care provision play a much greater role.

Newborn Infants

Surveys reviewing the incidence of both major and minor anomalies in newborn infants have been undertaken in many parts of the world. A *major* anomaly can be defined as one that has an adverse outcome on either the function or the social acceptability of the individual (Table 16.1). In contrast, minor abnormalities are of neither medical nor cosmetic importance (Box 16.1). However, the division between major and minor abnormalities is not always straightforward; for instance, an inguinal hernia occasionally leads to strangulation of bowel and always requires surgical correction, so there is a risk of serious sequelae.

These surveys have consistently shown that 2% to 3% of all newborns have at least one major abnormality apparent at birth. The true incidence, taking into account abnormalities that present later in life, such as brain malformations, is probably close to 5%. Minor abnormalities are found in

Table 16.1 Examples of Major Congenital Structural Abnormalities

System and Abnormality	Incidence per 1000 Births
Cardiovascular	10
Ventricular septal defect	2.5
Atrial septal defect	1
Patent ductus arteriosus	1
Tetralogy of Fallot	1
Central nervous system	10
Anencephaly	1
Hydrocephaly	1
Microcephaly	1
Lumbosacral spina bifida	2
Gastrointestinal	4
Cleft lip/palate	1.5
Diaphragmatic hernia	0.5
Esophageal atresia	0.3
Imperforate anus	0.2
Limb	2
Transverse amputation	0.2
Urogenital	4
Bilateral renal agenesis	2
Polycystic kidneys (infantile)	0.02
Bladder exstrophy	0.03

Table 16.2 Incidence of Structural Abnormalities

Incidence	(%)
Spontaneous Miscarriages	
First trimester	80–85
Second trimester	25
All Babies	
Major abnormality apparent at birth	2–3
Major abnormality apparent later	2
Minor abnormality	10
Death in perinatal period	25
Death in first year of life	25
Death at 1–9 years	20
Death at 10–14 years	7.5

approximately 10% of all newborns. If two or more minor abnormalities are present in a newborn, there is a 10% to 20% risk that the baby will also have a major malformation.

The long-term outlook for a baby with a major abnormality obviously depends on the nature of the specific birth defect and whether it can be treated. The overall prognosis for this group of newborns is relatively poor, with 25% dying in early infancy, 25% having subsequent mental or physical disability, and the remaining 50% having a fair or good outlook after treatment.

Childhood Mortality

Congenital abnormalities make a significant contribution to mortality throughout childhood. During infancy, approximately 25% of all deaths are the result of major structural abnormalities, falling to 20% between 1 to 10 years of age, and to ~7.5% between 10 to 15 years.

Box 16.1 Examples of Minor Congenital Structural Abnormalities

Preauricular pit or tag
Epicanthic folds
Lacrimal duct stenosis
Brushfield spots in the iris
Lip pits
Single palmar crease
Fifth finger clinodactyly
Syndactyly between second and third toes
Supernumerary nipple
Umbilical hernia
Hydrocele
Sacral pit or dimple

Collating the incidence data on abnormalities noted in early spontaneous miscarriages and newborns, at least 15% of all recognized human conceptions are structurally abnormal (Table 16.2), and genetic factors are probably implicated in at least 50% of these.

Definition and Classification of Birth Defects

So far in this chapter the terms *congenital abnormality* and *birth defect* have been used in a general sense to describe all types of structural abnormality that can occur in an embryo, fetus, or newborn infant. Although these terms are perfectly acceptable for the purpose of lumping together all these abnormalities when studying their overall incidence, they do not provide any insight into possible underlying mechanisms. More specific definitions have been devised that have the added advantage of providing a combined clinical and etiological classification.

Single Abnormalities

Single abnormalities may have a genetic or non-genetic basis. The system of terms used helps us to understand the different mechanisms that might be implicated, and these can be illustrated in schematic form (Figure 16.1).

Malformation

A **malformation** is a primary structural defect of an organ, or part of an organ, that results from an inherent abnormality in development. This used to be known as a primary or intrinsic malformation. The presence of a malformation implies that the early development of a particular tissue or organ has been arrested or misdirected. Common examples include congenital heart abnormalities such as ventricular or atrial septal defects, cleft lip and/or palate, or neural tube defects (Figure 16.2). Most malformations involving only a single organ show multifactorial inheritance, implying an interaction of gene(s) with other factors (p. 143). Multiple malformations are more likely to be due to chromosomal abnormalities but may be due to single gene mutations.

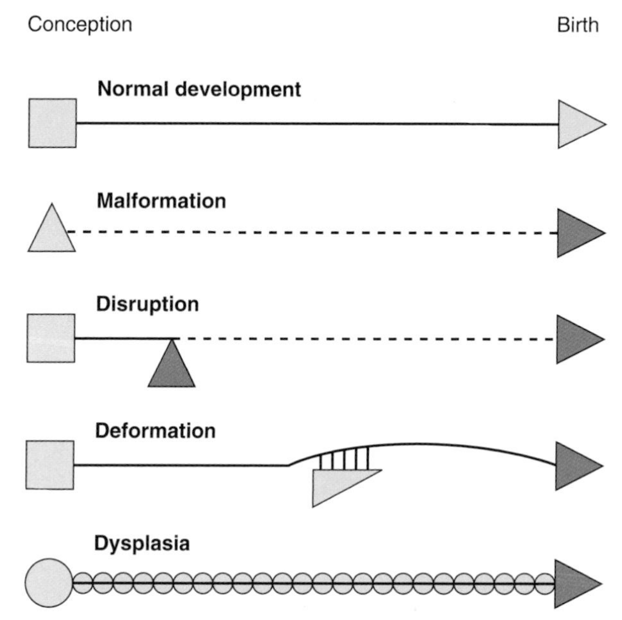

FIGURE 16.1 Schematic representation of the different mechanisms in morphogenesis. For malformation, disruption, and dysplasia, the broken line symbolizes developmental potential rather than timing of the manifestation of the defect, which might be late in embryogenesis. (Adapted from Spranger, et al 1982 Errors of morphogenesis: concepts and terms. J Pediatr 100:160–165.)

Disruption

The term **disruption** refers to an abnormal structure of an organ or tissue as a result of external factors disturbing the normal developmental process. This used to be known as a secondary or extrinsic malformation, and includes ischemia, infection, and trauma. An example of a disruption is the effect seen on limb development when a strand or band of amnion becomes entwined around a baby's forearm or digits (Figure 16.3). By definition a disruption is not genetic, although occasionally genetic factors can predispose to disruptive events. For example, a small proportion of amniotic bands are caused by an underlying genetically determined defect in collagen that weakens the amnion, making it more liable to tear or rupture spontaneously.

Deformation

A **deformation** is a defect resulting from an abnormal mechanical force that distorts an otherwise normal structure. Examples include dislocation of the hip and mild 'positional' talipes ('clubfoot') (Figure 16.4) resulting from reduced amniotic fluid (oligohydramnios) or intrauterine crowding from twinning or a structurally abnormal uterus. Deformations usually occur late in pregnancy and convey a good prognosis with appropriate treatment—for instance, gentle splinting for talipes, because the underlying organ is fundamentally normal in structure.

Dysplasia

A **dysplasia** is an abnormal organization of cells into tissue. The effects are usually seen wherever that particular tissue is present. For example, in a skeletal dysplasia such as thanatophoric dysplasia, which is caused by mutations in *FGFR3* (p. 93), almost all parts of the skeleton are affected (Figure 16.5). Similarly, in an ectodermal dysplasia, widely dispersed tissues of ectodermal origin, such as hair, teeth, skin, and nails, are involved (Figure 16.6). Most dysplasias are caused by single-gene defects and are associated with high recurrence risks for siblings and/or offspring.

Multiple Abnormalities

Sequence

This concept describes the findings that occur as a consequence of a cascade of events initiated by a single primary factor and may result in a single organ malformation. In the 'Potter' sequence, chronic leakage of amniotic fluid or defective fetal urinary output results in oligohydramnios (Figure 16.7). This, in turn, leads to fetal compression, resulting in squashed facial features, dislocation of the hips,

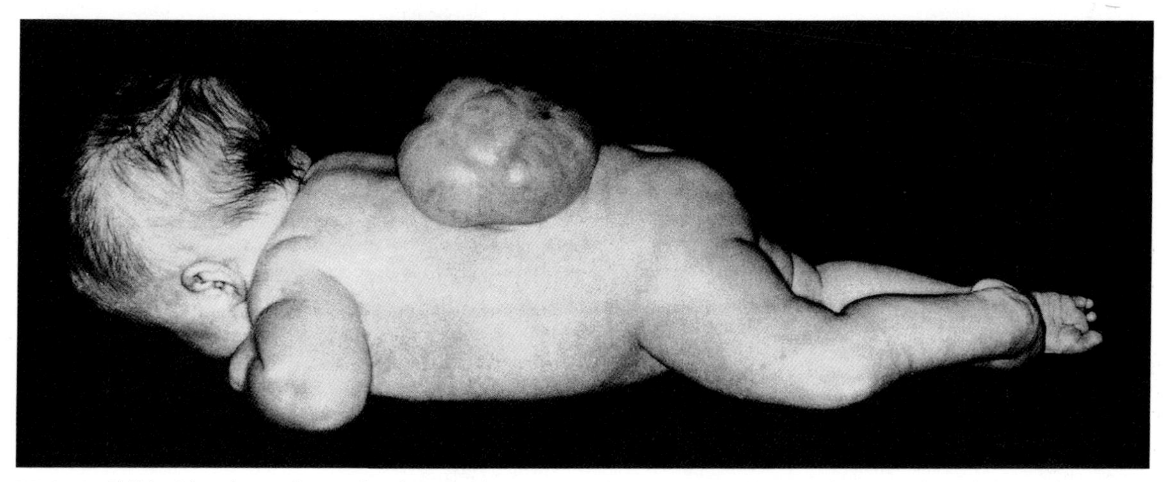

FIGURE 16.2 Child with a large thoracolumbar myelomeningocele consisting of protruding spinal cord covered by meninges.

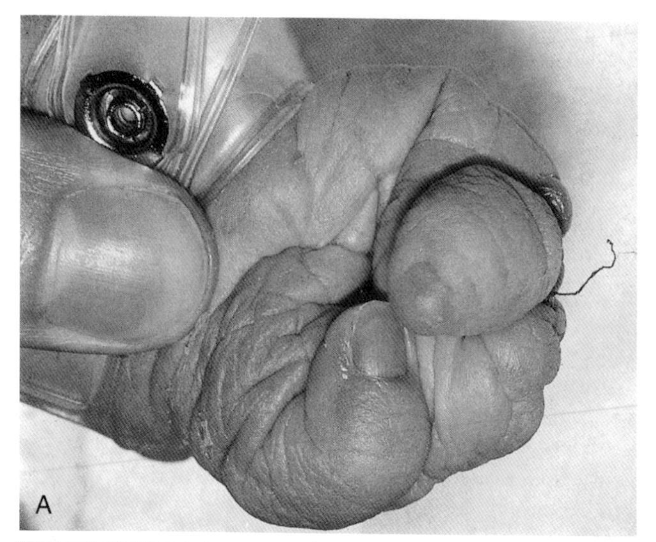

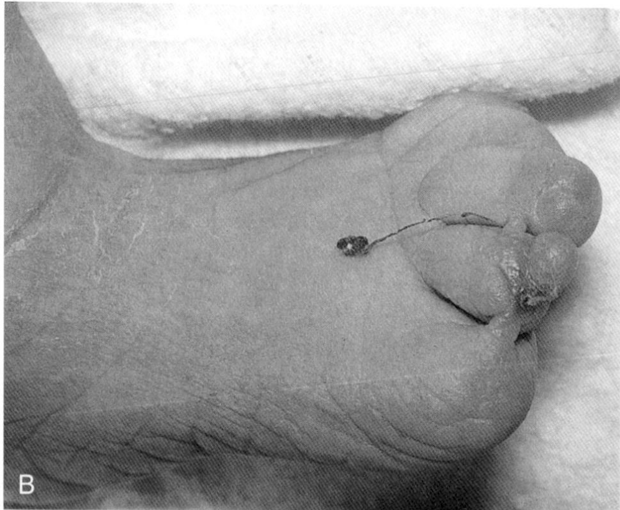

FIGURE 16.3 Hand (**A**) and foot (**B**) of a baby with digital amputations resulting from amniotic bands showing residual strands of amnion. (Courtesy Dr. Una MacFadyen, Leicester Royal Infirmary, UK.)

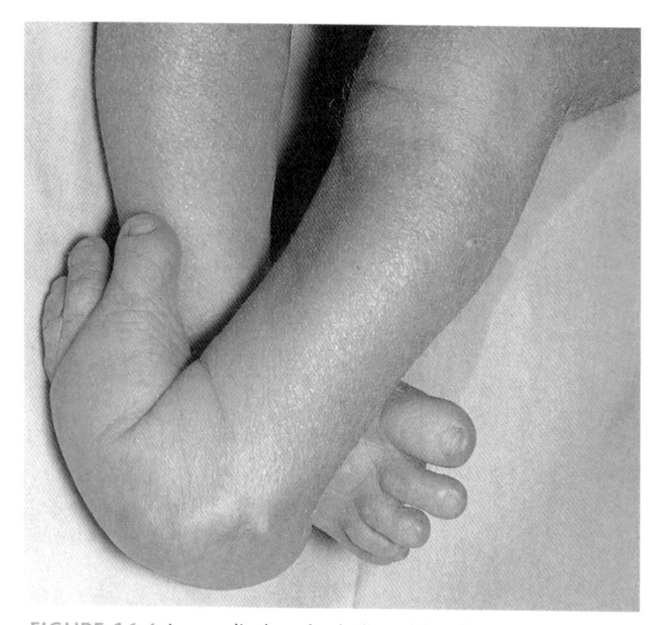

FIGURE 16.4 Lower limbs of a baby with talipes equinovarus.

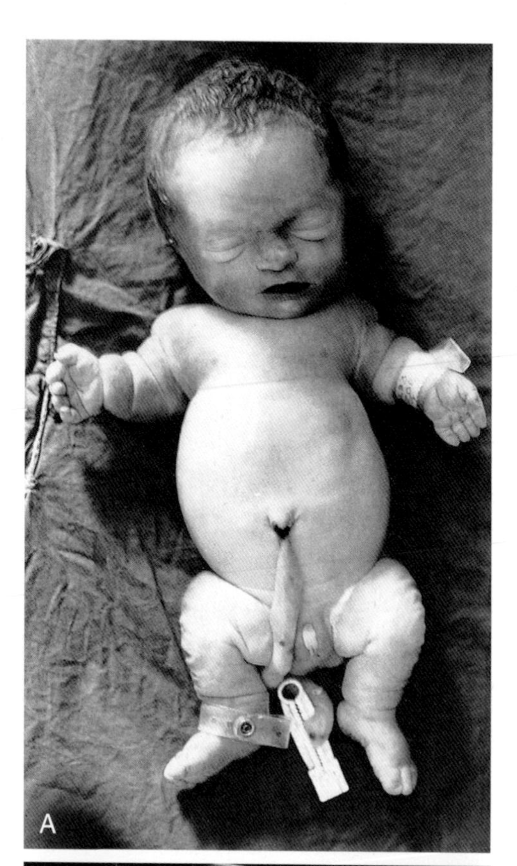

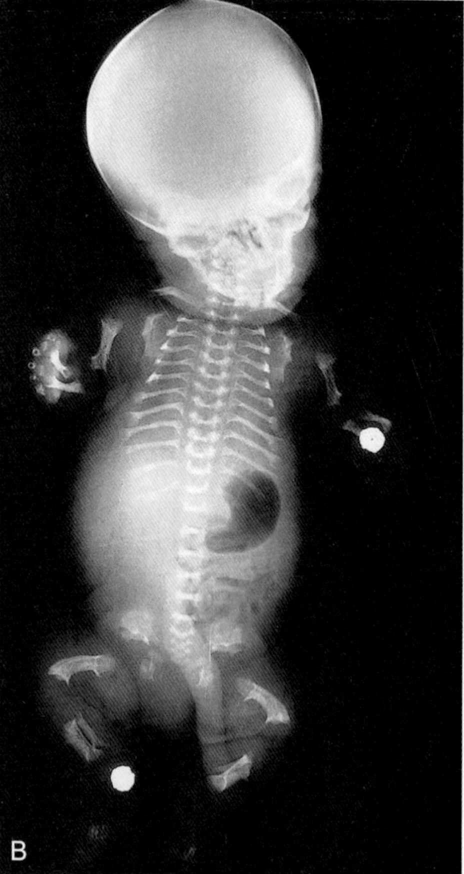

FIGURE 16.5 **A**, Infant with thanatophoric dysplasia. **B**, Radiograph of the infant showing short ribs, flat vertebral bodies, and curved femora.

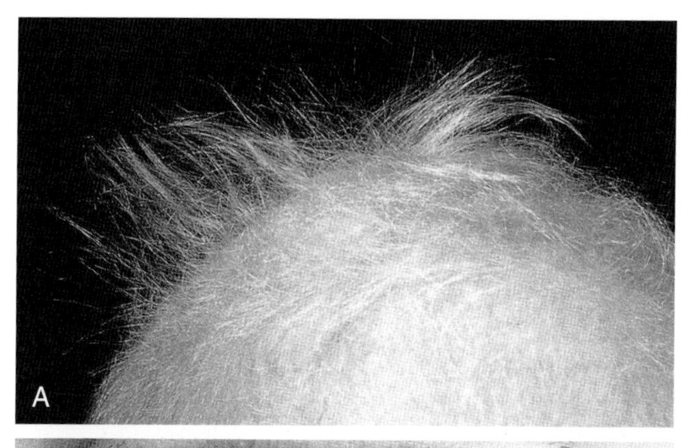

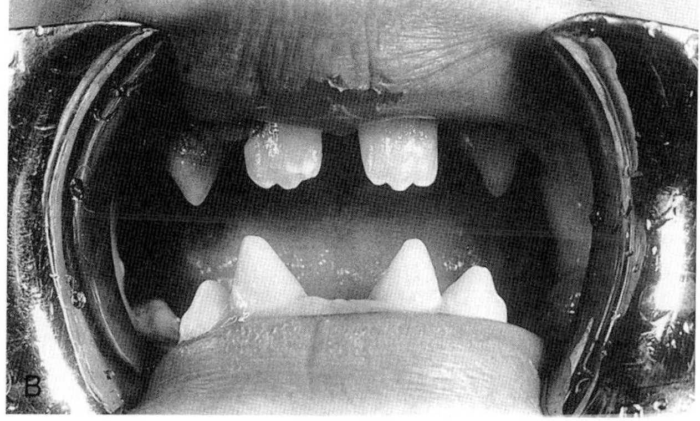

FIGURE 16.6 Hair (**A**) and teeth (**B**) of a male with ectodermal dysplasia.

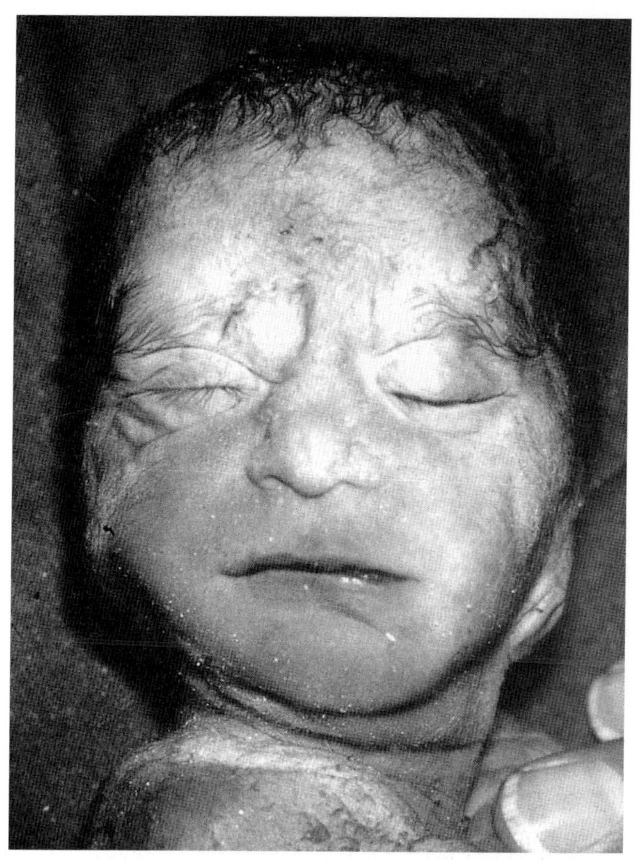

FIGURE 16.8 Facial appearance of a baby with Potter sequence from oligohydramnios as a consequence of bilateral renal agenesis. Note the squashed appearance caused by in utero compression.

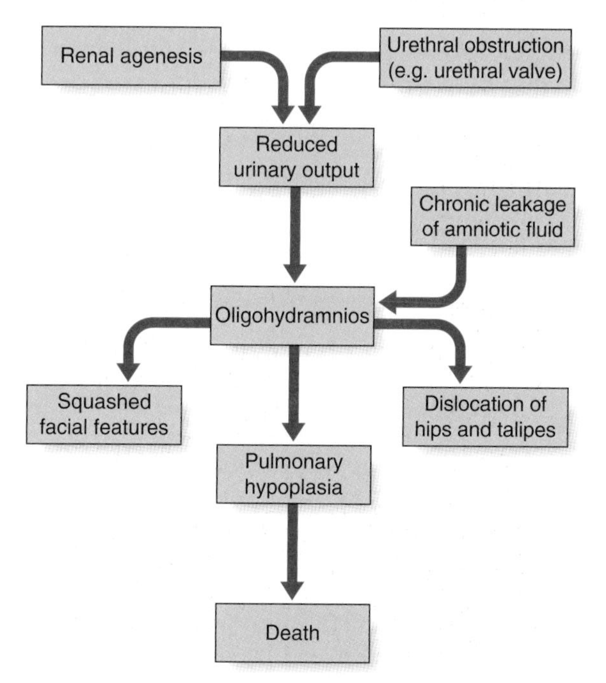

FIGURE 16.7 The 'Potter' sequence showing the cascade of events leading to and resulting from oligohydramnios (reduced volume of amniotic fluid).

talipes, and pulmonary hypoplasia (Figure 16.8), usually resulting in neonatal death from respiratory failure.

Syndrome

In practice the term **syndrome** is used very loosely (e.g., the amniotic band 'syndrome'), but in theory it should be reserved for consistent and recognizable patterns of abnormalities for which there will often be a known underlying cause. These underlying causes can include chromosome abnormalities, as in Down syndrome, or single gene defects, as in the Van der Woude syndrome, in which cleft lip and/or palate occurs in association with pits in the lower lip (Figure 16.9).

Several thousand multiple malformation syndromes are recognized, and their clinical study has been the discipline of **dysmorphology**. Clinical diagnosis has been greatly helped by the development of computerized databases (see Appendix) with a search facility based on key abnormal features. Even with the help of this extremely valuable diagnostic tool, there are many dysmorphic children for whom no diagnosis is reached, so that it can be very difficult to provide accurate information about the likely prognosis and recurrence risk (p. 266). The technique of microarray-CGH (p. 281) is making inroads into this large group of undiagnosed patients.

Association

The term **association** has been introduced in recognition of the fact that certain malformations tend to occur together more often than would be expected by chance, yet this non-random occurrence of abnormalities cannot be easily explained on the basis of a sequence or a syndrome. The main differences from a syndrome are the lack of consistency of abnormalities from one affected individual to another and the absence of a satisfactory underlying explanation. The names of associations are often acronyms; for example, the VATER association features ***v***ertebral, ***a***nal, ***t***racheo***e***sophageal, and ***r***enal abnormalities. Associations generally convey a low risk of recurrence and are generally thought not to be genetic. However, heterogeneity is likely and some cases are likely to be genetic.

This classification of birth defects is not perfect—it is far from being either fully comprehensive or mutually exclusive. For example, bladder outflow obstruction caused by a primary malformation such as a urethral valve will result in the oligohydramnios or Potter sequence, leading to secondary deformations such as dislocation of the hip and talipes. To complicate matters further, the absence of both kidneys, which will result in the same sequence of events, is usually erroneously referred to as Potter syndrome. Despite this semantic confusion, classifications can aid understanding of causes and recurrence risks (Chapter 17).

Genetic Causes of Malformations

There are many recognized causes of congenital abnormality, although it is notable that in up to 50% of all cases no clear explanation can be established (Table 16.3).

Table 16.3 Causes of Congenital Abnormalities

Cause	%
Genetic	30–40
Chromosomal	6
Single gene	7.5
Multifactorial	20–30
Environmental	5–10
Drugs and chemicals	2
Infections	2
Maternal illness	2
Physical agents	1
Unknown	50
Total	100

Chromosome Abnormalities

These account for approximately 6% of all recognized congenital abnormalities. As a general rule, any perceptible degree of autosomal imbalance, such as duplication, deletion, trisomy, or monosomy, will result in severe structural and developmental abnormality, which may lead to early miscarriage. Common chromosome syndromes are described in Chapter 18. It is not known whether malformations caused by a significant chromosome abnormality, such as a trisomy, are the result of dosage effects of the individual genes involved ('additive' model) or general developmental instability caused by a large number of abnormal developmental gene products ('interactive' model).

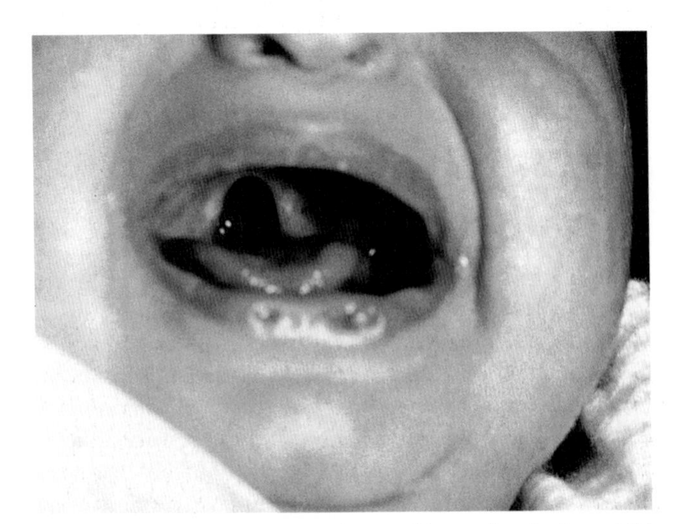

FIGURE 16.9 Posterior cleft palate and lower lip pits in a child with Van der Woude syndrome.

Single-Gene Defects

These account for 7% to 8% of all congenital abnormalities. Some of these are isolated—i.e., they involve only one organ or system (Table 16.4). Other single-gene defects result in multiple congenital abnormality syndromes involving many organs or systems that do not have any obvious underlying embryological relationship. For example, ectrodactyly (Figure 16.10) in isolation can be inherited as an autosomal dominant trait, occasionally autosomal recessive, and rarely X-linked. It can also occur as one manifestation of the EEC syndrome (***e***ctodermal dysplasia, ***e***ctrodactyly and ***c***left lip/palate), which follows autosomal dominant inheritance. Therefore different mutations, allelic or non-allelic, can cause similar or identical malformations.

The importance of determining a single-gene basis for birth defects lies in the need for accurate genetic counseling for the immediate and wider family. In addition, from a research perspective single gene causes can provide clues to susceptibility loci for similar malformations and phenotypes that appear to show multifactorial inheritance.

From the many examples of progress in identifying the genes that cause congenital abnormalities and dysmorphic syndromes, two are now illustrated from the field of pediatric genetics. In both the gene function in relation to widespread expression in many tissues has yet to be determined.

Noonan Syndrome

First described by Noonan and Ehmke in 1963, this well-known condition has incidence that may be as high as 1:2000 births, with equal sex ratio. The features resemble those of Turner syndrome in females—short stature, neck

Table 16.4 Congenital Abnormalities that Can Be Caused by Single-Gene Defects

	Inheritance	Abnormalities
Isolated		
CENTRAL NERVOUS SYSTEM		
Hydrocephalus	XR	
Megalencephaly	AD	
Microcephaly	AD/AR	
OCULAR		
Aniridia	AD	
Cataracts	AD/AR	
Microphthalmia	AD/AR	
LIMB		
Brachydactyly	AD	
Ectrodactyly	AD/AR/XR	
Polydactyly	AD	
OTHER		
Infantile polycystic	AR kidneys	
Syndromes		
Apert	AD	Craniosynostosis, syndactyly
EEC	AD	Ectodermal dysplasia, ectrodactyly, cleft lip/palate
Meckel	AR	Encephalocele, polydactyly, polycystic kidneys
Roberts	AR	Cleft lip/palate, phocomelia
Van der Woude	AD	Cleft lip/palate, lip pits

AD, Autosomal dominant; *AR*, autosomal recessive; *XR*, X-linked recessive.

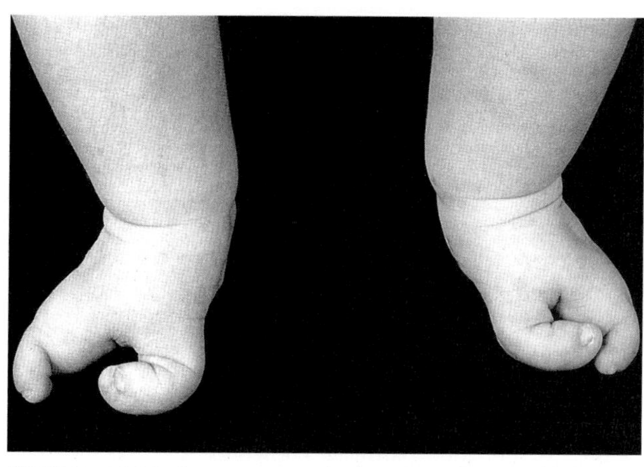

FIGURE 16.10 Appearance of the feet in a child with ectrodactyly.

webbing, increased carrying angle at the elbow and congenital heart disease. Pulmonary stenosis is the most common lesion but atrial septal defect, ventricular septal defect, and occasionally hypertrophic cardiomyopathy occur. A characteristic mild pectus deformity may be seen, and the face shows hypertelorism, down-slanting palpebral fissures, and low-set ears (Figure 16.11). Some patients have a mild bleeding diathesis, and learning difficulties occur in about one-quarter.

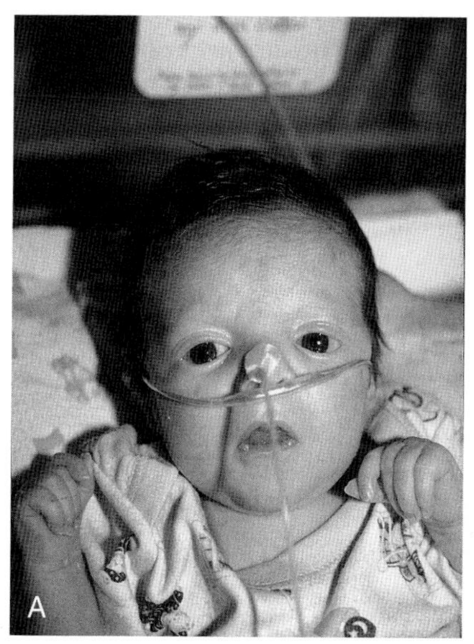

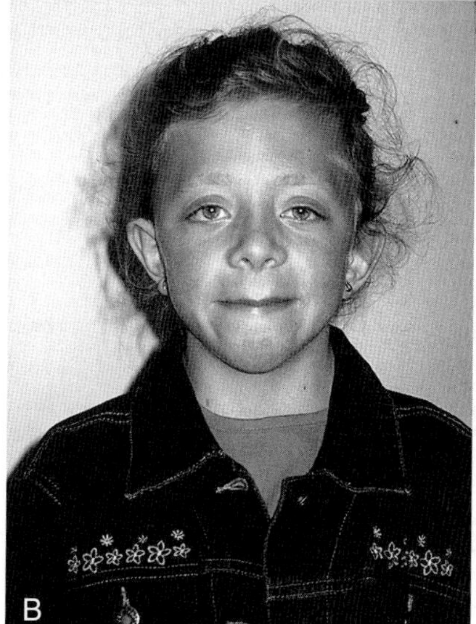

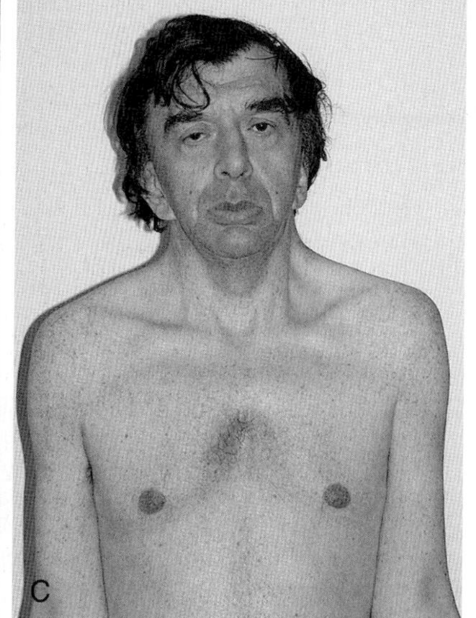

FIGURE 16.11 Noonan syndrome: (**A**) In a baby presenting with cardiomyopathy at birth (which later resolved); (**B**) in a child; and (**C**) in a 57-year-old man.

In a three-generation Dutch family, Noonan syndrome (NS) was mapped to 12q22 in 1994, but it was not until 2001 that mutations were identified in the protein tyrosine phosphatase, non-receptor-type, 11 *(PTPN11)* gene. Attention has turned rapidly to phenotype–genotype correlation, and mutation-positive cases have a much higher frequency of pulmonary stenosis than mutation-negative cases, and very few mutations have been found in patients with cardiomyopathy. However, facial features are similar, whether or not a mutation is found. Mutations in *PTPN11* account for about half of all cases of NS. Mutations in the *SOS1*, *SHOC2*, *KRAS*, and *MAPZK1* genes have been found in a small proportion of *PTPN11*-negative cases. These genes belong to the same pathway, known as RAS-MAPK. The protein product of *PTPN11* is SHP-2 and this, together with SOS1, positively transduces signals to Ras-GTP, a downstream effector (Figure 16.12). The *KRAS* mutations in NS appear to lead to K-ras proteins with impaired responsiveness to GTPase activating proteins (p. 213). Neurofibromatosis, the most common of this group, is dealt with in Chapter 18 (p. 298).

For years dysmorphologists recognized overlapping features between NS and the rarer conditions known as cardio-facio-cutaneous and Costello syndromes. These conditions are now recognized to form part of a spectrum of disorders explained by mutations in different components of the RAS-MAPK pathway, with each syndrome displaying considerable genetic heterogeneity (Table 16.5). Many of the mutations are gain-of-function missense mutations, which may explain the increase in solid tumors in Costello syndrome as well as cellular proliferation in some tissues in cardio-facio-cutaneous syndrome (e.g., hyperkeratosis). The effect is for RAS to bind GTP, which results in activation of the pathway (gain-of-function). Neurofibromin is a GTPase activating protein, and functions as a tumor-suppressor.

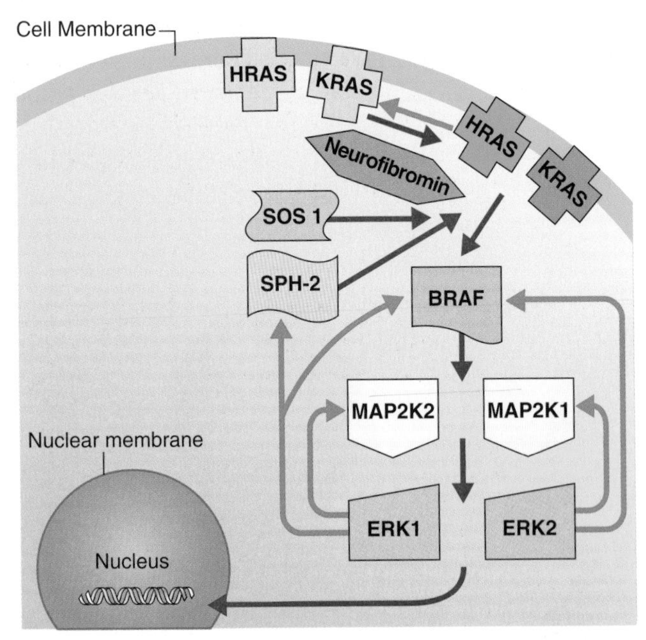

FIGURE 16.12 The RAS-MAPK pathway. HRAS and KRAS are activated by SPH-2 and SOS1 *(red arrows). Orange arrows* = inhibition. The pathway is dysregulated by mutations in key components, resulting in the distinct but related phenotypes of Noonan syndrome, CFC syndrome, Costello syndrome, and neurofibromatosis (see Table 16.5). Neurofibromin is a GAP (GTPase activating) protein that functions as a tumor-suppressor. Mutant RAS proteins display impaired GTPase activity and are resistant to GAPs. The effect is for RAS to bind GTP, which results in activation of the pathway (gain of function).

Sotos Syndrome

First described in 1964, this is one of the 'overgrowth' syndromes, previously known as cerebral gigantism. Birth weight is usually increased and macrocephaly noted. Early feeding difficulties and hypotonia may prompt many investigations and there is often motor delay and ataxia. Height progresses along the top of, or above, the normal centile lines, but final adult height is not necessarily markedly increased. Advanced bone age may be present, as well as large hands and feet, and the cerebral ventricles may be mildly dilated on imaging. The face is characteristic (Figure 16.13), with a high prominent forehead, hypertelorism with down-slanting palpebral fissures, a characteristic nose in early childhood, and a long pointed chin. Scoliosis develops in some cases during adolescence. Parent–child transmission is rare, probably because most patients have learning difficulties. However, the author has seen a three-generation family including individuals with above average intelligence.

Among patients with Sotos syndrome reported to have balanced chromosome translocations were two with breakpoints at 5q35. From these crucial patients a Japanese group in 2002 went on to identify a 2.2-Mb deletion in a series of Sotos syndrome cases. The deletion takes out a gene called *NSD1*, an androgen receptor-associated co-regulator with 23 exons. The Japanese found a small number of frameshift mutations in their patients but, interestingly, a study of European cases found that mutations were far more common

Table 16.5 Genes of the RAS-MAPK Pathway and Associated Syndromes

Gene	Noonan Syndrome	Cardio-Facio-Cutaneous Syndrome	Costello Syndrome
PTPN11	Common—≤50%	—	—
KRAS	Rare	Rare	Rare
HRAS	—	—	Common—>50%
SOS1	Rare	—	—
BRAF	—	Common—≤50%	Some
MAP2K1	Rare	Some	Some
MAP2K2	—	Rare	—

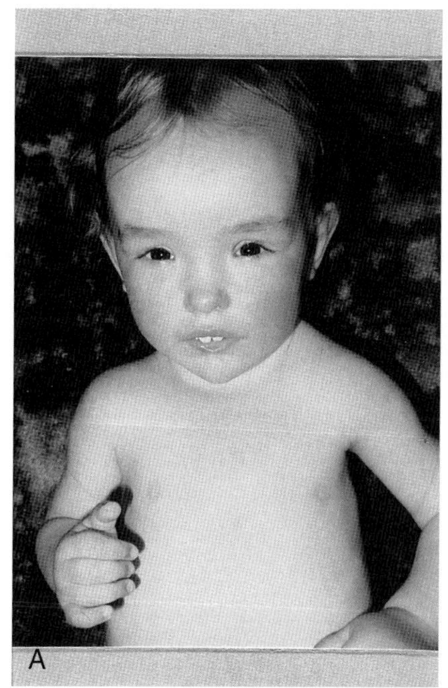

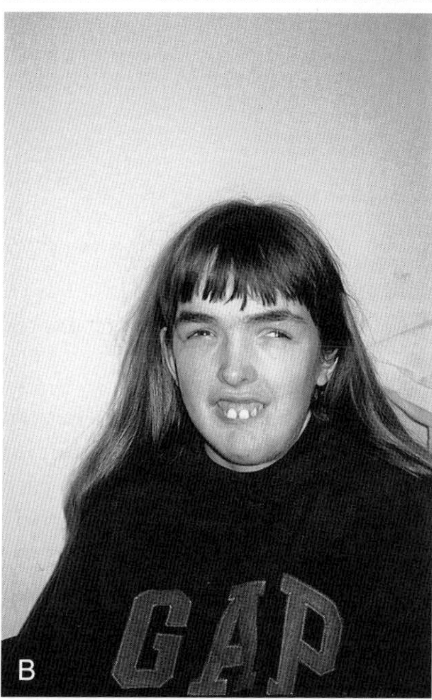

FIGURE 16.13 Sotos syndrome. **A**, In a young child who has the typical high forehead, large head, and characteristic tip to the nose. **B**, The same individual at age 18 years, with learning difficulties and a spinal curvature (scoliosis).

than deletions. For the large majority of cases the mutations and deletions occur de novo.

Multifactorial Inheritance

This accounts for the majority of congenital abnormalities in which genetic factors can clearly be implicated. These include most isolated ('non-syndromal') malformations involving the heart, central nervous system, and kidneys (Box 16.2). For many of these conditions, empirical risks have been derived (p. 346) based on large epidemiological family studies, so that it is usually possible to provide the parents of an affected child with a clear indication of the likelihood that a future child will be similarly affected. Risks to the offspring of patients who were themselves treated successfully in childhood are becoming available. These are usually similar to the risks that apply to siblings, as would be predicted by the multifactorial model (p. 143).

Genetic Heterogeneity

It has long been recognized that specific congenital malformations can have many different causes (p. 346), hence the importance of trying to distinguish between syndromal and isolated cases. This causal diversity has become increasingly apparent as developments in molecular biology have led to the identification of highly conserved families of genes that play crucial roles in early embryogenesis.

This subject is discussed at length in Chapter 6. In the current chapter, two specific malformations, holoprosencephaly and neural tube defects, will be considered to demonstrate the rate of progress in this field and the extent of the challenge that lies ahead.

Holoprosencephaly

This severe and often fatal malformation is caused by a failure of cleavage of the embryonic forebrain or prosencephalon. Normally this divides transversely into the telencephalon and the diencephalon. The telencephalon divides in the sagittal plane to form the cerebral hemispheres and the olfactory tracts and bulbs. The diencephalon develops to form the thalamic nuclei, the pineal gland, the optic chiasm, and the optic nerves. In holoprosencephaly, there is incomplete or partial failure of these developmental processes, and in the severe alobar form this results in an abnormal facial appearance (see Figure 6.7, p. 88) with profound neurodevelopmental impairment.

Box 16.2 Isolated (Non-Syndromal) Malformations that Show Multifactorial Inheritance

Cardiac
Atrial septal defect
Tetralogy of Fallot
Patent ductus arteriosus
Ventricular septal defect

Central Nervous System
Anencephaly
Encephalocele
Spina bifida

Genitourinary
Hypospadias
Renal agenesis
Renal dysgenesis

Other
Cleft lip/palate
Congenital dislocation of hips
Talipes

Etiologically, holoprosencephaly can be classified as chromosomal, syndromal, or isolated. Chromosomal causes account for around 30% to 40% of all cases, with the most common abnormality being trisomy 13 (p. 275). Other chromosomal causes include deletions of 18p, 2p21, 7q36, and 21q22.3, duplication of 3p24-pter, duplication or deletion of 13q, and triploidy (p. 276). Syndromal causes of holoprosencephaly are numerous and include relatively well known conditions such as the deletion 22q11 (DiGeorge) syndrome (p. 282) and a host of much rarer multiple malformation syndromes, some of which show autosomal recessive inheritance. One of these, Smith-Lemli-Opitz syndrome (p. 87), is associated with low levels of cholesterol; this is relevant in that it is known that cholesterol is necessary for normal functioning of the sonic hedgehog pathway (p. 86).

The third group, isolated holoprosencephaly, is sometimes explained by heterozygous mutations in three genes. The effects can be very variable, ranging from very mild with minimal features such as anosmia, to the full-blown, lethal, alobar form. The genes *Sonic hedgehog (SHH)* on chromosome 7q36, *ZIC2* on chromosome 13q32, and *SIX3* on chromosome 2p21. Of these *SHH* is thought to make the greatest contribution, accounting for up to 20% of all familial cases and between 1% and 10% of isolated cases. Some sibling recurrences of holoprosencephaly, not because of recessive Smith-Lemli-Opitz syndrome, have been shown to be due to germline mutations in these genes.

That so many familial cases remain unexplained indicates that more holoprosencephaly genes await identification. Causal heterogeneity is further illustrated by its association with poorly controlled maternal diabetes mellitus (p. 261).

Neural Tube Defects

Neural tube defects (NTDs), such as spina bifida and anencephaly, illustrate many of the underlying principles of multifactorial inheritance and emphasize the importance of trying to identify possible adverse environmental factors. These conditions result from defective closure of the developing neural tube during the first month of embryonic life. A defect occurring at the upper end of the developing neural tube results in either exencephaly/anencephaly or an encephalocele (Figure 16.14). A defect occurring at the lower end of the developing neural tube leads to a spinal lesion such as a lumbosacral myelocele or meningomyelocele (see Figure 16.2), and a defect involving the head plus cervical and thoracic spine leads to craniorachischisis. These different entities relate to the different embryological closure points of the neural tube. Most NTDs have serious consequences. Anencephaly and craniorachischisis are not compatible with survival for more than a few hours after birth. Large lumbosacral lesions usually cause partial or complete paralysis of the lower limbs with impaired bladder and bowel continence.

As with many malformations, NTDs can be classified etiologically under the headings of chromosomal, syndromal, and isolated. Chromosomal causes include trisomy 13 and trisomy 18, in both of which NTDs show an incidence of around 5% to 10%. Syndromal causes include the relatively rare autosomal recessive disorder Meckel-Gruber syndrome, characterized by encephalocele in association with polycystic kidneys and polydactyly. However, most NTDs represent isolated malformations in otherwise normal infants, and appear to show multifactorial inheritance.

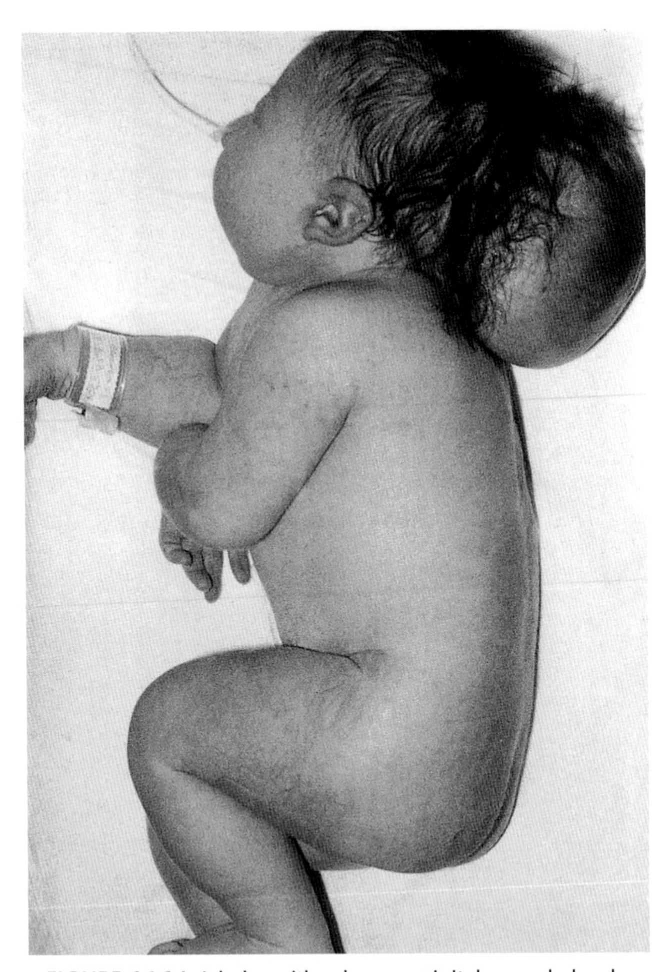

FIGURE 16.14 A baby with a large occipital encephalocele.

The empirical recurrence risks to first-degree relatives (siblings and offspring) vary according to the local population incidence and are as high as 4% to 5% in areas where NTDs are common. The incidence in the United Kingdom is highest in people of Celtic origin. If such individuals move from their country of origin to another part of the world, the incidence in their offspring declines but remains higher than among the indigenous population. These observations suggest a relatively high incidence of susceptibility genes in the Celtic populations.

No single NTD susceptibility genes have been identified in humans, although there is some evidence that the common 677C>T polymorphism in the *Methylenetetrahydrofolate reductase (MTHFR)* gene can be a susceptibility factor in some populations. Reduction in MTHFR activity results in decreased plasma folate levels, which are known to be causally associated with NTDs (see the following section). Research efforts have also focused on developmental genes, such as the *PAX* family (p. 91), which are expressed in the embryonic neural tube and vertebral column. In mouse models, about 80 genes have been linked to exencephaly, about 20 genes to lumbosacral meningomyelocele, and about 5 genes to craniorachischisis. One example is an interaction between mutations of *PAX1* and the *Platelet-derived growth factor α* gene *(PDGFRA)* that results in severe NTDs in 100% of double-mutant embryos. This rare example of digenic inheritance (p. 119) serves as a useful illustration of the difficulties posed by a search for susceptibility genes in a multifactorial disorder. However, to date there have been no equivalent breakthroughs in understanding the processes in human NTDs.

Environmental factors include poor socioeconomic status, multiparity, and valproic acid embryopathy (p. 261). Firm evidence has also emerged that periconceptional multivitamin supplementation reduces the risk of recurrence by a factor of 70% to 75% when a woman has had one affected child. Several studies have shown that folic acid is likely to be the effective constituent in multivitamin preparations. In both the United Kingdom and the United States, it is recommended that all women who have had a previous child with a NTD should take 4 to 5 mg folic acid daily both before and during the early stages of all subsequent pregnancies. Similarly, in the United Kingdom, it has been recommended that all women who are trying to conceive should take 0.4 mg folic acid daily. In the United States, where bread is fortified with folic acid, this recommendation applies to all women of reproductive age throughout their reproductive years. In the United Kingdom, this recommendation has not as yet resulted in a noticeable decline in the incidence of NTDs.

Environmental Agents (Teratogens)

An agent that can cause a birth defect by interfering with normal embryonic or fetal development is known as a teratogen. Many teratogens have been identified and exhaustive tests are now undertaken before any new drug is approved for use by pregnant women. The potential effects of any particular teratogen usually depend on the dosage and timing of administration during pregnancy, along with the susceptibility of both the mother and fetus.

An agent that conveys a high risk of teratogenesis, such as the rubella virus or thalidomide, can usually be identified relatively quickly. Unfortunately, it is much more difficult to detect a low-grade teratogen that causes an abnormality in only a small proportion of cases. This is because of the relatively high background incidence of congenital abnormalities, and also because many pregnant women take medication at some time in pregnancy, often for an ill-defined 'flulike' illness. Despite extensive study, controversy still surrounds the use of a number of drugs in pregnancy. The anti-nausea drug Debendox was the subject of successful litigation in the United States despite a lack of firm evidence to support a definite teratogenic effect.

Table 16.6 Drugs with a Proven Teratogenic Effect in Humans

Drug	Effects
ACE inhibitors	Renal dysplasia
Alcohol	Cardiac defects, microcephaly, characteristic facies
Chloroquine	Chorioretinitis, deafness
Diethylstilbestrol	Uterine malformations, vaginal adenocarcinoma
Lithium	Cardiac defects (Ebstein anomaly)
Phenytoin	Cardiac defects, cleft palate, digital hypoplasia
Retinoids	Ear and eye defects, hydrocephalus
Streptomycin	Deafness
Tetracycline	Dental enamel hypoplasia
Thalidomide	Phocomelia, cardiac and ear abnormalities
Valproic acid	Neural tube defects, clefting, limb defects, characteristic facies
Warfarin	Nasal hypoplasia, stippled epiphyses

ACE, Angiotensin-converting enzyme.

Drugs and Chemicals

Drugs and chemicals with a proven teratogenic effect in humans are listed in Table 16.6. These may account for approximately 2% of all congenital abnormalities.

Many other drugs have been proposed as possible teratogens, but the relatively small numbers of reported cases make it difficult to confirm that they are definitely teratogenic. This applies to many anticancer drugs, such as methotrexate and chlorambucil, and to anticonvulsants, such as sodium valproate, carbamazepine, and primidone. Exposure to environmental chemicals is also an area of widespread concern. Organic mercurials ingested in contaminated fish in Minamata, Japan, as a result of industrial pollution caused a 'cerebral palsy-like' syndrome in babies who had been exposed in utero. Controversy surrounds the use of agents used in warfare, such as dioxin (Agent Orange) in Vietnam and various nerve gases in the Gulf War.

The Thalidomide Tragedy

Thalidomide was used widely in Europe during 1958 to 1962 as a sedative. In 1961 an association with severe limb anomalies in babies whose mothers had taken the drug during the first trimester was recognized and the drug was subsequently withdrawn from use. It is possible that more than 10,000 babies were damaged over this period. Review of these babies' records indicated that the critical period for fetal damage was between 20 and 35 days postconception (i.e., 34 to 50 days after the beginning of the last menstrual period).

The most characteristic abnormality caused by thalidomide was phocomelia (Figure 16.15). This is the name given to a limb that is malformed due to absence of some or all of the long bones, with retention of digits giving a 'flipper' or 'seal-like' appearance. Other external abnormalities included ear defects, microphthalmia and cleft lip/palate. In addition, approximately 40% died in early infancy from

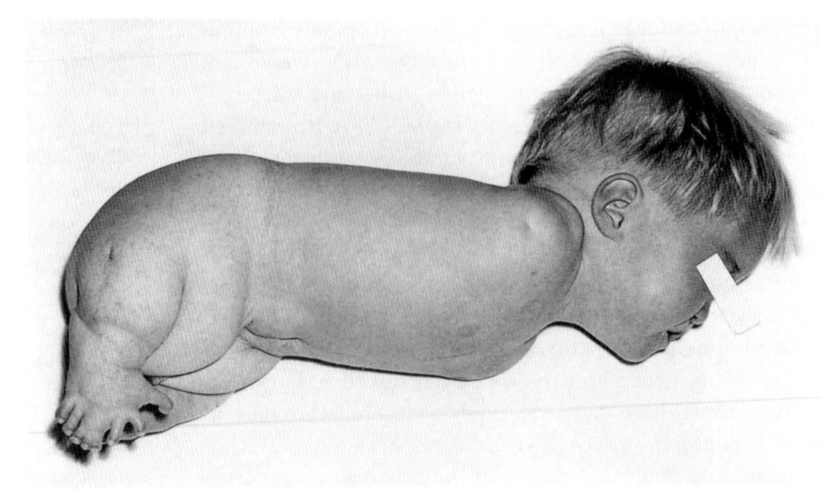

FIGURE 16.15 A child with thalidomide embryopathy. There is absence of the upper limbs (amelia). The lower limbs show phocomelia and polydactyly. (Courtesy Emeritus Professor R. W. Smithells, University of Leeds, UK)

severe internal abnormalities affecting the heart, kidneys, or gastrointestinal tract. Some 'thalidomide babies' have grown up and had children of their own, and in some cases these offspring have also had similar defects. It is therefore most likely, not surprisingly, that thalidomide was wrongly blamed in a proportion of cases that were in fact from single-gene conditions following autosomal dominant inheritance (e.g., *SALL4* mutations [see Figure 6.21, C, p. 99] in Okihiro syndrome).

The thalidomide tragedy focused attention on the importance of avoiding all drugs in pregnancy as far as is possible, unless absolute safety has been established. Drug manufacturers undertake extensive research trials before releasing a drug for general use, and invariably urge caution about the use of any new drug in pregnancy. Monitoring systems, in the form of congenital abnormality registers, have been set up in most Western countries so that it is unlikely that an 'epidemic' on the scale of the thalidomide tragedy could ever happen again.

Fetal Alcohol Syndrome

Children born to mothers who have consistently consumed large quantities of alcohol during pregnancy tend to show a degree of microcephaly, a distinctive facial appearance with short palpebral fissures, and a long smooth philtrum (Figure 16.16). They also show developmental delay with hyperactivity and clumsiness. This condition is referred to as fetal alcohol syndrome, but if the physical aspects are lacking the term 'alcohol-related neurodevelopmental defects' may be applied. There is uncertainty about the 'safe' level of alcohol consumption in pregnancy and there is evidence that mild-to-moderate ingestion can be harmful. Generally, total abstinence from alcohol is advised throughout pregnancy.

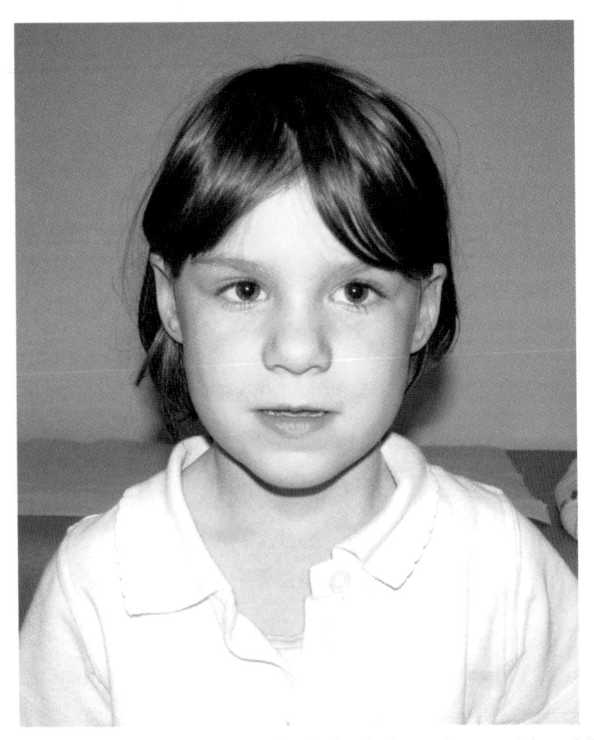

FIGURE 16.16 A child with fetal alcohol syndrome. The child has short palpebral fissures and a long, smooth philtrum.

Maternal Infections

Several infectious agents can interfere with embryogenesis and fetal development (Table 16.7). The developing brain, eyes, and ears are particularly susceptible to damage by infection.

Table 16.7 Infectious Teratogenic Agents

Infection	Effects
Viral	
Cytomegalovirus	Chorioretinitis, deafness, microcephaly
Herpes simplex	Microcephaly, microphthalmia
Rubella	Microcephaly, cataracts, retinitis, cardiac defects
Varicella zoster	Microcephaly, chorioretinitis, skin defects
Bacterial	
Syphilis	Hydrocephalus, osteitis, rhinitis
Parasitic	
Toxoplasmosis	Hydrocephalus, microcephaly, cataracts, chorioretinitis, deafness

Rubella

The rubella virus, which damages 15% to 25% of all babies infected during the first trimester, causes cardiovascular malformations such as patent ductus arteriosus and peripheral pulmonary artery stenosis. Congenital rubella infection can be prevented by the widespread use of immunization programs based on administration of either the measles, mumps, rubella vaccine in early childhood or rubella vaccine alone to young adult women.

Cytomegalovirus

At present no immunization is available against cytomegalovirus and naturally occurring infection does not always produce long-term immunity. The risk of abnormality is greatest when infection occurs during the first trimester. Overall this virus causes damage in about 5% of infected pregnancies.

Toxoplasmosis

Maternal infection with the parasite causing toxoplasmosis conveys a risk of 20% that the fetus will be infected during the first trimester, rising to 75% in the second and third trimesters. Vaccines against toxoplasmosis are not available.

Investigation for possible congenital infection can be made by sampling fetal blood to look for specific immunoglobulin-M antibodies. Fetal blood analysis can also reveal generalized evidence of infection, such as abnormal liver function and thrombocytopenia.

There is some evidence to suggest that maternal infection with *Listeria* can cause a miscarriage, and a definite association has been established between maternal infection with this agent and neonatal meningitis. Maternal infection with parvovirus can cause severe anemia in the fetus, resulting in hydrops fetalis and pregnancy loss.

Physical Agents

Women who have had babies with congenital abnormalities often scrutinize their own history in great detail and ask about exposure to agents such as radio waves, ultrasound, magnetic fields, and various chemicals, as well as minor trauma. It is invariably impossible to prove or disprove causal link but there is some evidence that two specific physical agents, ionizing radiation and prolonged hyperthermia, can have teratogenic effects.

Ionizing Radiation

Heavy doses of ionizing radiation, far in excess of those used in routine diagnostic radiography, can cause microcephaly and ocular defects in the developing fetus. The most sensitive time of exposure is 2 to 5 weeks postconception. Ionizing radiation can also have mutagenic (p. 26) and carcinogenic effects and, although the risks associated with low-dose diagnostic procedures are minimal, radiography should be avoided during pregnancy if possible.

Prolonged Hyperthermia

There is evidence that prolonged hyperthermia in early pregnancy can cause microcephaly and microphthalmia as well as neuronal migration defects. Consequently, it is recommended that care should be taken to avoid excessive use of hot baths and saunas during the first trimester.

Maternal Illness

Several maternal illnesses are associated with an increased risk of an untoward pregnancy outcome.

Diabetes Mellitus

Maternal diabetes mellitus is associated with a two- to threefold increase in the incidence of congenital abnormalities in offspring. Malformations that occur most commonly in such infants include congenital heart disease, neural tube defects, vertebral segmentation defects and sacral agenesis, femoral hypoplasia, holoprosencephaly, and sirenomelia ('mermaidism'). The likelihood of an abnormality is inversely related to the control of the mother's blood glucose levels during early pregnancy. This can be assessed by regular monitoring of blood glucose and glycosylated hemoglobin levels.

Phenylketonuria

Another maternal metabolic condition that conveys a risk to the fetus is untreated phenylketonuria (p. 167). A high serum level of phenylalanine in a pregnant woman with phenylketonuria will almost invariably result in serious damage (e.g., mental retardation). Structural abnormalities may include microcephaly and congenital heart defects. All women with phenylketonuria should be strongly advised to adhere to a strict and closely monitored low phenylalanine diet before and throughout pregnancy.

Maternal Epilepsy

There is a large body of literature devoted to the question of maternal epilepsy, the link with congenital abnormalities, and the teratogenic effects of antiepileptic drugs (AEDs). The largest and best controlled studies suggest that maternal epilepsy itself is not associated with an increased risk of congenital abnormalities. However, all studies have shown an increased incidence of birth defects in babies exposed to AEDs. The risks are in the region of 5% to 10%, which is two to five times the background population risk. These figures apply mainly to single drug therapy, and may be doubled if the fetus is exposed to more than one AED. Some drugs are more teratogenic than others, with the highest risks applying to sodium valproate. The range of abnormalities occurring in the 'fetal anticonvulsant syndromes' (FACS) are wide, including neural tube defect (about 2%), oral clefting, genitourinary abnormalities such as hypospadias, congenital heart disease, and limb defects. The abnormalities themselves are not specific to FACS, and making a diagnosis in an individual case can therefore be difficult. Sometimes characteristic

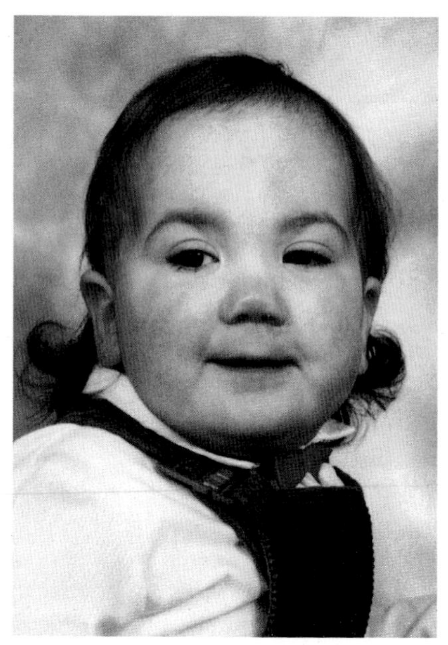

FIGURE 16.17 A child with fetal valproate syndrome. She has a broad nasal root, blunt nasal tip, and a thin upper lip.

facial features are seen, particularly in fetal valproate syndrome (Figure 16.17).

The most controversial aspect of AEDs and FACS is the risk of learning difficulties and behavioral problems. Controlled studies are problematic but evidence points to a higher incidence than in the general population—again, particularly in relation to sodium valproate. For practical purposes, potential risks to the fetus have to be weighed against the dangers of stopping AED treatment and risking seizures during pregnancy. If the patient has been seizure-free for at least 2 years, she can be offered withdrawal of anticonvulsant medication before proceeding with a pregnancy. If therapy is essential, then single-drug treatment is much preferred and sodium valproate should be avoided if possible.

Malformations of Unknown Cause

In up to 50% of all congenital abnormalities no clear cause can be established. This applies to many relatively common conditions such as isolated diaphragmatic hernia, tracheoesophageal fistula, anal atresia, and single-limb reduction defects. For an isolated limb defect, such as absence of a hand, it is reasonable to postulate that loss of vascular supply at a critical time during the development of the limb bud leads to developmental arrest, with the formation of only vestigial digits. It is more difficult to envisage how vascular occlusion could result in an abnormality such as esophageal atresia with an associated tracheoesophageal fistula.

Symmetry and Asymmetry

When trying to assess whether a birth defect is genetic or non-genetic, it may be helpful to consider aspects of symmetry. As a very broad generalization, symmetrical and midline abnormalities frequently have a genetic basis. Asymmetrical defects are less likely to have a genetic basis. In the examples shown in Figure 16.18, the child with cleidocranial dysplasia (Figure 16.18, A) has symmetrical

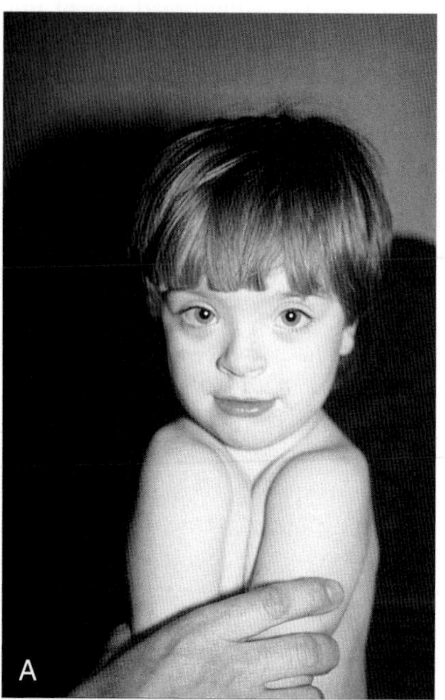

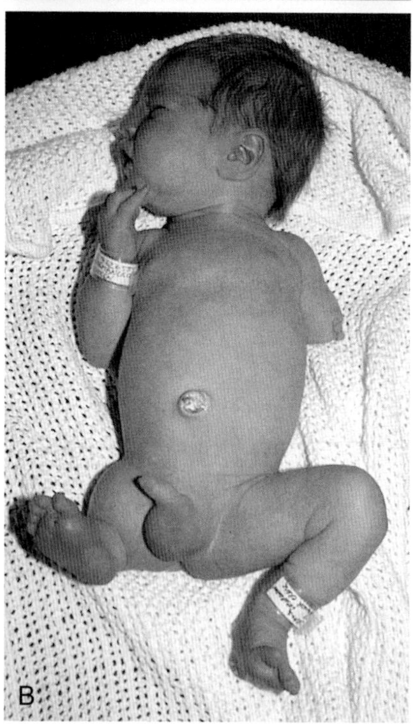

FIGURE 16.18 **A**, A boy with cleidocranial dysplasia in whom the clavicles have failed to develop, hence the remarkable mobility of his shoulders. He also has a relatively large head with widely spaced eyes (hypertelorism). He presented with ear problems—conductive deafness is a recognized feature. Skeletal dysplasias usually manifest in one main tissue and are symmetrical, suggesting a genetic basis. **B**, A child with congenital limb deformities from amniotic bands. The complete lack of symmetry suggests a non-genetic cause.

defects (absent or hypoplastic clavicles) and other features indicating a generalized tissue disorder that is overwhelmingly likely to have a genetic basis. The striking asymmetry of the limb deformities in Figure 16.18, B is likely to have a non-genetic basis.

Counseling

In cases where the precise diagnosis is uncertain, an assessment of symmetry and midline involvement may be helpful for genetic counseling. Although it may be very frustrating that no detailed explanation is possible, in many cases reassurance about a low recurrence risk in a future pregnancy can be given, based on empirical data. It is worth noting that this does not necessarily mean that genetic factors are irrelevant. Some 'unexplained' malformations and syndromes could well be due to new dominant mutations (p. 113), submicroscopic microdeletions (p. 280), or uniparental disomy (p. 121). All of these would convey negligible recurrence risks to future siblings, although those cases from new mutations or microdeletions would be associated with a significant risk to the offspring of affected individuals. There is optimism that molecular techniques will provide at least some of the answers to these many unresolved issues.

FURTHER READING

Aase J 1990 Diagnostic dysmorphology. London: Plenum

A detailed text of the art and science of dysmorphology.

Hanson JW 1997 Human teratology. In: Rimoin DL, Connor JM, Pyeritz RE eds. Principles and practice of medical genetics, 3rd edn, pp. 697–724. New York: Churchill Livingstone

A comprehensive, balanced overview of known and suspected human teratogens.

Jones KL 2006 Smith's recognizable patterns of human malformation, 6th edn. Philadelphia: Saunders

The standard pediatric textbook guide to syndromes.

Smithells RW, Newman CGH 1992 Recognition of thalidomide defects. J Med Genet 29:716–723

A comprehensive account of the spectrum of abnormalities caused by thalidomide.

Spranger J, Benirschke K, Hall JG, et al 1982 Errors of morphogenesis: concepts and terms. Recommendations of an international working group. J Pediatr 100:160–165

A short article providing a classification and clarification of the terms used to describe birth defects.

Stevenson RE, Hall JG, Goodman RM 1993 Human malformations and related anomalies. New York: Oxford University Press

The definitive guide, in two volumes, to human malformations.

ELEMENTS

1 Congenital abnormalities are apparent at birth in 1 in 40 of all newborn infants. They account for 20% to 25% of all deaths occurring in the perinatal period and in childhood up to the age of 10 years.

2 A single abnormality can be classified as a malformation, a deformation, a dysplasia, or a disruption. Multiple abnormalities can be classified as a sequence, a syndrome, or an association.

3 Congenital abnormalities can be caused by chromosome imbalance, single-gene defects, multifactorial inheritance, or non-genetic factors. Most isolated malformations, including isolated congenital heart defects and neural tube defects, show multifactorial inheritance, whereas most dysplasias have a single-gene etiology.

4 Many congenital malformations, including cleft lip/palate, congenital heart defects, and neural tube defects, show etiological heterogeneity, so that when counseling it is important to establish whether these malformations are isolated or are associated with other abnormalities.

5 Many environmental agents have been shown to have a teratogenic effect and, if at all possible, great care should be taken to avoid exposure to these agents during pregnancy.

CHAPTER 17

Genetic Counseling

Q. What's the difference between …
a doctor … and God?
A. God doesn't think He's a doctor.
ANON

Any couple that has had a child with a serious abnormality must inevitably reflect on why this happened and whether any child(ren) they choose to have in future might be similarly affected. Similarly, individuals with a family history of a serious disorder are likely to be concerned that they could either develop the disorder or transmit it to future generations. They are also very concerned about the risk that their normal children might transmit the condition to their offspring. For all those affected by a genetic condition that is serious to them, great sensitivity is needed in communication. Just a few words spoken with genuine caring concern can put patients at ease and allow a meaningful session to proceed; just a few careless words that make light of a serious situation can damage communication irrevocably. The importance of confidence and trust in the relationship between patient and health professional must never be underestimated. In the commercial world the same applies—confidence is a prerequisite for business contracts (between both parties).

Realization of the needs of individuals and couples, together with awareness of the importance of providing them with accurate and appropriate information, has led to the widespread introduction of genetic counseling clinics in parallel with the establishment of clinical genetics as a recognized medical specialty.

Definition

Since the first introduction of genetic counseling services approximately 40 years ago, many attempts have been made to devise a satisfactory and all-embracing definition. A theme common to all is the concept of genetic counseling being a process of communication and education that addresses concerns relating to the development and/or transmission of a hereditary disorder.

An individual who seeks genetic counseling is known as a **consultand**. During the genetic counseling process, it is widely agreed that the counselor should try to ensure that the consultand is provided with information that enables him or her to understand:

1. The medical diagnosis and its implications in terms of prognosis and possible treatment
2. The mode of inheritance of the disorder and the risk of developing and/or transmitting it
3. The choices or options available for dealing with the risks.

It is also agreed that genetic counseling should include a strong communicative and supportive element, so that those who seek information are able to reach their own fully informed decisions without undue pressure or stress (Box 17.1).

Establishing the Diagnosis

The most crucial step in any genetic consultation is that of establishing the diagnosis. If this is incorrect, then inappropriate and totally misleading information could be given, with potentially tragic consequences.

Reaching a diagnosis in clinical genetics usually involves the three fundamental steps of any medical consultation: taking a history, carrying out an examination, and undertaking appropriate investigations. Often, detailed information about the consultand's family history will have been obtained by a skilled genetic nurse counselor. A full and accurate family history is a cornerstone in the whole genetic assessment and counseling process. Further information about the family and personal medical history often emerges at the clinic, when a full examination can be undertaken and appropriate investigations initiated. These can include chromosome and molecular studies as well as referral on to specialists in other fields, such as neurology and ophthalmology. It cannot be overemphasized that the quality of genetic counseling is dependent upon the availability of facilities that ensure an accurate diagnosis can be made.

Even when a firm diagnosis has been made, problems can arise if the disorder in question shows etiological heterogeneity. Common examples include hearing loss and nonspecific mental retardation, both of which can be caused by either environmental or genetic factors. In these situations empirical risks can be used (p. 346), although these are not as satisfactory as risks based on a precise and specific diagnosis.

A disorder is said to show genetic **heterogeneity** if it can be caused by more than one genetic mechanism (p. 346). Many such disorders are recognized, and counseling can be extremely difficult if the heterogeneity extends to different modes of inheritance. Commonly encountered examples include the various forms of Ehlers-Danlos syndrome (Figure 17.1), Charcot-Marie-Tooth disease (p. 296), and

Box 17.1 Steps in Genetic Counseling

Diagnosis—based on accurate family history, medical history, examination, and investigations
Risk assessment
Communication
Discussion of options
Long-term contact and support

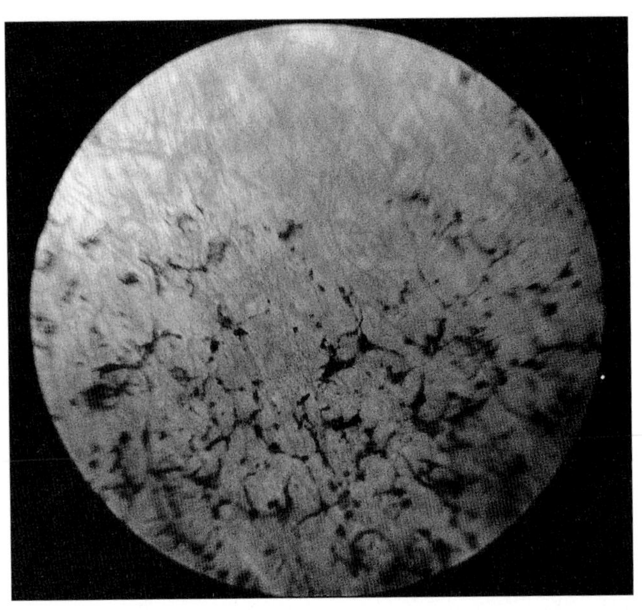

FIGURE 17.2 Fundus showing typical pigmentary changes of retinitis pigmentosa.

retinitis pigmentosa, all of which can show autosomal dominant, autosomal recessive, and X-linked recessive inheritance (Table 17.1). Fortunately, progress in molecular genetics is providing solutions to many of these problems. For example, mutations in the gene that codes for rhodopsin, a retinal pigment protein, are found in approximately 30% of families showing autosomal dominant inheritance of retinitis pigmentosa (Figure 17.2) and the molecular basis of the most common forms of Charcot-Marie-Tooth disease (type 1), also known as hereditary motor and sensory neuropathy, is now understood (p. 296).

Calculating and Presenting the Risk

In some counseling situations, calculation of the recurrence risk is relatively straightforward and requires little more than a reasonable knowledge of mendelian inheritance. However, many factors, such as delayed age of onset, reduced penetrance, and the use of linked DNA markers, can result in the calculation becoming much more complex. The theoretical aspects of risk calculation are considered in more detail in Chapter 22.

The provision of a recurrence risk does not simply involve conveying a stark risk figure in isolation. It is very important that the information provided is understood, and that parents are given as much background information as possible to help them reach their own decision. As a working rule of thumb, recurrence risks should be quantified, qualified, and placed in context.

Quantification—The Numerical Value of a Risk

Most prospective parents will be familiar to some degree with the concept of risk, but not everyone is comfortable with probability theory and the alternative ways of expressing risk, such as in the form of odds or as a percentage. Thus, for example, a risk of 1 in 4 can be presented as

FIGURE 17.1 Ehlers-Danlos syndrome. The inheritance pattern in this case is autosomal dominant because father and son are affected.

Table 17.1 Hereditary Disorders that Can Show Different Patterns of Inheritance

Disorder	Inheritance Patterns
Cerebellar ataxia	AD, AR
Charcot-Marie-Tooth disease	AD, AR, XR
Congenital cataract	AD, AR, XR
Ehlers-Danlos syndrome	AD, AR, XR
Ichthyosis	AD, AR, XR
Microcephaly	AD, AR
Polycystic kidney disease	AD, AR
Retinitis pigmentosa	AD, AR, XR, M
Sensorineural hearing loss	AD, AR, XR, M

AD, Autosomal dominant; *AR,* autosomal recessive; *XR,* X-linked recessive; *M,* mitochondrial.

an odds ratio of 3 to 1 against, or numerically as 25%. Consistency and clarity are important if confusion is to be avoided. It is also essential to emphasize that a risk applies to *each* pregnancy and that chance does not have a memory. For example, that parents have just had a child with an autosomal recessive disorder (recurrence risk of 1 in 4) does not mean that their next three children will be unaffected. A useful analogy is that of the tossed coin that cannot be expected to remember whether it landed heads or tails at the last throw and cannot therefore be expected to know what it should do at the next throw.

It is also important that genetic counselors should not be seen exclusively as prophets of doom. Continuing the penny analogy, the good side of the coin should also be emphasized, particularly if the odds strongly favor a successful outcome. For example, a couple faced with a probability of 1 in 25 that their next baby will have a neural tube defect should be reminded that there are 24 chances of 25 that their next baby will not be affected.

Qualification—The Nature of a Risk

Several studies have indicated that the most important factor determining the decision parents make about extending their family is the nature of the long-term *burden* (severity) of disease (or health care) associated with a risk, rather than its precise numerical value. Therefore a 'high' risk of 1 in 2 for a trivial problem such as an extra digit (polydactyly) will deter very few parents. In contrast a 'low' risk of 1 in 25 for a disabling condition such as a neural tube defect can have a very significant deterrent effect. A woman who grew up watching her brother develop Duchenne muscular dystrophy and subsequently die from the condition as a young adult may not risk having children even if there is only a 1% chance that she is a carrier. Other factors, such as whether a condition can be treated successfully, whether it is associated with pain and suffering, and whether prenatal diagnosis is available, may all be relevant to the decision-making process.

Placing Risks in Context

Prospective parents seen at a genetic counseling clinic should be provided with information that enables them to put their risks in context so as to be able to decide for themselves whether a risk is 'high' or 'low'. For example, it can be helpful (but also alarming) to point out that approximately 1 in 40 of all babies has a congenital malformation (often treatable) or handicapping disorder. Therefore, an additional quoted risk of 1 in 50, although initially alarming, might on reflection be perceived as relatively low. As an arbitrary guide, risks of 1 in 10 or greater tend to be regarded as high, 1 in 20 or less as low, and intermediate values as moderate.

Discussing the Options

Having established the diagnosis and discussed the risk of occurrence/recurrence, the counselor is then obliged to ensure that the consultands are provided with all of the information necessary for them to make their own informed decisions. This should include details of all the choices open to them. For example, if relevant, the availability of prenatal diagnosis should be discussed, together with details of the techniques, limitations and risks associated with the various methods employed (see Chapter 21). Mention will sometimes be made of other reproductive options. These can include alternative approaches to conception, such as artificial insemination using donor sperm, the use of donor ova and preimplantation genetic diagnosis (p. 335). These techniques can be used when one partner is infertile, as in the case of Klinefelter syndrome and Turner syndrome (see Chapter 18), or simply to bypass the possibility that one or other partner will transmit his or her disadvantageous gene(s) to the baby.

These are issues that should be broached with great care and sensitivity. For some couples, the prospect of prenatal diagnosis followed by selective termination of pregnancy is unacceptable, whereas others view this as their only means of ensuring that they have healthy children. Whatever the personal views of the counselor, the consultands are entitled to knowledge of prenatal diagnostic procedures that are technically feasible and legally permissible.

Communication and Support

The ability to communicate is essential in genetic counseling. Communication is a two-way process. Not only does the counselor provide information, he or she also has to be receptive to the fears and aspirations, expressed or unexpressed, of the consultand. A readiness to listen is a key attribute for anyone involved in genetic counseling, as is an ability to present information in a clear, sympathetic, and appropriate manner.

Often an individual or couple will be extremely upset when first made aware of a genetic diagnosis, and it is very common for guilt feelings to set in. The individual or couple may look back and scrutinize every event and happening, for example during a pregnancy. The delivery of potentially distressing information cannot be carried out in isolation. Genetic counselors need to take into account the complex psychological and emotional factors that can influence the counseling dialog. The setting should be agreeable, private, and quiet, with ample time for discussion and questions. When possible, technical terms should be avoided or, if used, fully explained. Questions should be answered openly and honestly, and if information is lacking it is certainly not a fault or sign of weakness to admit that this is so. Most couples respect and recognize the truth, and some parents of children whose condition cannot be diagnosed derive a curious pleasure from knowing that their child appears to be unique and has bamboozled the medical profession (unfortunately, this is not particularly difficult!).

Despite all of these measures, a counseling session can be so intense and intimidating that the amount and accuracy of information retained can be very limited. For this reason, a letter summarizing the topics discussed at a counseling session is usually sent to the family afterwards. In addition, they are sometimes contacted later by a member of the counseling team, thereby providing an opportunity for clarification of any confusing issues and for further questions to be answered.

It is poor practice to simply convey information of a distressing nature without offering an opportunity for further discussion and long-term support. Most genetic counseling centers maintain informal contact with relevant families through a network of genetic nurse counselors who are familiar with the family and their particular circumstances. This is especially valuable for prospective parents who subsequently request specific prenatal diagnostic investigations, and for presymptomatic adults who are shown to be at high risk of developing late-onset autosomal dominant disorders such as Alzheimer disease (p. 342) and Huntington disease (p. 293). Genetic registers (p. 332) provide a useful means of ensuring that effective contact can be maintained with all such relevant family members.

Patient Support Groups

Finally, mention should be made of the widespread network of support groups that now exists. These organizations have usually been established by highly motivated and well-informed parents or affected families who can provide enormous support and companionship for others affected by a particular genetic condition or syndrome. When confronted by a new diagnosis of a rare disorder, many families feel very isolated; if at all possible they should be given contact information so that they have the option of communicating with other affected families who have had similar experiences. Referral to an appropriate support group would now be viewed as an essential integral component of the genetic counseling process. As well as providing support and information for affected families, these groups have often been very successful, not only in fostering (and funding) research but also in developing new services.

Genetic Counseling—Directive or Non-Directive?

It has already been emphasized that genetic counseling should be viewed as a communication process that provides information. The ultimate goal is to ensure that an individual or couple can reach their own decisions based on full information about risks and options. There is universal agreement that genetic counseling should be non-directive, with no attempt being made to steer the consultand along a particular course of action. In the same spirit the genetic counselor should also strive to be non-judgmental, even if a decision is reached that seems ill advised or is contrary to the counselor's own beliefs. Thus the role of the genetic counselor is to facilitate and enhance individual autonomy rather than to give advice or recommend a particular course of action. This *person-centered* approach conforms most closely to the model of counseling theory developed by the American, Carl Rogers (1902–1987), rather the *psychodynamic* approach of Sigmund Freud (1856–1939).

Genetic counselors are sometimes asked what they themselves would do if placed in the consultand's position. Generally it is preferable to avoid being drawn into expressing an opinion, opting instead to suggest that the consultand try to imagine how he or she might feel in the future having pursued each of the available options. This approach, sometimes referred to as 'scenario-based decision counseling', provides individuals with an opportunity for careful reflection. This is particularly important if one of the options under consideration involves a potentially irreversible reproductive decision such as sterilization. There is a well-established maxim that it is the consultands and not the counselors who have to live with the consequences of their decisions and, indeed, consultands should be encouraged to make the decision that they can best live with—the one that they are least likely to regret.

Outcomes in Genetic Counseling

The issue of defining outcomes in genetic counseling is difficult and contentious, but also topical in today's climate of health economy where everything has to be justified. The difficulty arises because of the rather nebulous nature of genetic counseling, which, in contrast to most medical activities, does not have any easily quantified end points, such as rate of infection or survival after surgery. The issue is topical because of pressure from funding authorities, with emphasis on demonstrable quality and effectiveness. Finally, the issue is contentious because it raises serious questions about the purpose of genetic counseling and whether this should be viewed as simply the provision of information or whether there should be identified benefits for society in terms of a reduction in the incidence of genetic disease.

In practice the three main outcome measures that have been assessed are recall, impact on subsequent reproductive behavior, and patient satisfaction. Most studies have shown that the majority of individuals who have attended a genetic counseling clinic have a reasonable recall of the information given, particularly if this was reinforced by a personal letter or follow-up visit. Nevertheless, confusion can arise, and as many as 30% of counselees have difficulty in remembering a precise risk figure. Studies that have focused on the subsequent reproductive behavior of couples that have attended a genetic counseling clinic have shown that approximately 50% have been influenced to some extent. The factors that have been shown to be influential are the severity of the disorder, the desire of the parents to have children, and whether prenatal diagnosis and/or treatment are available. Finally, studies that have attempted to assess patient satisfaction have struggled to address the problem of how this should best be defined. For example, an individual could be very satisfied with the way in which they were counseled but remain very dissatisfied by lack of a precise diagnosis or the availability of subsequent prenatal diagnostic tests.

In an increasingly cost-conscious society, it is not surprising that the purchasers of health care, whether private or state funded, are keen to identify quantifiable targets with which they can assess the 'effectiveness' of genetics services, and genetic counseling in particular. Outcomes such as the number of abnormal pregnancies terminated or individuals screened can seem attractive to administrators and politicians who are preoccupied with balancing budgets and cost–benefit analyses. In contrast, clinical geneticists and non-medical counselor colleagues universally reject the use of these outcome criteria on the grounds that they hint at a eugenics philosophy that is totally unacceptable in a society in which patient autonomy is a guiding ethical principle. Instead, they emphasize the benefits of an educated informed community with enhanced individual autonomy. The goals of satisfaction as expressed by the users of genetic services and their ability to make informed decisions are seen as much more acceptable than a reduction in the financial and personal burdens caused by genetic disease.

In the future, it is possible that an outcome measure such as 'perceived personal control' will be developed. This will almost certainly incorporate both information and satisfaction criteria, together with an assessment of whether individuals have been able to understand the information and come to terms with their situation, thereby enabling them to make appropriate life decisions with which they are comfortable. The idea is also consistent with general government policy to encourage individuals to take more personal control of their own health agenda.

Special Problems in Genetic Counseling

There are a number of special problems that can arise in genetic counseling.

Consanguinity

A consanguineous relationship is one between blood relatives who have at least one common ancestor no more remote than a great-great-grandparent. Consanguineous marriage is widespread in many parts of the world (Table 17.2). In Arab populations, the most common type of consanguineous marriage occurs between first cousins who are the children of two brothers, whereas in the Indian subcontinent uncle–niece marriages are the most commonly encountered form of consanguineous relationship. Although there is in these communities some recognition of the potential disadvantageous genetic effects of consanguinity, there is also a strongly held view that these are greatly outweighed by social advantages such as greater family support and marital stability.

Many studies have shown that among the offspring of consanguineous marriages, there is an increased incidence of both congenital malformations and other conditions that will present later, such as hearing loss and mental retardation. For the offspring of first cousins, the incidence of congenital malformations is increased to approximately twice that seen in the offspring of unrelated parents. Almost all of this increase in morbidity and mortality is attributed to homozygosity for autosomal recessive disorders, a finding consistent with Garrod's original observation that 'the mating of first cousins gives exactly the conditions most likely to enable a rare, and usually recessive, character to show itself' (p. 113).

Table 17.2 Worldwide Incidence of Consanguineous Marriage

Country	Incidence (%)
Kuwait	54
Saudi Arabia	54
Jordan	50
Pakistan	40–50
India	5–60
Syria	33
Egypt	28
Lebanon	25
Algeria	23
Japan	2–4
France, UK, USA	2

Data adapted from various sources including Jaber L, Halpern GJ, Shohat M 1998 The impact of consanguinity worldwide. Commun Genet 1:12–17.

On the basis of studies of children born to consanguineous parents, it has been estimated that the average human carries between one and two genes for a harmful autosomal recessive disorder, together with several mutations for conditions that result in lethality before birth. Most prospective consanguineous parents are concerned primarily with the risk that they will have a handicapped child, and fortunately the overall risks are usually relatively small. When estimating a risk for a particular consanguineous relationship, it is generally assumed that each common ancestor carried one deleterious recessive mutation.

Therefore, for first cousins, the probability that their first child will be homozygous for their common grandfather's deleterious gene will be 1 in 64 (Figure 17.3). Similarly, the risk that this child will be homozygous for the common grandmother's recessive gene will also be 1 in 64. This gives a total probability that the child will be homozygous for one of the grandparent's deleterious genes of 1 in 32. This risk should be added to the general population risk of 1 in 40 that any baby will have a major congenital abnormality (p. 249), to give an overall risk of approximately 1 in 20 that a child born to first-cousin parents will be either malformed or handicapped in some way. Risks arising from consanguinity for more distant relatives are much lower.

For consanguineous marriages, there is also a slightly increased risk that a child will have a multifactorial disorder. In practice this risk is usually very small. In contrast, a close

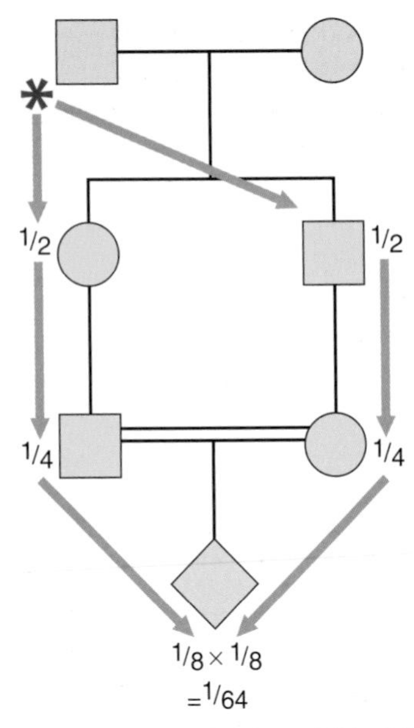

FIGURE 17.3 Probability that the first child of first cousins will be homozygous for the deleterious allele (*) carried by the common great-grandfather. A similar risk of 1 in 64 will apply to the deleterious allele belonging to the common great-grandmother, giving a total risk of 1 in 32.

family history of an autosomal recessive disorder can convey a relatively high risk that a consanguineous couple will have an affected child. For example, if the sibling of someone with an autosomal recessive disorder marries a first cousin, the risk that their first baby will be affected equals 1 in 24 (p. 342).

Incest

Incestuous relationships are those that occur between first-degree relatives—in other words, brother-sister or parent-child (Table 17.3). Marriage between first-degree relatives is forbidden, both on religious grounds and by legislation, in almost every culture. Incestuous relationships are associated with a very high risk of abnormality in offspring, with less than half the children of such unions being entirely healthy (Table 17.4).

Table 17.3 Genetic Relationship between Relatives and Risk of Abnormality in their Offspring

Genetic Relationship	Proportion of Shared Genes	Risk of Abnormality in Offspring (%)
First Degree	1/2	50
Parent–child		
Brother–sister		
Second Degree	1/4	5–10
Uncle–niece		
Aunt–nephew		
Double first cousins		
Third Degree	1/8	3–5
First cousins		

Table 17.4 Frequency of the Three Main Types of Abnormality in the Children of Incestuous Relationships

Abnormality	Frequency (%)
Intellectual impairment	
Severe	25
Mild	35
Autosomal recessive disorder	10–15
Congenital malformation	10

Adoption and Genetic Disorders

The issue of adoption can arise in several situations relating to genetics. First, parents at high risk of having a child with a serious abnormality often express interest in adopting rather than running the risk of having an affected baby. In genetic terms, this is a perfectly reasonable option, although in practice the number of couples wishing to adopt usually far exceeds the number of babies and children available for adoption.

The physician with a knowledge of genetics can also be called on to try to determine whether a child who is being placed for adoption will develop a genetic disorder. For the offspring of consanguineous or incestuous matings, risks can be given as outlined previously (see Tables 17.3 and 17.4). Adoption societies sometimes also wish to place a child with a known family history of a particular hereditary disorder. This raises the difficult ethical dilemma of predictive testing in childhood for conditions showing onset in adult life (p. 365). Increasingly it is felt that such testing should not be undertaken unless this will be of direct medical benefit to the child. In practice, even when a child is actually affected by a genetic disorder, suitable adoptive parents can usually be found.

Concern about the possible misuse of genetic testing in neonates and young children who are up for adoption has prompted the American Society of Human Genetics and the American College of Medical Genetics to issue joint recommendations. These are based on the best interests of the child. They can be summarized as supporting genetic testing in such children only when the testing would be appropriate for all children of that age and when the tests are undertaken for disorders that manifest during childhood, for which preventive measures can be undertaken during childhood. The joint statement does not support testing for untreatable disorders of adult onset or for detecting predispositions to 'physical, mental, or behavioral traits within the normal range'.

Disputed Paternity

This presents a difficult problem for which the help of a clinical geneticist is sometimes sought. Until recently paternity could never be proved with absolute certainty, although

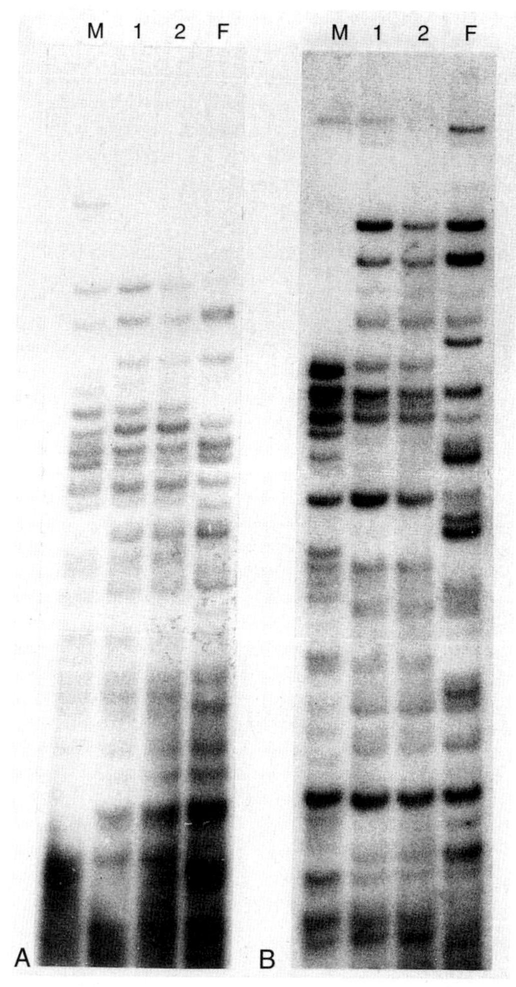

FIGURE 17.4 Genetic fingerprint obtained using two minisatellite probes with DNA from a mother (M), father (F) and their twins (1 and 2). The twins have an identical set of bands and each band in the twins originates from one of the parents. (Courtesy Dr. Raymond Dalgleish and Professor Sir Alec Jeffreys, University of Leicester. Reproduced from Young ID, Dalgleish R, Mackay EH, MacFadyen UM: Discordant expression of the G syndrome in monozygotic twins. Am J Med Genet 1988; 29:863–869, with permission from the *American Journal of Medical Genetics*.)

it could be disproved or excluded in two ways. If a child was found to possess a blood group or other polymorphism not present in either the mother or the putative father, then paternity could be confidently excluded. For example, if the mother and putative father both lacked blood group B, but this was present in the child, the putative father could be excluded. Similarly, if a child lacked a marker that the putative father would have had to transmit to all of his children, then once again paternity could be excluded. As an example, a putative father with blood group AB could not have a child with blood group O.

Early attempts to establish paternity were based on analysis of several different polymorphic systems, such as blood groups, isoenzymes and human leukocyte antigen haplotypes. The results of these studies can be consistent with paternity but cannot give absolute proof of it. Depending on the number of polymorphic systems analyzed and their frequencies in the general population, it is possible to calculate the relative probability that a particular male is the father compared with any male taken at random from the general population.

The limitations of these approaches have been overcome by the development of genetic fingerprinting using minisatellite repeat sequence probes (pp. 17, 69) and single nucleotide polymorphisms (SNPs) (p. 67). The pattern of DNA fragments generated by these probes, and SNP variants, is so highly polymorphic that the restriction map obtained is unique to each individual, with the exception of identical twins (Figure 17.4). If DNA from the child and the mother is analyzed, then the bands inherited from the biological father can be analyzed and compared with those present in DNA from the putative father(s). If these match, this gives an extremely high mathematical probability that the putative and biological fathers are the same individual.

FURTHER READING

ASHG/ACMG: Statement 2000 Genetic testing in adoption. Am J Hum Genet 66:761–767

The joint recommendations of the American Society of Human Genetics and the American College of Medical Genetics on genetic testing in young children who are being placed for adoption.

Clarke A (ed) 1994 Genetic counselling. Practice and principles. London: Routledge

A thoughtful and provocative multi-author text that addresses difficult issues such as predictive testing, screening, prenatal diagnosis, and confidentiality.

Clarke A, Parsons E, Williams A 1996 Outcomes and process in genetic counseling. Clin Genet 50:462–469

A critical review of previous studies of the outcomes of genetic counseling.

Frets PG, Niermeijer MF 1990 Reproductive planning after genetic counselling: a perspective from the last decade. Clin Genet 38:295–306

A review of studies undertaken between 1980 and 1989 to determine which factors are most important in influencing reproductive decisions.

Harper PS 1998 Practical genetic counseling, 5th ed. Oxford: Butterworth-Heinemann

An extremely useful practical guide to all aspects of genetic counseling.

Jaber L, Halpern GJ, Shohat M 1998 The impact of consanguinity worldwide. Community Genet 1:12–17

A review of the incidence and consequences of consanguinity in various parts of the world.

Jeffreys AJ, Brookfield JFY, Semeonoff R 1985 Positive identification of an immigration test-case using human DNA fingerprints. Nature 317:818–819

A clever demonstration of the value of genetic fingerprinting in analyzing alleged family relationships.

Turnpenny P (ed) 1995 Secrets in the genes: adoption, inheritance and genetic disease. London: British Agencies for Adoption and Fostering

A multi-author basic text covering aspects of genetics relevant to the adoption process.

ELEMENTS

1. Genetic counseling may be defined as a communication process that deals with the risk of developing or transmitting a genetic disorder.

2. The most important steps in genetic counseling are the establishment of a diagnosis, estimation of a recurrence risk, communication of relevant information, and provision of long-term support.

3. The pertinent counseling theory is person-centered, non-directive, and non-judgmental. The goal of genetic counseling is to provide accurate information that enables counselees to make their own fully informed decisions.

4. Marriage between blood relatives conveys an increased risk for an autosomal recessive disorder in future offspring. The probability that first cousins will have a child with an autosomal recessive condition is approximately 3%, although this risk can be greater if there is a family history of a specific genetic disorder.

5. The most sensitive technique for paternity testing is genetic fingerprinting. Other polymorphic systems can disprove paternity but cannot establish with such a high statistical probability that a particular male is the biological father.

CHAPTER 18

Chromosome Disorders

The development of a reliable technique for chromosome analysis in 1956 soon led to the discovery that several previously described conditions were due to an abnormality in chromosome number. Within 3 years, the causes of Down syndrome (47,XX/XY, +21), Klinefelter syndrome (47,XXY), and Turner syndrome (45,X) had been established. Shortly after, other autosomal trisomy syndromes were recognized, and over the ensuing years many other multiple malformation syndromes were described in which there was loss or gain of chromosome material.

To date, at least 20,000 chromosomal abnormalities have been registered on laboratory databases. On an individual basis, most of these are very rare, but together they make a major contribution to human morbidity and mortality. Chromosome abnormalities account for a large proportion of spontaneous pregnancy loss and childhood disability, and also contribute to malignancy throughout life as a consequence of acquired translocations and other aberrations.

In Chapter 3, the basic principles of chromosome structure, function, and behavior during cell division were described, together with an account of chromosome abnormalities and how they can arise and be transmitted in families. In this chapter, the medical aspects of chromosome abnormalities, and some of their specific syndromes, are described.

Incidence of Chromosome Abnormalities

Chromosome abnormalities are present in at least 10% of all spermatozoa and 25% of mature oocytes. Some 15% to 20% of all recognized pregnancies end in spontaneous miscarriage, and many more zygotes and embryos are so abnormal that survival beyond the first few days or weeks after fertilization is not possible. Approximately 50% of all spontaneous miscarriages have a chromosome abnormality (Table 18.1) and the incidence of chromosomal abnormalities in morphologically normal embryos is around 20%. Chromosome abnormalities therefore account for the spontaneous loss of a very high proportion of all human conceptions.

From conception onward, the incidence of chromosome abnormalities falls rapidly. By birth it has declined to a level of 0.5% to 1%, although the total is higher (5%) in stillborn infants. Table 18.2 lists the incidence figures for chromosome abnormalities based on newborn surveys. It is notable that among the commonly recognized aneuploidy syndromes, there is also a high proportion of spontaneous pregnancy loss (Table 18.3). This is illustrated by comparison of the incidence of conditions such as Down syndrome at the time of chorionic villus sampling (11 to 12 weeks), amniocentesis (16 weeks), and term (Figure 18.1).

Down Syndrome (Trisomy 21)

This condition derives its name from Dr Langdon Down, who first described it in the Clinical Lecture Reports of the London Hospital in 1866. The chromosomal basis of Down syndrome was not established until 1959 by Lejeune and his colleagues in Paris.

Incidence

The overall birth incidence, when adjusted for the increasingly widespread impact of antenatal screening, is approximately 1 : 1000 in the United Kingdom, which has a national register. In the United States, the birth incidence has been estimated at approximately 1 : 800. In the United Kingdom, approximately 60% of Down syndrome cases are detected prenatally. There is a strong association between the incidence of Down syndrome and advancing maternal age (Table 18.4).

Clinical Features

These are summarized in Box 18.1. The most common finding in the newborn period is severe hypotonia. Usually the facial characteristics of upward sloping palpebral fissures, small ears, and protruding tongue (Figures 18.2 and 18.3) prompt rapid suspicion of the diagnosis, although this can be delayed in very small or premature babies. Single palmar creases are found in 50% of children with Down syndrome (Figure 18.4), in contrast to 2% to 3% of the general population. Congenital cardiac abnormalities are present in 40% to 45% of babies with Down syndrome, with the three most common lesions being atrioventricular canal defects, ventricular septal defects, and patent ductus arteriosus.

Natural History

Affected children show a broad range of intellectual ability with IQ scores ranging from 25 to 75. The average IQ of young adults is around 40 to 45. Social skills are relatively well-advanced and most children are happy and very

Table 18.1 Chromosome Abnormalities in Spontaneous Abortions (Percentage Values Relate to Total of Chromosomally Abnormal Abortuses)

Abnormality	Incidence (%)
Trisomy 13	2
Trisomy 16	15
Trisomy 18	3
Trisomy 21	5
Trisomy other	25
Monosomy X	20
Triploidy	15
Tetraploidy	5
Other	10

Table 18.2 Incidence of Chromosome Abnormalities in the Newborn

Abnormality	Incidence per 10,000 Births
Autosomes	
Trisomy 13	2
Trisomy 18	3
Trisomy 21	15
Sex Chromosomes	
FEMALE BIRTHS	
45,X	1–2
47,XXX	10
MALE BIRTHS	
47,XXY	10
47,XYY	10
Other unbalanced rearrangements	10
Balanced rearrangements	30
Total	90

affectionate. Adult height is usually around 150 cm. In the absence of a severe cardiac anomaly, which leads to early death in 15% to 20% of cases, average life expectancy is 50 to 60 years. Most affected adults develop Alzheimer disease in later life, possibly because of a gene dosage effect—the amyloid precursor protein gene is on chromosome 21. This gene is known to be implicated in some familial cases of Alzheimer disease (p. 243).

Table 18.3 Spontaneous Pregnancy Loss in Commonly Recognized Aneuploidy Syndromes

Disorder	Proportion Undergoing Spontaneous Pregnancy Loss (%)
Trisomy 13	95
Trisomy 18	95
Trisomy 21	80
Monosomy X	98

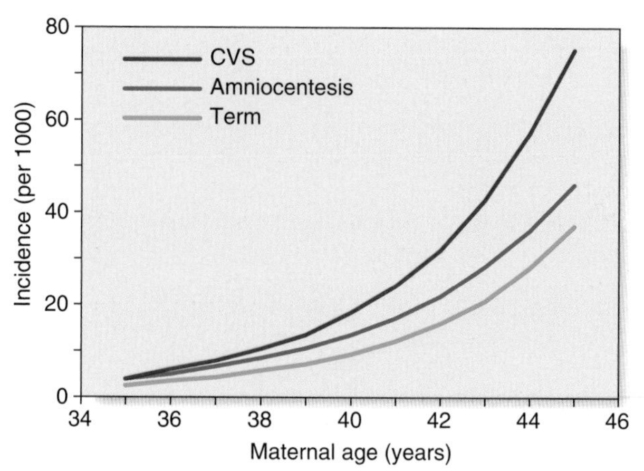

FIGURE 18.1 Approximate incidence of trisomy 21 at the time of chorionic villus sampling (CVS) (11–12 weeks), amniocentesis (16 weeks), and delivery. (Data from Hook EB, Cross PK, Jackson L, Pergament E, Brambati B 1988 Maternal age-specific rates of 47, 121 and other cytogenetic abnormalities diagnosed in the first trimester of pregnancy in chorionic villus biopsy specimens. Am J Hum Genet 42:797–807; and Cuckle HS, Wald NJ, Thompson SG 1987 Estimating a woman's risk of having a pregnancy associated with Down syndrome using her age and serum alpha-fetoprotein level. Br J Obstet Gynaecol 94:387–402.)

Chromosome Findings

These are listed in Table 18.5. In cases resulting from trisomy 21, the additional chromosome is maternal in origin in more than 90% of cases, and DNA studies have shown that this arises most commonly as a result of non-disjunction in maternal meiosis I (p. 39). Robertsonian translocations (p. 47) account for approximately 4% of all cases, in roughly one-third of which a parent is found to be a carrier. Children with mosaicism are often less severely affected than those with the full syndrome.

Table 18.4 Incidence of Down Syndrome in Relation to Maternal Age

Maternal Age at Delivery (Years)	Incidence of Down Syndrome
20	1 in 1500
25	1 in 1350
30	1 in 900
35	1 in 400
36	1 in 300
37	1 in 250
38	1 in 200
39	1 in 150
40	1 in 100
41	1 in 85
42	1 in 65
43	1 in 50
44	1 in 40
45	1 in 30

Adapted from Cuckle HS, Wald NJ, Thompson SG 1987 Estimating a woman's risk of having a pregnancy associated with Down syndrome using her age and serum alpha-fetoprotein level. Br J Obstet Gynaecol 94:387–402.

Box 18.1 Common Findings in Down Syndrome

Newborn period
Hypotonia, sleepy, excess nuchal skin

Craniofacial
Brachycephaly, epicanthic folds, protruding tongue, small ears, upward sloping palpebral fissures

Limbs
Single palmar crease, small middle phalanx of fifth finger, wide gap between first and second toes

Cardiac
Atrial and ventricular septal defects, common atrioventricular canal, patent ductus arteriosus

Other
Anal atresia, duodenal atresia, Hirschsprung disease, short stature, strabismus

Efforts have been made to correlate the various clinical features in trisomy Down syndrome with specific regions of chromosome 21, by studying children with partial trisomy for different regions. There is some support for a Down syndrome 'critical region' at the distal end of the long arm (21q22), because children with trisomy for this region

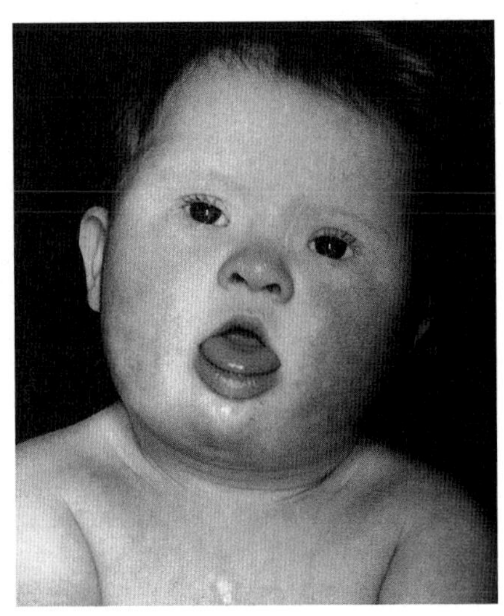

FIGURE 18.2 A child with Down syndrome.

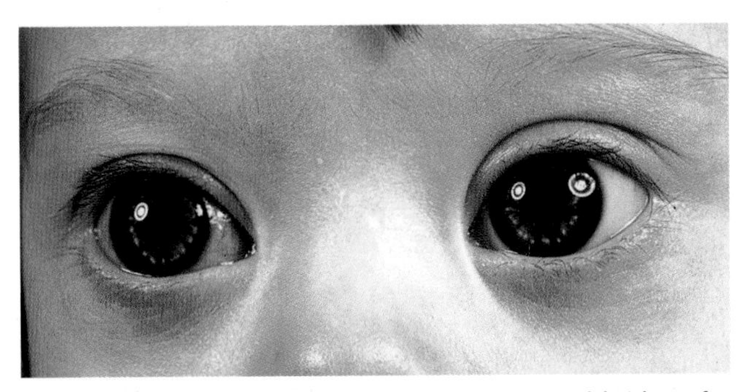

FIGURE 18.3 Close-up view of the eyes and nasal bridge of a child with Down syndrome showing upward sloping palpebral fissures, Brushfield spots, and bilateral epicanthic folds.

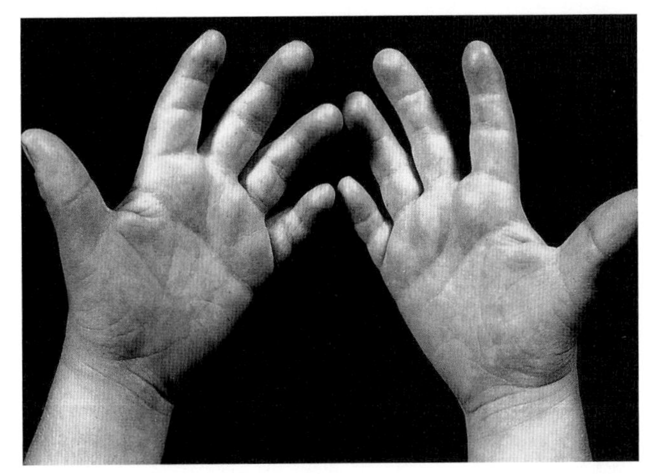

FIGURE 18.4 The hands of an adult with Down syndrome. Note the single palmar crease in the left hand plus bilateral short curved fifth fingers (clinodactyly).

alone usually have typical Down syndrome facial features. Chromosome 21 is a 'gene-poor' chromosome with a high ratio of AT to GC sequences (p. 69). At present the only reasonably well-established genotype-phenotype correlation in trisomy 21 is the high incidence of Alzheimer disease.

Recurrence Risk

For straightforward trisomy 21, the recurrence risk is related to maternal age (variable) and the simple fact that trisomy has already occurred (~1%). The combined recurrence risk is usually between 1 : 200 and 1 : 100. In translocation cases, similar figures apply if neither parent is a carrier. In familial translocation cases, the recurrence risks vary from around 1% to 3% for male carriers up to 10% to 15% for female carriers, with the exception of very rare carriers of a 21q21q translocation, for whom the recurrence risk is 100% (p. 48).

Prenatal diagnosis can be offered based on analysis of chorionic villi or cultured amniotic cells. Prenatal screening programs have been introduced based on the so-called triple or quadruple tests of maternal serum at 16 weeks' gestation (p. 328).

Patau Syndrome (Trisomy 13) and Edwards Syndrome (Trisomy 18)

These very severe conditions were first described in 1960 and share many features in common (Figures 18.5 and 18.6). The incidence for both is approximately 1 : 5000 and prognosis is very poor, with most infants dying during the

Table 18.5 Chromosome Abnormalities in Down Syndrome

Abnormality	Frequency (%)
Trisomy	95
Translocation	4
Mosaicism	1

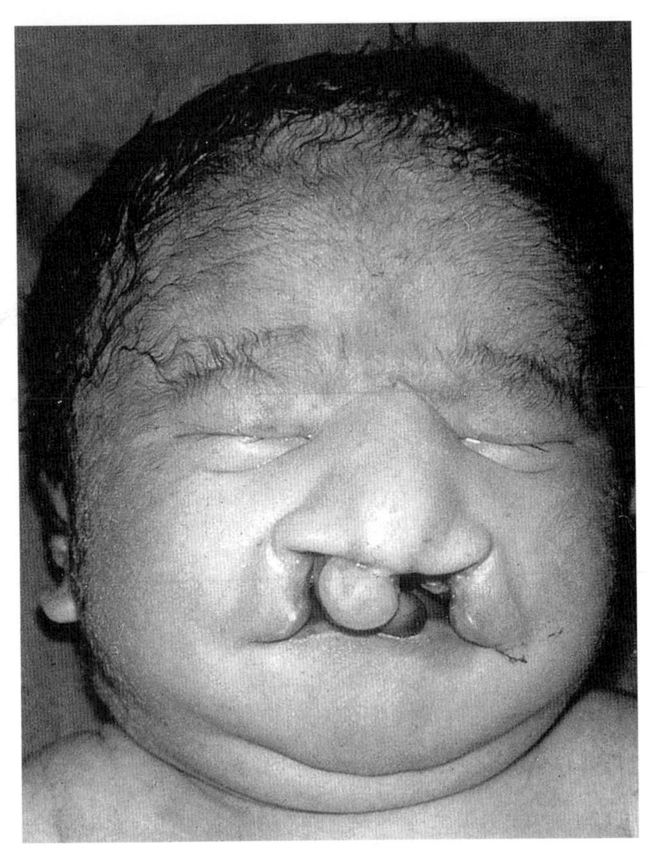

FIGURE 18.5 Facial view of a child with trisomy 13 showing severe bilateral cleft lip and palate.

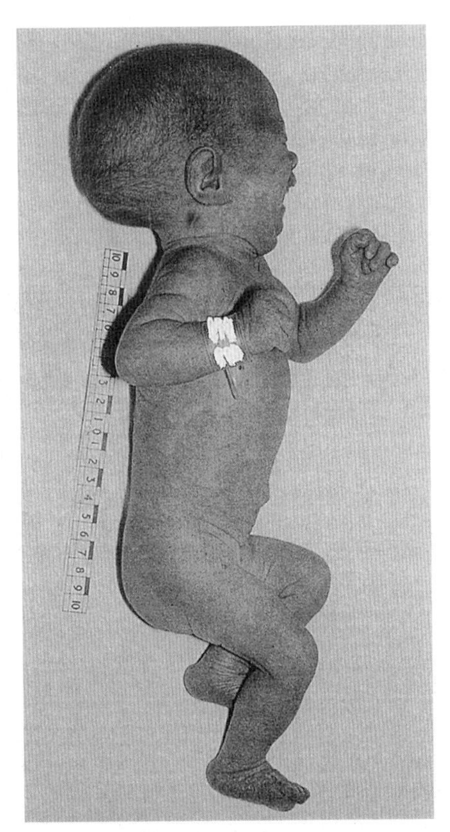

FIGURE 18.6 A baby with trisomy 18. Note the prominent occiput and tightly clenched hands.

first days or weeks of life, though most cases are detected prenatally, often leading to termination. In the unusual event of longer term survival, there are severe learning difficulties. Cardiac abnormalities occur in at least 90% of cases.

Chromosome analysis usually reveals straightforward trisomy. Both disorders occur more frequently with advanced maternal age, the additional chromosome being of maternal origin (see Table 3.4, p. 43). Approximately 10% of cases are caused by mosaicism or unbalanced rearrangements, particularly Robertsonian translocations in Patau syndrome.

Triploidy

Triploidy (69,XXX, 69,XXY, 69,XYY) is a relatively common finding in material cultured from spontaneous abortions, but is seen only rarely in a liveborn infant. Such a child almost always shows severe intrauterine growth retardation with relative preservation of head growth at the expense of a small thin trunk. Syndactyly involving the third and fourth fingers and/or the second and third toes is a common finding. Cases of triploidy resulting from a double paternal contribution usually miscarry in early to mid-pregnancy and are associated with partial hydatidiform changes in the placenta (p. 101). Cases with a double maternal contribution survive for longer but rarely beyond the early neonatal period.

Hypomelanosis of Ito

Several children with mosaicism for diploidy/triploidy have been identified. These can demonstrate the clinical picture seen in full triploidy but in a milder form. An alternative presentation occurs as the condition known as hypomelanosis of Ito. In this curious disorder, the skin shows alternating patterns of normally pigmented and depigmented streaks that correspond to the embryological developmental lines of the skin known as Blaschko's lines (see Figure 18.7). Most children with hypomelanosis of Ito have moderate learning difficulties and convulsions that can be particularly difficult to treat. There is increasing evidence that this clinical picture represents a non-specific embryological response to cell or tissue mosaicism. A similar pattern of skin pigmentation is sometimes seen in women with one of the rare X-linked dominant disorders (p. 117) with skin involvement, such as incontinentia pigmenti (see Figure 7.18, p. 118). Such women can be considered as being mosaic, as some cells express the normal gene, whereas others express only the mutant gene.

Disorders of the Sex Chromosomes

Klinefelter Syndrome (47,XXY)

First described clinically in 1942, this relatively common condition with an incidence of 1:1000 male live births was shown in 1959 to be due to the presence of an additional X chromosome.

FIGURE 18.7 Mosaic pattern of skin pigmentation on the arm of a child with hypomelanosis of Ito. (Reproduced with permission from Jenkins D, Martin K, Young ID 1993 Hypomelanosis of Ito associated with mosaicism for trisomy 7 and apparent 'pseudomosaicism' at amniocentesis. J Med Genet 1993; 30:783–784.)

Clinical Features

In childhood the presentation may be with clumsiness or mild learning difficulties, particularly in relation to verbal skills. The overall verbal IQ is reduced by 10 to 20 points below unaffected siblings and controls, and children can be rather self-obsessed in their behavior. Adults tend to be slightly taller than average with long lower limbs. Approximately 30% show moderately severe gynecomastia (breast enlargement) and all are infertile because of the absence of sperm in their semen *(azoospermia)*, with small, soft testes. Fertility has been achieved for a small number of affected males using the techniques of testicular sperm aspiration and intracytoplasmic sperm injection. There is an increased incidence of leg ulcers, osteoporosis, and carcinoma of the breast in adult life. Treatment with testosterone from puberty onward is beneficial for the development of secondary sexual characteristics and the long-term prevention of osteoporosis.

Chromosome Findings

Usually the karyotype shows an additional X chromosome. Molecular studies have shown that there is a roughly equal chance that this will have been inherited from the mother or from the father. The maternally derived cases are associated with advanced maternal age. A small proportion of cases show mosaicism (e.g., 46,XY/47,XXY). Rarely, a male with more than two X chromosomes can be encountered, for example 48,XXXY or 49,XXXXY. These individuals are usually quite severely retarded and also share physical characteristics with Klinefelter men, often to a more marked degree.

Turner Syndrome (45,X)

This condition was first described clinically in 1938. The absence of a Barr body, consistent with the presence of only one X chromosome, was noted in 1954 and cytogenetic confirmation was forthcoming in 1959. Although common at conception and in spontaneous abortions (see Table 18.1), the incidence in liveborn female infants is low, with estimates ranging from 1 : 5000 to 1 : 10,000.

Clinical Features

Presentation can be at any time from pregnancy to adult life. Increasingly, Turner syndrome is being detected during the second trimester as a result of routine ultrasonography, showing either generalized edema (hydrops) or swelling localized to the neck (nuchal cyst or thickened nuchal pad) (Figure 18.8). At birth many babies with Turner syndrome look entirely normal. Others show the residue of intrauterine edema with puffy extremities (Figure 18.9) and neck webbing. Other findings may include a low posterior hairline, increased carrying angles at the elbows, short fourth metacarpals, widely spaced nipples, and coarctation of the aorta, which is present in 15% of cases.

Intelligence in Turner syndrome is normal. However, studies have shown some differences in social cognition and higher order executive function skills according to whether the X chromosome was paternal or maternal in origin (p. 105). The two main medical problems are short stature

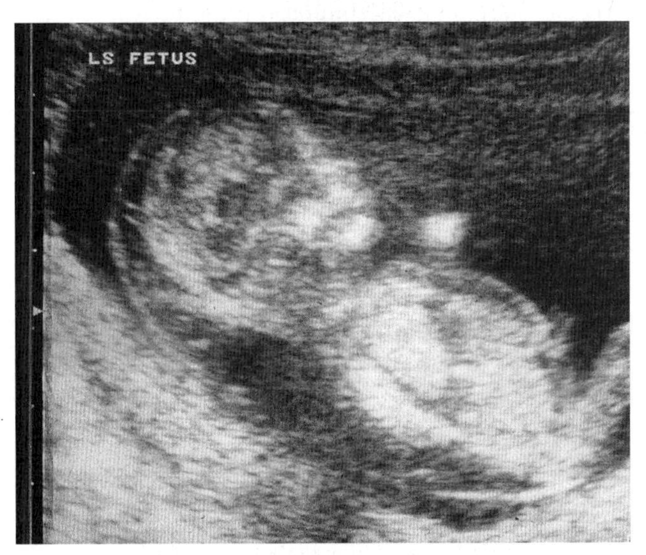

FIGURE 18.8 Ultrasonographic scan at 18 weeks' gestation showing hydrops fetalis. Note the halo of fluid surrounding the fetus. (Courtesy Dr. D. Rose, City Hospital, Nottingham, UK.)

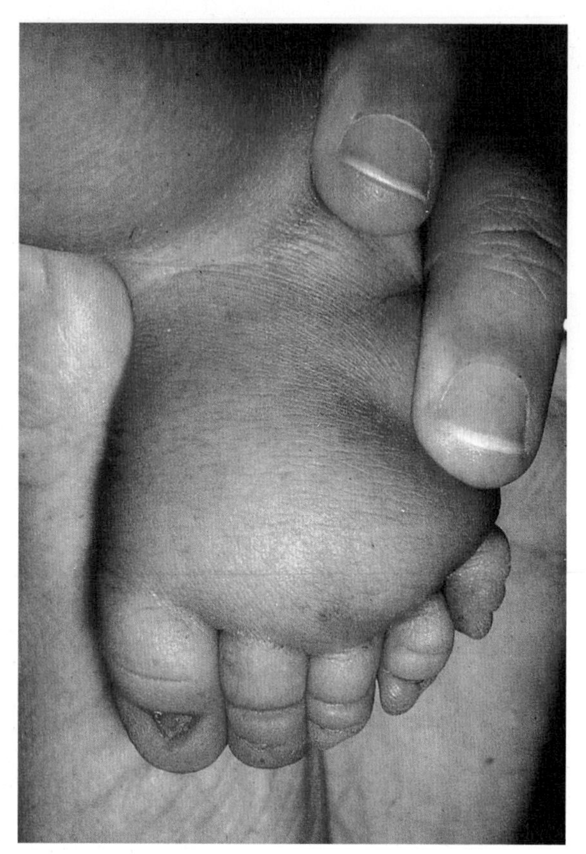

FIGURE 18.9 The foot of an infant with Turner syndrome showing edema and small nails.

and ovarian failure. The short stature becomes apparent by mid-childhood, and without growth hormone treatment the average adult height is 145 cm. This short stature is due, at least in part, to haploinsufficiency for the *SHOX* gene, which is located in the pseudoautosomal region (p. 118). Ovarian failure commences during the second half of intrauterine life and almost invariably leads to primary amenorrhea and infertility. Estrogen replacement therapy should be initiated at adolescence for the development of secondary sexual characteristics and long-term prevention of osteoporosis. In vitro fertilization using donor eggs offers the prospect of pregnancy for women with Turner syndrome.

Chromosome Findings

These are summarized in Table 18.6. The most common finding is 45,X (sometimes erroneously referred to as 45,XO). In 80% of cases, it arises through loss of a sex chromosome (X or Y) *paternal* meiosis. In a significant proportion of cases, there is chromosome mosaicism and those with a normal cell line (46,XX) have a chance of being fertile. Some cases with a 46,XY cell line are phenotypically male, and all cases with some Y-chromosome material in their second cell line must be investigated for possible gonadal dysgenesis—intracellular male gonads can occasionally become malignant and require surgical removal.

XXX Females

Birth surveys have shown that approximately 0.1% of all females have a 47,XXX karyotype. These women usually have no physical abnormalities, but can show a mild reduction of between 10 and 20 points in intellectual skills and sometimes quite oppositional behavior. This is rarely of sufficient severity to require special education. Studies have shown that the additional X chromosome is of *maternal* origin in 95% of cases and usually arises from an error in meiosis I. Adults are usually fertile and have children with normal karyotypes.

As with males who have more than two X chromosomes, women with more than three X chromosomes show a high incidence of learning difficulties, the severity being directly related to the number of X chromosomes.

XYY Males

This condition shows an incidence of about 1 : 1000 in males in newborn surveys but is found in 2% to 3% of males who are in institutions because of learning difficulties or antisocial criminal behavior. However, it is important to stress that most 47,XYY men have neither learning difficulty nor a criminal record, although they can show emotional immaturity and impulsive behavior. Fertility is normal.

Physical appearance is normal and stature is usually above average. Intelligence is mildly impaired, with an overall IQ score of 10 to 20 points below a control sample. The additional Y chromosome must arise either as a result of non-disjunction in *paternal* meiosis II or as a post-zygotic event.

Fragile X Syndrome

This condition, which could equally well be classified as a single gene disorder rather than a chromosome abnormality, has the unique distinction of being both the most common inherited cause of learning difficulties and the first disorder in which a dynamic mutation (triplet repeat expansion) was identified (p. 24) in 1991. It affects approximately 1 : 5000 males and accounts for 4% to 8% of all males with learning difficulties. Martin and Bell described the condition in the 1940s before the chromosome era, and it has also been known as Martin-Bell syndrome. The chromosomal abnormality was first described in 1969 but the significance not fully realized until 1977.

Table 18.6 Chromosome Findings in Turner Syndrome

Karyotype	Frequency (%)
Monosomy X—45,X	50
Mosaicism (e.g., 45,X/46,XX)	20
Isochromosome—46,X,i(Xq)	15
Ring—46,X,r(X)	5
Deletion—46,X,del(Xp)	5
Other	5

FIGURE 18.10 A family affected by fragile X syndrome. Two sisters, both carriers of a small *FRAXA* mutation inherited from their father, have had affected sons with different degrees of learning difficulty.

Clinical Features

Older boys and adult males usually have a recognizable facial appearance with high forehead, large ears, long face, and prominent jaw (Figure 18.10). After puberty most affected males have large testes (macro-orchidism). There may also be evidence of connective tissue weakness, with hyperextensible joints, stretch marks on the skin (striae) and mitral valve prolapse. The learning difficulties are moderate to severe and many show autistic features and/or hyperactive behavior. Speech tends to be halting and repetitive. Female carriers can show some of the facial features, and approximately 50% of women with the full mutation show mild-to-moderate learning difficulties.

The Fragile X Chromosome

The fragile X syndrome takes its name from the appearance of the X chromosome, which shows a *fragile site* close to the telomere at the end of the long arm at Xq27.3 (Figure 18.11). A fragile site is a non-staining gap usually involving both chromatids at a point at which the chromosome is liable to break. In this condition, detection of the fragile site involves the use of special culture techniques such as folate or thymidine depletion, which can result in the fragile site being detectable in up to 50% of cells from affected males. Demonstration of the fragile site in female carriers is much more difficult and cytogenetic studies alone are not a reliable means of carrier detection, in that, although a positive result confirms carrier status, the absence of the fragile site does not exclude a woman from being a carrier.

The Molecular Defect

The fragile X locus is known as *FRAXA* and the mutation consists of an increase in the size of a region in the 5′-untranslated region of the fragile X learning difficulties (*FMR-1*) gene. This region contains a long CGG trinucleotide repeat sequence. In the DNA of a normal person, there are between 10 and 50 copies of this triplet repeat and these are inherited in a stable fashion. However, a small increase to between 59 and 200 renders this repeat sequence unstable, a condition in which it is referred to as a **premutation**. Alleles of 51 to 58 are referred to as **intermediate**.

A man who carries a premutation is known as a 'normal transmitting male', although it has been recognized that these premutation carriers are at increased risk of a late-onset neurological condition named 'fragile X tremor/ataxia syndrome' (FXTAS). All of his daughters will inherit the premutation and have normal intelligence, but they are also at small risk of FXTAS in later years. When they have sons, there is a significant risk that the premutation will undergo a further increase in size during meiosis, and if this exceeds 200 CGG triplets, it becomes a full mutation.

The full mutation is unstable not only during female meiosis but also in somatic mitotic divisions. Consequently, in an affected male gel electrophoresis shows a 'smear' of DNA consisting of many different-sized alleles rather than a single band (Figure 18.12). Note that a normal allele and premutation can be identified by polymerase chain reaction (PCR), whereas Southern blotting is necessary to detect full mutations as the long GCC expansion is often refractory to PCR amplification. At the molecular level, a full mutation suppresses transcription of the *FMR-1* gene by hypermethylation, and this in turn is thought to be responsible for the

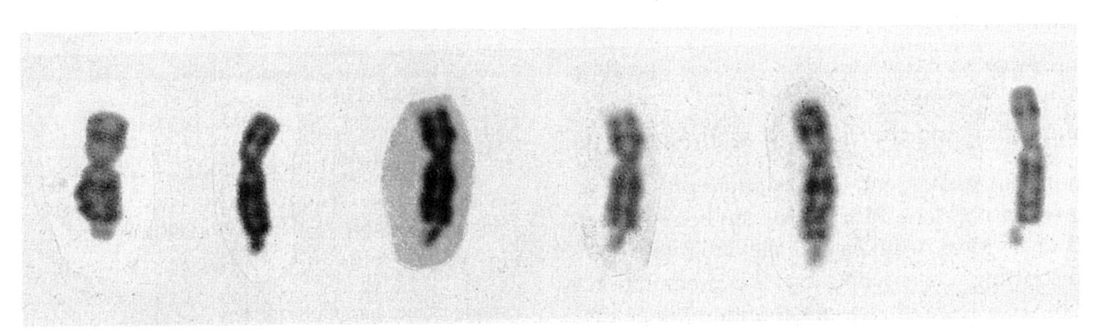

FIGURE 18.11 X chromosome from several males with fragile X syndrome. (Courtesy Ashley Wilkinson, City Hospital, Nottingham, UK.)

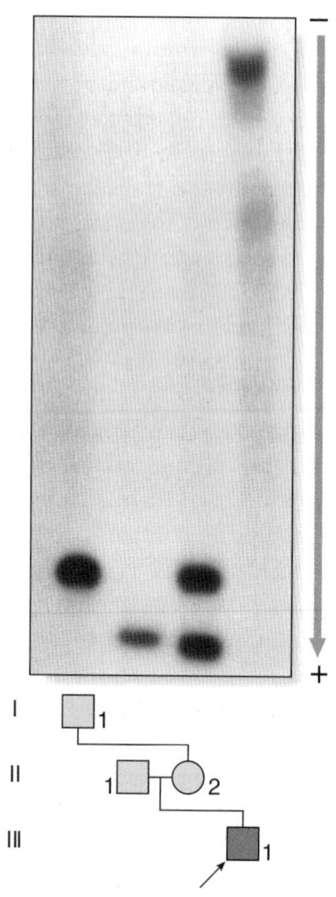

FIGURE 18.12 Southern blot of DNA from a family showing expansion of the CGG triplet repeat being passed from a normal transmitting male through his obligate carrier daughter to her son with fragile X learning difficulties. (Courtesy Dr. G. Taylor, St. James's Hospital, Leeds, UK.)

clinical features seen in males, and in some females with a large expansion (Table 18.7). The *FMR-1* gene contains 17 exons encoding a cytoplasmic protein that plays a crucial role in the development and function of cerebral neurons. The FMR-1 protein can be detected in blood using specific monoclonal antibodies.

Another fragile site adjacent to *FRAXA* has been identified at Xq28. This is known as *FRAXE*. The expansion mutations at *FRAXE* also involve CGG triplet repeats and occur much less frequently than *FRAXA* mutations. Some males with these mutations have mild learning difficulties, whereas others are just as severely affected as men with *FRAXA*. *FRAXE* may show up as a fragile site cytogenetically but the PCR test is separate. A third fragile site, *FRAXF*, has been identified close to *FRAXA* and *FRAXE*. This does not seem to cause any clinical abnormality.

Genetic Counseling and the Fragile X Syndrome

This common cause of learning difficulties presents a major counseling problem. Inheritance can be regarded as modified or atypical X-linked. All of the daughters of a normal transmitting male will carry the premutation. Their male offspring are at risk of inheriting either the premutation or a full mutation. This risk is dependent on the size of the premutation in the mother, with mutations greater than 100 CGG repeats almost invariably increasing in size to become full mutations.

For a woman who carries a full mutation there is a 50% risk that each of her sons will be affected with the full syndrome and that each of her daughters will inherit the full mutation. Because approximately 50% of females with the full mutation have mild learning difficulties, the risk that a female carrier of a full mutation will have a daughter with learning difficulties equals $\frac{1}{2} \times \frac{1}{2}$ (i.e., $\frac{1}{4}$). Prenatal diagnosis can be offered based on analysis of DNA from chorionic villi, but in the event of a female fetus with a full mutation accurate prediction of phenotype cannot be made.

The fragile X syndrome is a condition for which population screening could be offered, either among selected high-risk groups such as males with learning difficulties or on a widespread general population basis. Such programs will have to surmount major ethical, financial, and logistical concerns if they are to achieve widespread acceptance (p. 317).

Chromosome Deletion and Microdeletion Syndromes

Microscopically visible deletions of the terminal portions of chromosomes 4 and 5 cause the Wolf-Hirschhorn (4p–) (Figure 18.13) and cri-du-chat (5p–) (Figure 18.14) syndromes respectively. In both conditions severe learning

Table 18.7 Fragile X Syndrome: Genotype-Phenotype Correlations

Number of Triplet Repeats (Normal Range 10–50)	Fragile Site	Intelligence Detectable
Males		
51–58 (intermediate alleles)		
59–200 (premutation)	No	Normal (normal transmitting male)
200–2000 (full mutation)	Yes (in up to 50% of cells)	Moderate-to-severe learning difficulties
Females		
51–58 (intermediate alleles)		
59–200 (premutation)	No	Normal
200–2000 (full mutation)	Yes (usually <10% of cells)	50% normal, 50% mild learning difficulties

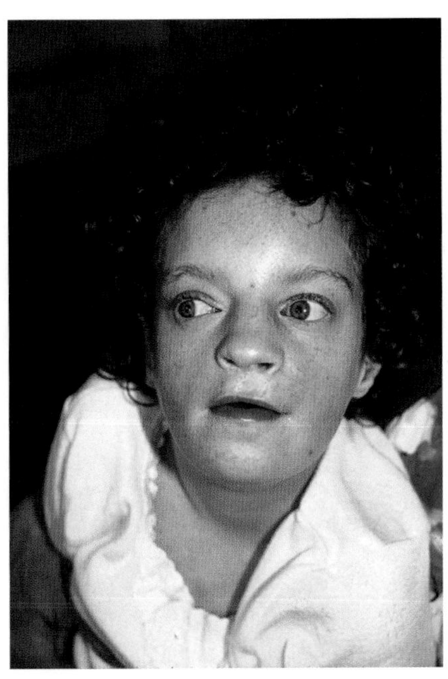

FIGURE 18.13 A child with deletion 4p syndrome—Wolf-Hirschhorn syndrome.

difficulties is usual, often with failure to thrive. However, there is considerable variability, particularly in Wolf-Hirschhorn syndrome, and no clear correlation of the phenotype with the precise loss of chromosomal material. Cri-du-chat syndrome derives its name from the characteristic cat-like cry of affected neonates—a consequence of underdevelopment of the larynx. Both conditions are rare, with estimated incidences of approximately 1:50,000 births. Not all cases have cytogenetically visible chromosome deletions, and clinicians can ask for specific fluorescent in-situ hybridization (FISH; see p. 34) analysis if microarray-CGH (pp. 36, 281) is not available.

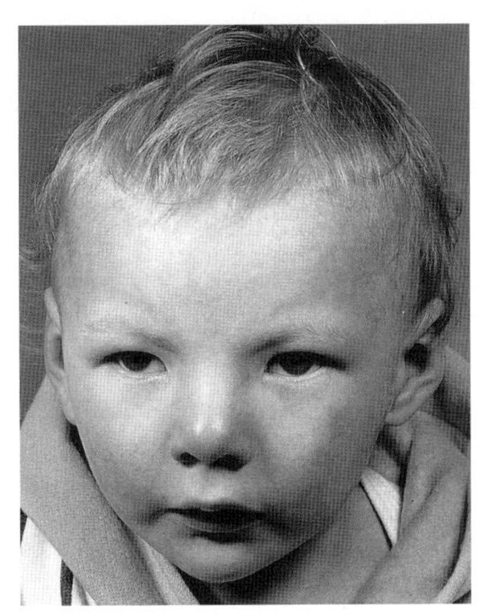

FIGURE 18.14 Facial view of a 2-year-old boy with cri-du-chat syndrome.

Microdeletions

Through high-resolution prometaphase banding (p. 33) and FISH (p. 34), several previously unexplained syndromes have been shown to be due to submicroscopic or *micro*-deletions. Some microdeletions involve loss of only a few genes at closely adjacent loci, resulting in **contiguous gene syndromes**. For example, several boys with Duchenne muscular dystrophy (DMD) have been described who also have other X-linked disorders, such as retinitis pigmentosa and glycerol kinase deficiency. The loci for these disorders are very close to the DMD locus on Xp21. Examples of well known microdeletion syndromes are given in Table 18.8, all of them relatively rare.

Microarray-CGH

The 1990s witnessed the development of FISH-based analysis of all chromosome telomeres using subtelomeric probes. This led to the diagnosis of some cases of learning difficulty/dysmorphic patients, not detected by multiplex ligation-dependent probe amplification (p. 66). In recent few years this has been rapidly superseded by extensive microarray comparative genomic hybridization (CGH) testing (p. 36) and a number of new microdeletion (and to a lesser extent microduplication) syndromes have emerged. At high resolution, this new technology is yielding significant results in about 20% of cases of well selected, previously unknown dysmorphic patients with developmental delay/learning disability. This compares to a positive pick-up rate of 4% to 5% from standard karyotyping on patients considered likely to have a chromosome disorder. Examples of these new and emerging syndromes are shown in the following section.

Lessons from Microdeletion Syndromes

Retinoblastoma

It was originally observed that approximately 5% of children presenting with retinoblastoma had other abnormalities, including learning difficulties. In some of these, a constitutional interstitial deletion of a region of chromosome 13q

Table 18.8 Microdeletion Syndromes

Syndrome	Chromosome
Deletion 1p36	1
Williams	7
Langer-Giedion	8
WAGR	11
Angelman	15
Prader-Willi	15
Rubinstein-Taybi	16
Miller-Dieker	17
Smith-Magenis	17
DiGeorge/Sedláčková/velocardiofacial	22

WAGR, Wilms' tumor, aniridia, genitourinary malformations, and retardation of growth and development.

was identified. The smallest region of overlap was 13q14, which was subsequently shown to be the position of the locus for the autosomal dominant form of retinoblastoma due to mutations in the *RB1* gene (p. 216).

Wilms' Tumor

Some children with the rare renal embryonal neoplasm known as Wilms' tumor (or hypernephroma) also have aniridia, genitourinary abnormalities, and retardation of growth and development. This combination is referred to as the WAGR syndrome. Chromosome analysis in these children often reveals an interstitial deletion of 11p13 (Figure 18.15). The deletion genes include *PAX6*, which is responsible for the aniridia (Figure 18.16). Confirmation is made by FISH probe analysis (or direct gene mutation analysis in cases of pure aniridia). Loss of the *WT1* gene causes the development of Wilms' tumor (see also p. 219). Knowing this, it can now be predicted whether a newly diagnosed child with deletion 11p13 is at high risk of developing a Wilms' tumor, using a separate FISH analysis at the *WT1* locus. It is important to note that, however, that the genetics of Wilms' tumor is complex, with other loci sometimes involved.

Angelman and Prader-Willi Syndromes

These two conditions have special place in medical genetics as paradigms for genomic imprinting. Children with Angelman syndrome (see Figure 7.24; p. 123) have inappropriate laughter, convulsions, poor coordination (ataxia), and severe learning difficulties. Children with Prader-Willi syndrome (see Figure 7.22; p. 122) are extremely floppy (hypotonic) with poor feeding in infancy, and later develop hyperphagia and obesity, with mild-to-moderate learning difficulties. A large proportion of children with these disorders have a microdeletion involving 15q11-13.

In contrast, a deletion occurring at the same region on the *maternally* inherited chromosome 15 causes Angelman syndrome. Non-deletion cases also exist and are often due to uniparental disomy (p. 121), with both number 15 chromosomes being *paternal* in origin in Angelman syndrome, and *maternal* in origin in Prader-Willi syndrome. These parent-of-origin effects are explained by imprinting (see Figure 7.23; p. 123).

DiGeorge/Sedláčková/Velocardiofacial Syndrome

DiGeorge syndrome affects approximately 1:4000 births, is usually sporadic, and is characterized by heart malformations (particularly those involving the cardiac outflow tract), thymic and parathyroid hypoplasia, cleft palate and typical facies. The molecular defect is a 3-Mb microdeletion on chromosome 22 (22q11.2). Dr Eva Sedláčková from Prague reported a large series of children with a congenitally short palate in 1955, 10 years earlier than DiGeorge, and these patients clearly had the same condition. A similar phenotype was described by Shprintzen and referred to as velocardiofacial syndrome. Because of the confusion of eponyms and other terms given to this condition over the years, 'deletion 22q11 syndrome' now has the most widespread acceptance (at the molecular level the deleted DNA segment is still called the DiGeorge Critical Region). Figure 18.17 shows individuals with deletion 22q11.2 at different ages. Because it is the most common of the microdeletion

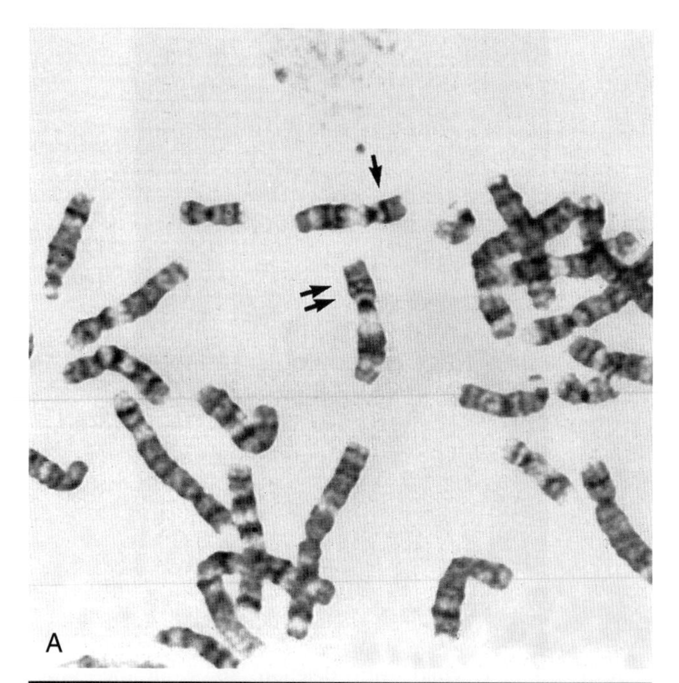

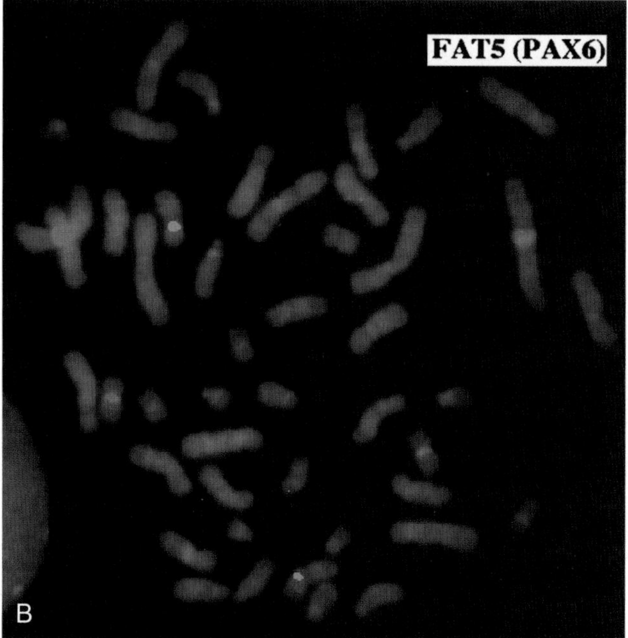

FIGURE 18.15 **A**, Metaphase spread showing the number 11 chromosomes (*double arrows*). The chromosome indicated by the *single arrow* has an interstitial deletion in the short arm. See Figures 18.11 and 18.12. (Courtesy Meg Heath, City Hospital, Nottingham, UK) **B**, FISH showing failure of a *PAX6* locus specific probe (*red*) to hybridize to the deleted number 11 chromosome shown in **A** from a child with WAGR syndrome. The *green* probe acts as a marker for the centromere of each number 11 chromosome. (Courtesy Dr. John Crolla, Salisbury and Dr. Veronica van Heyningen, Edinburgh, UK.)

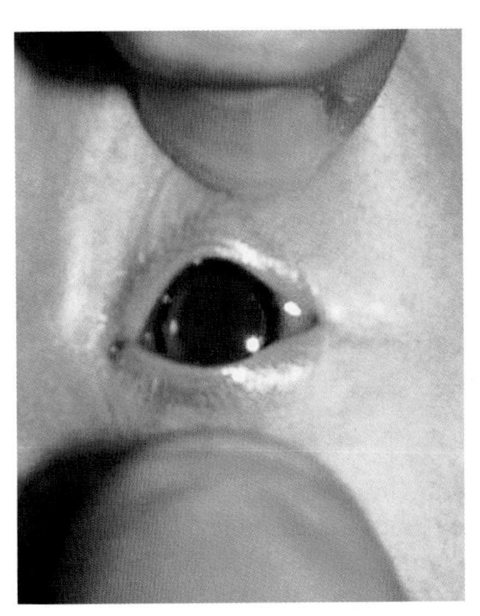

FIGURE 18.16 A baby with deletion 11p13 presenting with aniridia on routine neonatal examination.

syndromes, it has been intensely researched. It is variable and many affected individuals are able to reproduce, so the condition follows autosomal dominant inheritance in some families. The 3 Mb deletion occurs because this region is flanked by two identical sequences of DNA, known as low-copy repeats (LCRs), of the type that occur frequently throughout the genome. At meiosis the chromosomes can be 'confused' when they align, such that the downstream DNA sequence aligns with the upstream. If recombination occurs between these two flanking regions, a deletion of 3 Mb results on one chromosome 22. It is possible that the phenotypic features may be due largely to haploinsufficiency for the *TBX1* gene that lies within the region.

Those diagnosed should be investigated for cardiac malformations, calcium and parathyroid status, immune function, and renal anomalies. About half have short stature and a small proportion of these have partial growth hormone deficiency. About 25% have schizophrenia-like episodes in adult life.

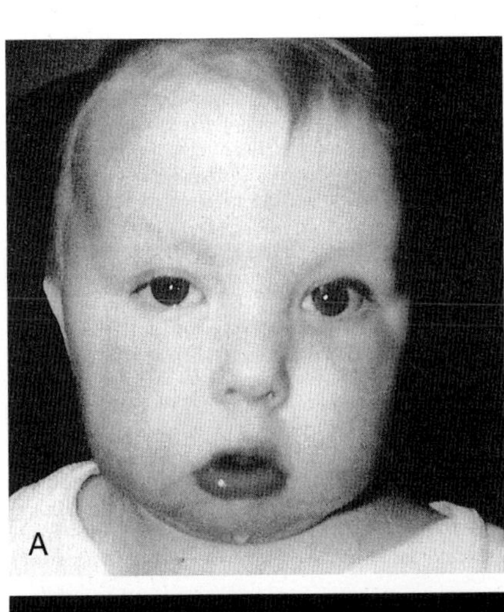

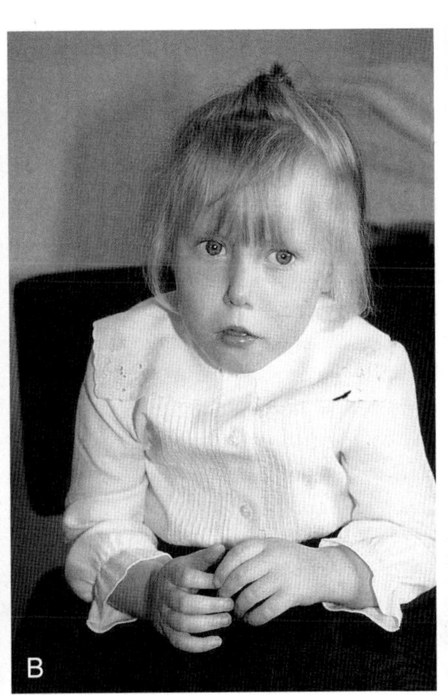

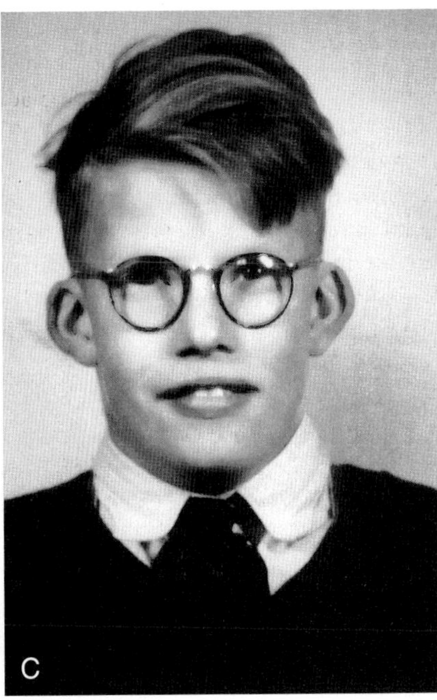

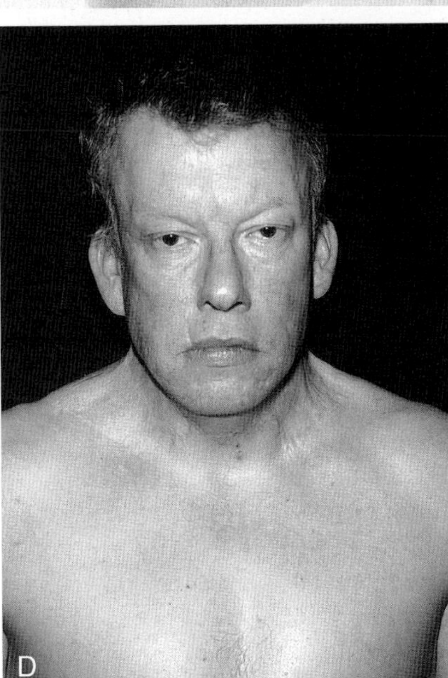

FIGURE 18.17 Deletion 22q11 (DiGeorge/Sedláčková/velocardiofacial) syndrome. **A,** A young infant. **B,** A young child. **C,** An older child. **D,** The same individual (shown in C) as an adult age 49 years.

Duplication 22q11.2

The misaligned pairing at meiosis, of the LCRs that flank the 3 Mb region at 22q11.2 that causes DiGeorge syndrome, predicts that gametes *duplicated* for this DNA segment would be present in equal numbers. However, duplication 22q11.2 syndrome is seldom seen in clinical practice, suggesting that it might be subclinical in its effects.

Some patients have been reported but there is no consistent phenotype. Some cases bear similarity to the *deletion* 22q11.2 phenotype, but marked variability occurs. The problems range from isolated mild learning difficulties to multiple abnormalities with non-specific dysmorphic features, congenital heart disease, cleft palate, hearing loss, and postnatal growth deficiency. More cases may be diagnosed using microarray-CGH.

Williams Syndrome

Williams syndrome occurs because of a microdeletion at chromosome 7q11 and diagnosis can be confirmed by FISH. The clinical phenotype was first reported by Williams in 1961 and later expanded by Beuren (hence, sometimes, Williams-Beuren syndrome). Hypocalcaemia is a variable feature in childhood and sometimes persists, whilst supravalvular aortic stenosis (SVAS) and peripheral pulmonary artery stenosis are congenital abnormalities of the great vessels. Haploinsufficiency at 7q11 leads to loss of one copy of the gene that encodes elastin, a component of connective tissue. This is probably the key factor causing SVAS and the vascular problems that are more common in later life. Individuals have a characteristic appearance (Figure 18.18) with mild short stature, a full lower lip, and sloping shoulders. Equally characteristic is their behavior. They are typically very outgoing in childhood—having a 'cocktail party manner'—but become withdrawn and sensitive as adults. All are intellectually impaired to the extent that they cannot lead independent lives, and the majority do not reproduce, although parent-child transmission has been reported.

Smith-Magenis Syndrome

This microdeletion syndrome is due to loss of chromosome material at 17p.11.2, often visible cytogenetically. As with DiGeorge syndrome, the mechanism of deletion in many cases involves homologous recombination between flanking LCRs. The physical characteristics are not highly distinctive (Figure 18.19), but congenital heart disease occurs in one-third, scoliosis develops in late childhood in more than half, and hearing impairment in about two-thirds. The syndrome is most likely to be recognized by the behavioral characteristics: as children, patients exhibit self-harming (head-banging, pulling out nails, and inserting objects into orifices), a persistently disturbed sleep pattern, and characteristic 'self-hugging'. Some degree of learning difficulty is the norm. The sleep pattern can often be managed by judicious use of melatonin.

Deletion 1p36 Syndrome

A new microdeletion syndrome that emerged through improved cytogenetic techniques and the use of FISH in the 1990s is 1p36 syndrome which, in keeping with today's approach to nomenclature, has no eponym. The features are hypotonia, microcephaly, growth delay, severe learning difficulties, epilepsy (including infantile spasms), characteristically straight eyebrows with slightly deep-set eyes, and midface hypoplasia (Figure 18.20). Some cases developed dilated cardiomyopathy.

Deletion 9q34 Syndrome

Another of the relative new microdeletion syndromes, this was first reported as a condition featuring significant learning difficulties, hypotonia, obesity, brachycephaly, arched eyebrows, synophrys, anteverted nostrils, prognathism, sleep disturbances, and behavioural problems. Many patients have severe speech delay and not all manifest obesity. The case pictured here (Figure 18.21) bears a passing resemblance to Angelman syndrome. Some patients with the phenotypic features but no microdeletion have been shown to have mutations in the Euchromatin histone methyl transferase 1 *(EHMT1)* gene, which lies within the region. The syndrome might therefore be mainly due to haploinsufficiency for this gene.

Deletion 17q21.31 Syndrome

This new condition has a prevalence of approximately 1:16,000 and is probably significantly underdiagnosed. The main features are severe developmental delay, hypotonia, and characteristic facial dysmorphisms including a long face with a high forehead and tubular or pear-shaped nose, a bulbous nasal tip, large ears, and everted lower lip (Figure 18.22). Individuals tend to be friendly. Other clinically important features include epilepsy, heart defects, kidney anomalies, and long slender fingers.

Deletion 1q21.1 Syndrome

This condition was first identified in three individuals from a cohort of 505 with congenital heart disease. The phenotype is broad and includes mild-moderate mental retardation, small head size, growth retardation, heart defects, cataracts, hand deformities and skeletal problems, learning disabilities, seizures, and autism. On the other hand, however, some individuals with the aberration are only slightly affected, or apparently unaffected. A mother and her child, both with the deletion, are shown in Figure 18.23. Variable penetrance and lack of highly distinctive features make genetic counseling for this aberration highly problematic. Some experts regard it as a susceptibility locus rather than a clinically distinct syndrome.

Chromosome disorders and behavioral phenotypes. The distinctive behavior of children with Williams syndrome—their outgoing 'cocktail party manner'—has

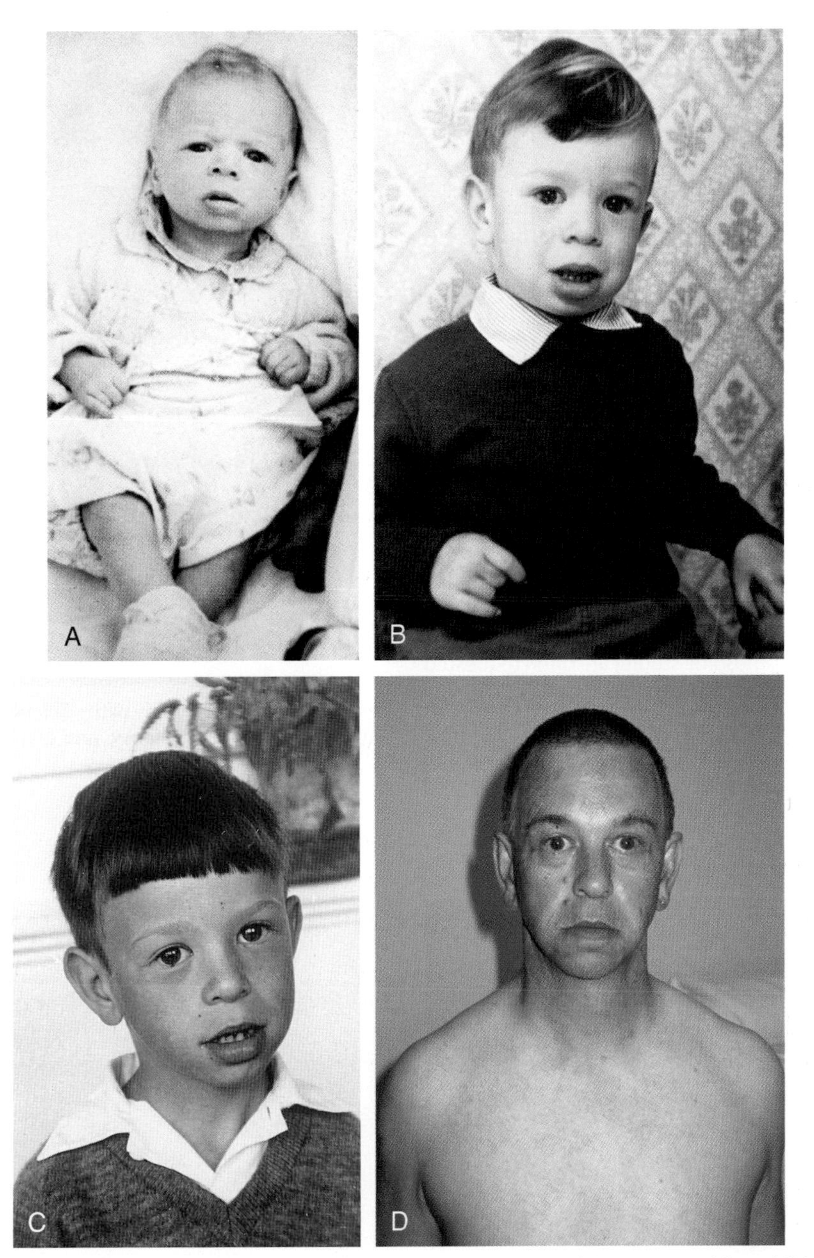

FIGURE 18.18 A person with Williams syndrome as a baby **(A)**, a young child **(B)**, an older child **(C)**, and in his early forties **(D)**.

been recognized as part of the condition for a long time. As the microdeletion conditions have emerged, it has been increasingly clear that patterns of behavior can reliably be attributed to certain disorders. This is very striking in Smith-Magenis syndrome, but also apparent to a lesser extent in deletion 22q11, cri-du-chat, and Angelman and Prader-Willi syndromes. It is also apparent in the aneuploidies (Down and Klinefelter syndromes), as well as in 47,XXX and 47,XYY and fragile X syndromes. Behavioral phenotypes have therefore become an area of considerable interest to clinical scientists and the observations lend support to the belief that behavior, to some extent at least, is genetically determined. In studying chromosome disorders we are of course looking at genetically abnormal situations, and from this we cannot necessarily extrapolate directly to 'normal' situations. For the latter, twin studies have provided substantial and valuable information. This field of study remains complex and understandably controversial. However, most now accept that behavior is a complex interaction of genetic background, physical influences during early development (e.g., fetal well-being), nurturing experiences, family size, culture, and belief systems.

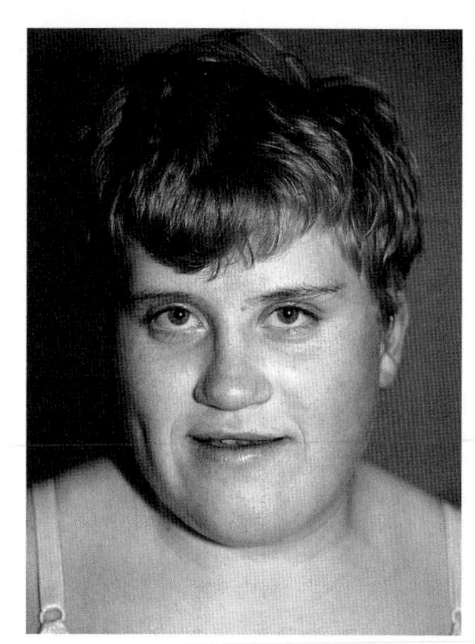

FIGURE 18.19 A young person with Smith-Magenis syndrome; the facial features are not highly distinctive, but the philtrum is usually short. As babies, chromosome studies are often requested because the possibility of Down syndrome is raised.

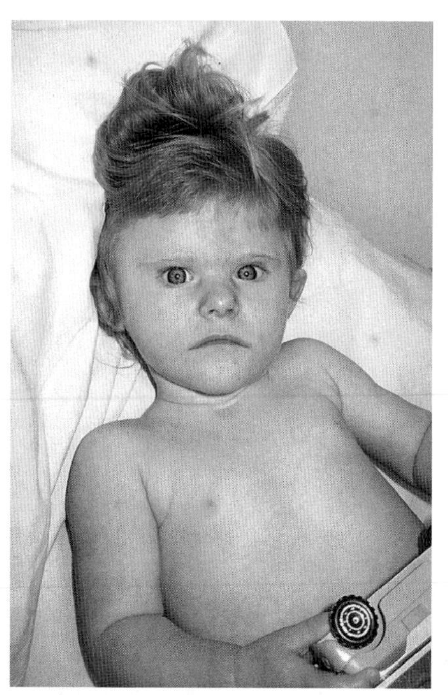

FIGURE 18.20 A child with deletion 1p36 syndrome—very straight eyebrows, epilepsy, and learning difficulties.

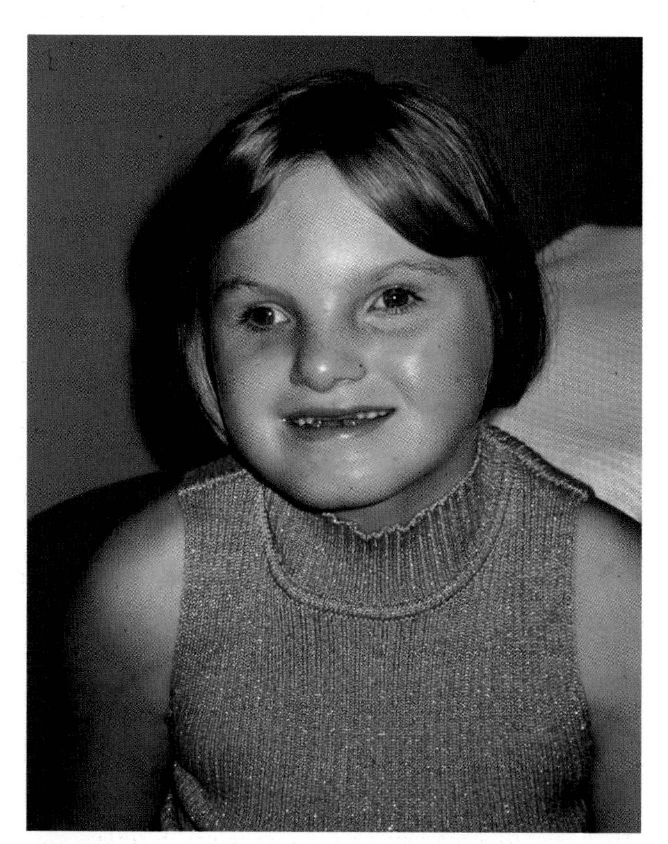

FIGURE 18.21 A child with deletion 9q34 syndrome. She has arched eyebrows, narrow upslanting palpebral fissures, brachycephaly, prognathism, and severe learning difficulties. She was initially investigated for possible Angelman syndrome.

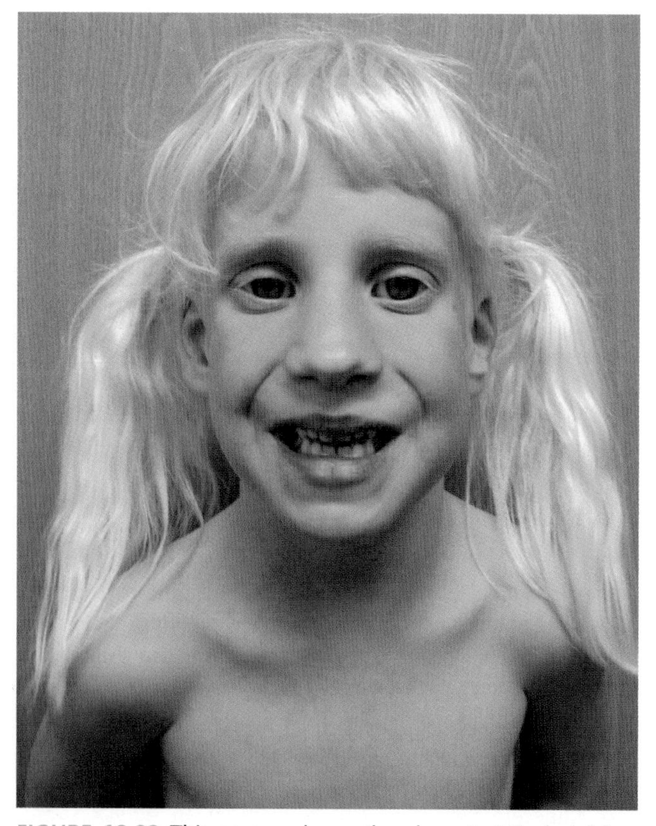

FIGURE 18.22 This person shows the characteristic facial features of deletion 17q21.31 syndrome. The face is long and the nose somewhat tubular or pear-shaped, and the nasal tip bulbous. There is developmental delay. (Courtesy Dr. David Koolen, Nijmegen, Netherlands.)

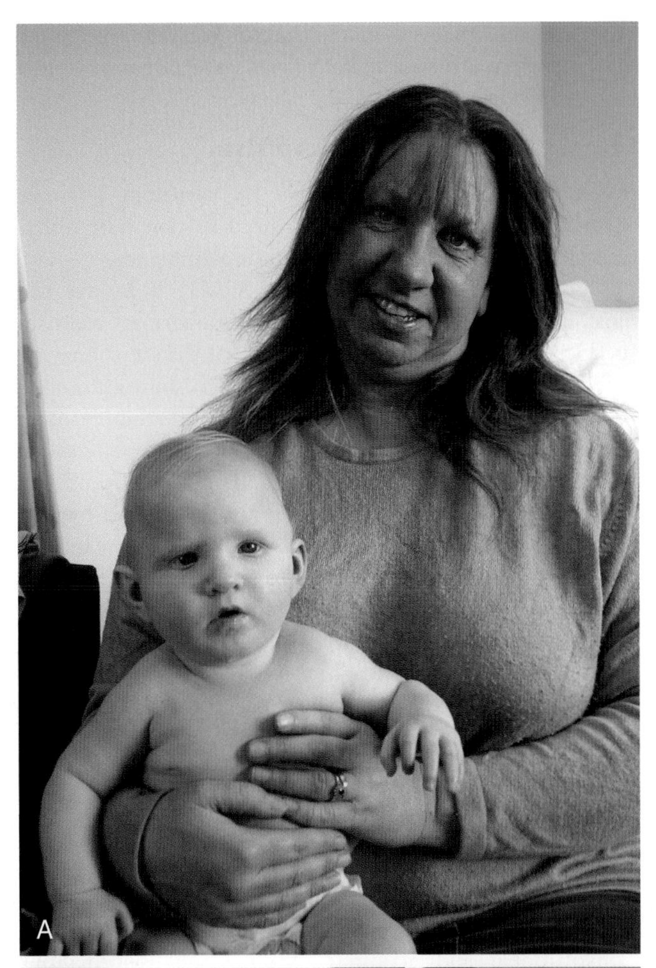

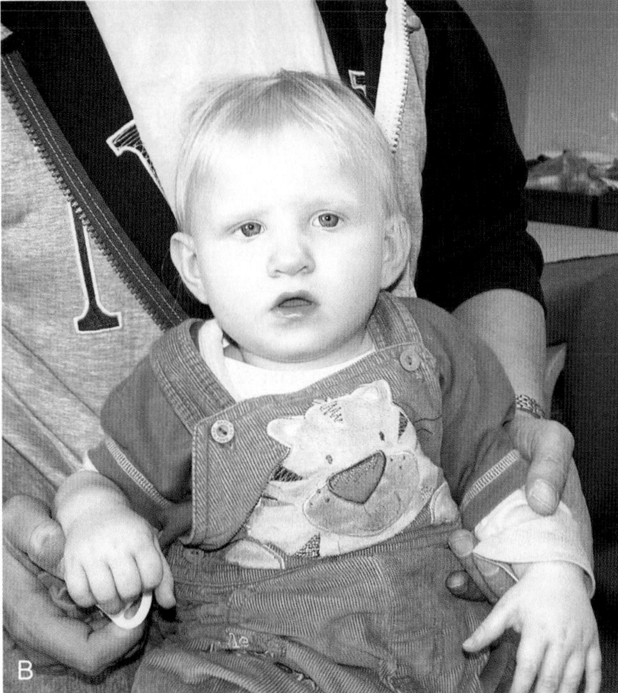

FIGURE 18.23 **A,** This mother and child have deletion 1p21.1 syndrome. They bear a resemblance to each other and there is evidence of mild development delay and small head size. **B,** The same child nearly a year after the first picture.

Box 18.2 Disorders of Sexual Differentiation and Development

Seminiferous tubule dysgenesis (Klinefelter syndrome)
47,XXY, 48,XXXY, 48,XXYY, 49,XXXXY
Ovarian dysgenesis (Turner syndrome)
45,X, 46,X,i(Xq), 46,X,del(Xp), 46,X,r(X)
True hermaphroditism
46,XX with Y-derived sequences
46,XX/46,XY chimerism
Male pseudohermaphroditism
Androgen insensitivity
Complete—testicular feminization
Incomplete—Reifenstein syndrome
Inborn errors of testosterone biosynthesis
e.g., 5α-Reductase deficiency
45,X/46,XY mosaicism
Female pseudohermaphroditism
Congenital adrenal hyperplasia
Maternal androgen ingestion or androgen-secreting tumor

Disorders of Sexual Differentiation

The process of sexual differentiation has been described in Chapter 6 (p. 101). Given the complexity of the sequential cascade of events that takes place between 6 and 14 weeks of embryonic life, it is not surprising that errors can occur. Many of these errors can lead to sexual ambiguity or to discordance between the chromosomal sex and the appearance of the external genitalia. These disorders are also sometimes referred to as various forms of intersex (Box 18.2).

True Hermaphroditism

In this extremely rare condition, an individual has both testicular and ovarian tissue, often in association with ambiguous genitalia. When an exploratory operation is carried out in these patients, an ovary can be found on one side and a testis on the other. Alternatively, there can be a mixture of ovarian and testicular tissue in the gonad, which is known as an **ovotestis**. Most patients with true hermaphroditism have a 46,XX karyotype, and in many of these individuals the paternally derived X chromosome carries Y chromosome–specific DNA sequences as a result of illegitimate crossing over between the X and Y chromosomes during meiosis I in spermatogenesis (Figure 18.24).

A small proportion of patients with true hermaphroditism are found to be chimeras with both 46,XX and 46,XY cell lines, a situation analogous to freemartins in cattle (p. 51).

Male Pseudohermaphroditism

In pseudohermaphroditism there is gonadal tissue of only one sex. The external genitalia can be ambiguous or of the sex opposite to that of the chromosomes. Thus in male pseudohermaphroditism there is a 46,XY karyotype with ambiguous or female genitalia.

The most widely recognized cause of male pseudohermaphroditism is androgen insensitivity (p. 175). In this condition, which is also known as **testicular feminization**

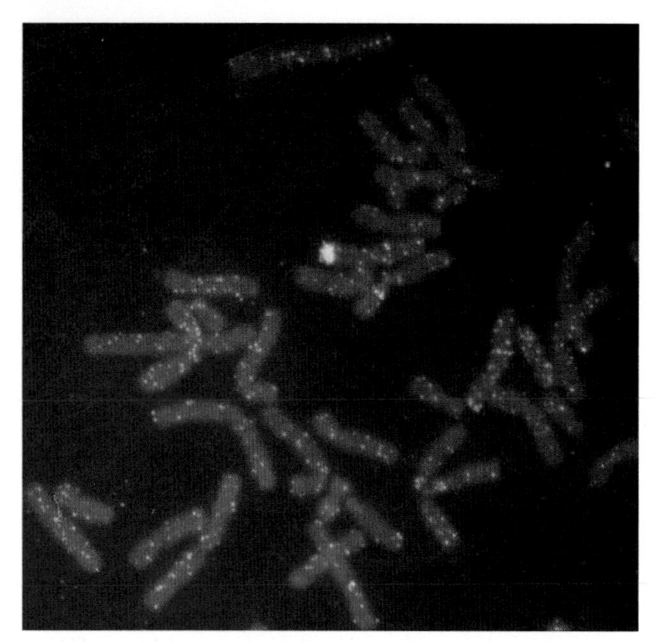

FIGURE 18.24 FISH showing hybridization of a Y chromosome paint to the short arm of an X chromosome in a 46,XX male. (Courtesy Nigel Smith, City Hospital, Nottingham, UK.)

syndrome, the karyotype is normal male and the external phenotype is essentially that of a normal female. Internally the vagina ends blindly and the uterus and fallopian tubes are absent. Testes are located in the abdomen or in the inguinal canal, where they can be mistaken for inguinal herniae. This condition is caused by the absence of androgen receptors in the target organs, so that, although testosterone is formed normally, its peripheral masculinizing effects are blocked. Androgen receptors are encoded by a gene on the X chromosome in which both deletions and point mutations have been identified. Curiously, expansion of a CAG repeat in the first exon of this androgen receptor gene causes a neurological disorder known as **Kennedy disease**, or **spinobulbar muscular atrophy**. This is a rare example of the phenomenon sometimes called a 'gene within a gene'.

Other causes of male pseudohermaphroditism include:

1. An incomplete form of androgen insensitivity known as Reifenstein syndrome in which affected males have hypospadias, small testes, and gynecomastia.
2. Enzyme defects in testosterone synthesis such as 5α-reductase deficiency (see Figure 11.5; p. 174), in which the external genitalia are ambiguous at birth but undergo masculinization (virilization) at puberty.
3. Chromosome mosaicism (45,X/46,XY), in which most individuals are normal males but a small proportion have ambiguous or female external genitalia.
4. Campomelic dysplasia, which is caused by mutations in the *SOX9* gene on chromosome 17. *SOX9* is believed to be an important gene in the regulatory pathway by which *SRY* causes masculinization of the undifferentiated fetal gonads (p. 92).
5. The Smith-Lemli-Opitz syndrome, which is caused by deficiency of 7-dehydrocholesterol reductase, an enzyme involved in cholesterol biosynthesis. Some severely affected male infants have female external genitalia.

Female Pseudohermaphroditism

In female pseudohermaphroditism, the karyotype is female and the external genitalia are virilized so that they either resemble those of a normal male or are ambiguous.

Congenital adrenal hyperplasia (CAH) is by far the most important cause of female pseudohermaphroditism (p. 174). This can be caused by several different enzyme defects in the adrenal cortex, all of which show autosomal recessive inheritance. Reduced cortisol production leads to an increase in adrenocorticotrophic hormone secretion, which in turn causes hyperplasia of the adrenal glands. In the most common form of CAH, because of a 21-hydroxylase deficiency, hormone synthesis switches from the manufacture of cortisol and aldosterone to the androgen pathway (see Figure 11.5; p. 174), leading to striking virilization of a female fetus (see Figure 11.6; p. 175). The lack of cortisol and aldosterone usually leads to rapid collapse within days of birth, which can prove fatal unless appropriate hormone and electrolyte supplementation is initiated.

Rarer causes of female pseudohermaphroditism include an androgen-secreting tumor and maternal androgen ingestion during pregnancy.

Chromosomal Breakage Syndromes

Constitutional and acquired chromosome abnormalities that predispose to malignancy are considered in Chapter 14. In addition to these conditions, it is recognized that a small number of hereditary disorders are characterized by an excess of chromosome breaks and gaps as well as an increased susceptibility to neoplasia.

Ataxia Telangiectasia

This is an autosomal recessive disorder that presents in early childhood with ataxia, oculocutaneous telangiectasia (Figure 18.25), radiation sensitivity, and susceptibility to sinus and pulmonary infection (p. 204). There is a 10% to 20% risk of leukemia or lymphoma. Cells from patients show an increase in spontaneous chromosome abnormalities, such as chromatid gaps and breaks, which are enhanced by radiation. The gene for ataxia telangiectasia is called *ATM* and maps to chromosome 11q23. The protein product is thought to act as a 'checkpoint' protein kinase, which interacts with the *TP53* and *BRCA1* gene products to arrest cell division and thereby allow repair of radiation-induced chromosome breaks before the S phase in the cell cycle (p. 39).

Bloom Syndrome

Children with this autosomal recessive disorder are small with a light-sensitive facial rash and reduced immunoglobulin levels (IgA and IgM). The risk of lymphoreticular malignancy is approximately 20%. Cultured cells show an increased frequency of chromosome breaks, particularly if

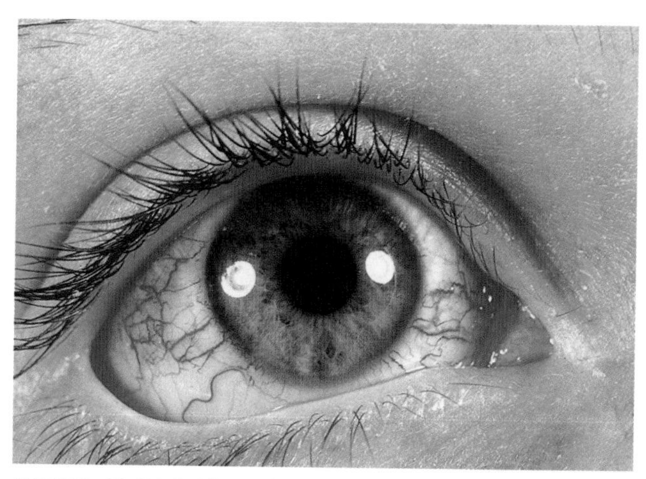

FIGURE 18.25 Ocular telangiectasia in a child with ataxia telangiectasia.

they are exposed in vitro to ultraviolet light. The gene for Bloom syndrome maps to chromosome 15q26, where it encodes one member of a group of enzymes called the DNA helicases (p. 14). These are responsible for unwinding double-stranded DNA before replication, repair, and recombination. Normally the Bloom syndrome gene plays a major role in maintaining genome stability. When defective in the homozygous state, DNA repair is impaired and the rate of recombination between sister chromatids is increased dramatically. This can be demonstrated by looking for sister chromatid exchanges (see the following section).

Fanconi Anemia

This autosomal recessive disorder is associated with upper limb abnormalities involving the radius and thumb (Figure 18.26), increased pigmentation, and bone marrow failure leading to deficiency of all types of blood cells (i.e., pancytopenia). There is also an increased risk of neoplasia, particularly leukemia, lymphoma, and hepatic carcinoma. Multiple chromosomal breaks are observed in cultured cells (Figure 18.27) and the basic defect lies in the repair of DNA strand cross-links. There are five known subtypes of Fanconi anemia, each caused by recessive mutations at different autosomal loci. The most common, type A, maps to chromosome 16q24.

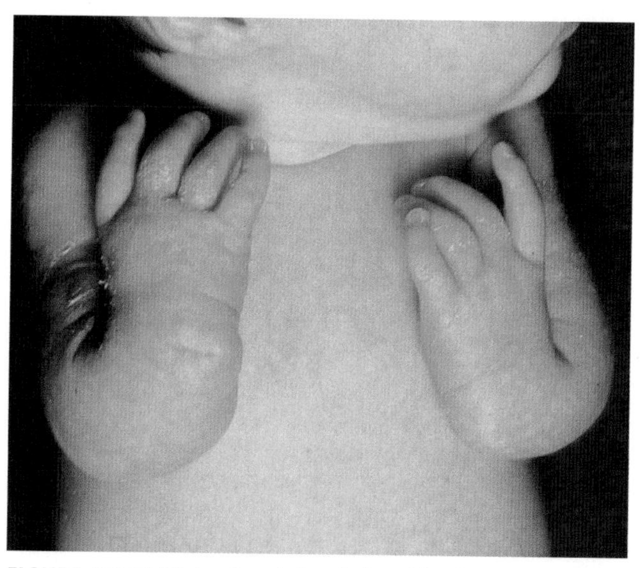

FIGURE 18.26 Bilateral radial aplasia with absent thumbs in an infant with Fanconi anemia.

Xeroderma Pigmentosa

This exists in at least seven different forms, all of which show autosomal recessive inheritance. Patients present with a light-sensitive pigmented rash and usually die from skin malignancy in sun-exposed areas before the age of 20 years. Cells cultured from these patients show chromosome abnormalities only after exposure to ultraviolet light. These disorders are due to defects in the nucleotide excision repair pathway. This involves endonuclease cleavage 5′ and 3′ to each damaged nucleotide, excision of the damaged nucleotide(s), and finally restoration of the damaged strand using the intact opposite strand as a template.

Chromosome Breakage and Sister Chromatid Exchange

Strong evidence of increased **chromosome instability** is provided by the demonstration of an increased number of **sister chromatid exchanges** (SCEs) in cultured cells. An SCE is an exchange (crossing over) of genetic material between the two chromatids of a chromosome in mitosis, in contrast to recombination in meiosis I, which is between homologous chromatids. SCEs can be demonstrated by differences in the uptake of certain stains by the two chromatids of each metaphase chromosome after two rounds of cell division in the presence of the thymidine analogue 5-bromodeoxyuridine (BUdR), which becomes incorporated in the newly synthesized DNA (Figure 18.28). There are normally about 10 SCEs per cell, but the number is greatly increased in cells from patients with Bloom syndrome and xeroderma pigmentosa. In the latter condition, this is apparent only after the cells have been exposed to ultraviolet light.

It is not clear how SCEs relate to the increased chromosome breakage observed in these two disorders, but it is thought that the explanation could involve one of the steps in DNA replication. It is also of interest that the number of SCEs in normal cells is increased on exposure to certain carcinogens and chemical mutagens. For this reason the frequency of SCEs in cells in culture has been suggested as a useful in vitro test of the carcinogenicity and mutagenicity of chemical compounds (p. 26).

Indications for Chromosomal/ Microarray-CGH Analysis

It should be apparent from the contents of this chapter that chromosome abnormalities can present in many different ways. Consequently it is appropriate to consider the

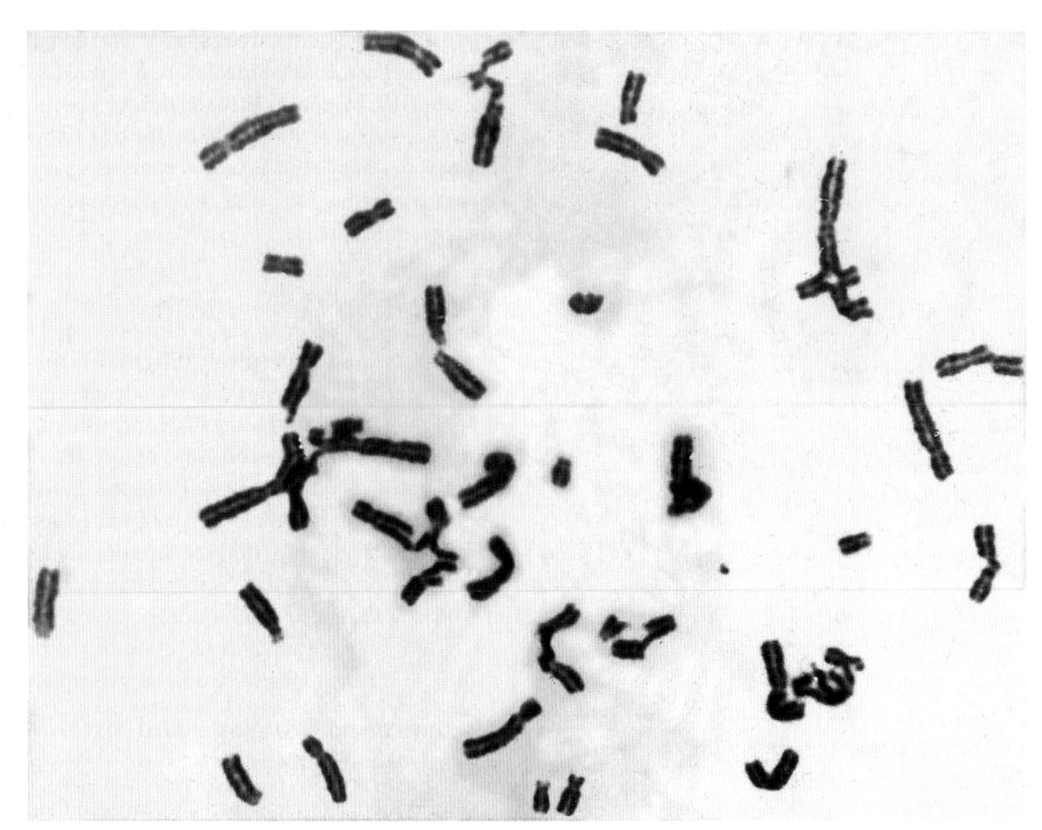

FIGURE 18.27 Multiple chromosome breaks and gaps in a metaphase spread prepared from a child with Fanconi anemia.

indications for chromosome analysis, which increasingly means **microarray-CGH**, under a number of different headings (Box 18.3).

Multiple Congenital Abnormalities

Every child with multiple congenital abnormalities should have chromosome studies undertaken. This is important for several reasons:

1. Establishing a chromosomal diagnosis will prevent further potentially unpleasant investigations being undertaken.
2. Information about the prognosis can be provided, along with details of the relevant support group and an offer of contact with other families.
3. A chromosomal diagnosis should facilitate accurate genetic risk counseling.

Although it can be very distressing for parents to be told that their child has a chromosome abnormality, they will often be relieved that an explanation for their child's problems has been found.

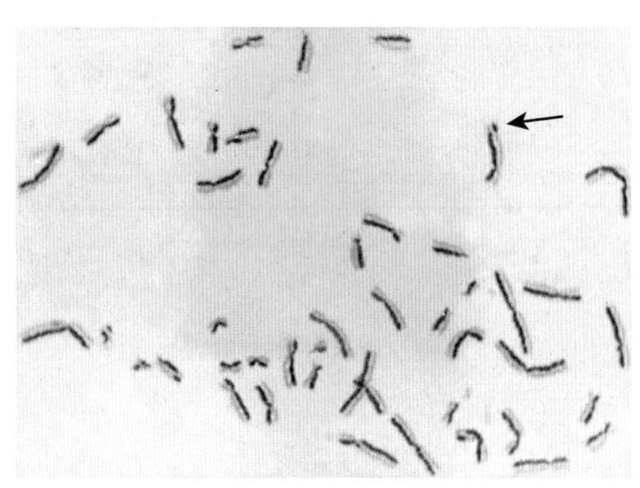

FIGURE 18.28 Chromosome preparation showing sister chromatid exchanges (*arrow*).

Unexplained Learning Difficulties

Chromosome abnormalities cause at least one-third of the 50% of learning difficulties that are attributable to genetic factors. Although most children with a chromosome abnormality have other features such as growth retardation and physical anomalies, this is not always so. If the fragile X

Box 18.3 Indications for Chromosome Analysis

Multiple congenital abnormalities
Unexplained mental retardation
Sexual ambiguity or abnormality in sexual development
Infertility
Recurrent miscarriage
Unexplained stillbirth
Malignancy and chromosome breakage syndromes

syndrome is a possibility, it is important that the cytogenetics laboratory is informed so that the correct culture conditions are used (p. 279), although in most centers the fragile X syndrome is now diagnosed by molecular methods rather than chromosome analysis. If standard karyotyping and fragile X syndrome testing is negative, great reliance will increasingly be placed on microarray CGH.

Sexual Ambiguity

The birth of a child with ambiguous genitalia should be regarded as a medical emergency, not only because of the inevitable parental anxiety, but also because of the importance of ruling out the potentially life-threatening diagnosis of salt-losing congenital adrenal hyperplasia (p. 174). A chromosome analysis should be among the first investigations undertaken.

Disorders of sexual development presenting in later life with problems such as delayed puberty, primary amenorrhea or male gynecomastia are also strong indications for chromosome analysis as a first-line investigation. This can provide a diagnosis such as Turner syndrome (45,X) or Klinefelter syndrome (47,XXY). Alternatively a normal karyotype will stimulate a search for other possible explanations, such as an endocrine abnormality.

Infertility and Recurrent Miscarriage

Unexplained involuntary infertility should prompt a request for chromosome studies, particularly if investigations reveal evidence of azoospermia in the male partner. At least 5% of such men are found to have Klinefelter syndrome. More rarely a complex chromosome rearrangement such as a translocation can cause such severe mechanical disruption in meiosis that complete failure of gametogenesis ensues.

At least 15% of all recognized pregnancies end in spontaneous miscarriage; in 50% of cases this is because of a chromosome abnormality (p. 273). Unfortunately, some couples experience recurrent pregnancy loss—usually defined as more than three spontaneous miscarriages. Often no explanation is found and many such couples go on to have successful pregnancies. However, in 3% to 6% one partner is found to carry a chromosome rearrangement that predisposes to severe imbalance through malsegregation at meiosis (p. 39). Consequently it is now standard practice to offer chromosome analysis to all such couples.

Unexplained Stillbirth/Neonatal Death

The presence of growth retardation and at least one congenital abnormality in a stillbirth or neonatal death would be an indication for chromosome studies based on analysis of blood or skin collected from the baby before or as soon after death as possible. Skin fibroblasts continue to be viable for several days after death. Chromosome abnormalities account for 5% of all stillbirths and neonatal deaths, and not all of these babies have multiple abnormalities that would immediately suggest a chromosomal cause.

Malignancy and Chromosome Breakage Syndromes

Certain types of leukemia and many solid tumors, such as retinoblastoma (p. 215) and Wilms' tumor (p. 282), are associated with specific chromosomal abnormalities that can be of both diagnostic and prognostic value. Clinical features suggestive of a chromosome breakage syndrome (p. 288), such as a combination of photosensitivity and short stature, should also lead to appropriate chromosome fragility studies, such as analysis of sister chromatid exchanges.

FURTHER READING

De Grouchy J, Turleau C 1984 Clinical atlas of human chromosomes, 2nd ed. Chichester: John Wiley

A lavishly illustrated atlas of known chromosomal syndromes.

Donnai D, Karmiloff-Smith A 2000 Williams syndrome: from genotype through to the cognitive phenotype. Am J Med Genet (Semin Med Genet) 97:164–171

A review dealing with one microdeletion syndrome in detail and the efforts to understand how the phenotype can be explained by the molecular findings.

Gardner RJM, Sutherland GR 1996 Chromosome abnormalities and genetic counseling, 2nd ed. Oxford: Oxford University Press

A useful updated guide to genetic counseling in families with a chromosome disorder.

Hagerman RJ, Silverman AC (eds) 1991 Fragile X syndrome. Diagnosis, treatment and research. Baltimore: Johns Hopkins University Press

A detailed account of the clinical and genetic aspects of the fragile X syndrome.

Jacobs PA, Browne C, Gregson N, Joyce C, White H 1992 Estimates of the frequency of chromosome abnormalities detectable in unselected newborns using moderate levels of banding. J Med Genet 29:103–108

A review of the results of more than 14,000 prenatal diagnoses with estimates of the incidence of chromosome abnormalities in term infants.

Ratcliffe S 1999 Long term outcome of children of sex chromosome abnormalities. Arch Dis Child 80:192–195

A very useful and clear description of the cognitive and social outcomes of long-term follow-up studies of sex chromosome aneuploidies.

Schinzel A 1994 Human cytogenetics database. Oxford: Oxford University Press

A regularly updated computerized database of all known chromosome abnormalities. This is an invaluable aid to diagnosis and counseling.

ELEMENTS

1 Chromosome abnormalities account for 50% of all spontaneous miscarriages and are present in 0.5–1.0% of all newborn infants.

2 Down syndrome is the most common autosomal chromosomal syndrome and shows a strong association between increasing incidence and advancing maternal age. Some 95% of all cases are caused by trisomy 21. Chromosome studies are necessary in all cases so that the rare but important cases due to unbalanced familial robertsonian translocations can be identified.

3 An increasing number of chromosome microdeletion syndromes are being recognized. These have helped in gene mapping and in enhancing understanding of underlying genetic mechanisms such as imprinting. Microdeletions of chromosome 15 are found in both Angelman and Prader-Willi syndromes, and are maternally and paternally derived, respectively.

4 Triploidy is a common finding in spontaneously aborted products of conception but rare in a live-born infant. Some children with diploidy/triploidy mosaicism present with learning difficulties and areas of depigmentation, a condition known as hypomelanosis of Ito.

5 Sex chromosome abnormalities include Klinefelter syndrome (47,XXY), Turner syndrome (45,X), the XYY syndrome (47,XYY), and the triple X syndrome (47,XXX). In all of these conditions, intelligence is either normal or only mildly impaired. Infertility is the rule in Klinefelter and Turner syndromes. Fertility is normal in the XYY and the triple X syndrome.

6 The fragile X syndrome is the most common inherited cause of learning difficulties. It is associated with a fragile site on the long arm of the X chromosome and shows modified X-linked inheritance. Affected males have moderate-to-severe learning difficulties; carrier females can show mild learning difficulties. At the molecular level there is expansion of a CGG triplet repeat, which can exist as a premutation or a full mutation.

7 Disorders of sexual differentiation include true and pseudohermaphroditism. True hermaphroditism is extremely rare. Male pseudohermaphroditism is caused most commonly by androgen insensitivity, an X-linked disorder involving the formation of androgen receptors. The most common cause of female pseudohermaphroditism is congenital adrenal hyperplasia, in which virilized infants can collapse with adrenal failure during the first week of life.

8 The chromosome breakage syndromes are rare autosomal recessive disorders characterized by increased chromosome breakage in cultured cells and an increased tendency to neoplasia, such as leukemia and lymphoma. They are caused by underlying defects in DNA repair.

CHAPTER 19

Single-Gene Disorders

To date, more than 10,000 single-gene traits and disorders have been identified. Most of these are individually rare, but together they affect between 1% and 2% of the general population at any one time. The management of these disorders in affected individuals and in their extended families presents the major workload challenge in clinical genetics.

A wide variety of single-gene disorders has been mentioned throughout this book. In this chapter some of the more common and important single gene disorders are described, as well as a small number that hold particular interest for clinicians, with emphasis on their molecular defects. Each one illustrates important genetic principles and, for many, the identification of the mutational basis and associated protein product represent major scientific achievements in the last two decades.

Huntington Disease

Huntington disease (HD) derives its eponymous title from Dr George Huntington, who described multiple affected individuals in a large North American kindred in 1872. His paper, published in the Philadelphia journal *The Medical and Surgical Reporter*, gave a graphic description of the progressive neurological disability that has endowed HD with the unenviable reputation of being one of the most feared and unpleasant hereditary disorders in man. The natural history is characterized by slowly progressive selective cell death in the central nervous system, and there is no effective treatment or cure. The prevalence in most parts of the world is approximately 1 : 10,000, although higher in some areas, such as Tasmania and the Lake Maracaibo region of Venezuela. The onset is mostly between 30 and 50 years, but it can start at virtually any age, including a rare juvenile form with different clinical features. The variable age of onset has been explained, at least in part, by the discovery of the underlying molecular defect.

Clinical Features

The usual pattern of disease is characterized by a slowly progressive movement disorder and insidious impairment of intellectual function with psychiatric disturbance and eventual dementia. The mean duration of the illness is approximately 15 to 20 years and chorea is the most common movement abnormality. This takes the form of subtle involuntary movements such as facial grimacing, twitching of the face and limbs, folding of the arms, and crossing of the legs. As the disease progresses the gait becomes very unsteady and speech unclear.

Intellectual changes in the early stages of HD include memory impairment and poor concentration span. Anxiety and panic attacks, mood changes and depression, aggressive behavior, paranoia, irrationality, increased libido, and alcohol abuse can also occur. There is a gradual deterioration in intellectual function, leading eventually to total incapacitation and dementia.

Juvenile HD

Up to 5% of HD cases present before the age of 20 years and instead of chorea there is rigidity, with slowing of voluntary movement and clumsiness. A decline in school performance heralds the onset of a severe progressive dementia, often in association with epileptic seizures. The average duration of the illness is 10 to 15 years.

Genetics

Traditionally HD has been said to show autosomal dominant inheritance with a variable age of onset, close to complete penetrance, and a very low mutation rate. In addition, it has been noted that the disorder often shows anticipation, whereby the onset is at a younger age in succeeding generations, particularly when transmitted by a male. The discovery of the HD gene in 1993 provided an explanation for some of these observations.

Mapping and Isolation of the HD Gene

HD was one of the first disorders to be mapped by linkage analysis using polymorphic DNA markers when, in 1983, the disorder was found to show close linkage with a probe known as G8 on the short arm of chromosome 4, greatly aided by collection of blood samples from the huge pedigree containing over 100 affected subjects living on the shores of Lake Maracaibo in Venezuela. As well as providing the first means of predictive testing for HD, this work also revealed that HD homozygotes are no more severely affected than heterozygotes. This is in contrast to many other autosomal dominant disorders (p. 109). When the gene itself was isolated in 1993, it was found to contain a highly polymorphic CAG (polyglutamine) repeat sequence located in the 5′ region. The messenger RNA (mRNA)

Table 19.1 Comparison of Genetic Aspects of Huntington Disease and Myotonic Dystrophy

	Huntington Disease	Myotonic Dystrophy
Inheritance	Autosomal dominant	Autosomal dominant
Chromosome locus	4p16.3	19q13.3
Trinucleotide repeat	CAG in 5′ translated region	CTG in 3′ untranslated region
Repeat sizes	Normal ≤26	Normal <37
	Mutable 27–35	
	Reduced penetrance 36–39	Full mutation 50–2000+
	Fully penetrant ≥40	
Protein product	Huntingtin	MD protein kinase (DMPK)
Early-onset form	Juvenile	Congenital
	Usually paternally transmitted	Usually maternally transmitted

codes for a protein of approximately 350 kDa, known as *huntingtin* (also IT15). Huntingtin is expressed in many different cells throughout the central nervous system, as well as other tissues, although its function remains unclear.

The Mutation in HD

Almost all individuals with HD possess an expansion of a CAG polyglutamine (triplet) repeat sequence located in the 5′ region of the HD gene, a mutational mechanism first identified in humans in contrast to almost all other types of mutation that were first reported in other species such as *Drosophila* and mice. A joint working party of the American College of Medical Genetics and the American Society of Human Genetics recommended that HD genes should be categorized under four headings on the basis of CAG repeat length (Table 19.1).

Normal Alleles

Alleles containing 26 or fewer CAG repeats are not associated with disease manifestations and are stable in meiosis.

Mutable Alleles

Allele sizes of 27 to 35 CAG repeats do not cause disease but may show meiotic instability with a potential to increase or decrease in size. These 'mutable' alleles thus constitute a reservoir for new mutations. When an affected individual presents with what appears to be a new mutation, it usually emerges that the father carries a mutable allele. Furthermore, there is evidence that mutable alleles that expand are associated with a particular haplotype, as identified by intragenic and flanking DNA markers. This implies that certain haplotypes are more mutable than others.

Reduced Penetrance Alleles

This third category consists of alleles containing 36 to 39 CAG repeats. These are associated with either late-onset disease or complete absence of disease expression (i.e., non-penetrance).

Disease Alleles

The final group of HD genes contains 40 or more CAG repeats. These are invariably associated with disease, although sometimes this may not develop until the seventh or eighth decade. There is a direct relationship between length of repeat and disease expression, with the average age of onset for repeat sizes of 40, 45, and 50 being 57, 37, and 26 years, respectively. Most affected adults have repeat sizes of between 36 and 50, whereas juvenile cases often have an expansion greater than 55 repeats.

Parent of Origin Effect in Disease Transmission

The risk to offspring is 50% regardless of whether the affected parent is male or female, according to autosomal dominant inheritance. However, for reasons that are not clear, meiotic instability is greater in spermatogenesis than oogenesis. This is reflected in anticipation, occurring mainly when the mutant allele is transmitted by a male. Juveniles with the rigid form of HD have almost always inherited the mutant allele from their more mildly affected father.

Explanations for this include the possibility that expansion is caused by **slippage** (p. 24) of DNA polymerase, simply reflecting the number of mitoses undergone during gametogenesis (p. 41). An alternative possibility is based on the observation that *huntingtin* is expressed in oocytes, so that there could be selection against oocytes with large expansions as a consequence of preferential apoptosis.

Clinical Applications and Future Prospects

Predictive genetic testing is part of routine clinical genetic practice, but there is universal agreement that this should be offered only as part of a careful counseling package. Experience to date indicates that more women than men come forward for this, and the psychological disturbance in those given positive results is low. Some 60% of candidates test negative (i.e., they receive good news), and the reasons for this departure from the expected 50% are not clear.

Prenatal diagnosis is possible for those couples who find this acceptable, although only about 25 such tests are performed in the United Kingdom annually. Obviously there are considerable emotional and ethical issues associated with termination of pregnancy; the condition is late in onset and the couple must consider the possibility of effective therapy being available in the foreseeable future. One appealing therapeutic approach is based on the observation that large CAG repeats result in intracellular accumulation of *huntingtin* 'aggregates', which are cleaved by a protease known as **caspase** to form a toxic product that causes cell death (apoptosis). Caspase inhibitors have been shown to have a beneficial effect in a HD mouse model. Another therapeutic approach under consideration is fetal neuronal cell transfer into regions of the brain, such as the caudate nucleus and putamen, which become atrophic in the early stages of the disease. This approach carries ethical considerations that will be difficult for some couples.

Myotonic Dystrophy

Myotonic dystrophy (MD) is the most common form of muscular dystrophy seen in adults, with an overall incidence of approximately 1:8000. It shares many features in common with HD (see Table 19.1)—both show autosomal dominant inheritance with anticipation, and an early-onset form with different clinical features. However, in MD the early-onset form is transmitted almost exclusively by the mother and presents at birth, in contrast to juvenile HD, which is generally paternally transmitted with an age of onset in the teens.

Clinical Features

In contrast to most forms of muscular dystrophy, clinical features in MD are not limited exclusively to the neuromuscular system. Individuals with MD usually present in adult life with slowly progressive weakness and myotonia. This latter term refers to tonic muscle spasm with prolonged relaxation, which can manifest as a delay in releasing the grip on shaking hands. Other clinical features include cataracts (Figure 19.1), cardiac conduction defects, disturbed gastrointestinal peristalsis (dysphagia, constipation, diarrhea), weak sphincters, increased risk of diabetes mellitus and gallstones, somnolence, frontal balding, and testicular atrophy. The age of onset is very variable and in its mildest form usually runs a relatively benign course. However, as the age of onset becomes earlier, so the clinical symptoms increase in severity and more body systems are involved. In the 'congenital' form, affected babies present at birth with hypotonia, talipes, and respiratory distress that can prove life threatening (see Figure 7.19). Children who survive tend to show a lack of facial expression ('myopathic facies') with delayed motor development and learning difficulties (Figure 19.2).

The diagnosis of MD used to be based on the myotonic discharges seen on electromyography but mutation analysis is more reliable and much less painful.

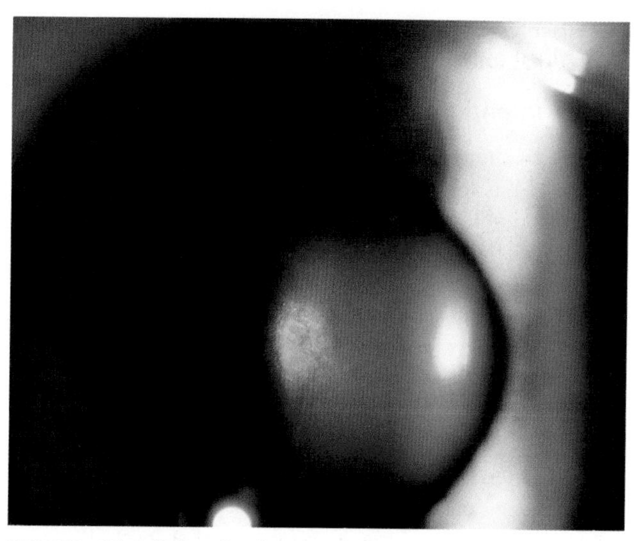

FIGURE 19.1 Refractile lens opacities in an asymptomatic person with myotonic dystrophy. (Courtesy Mr. R. Doran and Mr. M. Geall, Department of Ophthalmology, General Infirmary, Leeds, UK.)

Genetics

It follows autosomal dominant inheritance with increasing severity in succeeding generations—**anticipation** (p. 120). It used to be thought that this reflected ascertainment bias, caused by the greater likelihood of detecting a mildly affected parent with a severely affected child, rather than

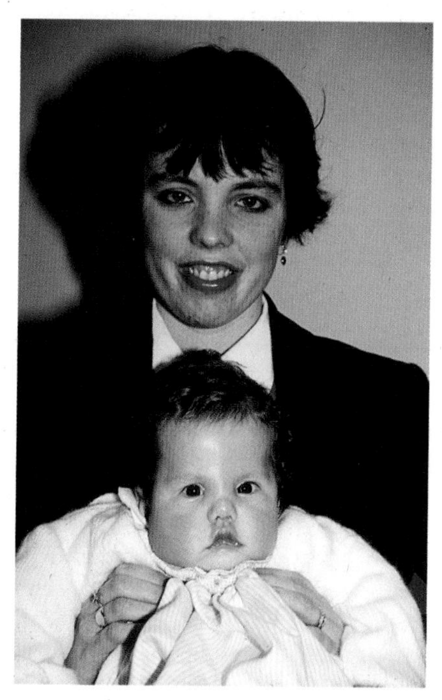

FIGURE 19.2 A mother and child with myotonic dystrophy. The child has clear features of facial myopathy and suffers from the congenital form; the mother has only mild facial myopathy. The marked generational difference in the severity of disease illustrates the phenomenon of anticipation.

the other way round. However, studies in the 1980s confirmed that anticipation is a real phenomenon.

Mapping and Isolating the MD Gene

Linkage to the secretor and Lutheran blood group loci was established as long ago as 1971, with the locus mapped to chromosome 19 in 1982. The relevant region was cloned with great effort and the mutational basis shown in 1992 to be instability in a CTG repeat sequence, which is present in the 3′ untranslated region of a protein kinase gene—dystrophia myotonica protein kinase *(DMPK)*.

Genotype–Phenotype Correlation in Myotonic Dystrophy

In unaffected persons the CTG sequence lying 3′ to the *DMPK* gene consists of up to 37 repeats (see Table 19.1). Affected individuals have an expansion of at least 50 copies of the CTG sequence. There is a close correlation between disease severity and the size of the expansion, which can exceed 2000 repeats. The severe congenital cases show the largest repeat copy number, with almost invariable inheritance from the mother. Thus, meiotic or germline instability is greater in the female for alleles containing large sequences. Curiously, expansion of a relatively small number of repeats appears to occur more commonly in the male, and most MD mutations are thought to have occurred originally during meiosis in the male. One possible explanation for these observations is that mature spermatozoa can carry only small expansions, whereas ova can accommodate much larger expansions.

Another puzzling feature of MD is the reported tendency for healthy individuals who are heterozygous for MD alleles in the normal size range to preferentially transmit alleles greater than 19 CTG repeats in size. This possible example of meiotic drive (p. 135) could explain the relatively high frequency of MD with constant replenishment of a reservoir of potential MD mutations.

The MD Protein Kinase

Perhaps surprisingly, it may be that *DMPK* is not directly responsible for muscle symptoms—mice with both overexpression and underexpression of *Dmpk* show neither myotonia nor other typical clinical features of MD. It is now appears that the RNA produced by expanded *DMPK* alleles interferes with the cellular processing of RNA produced by a variety of other genes. Expanded *DMPK* transcripts accumulate in the cell nuclei, and this is believed to have a gain-of-function effect through its binding with a CUG RNA-binding protein (CUG-BP) that has been identified. Excess CUG-BP has been shown to interfere with a number of genes relevant to MD; this is not surprising because CUG repeats are known to exist in various alternately spliced muscle-specific enzymes.

Clinical Applications and Future Prospects

Presymptomatic genetic testing and prenatal diagnosis can be offered to those families for whom it is appropriate and acceptable. This is particularly relevant for couples who have had a child with the severe congenital form, for whom the risk of recurrence is relatively high. As in HD, presymptomatic testing should not be undertaken without an offer of long-term support and medical care, and a discussion about possible difficulty in obtaining life and health insurance (p. 366).

Important components of the management of MD include regular surveillance for cardiac conduction defects and the provision of information about risks associated with general anesthesia. Logical approaches to gene therapy will almost certainly have to await a better understanding of the mechanism whereby expansion of the repeat sequence in the 3′ untranslated region of the *DMPK* gene causes such diverse and variable clinical abnormalities.

Type 2 MD

Some families with a variable presentation of similar features to MD, but without the $(CTG)_n$ expansion of *DMPK*, have been shown to link to 3q21. Originally referred to as **proximal myotonic myopathy** (PROMM), these cases are designated type 2 MD to distinguish them from the more common type 1 MD. The molecular defect has been shown to be a $(CCTG)_n$ expansion mutation in intron 1 of a gene called *ZNF9* and its protein is thought to bind RNA. Most families are of German descent, and haplotype studies suggest a single founder mutation occurring between 200 and 500 generations ago.

Hereditary Motor and Sensory Neuropathy

Hereditary motor and sensory neuropathy (HMSN) comprises a group of clinically and genetically heterogeneous disorders characterized by slowly progressive distal muscle weakness and wasting. Other names for these disorders include **Charcot-Marie-Tooth disease** and **peroneal muscular atrophy**. Their overall incidence is approximately 1:2500.

HMSN can be classified on the basis of the results of motor nerve conduction velocity (MNCV) studies. In HMSN type I, MNCV is reduced and nerve biopsies from patients show segmental demyelination accompanied by hypertrophic changes with ‘onion bulb’ formation. In HMSN type II, MNCV is normal or only slightly reduced and nerve biopsies show axonal degeneration.

Clinical Features

In autosomal dominant HMSN-I—the most common form—there is onset of slowly progressive distal muscle weakness and wasting in the lower limbs between the ages of 10 and 30 years, followed later by the upper limbs in many patients, and associated ataxia and tremor. The appearance of the lower limbs has been likened to that of an ‘inverted champagne bottle’ (Figure 19.3). With age, locomotion becomes more difficult and the feet tend to

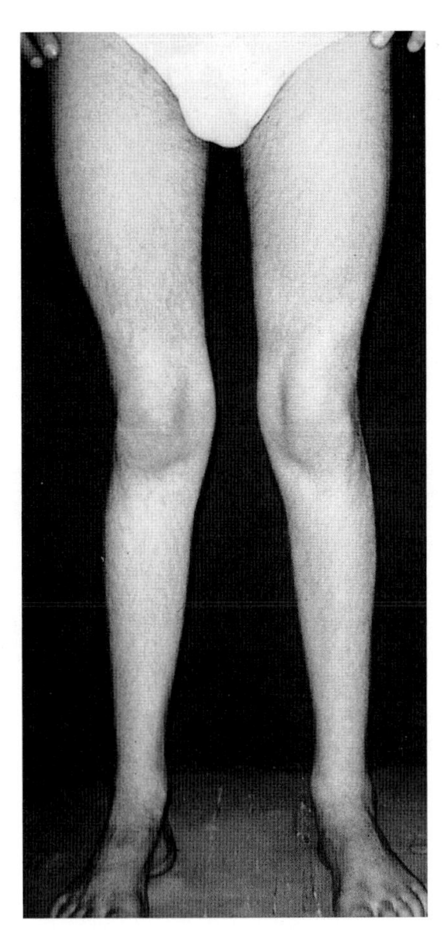

FIGURE 19.3 Lower limbs of a male with hereditary motor and sensory neuropathy (HMSN) showing severe muscle wasting below the knees.

show exaggeration of their normal arch, known as 'pes cavus'. Despite these quite striking changes, many patients retain reasonable muscle strength and are not too seriously disabled. Other faculties such as vision, hearing, and intellect are not impaired. Palpable thickening of peripheral nerves can sometimes be detected.

The clinical features in other forms of HMSN are similar but differ in the age of onset, rate of progression, and presence of other neurological involvement. For example, in HMSN-II onset is usually later than in HMSN-I, and the disease course is milder to the extent that some affected individuals are asymptomatic. By contrast, in HMSN-III, which is very rare, onset is in early childhood and there is severe delay in achieving motor milestones.

Genetics

HMSN can show autosomal dominant, autosomal recessive or X-linked inheritance, although autosomal dominant forms are by far the most common. More than 70% of cases of HMSN-I are due to a DNA **duplication** of 1.5 Mb chromosome 17p that encompasses the peripheral myelin protein-22 (*PMP22*) gene. The glycoprotein product is present in the myelin membranes of peripheral nerves, where it helps to arrest Schwann cell division. HMSN-I in humans is therefore thought to be the result of a *PMP22* dosage effect, though point mutations can be found some patients. The duplication is generated by misalignment and subsequent recombination between homologous sequences that flank the *PMP22* gene (Figure 19.4); this event usually occurs in male gametogenesis (rather than in the female, which is the case in Duchenne muscular dystrophy [p. 307]). The reciprocal **deletion** product of this misaligned recombination event, giving rise to haploinsufficiency, causes a relatively mild disorder known as **hereditary neuropathy with liability to pressure palsies**. Minor nerve trauma, such as pressure from prolonged sitting on a long-haul flight, causes focal numbness and weakness. The same misalignment recombination mechanism occurs in Hb Lepore and anti-Lepore (see Figure 10.3; p. 157), congenital adrenal hyperplasia (p. 174), and deletion 22q11 syndrome (p. 282), to name but a few.

Other Forms of Hereditary Motor and Sensory Neuropathy

A second HMSN-I locus is found on chromosome 1, which has led to the duplication chromosome 17 cases being referred to as HMSN-Ia and chromosome 1 cases as HMSN-Ib. The latter is caused by mutations in the gene that codes for another major myelin protein, known as **myelin protein zero** *(MPZ)*. This plays a crucial role as an adhesion molecule in the compaction of myelin in peripheral nerves.

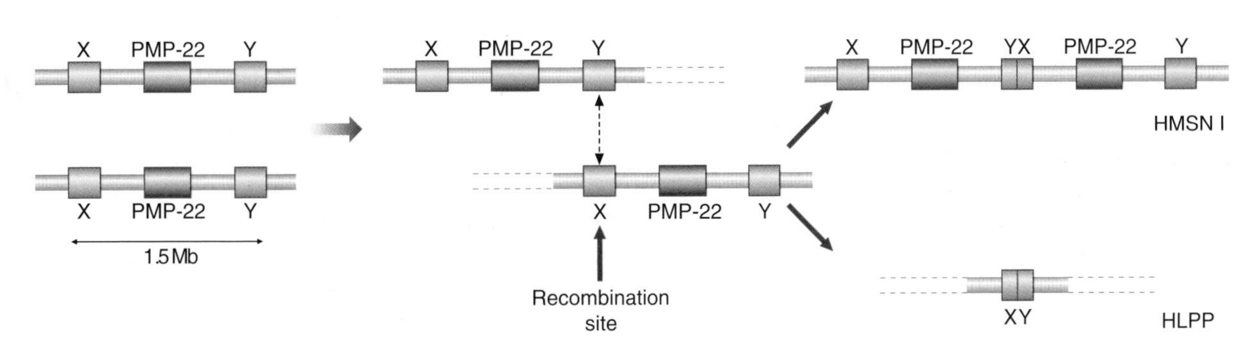

FIGURE 19.4 Mechanism by which misalignment and recombination with unequal crossing over lead to formation of the duplication and deletion that cause hereditary motor and sensory neuropathy type I (HMSN-I) and hereditary neuropathy with liability to pressure palsies (HNPP). X and Y represent homologous sequences flanking the *PMP22* gene.

A more rare form of HMSN shows X-linked dominant inheritance, with males having typical HMSN-I features and females being more mildly affected—sometimes with HMSN-II characteristics. This is due to mutations in the gene that encodes a gap junction protein called *GJB1* (previously *Connexin 32*).

HMSN-II is genetically heterogeneous and progress with clinical mutation testing has been considerably slower. However, a range of genes can now be analysed, including Mitofusin 2 *(MFN2)* and Neurofilament Protein, Light Peptide *(NEFL)*.

Future Prospects

Genetic testing in HMSN has greatly improved diagnostic precision, though there is still a place for nerve conduction studies in clinical assessment, especially when genetic testing fails to yield a positive result. Treatment remains elusive but in HMSN-Ia efforts at gene therapy may be directed at trying to achieve a reduction in gene dosage by switching off, or 'downregulating', *PMP22* expression.

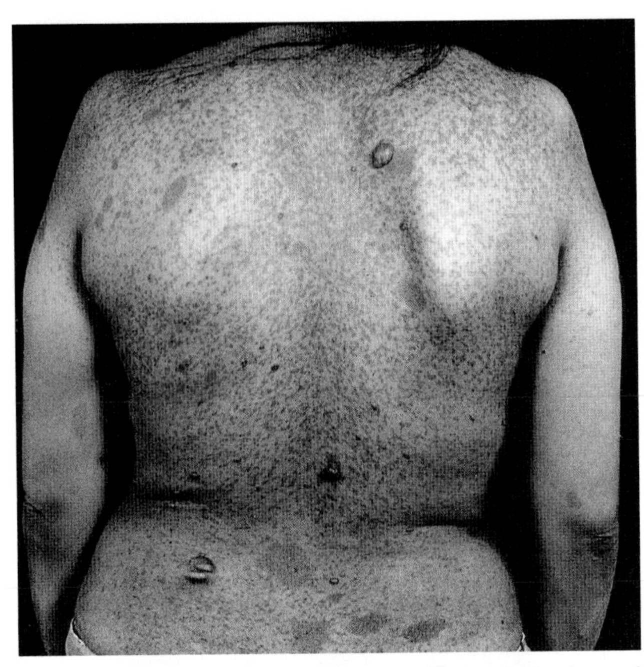

FIGURE 19.5 A patient with neurofibromatosis type I showing truncal freckling, café-au-lait spots, and multiple neurofibromata.

Neurofibromatosis

References to the clinical features of neurofibromatosis (NF) first appeared in the eighteenth-century medical literature, but historically the disorder is most commonly associated with the name Von Recklinghausen, a German pathologist who coined the term 'neurofibroma' in 1882.

This is one of the most common genetic disorders in humans and gained public notoriety when it was suggested that Joseph Merrick, the 'Elephant Man', was probably affected. However, many now think he had the much rarer disorder known as **Proteus syndrome**.

There are two main types of neurofibromatosis, NF1 and NF2. Both conditions, especially NF2, could be included under familial cancer syndromes (see Chapter 14) but are covered in more detail here. NF1 has a birth incidence of approximately 1:3000; NF2 approximately 1:35,000 and a prevalence of around 1:200,000.

Clinical Features

The most notable features of NF1 are small pigmented skin lesions, known as **café-au-lait** (CAL) spots, and small soft fleshy growths known as **neurofibromata** (Figure 19.5). CAL spots first appear in early childhood and continue to increase in both size and number until puberty. A minimum of six CAL spots at least 5 mm in diameter is required to support the diagnosis in childhood, and axillary and/or inguinal freckling should be present. Neurofibromata are benign tumors that arise most commonly in the skin, usually appearing in adolescence or adult life, and increasing in number with age.

Other clinical findings include relative macrocephaly (large head) and Lisch nodules. These are small harmless raised pigmented hamartomata of the iris (Figure 19.6). The most common complication, occurring in a third of childhood cases, is mild developmental delay characterized by a non-verbal learning disorder. For many, significant improvement is seen through the school years. Most individuals with NF1 enjoy a normal life and are not unduly inconvenienced by their condition. However, a small number of patients develop one or more major complications, such as epilepsy, a central nervous system tumor, or scoliosis.

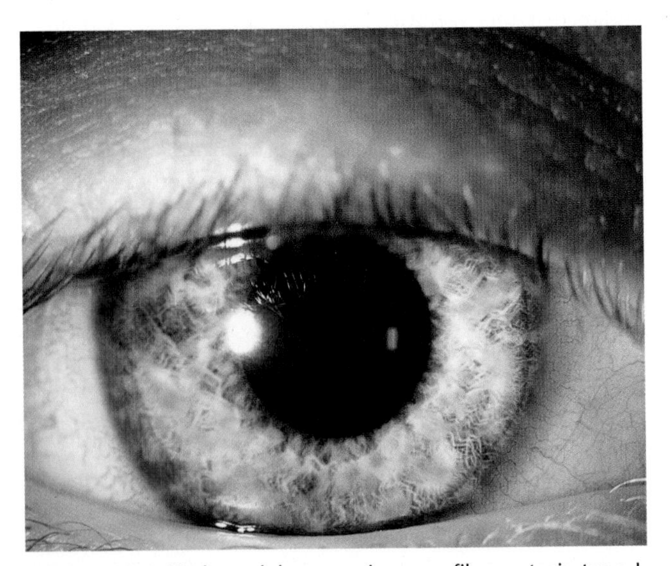

FIGURE 19.6 Lisch nodules seen in neurofibromatosis type I. (Courtesy Mr. R. Doran, Department of Ophthalmology, General Infirmary, Leeds, UK.)

Genetics

NF1 shows autosomal dominant inheritance with virtually 100% penetrance by the age of 5 years. The effects are very variable and affected members of the same family can show striking differences in disease severity. The features in affected MZ twins are usually very similar, so the variability in family members with the same mutation may be due to modifying genes at other loci. Approximately 50% of cases of NF1 are due to new mutations, with the estimated mutation rate being approximately 1 per 10,000 gametes. This is around 100 times greater than the average mutation rate per generation per locus in humans.

There are a few reports of more than one affected child born to unaffected parents—the result of gonadal mosaicism (p. 121), usually paternal in origin. Somatic mosaicism in NF1 can manifest with features limited to a particular part of the body. This is referred to as **segmental NF**.

The Neurofibromatosis Type 1 Gene and Its Product

The NF1 gene, *neurofibromin*, was successfully mapped to chromosome 17, adjacent to the centromere, in 1987. Its isolation was aided by the identification of two patients who both had a balanced translocation with a breakpoint at 17q11.2. A cosmid clone was identified containing both translocation breakpoints and a search for transcripts from this region yielded four genes, one of which was shown to be *neurofibromin*. It is large, spanning over 350 kilobases (kb) of genomic DNA and comprising at least 59 exons. The other three genes identified in this region were found to lie *within* a single intron of the *neurofibromin* gene, where they are transcribed in the opposite direction from the complementary strand (p. 14).

The neurofibromin protein encoded by this gene shows structural homology to the guanosine triphosphatase (GTPase)-activating protein (GAP), which is important in signal transduction (p. 214) by downregulating *RAS* activity. The place of neurofibromin in the RAS-MAPK pathway is shown in Figure 16.12, highlighting the link with Noonan syndrome (p. 254). Loss of heterozygosity (p. 215) for chromosome 17 markers has been observed in several malignant tumors in patients with NF1, as well as in a small number of benign neurofibromata. These observations indicate that the *neurofibromin* gene functions as a tumor suppressor (p. 214). It has been shown to contain a GAP-related domain (GRD), which interacts with the *RAS* proto-oncogene product. A mRNA editing site exists in the *neurofibromin* gene and edited transcript causes GRD protein truncation, which inactivates the tumor suppressor function. A higher range of editing is seen in more malignant tumors.

Other genes, including *TP53* (p. 218) on the short arm of chromosome 17, are also involved in tumor development and progression in NF1. Conversely, it is also known that the *neurofibromin* gene is implicated in the development of sporadic tumors not associated with NF, including carcinoma of the colon, neuroblastoma, and malignant melanoma. These observations confirm that the *neurofibromin* gene plays an important role in cell growth and differentiation.

Genotype–Phenotype Correlation

Many different mutations have been identified in the *neurofibromin* gene, which include deletions, insertions, duplications, and point substitutions (p. 23). Most lead to severe truncation of the protein or complete absence of gene expression. To date, there is little evidence for a clear genotype-phenotype relationship with the exception of one specific mutation, a 3-bp inframe deletion in exon 17, which has recurred in different cases and families, and affected individuals do not appear to develop cutaneous neurofibromata. Generally, NF1 shows quite striking intrafamilial variation, suggesting the possibility of modifier genes. Patients with large deletions that include the entire *neurofibromin* gene tend to be more severely affected, with significant intellectual impairment, a somewhat marfanoid habitus, and a larger than average number of cutaneous neurofibromata.

Neurofibromatosis Type 2

In NF2 both CAL spots and **neurofibromata** can occur, but these are much less common than in NF1. The most characteristic feature is the development in early adult life of tumors involving the eighth cranial nerves—**vestibular schwannomas** (still sometimes called **acoustic neuromas**). Several other central nervous system tumors occur frequently, although more than half remain asymptomatic. Autosomal dominant spinal and peripheral schwannomas without vestibular schwannomas is an entity known as **schwannomatosis**. An ophthalmic feature seen in NF2, but not NF1, is cataracts, which are frequent but often subclinical.

The NF2 locus was mapped to chromosome 22q by linkage analysis in 1987. The gene, called **schwannomin**, was cloned in 1993, found to span 110 kb with 17 exons, and is thought to be a cytoskeleton protein that acts as a tumor suppressor.

Clinical Applications and Future Prospects

Mapping of the *neurofibromin* gene has provided a means of offering both presymptomatic and prenatal diagnosis, usually by direct mutation analysis but linkage if no mutation has been found. In practice very few families pursue either of these options, partly because NF1 is not perceived as a serious illness and also because mutation analysis does not help in predicting disease severity.

At present there is no cure for NF1. Drug therapy aimed at upregulating neurofibromin GAP activity or downregulating *RAS* activity could prove beneficial in the absence of effective gene therapy. However, it is difficult to envisage how this could be applied to diverse target tissues, including

Table 19.2 Revised (Gent) Criteria for Making a Diagnosis of Marfan Syndrome

System	Major Criteria	Minor Criteria
Skeletal	*Four* of these should be present: Pectus carinatum Pectus excavatum requiring surgery Reduced upper to lower segment body ratio or span : height ratio >1.05 Hypermobility of wrist and thumbs Medial displacement of medial malleolus Radiological protrusio acetabulae	Pectus excavatum Joint hypermobility High arched palate with dental crowding Facial features, including down-slanting palpebral fissures causing pes planus
Ocular	Ectopia lentis	Flat cornea Increased axial length of the globe Hypoplastic iris
Cardiovascular	Dilatation of the ascending aorta Dissection of the ascending aorta	Mitral valve prolapse Dilatation or dissection of descending thoracic or abdominal aorta under 50 years
Pulmonary	None	Spontaneous pneumothorax Apical blebs
Skin/connective tissue		None
Dura	Lumbosacral dural ectasia	None
Family history/genetics	First-degree relative who meets criteria	None
	Presence of *FBN1* mutation, or high-risk haplotype in MFS family	None

the central nervous system. Nevertheless, there is great interest in the whole RAS-MAPK pathway (see Figure 16.12, p. 256), especially its role in tumor formation, and whether drugs can effectively modify its activity.

Marfan Syndrome

The original patient described by the French pediatrician Bernard Marfan, in 1896, probably had the similar but rarer condition now known as **Beal syndrome**, or **congenital contractual arachnodactyly** (p. 301). In clinical practice physicians often consider the diagnosis of Marfan syndrome (MFS) for any patient who is tall with subjective features of long limbs and fingers. However, it is essential to be objective in clinical assessment because a number of conditions have 'marfanoid' features, and many tall, thin people are entirely normal. Detailed diagnostic criteria, referred to as the **Gent criteria**, are in general use by geneticists (Table 19.2).

Clinical Features

MFS is a disorder of fibrous connective tissue, specifically a defect in type 1 fibrillin, a glycoprotein encoded by the *FBN1* gene. In the classic presentation affected individuals are tall compared with unaffected family members, have joint laxity, a span : height ratio greater than 1.05, a reduced upper to lower segment body ratio, pectus deformity, and scoliosis (Figure 19.7). The connective tissue defect gives rise to ectopia lentis (lens subluxation) in a proportion of (but not all) families and, very importantly, dilatation of the ascending aorta, which can lead to dissection. The latter complication is obviously life threatening, and for this reason alone care must be taken over the diagnosis. Aortic dilatation may be progressive but the rate of change can be reduced by β-adrenergic blockade (if tolerated) and there is great hope for angiotensin-II receptor antagonists (similar properties to angiotensin-converting enzyme inhibitors), whose trials are under way. Surgical replacement should be undertaken if the diameter reaches 50 to 55 mm. Pregnancy is a risk factor for a woman with MFS who already has some dilatation of the aorta, and monitoring is very important.

A diagnosis of MFS requires careful clinical assessment, body measurements looking for evidence of disproportion, echocardiography, ophthalmic evaluation, and, in some doubtful cases, lumbar magnetic resonance imaging to look for evidence of dural ectasia (see Table 19.2). The metacarpophalangeal index, a radiological measurement of the ratio of these hand bone lengths, does not feature in the revised criteria. Where the family history is non-contributory, a positive diagnosis is made when the patient has a minimum of two major criteria plus involvement of a third organ system; for a person with a close relative who is definitely affected, it is sufficient to have one major criterion plus involvement of a second organ system.

Genetics

MFS follows autosomal dominant inheritance and the majority of cases are linked to the large *FBN1* gene on 15q21, with 65 exons spanning 200 kb and containing five distinct domains. The largest of these, occupying about 75% of the gene, comprises about 46 epidermal growth factor repeats (see p. 190). Finding the causative mutations in affected patients was initially very difficult, but hundreds have now been reported. Most are missense and have a dominant-negative effect, resulting in less than 35% of the expected amount of fibrillin in the extracellular matrix.

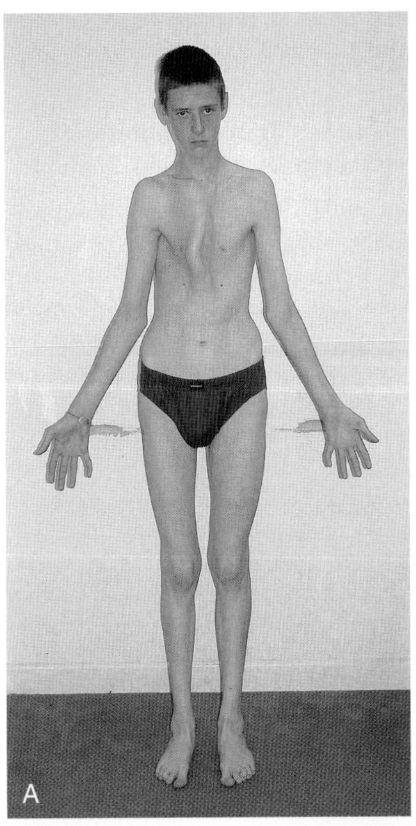

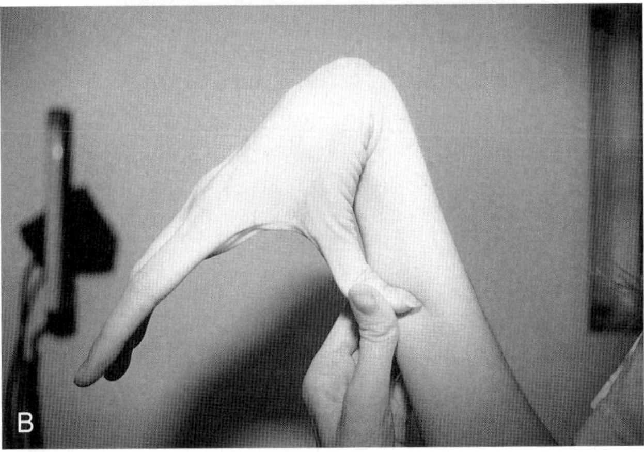

FIGURE 19.7 **A**, An adolescent with Marfan syndrome showing disproportionately long limbs (arachnodactyly) and a very extreme example of chest bone deformity; he also has a dilated aortic root. **B**, Joint hypermobility at the wrist in a woman with Marfan syndrome; this appearance might also be seen in other joint-laxity conditions, such as Ehlers-Danlos syndrome.

Mutations have also occasionally been found in related phenotypes such as neonatal MFS, familial ectopia lentis, Shrintzen-Goldberg syndrome, and the MASS phenotype (mitral valve prolapse, myopia, borderline aortic enlargement, non-specific skin and skeletal findings).

Loeys-Dietz Syndrome

Familial aortic aneurysm is not confined to MFS and a new, distinct, condition has been delineated. This also follows autosomal dominant inheritance and aneurysms can be aggressive and occur before major aortic dilatation. Additional findings include cleft palate or bifid uvula, craniosynostosis, mental retardation, and generalized arterial tortuosity with aneurysms occurring elsewhere in the circulation. Some individuals have features overlapping with MFS but they do not fulfill the accepted Gent diagnostic criteria. The condition is now known as **Loeys-Dietz syndrome**, and the gene was identified through a candidate approach. Transforming growth factor (TGF) signaling (p. 85) had been shown to be important in vascular and craniofacial development in mouse models; this led Loeys and colleagues to sequence the TGF-β receptor 2 *(TGFBR2)* gene in a series of families. Heterozygous mutations were found in most of these, and in the others missense mutations were found in the related gene, *TGFBR1*.

Congenital Contractural Arachnodactyly

Also known as **Beal syndrome**, this is probably the condition originally described by Marfan in 1896. Many features overlap with MFS, but there is less tendency to aortic dilatation and its catastrophic consequences. Individuals have congenital contractures of their digits, a crumpled ear helix, and sometimes marked scoliosis. It is due to mutated **type 2 fibrillin**, which shares the same organizational structure as fibrillin-1 and maps to 5q23.

Cystic Fibrosis

Cystic fibrosis (CF) was first recognized as a discrete entity in 1936 and used to be known as 'mucoviscidosis' because of the accumulation of thick mucous secretions that lead to blockage of the airways and secondary infection. Although antibiotics and physiotherapy have been very effective in increasing the average life expectancy of a child with CF from less than 5 years in 1955 to at least 30 years, CF remains a significant cause of chronic ill health and death in childhood and early adult life.

CF is one of the most common autosomal recessive disorders encountered in individuals of western European origin, in whom the incidence varies from 1 in 2000 to 1 in 3000. The incidence is slightly lower in eastern and southern European populations, and much lower in African Americans (1 in 15,000) and Asian Americans (1 in 31,000).

Clinical Features

The organs most commonly affected in CF are the lungs and the pancreas. Chronic lung disease caused by recurrent infection eventually leads to fibrotic changes in the lungs with secondary cardiac failure, a condition known as **cor pulmonale**. When this complication occurs, the only hope for long-term survival rests in a successful heart–lung transplant.

In 85% of people with CF, pancreatic function is impaired, with reduced enzyme secretion from blockage of the pancreatic ducts by inspissated secretions. This leads to malabsorption with an increase in the fat content of the stools but is satisfactorily treated with oral supplements of pancreatic enzymes.

Other problems commonly encountered in CF include nasal polyps, rectal prolapse, cirrhosis, and diabetes mellitus. Around 10% of children with CF present in the newborn period with obstruction of the small bowel from thickened meconium, known as **meconium ileus**. Almost all males with CF are sterile because of congenital bilateral absence of the vas deferens (CBAVD). There are some males with CBAVD, sometimes with rare mutations in the CF gene, who have virtually no other features of CF. Other rare presentations include chronic pancreatitis, diffuse bronchiectasis, and bronchopulmonary allergic aspergillosis.

Genetics

CF shows autosomal recessive inheritance. Other autosomal recessive disorders, such as hemochromatosis, which causes tissue iron overload, have a higher carrier frequency, but CF is by far the most serious autosomal recessive disorder encountered in children of western European origin. Possible explanations for this high incidence include a high mutation rate, meiotic drive, and heterozygote advantage. The latter explanation, possibly mediated by increased heterozygote resistance to chloride-secreting bacterially induced diarrhea, is often favored but does not explain why CF is rare in tropical regions where diarrheal diseases are common.

Mapping and Isolation of the Cystic Fibrosis Gene

The mapping and isolation of the CF was a celebrated milestone in the history of human molecular genetics and it is easy to forget how very difficult and time consuming such research was just 25 years ago. The CF locus was mapped to chromosome 7q31 in 1985 by the demonstration of linkage to the gene for a polymorphic enzyme known as **paraoxonase**. Shortly afterward, two polymorphic DNA marker loci, known as MET and D7S8, were shown to be closely linked flanking markers. The region between these markers was scrutinized for the presence of HTF or CpG islands, which are known to be present close to the 5′ end of many genes (p. 75). This led to the identification of several new DNA markers that were shown to be very tightly linked to the CF locus with recombination frequencies of less than 1%. These loci were found to be in linkage disequilibrium (p. 138) with the CF locus, and one CF mutation was found to be associated with one particular haplotype in 84% of cases, consistent with the concept of a single original mutation being responsible for a large proportion of all CF genes. The identification of loci tightly linked to the CF locus narrowed its location down to a region of approximately 500 kb. The CF gene was eventually cloned by two groups of scientists in North America in 1989 by a combination of chromosome jumping, physical mapping, isolation of exon sequences and mutation analysis. It was named the CF transmembrane conductance regulator *(CFTR)* gene, spans a genomic region of approximately 250 kb, and contains 27 exons.

The Cystic Fibrosis Transmembrane Conductance Regulator Protein

The structure of CFTR is consistent with a protein product containing 1480 amino acids with a molecular weight of 168 kDa. It is thought to consist of two transmembrane (TM) domains that anchor it to the cell membrane, two nucleotide binding folds (NBFs) that bind ATP, and a regulatory (R) domain, which is phosphorylated by protein kinase-A (Figure 19.8).

The primary role of the *CFTR* protein is to act as a chloride channel. Activation by phosphorylation of the regulatory domain, followed by binding of ATP to the NBF domains, opens the outwardly rectifying chloride channel and exerts a negative effect on intracellular sodium absorption by closure of the epithelial sodium channel. The net effect is to reduce the level of intracellular sodium chloride, which improves the quality of cellular mucous secretions.

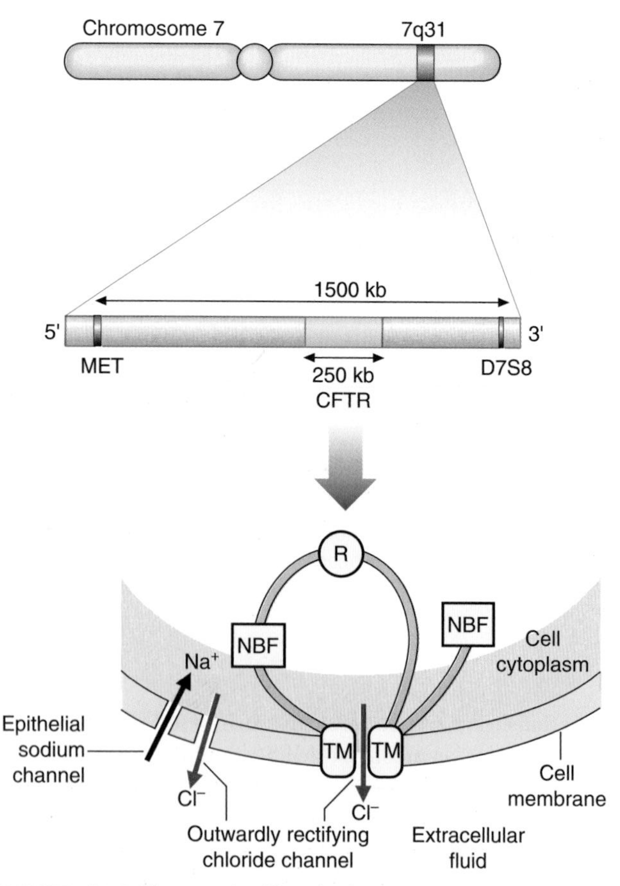

FIGURE 19.8 The cystic fibrosis locus, gene, and protein product, which influences closely adjacent epithelial sodium and outwardly rectifying chloride channels. *R,* Regulatory domain; *NBF,* nucleotide binding fold; *TM,* transmembrane domain.

Table 19.3 Contribution of Phe508del Mutation to All CF Mutations

Country	%
Denmark	88
Netherlands	79
UK	78
Ireland	75
France	75
USA	66
Germany	65
Poland	55
Italy	50
Turkey	30

Data from European Working Group on CF Genetics (EWGCFG) gradient of distribution in Europe of the major CF mutation and of its associated haplotype. Hum Genet 1990; 85:436–441, and worldwide survey of the Phe508del mutation—report from the Cystic Fibrosis Genetic Analysis Consortium. Am J Hum Genet 1990; 47:354–359

Mutations in the Cystic Fibrosis Transmembrane Conductance Regulator Gene

The first mutation to be identified in *CFTR* was a deletion of three adjacent base pairs at the 508th codon which results in the loss of a phenylalanine residue. This mutation is now known as Phe508del (previously deltaF508) and accounts for approximately 70% of all mutations in *CFTR*, the highest incidence of 88% being in Denmark (Table 19.3). The mutation can be demonstrated very simply by PCR using primers that flank the 508th codon (Figure 19.9).

More than 1500 other mutations in the *CFTR* gene have been identified. These include missense, frameshift, splice-site, nonsense, and deletion mutations (p. 23). Most of these are extremely uncommon, although a few can account for a small but significant proportion of mutations in a particular population. For example, the G542X and G551D mutations account for 12% and 3%, respectively, of all CF mutations in the Ashkenazi Jewish and North American Caucasian populations. Commercial multiplex PCR-based kits have been developed that detect approximately 90% of all carriers. Using these it is possible to reduce the carrier risk for a healthy individual from a population risk of 1 in 25 to less than 1 in 200.

Genotype-Phenotype Correlation

Mutations in *CFTR* can influence the function of the protein product by:

1. Causing a complete or partial reduction in its synthesis—e.g., G542X and IVS8-6(5T)
2. Preventing it from reaching the epithelial membrane—e.g., Phe508del
3. Causing it to function incorrectly when it reaches its final location—e.g., G551D and R117H.

The net effect of all these mutations is to reduce the normal functional activity of the CFTR protein. The extent to which normal CFTR protein activity is reduced correlates well with the clinical phenotype. Levels of less than 3% are associated with severe 'classic' CF, sometimes referred to as the **PI type** because of associated pancreatic insufficiency. Levels of activity between 3% and 8% cause a milder 'atypical' form of CF in which there is respiratory disease but relatively normal pancreatic function. This is referred to as the **PS** (pancreatic sufficient) **form**. Finally, levels of activity between 8% and 12% cause the mildest CF phenotype, in which the only clinical abnormality is CBAVD in males.

The relationship between genotype and phenotype is complex. Homozygotes for Phe508del almost always have severe classical CF, as do compound heterozygotes with Phe508del and G551D or G542X. The outcome for other compound heterozygote combinations can be much more difficult to predict. The complexity of the interaction between *CFTR* alleles is illustrated by the IVS8-6 poly T variant. This contains a polythymidine tract in intron 8 that influences the splicing efficiency of exon 9, resulting in reduced synthesis of normal CFTR protein. Three variants consisting of 5T, 7T, and 9T have been identified. The 9T variant is associated with normal activity but the 5T allele leads to a reduction in the number of transcripts containing exon 9. The 5T variant has a population frequency of approximately 5%, but is more often found in patients with CBAVD (40–50%) or disseminated bronchiectasis (30%).

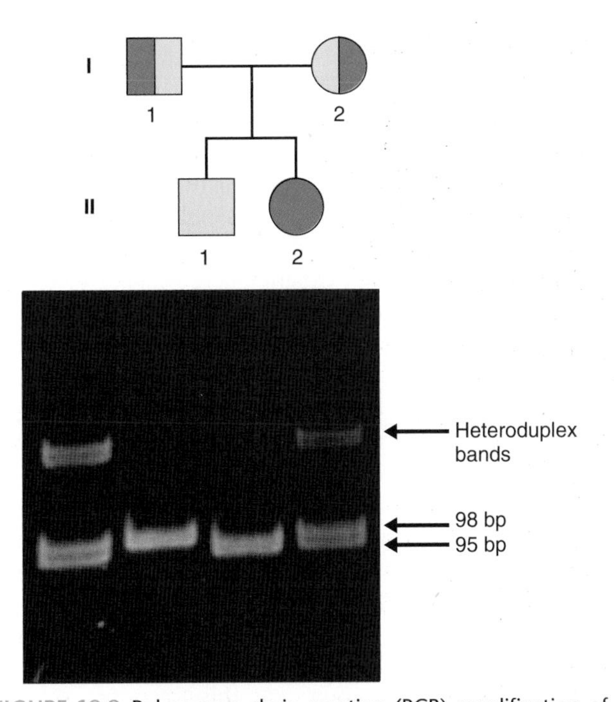

FIGURE 19.9 Polymerase chain reaction (PCR) amplification of 98- and 95-bp DNA fragments surrounding the Phe508del mutation site in the *CFTR* gene from a child with cystic fibrosis and her parents. The child, II_2, is homozygous for Phe508del. Her parents, I_1 and I_2, are heterozygous, and her brother, II_1, is homozygous for the normal allele. Heterozygotes are readily identified by the presence of heteroduplex bands formed between 95- and 98-bp products and electrophoresed on a non-denaturing gel.

Curiously, it has been shown that the number of thymidine residues influences the effect of another mutation, R117H. When R117H is in *cis* with 5T (i.e., in the same allele) it causes the PS (pancreatic sufficiency) form of CF when another CF mutation is present on the other allele. However, in compound heterozygotes (e.g., Phe508del /R117H) where R117H is in *cis* with 7T, it can result in a milder but variable phenotype, ranging from CBAVD to PS CF. The milder phenotype is likely to result from the expression of higher levels of full-length R117H protein with some residual activity. The increasing number of *CFTR* mutations and variability of the associated phenotypes has led some authors to propose a spectrum of 'CFTR disease', recognizing that a label of CF may be inappropriate for patients with milder symptoms.

Clinical Applications and Future Prospects

Before the mapping of the CF locus and the subsequent isolation of *CFTR*, it was not possible to offer either carrier detection or reliable prenatal diagnosis. Now parents of an affected child can almost always be offered prenatal diagnosis by direct mutation analysis. Similarly, knowledge of one or both of the mutations in an affected child now permits the offer of carrier detection to close family relatives. In many parts of the world, it is now standard practice to offer **cascade screening** to all families in which a mutation has been identified. Population screening for carriers of CF (p. 318) and neonatal screening for CF homozygotes (p. 320) have been widely implemented.

CF is a prime candidate for gene therapy because of the relative accessibility of the crucial target organs (i.e., the lungs). Gene transfer studies carried out using adenoviruses and *CFTR* complementary DNA (cDNA)–liposome complexes have resulted in the restoration of chloride secretion in CF transgenic mice. Several clinical trials have been undertaken in small groups of volunteer patients with CF. Although there has been experimental evidence of *CFTR* expression in the treated patients, this has generally been transient. Problems have been encountered with poor vector efficiency and inflammatory reactions, particularly when adenoviruses have been used as the vector. Despite these initial problems, there is cautious optimism that effective gene therapy for CF will be developed eventually.

Inherited Cardiac Arrhythmias and Cardiomyopathies

In about 4% of sudden cardiac death in persons ages 16 to 64 years, no explanation is evident; this is enormously traumatic for the family left behind. In England this equates to about 200 such deaths annually. Understandably, there can be great anxiety when this is familial and affects young adults. Over the past few years, the term **sudden adult death syndrome** (**SADS**) has been applied, but also in use are **sudden cardiac death** (**SCD**) and **inherited cardiac condition** (**ICC**). This group of conditions includes the **long QT syndromes** (**LQTS**), **Brugada syndrome**, **catecholaminergic** (stress-induced) **polymorphic ventricular tachycardia** (**CPVT**), and **arrhythmogenic right ventricular cardiomyopathy** (**ARVC**). LQTS and Brugada syndrome are sodium and potassium **ion channelopathies**. CPVT and ARVC demonstrate overlap with the inherited cardiomyopathies and some cases are due to molecular defects affecting the calcium channel. In ARVC there is often pathological evidence of either a hypertrophic or a dilated myocardium.

Inherited Arrhythmias

Clinical Features

When sudden unexplained death occurs, a careful review of the post-mortem findings and an exploration of the deceased's history, as well as the family history, are indicated. Most of those who die are young males, and death occurs during sleep or while inactive. In a proportion of cases, death occurs while swimming, especially in LQT1. Emotional stress can be a trigger, especially in LQT2, and cardiac events are more likely in sleep for LQT2 and LQT3. Careful investigation and questioning may reveal an antecedent history of episodes of syncope, palpitation, chest discomfort, and dyspnea, and these symptoms should be explored in the relatives in relation to possible triggers. If the deceased had a 12-lead electrocardiogram (ECG), this may hold some key evidence; however, a normal ECG is present in about 30% of proven LQTS and possibly a higher proportion of Brugada syndrome cases.

In LQTS, also known as **Romano-Ward syndrome**, the ECG findings are dominated by, as the name suggests, a QT interval outside the normal limits, remaining long when the heart rate increases. They are classified according to the gene involved (Table 19.4). The inheritance is overwhelmingly autosomal dominant but a rare recessive form exists, combined with sensorineural deafness, which is known as **Jervell and Lange-Nielsen syndrome**. The ECG changes may be evident from a young age and a cardiac event occurs by age 10 years in about 50%, and by age 20 years in 90%. First cardiac events tend to be later in LQT2 and LQT3. Predictive genetic testing, where possible, is helpful to identify those at risk in affected families, and decisions about prophylactic β-blockade can be made. β-Blockers are particularly useful in LQT1 but less so in LQT2 and LQT3; indeed, it is possible that β-blockers may be harmful in LQT3.

Brugada syndrome also follows autosomal dominant inheritance and was first described in 1992. The cardiac event is characterized by a proneness to idiopathic ventricular tachycardia (VT), and there may be abnormal ST-wave elevation in the right chest leads with incomplete right bundle branch block. In at-risk family members with a normal ECG, the characteristic abnormalities can usually be unmasked by the administration of potent sodium channel blockers such as flecainide. The condition is relatively common in Southeast Asia; there is a male

Table 19.4 The Inherited Cardiac Arrhythmias

Arrhythmia	Onset	Triggers	Gene	Locus
LQT1 (Romano-Ward)	90% by age 20 years	Exercise (swimming)	*KCNQ1*	11p15
LQT2	Early adult life	Stress/sleep	*KCNH2* (HERG)	7q35
LQT3	Early adult life	Stress/sleep	*SCN5A*	3p21
LQT4	Adulthood		*Ankyrin-B*	4q25
LQT5	Childhood		*KCNE1*	21q22
LQT6	Adulthood		*KCNE2*	21q22
LQT7 (Andersen syndrome)	Adulthood		*KCNJ2*	17q23
Brugada syndrome	Adulthood		*SCN5A*	3p21
CPVT	Childhood/adolescence	Stress	*RYR2*	1q42
ARVC1	Childhood/adolescence		*TGFB3*	14q23
ARVC2	Childhood/adolescence		*RYR2*	1q42
ARVC3, 4, 5, 6, 7	Childhood/adolescence			14q12, 2q32, 10p14, 10q22
ARVC8	Childhood/adolescence		*Desmoplakin*	6p24
ARVC9	Childhood/adolescence		*PKP2—plakophilin-2*	12p11
Naxos disease (autosomal recessive)	Childhood		*JUP—plakoglobin*	17q21

predominance of 8:1, and the average age of arrhythmic events is 40 years. The definitive treatment is an implantable defibrillator and exercise is not a particular risk factor. Mutations in the *SCN5A* gene are found in about 20% of Brugada syndrome patients, as well as some cases of LQT3 (see Table 19.4). In some families both arrhythmias occur.

ARVC, which follows mainly dominant inheritance, is characterized by localized or diffuse atrophy and fatty infiltration of the right ventricular myocardium. It can lead to VT and sudden cardiac death in young people, especially athletes with apparently normal hearts. The ECG shows right precordial T-wave inversion and prolongation of the QRS complex. ARVC appears to demonstrate substantial genetic heterogeneity (see Table 19.4) with five genes identified, one of which, encoding *plakoglobin*, is implicated in the rare recessive form found on the island of Naxos. The *RYR2* gene, for ARVC2, is also mutated in **catecholaminergic polymorphic ventricular tachycardia (CPVT)**, also known as **Coumel's VT**. Individuals with CPVT present with syncopal events, sometimes in childhood or adolescence, and reproducible stress-induced ventricular tachycardia, without a prolonged QT interval; the heart is structurally normal.

Genetics

These are genetically heterogeneous conditions. Nearly all follow autosomal dominant inheritance; the genes and their loci are summarized in Table 19.4. In some cases, however, there is evidence for biallelic inheritance (p. 119)—i.e., patients require mutations at two different loci to have clinical symptoms and ECG changes. This poses very significant difficulties in relation to genetic testing strategies, interpretation of mutation tests, and the usefulness of predictive genetic testing based on the findings at one locus. The same problem may also apply to some cases of cardiomyopathy.

Inherited Cardiomyopathies

Dilated cardiomyopathy is characterized by cardiac dilatation and reduced systolic function. Causes include myocarditis, coronary artery disease, systemic and metabolic diseases, and toxins. When these are excluded the prevalence of idiopathic dilated cardiomyopathy is 35 to 40 per 100,000 and familial cases account for about 25%. As with the inherited cardiac arrhythmias, they are genetically heterogeneous but nearly always follow autosomal dominant inheritance. They are also very variable, and within the same family affected members may show symptoms in childhood at one end of the spectrum, whereas in other individuals the onset of cardiac symptoms may not occur until late in adult life. At least 10 different loci have been mapped in different family studies. One cause is the result of mutations in the *LMNA* gene (which encodes lamin A/C), noted for its pleiotropic effects (p. 112), of which dilated cardiomyopathy is one and may occasionally be isolated.

Hypertrophic cardiomyopathy is similarly genetically heterogeneous but the large majority follow autosomal dominant inheritance. The group includes asymmetric septal hypertrophy, hypertrophic subaortic stenosis, and ventricular hypertrophy. The most common single gene involved appears to be that which encodes the cardiac β-myosin heavy chain (*MYH7*) on chromosome 14q but, again, there are at least a further eight loci mapped for genes encoding different cardiac muscle proteins. Sudden death can occur, especially in young athletes. Notable is cardiomyopathy due to mutations in the gene encoding the 'T' isoform of cardiac troponin *(TNNT2)*, located on chromosome 1q32. This isoform is not expressed in skeletal muscle but, when mutated, a mild and sometimes subclinical hypertrophy results. Unfortunately, there is a high incidence of sudden death.

Genetic testing is now available within clinical services, but the vast genetic heterogeneity means that the pick-up rate for mutations is low. After a diagnosis has been made in an index case, a detailed family history is indicated and investigation by ECG and echocardiogram should be offered. Screening may need to continue well into adult life. Among the causes of cardiomyopathy that can be detected relatively easily by a biochemical test is X-linked Fabry disease, for which enzyme replacement is available (see Table 23.1; p. 350).

Spinal Muscular Atrophy

Spinal muscular atrophy (SMA) is the term used to describe a clinically and genetically heterogeneous group of disorders that are among the most common genetic causes of death in childhood. The disease is characterized by degeneration of the anterior horn cells of the spinal cord leading to progressive muscle weakness and ultimately death.

Three common childhood forms of SMA are recognized with a collective incidence of approximately 1:10,000 and carrier frequency of about 1:50. Of these, SMA type I is the most common and the most severe. Although three types are delineated here, it is now clear that the disease spectrum forms a continuum.

Clinical Features

Spinal Muscular Atrophy Type I (Werdnig-Hoffmann Disease)

This form of SMA presents at birth or in the first 6 months of life with severe hypotonia and lack of spontaneous movement. Sometimes the mother will have noticed that intrauterine fetal movements were reduced in both strength and frequency. These children show normal intellectual activity but their profound muscle weakness, which affects swallowing and respiratory function, leads to death within the first 2 years of life. The diagnosis is confirmed by electromyography and there is no effective means of treating the disorder or even delaying its rate of progression.

Spinal Muscular Atrophy Type II

This is less severe than type I, with an age of onset between 6 and 18 months. As with type I, muscle weakness and hypotonia are the main presenting features. These children can sit unaided but are never able to achieve independent locomotion. The rate of progression is slow and most affected children survive into early adult life.

Spinal Muscular Atrophy Type III (Kugelberg-Welander Disease)

This relatively mild form of childhood or juvenile-onset SMA presents after 18 months and all patients are able to walk without support. Slowly progressive muscle weakness results in many affected individuals having to use a wheelchair by early adult life. Long-term survival can be jeopardized by recurrent respiratory infection and the development of a scoliosis caused by weakness of the spinal muscles.

Genetics

All three types of childhood-onset SMA show autosomal recessive inheritance. Several other much rarer forms of SMA have been described, and the late, adult onset forms may follow autosomal dominant inheritance. SMA type I generally shows a high degree of intrafamilial concordance, with affected siblings showing an almost identical clinical course. In types II and III, intrafamilial variation can be quite marked. In all childhood-onset SMA, the predominant gene involved is *SMN1*.

Mapping and Isolating the Spinal Muscular Atrophy Gene

All three childhood forms of SMA were mapped to chromosome 5q in 1990 using linkage. Detailed mapping narrowed the locus to a 1000 kb consisting of a 500-kb inverted duplication (Figure 19.10). This region is noted for its high rate of instability, with several DNA duplications and a relatively large number of pseudogenes (p. 17).

Within the candidate region two distinct genes were isolated that show a high incidence of deletion in patients with SMA—*SMN* and *NAIP*—each present as two almost identical copies. The *SMN* genes are now referred to as *SMN1* and *SMN2* (the pseudogene of *SMN1* that shares ~99% homology). *SMN1* shows homozygous deletion of exons 7–8 in 95% to 98% of all patients with childhood-onset SMA. Point mutations in *SMN1* have been identified in 1% to 2% of patients with childhood SMA who do not show the exons 7–8 deletion on one allele. The number of copies of *SMN2*, arranged in tandem in *cis* configuration on each chromosome, varies between zero and five. It produces a similar transcript to *SMN1* but this is not sufficient to fully

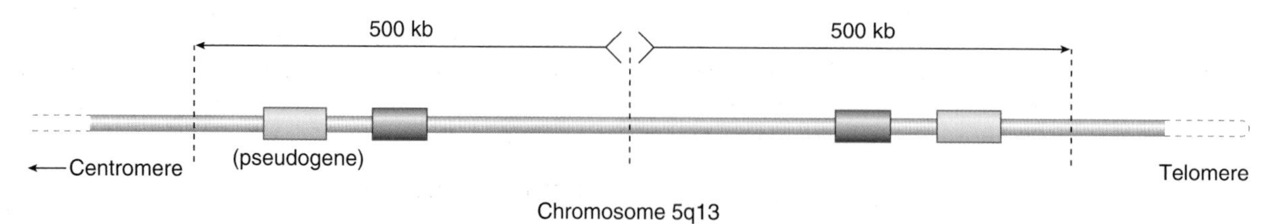

FIGURE 19.10 The inverted duplication with the *SMN* and *NAIP* genes. SMA occurs when both copies of the *SMN1* gene are mutated (autosomal recessive inheritance); in 95% to 98% this is a deletion of exons 7–8, and point mutations in the remainder. *SMN*, Survival motor neuron; *NAIP*, neuronal apoptosis-inhibitory protein.

compensate. Nevertheless, the presence of copies of *SMN2* modifies the phenotype, causing milder forms of SMA.

The other gene originally isolated, *NAIP*, codes for the neuronal apoptosis-inhibitory protein, and is deleted in ~45% of individuals with SMA type I and ~20% of those with SMA types II and III. However, for the purposes of clinical molecular genetics, it is no longer considered relevant.

Clinical Applications and Future Prospects

SMN1 is always mutated in SMA, in the vast majority by deletion of exons 7–8, and in the remainder by point mutation. Diagnostic testing is therefore very reliable and prenatal testing is an option for those couples who request it, assuming both parents are carriers. Carrier detection is based on determining the number of exon 7–containing *SMN1* gene copies present in an individual. However, results can be difficult to interpret because some carriers have the normal number of *SMN1* gene copies caused by the presence either of two *SMN1* gene copies in *cis* configuration on one chromosome, or of a *SMN1* point mutation. About 4% of the general population has two copies of *SMN1* on a single chromosome. Furthermore, 2% of individuals with SMA have one de novo mutation, meaning that only one parent is a carrier. Because of these difficulties, SMA carrier testing should be provided in the context of formal, expert genetic counseling.

Duchenne Muscular Dystrophy

Duchenne muscular dystrophy (DMD) is the most common and most severe form of muscular dystrophy. The eponymous title is derived from the French neurologist Guillaume Duchenne, who described a case in 1861. A similar but milder condition, Becker muscular dystrophy (BMD), is caused by mutations in the same gene. The incidences of DMD and BMD are approximately 1 : 3500 males and 1 : 20,000 males, respectively. Great efforts are being directed at devising novel treatments for these serious disorders.

Clinical Features

Males with DMD usually present between the ages of 3 and 5 years with slowly progressive muscle weakness resulting in an awkward gait, inability to run quickly, and difficulty in rising from the floor, which can be achieved only by pushing on, or 'climbing up', the legs and thighs (Gowers' sign). Most affected boys have to use a wheelchair by the age of 11 years because of severe proximal leg muscle weakness. Subsequent deterioration leads to lumbar lordosis, joint contractures, and cardiorespiratory failure, resulting in death at a mean age of 18 years without aggressive supportive measures.

On examination, boys with DMD show an apparent increase in the size of the calf muscles, which is actually due to replacement of muscle fibers by fat and connective tissue—referred to as **pseudohypertrophy** (Figure 19.11).

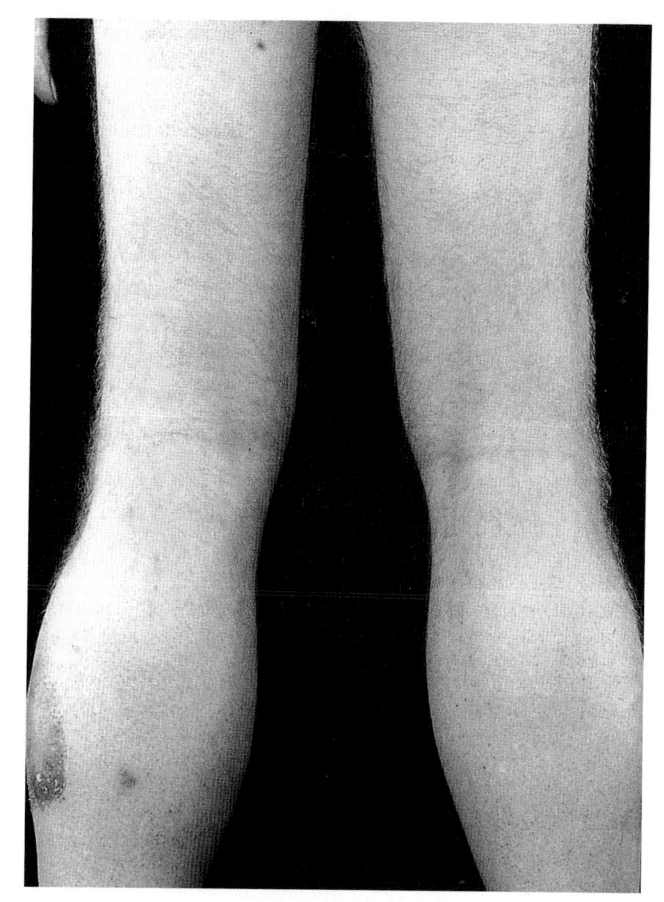

FIGURE 19.11 Lower limbs of an adult male with Becker muscular dystrophy showing proximal wasting and calf pseudohypertrophy.

DMD is sometimes known as **pseudohypertrophic muscular dystrophy**. In addition, approximately one-third of boys with DMD show mild-moderate intellectual impairment, with the mean IQ being 83.

In BMD the clinical picture is very similar but the disease process runs a much less aggressive course. The mean age of onset is 11 years and many patients remain ambulant until well into adult life. Overall life expectancy is only slightly reduced. A few patients with proven mutations in the DMD/BMD gene have been asymptomatic in their fifth or sixth decade.

Genetics

Both DMD and BMD show X-linked recessive inheritance. Males with DMD rarely, if ever, reproduce. Therefore, as genetic fitness equals zero, the mutation rate equals the incidence in affected males divided by 3 (p. 133), which approximates to 1 : 10,000—one of the highest known mutation rates in humans.

Isolation of the Gene for DMD

The isolation of the gene for DMD—the *dystrophin gene*—represented a major scientific achievement at the time, because of a successfully applied positional cloning strategy.

The initial clue to the site of the DMD locus was provided by reports of several females affected with DMD who had a balanced X-autosome translocation with a common X-chromosome breakpoint at Xp21. In these women, those cells in which the derivative X chromosome is randomly inactivated are at a major disadvantage because of inactivation of the autosomal segment (Figure 7.16; p. 117). Consequently, cells in which the normal X chromosome has been randomly inactivated are more likely to survive. The net result is that the derivative X autosome is active in most cell lines, and if the breakpoint has damaged an important gene, in this case dystrophin, the woman will be affected by the disease in question.

Further support for mapping came from affected males with visible microdeletions involving Xp21 and the next phase of gene isolation was the identification of conserved sequences in muscle cDNA libraries that were shown to be exons from the gene itself. The task was completed in 1987.

The *dystrophin* gene is huge in molecular terms, consisting of 79 exons and 2.3 Mb of genomic DNA, though only 14 kb are transcribed into mRNA. It is transcribed in brain as well as muscle, which explains why some boys with DMD show learning difficulties. The large size may explain the high mutation rate.

Mutations in the Dystrophin Gene

Deletions, which can be almost any size and location, account for two-thirds of all *dystrophin* gene mutations and arise almost exclusively in maternal meiosis, probably due to unequal crossing over. A smaller number of affected males with **duplications** have been described. Deletion 'hotspots' are found in the first 20 exons as well as around exons 45 through 53. One of the deletion breakpoint 'hotspots' in intron 7 contains a cluster of transposon-like repetitive DNA sequences that could facilitate misalignment in meiosis, with a subsequent crossover leading to deletion and duplication products.

Deletions that cause DMD usually disturb the translational reading frame (p. 20), but those seen in males with BMD usually do not alter the reading frame (i.e., they are 'in-frame'). This means that the amino-acid sequence of the protein product downstream of the deletion is normal, explaining the relatively mild features in BMD. Mutations in the remaining one-third of boys with DMD include stop codons, frameshift mutations, altered splicing signals and promoter mutations. Most lead to premature translational termination, resulting in the production of little, if any, protein product. In contrast to deletions, point mutations in the *dystrophin* gene often arise in paternal meiosis, most probably because of a copy error in DNA replication. Full sequencing of the *dystrophin* gene is now available as a service, which has transformed molecular diagnosis of DMD and carrier detection.

The Gene Product Dystrophin

The 427-kDa *dystrophin* protein is located close to the muscle membrane, where it links intracellular actin with extracellular laminin. Absence of dystrophin, as in DMD, leads gradually to muscle cell degeneration. The presence of dystrophin in a muscle biopsy sample can be assessed by immunofluorescence. Levels less than 3% are diagnostic. In muscle biopsies from males with BMD, the dystrophin shows qualitative rather than gross quantitative abnormalities.

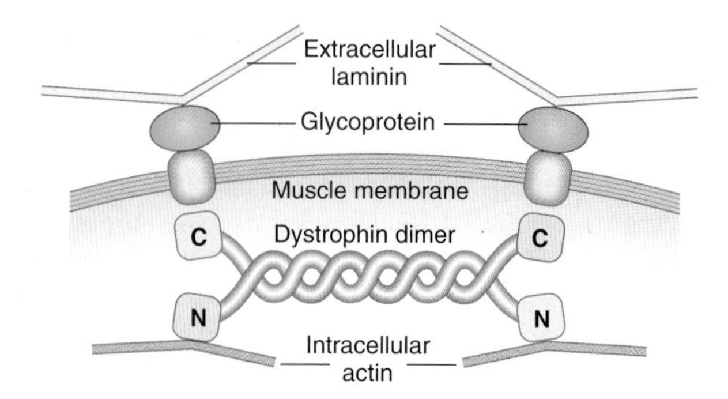

FIGURE 19.12 Probable structure of the dystrophin protein molecule, depicted as a dimer linking intracellular actin with extracellular laminin. (Adapted from Ervasti JM, Campbell KP 1991 Membrane organization of the dystrophin–glycoprotein complex. Cell 66:1121–1131.)

Dystrophin binds to a glycoprotein complex in the muscle membrane through its C-terminal domain (Figure 19.12). This glycoprotein complex consists of several subunits, abnormalities of which cause other rare genetic muscle disorders, including several different types of autosomal recessive limb girdle muscular dystrophy, as well as congenital muscular dystrophy.

Carrier Detection

Before DNA analysis, carrier detection was based on pedigree information combined with serum creatine kinase (CK) assay (p. 314). CK levels are grossly increased in boys with DMD, and marginally raised in approximately two-thirds of all carriers (see Figure 20.1; p. 314). CK levels are only occasionally useful today, as DNA testing has become increasingly sophisticated. Linkage studies may still be useful in circumstances where no DNA is available from an affected male, but perhaps available from normal males in the same family; each situation has to be assessed individually. Care has to be taken when using linkage for carrier detection because of the high recombination rate of 12% across the DMD gene.

Prospects for Treatment

At present, there is no cure for DMD or BMD, although physiotherapy is beneficial for maintaining mobility and preventing muscle spasm and joint contractures.

Gene therapy offers the only realistic hope of a cure in the short to medium term. Several approaches have been tried experimentally in transgenic and naturally occurring mutant mice with dystrophin-negative muscular dystrophy.

These include direct injection of recombinant DNA, myoblast implantation, and transfection with retroviral or adenoviral vectors carrying a dystrophin minigene containing only those sequences that code for the important functional domains. One approach is antisense technology to block an exon splicing enhancer sequence and generate a protein with an in-frame deletion that encodes a protein with some residual function (i.e., a BMD rather than a DMD phenotype). That mice with dystrophin-negative muscular dystrophy can show spontaneous muscle repair indicates that there could be a way of switching on an alternative compensatory protein such as *utrophin*. This is expressed in the fetus rather than dystrophin, with which it shares a large degree of homology. Genetically engineered mice deficient for both dystrophin and utrophin develop a typical DMD dystrophy. If the utrophin gene could be reactivated as in the dystrophin-negative mouse, this may have a therapeutic benefit.

Hemophilia

There are two forms of hemophilia: A and B. **Hemophilia A** is the most common severe inherited coagulation disorder, with an incidence of 1:5000 males. It is caused by a deficiency of factor VIII, which, together with factor IX, plays a critical role in the intrinsic pathway activation of prothrombin to thrombin. Thrombin then converts fibrinogen to fibrin, which forms the structural framework of clotted blood. The existence of hemophilia was recognized in the Talmud, and the tendency for males to be affected much more often than females was acknowledged by the Jewish authorities 2000 years ago when they excused from circumcising the sons of the sisters of a mother who had an affected son. Queen Victoria was a carrier and, as well as having an affected son—Leopold Duke of Albany—she transmitted the disorder through two of her daughters to most of the royal families of Europe (Figure 19.13).

Hemophilia B affects approximately 1:40,000 males and is caused by deficiency of factor IX. It is also known as **Christmas disease**, whereas hemophilia A is sometimes referred to as 'classic hemophilia'.

Clinical Features

These are similar in both forms of hemophilia and vary from mild bleeding following major trauma or surgery to spontaneous hemorrhage into muscles and joints. The degree of severity shows a close correlation with the reduction in factor VIII or IX activity. Levels below 1% are usually associated with a severe hemorrhagic tendency from birth. Hemorrhage into joints causes severe pain and swelling which, if recurrent, causes a progressive arthropathy with severe disability (Figure 19.14). Affected family members generally show the same degree of severity.

Genetics

Both forms of hemophilia show X-linked recessive inheritance. The loci lie close together near the distal end of Xq.

Hemophilia A

The factor VIII comprises 26 exons and spans 186 kb with a 9-kb mRNA transcript. Deletions account for 5% of all cases and usually cause complete absence of factor VIII expression. In addition, hundreds of frameshift, nonsense, and missense mutations have been described, besides insertions and a 'flip' inversion, which represented a new form of mutation when first identified in hemophilia A in 1993. Inversions account for 50% of all severe cases with <1% factor VIII activity. They are caused by recombination between a small gene called *A* located within intron 22 of the factor VIII gene and other copies of the *A* gene, which are located upstream near the telomere (Figure 19.15). The inversion disrupts the *factor VIII* gene, resulting in very low factor VIII activity. The genetic test is straightforward but

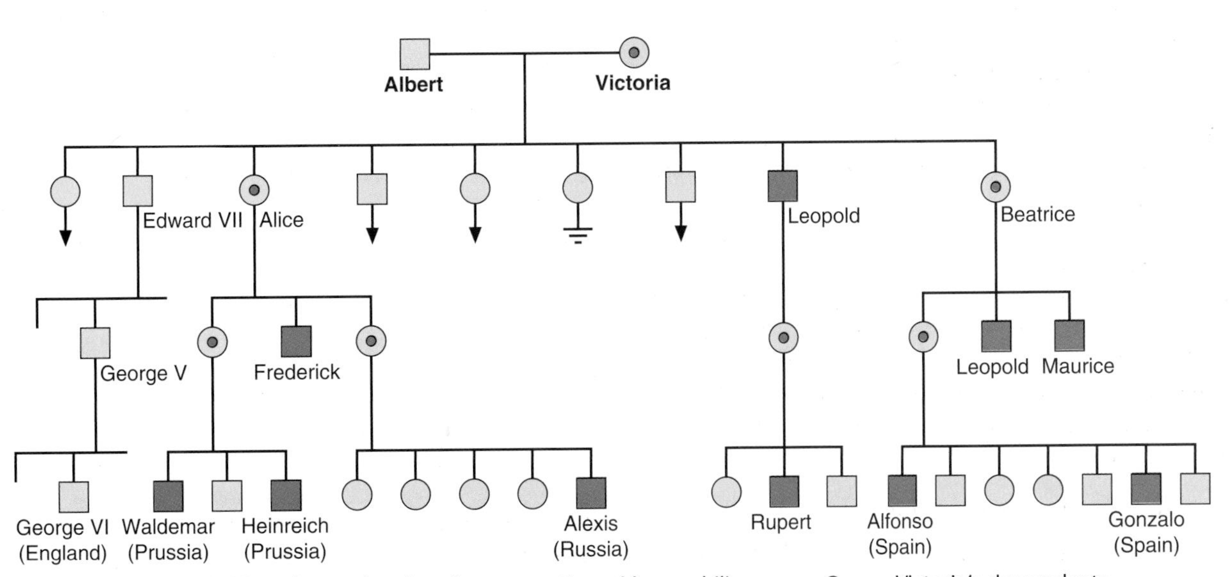

FIGURE 19.13 Pedigree showing the segregation of hemophilia among Queen Victoria's descendants.

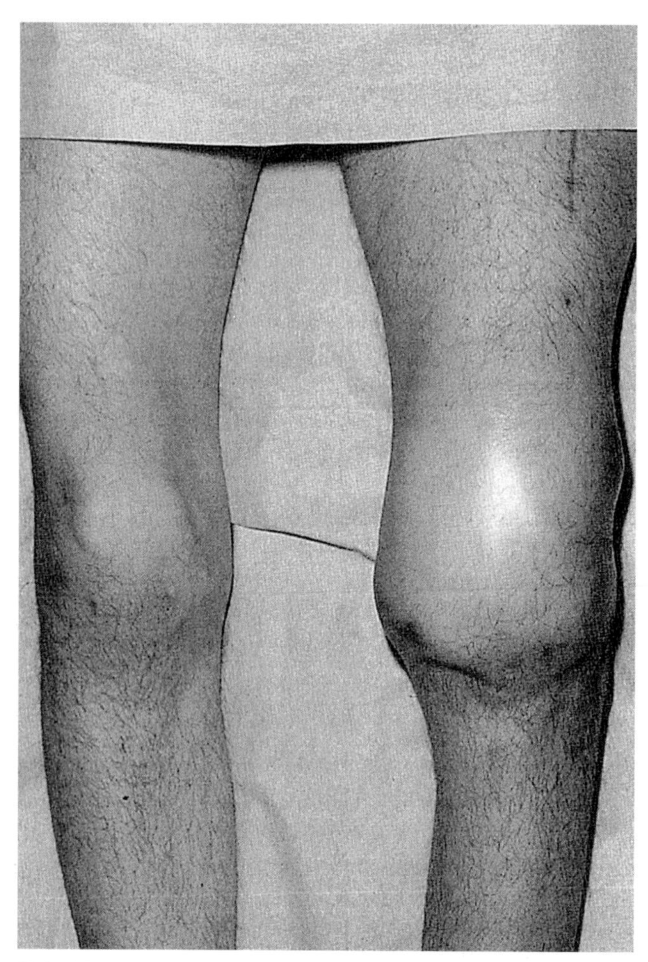

FIGURE 19.14 Lower limbs of a male with hemophilia showing the effect of recurrent hemorrhage into the knees. (Courtesy Dr. G. Dolan, University Hospital, Nottingham, UK.)

detection of the numerous other mutations may require direct sequencing.

As in DMD, point mutations usually originate in male germ cells whereas deletions arise mainly in the female. The 'flip' inversions show a greater than 10-fold higher mutation rate in male compared with female germ cells, probably because Xq does not pair with a homologous chromosome in male meiosis—so that there is much greater opportunity for **intrachromosomal** recombination to occur via looping of the distal long arm (see Figure 19.15).

Factor VIII levels are about half normal in carrier females and many are predisposed to a bleeding tendency. Carrier detection used to be based on assay of the ratio of factor VIII coagulant activity to the level of factor VIII antigen but, as with CK assay in DMD, this is not always discriminatory, and direct mutation analysis is now routine. Linkage analysis may sometimes be necessary to resolve carrier status.

Hemophilia B

The factor IX gene comprised 8 exons and is 34 kb long. More than 800 different point mutations, deletions, and insertions have been reported but analysis of only 2.2 kb of the gene detects the mutation in 96% of all patients. A rare variant form known as **hemophilia B Leyden** shows the extremely unusual characteristic of age-dependent expression. During childhood the disease is very severe, with factor IX levels of less than 1%. After puberty the levels rise to around 50% of normal and the condition resolves to become asymptomatic. Hemophilia B Leyden has been shown to be caused by mutations in the promoter region.

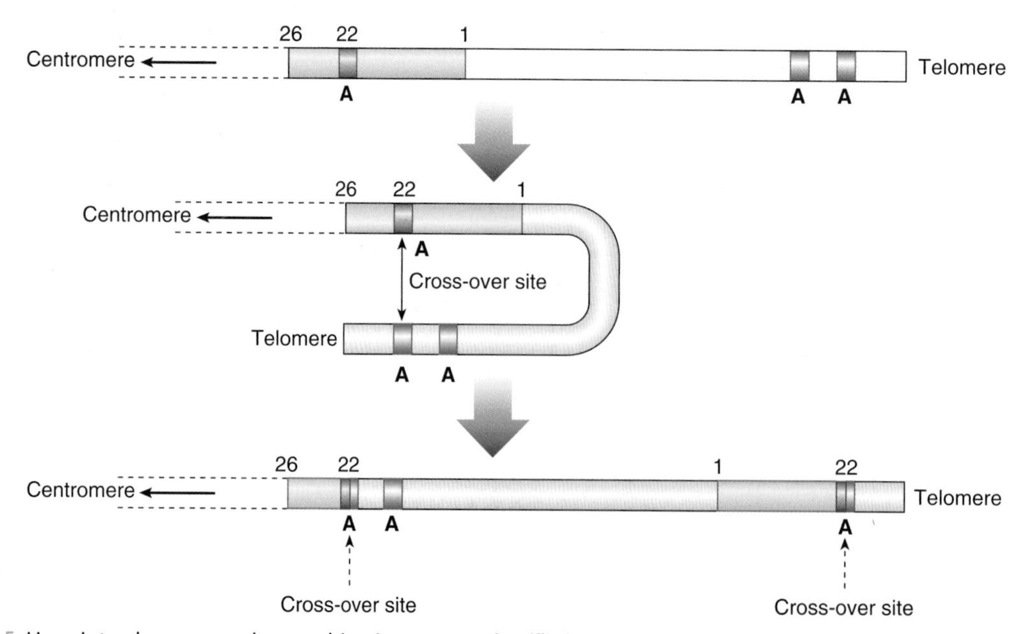

FIGURE 19.15 How intrachromosomal recombination causes the 'flip' inversion, which is the most common mutation found in severe hemophilia A. (Adapted from Lakich D, Kazazian HH, Antonarakis SE, Gitschier J 1993 Inversions disrupting the factor VIII gene are a common cause of severe hemophilia A. Nat Genet 5:236–241.)

Factor IX levels are now rarely used in carrier detection and prenatal diagnosis, both of which are achieved much more reliably by direct mutation analysis.

Treatment

Protein Substitution

Both forms of hemophilia have been treated successfully for many years using plasma-derived factor VIII or factor IX. Factor VIII is concentrated in the cryoprecipitate fraction of plasma and has been used widely for replacement therapy. It has a half-life of 8 h, so repeated infusions are necessary for elective surgery or major trauma.

Two major disadvantages emerged using this approach. The first was that the purification process for preparation of cryoprecipitate did not prevent transmission of viral infection such as hepatitis B and HIV, with the inevitable disastrous consequence that many males with hemophilia developed acquired immune deficiency syndrome (AIDS). In due course, better purification and screening processes were developed. Also, recombinant factor VIII became available in 1994, though expensive. The second disadvantage was that ~10% of patients with both forms of hemophilia developed inhibitory antibodies to the relevant factor, which their immune systems recognized as a foreign agent. This problem can sometimes be overcome by using porcine factor VIII or by immunosuppression.

Gene Therapy

Hemophilia A and B are excellent candidates for gene therapy as only a slight increase in the plasma level of the relevant factor is of major clinical benefit. Trials in animal models (dogs and mice) for factor VIII, using adenoviral systems, have shown a decrease in severity for hemophilia A. The effect lasted for a number of months. Similar results have been shown in dogs and mice with hemophilia B, again using an adeno-associated viral vector expressing factor IX, injected into skeletal muscle.

The early results of similar studies in severely affected humans have been encouraging, with evidence that severe disease might be converted to a milder form. However, the small increases seen in factor levels was not sustained in the long term and there were transient side effects such as fever and mild thrombocytopenia from the high adenoviral load; this has led to the trials being discontinued. As well as these in vivo therapies, treatment attempts have also been made with ex vivo systems. Non-viral vectors and retroviral systems have been used in a mouse model with some success, as well as autologous fibroblasts in human subjects, which gave transient rises in factor VIII levels for a year without side effects. Despite these setbacks, hemophilia will remain a prime target for gene therapy.

FURTHER READING

Biros I, Forrest S 1999 Spinal muscular atrophy: untangling the knot? J Med Genet 36:1–8

A contemporary account of current understanding of the genetic basis of childhood spinal muscular atrophy.

Bolton-Maggs PHB, Pasi KJ 2003 Haemophilias A and B. Lancet 361:1801–1809

An excellent recent review.

Brown T, Schwind EL 1999 Update and review: cystic fibrosis. J Genet Counseling 8:137–162

A useful review of recent genetic developments in cystic fibrosis.

Collinge J 1997 Human prion diseases and bovine spongiform encephalopathy (BSE). Hum Mol Genet 6:1699–1705

A clear account of human prion diseases and their known causes with particular reference to BSE.

De Paepe A, Devereux RB, Hennekam RCM, et al 1996 Revised diagnostic criteria for the Marfan syndrome. Am J Med Genet 62:417–426

Essential reading for those required to make a diagnosis of Marfan syndrome.

Emery AEH 1993 Duchenne muscular dystrophy, 2nd edn. Oxford, UK: Oxford University Press

A detailed monograph reviewing the history, clinical features and genetics of Duchenne and Becker muscular dystrophy.

Harper PS 1996 Huntington's disease, 2nd ed. London: WB Saunders

A comprehensive review of the clinical and genetic aspects of Huntington disease.

Harper PS: Myotonic Dystrophy, 3rd ed. WB Saunders, London, 2001.

A comprehensive review of the clinical and genetic aspects of myotonic dystrophy.

Huson SM, Hughes RA C (eds) 1994 The neurofibromatoses. London: Chapman & Hall,

A very thorough description of the different types of neurofibromatosis. Includes a chapter on the 'Elephant Man'.

Karpati G, Pari G, Molnar MJ 1999 Molecular therapy for genetic muscle diseases—status 1999. Clin Genet 55:1–8

An optimistic review of possible approaches to gene therapy for inherited muscle disorders such as Duchenne muscular dystrophy.

Kay MA, Manno CS, Ragni MV, et al 2000 Evidence for gene transfer and expression of factor IX in haemophilia B patients treated with an AV vector. Nature Genet 24:257–261

Report of provisional encouraging results of gene therapy in patients with hemophilia B.

Lakich D, Kazazian HH, Antonarakis SE, Gitschier J 1993 Inversions disrupting the factor VIII gene are a common cause of severe haemophilia A. Nature Genet 5:236–241

The first report showing how the common 'flip' inversion is generated.

ELEMENTS

1. Huntington disease is an autosomal dominant disorder characterized by choreiform movements and progressive dementia. The disease locus has been mapped to the short arm of chromosome 4 and the mutational basis involves expansion of a CAG triple repeat sequence. Meiotic instability is greater in the male than in the female, which probably explains why the severe 'juvenile'-onset form is almost always inherited from a more mildly affected father.

2. Myotonic dystrophy shows autosomal dominant inheritance and is characterized by slowly progressive weakness and myotonia. The disease locus lies on chromosome 19q and the mutation is the expansion of an unstable CTG triple repeat sequence. The range of meiotic expansion is greater in females, almost certainly accounting for the almost exclusive maternal inheritance of the severe 'congenital' form.

3. Hereditary motor and sensory neuropathy (HMSN) includes several clinically and genetically heterogeneous disorders characterized by slowly progressive distal muscle weakness and wasting. HMSN-Ia, the most common form, is due to duplication of the *PMP22* gene on chromosome 17p, which encodes a protein present in the myelin membrane of peripheral nerve. The reciprocal deletion product of the unequal crossover leads to a mild disorder known as hereditary liability to pressure palsies.

4. Neurofibromatosis type I (NF1) shows autosomal dominant inheritance with complete penetrance and variable expression. The NF1 gene is located on chromosome 17q and encodes a protein known as neurofibromin. This normally acts as a tumor suppressor by inactivating the RAS-mediated signal transduction of mitogenic signaling.

5. Cystic fibrosis (CF) shows autosomal recessive inheritance and is characterized by recurrent chest infection and malabsorption. The CF locus lies on chromosome 7, where the gene (*CFTR*) encodes the CF transmembrane receptor protein. This acts as a chloride channel and controls the level of intracellular sodium chloride, which in turn influences the viscosity of mucous secretions.

6. The childhood forms of spinal muscular atrophy (SMA) are characterized by hypotonia and progressive muscle weakness. They show autosomal recessive inheritance and the disease locus has been mapped to chromosome 5q13. This region shows a high incidence of instability, with duplication of a 500-kb fragment containing *SMN* genes and a characteristic deletion in most patients.

7. Duchenne muscular dystrophy (DMD) shows X-linked recessive inheritance, with most carriers being entirely healthy. The DMD locus lies at chromosome Xp21 and is the largest known in humans. The gene product, dystrophin, links intracellular actin with extracellular laminin. The commonest mutational mechanism is a deletion that disturbs the translational reading frame. Deletions that maintain the reading frame cause the milder Becker form of muscular dystrophy.

8. Hemophilia A is the most common severe inherited coagulation disorder in humans. It shows X-linked recessive inheritance and is caused by a deficiency of factor VIII. The most common mutation in severe hemophilia A is caused by a 'flip' inversion that disrupts the factor VIII gene at intron 22. Treatment with factor VIII replacement therapy is very effective and the results of gene therapy in animal models offer hope that this will soon be possible in humans.

CHAPTER 20

Screening for Genetic Disease

Genetic disease affects individuals and their families dramatically but every person, and every couple having children, is at some risk of seeing a disorder with a genetic component suddenly appear. Our concepts and approaches to screening reflect the different burdens that these two realities impose. First, there is screening of individuals and couples known to be at significant or high risk because of a positive family history—sometimes referred to as **targeted**, or **family, screening**. This includes **carrier**, or **heterozygote, screening**, as well as **presymptomatic testing**. Second, there is the screening offered to the general population, who are at low risk—sometimes referred to as **community genetics**. Population screening involves the offer of genetic testing on an equitable basis to all relevant individuals in a defined population. Its primary objective is to enhance autonomy by enabling individuals to be better informed about genetic risks and reproductive options. A secondary goal is the prevention of morbidity resulting from genetic disease and alleviation of the suffering this would impose.

Screening Those at High Risk

Here we focus on the very wide range of general genetic disease as opposed to screening in the field of cancer genetics, which is addressed in Chapter 14. Prenatal screening is also covered in more detail in the next chapter. If it were easy to recognize carriers of autosomal and X-linked recessive disorders and persons who are heterozygous for autosomal dominant disorders that show reduced penetrance or a late age of onset, much doubt and uncertainty would be removed when providing information in genetic counseling. Increasingly, mutation analysis in genes that cause these disorders is indeed making the task easier. Where this is not possible, either because no gene test is available or the molecular pathology cannot be easily detected in a gene known to be associated with the disorder in question, a number of strategies and types of analysis is available to detect carriers for autosomal and X-linked recessive disorders, and for presymptomatic diagnosis of heterozygotes for autosomal dominant disorders.

Carrier Testing for Autosomal Recessive and X-Linked Disorders

In a number of autosomal recessive disorders, such as some of the inborn errors of metabolism (e.g., Tay-Sachs disease; p. 178) and the hemoglobinopathies (e.g., sickle cell disease; p. 159), carriers can be recognized with a high degree of certainty using biochemical or hematological techniques such that DNA analysis is not necessary. In other single-gene disorders, it is possible to detect or confirm carrier status by biochemical means in only a proportion of carriers; for example, mildly abnormal coagulation study results in a woman at risk of being a carrier for hemophilia (p. 309). A significant proportion of obligate carriers of hemophilia will have normal coagulation, however, so a normal result does not exclude a woman at risk from being a carrier.

There are several possible ways in which carriers of genetic diseases can be recognized.

Clinical Manifestations in Carriers

Occasionally, carriers for certain disorders can have mild clinical manifestations of the disease (Table 20.1), particularly with some of the X-linked disorders. These manifestations are usually so slight that they are apparent only on careful clinical examination. Such manifestations, even though minimal, are unmistakably pathological; for instance, the mosaic pattern of retinal pigmentation seen in manifesting female carriers of X-linked ocular albinism. Unfortunately, in most autosomal and X-linked recessive disorders there are either no manifestations at all in carriers, or they overlap with the variation seen in the general population. An example would be female carriers of hemophilia, who have a tendency to bruise easily—this is not a reliable sign of carrier status, as this is seen in a significant proportion of the general population, for different reasons. In X-linked adrenoleukodystrophy, a proportion of carrier females manifest neurological features, sometimes relatively late in life when the signs might easily be confused with the problems of aging. Thus clinical manifestations are helpful in detecting carriers only when they are unmistakably pathological, which is the exception rather than the rule with most single-gene disorders.

Biochemical Abnormalities in Carriers

By far the most important approach to determining the carrier status for autosomal recessive and X-linked disorders has been the demonstration of detectable biochemical abnormalities in carriers of certain diseases. In some disorders, the biochemical abnormality seen is a direct product of the gene and the carrier status can be tested for with confidence. For example, in carriers of Tay-Sachs disease the range of enzyme activity (**hexosaminidase**) is intermediate

Table 20.1 Clinical and Biochemical Abnormalities Used in Carrier Detection of X-Linked Disorders

Disorder	Abnormality
Clinical	
Ocular albinism	Mosaic retinal pigmentary pattern
Retinitis pigmentosa	Mosaic retinal pigmentation, abnormal electroretinographic findings
Anhidrotic ectodermal dysplasia	Sweat pore counts reduced, dental anomalies
Lowe syndrome	Lens opacities
Alport syndrome	Hematuria
Biochemical	
Hemophilia A	Reduced factor VIII activity : antigen ratio
Hemophilia B	Reduced levels of factor IX
G6PD deficiency	Erythrocyte G6PD activity reduced
Lesch-Nyhan syndrome fibroblasts	Reduced hypoxanthine-guanine phosphoribosyl transferase activity in skin
Hunter syndrome	Reduced sulfoiduronate sulfatase activity in skin fibroblasts
Vitamin D–resistant rickets	Serum phosphate level reduced
Duchenne muscular dystrophy	Raised serum creatine kinase level
Becker muscular dystrophy	Raised serum creatine kinase level
Fabry disease	Reduced α-galactosidase activity in hair root follicles

G6PD, Glucose 6-phosphate dehydrogenase.

between levels found in normal and affected people. In many inborn errors of metabolism, however, the enzyme activity levels in carriers overlap with those in the normal range, so that it is not possible to distinguish reliably between heterozygote carriers and those who are homozygous normal.

Carrier testing for Tay-Sachs disease in many orthodox Jewish communities, which are at significantly increased risk of the disorder, is highly developed. Because of faith-based objections to termination of pregnancy, carrier testing may be crucial in the selection of life partners. A couple considering betrothal will first see their rabbi. In addition to receiving spiritual advice, they will undergo carrier testing for Tay-Sachs disease. If both prove to be carriers, the proposed engagement will be called off, leaving them free to look for a new partner. If only one proves to be a carrier the engagement can proceed, although the rabbi does not disclose which one is the carrier. Although such a strategy to prevent genetic disease may be possible in many communities where inbreeding is the norm, and their 'private' diseases have been well characterized either biochemically or by molecular genetics, in practice this is very rare.

In many single-gene disorders, the biochemical abnormality used in the diagnosis of the disorder in the affected individual is not a direct result of action of the gene product but the consequence of a secondary or downstream process. Such abnormalities are further from the primary action of the gene and, consequently, are usually even less likely to be useful in identifying carriers. For example, in Duchenne muscular dystrophy (DMD) there is an increased permeability of the muscle membrane, resulting in the escape of muscle enzymes into the blood. A grossly raised serum creatine kinase (CK) level often confirms the diagnosis of DMD in a boy presenting with features of the disorder (p. 307). Obligate female carriers of DMD have, on average, serum CK levels that are increased compared with those of the general female population (Figure 20.1). There is, however, a substantial overlap of CK values between normal and obligate carrier females. Nevertheless, this information can be used in conjunction with pedigree risk information (p. 344) and the results of linked DNA markers (p. 345) to help calculate the likelihood of a woman being a carrier for this disorder.

There is another reason for difficulty with carrier testing in the case of X-linked recessive disorders. Random inactivation of the X chromosome in females (p. 103) means that many, often the majority, of female carriers of X-linked disorders cannot be detected reliably by biochemical methods. An exception to this involves analysis of individual clones to look for evidence of two populations of cells, as with peripheral blood lymphocytes in female carriers

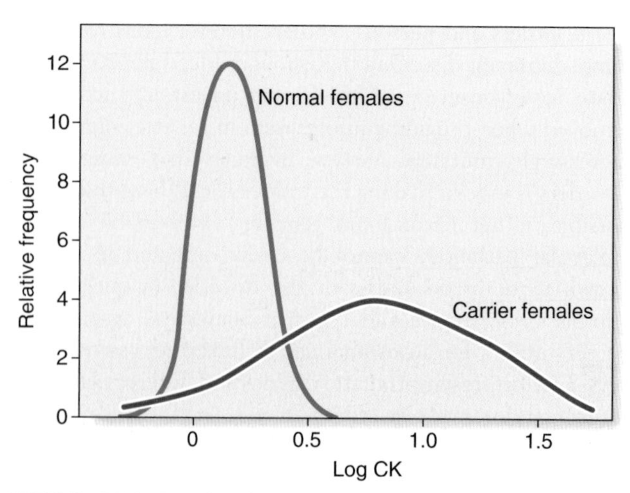

FIGURE 20.1 Creatine kinase (CK) levels in obligate carrier females of Duchenne muscular dystrophy and women from the general population. (Adapted from Tippett PA, Dennis NR, Machin D, et al 1982 Creatine kinase activity in the detection of carriers of Duchenne muscular dystrophy: comparison of two methods. Clin Chim Acta 121:345–359.)

of some of the X-linked immunodeficiency syndromes (p. 204). In a clinical setting, this is referred to as 'X-inactivation studies'.

Linkage between a Disease Locus and a Polymorphic Marker

DNA Polymorphic Markers

The advent of recombinant DNA technology revolutionized the approach to carrier detection. Relatively seldom, nowadays, do conventional biochemical and blood group polymorphisms have a role because there are few that are sufficiently informative to be of practical clinical value. The large number of different types of DNA sequence variants (p. 17) in the human genome means that, if sufficient numbers of families are available, linkage of any disease with a polymorphic DNA marker is possible, providing the disease is not genetically heterogeneous (see *locus heterogeneity* below). After this is achieved, the information can be applied to smaller families. The demonstration of linkage between a DNA sequence variant and a disease locus overcomes the need to identify a biochemical defect or protein marker and the necessity for it to be expressed in accessible tissues. In addition, use of markers at the DNA level also overcomes the difficulties that occur in carrier detection due to X-inactivation for women at risk for X-linked disorders (p. 103).

Linked polymorphic DNA markers were frequently used in determining the carrier status of females in families where DMD has occurred; nowadays, direct gene sequencing is more likely to be used. An example is shown in Figure 20.2; individual III_3 wants to know whether she is a carrier and therefore at risk for having sons affected with DMD.

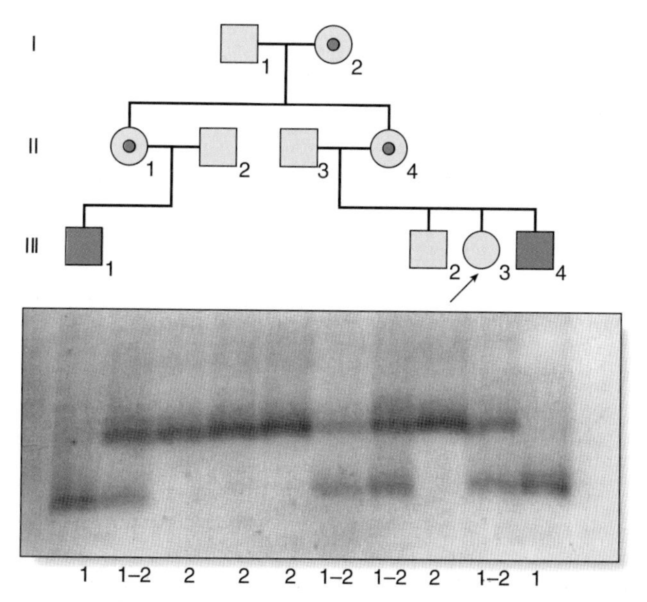

FIGURE 20.2 Family with Duchenne muscular dystrophy showing segregation of the CA repeat 5′ to the dystrophin gene known as Dys 5′ II. (Courtesy J. Rowland, Yorkshire Regional DNA Laboratory, St. James's Hospital, Leeds, UK.)

Analysis of the pedigree reveals that her mother, II_4, along with her sister, II_1, and their mother, I_2, are all obligate carriers of DMD. The family is informative for a polymorphic CA dinucleotide repeat (p. 69) in the closely flanking region 5′ to the dystrophin gene known as Dys 5′ II, which can be demonstrated by polymerase chain reaction (PCR) (p. 56). The mutation in the dystrophin gene in the family is segregating with allele 1 and, because individual III_3 has inherited this allele from her mother, she is likely to be a carrier. Linked polymorphic DNA markers can be used for prenatal diagnosis to predict whether a male fetus is likely to be affected with DMD, even without knowing the specific dystrophin gene mutation in the affected male(s) (p. 308).

Potential Pitfalls with Linked Polymorphic DNA Markers

A number of potential pitfalls should be kept in mind with the use of linked polymorphic DNA markers.

Recombination

The first potential pitfall is the chance of recombination occurring between the polymorphic DNA marker and the disease locus to which linkage has been shown. The risk of a recombination can be minimized, in most instances, by the identification of either intragenic or closely linked markers on *either side* of the disease locus—termed **flanking markers**. In some instances, for example the dystrophin locus, there is a 'hotspot' for recombination (p. 308). Even with closely flanking or intragenic markers there appears to be a minimal chance of approximately 12% that recombination will occur in any meiosis in a female. The uncertainty introduced by this possibility needs to be accounted for when combining the results of linked polymorphic DNA markers with pedigree risks and the results of CK testing for women at risk of being carriers of DMD (p. 308).

Sample Availability

The use of linked polymorphic DNA markers means that samples from the appropriate family members are required, and therefore their cooperation is essential. This can prove difficult, depending on relationships within the family and the need to maintain confidentiality. In addition, families (and their physicians) with inherited disorders that cause early demise; for example, the autosomal recessive disorder spinal muscular atrophy type 1 (Werdnig-Hoffmann disease), need foresight to arrange for DNA to be banked from the affected individual(s). In DMD it is not unusual that the affected male has passed away by the time a younger sister seeks advice about her carrier status. If the family structure is suitable (e.g., there are one or more *unaffected* males in the pedigree), it is often possible to 'reconstruct' the likely alleles of the linked polymorphic DNA markers in the affected male. This will, however, affect the risk estimation in the pedigree as the phase of the marker in the affected male cannot be known with certainty.

Polymorphic Variation

Another problem encountered in the use of linked DNA markers is whether the family possesses the necessary variation in a linked marker to be what is known as **informative**. This is increasingly rare due to the availability of large numbers of different DNA sequence variants in the human genome, for example tandem repeat sequences such as CA dinucleotide repeats (p. 69), and single nucleotide polymorphisms (p. 67). It has become less problematic as specific mutation analysis becomes available for an increasing number of single-gene disorders.

Locus Heterogeneity

Polymorphic DNA markers can be extremely reliable if the disease in question is caused by mutations in only one gene in the entire genome. In infantile polycystic kidney disease, for example, an autosomal recessive disorder, there is no evidence for a locus other than that on chromosome 6p21 (the gene, *PKHD1*, encodes fibrocystin). In the overwhelming majority of cases, it is a fatal condition in early postnatal life and the pathology is characteristic. For couples who have had one affected child in whom the diagnosis is confidently made, and from whom DNA is available, linked DNA markers can be used for prenatal diagnosis in subsequent pregnancies, provided the markers are informative. In many other conditions, however, mutations in more than one gene can give rise to the same basic phenotype. This is true for sensorineural hearing loss, retinitis pigmentosa, limb girdle muscular dystrophy, Bardet-Biedl syndrome, and many others that demonstrate enormous **locus heterogeneity**. It is usually not practical to use DNA polymorphic markers in conditions such as these because of the high chance of false-positive results.

Presymptomatic Diagnosis of Autosomal Dominant Disorders

Many autosomal dominant single-gene disorders either have a delayed age of onset (p. 341) or exhibit reduced penetrance (p. 340). The results of clinical examination, specialist investigations, biochemical studies, and family DNA studies can enable the genetic status of the person at risk to be determined before the onset of symptoms or signs. This is known as **presymptomatic**, or **predictive, testing**.

Clinical Examination

In some dominantly inherited disorders, simple clinical means can be used for presymptomatic diagnosis, taking into account possible pleiotropic effects of a gene (p. 109). For example, individuals with neurofibromatosis type I (NF1) can have a variety of clinical features (p. 298). It is not unusual to examine an apparently unaffected relative of someone with NF1, who has had no medical problems, only to discover that they have sufficient numbers of café-au-lait spots or cutaneous **neurofibromas** to confirm that they are affected. However, NF1 is a relatively rare example of a dominantly inherited disorder that is virtually 100% penetrant by the age of 5 or 6 years, with visible external features. With many other disorders, clinical examination is less helpful.

In tuberous sclerosis (TSC) a number of body systems may be involved and the external manifestations, such as the facial rash of angiokeratoma (Chapter 7; see Figure 7.5, A, p. 111) may not be present. Similarly, seizures and learning difficulties are not inevitable. In autosomal dominant polycystic kidney disease, which is extremely variable and may have a delayed age of onset, there may be no suspicion of the condition from routine examination, and hypertension may be borderline without raising suspicions of an underlying problem. Reaching a diagnosis in Marfan syndrome (p. 300) can be notoriously difficult because of the variable features and overlap with other joint hypermobility disorders, even though very detailed diagnostic criteria have been established.

Specialist Investigation

In conditions where clinical assessment leaves diagnostic doubt or ambiguity, special investigations of relevant body systems can serve to clarify status and presymptomatic diagnosis. In TSC imaging, studies of the brain by computed tomography to look for intracranial calcification (Figure 20.3) is more or less routine, as well renal ultrasonography to identify the cysts known as angiomyolipoma(ta) (Figure

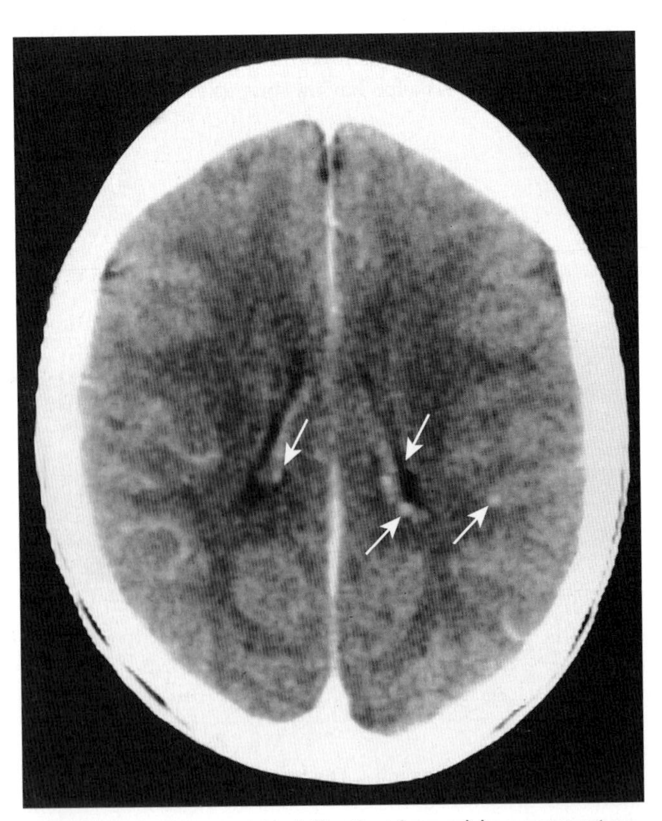

FIGURE 20.3 Intracranial calcification (*arrows*) in an asymptomatic person with tuberous sclerosis.

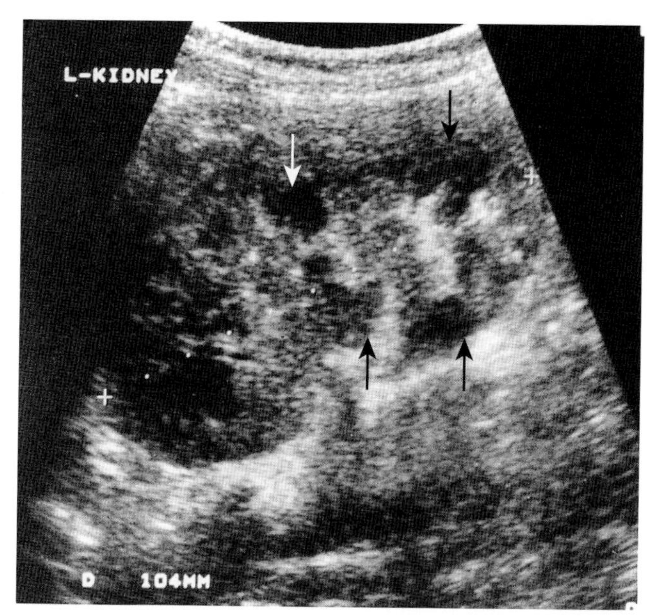

FIGURE 20.4 Renal ultrasonogram of an asymptomatic person with tuberous sclerosis showing abnormal echogenicity due to presumed angiomyolipomata (*arrows*).

20.4). Use of these relatively non-invasive tests in relatives of persons with TSC can often detect evidence of the condition in asymptomatic persons.

Similarly, assessment for Marfan syndrome involves ophthalmic examination for evidence of ectopia lentis, echocardiography for measurement of the aortic root diameter, and sometimes magnetic resonance imaging of the lumbar spine to look for evidence of dural ectasia—all of these features count as major criteria in the disorder.

It is important to point out, however, that the absence of these findings on clinical or specialist investigation does not always exclude the diagnosis, although it does reduce the likelihood of the person having inherited the gene. If the relative frequencies of positive findings in persons with the disorder and in persons from the general population are known, it is possible to calculate a residual relative likelihood of having inherited the gene. This information can be used in conjunction with other information, such as the pedigree risk, in genetic counseling.

Biochemical Tests

For a number of autosomal dominant disorders, biochemical tests can determine whether or not a person at risk has inherited a gene. Examples include the use of serum cholesterol levels in those at risk for familial hypercholesterolemia (p. 175), currently more straightforward than genetic testing, and assay of the appropriate urinary porphyrins or the specific enzyme deficiency in the various dominant porphyrias (p. 179).

Linked DNA Markers

Linked DNA polymorphic markers can be used in presymptomatic diagnosis of dominantly inherited disorders but the principles and pitfalls discussed earlier apply. A common difficulty in dominantly inherited conditions is that of informativity, because often a key individual is deceased or unavailable for some other reason. Despite this, the availability of linked DNA markers has found widespread use in presymptomatic or predictive testing for a number of single-gene disorders inherited in an autosomal dominant manner. Increasingly, however, direct mutation analysis is possible and has replaced the use of linked markers—for example, in Huntington disease (HD). In NF1, TSC, and Marfan syndrome, mutation analysis is expensive and used more discerningly because it is not guaranteed to identify the causative mutation even where there is confidence about the diagnosis. Judicious use of linked DNA markers can be very helpful. Box 20.1 lists some of the more common conditions in which DNA analysis is regularly used to offer presymptomatic diagnosis, but there are of course many more.

Ethical Considerations in Carrier Detection and Predictive Testing

Medically, there are often advantages in being able to determine the carrier status of a person at risk an autosomal or X-linked recessive disorder, mainly relating to a couple being able to make an informed choice when having children. For some individuals and couples, however, the knowledge that there is a significant risk of having an affected child may present options and choices that they would rather not have. The attendant risk, and the awareness that prenatal diagnosis is available, may create a sense of guilt whichever decision is taken—either to have a child knowing it could be affected, or to have prenatal testing and possible termination of pregnancy. The latter option is especially difficult when the prognosis of the disease in question

Box 20.1 Autosomal Disorders that Show a Delayed Age of Onset or Exhibit Reduced Penetrance in Which Linked DNA Markers or Specific Mutational Analysis Can be Used to Offer Presymptomatic Diagnosis

Breast cancer
Familial adenomatous polyposis
Hereditary motor and sensory neuropathy type I
Hereditary non-polyposis colonic cancer
Huntington disease
Inherited cardiac arrhythmias
Marfan syndrome
Myotonic dystrophy
Neurofibromatosis type I
Neurofibromatosis type II
Tuberous sclerosis
Von Hippel–Lindau disease

cannot be stated with any certainty because of variability or reduced penetrance, or if there is hope that treatment may be developed in time to help the child. Because of these difficulties and potential dilemmas it is normal practice to suggest that information is passed on *within* families, rather than by professionals. In general this approach works well, but professional dilemmas can arise if family members refuse to communicate with one another when the disease in question carries significant morbidity and the risk may be high, particularly with X-linked conditions.

In those at risk for late-onset autosomal dominant disorders, many of which have neurological features, there can in some instances be a clear advantage in presymptomatic diagnosis. For example, in those at risk for familial adenomatous polyposis (p. 221), colonoscopy looking for the presence of colonic polyps can be offered as a regular screening procedure to those who have been shown to be at high risk of developing colonic cancer by molecular studies. Conversely, individuals who have been shown to have not inherited a mutation in the *APC* gene do not need to be screened.

In contrast, for persons at risk for HD, in which there is not yet any effective treatment to delay the onset or progression of the disorder, the benefit of predictive testing is not immediately obvious. The same is true for familial forms of Alzheimer disease, motor neurone disease, CADASIL (cerebral autosomal dominant arteriopathy with subcortical infarcts and leukoencephalopathy), and the spinocerebellar ataxias. Although choice is often considered to be of paramount importance in genetic counseling for those at risk for inherited disorders, it is important to remember that those considering presymptomatic or predictive testing should proceed only if they can give truly informed consent and are free from coercion from any outside influence. It is possible that employers, life insurance companies, and society in general will put indirect, and on occasion direct, pressure on those at high at risk to be tested (p. 366). Indeed, there are examples in which individuals at risk of HD have received prejudicial treatment in relation to employment, and higher than average insurance premiums can be expected on the basis of the family history alone. Here again, the very knowledge itself may torment and frustrate an individual who wants to move on in their life but, once acquired, the knowledge and awareness cannot be removed.

A significant problem raised by predictive testing for late-onset disorders is that it can, in theory, be used for children and minors. This issue can be very contentious, with parents sometimes arguing that it is their right to know the status of their child(ren). However, this conflicts with the high ideal of upholding the principle of individual autonomy wherever possible. Presymptomatic testing of children is therefore usually discouraged unless an early medical intervention or screening is beneficial for the disorder in question. The latter is certainly true for a number of the familial cancer conditions, but also applies to Marfan syndrome and autosomal dominant polycystic kidney disease, among others. The issue of genetic testing of children is addressed more fully in Chapter 24 (p. 365).

Box 20.2 Current Nationally Managed Screening Programs in the UK

Antenatal
Down syndrome
Sickle cell disease
Thalassemia
Structural abnormalities (fetal anomaly scanning at 18–20 weeks' gestation)

Newborn
Phenylketonuria (PKU)
Congenital hypothyroidism
Sickle cell disease
Thalassemia
Cystic fibrosis
Medium chain acyl-CoA dehydrogenase deficiency (MCADD)
Hearing impairment

Adult
Breast cancer
Cervical cancer
Sight-threatening diabetic retinopathy

Population Screening

One definition of population screening is: 'The systematic application of a test or inquiry, to identify individuals at sufficient risk of a specific disorder to warrant further investigation or treatment, amongst persons who have not sought medical attention on account of symptoms of that disorder'. Neonatal screening for phenylketonuria is the paradigm of a good screening program and has been available for more than 40 years, with screening for congenital hypothyroidism not far behind. In the United Kingdom, since 1996, population screening has been overseen by the UK National Screening Committee, which advises the government. Under their auspices a National Programme Centre was established in 2002. The current, nationally managed, screening programs are listed in Box 20.2. The implementation of a screening program is a huge logistical exercise requiring financial, expertise, and technology resources, as well as setting up practical mechanisms to introduce the program and monitor outcomes and quality assurance.

Criteria for a Screening Program

These can be considered under the headings of the disease, the test, and the practical aspects of the program (Box 20.3). These criteria apply equally to prenatal screening, which is addressed in the next chapter.

The Disease

To justify the applied effort and resources allocated to screening, the disease should be sufficiently common and have potentially serious effects that are amenable to

Box 20.3 Criteria for a Screening Program

Disease
High incidence in target population
Serious effect on health
Treatable or preventable

Test
Non-invasive and easily carried out
Accurate and reliable (high sensitivity and specificity)
Inexpensive

Program
Widespread and equitable availability
Voluntary participation
Acceptable to the target population
Full information and counseling provided

Table 20.3 In this Hypothetical Scenario a Screening Test for Congenital Adrenal Hyperplasia (CAH) Has Been Implemented, with the Following Results

CAH Present		CAH Absent	
Positive	Negative	Positive	Negative
96	4	4980	510,100

Positive predictive value: 96/(96 + 4980) ≅ 2%
Sensitivity: 96/(96 + 4) = 96%
Specificity: 510,100/(510,100 + 4980) ≅ 99%

prevention or amelioration. This may involve early treatment, as in phenylketonuria diagnosed in the neonatal period (p. 167), or the offer of termination of pregnancy for disorders that cannot be treated effectively and are associated with serious morbidity and/or mortality.

The Test

The test should be accurate and reliable with high **sensitivity** and **specificity**. Sensitivity refers to the proportion of cases that are detected. A measure of sensitivity can be made by determining the proportion of false-negative results (i.e., how many cases are missed). Thus, if a test detects only 70 of 100 cases, it shows a sensitivity of 70%. Specificity refers to the extent to which the test detects only affected individuals. If unaffected people test positive, these are referred to as false positives. Thus, if 10 of 100 unaffected individuals have a false-positive test result, the test shows a **specificity** of 90%. Table 20.2 explains this further. Of great interest too is the **positive predictive value** of a screening test, which is the proportion of positive tests that are true positives; this is illustrated in Table 20.3.

The Program

The program should be offered in a fair and equitable manner, and should be widely available. It must also be morally acceptable to a substantial proportion of the population to which it is offered. Participation must be entirely voluntary in the case of prenatal programs, but the ethical principles are more complex in neonatal screening for conditions where early treatment is essential and effective in preventing morbidity. In these situations, the principles of **beneficence** (doing good) and **non-maleficence** (not doing harm) are relevant. Easily understood information and well-informed counseling should both be readily available.

It is often stated that the cost of a screening program should be reasonable and affordable. This does not mean that the potential savings gained through a reduction in the number of affected cases requiring treatment should exceed or even balance the cost of screening, although this argument is popular with health economists who have to fund the program. It is reasonable to point out that cost–benefit analyses also have to take into account non-tangible factors such as the emotional costs of human suffering borne by both the affected individuals and those who care for them.

Table 20.2 Sensitivity and Specificity

	Disease Status	
	Affected	Unaffected
Screening test result		
Positive	a (true positive)	b (false positive)
Negative	c (false negative)	d (true negative)

Sensitivity: a/(a + c) – proportion of true positives
Specificity: d/(d + b) – proportion of true negatives

Neonatal Screening

Newborn screening programs have been introduced on a widespread basis for phenylketonuria, galactosemia, and congenital hypothyroidism. In all of these disorders early treatment can dramatically prevent the development of learning disability. To this group have been added hemoglobinopathies, cystic fibrosis, and medium-chain acyl-CoA dehydrogenase more recently, where knowledge of the condition and intervention is important but less dramatic in its benefits. Screening for several other disorders is carried out more selectively in different centers (Table 20.4). In the United States, for example, all 50 states have a legal duty to provide screening to all newborns, at least for the first three conditions mentioned. In some states, (West Virginia, Montana, and South Dakota), screening is limited to these three conditions but at the other extreme up to 30 diseases are screened for (North Carolina and Oregon). In the United Kingdom, consideration is being given to screening for Pompe disease, maple syrup urine disease, tyrosinemia, congenital adrenal hyperplasia, isovaleric academia, glutaric aciduria type 1, and homocystinuria.

For most of these conditions, the rationale is to prevent subsequent morbidity and in most countries is either mandatory or pursued under the auspices of implied consent.

Table 20.4 Conditions for Which Neonatal Screening Can Be Undertaken

Disorder	Test/Method
Widely Applied	
Phenylketonuria	Guthrie test[a] or automated fluorometric assay
Congenital hypothyroidism	Thyroxine or thyroid-stimulating hormone
Other Inborn Errors	
Biotinidase deficiency	Specific enzyme assay
Galactosemia	Modified Guthrie test
Homocystinuria	Modified Guthrie test
Maple syrup urine disease	Modified Guthrie test
Tyrosinemia	Modified Guthrie test
Miscellaneous	
Congenital adrenal hyperplasia	17-Hydroxyprogesterone assay
Cystic fibrosis	Immunoreactive trypsin and DNA analysis
Duchenne muscular dystrophy	Creatine kinase
Sickle-cell disease	Hemoglobin electrophoresis

[a]The Guthrie test is based on reversal of bacterial growth inhibition by a high level of phenylalanine.

The Netherlands, by contrast, have adopted a more explicit consent process, whereas the screening remains highly recommended. The importance of adhering to the principle of screening for a disorder that needs to be treated early is illustrated by the Swedish experience of neonatal screening for α_1-antitrypsin deficiency. In this condition neonatal complications occur in up to 10%, but for most cases the morbidity is seen in adult life, and the main message on diagnosing the disorder is avoidance of smoking. Between 1972 and 1974, 200,000 newborns were screened and follow-up studies showed that considerable anxiety was generated when the information was conveyed to parents, who perceived their children to be at risk of a serious, life-threatening disorder. The case of newborn screening for Duchenne muscular dystrophy also deviates from the screening paradigm because, thus far, no early intervention is helpful. Here, the parents (or mother) can be counseled before having more children and, in the wider family, identification of female carriers (of reproductive age) may be possible. However, here again, parental reaction has not been uniformly favorable.

Phenylketonuria

Routine biochemical screening of newborn infants for phenylketonuria was recommended by the Ministry of Health in the United Kingdom in 1969, after it had been shown that a low-phenylalanine diet could prevent the severe learning disabilities that previously had been a hallmark of this condition (p. 167). The screening test, which is sometimes known as the **Guthrie test**, is carried out on a small sample of blood obtained by heel-prick at age 7 days. An abnormal test result is further investigated by repeat analysis of phenylalanine levels in a venous blood sample. A low-phenylalanine diet is extremely effective in preventing learning disabilities, and, although it is not particularly palatable, most affected children can be persuaded to adhere to it until early adult life when it can be relaxed. However, because high phenylalanine levels are toxic to the developing brain, a woman with phenylketonuria who is contemplating pregnancy should adhere to a strict low-phenylalanine diet both before and during pregnancy (p. 261).

Galactosemia

Classic galactosemia affects approximately 1 in 50,000 newborn infants and usually presents with vomiting, lethargy, and severe metabolic collapse within the first 2 or 3 weeks of life. Newborn screening is based on a modification of the Guthrie test with subsequent confirmation by specific enzyme assay. The early introduction of appropriate dietary restriction can prevent the development of serious complications such as cataracts, liver failure, and learning disability.

Congenital Hypothyroidism

Screening for congenital hypothyroidism was first introduced in the United States in 1974 and is now undertaken in most parts of the developed world. The test is based on assay of either **thyroxine** or **thyroid-stimulating hormone**. This disorder is particularly suitable for screening as it is relatively common, with an incidence of approximately 1 in 4000, and treatment with lifelong thyroxine replacement is extremely effective in preventing the severe developmental problems associated with the classic picture of 'cretinism'. The most common cause of congenital hypothyroidism is absence of the thyroid gland rather than an inborn error of metabolism (p. 167). Congenital absence of the thyroid gland is usually not caused by genetic factors but on rare occasion is part of a wider syndrome.

Cystic Fibrosis

Newborn screening for cystic fibrosis has been introduced in several countries with a significant population of northern European origin. It is based on the detection of a raised blood level of **immunoreactive trypsin**, which is a consequence of blockage of pancreatic ducts in utero, supplemented by DNA analysis. The rationale for screening is that early treatment with physiotherapy and antibiotics improve the long-term prognosis. Initial results are encouraging but absolute confirmation of long-term benefit is awaited. In 2006 it became policy in England to introduce screening for all newborns.

Sickle Cell Disease and Thalassemia

Newborn screening based on **hemoglobin electrophoresis** is undertaken in many countries with a significant Afro-Caribbean community. As with cystic fibrosis, it is hoped that early prophylaxis will reduce morbidity and mortality, thereby improving the long-term outlook. In the case of sickle cell disease, treatment involves the use of oral penicillin to reduce the risk of pneumococcal infection resulting from immune deficiency secondary to splenic infarction

(p. 152). Even in Western countries with good medical facilities, a significant proportion of sickle cell homozygotes, possibly as many as 15%, die as a result of infection in early childhood. In the case of thalassemia, early diagnosis makes it possible to optimize transfusion regimens and iron-chelation therapy from an early stage. Neonatal screening programs for both of these hemoglobinopathies were implemented in the United Kingdom in 2005, and antenatal screening (the mother, followed by the father if necessary) is under way in some areas of high risk. In some low-risk areas, there is a preference for antenatal screening to be targeted to high-risk couples after completion of an ethnicity questionnaire (p. 164).

Population Carrier Screening

Widespread screening for carriers of autosomal recessive disorders in high-incidence populations was first introduced for the hemoglobinopathies (see Chapter 10) and has been extended to several other disorders (Table 20.5). The rationale behind these programs is that carrier detection can be supported by genetic counseling so that carrier couples can be forewarned of the 1 in 4 risk that each of their children could be affected. The example of Tay-Sachs disease in orthodox Jewish communities has been discussed previously (pp. 313–314); this does not amount to 'population' screening.

Experience with the two common hemoglobinopathies, thalassemia, and sickle cell disease, illustrates the extremes of success and failure that can result from well or poorly planned screening programs.

Thalassemia

α- and β-thalassemia are caused by abnormal globin chain synthesis because of mutations involving the α- and β-globin genes or their promoter regions (p. 160). Both disorders show autosomal recessive inheritance and are extremely common in certain parts of the world, notably South-East Asia (α-thalassemia), Cyprus and the Mediterranean region, Italy, and the Indian subcontinent (β-thalassemia).

In Cyprus in 1974 the birth incidence of β-thalassemia was 1 in 250 (carrier frequency 1 in 8). After the introduction of a comprehensive screening program to determine the carrier status of young adults, which had the support of the Greek Orthodox Church, the incidence of affected babies declined by more than 95% within 10 years. Similar programs in Greece and Italy have seen a drop in the incidence of affected homozygotes of more than 50%.

If it is acceptable to judge the outcome of these screening programs on the basis of a reduction in the births of affected babies, then they have been very successful, due largely to the efforts of highly motivated staff interacting with a well-informed target population that has usually opted not to have affected children.

Sickle Cell Disease

In contrast to the Cypriot response to β-thalassemia screening, early attempts to introduce sickle cell carrier detection in the black population of North America were disastrous. Information pamphlets tended to confuse the sickle cell carrier state, or trait, which is usually harmless, with the homozygous disease, which conveys significant morbidity (p. 159). Several US states passed legislation making sickle cell screening in black people mandatory, and carriers suffered discrimination by employers and insurance companies. It is not surprising that public criticism was aroused, leading to abandonment of the screening programs and amendment of the ill-conceived legislation.

This experience emphasizes the importance of ensuring voluntary participation and providing adequate and appropriate information and counseling. Later pilot studies in the United States and in Cuba have shown that individuals of Afro-Caribbean origin are perfectly receptive to well planned, non-directive sickle cell screening programs.

Cystic Fibrosis

The discovery in 1989 that the Phe508del deletion/mutation (formerly deltaF508, or ΔF508) accounts for a high proportion of all cystic fibrosis heterozygotes soon led to the suggestion that screening programs could be implemented for carrier detection on a population basis. In the white population of the United Kingdom, the cystic fibrosis carrier frequency is approximately 1 in 25 and the Phe508del mutation accounts for 75% to 80% of all heterozygotes. A further 10% to 15% of carriers can be detected relatively easily and cheaply using a multiplex PCR analytical procedure.

Initial studies of attitudes to cystic fibrosis carrier detection yielded quite divergent results. A casual, written invitation generates a poor take-up response of around 10%, whereas personal contact during early pregnancy, whether mediated through general practice or the antenatal clinic,

Table 20.5 Autosomal Recessive Disorders Suitable for Population Carrier Screening

Disorder	Ethnic Group or Community	Test
α-Thalassemia electrophoresis	China and eastern Asia	Mean corpuscular hemoglobin and hemoglobin
β-Thalassemia electrophoresis	Indian subcontinent and Mediterranean countries	Mean corpuscular hemoglobin and hemoglobin
Sickle cell disease	Afro-Caribbean	Sickle test and hemoglobin electrophoresis
Cystic fibrosis	Western European whites	Common mutation analysis
Tay-Sachs disease	Ashkenazi Jews	Hexosaminidase A

results in uptake rates of more than 80%. Studies have been undertaken to explore attitudes to cystic fibrosis screening among specific groups, such as school leavers and women in early pregnancy.

Two approaches for screening pregnant women have been considered. The first is referred to as **two-step** and involves testing pregnant mothers at the antenatal clinic. Those who test positive for a common mutation (approximately 85% of all cystic fibrosis carriers) are informed of the result and invited to bring their partners for testing—hence 'two-step' testing. If both partners are found to be carriers, an offer of prenatal diagnosis is made. This approach has the advantage that all carriers detected are informed of their result and further family studies—**cascade screening**—can be initiated. The disadvantage is that women whose partners are subsequently found to have a normal result experience considerable unnecessary anxiety, which creates a need for counseling and support.

The second approach is referred to as **couple screening**. This involves testing both partners simultaneously and disclosing positive results only if both partners are found to be carriers. In this way much less anxiety is generated, but the opportunity for offering tests to the extended family when only one partner is a carrier is lost. The results of pilot studies indicated that these 'two-step' and 'couple screening' approaches are equally acceptable to pregnant women, with take-up rates of approximately 70%. However, there is no publicly available cystic fibrosis screening for adults in the United Kingdom.

Positive and Negative Aspects of Population Screening

Well-planned population screening enhances informed choice and offers the prospect of a significant reduction in the incidence of disabling genetic disorders. These potential advantages have to be weighed against the potential disadvantages that can arise from the overenthusiastic pursuit of a poorly planned or ill-judged screening program (Box 20.4). Experience to date indicates that in relatively small, well-informed groups, such as the Greek Cypriots and American Ashkenazi Jews, community screening is welcomed. When screening is offered to larger populations the outcome is less certain.

Box 20.4 Potential Advantages and Disadvantages of Genetic Screening

Advantages
Informed choice
Improved understanding
Early treatment when available
Reduction in births of affected homozygotes

Disadvantages and hazards
Pressure to participate causing mistrust and suspicion
Stigmatization of carriers (social, insurance, and employment)
Inappropriate anxiety in carriers
Inappropriate reassurance if test is not 100% sensitive

Box 20.5 Roles of a Genetic Register

- To maintain an informal two-way communication process between the family and the genetics unit
- To offer carrier detection to relevant family members as they reach adult life
- To coordinate presymptomatic and prenatal diagnosis when requested
- To coordinate multidisciplinary management of patients with complex hereditary conditions such as the familial cancer syndromes
- To ensure effective implementation of new technology and treatment
- To provide a long-term source of information and support

A 3-year follow-up of almost 750 individuals screened for cystic fibrosis carrier status in the United Kingdom revealed that a positive test result did not cause undue anxiety, although some carriers had a relatively poor perception of their own general health. A more worrying outcome was that almost 50% of the individuals tested could not accurately recall or interpret their results. This emphasizes the importance of pretest counseling and the provision of accurate information that is easily processed and understood.

Genetic Registers

Regional genetic centres maintain a **genetic register** of families and individuals who are either affected by, or at risk of developing, a serious hereditary disorder. The main difference with a conventional digital medical record is the facility to link biological relatives, some affected by a disorder, some unaffected, and others at risk. This makes it possible to manage, support, and inform the family (Box 20.5), as well as schedule review, screening, and the offer of appropriate and timely genetic tests. Ideally, patients should be aware that their information is held in this way and the genetics unit is responsible for security and confidentiality. Unlike other medical records, the destruction of these records at a given time after patients' death is strongly resisted by geneticists because the information may be crucially important to the generations that follow.

Genetic registers are especially appropriate for conditions that are relatively common, have potentially serious effects, convey high risks to other family members, and in which complications can be treated or prevented (Box 20.6). They are particularly valuable for conditions that show a delayed age of onset or those for which unaffected carriers could be at high risk of having seriously affected children.

Well-organized registers can facilitate the coordination of a multidisciplinary approach to the management of families with familial cancer-predisposing syndromes (p. 229). This often involves the interpretation of molecular investigations and the organization of hospital appointments and screening (see Table 14.9, p. 228). This is an important part

Box 20.6 Disorders Suitable for a Genetics Register

Autosomal dominant
Adult-onset polycystic kidney disease
Cardiomyopathies
Familial adenomatous polyposis
Familial common cancers—breast, ovarian, colorectal
Huntington disease
Inherited cardiac arrhythmias
Marfan syndrome
Multiple endocrine neoplasia types 1 and 2
Myotonic dystrophy
Neurofibromatosis types I and II
Retinoblastoma
Tuberous sclerosis
Von Hippel–Lindau syndrome

Autosomal recessive
Congenital adrenal hyperplasia
Cystic fibrosis
Sickle cell disease
Spinal muscular atrophy
Thalassemia

X-linked
Duchenne/Becker muscular dystrophy
Fragile X syndrome
Hemophilia
Retinitis pigmentosa

Chromosomal
Deletions/insertions
Inversions
Translocations

of patient care and registers will continue to be necessary as more genetic health information is generated in the future.

FURTHER READING

Axworthy D, Brock DJH, Bobrow M, Marteau TM 1996 Psychological impact of population-based carrier testing for cystic fibrosis: 3-year follow-up. Lancet 1996; 347:1443–1446

A review of the impact of carrier testing for cystic fibrosis on over 700 individuals.

Baily MA, Murray TH (eds) 2009 Ethics and newborn genetic screening: new technologies, new challenges. Baltimore: Johns Hopkins University Press

A multi-author volume with a focus on the health economics of newborn screening and distributive justice.

Brock DJH, Rodeck CH, Ferguson-Smith MA (eds) 1992 Prenatal diagnosis and screening. Edinburgh: Churchill Livingstone

A huge multiauthor textbook with excellent chapters on all aspects of genetic screening.

Cunningham GC, Tompkinson DG 1999 Cost and effectiveness of the California triple marker prenatal screening program. Genet Med 1:199–206

A detailed review of the impact of triple test screening over a 10-year period in California. The report of a working party on genetic screening on the associated ethical and societal issues.

Harper PS 2004 Practical genetic counselling, 6th ed. London: Arnold

As the title suggests, a practical book that serves as a good starting point in almost every aspect of genetic counseling, including carrier testing.

Marteau T, Richards M (eds) 1996 The troubled helix. Cambridge: Cambridge University Press

Perspectives on the social and psychological implications of genetic testing and screening.

Modell B, Modell M 1992 Towards a healthy baby. Oxford: Oxford University Press

A clearly written and easily understood guide to genetic counseling and community genetics.

Nuffield Council on Bioethics 1993 Genetic screening: ethical issues. London: Nuffield Council on Bioethics

Pauli R, Motulsky AG 1981 Risk counselling in autosomal dominant disorders with undetermined penetrance. J Med Genet 18:340–343

A paper that considers the problem of counseling for autosomal dominant disorders with reduced penetrance.

Pembrey ME, Davies KE, Winter RM, et al 1984 Clinical use of DNA markers linked to the gene for Duchenne muscular dystrophy. Arch Dis Child 59:208–216

A useful discussion of how DNA markers can be used for carrier detection in Duchenne muscular dystrophy.

ELEMENTS

1. Determination of carrier status for autosomal recessive and X-linked disorders, if not possible by direct gene testing, can involve detailed clinical examination looking for specific minor features, specialist clinical investigations, biochemical tests, or family studies using linked DNA polymorphic markers.
2. The same principles of assessment and investigation apply to presymptomatic or predictive testing for those at risk of autosomal dominant disorders with reduced penetrance or a delayed age of onset.
3. Consideration should be given to the advantages and disadvantages of presymptomatic or predictive testing from both a practical and an ethical point of view.
4. Population screening involves the offer of genetic testing to all members of a particular population, with the objectives of preventing later ill-health and providing informed personal choice.
5. Participation should be voluntary and each program should be widely available, equitably distributed, acceptable to the target population and supported by full information and counseling.
6. Prenatal screening is routinely available for neural tube defects and Down syndrome using a combination of fetal ultrasonography and assay of biochemical markers in maternal serum, which may lead to the offer of amniocentesis and fetal karyotyping.
7. Neonatal screening is widely available for phenylketonuria, galactosemia, congenital hypothyroidism, and cystic fibrosis, as well as other conditions in specific populations.
8. Population screening programs for carriers of β-thalassemia have resulted in a major fall in the incidence of births of affected homozygotes. This has provided the paradigm for the introduction of screening for other disorders with serious long-term morbidity.
9. Well-organized genetic registers provide an effective means of maintaining two-way contact between genetics centers and families with hereditary disease.

CHAPTER 21

Prenatal Testing and Reproductive Genetics

The more alternatives, the more difficult the choice.

ABBE D'ALLAINVAL

Until recently, couples at high risk of having a child with a genetic disorder had to choose between taking the risk or considering other reproductive options, such as long-term contraception, sterilization, and termination of pregnancy. Other alternatives included adoption, long-term fostering, and donor insemination (DI).

Over the past three decades, prenatal diagnosis—the ability to detect abnormalities in an unborn child—has been widely used. Although it may be very difficult for a couple to decide to pursue prenatal diagnosis because of the possibility that this will lead to termination of pregnancy, prenatal diagnosis is an option that is chosen by many couples at high risk of having a child with a serious genetic disorder or birth defect.

The ethical issues surrounding prenatal diagnosis and selective termination of pregnancy are both complex and emotive, and are considered more fully in Chapter 24 (p. 363). In this chapter, we focus on the practical aspects of prenatal testing and diagnosis, including prenatal screening, as well as some aspects of reproductive genetics.

Techniques Used in Prenatal Diagnosis

There are several techniques that can be used for the prenatal diagnosis of hereditary disorders and structural abnormalities (Table 21.1).

Amniocentesis

Amniocentesis involves the aspiration of 10 to 20 ml of amniotic fluid through the abdominal wall under ultrasonographic guidance (Figure 21.1). This is usually performed around the 16th week of gestation. The sample is spun down to yield a pellet of cells and supernatant fluid. The fluid can be used in the prenatal diagnosis of neural tube defects by assay of α-fetoprotein (p. 328). The cell pellet is resuspended in culture medium with fetal calf serum, which stimulates cell growth. Most of these cells in the amniotic fluid, that have been shed from the amnion, fetal skin, and urinary tract epithelium, are non-viable, but a small proportion will grow. After approximately 14 days, there are usually sufficient cells for chromosome and DNA analysis, although a longer period may be required before enough cells are obtained for biochemical assays. Increasingly, sensitive polymerase chain reaction (PCR) techniques make direct DNA analysis possible without the need for culture.

When a couple is considering amniocentesis, they should be informed of the 0.5% to 1% risk of miscarriage associated with the procedure, and that if the result is abnormal they will be facing the possibility of having to consider a mid-trimester termination of pregnancy that involves an induction of labor.

Trials of amniocentesis earlier in pregnancy, at 12 to 14 weeks' gestation, yielded comparable rates of success in obtaining results with a similar risk of miscarriage. However, concerns have been expressed regarding the reduction in amniotic fluid at this early stage of pregnancy, and early amniocentesis is not widely practiced. Although it has the advantage of allowing a result to be given earlier in the pregnancy, a mid-trimester termination of pregnancy is still usually required if the fetus is found to be affected.

Chorionic Villus Sampling

In contrast to amniocentesis, chorionic villus sampling (CVS), first developed in China, enables prenatal diagnosis to be undertaken during the first trimester. This procedure is usually carried out at 11 to 12 weeks' gestation under ultrasonographic guidance by either transcervical or, more usually, transabdominal aspiration of chorionic villus (CV) tissue (Figure 21.2). This tissue is fetal in origin, being derived from the outer cell layer of the blastocyst (i.e., the trophoblast). Maternal decidua, normally present in the biopsy sample, must be removed before the sample is analyzed. **Placental biopsy** is the term used when the procedure is carried out at later stages of pregnancy.

Chromosome analysis can be undertaken on CV tissue either directly, looking at metaphase spreads from actively dividing cells, or after culture. Direct chromosomal analysis of CV tissue usually allows a provisional result to be given within 24 hours. Nowadays rapid, direct fluorescent in-situ hybridization (FISH) probing (p. 34), or DNA analysis by the multiplex ligation-dependent probe amplification

Table 21.1 Standard Techniques Used in Prenatal Diagnosis

Technique	Optimal Time (Weeks)	Disorders Diagnosed
Non-Invasive		
MATERNAL SERUM SCREENING		
α-Fetoprotein	16	Neural tube defects
Triple test	16	Down syndrome
Ultrasound	18	Structural abnormalities (e.g., central nervous system, heart, kidneys, limbs)
Invasive		
Amniocentesis	16	
Fluid		Neural tube defects
Cells		Chromosome abnormalities, metabolic disorders, molecular defects
Chorionic villus sampling	10–12	Chromosome abnormalities, metabolic disorders, molecular defects
Fetoscopy		
Blood (cordocentesis)		Chromosome abnormalities, hematological disorders, congenital infection
Liver		Metabolic disorders (e.g., ornithine transcarbamylase deficiency)
Skin		Hereditary skin disorders (e.g., epidermolysis bullosa)

technique (p. 66), is used to test for common chromosome aneuploidies prior to a standard karyotype following culture of CV tissue; this may also detect other chromosome abnormalities and balanced rearrangements. For single-gene disorders, sufficient CV tissue is usually obtained to allow prenatal diagnosis by immediate biochemical assay or DNA analysis using uncultured CV tissue.

The major advantage of CV sampling is that it offers first-trimester prenatal diagnosis, although it has the disadvantage that even in experienced hands the procedure conveys a 1% to 2% risk of causing miscarriage. There is also evidence that this technique can cause limb abnormalities in the embryo if carried out before 9 to 10 weeks' gestation; for this reason is not performed before 11 weeks' gestation.

Ultrasonography

Ultrasonography offers a valuable means of prenatal diagnosis. It can be used not only for obstetric indications, such as placental localization and the diagnosis of multiple pregnancies, but also for prenatal diagnosis of structural abnormalities not associated with known chromosomal, biochemical or molecular defects. Ultrasonography is particularly valuable because it is non-invasive and conveys no known risk to the fetus or mother. It does, however, require expensive equipment and a skilled, experienced operator. For example, a search can be made for polydactyly as a diagnostic feature of a multiple abnormality syndrome, such as one of the autosomal recessive short-limb polydactyly syndromes that are associated with severe pulmonary

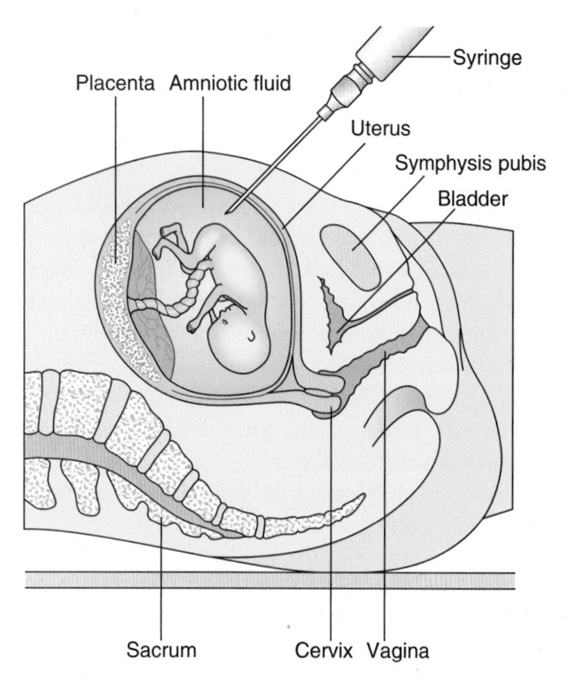

FIGURE 21.1 Diagram of the technique of amniocentesis.

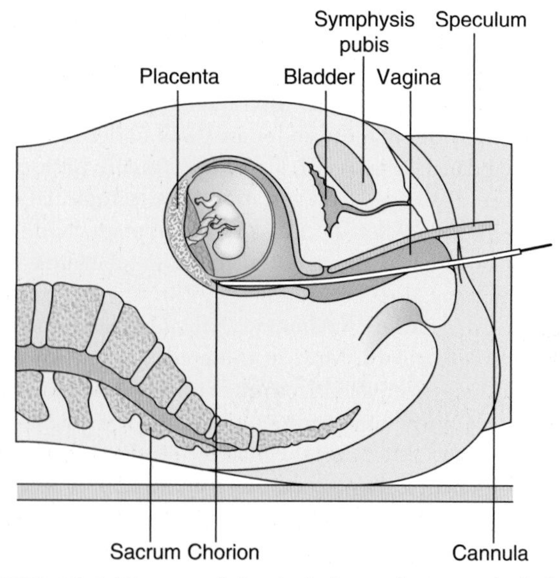

FIGURE 21.2 Diagram of the technique of transvaginal chorionic villus sampling.

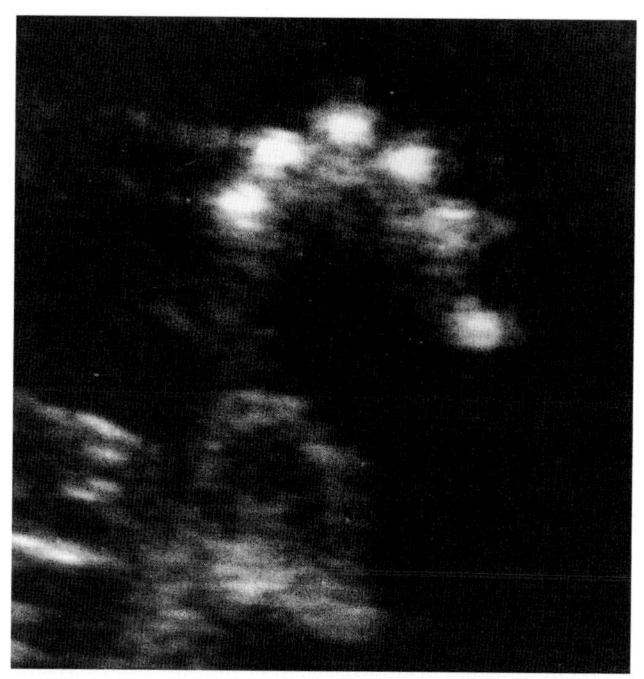

FIGURE 21.3 Ultrasonographic image of a transverse section of the hand of a fetus showing polydactyly.

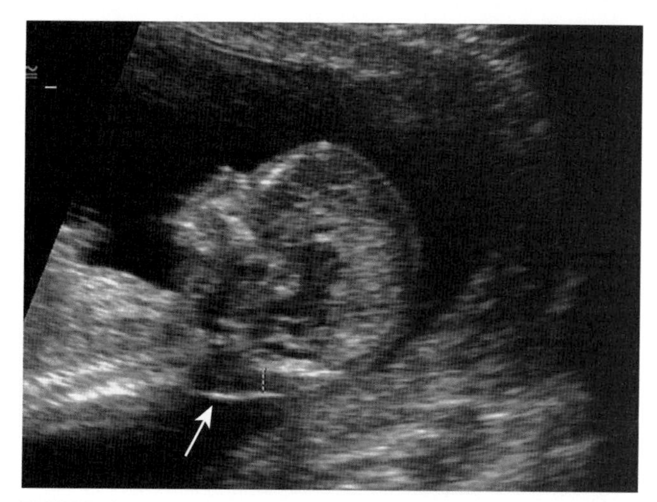

FIGURE 21.5 Nuchal thickening—an accumulation of fluid at the back of the neck. The greater the thickness, the more likely there will be a chromosomal abnormality (e.g., Down syndrome) and/or cardiac anomaly. This finding leads to detailed fetal heart scanning and, usually, fetal karyotyping. (Courtesy Dr. Helen Liversedge, Exeter.)

hypoplasia—invariably lethal (Figure 21.3). Similarly, a scan can reveal that the fetus has a small jaw, which can be associated with a posterior cleft palate and other more serious abnormalities in several single-gene syndromes (Figure 21.4).

Until a few years ago, detailed ultrasonography for structural abnormalities was offered only to couples who had a child with a genetic disorder or syndrome for which there was no chromosomal, biochemical, or molecular marker. Increasingly, however, detailed 'fetal anomaly' scanning is being offered routinely to all pregnant women at around 18 weeks' gestation as screening for structural abnormalities such as neural tube defects or cardiac anomalies. This technique can also identify features that suggest the presence of an underlying chromosomal abnormality. Such a finding would lead to an offer of amniocentesis or placental biopsy for definitive chromosome analysis.

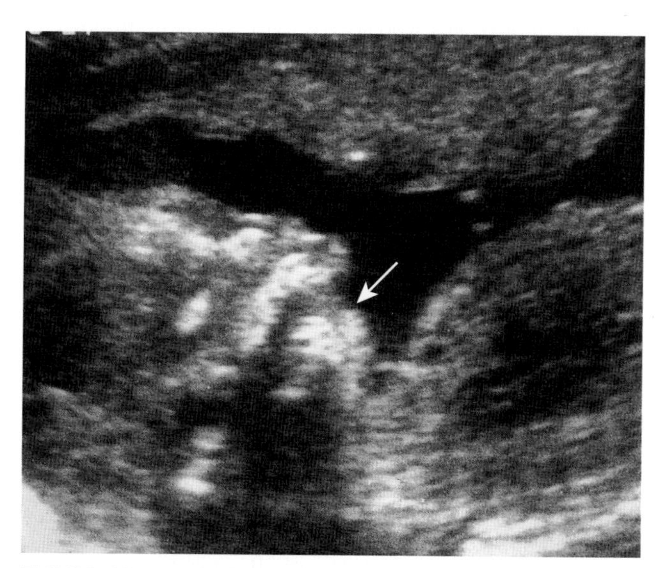

FIGURE 21.4 Longitudinal sagittal ultrasonographic image of the head and upper chest of a fetus showing micrognathia (small jaw) (*arrow*).

The future of fetal scanning holds the prospect of three-dimensional imaging and magnetic resonance imaging being used more widely and routinely. Although this will clearly enable the unborn baby to be visualized in far greater detail, it will also generate bigger challenges for the dysmorphologist, who might be expected to diagnose serious disorders on the basis of very subtle features.

Nuchal Translucency

In addition, the observation that increased nuchal translucency (NT) is seen in fetuses who are subsequently born with Down syndrome has resulted in the introduction of measurements of nuchal pad thickness (Figure 21.5) in the first and second trimesters as part of screening for Down syndrome (p. 273). In fact, the finding is not specific and may be seen in various chromosomal anomalies as well as isolated congenital heart disease.

Fetoscopy

Fetoscopy involves visualization of the fetus by means of an endoscope. Increasingly, this technique is being superseded by detailed ultrasonography, although occasionally fetoscopy is still undertaken during the second trimester to try to detect the presence of subtle structural abnormalities that would point to a serious underlying diagnosis.

Fetoscopy has also been used to obtain samples of tissue from the fetus that can be analyzed as a means of achieving the prenatal diagnosis of several rare disorders. These have included inherited skin disorders such as epidermolysis bullosa and, before DNA testing became available, metabolic disorders in which the enzyme is expressed only in certain tissues or organs, such as the liver—e.g., ornithine transcarbamylase deficiency (p. 172).

Unfortunately, fetoscopy is associated with a 3% to 5% risk of miscarriage. This relatively high risk, coupled with the increasing sensitivity of ultrasonography and the availability of either linked DNA markers or specific mutation analysis, means that fetoscopy is used only infrequently and in highly specialized prenatal diagnostic centers.

Cordocentesis

Although fetoscopy can also be used to obtain a small sample of fetal blood from one of the umbilical cord vessels in the procedure known as cordocentesis, improvements in ultrasonography have enabled visualization of the vessels in the umbilical cord, allowing transabdominal percutaneous fetal blood sampling. Fetal blood sampling is used routinely in the management of rhesus iso-immunization (p. 205) and can be used to obtain samples for chromosome analysis to resolve problems associated with possible chromosomal mosaicism in CV or amniocentesis samples.

Radiography

The fetal skeleton can be visualized by radiography from 10 weeks onwards, and this technique has been used in the past to diagnose inherited skeletal dysplasias. It is now employed only occasionally because of the dangers of radiography to the fetus (p. 26) and the widespread availability of detailed ultrasonography.

Prenatal Screening

The history of widespread prenatal (antenatal) screening really began with the finding, in the early 1970s, of an association between raised maternal serum α-fetoprotein (AFP) and neural tube defects (NTDs). Estimation of AFP levels was gradually introduced into clinical service, and the next significant development was ultrasonography, followed, in the 1980s, by the identification of maternal serum biochemical markers for Down syndrome. These are discussed in more detail below. Where the incidence of a genetic condition was high, for instance thalassemia in Cyprus, prenatal screening came into practice, as described in Chapter 20 (p. 321). However, molecular genetic advances, rather than biochemical, mean that the range of prenatal screening is continuing to evolve.

Testing for cystic fibrosis and fragile X syndrome are available in the UK, mainly for those willing to pay privately, and in Israel, for example, a wide range of relatively rare diseases can be screened for on the basis that they are more common in specific population groups that were originally isolates with multiple inbreeding, and therefore certain mutations are prevalent. Besides Tay-Sachs disease (carrier testing in this case is biochemical; see Chapter 20), familial dysautonomia, Canavan disease, Bloom syndrome, ataxia telangiectasia (North African Jews), limb-girdle muscular dystrophy (Libyan Jews) and Costeff syndrome (Iraqi Jews) are among the conditions for which screening is available. It does not come free of charge but the level of uptake of this screening is high, revealing the lengths to which some societies will go in order to avoid having children with serious genetic conditions. As DNA testing becomes more automated, rapid, and affordable, there will be pressure from some quarters to screen for many conditions, even though they are individually very rare. This challenge is already emerging with the potential use of microarray CGH in prenatal testing. If the use of microarray becomes routine there may be great difficulty in interpreting the consequences of rare or unique copy number variants. This ethical challenge is discussed more fully in Chapter 24.

Maternal Serum Screening

It has been government policy in the UK since 2001 that antenatal Down syndrome screening be available to all women, though it was introduced in the late 1980s. Where it is standard practice, maternal serum screening is offered for NTDs and Down syndrome using a blood sample obtained from the mother at 16 weeks' gestation. In this way up to 75% of all cases of open NTDs and 60% to 70% of all cases of Down syndrome can be detected.

Neural Tube Defects

In 1972 it was recognized that many pregnancies in which the baby had an open NTD (p. 258) could be detected at 16 weeks' gestation by assay of AFP in maternal serum. AFP is the fetal equivalent of albumin and is the major protein in fetal blood. If the fetus has an open NTD, the level of AFP is raised in both the amniotic fluid and maternal serum as a result of leakage from the open defect. Open NTDs fulfil the criterion of being serious disorders, as anencephaly is invariably fatal, and between 80% and 90% of the small proportion of babies who survive with an open lumbosacral lesion are severely handicapped.

Unfortunately maternal serum AFP screening for NTDs is neither 100% sensitive nor 100% specific (p. 319). The curves for the levels of maternal serum AFP in normal and affected pregnancies overlap (Figure 21.6), so that in practice an arbitrary cut-off level has to be introduced below which no further action is taken. This is usually either the 95th centile, or 2.5 multiples of the median (MoM); as a result around 75% of screened open spina bifida cases are detected. Those pregnant women with results that lie above this arbitrary cut-off level are offered detailed ultrasonography; which is usually sufficient to diagnose NTD. In fact, ultrasonography has more or less superceded maternal serum screening as a means of diagnosing NTD. Anencephaly shows a dramatic deficiency in the cranium (Figure 21.7) and an open myelomeningocele is almost invariably associated with herniation of the cerebellar tonsils through the foramen magnum. This deforms the cerebellar hemispheres, which then have a curved appearance known as the 'banana

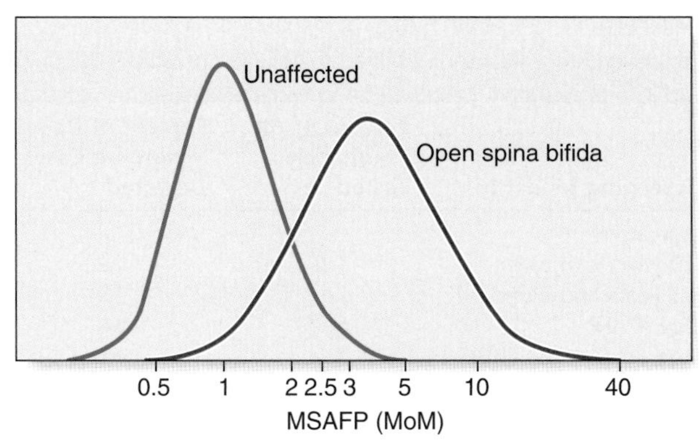

FIGURE 21.6 Maternal serum α-fetoprotein (AFP) levels at 16 weeks' gestation plotted on a logarithmic scale as multiples of the median (MoMs). Women with a value off or above 2.5 multiples of the median are offered further investigations. (Adapted from Brock DJH, Rodeck CH, Ferguson-Smith MA [eds] 1992 Prenatal diagnosis and screening. Edinburgh, UK: Churchill Livingstone.)

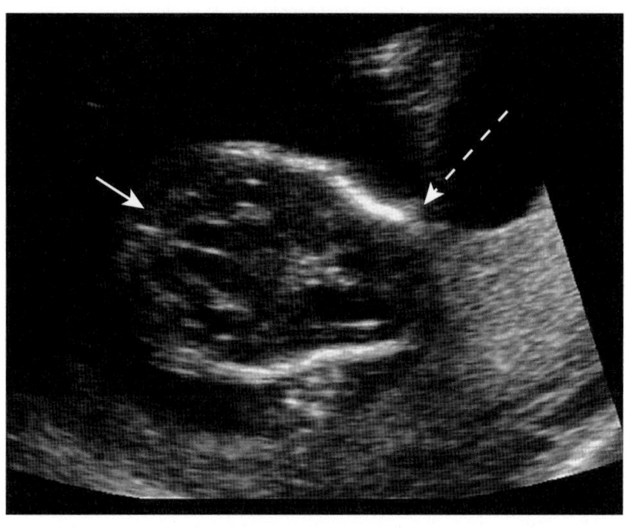

FIGURE 21.8 The so-called banana sign showing the distortion of the cerebellar hemispheres into a curved structure *(solid arrow)*. The forehead is also distorted into a shape referred to as the 'lemon sign' *(broken arrow)*. (Courtesy Dr. Helen Liversedge, Exeter, UK.)

sign'; the forehead is also distorted, giving rise to a shape referred to as the 'lemon sign' (Figure 21.8). A posterior encephalocele is readily visualized as a sac in the occipital region (Figure 21.9) and always prompts a search for additional anomalies that might help diagnose a recognizable condition such as Meckel-Gruber syndrome.

A raised maternal serum AFP concentration is not specific for open NTDs (Box 21.1). Other causes include threatened miscarriage, twin pregnancy and a fetal abnormality such as exomphalos, in which there is a protrusion of abdominal contents through the umbilicus.

As a result of these screening modalities there has been a striking decline in the incidence of open NTDs in liveborn and stillborn babies. Other contributory factors are a general

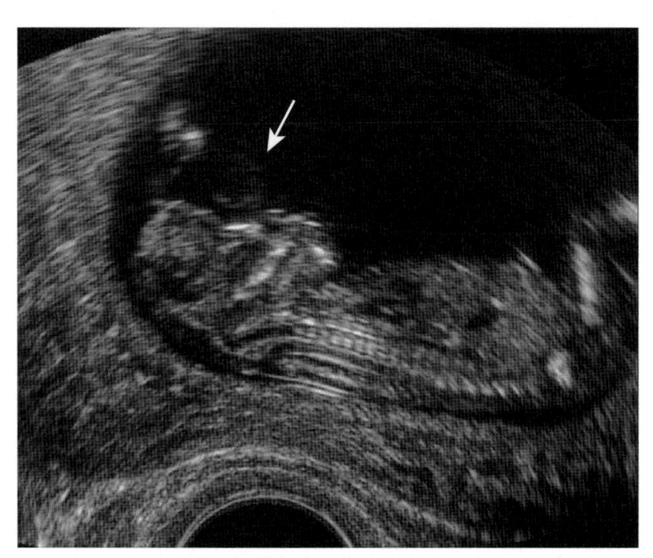

FIGURE 21.7 Anencephaly (*arrow*). There is no cranium and this form of neural tube defect is incompatible with life. (Courtesy Dr. Helen Liversedge, Exeter, UK.)

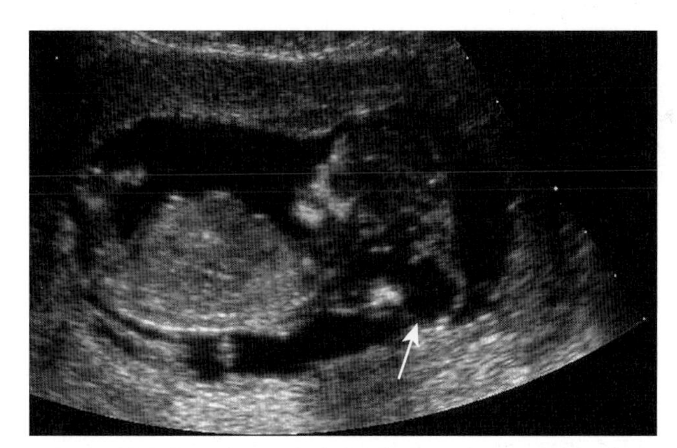

FIGURE 21.9 Posterior encephalocele (*arrow*), a more rare form of neural tube defect. This may be an isolated finding or associated with polydactyly and cystic renal changes in Meckel-Gruber syndrome. (Courtesy Dr. Helen Liversedge, Exeter, UK.)

Box 21.1 Causes of Raised Maternal Serum AFP Level

Anencephaly
Open spina bifida
Incorrect gestational age
Intrauterine fetal bleed
Threatened miscarriage
Multiple pregnancy
Congenital nephrotic syndrome
Abdominal wall defect

improvement in diet and the introduction of periconceptional folic acid supplementation (p. 258). In England and Wales the combined incidence of anencephaly and spina bifida in liveborn and stillborn babies fell from 1 in 250 in 1973 to 1 in 6250 in 1993.

Down Syndrome and Other Chromosome Abnormalities

The Triple Test

Confirmation of a chromosome abnormality in an unborn baby requires cytogenetic or molecular studies using material obtained by an invasive procedure such as CVS or amniocentesis (p. 325). However, chromosome abnormalities, and in particular Down syndrome, can be screened for in pregnancy by taking into account risk factors such as maternal age and the levels of three biochemical markers in maternal serum (Table 21.2).

This latter approach is based on the discovery that, at 16 weeks' gestation, maternal serum AFP and unconjugated estriol levels tend to be *lower* in Down syndrome pregnancies than in normal pregnancies, whereas the level of maternal serum human chorionic gonadotropin (hCG) is usually raised. None of these parameters gives absolute discrimination, but taken together they provide a means of modifying a woman's prior age-related risk to give an overall probability that the unborn baby is affected. When this probability exceeds 1 in 250, invasive testing in the form of amniocentesis or placental biopsy is offered.

Using age alone as a screening parameter, if all pregnant women aged 35 years and over opt for fetal chromosome analysis approximately 35% of all Down syndrome pregnancies will be detected (Table 21.3). If three biochemical markers are also included (this being the so-called triple test), 60% of all Down syndrome pregnancies will be detected when a risk of 1 in 250 or greater is the cut-off for offering amniocentesis. This approach will also result in the detection of approximately 50% of all cases of trisomy 18 (p. 275). In the latter condition *all* the biochemical parameters are *low*, including hCG.

It has recently been shown that another biochemical marker, inhibin-A, is also increased in maternal serum in Down syndrome pregnancies. If this fourth marker is used as part of a 'quadruple' serum screening test, the proportion of Down syndrome pregnancies detected rises from 60% to 75% when amniocentesis is offered to the 5% of mothers with the highest risk.

Table 21.2 Maternal Risk Factors for Down Syndrome

Advanced age (35 years or older)	MoM*
Maternal serum	
α-Fetoprotein	(0.75)
Unconjugated estriol	(0.73)
Human chorionic gonadotrophin	(2.05)
Inhibin-A	(2.10)

*Values in parentheses refer to the mean values in affected pregnancies, expressed as multiples of the median (MoMs) in normal pregnancies.

Table 21.3 Detection Rates Using Different Down Syndrome Screening Strategies

Screening Modality	Percent of All Pregnancies Tested	Percent of Down Syndrome Cases Detected
Age alone		
40 years and older	1.5	15
35 years and older	7	35
Age + AFP	5	34
Age + AFP, μE3 + hCG	5	61
Age + AFP, μE3, hCG + inhibin-A	5	75
NT alone	5	61
NT + age	5	69
hCG, AFP + age	5	73
NT + AFP, hCG + age	5	86

AFP, α-fetoprotein; *μE3*, unconjugated estriol; *hCG*, human chorionic gonadotrophin; *NT*, nuchal translucency.

Published results from California provide a useful indication of the outcome of a triple-test prenatal screening program. In a population of 32 million, all pregnant women were offered the triple-test. This was accepted by 67% of all eligible women, of whom 2.6% went on to have amniocentesis, resulting in the detection of 41% of all cases of Down syndrome. These figures are similar to those observed in other studies and serve to illustrate the discrepancy between what is possible in theory (i.e., a detection rate of 60%) and what actually happens in practice.

Ultrasonography

Almost all pregnant women are routinely offered a 'dating' scan at around 12 weeks' gestation. At around this time there is a strong association between chromosome abnormalities and the abnormal accumulation of fluid behind the baby's neck—increased fetal nuchal translucency (NT) (see Figure 21.5). This applies to Down syndrome, the other autosomal trisomy syndromes (trisomies 13 and 18; p. 275), Turner syndrome, and triploidy, as well as a wide range of other fetal abnormalities and rare syndromes. The risk for Down syndrome correlates with absolute values of NT as well as maternal age (Figure 21.10) but, because NT also increases with gestational age, it is more usual now to relate the risk to the percentile value for any given gestational age. In one study, for example, 80% of Down syndrome fetuses had NT above the 95th percentile. By combining information on maternal age with the results of fetal NT thickness measurements, together with maternal serum markers, it is possible to detect more than 80% of fetuses with trisomy 21 if invasive testing is offered to the 5% of pregnant women with the highest risk (see Table 21.3). Some babies with Down syndrome have duodenal atresia, which shows up as a 'double bubble sign' on ultrasonography of the fetal abdomen (Figure 21.11).

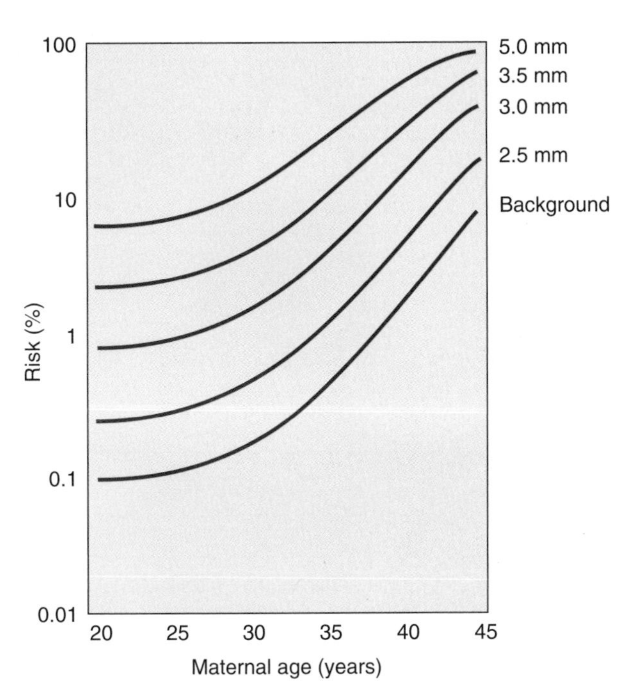

FIGURE 21.10 Risk for trisomy 21 (Down syndrome) by maternal age, for different absolute values of nuchal translucency (NT) at 12 weeks' gestation.

In many centers, it is also standard practice to offer a detailed 'fetal anomaly' scan to all pregnant women at 18 weeks. Although chromosome abnormalities cannot be diagnosed directly, their presence can be suspected by the detection of an abnormality, such as exomphalos (Figure 21.12) or a rocker-bottom foot (Figure 21.13) (Table 21.4). A chromosome abnormality is found in 50% of fetuses with exomphalos identified at 18 weeks, and a rocker-bottom foot is a very characteristic, though not specific, finding in babies with trisomy 18 (p. 275), who are invariably growth

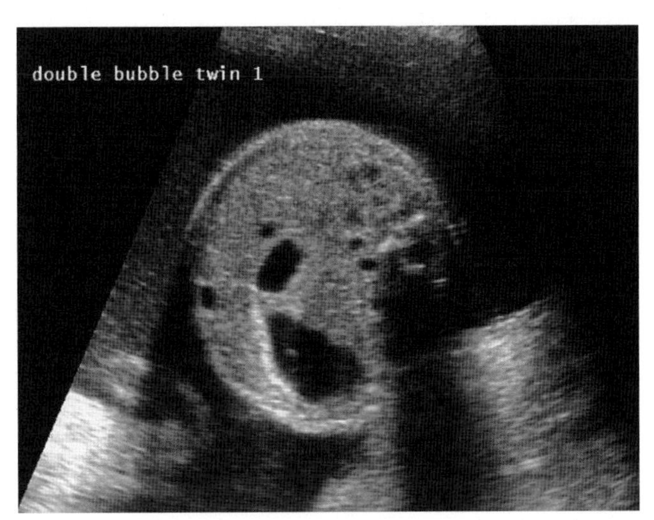

FIGURE 21.11 The 'double bubble sign', suggestive of duodenal atresia, sometimes associated with Down syndrome. (Courtesy Dr. Helen Liversedge, Exeter, UK.)

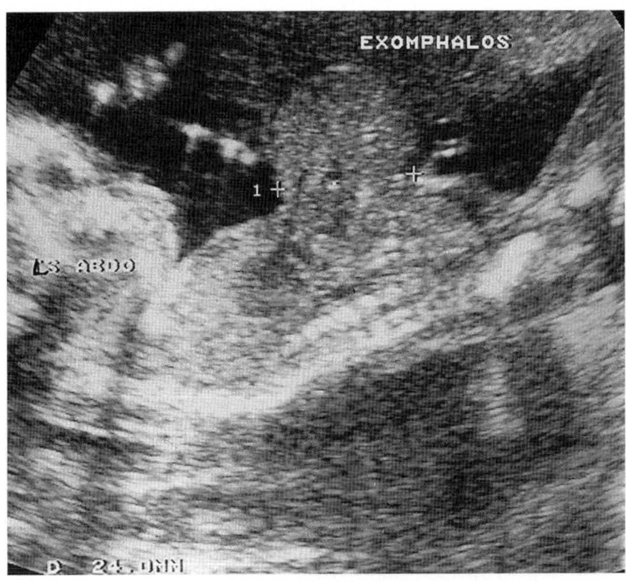

FIGURE 21.12 Ultrasonogram at 18 weeks showing exomphalos. (Courtesy Dr. D. Rose, City Hospital, Nottingham, UK.)

retarded. The use of other ultrasonographic 'soft markers' in identifying chromosome abnormalities in pregnancy is discussed in the following section (p. 335).

Indications for Prenatal Diagnosis

There are numerous indications for offering prenatal diagnosis. Ideally, couples at high prior risk of having a baby with an abnormality should be identified and assessed before embarking on a pregnancy so that, in an unrushed manner, they can be counseled and come to a decision about which option they wish to pursue. Certain orthodox Jewish communities are extremely well organized in this respect vis-à-vis Tay-Sachs disease, as described in Chapter 20 (pp. 313–314). A less satisfactory alternative is that couples are identified early in pregnancy so that they still have an opportunity to consider prenatal diagnostic options. Unfortunately, many couples at increased risk because of their family history or previous reproductive history are still not

Table 21.4 Prenatal Ultrasonographic Findings Suggestive of a Chromosome Abnormality

Feature	Chromosome Abnormality
Cardiac defect (especially common atrioventricular canal)	Trisomy 13, 18, 21
Clenched overlapping fingers	Trisomy 18
Cystic hygroma or fetal hydrops	Trisomy 13, 18, 21
Duodenal atresia	45,X (Turner syndrome) Trisomy 21
Exomphalos	Trisomy 13, 18
Rocker-bottom foot	Trisomy 18

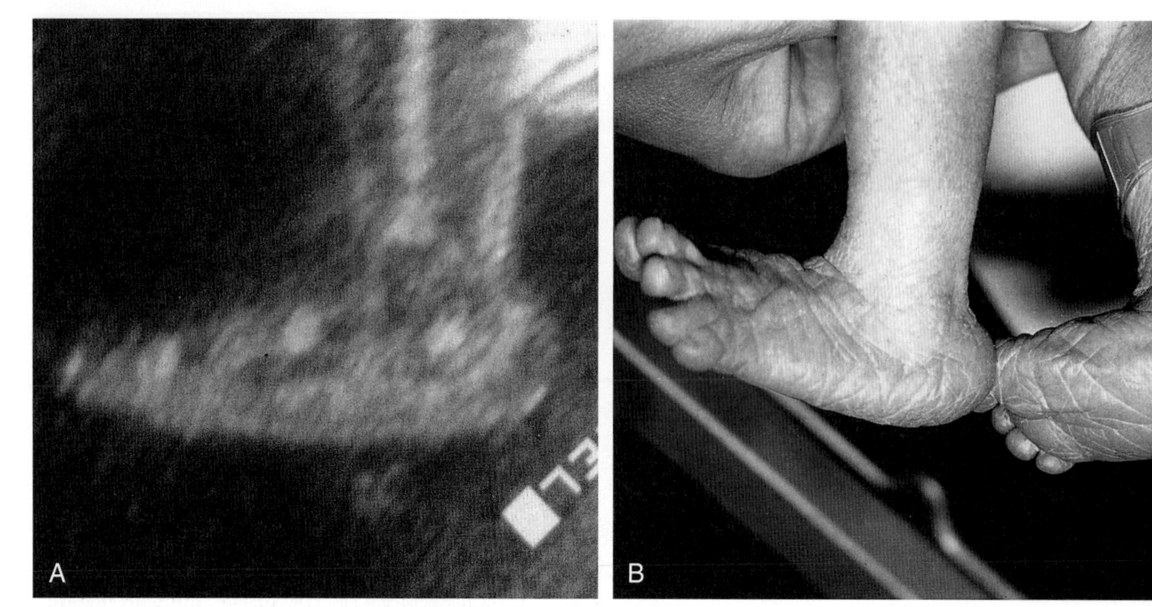

FIGURE 21.13 **A**, Ultrasonogram at 18 weeks showing a rocker-bottom foot in a fetus subsequently found to have trisomy 18. **B**, Photograph of the feet of a newborn with trisomy 18. (Courtesy Dr. D. Rose, City Hospital, Nottingham, UK.)

referred until mid-pregnancy, when it may be too late to undertake the most thorough clinical and laboratory work-up in preparation for prenatal diagnosis.

Advanced Maternal Age

This has been the most common indication for offering prenatal diagnosis. There is a well-recognized association of advanced maternal age with increased risk of having a child with Down syndrome (see Table 18.4; p. 274) and the other autosomal trisomy syndromes. No standard criterion exists for determining at what age a mother should be offered the option of an invasive prenatal diagnostic procedure for fetal chromosome analysis. Most centers routinely offer amniocentesis or CVS to women age 37 years or older, and the option is often discussed with women from the age of 35 years onward. These risk figures relate to the maternal age at the expected date of delivery. The risk figures for Down syndrome at the time of CVS, amniocentesis, and delivery differ (see Figure 18.1; p. 274) because a proportion of pregnancies with trisomy 21 are lost spontaneously during the first and second trimesters. Interestingly, despite industrial-scale efforts to screen for Down syndrome, there has been a slight rise in the numbers of live births in the United Kingdom since 2000, following a steady decline from the widespread introduction in screening in 1989 (National Down Syndrome Cytogenetic Register). However, the numbers of prenatal diagnoses and terminations for Down syndrome has also increased over this period. Both observations are attributed to the slightly older age at which women are now having children, and there may also be an increasing willingness to raise a child with the condition.

Previous Child with a Chromosome Abnormality

Although there are a number of series with slightly different recurrence risk figures, for couples who have had a child with Down syndrome because of non-disjunction, or a de novo unbalanced Robertsonian translocation, the risk in a subsequent pregnancy is usually given as the mother's age-related risk plus approximately 1%. If one of the parents has been found to carry a balanced chromosomal rearrangement, such as a chromosomal translocation (p. 44) or pericentric inversion (p. 48), that has caused a previous child to be born with serious problems due to an unbalanced chromosome abnormality, the recurrence risk is likely to be between 1% to 2% and 15% to 20%. The precise risk will depend on the nature of the parental rearrangement and the specific segments of the individual chromosomes involved (p. 48).

Family History of a Chromosome Abnormality

Couples may be referred because of a family history of a chromosome abnormality, most commonly Down syndrome. For most couples, there will usually be no increase in risk compared with the general population, as most cases of trisomy 21 and other chromosomal disorders will have arisen as a result of non-disjunction rather than as a result of a familial translocation, or other rearrangement. However, each situation should be evaluated carefully, either by confirming the nature of the chromosome abnormality in the affected individual or, if this is not possible, by urgent chromosome analysis of blood from the relevant parent at risk.

The results of parental chromosomal analysis can usually be obtained within 3 to 4 days; if normal, an invasive prenatal diagnostic procedure is not then appropriate as the risk is no greater than that for the general population.

Family History of a Single-Gene Disorder

If prospective parents have already had an affected child, or if one of the parents is affected or has a positive family history of a single-gene disorder that conveys a significant risk to offspring, then the option of prenatal diagnosis should be discussed with them. Prenatal diagnosis is available for a large and ever-increasing number of single-gene disorders by either biochemical or DNA analysis.

Family History of a Neural Tube Defect

Careful evaluation of the pedigree is necessary to determine the risk that applies to each pregnancy. Risks can be determined based on empiric data (p. 346). In high-risk situations, ultrasonographic examination of the fetus, possibly in conjunction with assay of maternal serum AFP, can be offered. However, even with good equipment and an experienced ultrasonographer, small closed NTDs can still be missed. Fortunately, the latter types of NTD are not usually associated with the serious problems seen with large open NTDs (p. 258).

Family History of Other Congenital Structural Abnormalities

As with NTDs, evaluation of the family pedigree should enable the provision of a risk derived from the results of empiric studies. If the risk to a pregnancy is increased, detailed ultrasonographic examination looking for the specific structural abnormality can be offered at around 16 to 18 weeks' gestation. Mid-trimester ultrasonography will detect most serious cranial, cardiac, renal and limb malformations. Some couples request detailed ultrasonographic scanning, not because they wish to pursue the option of termination of pregnancy but because they wish to prepare themselves if the baby is found to be affected.

Family History of Undiagnosed Learning Difficulty

An increasingly common scenario is the urgent referral of a pregnant couple who already has a child, or close relative, with an undiagnosed learning difficulty, with or without dysmorphic features. Whereas in the past a standard karyotype and fragile X syndrome test might be carried out, today it is incumbent on geneticists to use the latest technique of microarray-CGH (p. 281). This must first be undertaken on the affected child or individual, of course, and producing an urgent report (2 weeks) is a challenge for the laboratories.

Abnormalities Identified In Pregnancy

The widespread introduction of prenatal diagnostic screening procedures, such as triple testing and fetal anomaly scanning, has meant that many couples unexpectedly present with diagnostic uncertainty during the pregnancy that can be resolved only by an invasive procedure such as amniocentesis or CVS. Other factors, such as poor fetal growth, can also be an indication for prenatal chromosome analysis, as confirmation of a serious and non-viable chromosome abnormality, such as trisomy 18 or triploidy (p. 276), can influence subsequent management of the pregnancy and mode of delivery.

Other High-Risk Factors

These factors include parental consanguinity, a poor obstetric history, and certain maternal illnesses. Parental consanguinity increases the risk that a child will have a hereditary disorder or congenital abnormality (pp. 113). Consequently, if the parents are concerned, it is appropriate to offer detailed ultrasonography to try to exclude a serious structural abnormality. It may also be appropriate to offer to test the couple for cystic fibrosis and spinal muscular atrophy carrier status, and possibly other conditions depending on ethnicity. A poor obstetric history, such as recurrent miscarriages or a previous unexplained stillbirth, could indicate an increased risk of problems in a future pregnancy and detailed ultrasonographic monitoring. A history of three or more unexplained miscarriages should be investigated by parental chromosome studies to exclude a chromosomal rearrangement such as a translocation or inversion (pp. 44, 48). Maternal illnesses, such as poorly controlled diabetes mellitus (p. 235) or epilepsy treated with anticonvulsant medications such as sodium valproate (p. 261), would also be indications for detailed ultrasonography. Both of these factors convey an increased risk of structural abnormality in a fetus.

Special Problems in Prenatal Diagnosis

The significance of the result of a prenatal diagnostic investigation is usually clear-cut, but situations can arise that pose major problems of interpretation. Problems also occur when the diagnostic investigation is unsuccessful or an unexpected result is obtained.

Failure to Obtain a Sample or Culture Failure

It is important that every woman undergoing one of these invasive procedures is alerted to the possibility that, on occasion, it can prove impossible to obtain a suitable sample or the cells obtained subsequently fail to grow. Fortunately, the risk of either of these events occurring is less than 1%.

An Ambiguous Chromosome Result

In approximately 1% of cases, CVS shows evidence of apparent chromosome mosaicism—i.e., the presence of two or more cell lines with different chromosome constitutions (p. 50). This can occur for several reasons:

1. The sample is *contaminated* by maternal cells. This is more likely to be seen when using cultured cells than with direct preparations.

2. The mosaicism is a **culture artifact**. Usually, several separate cell cultures are routinely established at the time of the procedure in order to help resolve this problem rapidly. If mosaicism is present in only one culture then it is probably an artifact and does not reflect the true fetal karyotype.
3. The mosaicism is limited to a portion of the placenta, or what is known as **confined placental mosaicism**. This arises due to an error in mitosis during the formation and development of the trophoblast and is of no consequence to the fetus.
4. There is **true fetal mosaicism**.

In the case of amniocentesis, in most laboratories it is routine for the sample to be split and for two or three separate cultures to be established. If a single abnormal cell is identified in only one culture, this is assumed to be a culture artifact, or what is termed **level 1 mosaicism**, or **pseudomosaicism**. If the mosaicism extends to two or more cells in two or more cultures this is taken as evidence of true mosaicism, or what is known as **level 3 mosaicism**. The most difficult situation to interpret is when mosaicism is present in two or more cells in only one culture, termed **level 2 mosaicism**. This is most likely to represent a culture artifact, but there is up to a 20% chance that the mosaicism is real and will be present in the fetus.

To resolve the uncertainty of chromosomal mosaicism in cultured CV tissue it may be necessary to proceed to amniocentesis. If the latter test yields a normal chromosomal result, then it is usually concluded that the earlier result was not a true indication of the fetal karyotype.

Counseling couples in this situation may be extremely difficult. If true mosaicism is confirmed, it is often impossible to predict the phenotypic outcome for the baby. An attempt can be made to resolve ambiguous findings by fetal blood sampling for urgent karyotype analysis, but this too is limited in terms of the information it yields about the phenotype. Whatever option the parents choose, it is important that tissue (blood, skin, or placenta) is obtained at the time of delivery, whether the couple elects to terminate or continue with the pregnancy, to resolve the significance of the prenatal findings.

An Unexpected Chromosome Result

Three different types of unexpected chromosome results may occur, each of which usually necessitates specialized detailed genetic counseling.

A Different Numerical Chromosomal Abnormality

Although most invasive prenatal diagnostic procedures, such as CVS and amniocentesis, are carried out because of an increased risk of trisomy 21 through increased maternal age, or as a result of increased risk through the triple test or NT screening, a chromosomal abnormality other than trisomy 21 may be found, for example another autosomal trisomy (13 or 18) or a sex chromosome aneuploidy (45,X, 47,XXX, 47,XXY, or 47,XYY). The implications for the outcome of a pregnancy with an autosomal trisomy are reasonably straightforward but not so clear-cut with the sex chromosome aneuploidies. Ideally, all women undergoing amniocentesis should be alerted to this possibility before the test is carried out, although in practice this is rarely done. When a diagnosis such as Turner syndrome (45,X) or Klinefelter syndrome (47,XXY) is obtained, it is essential that the parents be given full details of the nature and consequences of the diagnosis. When objective and informed counseling is available, less than 50% of the parents of a fetus with an 'incidental' diagnosis of a sex chromosome abnormality opt for termination of the pregnancy.

A Structural Chromosomal Rearrangement

A second difficult situation is the discovery of an apparently balanced chromosome rearrangement in the fetus, such as an inversion or translocation. If analysis of parental chromosomes shows that one of the parents has the same structural chromosomal rearrangement, the parents can be reassured that this is very unlikely to cause any problems in the fetus. If, however, the apparently balanced chromosomal rearrangement has occurred as a de novo event in the fetus, there is a 5% to 10% chance that the fetus will have physical abnormalities and/or subsequently show developmental delay. This is likely to reflect a subtle chromosomal imbalance that cannot be detected using conventional cytogenetic techniques, though may be resolved by microarray-CGH if this is available. It is also possible that damage to a critical gene at one or both of the rearrangement breakpoints has occurred, which would not be picked up by microarray. It is not surprising that couples in this situation can have great difficulty in deciding what to do. Detailed ultrasonography, if normal, can provide some, but not complete, reassurance. Later on, the extended family should be investigated if the rearrangement is found to be present in one of the parents.

The Presence of a Marker Chromosome

A third difficult situation is the finding of a small additional chromosome known as a marker chromosome, that is, a small chromosomal fragment the specific identity of which cannot be determined by conventional cytogenetic techniques (p. 33). If this is found to be present in one of the parents, then it is unlikely to be of any significance to the fetus. If, on the other hand, it is a de novo finding, there is up to a 15% chance that the fetus will be phenotypically abnormal. The risk is lower when the marker chromosome contains satellite material (p. 17), or is made up largely of heterochromatin (p. 32), than when it does not have satellites and is mostly made up of euchromatin (p. 32). The availability of FISH (p. 34) means that the origin of the marker chromosome can often be determined more specifically, so that it is possible to give more precise prognostic information. The most common single abnormality of this kind is a marker chromosome 15.

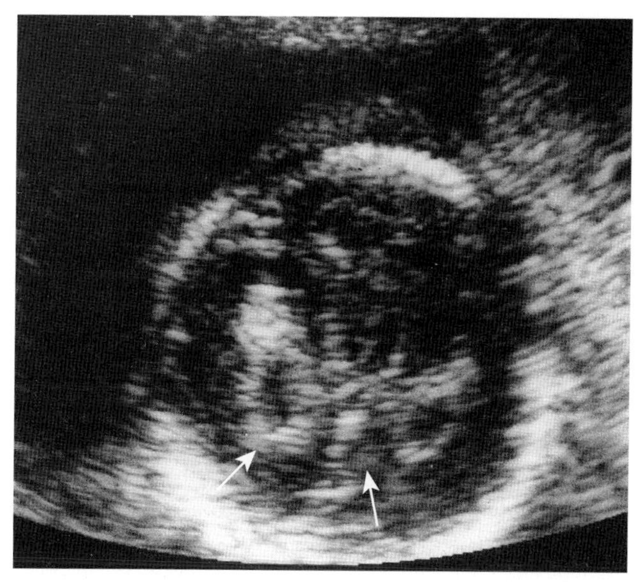

FIGURE 21.14 Ultrasonogram of a fetal brain showing bilateral choroid plexus cysts (*arrows*).

Ultrasonographic 'Soft' Markers

Sophisticated ultrasonography has resulted in the identification of subtle anomalies in the fetus, the significance of which is not always clear. For example, choroid plexus cysts are sometimes seen in the developing cerebral ventricles in mid-trimester (Figure 21.14). Initially, it was thought that these were invariably associated with the fetus having trisomy 18 but in fact they occur frequently in normal fetuses, although if they are very large and do not disappear spontaneously they can be indicative of a chromosome abnormality.

Increased echogenicity of the fetal bowel (Figure 21.15) has been reported in association with cystic fibrosis—the prenatal equivalent of meconium ileus (p. 301). Initial reports suggested this finding could convey a risk as high as 10% for the fetus having cystic fibrosis, but it is now clear that this risk is probably no greater than 1% to 2%. Novel ultrasonographic findings of this kind are often called **soft markers**, and their interpretation must be approached cautiously in the effort to distinguish normal from abnormal variation.

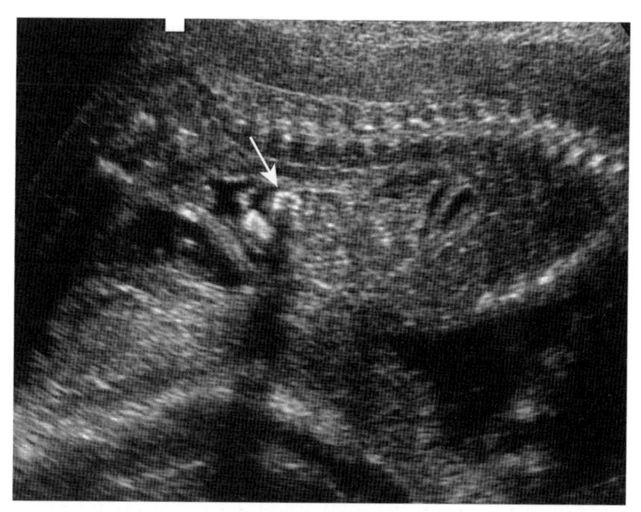

FIGURE 21.15 Echogenic bowel. Regions of the bowel showing unusually high signal (*arrow*). This is occasionally a sign of meconium ileus seen in cystic fibrosis. (Courtesy Dr. Helen Liversedge, Exeter, UK)

Termination of Pregnancy

The presence of a serious abnormality in a fetus in the majority of developed countries is an acceptable legal indication for termination of pregnancy (TOP). This does not mean, however, that this is an easy choice for a couple to make. It is essential that all couples undergoing any form of prenatal diagnostic investigation, whether invasive or noninvasive, be provided with information about the practical aspects of TOP before the prenatal diagnostic procedure is carried out. This should include a practical explanation that termination in the first trimester is carried out by surgical means under general anesthesia, whereas a woman undergoing a mid-trimester termination will have to experience labor and delivery.

Preimplantation Genetic Diagnosis

For many couples prenatal diagnosis on an established pregnancy, with a view to possible termination, is too difficult to contemplate. For some of these couples **preimplantation genetic diagnosis** (PGD) provides an acceptable alternative. The second largest group of PGD users are those with subfertility or infertility who wish to combine assisted reproduction with genetic testing of the early embryo. In the procedure, the female partner is given hormones to induce hyperovulation, and oocytes are then harvested transcervically, under sedation and ultrasonographic guidance. Motile sperm from a semen sample are added to the oocytes in culture (**in vitro fertilization** [IVF]—the same technique as developed for infertility) and incubated to allow fertilization to occur. If genetic analysis is to be undertaken on DNA from a single cell (blastomere) from the early embryo (blastocyst) at the eight-cell stage on the third day, fertilization is achieved using **intracytoplasmic sperm injection** (ICSI) of a single sperm to avoid the presence of extraneous sperm.

At the eight-cell stage, the early embryo is biopsied and one, or sometimes two, cells are removed for analysis. Whatever genetic analysis is undertaken, it is essential that this is a practical possibility on genomic material from a single cell. From the embryos tested, two that are both healthy and unaffected by the disorder from which they are at risk are reintroduced into the mother's uterus. Implantation must then occur for a successful pregnancy and this is a major hurdle—the success rate for the procedure is only about 25% to 30% per cycle of treatment, even in the best centers. A variation of the technique is removal of the first, and often second, polar bodies from the unfertilized oocyte, which lie under the zona pellucida. Because the first polar body degenerates quite rapidly, analysis is necessary within 6 hours of retrieval. Analysis of polar bodies is an indirect method of genotyping because the oocyte and first polar body divide from each other during meiosis I and therefore contain different members of each pair of homologous chromosomes.

In the United Kingdom, centers must be licensed to practice PGD and are regulated by the Human Fertilization and Embryology Authority (HFEA), though this body is due to be abolished. In numerical terms, the impact of PGD has been small to date, but a wide and increasing range of genetic conditions has now been tested (Table 21.5). The most common referral reasons for single-gene disorders are cystic fibrosis, myotonic dystrophy, Huntington disease, β-thalassemia, spinal muscular atrophy, and fragile X syndrome. The technique for identifying normal and abnormal alleles in these conditions, and DNA linkage analysis where appropriate, is PCR (p. 56). Sex selection in the case of serious X-linked conditions is available where single-gene analysis is not possible. The biggest group of referrals for PGD, however, is chromosome abnormalities—reciprocal and robertsonian translocations in particular (pp. 45, 47). Genetic analysis in these cases uses FISH technology (p. 34) and substantial work has to be undertaken for the couple prior to treatment because of the unique nature of many translocations.

In recent years, PGD has occasionally been used not only to select embryos unaffected for the genetic disorder for which the pregnancy is at risk, but also to provide a human leukocyte antigen tissue-type match so that the new child can act as a bone marrow donor for an older sibling affected by, for example, Fanconi anemia. The ethical debate surrounding these so-called savior sibling cases is discussed further in Chapter 24.

A further development using micromanipulation methods has attracted a lot of attention. To circumvent the problem of genetic disease resulting from mutation in the mitochondrial genome, the nucleus of the oocyte from the genetic mother (who carried the mitochondrial mutation) was removed and inserted into a donor oocyte from which the nucleus had been removed. This is cell nuclear replacement technology, similar to that used in reproductive cloning experiments in animals ('Dolly' the sheep; see p. 369). The resulting fertilization led to the headline that the fetus had three genetic parents. The technique has also been used in other situations where the oocytes are generally of poor quality, but its use is extremely limited.

Table 21.5 Some of the Conditions for Which Preimplantation Genetic Diagnosis Has Been Used and Is Available

Mode of Inheritance	Disease
Autosomal dominant	Charcot-Marie-Tooth
	Familial adenomatous polyposis
	Huntington disease
	Marfan syndrome
	Myotonic dystrophy
	Neurofibromatosis
	Osteogenesis imperfecta
	Tuberous sclerosis
Autosomal recessive	β-Thalassemia
	Cystic fibrosis
	Epidermolysis bullosa
	Gaucher disease
	Sickle cell disease
Spinal muscular atrophy	Tay-Sachs disease
X-linked	Alport syndrome
	Duchenne muscular dystrophy (DMD)
	Hunter syndrome
	Kennedy syndrome
	Fragile X syndrome
X-linked—sexing only	DMD
	Ornithine transcarbamylase deficiency
	Incontinentia pigmenti
	Other serious disorders
Mitochondrial	MELAS
Chromosomal	Robertsonian translocations
	Reciprocal translocations
	Aneuploidy screening
	Inversions, deletions

MELAS, Mitochondrial myopathy encephalopathy, lactic acidosis, stroke.

Assisted Conception and Implications for Genetic Disease

In Vitro Fertilization

Many thousands of babies worldwide have been born by IVF over the past 30 years, when the technique was first successful. The indication for the treatment in most cases is subfertility, which now affects one in seven couples. In some Western countries, 1% to 3% of all births are the result of assisted reproductive technologies (ARTs). The cohort of offspring conceived in this way is therefore very large, and evidence is gathering that the risk of birth defects is increased by 30% to 40% compared with the general population conceived in the normal way and about 50% more children are likely to be small for gestational age (SGA). Specifically, a small increase in certain epigenetic conditions due to defective genomic imprinting (p. 121) has been observed—Beckwith-Wiedemann (p. 124) and Angelman (p. 123) syndromes, and 'hypomethylation' syndrome, though the possible mechanisms are unclear. In cases studied, loss of imprinting (LOI) was observed at the *KCNQ1OT1* locus (see Figure 7.27; p. 125) in the case of Beckwith-Wiedemann syndrome, and at the *SNRPN* locus (Figure 7.23; p. 123) in the case of Angelman syndrome. No apparent imprinting differences explain the increase in SGA babies conceived by ICSI.

Epigenetic events around the time of fertilization and implantation are crucial for normal development (p. 103). If there is a definite increased risk of conditions from abnormal imprinting after ARTs, this may relate, in part, to the extended culture time of embryos, which has become a trend in infertility clinics. Instead of transferring cleavage-stage embryos, it is now more routine to transfer blastocysts, which allows the healthier looking embryos to be selected. However, in animal models it has been shown that in vitro culture affects the extent of imprinting, gene expression, and therefore the potential for normal development.

Intracytoplasmic Sperm Injection

As mentioned, this technique is employed as part of IVF when combined with PGD, although the main indication for directly injecting the sperm into the egg is male subfertility because of low sperm count, poor sperm motility, abnormal sperm morphology, or mechanical blockage to the passage of sperm along the vas deferens. Chromosomal abnormalities or rearrangements have been found in about 5% of men for whom ICSI is suitable, and 10% to 12% in those with azoospermia or severe oligospermia. Examples include the robertsonian 13 : 14 translocation and Y-chromosome deletions. For men with azoospermia or severe oligospermia the karyotype should be checked, including the application of molecular techniques looking for submicroscopic Y deletions. In those with mechanical blockage due to congenital bilateral absence of the vas deferens (CBAVD), a significant proportion has cystic fibrosis mutations. ICSI offers hope to men with CBAVD, as well as those with Klinefelter syndrome, following testicular aspiration of sperm.

Some of the chromosomal abnormalities in the men may be heritable—especially those involving the sex chromosomes—and there is a small but definite increase in chromosomal abnormalities in the offspring (1.6%).

Donor Insemination

As a means of assisted conception to treat male infertility, or circumvent the risk of a genetic disease, **donor insemination** (DI) has been used since the 1950s. Only relatively recently, however, has awareness of medical genetic issues been incorporated into practice. Following the cases of children conceived by DI who were subsequently discovered to have balanced or unbalanced chromosome disorders, or in some cases cystic fibrosis (indicating that the sperm donor was a carrier for cystic fibrosis), screening of sperm donors for cystic fibrosis mutations and chromosome rearrangements has become routine practice in many countries. This was recommended only as recently as 2000 by the British Andrology Society. In the Netherlands, a donor whose sperm was used to father 18 offspring developed an autosomal dominant late-onset neurodegenerative disorder (one of the spinocerebellar ataxias), thus indicating that all 18 offspring were conceived at 50% risk. This led to a ruling that the sperm from one donor should be used no more than 10 times, as against 25 before this experience. In the United Kingdom, men older than age 40 years cannot be donors because of the small but increasing risk of new germline mutations arising in sperm with advancing paternal age.

Box 21.2 Assisted Conception Treatments Requiring a Licence from the Human Fertilization and Embryology Authority

In vitro fertilization
Intracytoplasmic sperm injection
Preimplantation genetic diagnosis
Sperm donation
Egg donation
Embryo donation
Surrogacy

Of course, it is not possible to screen the donor for all eventualities, but these cases have served to highlight the potential conflict between treating infertility (or genetic disease) by DI and maintaining a high level of concern for the welfare of the child conceived. More high profile in this respect is the ongoing debate about how much information DI children should be allowed about their genetic fathers, and the law varies across the world.

Naturally, all of these issues apply in an equivalent way to women who wish to be egg donors.

Assisted Conception and the Law

In the United States, no federal law exists to regulate the practice of assisted conception other than the requirement that outcomes of IVF and ICSI must be reported. In the United Kingdom, strict regulation operates through the HFEA based on the Human Fertilization and Embryology Act of 1990. The HFEA reports to the Secretary of State for Health, issues licences, and arranges inspections of registered centers. The different licences granted are for *treatment* (Box 21.2), *storage* (gametes and embryos), and *research* (on human embryos in vitro). A register of all treatment cycles, the children born by IVF, and the use of donated gametes, must be kept. The research permitted under licence covers treatment of infertility, increase in knowledge regarding birth defects, miscarriage, genetic testing in embryos, the development of the early embryo, and potential treatment of serious disease. At the time of writing it is not clear what arrangements will be put in place when the HFEA is abolished.

Non-Invasive Prenatal Diagnosis

At the turn of the 19th century, it was discovered that fetal cells reach the maternal circulation, but confirmation that **cell-free** DNA of fetal origin (placentally derived) is present in the plasma of pregnant women was not made until 1997. This fact has now been exploited in clinical practice as early as 6 to 7 weeks of pregnancy to determine fetal sex by detection of Y-chromosome DNA, as well as fetal Rhesus D gene. Early determination of fetal sex is clinically useful in a pregnancy at risk of an X-linked recessive disorder, and also in congenital adrenal hyperplasia (see the following section). The problem with analyzing cell-free fetal DNA is one of isolation because maternal cell-free DNA constitutes about 95% of all the cell-free DNA in the maternal circulation. The absence of Y-chromosome DNA might indicate the fetus is female, or that the quantity of fetal DNA is very low. This is resolved by using real-time PCR to quantify the amount of fetal or total DNA present in plasma.

Much effort is now focused on enriching fetal cells from maternal blood, which would make it possible to analyse the pure fetal genome because fetal red cells are nucleated; however, only about one fetal cell is present in 1 ml of maternal blood, and consequently current techniques are limited because of the scarcity of cells. If whole cells could be efficiently enriched, in theory fetal aneuploids could be detected using FISH, or other techniques based on allele ratios. Advances are taking place rapidly and non-invasive techniques are likely to become a reality in due course, and therefore change perceptions of prenatal diagnosis dramatically.

Prenatal Treatment

So far this chapter has focused on prenatal diagnosis and screening for abnormalities with the subsequent option of termination of pregnancy, as well as other techniques designed to prevent genetic disease. Although these are the only options in most situations, there is cautious optimism that prenatal diagnosis will, in time, lead to the possibility of effective treatment in utero, at least for some conditions.

A possible model for successful prenatal treatment is provided by the autosomal recessive disorder congenital adrenal hyperplasia (CAH) (p. 174). Affected female infants are born with virilization of the external genitalia. There is evidence that in a proportion of cases the virilization can be prevented if the mother takes a powerful steroid known as dexamethasone in a very small dose from 4 to 5 weeks' gestation onward. Specific prenatal diagnosis of CAH can be achieved by DNA analysis of CV tissue. If this procedure confirms that the fetus is both female and affected, the mother continues to take low doses of dexamethasone throughout pregnancy, which suppresses the fetal pituitary–adrenal axis and can prevent virilization of the female fetus. If the fetus is male and either affected or unaffected, the mother ceases to take dexamethasone and the pregnancy can proceed uneventfully.

Treatment of a fetus affected with severe combined immunodeficiency (p. 203) has also been reported. The immunological tolerance of the fetus to foreign antigens introduced in utero means that the transfused stem cells are recognized as 'self', with the prospect of good long-term results.

When gene therapy (p. 350) has been proved to be both safe and effective, the immunological tolerance of the fetus should make it easier to commence such therapy before birth rather than afterward. This will have the added advantage of reducing the period in which irreversible damage can occur in organs such as the central nervous system, which can be affected by progressive neurodegenerative disorders.

FURTHER READING

Abramsky L, Chapple J (eds) 2003 Prenatal diagnosis: the human side. Cheltenham, UK: Nelson Thornes

Dealing with the legal, emotional, and ethical issues, this nevertheless contains a lot of medical information in a very readable format with interesting case studies.

Brock DJH, Rodeck CH, Ferguson Smith MA (eds) 1992 Prenatal diagnosis and screening. Edinburgh, UK: Churchill Livingstone

A comprehensive multiauthor textbook covering all aspects of prenatal diagnosis.

Drife JO, Donnai D (eds) 1991 Antenatal diagnosis of fetal abnormalities. London, UK: Springer

The proceedings of a workshop on the practical aspects of prenatal diagnosis.

European Society for Human Reproduction and Embryology PGD Steering Committee 2002 ESHRE Preimplantation Genetic Diagnosis Consortium: data collection III (May 2001). Hum Reprod 17:233–246

An up-to-date appraisal of the use of PGD.

Lilford RJ (ed) 1990 Prenatal diagnosis and prognosis. Oxford, UK: Butterworth-Heinemann

Provides useful information on recurrence risks for Down syndrome, the prognosis for abnormalities detected by ultrasonography, and decision analysis.

Stranc LC, Evans JA, Hamerton JL 1997 Chorionic villus sampling and amniocentesis for prenatal diagnosis. Lancet 349:711–714

A good review of the practical and ethical aspects of the two main prenatal invasive diagnostic techniques.

Whittle MJ, Connor JM (eds) 1989 Prenatal diagnosis in obstetric practice. Oxford, UK: Blackwell

Describes prenatal diagnostic techniques and the types of abnormalities identified.

ELEMENTS

1. Prenatal diagnosis can be carried out by non-invasive procedures such as maternal serum α-fetoprotein screening for neural tube defects, the triple test, and nuchal pad screening for Down syndrome, ultrasonography for structural abnormalities, and in the future analysis of fetal DNA in the maternal circulation.

2. Specific prenatal diagnosis of chromosome and single-gene disorders usually requires an invasive technique, such as amniocentesis or chorionic villus sampling, by which material of fetal origin can be obtained for analysis.

3. Invasive prenatal diagnostic procedures convey small risks for causing miscarriage (e.g., amniocentesis 0.5% to 1%, chorionic villus sampling 1% to 2%, cordocentesis 1% to 2%, fetoscopy 3% to 5%).

4. The most common indication for prenatal diagnosis is advanced maternal age. Other indications include a family history of a chromosome, single-gene, or structural abnormality or an increased risk predicted from the result of a screening test.

5. Although the significance of most prenatal diagnostic findings is clear, situations can arise in which the implications for the fetus are very difficult to predict. When this occurs, the parents should be offered specialized genetic counseling.

CHAPTER 22

Risk Calculation

As far as the laws of mathematics refer to reality, they are not certain; and as far as they are certain, they do not refer to reality.

ALBERT EINSTEIN

One of the most important aspects of genetic counseling is the provision of a risk figure. This is often referred to as a **recurrence** risk. Estimation of the recurrence risk usually requires careful consideration and takes into account:

1. The diagnosis and its mode of inheritance
2. Analysis of the family pedigree
3. The results of tests that can include linkage studies using DNA markers, but may also include clinical data from standard investigation.

Sometimes the provision of a risk figure can be quite easy, but in a surprisingly large number of complicating factors arise that make the calculation very difficult. For example, the mother of a boy who is an isolated case of a sex-linked recessive disorder could very reasonably wish to know the recurrence risk for her next child. This is a very simple question, but the solution may be far from straightforward, as will become clear later in this chapter.

Before proceeding any further, it is necessary to clarify what we mean by **probability** and review the different ways in which it can be expressed. The probability of an outcome can be defined as the number or, more correctly, the **proportion** of times it occurs in a large series of events. Conventionally, probability is indicated as a proportion of 1, so that a probability of 0 implies that an outcome will never be observed, whereas a probability of 1 implies that it will always be observed. Therefore, a probability of 0.25 indicates that, on average, a particular outcome or event will be observed on 1 in 4 occasions, or 25%. The probability that the outcome will not occur is 0.75, which can also be expressed as 3 chances out of 4, or 75%. Alternatively, this probability could be expressed as odds of 3 to 1 against, or 1 to 3 in favor of the particular outcome being observed. In this chapter, fractions are used where possible as these tend to be more easily understood than proportions of 1 expressed as decimals.

Probability Theory

To calculate genetic risks it is necessary to have a basic understanding of probability theory. This will be discussed in so far as it is relevant to the skills required for genetic counseling.

Laws of Addition and Multiplication

When considering the probability of two different events or outcomes, it is essential to clarify whether they are mutually exclusive or independent. If the events are mutually exclusive, then the probability that *either* one or the other will occur equals the sum of their individual probabilities. This is known as the **law of addition**.

If, however, two or more events or outcomes are independent, then the probability that *both* the first and the second will occur equals the **product** of their individual probabilities. This is known as the **law of multiplication**.

As a simple illustration of these laws, consider parents who have embarked upon their first pregnancy. The probability that the baby will be *either* a boy *or* a girl equals 1—i.e., 1/2 + 1/2. If the mother is found on ultrasonography to be carrying twins who are non-identical, then the probability that *both* the first *and* the second twin will be boys equals 1/4—i.e., 1/2 × 1/2.

Bayes' Theorem

Bayes' theorem, which was first devised by the Reverend Thomas Bayes (1702–1761) and published after his death in 1763, is widely used in genetic counseling. Essentially it provides a very valuable method for determining the overall probability of an event or outcome, such as carrier status, by considering all initial possibilities (e.g., carrier or non-carrier) and then modifying or 'conditioning' these by incorporating information, such as test results or pedigree information, that indicates which is the more likely. Thus, the theorem combines the probability that an event *will* occur with the probability that it *will not* occur. The theorem lay fairly dormant for a long time, but has been enthusiastically employed by geneticists. In recent years its beauty, simplicity and usefulness have been recognized in many other fields—for example, legal work, computing, and statistical analysis—such that it has truly come of age.

The initial probability of each event is known as its **prior probability**, and is based on ancestral or **anterior information**. The observations that modify these prior probabilities allow **conditional probabilities** to be determined. In genetic counseling these are usually based on numbers of offspring and/or the results of tests. This is **posterior information**. The resulting probability for each event or outcome is known as its **joint probability**. The final probability for each

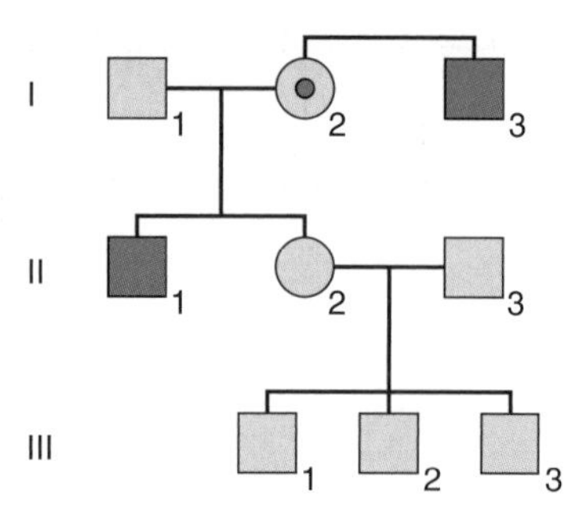

FIGURE 22.1 Pedigree showing sex-linked recessive inheritance. When calculating the probability that II_2 is a carrier, it is necessary to take into account her three unaffected sons.

event is known as its **posterior** or **relative probability** and is obtained by dividing the joint probability for that event by the sum of all the joint probabilities.

This is not an easy concept to grasp! To try to make it a little more comprehensible, consider a pedigree with two males, I_3 and II_1, who have a sex-linked recessive disorder (Figure 22.1). The sister, II_2, of one of these men wishes to know the probability that she is a carrier. Her mother, I_2, must be a carrier because she has both an affected brother and an affected son (i.e., she is an **obligate** carrier). Therefore, the prior probability that II_2 is a carrier equals 1/2. Similarly, the prior probability that II_2 is not a carrier equals 1/2.

The fact that II_2 already has three healthy sons must be taken into consideration, as intuitively this makes it rather unlikely that she is a carrier. Bayes' theorem provides a way to quantify this intuition. These three healthy sons provide posterior information. The conditional probability that II_2 will have three healthy sons if she is a carrier is $1/2 \times 1/2 \times 1/2$, which equals 1/8. These values are multiplied as they are independent events, in that the health of one son is not influenced by the health of his brother(s). The conditional probability that II_2 will have three healthy sons if she is not a carrier equals 1.

This information is now incorporated into a bayesian calculation (Table 22.1). From this table, the posterior probability that II_2 is a carrier equals $1/16/(1/16 + 1/2)$, which reduces to 1/9. Similarly the posterior probability that II_2 is not a carrier equals $1/2/(1/16 + 1/2)$, which reduces to 8/9. Another way to obtain these results is to consider that the odds for II_2 being a carrier versus not being a carrier are 1/16 to 1/2 (i.e., 1 to 8, which equals 1 in 9). Thus, by taking into account the fact that II_2 has three healthy sons, we have been able to reduce her risk of being a carrier from 1 in 2 to 1 in 9.

Table 22.1 Bayesian Calculation for II_2 in Figure 22.1

Probability	II_2 Is a Carrier	II_2 Is Not a Carrier
Prior	1/2	1/2
Conditional		
Three healthy sons	$(1/2)^3 = 1/8$	$(1)^3 = 1$
Joint	1/6	1/2 (= 8/16)
Expressed as odds	1 to	8
Posterior	1/9	8/9

Perhaps by now the use of Bayes' theorem will be a little clearer. Try to remember that the basic approach is to draw up a table showing all of the possibilities (e.g., carrier, not a carrier), then establish the background (prior) risk for each possibility, next determine the chance (conditional possibility) that certain observed events (e.g., healthy children) would have happened if each possibility were true, then work out the combined (joint) likelihood for each possibility, and finally weigh up each of the joint probabilities to calculate the exact (posterior) probability for each of the original possibilities. If this is still confusing, some of the following examples may bring more clarity.

Autosomal Dominant Inheritance

For someone with an autosomal dominant disorder, the risk that each of his or her children will inherit the mutant gene equals 1 in 2. This will apply whether the affected individual inherited the disorder from a parent or developed the condition as the result of a new mutation. Therefore the provision of risks for disorders showing autosomal dominant inheritance is usually straightforward as long as there is a clear family history, the condition is characterized by being fully penetrant, and there is a reliable means of diagnosing heterozygotes. However, if penetrance is incomplete or there is a delay in the age of onset so that heterozygotes cannot always be diagnosed, the risk calculation becomes more complicated. Two examples will be discussed to illustrate the sorts of problem that can arise.

Reduced Penetrance

A disorder is said to show **reduced penetrance** when it has clearly been demonstrated that individuals who must possess the abnormal gene, who by pedigree analysis must be obligate heterozygotes, show absolutely no manifestations of the condition. For example, if someone who was completely unaffected had both a parent and a child with the same autosomal dominant disorder, this would be an example of **non-penetrance**. Penetrance is usually quoted as a percentage (e.g., 80%) or as a proportion of 1 (e.g., 0.8). This would imply that 80% of all heterozygotes express the condition in some way.

For a condition showing reduced penetrance, the risk that the child of an affected individual will be affected equals 1/2—i.e., the probability that the child will inherit the mutant allele, × P, the proportion of heterozygotes who are affected. Therefore, for a disorder such as hereditary retinoblastoma, an embryonic eye tumor (p. 215), which shows dominant inheritance in some families with a penetrance of $P = 0.8$, the risk that the child of an affected parent will develop a tumor equals $1/2 \times 0.8$, which equals 0.4.

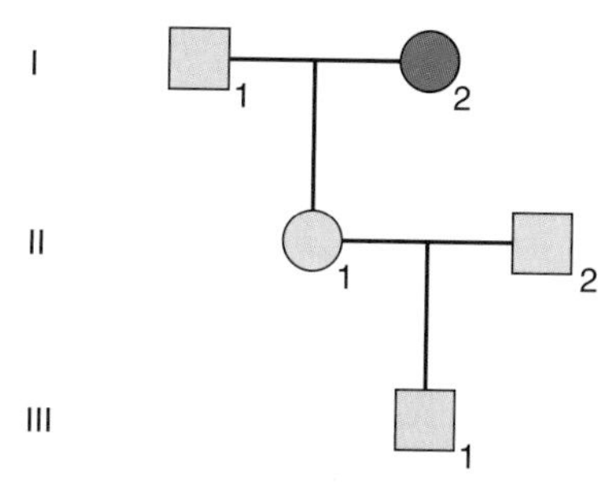

FIGURE 22.2 I_2 has an autosomal dominant disorder that shows reduced penetrance. The probability that III_1 will be affected has to take into account the possibility that his mother (II_1) is a nonpenetrant heterozygote.

A more difficult calculation arises when a risk is sought for the future child of someone who is healthy but whose parent has, or had, an autosomal dominant disorder showing reduced penetrance (Figure 22.2).

Let us assume that the penetrance, P, equals 0.8. Calculation of the risk that III_1 will be affected can be approached in two ways. The first simply involves a little logic. The second uses Bayes' theorem.

1. Imagine that I_2 has 10 children. On average, five children will inherit the gene, but because P = 0.8, only four will be affected (Figure 22.3). Therefore, six of the 10 children will be unaffected, one of whom has the mutant allele, with the remaining five having the normal allele. II_1 is unaffected, so there is a probability of 1 in 6 that she is, in fact, a heterozygote. Consequently, the probability that III_1 will both inherit the mutant gene and be affected equals $1/6 \times 1/2 \times P$, which equals 1/15 if P is 0.8.
2. Now consider II_1 in Figure 22.2. The prior probability that she is a heterozygote equals 1/2. Similarly, the prior probability that she is not a heterozygote equals 1/2. Now a bayesian table can be constructed to determine how these prior probabilities are modified by the fact that II_1 is not affected (Table 22.2).

The posterior probability that II_1 is a heterozygote equals $1/2(1 - P)/[1/2(1 - P) + 1/2]$, which reduces to $\{1 - P/2 - P\}$. Therefore, the risk that III_1 will both inherit the mutant allele and be affected equals $(\{1 - P/2 - P\}) \times 1/2 \times P$, which reduces to $\{(P - P^2)/(4 - 2P)\}$. If P equals 0.8, this expression equals 1/15 or 0.067.

Table 22.2 Bayesian Calculation for II_1 in Figure 22.2

Probability	II_1 Is Heterozygous	II_1 Is Not Heterozygous
Prior	1/2	1/2
Conditional		
Not affected	1 – P	1
Joint	1/2 (1 – P)	1/2

By substituting different values of P in the above expression, it can be shown that the maximum risk for III_1 being affected equals 0.086, approximately 1/12, which is obtained when P equals 0.6. This maximal risk figure can be used when counseling people at risk for late-onset autosomal disorders with reduced penetrance and who have an affected grandparent and unaffected parents.

Delayed Age of Onset

Many autosomal dominant disorders do not present until well into adult life. Healthy members of families in which these disorders are segregating often wish to know whether they themselves will develop the condition or pass it on to their children. Risks for these individuals can be calculated in the following way.

Consider someone who has died with a confirmed diagnosis of Huntington disease (Figure 22.4). This is a late-onset autosomal dominant disorder. The son of I_2 is entirely healthy at age 50 years and wishes to know the probability that his 10-year-old daughter, III_1, will develop Huntington disease in later life. In this condition, the first signs usually appear between the ages of 30 and 60 years, and approximately 50% of all heterozygotes have shown signs by the age of 50 years (Figure 22.5).

To answer the question about the risk to III_1, it is first necessary to calculate the risk for II_1 (if III_1 was asking about

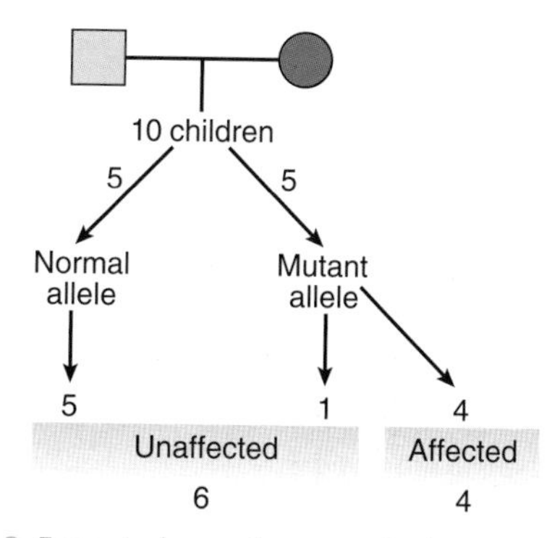

FIGURE 22.3 Expected genotypes and phenotypes in 10 children born to an individual with an autosomal dominant disorder with penetrance equal to 0.8.

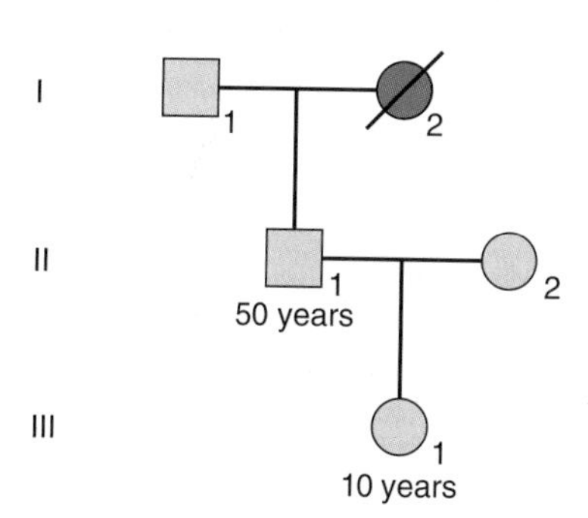

FIGURE 22.4 I_2 had an autosomal dominant disorder showing delayed age of onset. When calculating the probability that III_1 will develop the disorder, it is necessary to determine the probability that II_1 is a heterozygote who is not yet clinically affected.

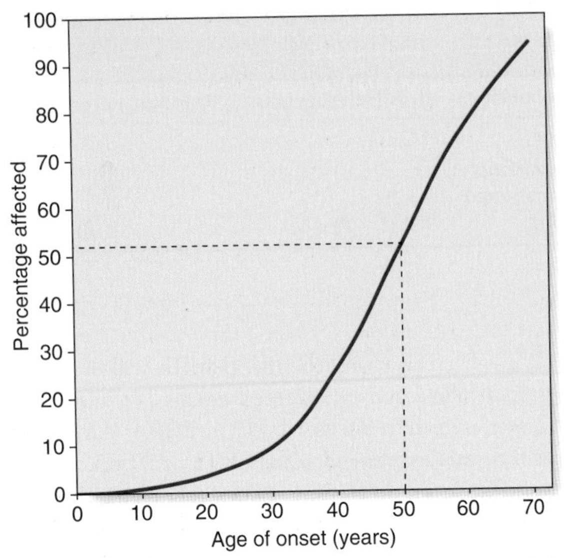

FIGURE 22.5 Graph showing age of onset in years of clinical expression in Huntington disease heterozygotes. Approximately 50% show clinical signs or symptoms by age 50 years. (Data from Newcombe RG 1981 A life table for onset of Huntington's chorea. Ann Hum Genet 45:375–385.)

her own risk, her father might be referred to as the **dummy consultand**). The probability that II_1 has inherited the gene, given that he shows no signs of the condition, can be determined by a simple bayesian calculation (Table 22.3).

The posterior probability that II_1 is heterozygous equals $1/4/(1/4 + 1/2)$, which equals 1/3. Therefore the prior probability that his daughter III_1 will have inherited the disorder equals $1/3 \times 1/2$, or 1/6.

There is a temptation when doing calculations such as these to conclude that the overall risk for II_1 being a heterozygote simply equals $1/2 \times 1/2$—i.e., the prior probability that he will have inherited the mutant gene times the probability that a heterozygote will be unaffected at age 50 years, giving a risk of 1/4. This is correct in as much as it gives the joint probability for this possible outcome, but it does not take into account the possibility that II_1 is not a heterozygote. Consider the possibility that I_2 has four children. On average, two will inherit the mutant allele, one of whom will be affected by the age of 50 years. The remaining two children will not inherit the mutant allele. By the time these children have grown up and reached the age of 50 years, on average one will be affected and three will not. Therefore, on average, one-third of the healthy 50-year-old offspring of I_2 will be heterozygotes. Hence the correct risk for II_1 is 1/3 and not 1/4.

Table 22.3 Bayesian Calculation for II_1 in Figure 22.4

Probability	II_1 Is Heterozygous	II_1 Is Not Heterozygous
Prior	1/2	1/2
Conditional		
Unaffected at age 50 years	1/2	1
Joint	1/4	1/2

Autosomal Recessive Inheritance

With an autosomal recessive condition, the **biological** parents of an affected child are both heterozygotes. Apart from undisclosed non-paternity and donor insemination, there are two possible exceptions, both of which are very rare. These arise when only one parent is a heterozygote, in which case a child can be affected if either a new mutation occurs on the gamete inherited from the other parent, or uniparental disomy occurs resulting in the child inheriting two copies of the heterozygous parent's mutant allele (p. 113). For practical purposes, it is usually assumed that both parents of an affected child are carriers.

Carrier Risks for the Extended Family

When both parents are heterozygotes, the risk that each of their children will be affected is 1 in 4. On average three of their four children will be unaffected, of whom, on average, two will be carriers (Figure 22.6). Therefore the probability that the healthy sibling of someone with an autosomal recessive disorder will be a carrier equals 2/3. Carrier risks can be derived for other family members, starting with the assumption that both parents of an affected child are carriers (Figure 22.7).

When calculating risks in autosomal recessive inheritance the underlying principle is to establish the probability that each prospective parent is a carrier, and then multiply the product of these probabilities by 1/4, this being the risk that any child born to two carriers will be affected. Therefore, in Figure 22.7, if the sister, III_3, of the affected boy was to marry her first cousin, III_4, the probability that their first baby would be affected would equal $2/3 \times 1/4 \times 1/4$—i.e., the probability that III_3 is a carrier times the probability

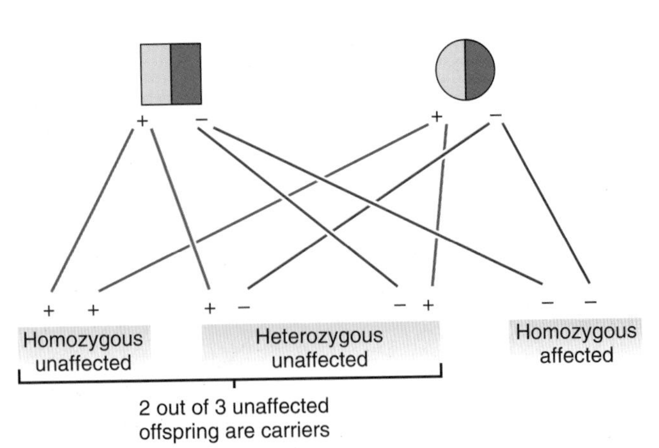

FIGURE 22.6 Possible genotypes and phenotypes in the offspring of parents who are both carriers of an autosomal recessive disorder. On average, two of three healthy offspring are carriers.

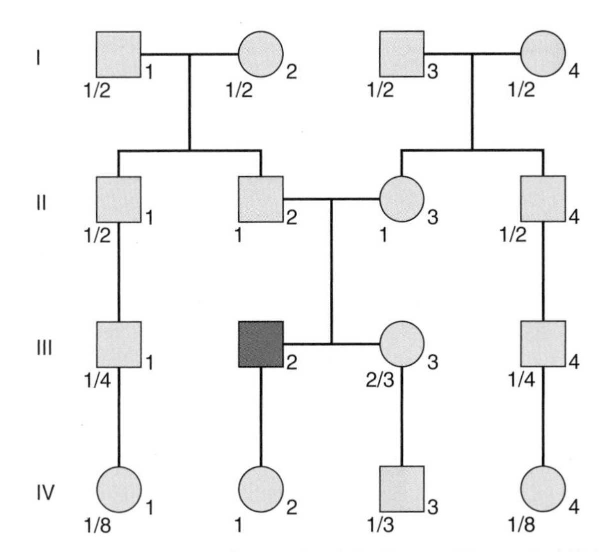

FIGURE 22.7 Autosomal recessive inheritance. The probabilities that various family members are carriers are indicated as fractions.

that III_4 is a carrier times the probability that a child of two carriers will be affected. This gives a total risk of 1/24.

If this same sister, III_3, was to marry a healthy unrelated individual, the probability that their first child would be affected would equal $2/3 \times 2pq \times 1/4$—i.e., the probability that III_3 is a carrier times the carrier frequency in the general population (p. 132) times the probability that a child of two carriers will be affected. For a condition such as cystic fibrosis, with a disease incidence of approximately 1 in 5000, $q^2 = 1/2500$ and therefore $q = 1/50$ and thus $2pq = 1/25$. Therefore the final risk would be $2/3 \times 1/25 \times 1/4$, or 1 in 150.

Modifying a Carrier Risk by Mutation Analysis

Population screening for cystic fibrosis has been introduced in the United Kingdom after pilot studies (p. 321). More than 1500 different mutations have been identified in the cystic fibrosis gene, so that carrier detection by DNA mutation analysis is not straightforward. However, a relatively simple test has been developed for the most common mutations, which enables about 90% of all carriers of western European origin to be detected. What is the probability that a healthy individual who has no family history of cystic fibrosis, and who tests negative on the common mutation screen, is a carrier?

The answer is obtained, once again, by drawing up a simple bayesian table (Table 22.4). The prior probability that this healthy member of the general population is a carrier equals 1/25; therefore the prior probability that he or she is not a carrier equals 24/25. If this individual is a carrier, then the probability that the common mutation test will be normal is 0.10 as only 10% of carriers do not have a common mutation. The probability that someone who is not a carrier will have a normal common mutation test result is 1.

This gives a joint probability for being a carrier of 1/250 and for not being a carrier of 24/25. Therefore the posterior probability that this individual is a carrier equals 1/250/(1/250 + 24/25), which equals 1/241. Thus, the normal result on common mutation testing has reduced the carrier risk from 1/25 to 1/241.

Table 22.4 Bayesian Table for Cystic Fibrosis Carrier Risk if Common Mutation Screen Is Negative

Probability	Carrier	Not a Carrier
Prior	1/25	24/25
Conditional		
Normal result on common mutation screening	0.10	1
Joint	1/250	24/25

Sex-Linked Recessive Inheritance

This pattern of inheritance tends to generate the most complicated risk calculations when counseling for mendelian disorders. In severe sex-linked conditions, affected males are often unable to have their own children. Consequently, these conditions are usually transmitted only by healthy female carriers. The carrier of a sex-linked recessive disorder transmits the gene on average to half of her daughters, who are therefore carriers, and to half of her sons who will thus be affected. If an affected male does have children, he will transmit his Y chromosome to all of his sons, who will be unaffected, and his X chromosome to all of his daughters, who will be carriers (Figure 22.8).

An example of how the birth of unaffected sons to a possible carrier of a sex-linked disorder results in a reduction of her carrier risk has already been discussed in the introductory section on Bayes' theorem (p. 339). In this section, we consider two further factors that can complicate risk calculation in sex-linked recessive disorders.

The Isolated Case

If a woman has only one affected son, then in the absence of a positive family history there are three possible ways in which this can have occurred.

1. The woman is a carrier of the mutant allele, in which case there is a risk of 1/2 that any future son will be affected.
2. The disorder in the son arose because of a new mutation that occurred during meiosis in the gamete that led to his conception. The recurrence risk in this situation is negligible.
3. The woman is a **gonadal mosaic** (p. 121) for the mutation that occurred in an early mitotic division during her own embryonic development. The recurrence risk will be equal to the proportion of ova that carry the mutant allele (i.e., between 0% and 50%).

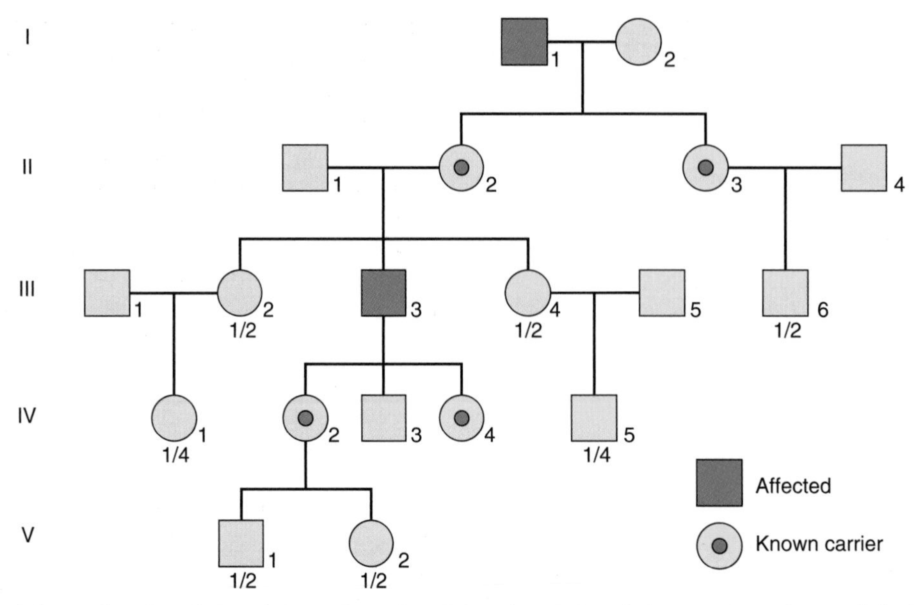

FIGURE 22.8 Probabilities of male relatives being affected and female relatives being carriers of an X-linked recessive disorder. All the daughters of an affected male are obligate carriers.

In practice it is often very difficult to distinguish among these three possibilities unless reliable tests are available for carrier detection. If a woman is found to be a carrier, then risk calculation is straightforward. If the tests indicate that she is not a carrier, the recurrence risk is probably low, but not negligible because of the possibility of **gonadal mosaicism**.

For example, in Duchenne muscular dystrophy (DMD), it has been estimated that among the mothers of isolated cases approximately two-thirds are carriers, 5% to 10% are gonadal mosaics, and in the remaining 25% to 30% the disorder has arisen as a new mutation in meiosis.

Leaving aside the complicating factor of gonadal mosaicism, risk calculation in the context of an isolated case (Figure 22.9) is possible, but may require calculation of the risk for a **dummy consultand** within the pedigree as well as taking account of the **mutation rate**, or **μ**. For a fuller understanding of μ, the student is referred to one of the more detailed texts listed at the end of the chapter.

Incorporating Carrier Test Results

Several biochemical tests are available for detecting carriers of sex-linked recessive disorders. Unfortunately, there is often overlap in the values obtained for controls and women known to be carriers (i.e., obligate carriers). Although an abnormal result in a potential carrier would suggest that she is likely to be a carrier, a normal test result does not exclude a woman from being a carrier. Although for many sex-linked recessive disorders this problem can be overcome by using linked DNA markers, the difficulties presented by overlapping biochemical test results arise sufficiently often to justify further consideration.

For example, in DMD, the serum creatine kinase level is raised in approximately two out of three obligate carriers (see Figure 20.1; p. 314). Therefore, if a possible carrier such as II_2 in Figure 22.1 is found to have a normal level of creatine kinase, this would provide further support for her not being a carrier. The test result therefore provides a conditional probability, which is included in a new bayesian calculation (Table 22.5).

The posterior probability that II_2 is a carrier equals $1/48/(1/48 + 1/2)$, or 1/25. Consequently, by first taking into account this woman's three healthy sons, and second her normal creatine kinase test result, it has been possible to reduce her carrier risk from 1 in 2 to 1 in 9 and then to 1 in 25.

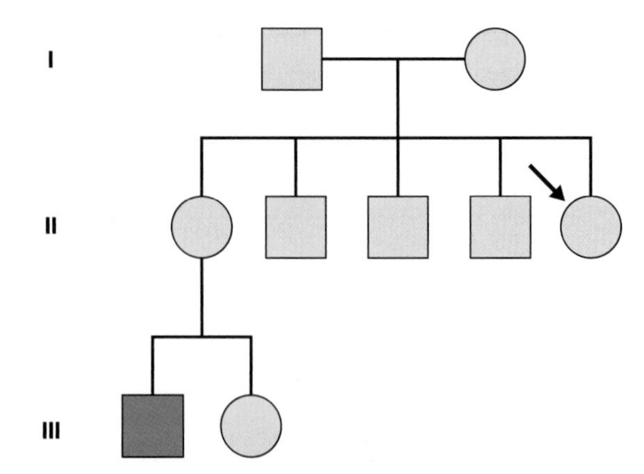

FIGURE 22.9 In this pedigree III_1 is affected by Duchenne muscular dystrophy and is an isolated case (i.e., there is no history of the condition in the wider family). The consultand, II_5 (*arrow*), wishes to know whether she is at risk of having affected sons. To calculate her risk, the risk that her mother, I_2, is a carrier is first calculated; this requires consideration of the mutation rate, μ. I_2 is the *dummy consultand* in this scenario.

Table 22.5 Bayesian Calculation for II_2 in Figure 22.1

Probability	II_2 Is a Carrier	II_2 Is Not a Carrier
Prior	1/2	1/2
Conditional		
Three healthy sons	1/8	1
Normal creatine kinase	1/3	1
Joint	1/48	1/2

The Use of Linked Markers

For many single-gene disorders the genomic location is known and sequence analysis possible. Linked DNA markers are therefore used less often today than formerly but still sometimes have a role in clarifying the genetic status of an individual in a pedigree, providing there is certainty that the disease in question is caused by mutations at one particular gene locus (i.e., the disease is not **genetically** heterogeneous). Take the example of DMD (p. 307), in which each family usually has its own unique mutation. If there are no surviving affected males, linked markers may be employed to help determine carrier detection.

As an illustration, consider the sister of a boy affected with DMD, whose mother is an obligate carrier as she herself had an affected brother (Figure 22.10). A DNA marker with alleles A and B is available and is known to be closely linked to the DMD disease locus with a recombination fraction (θ) equal to 0.05. The disease allele must be in coupling with the A marker allele in II_2 as this woman has inherited both the A allele and the DMD allele on the X chromosome from her mother (she must have inherited the B allele from her father, so the A allele must have come from her mother). Therefore, if III_3 inherits this A allele from her mother, the probability that she will also inherit the disease allele and be a carrier equals $1 - \theta$—i.e., the probability that a crossover will not have occurred between the disease and marker loci in the meiosis of the ova that resulted in her conception. For a value of θ equal to 0.05, this gives a carrier risk of 0.95 or 95%. Similarly, the probability that III_3 will be a carrier if she inherits the B allele from her mother equals 0.05 or 5%.

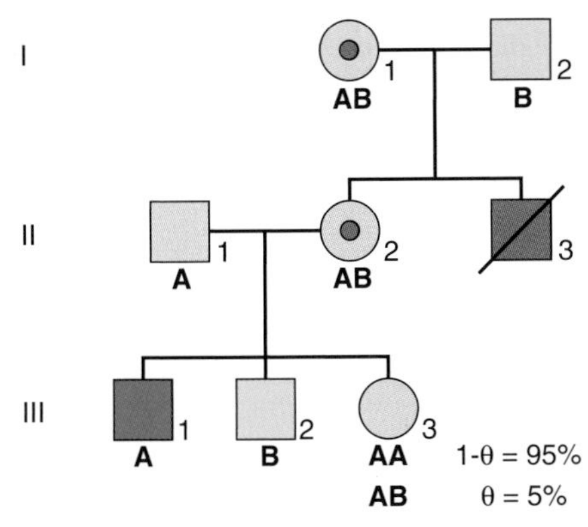

FIGURE 22.10 Pedigree showing sex-linked recessive inheritance. **A** and **B** represent alleles at a locus closely linked to the disease locus.

Closely linked DNA markers also still have a role in prenatal diagnosis when direct mutation testing is not available. The smaller the value of θ, the smaller the likelihood of a predictive error. If DNA markers are available that 'bridge' or 'flank' the disease locus, this greatly reduces the risk of a predictive error as only a double crossover will go undetected, and the probability of a double crossover is extremely low.

Bayes' Theorem and Prenatal Screening

As a further illustration of the potential value of Bayes' theorem in risk calculation and genetic counseling, an example from prenatal screening is given. Consider the situation that arises when a woman age 20 years presents at 13 weeks' gestation with a fetus that has been shown on ultrasonography to have significant nuchal translucency (NT) (see Figure 21.5). NT may be present in about 75% of fetuses with Down syndrome (p. 327). In contrast, the incidence in babies not affected with Down syndrome is approximately 5%. In other words, NT is 15 times more common in Down syndrome than in unaffected babies.

Question: Does this mean that the odds are 15 to 1 that this unborn baby has Down syndrome? No! This risk, or more precisely **odds ratio**, would be correct only if the prior probabilities that the baby would be affected or unaffected were equal. In reality the prior probability that the baby will be unaffected is much greater than the prior probability that it will have Down syndrome.

Actual values for these prior probabilities can be obtained by reference to a table showing maternal age-specific risks for Down syndrome (see Table 18.4; p. 274). For a woman age 20 years, the incidence of Down syndrome is approximately 1 in 1500; hence, the prior probability that the baby will be unaffected equals 1499/1500. If these prior probability values are used in a bayesian calculation, it can be shown that the posterior probability risk that the unborn baby will have Down syndrome is approximately 1 in 100 (Table 22.6). Obviously, this is much lower than the

Table 22.6 Bayesian Calculation to Show the Posterior Probability that a Fetus with Nuchal Translucency Conceived by a 20-Year-Old Mother Will Have Down Syndrome

Probability	Fetus Unaffected	Fetus Affected
Prior	1499/1500	1/1500
Conditional		
Nuchal translucency	1	15
Joint	1499/1500 = 1	1/100
Expressed as odds	100 to	1
Posterior	100/101	1/101

conditional odds of 15 to 1 in favor of the baby being affected.

In practice, the demonstration of NT on ultrasonography in a fetus would usually prompt an offer of definitive chromosome analysis by placental biopsy, amniocentesis, or fetal blood sampling (see Chapter 21). This example of NT has been used to emphasize that an observed conditional probability ratio should always be combined with prior probability information to obtain a correct indication of the actual risk.

Empiric Risks

Up to this point, risks have been calculated for single-gene disorders using knowledge of basic mendelian genetics and applied probability theory. In many counseling situations, it is not possible to arrive at an accurate risk figure in this way, either because the disorder in question does not show single-gene inheritance or because the clinical diagnosis with which the family has been referred shows causal heterogeneity (see below). In these situations, it is usually necessary to resort to the use of observed or **empiric risks**. These are based on observations derived from family and population studies rather than theoretical calculations.

Multifactorial Disorders

One of the basic principles of multifactorial inheritance is that the risk of recurrence in first-degree relatives, siblings and offspring, equals the square root of the incidence of the disease in the general population (p. 146)—i.e., $P^{1/2}$, where P equals the general population incidence. For example, if the general population incidence equals 1/1000, then the theoretical risk to a first-degree relative equals the square root of 1/1000, which approximates to 1 in 32 or 3%. The theoretical risks for second- and third-degree relatives can be shown to approximate to $P^{3/4}$ and $P^{7/8}$, respectively. Therefore, if there is strong support for multifactorial inheritance, it is reasonable to use these theoretical risks when counseling close family relatives.

However, when using this approach it is important to remember that the confirmation of multifactorial inheritance will often have been based on the study of observed recurrence risks. Consequently, it is generally more appropriate to refer back to the original family studies and counsel on the basis of the risks derived in these (Table 22.7).

Ideally, reference should be made to local studies as recurrence risks can differ quite substantially in different communities, ethnic groups, and geographical locations. For example, the recurrence risk for neural tube defects in siblings used to be quoted as 4% (before the promotion of periconceptional maternal folate intake). This, essentially, was an average risk. The actual risk varied from 2% to 3% in southeast England up to 8% in Northern Ireland, and also showed an inverse relationship with the family's socioeconomic status, being greatest for mothers in poorest circumstances.

Unfortunately, empiric risks are rarely available for families in which there are several affected family members, or for disorders with variable severity or different sex incidences. For example, in a family where several members have been affected by cleft lip/palate, the empiric risks based on population data may not apply—the condition may appear to be segregating as an autosomal dominant trait with a high penetrance. In the absence of a syndrome diagnosis being made and genetic testing being possible, the clinical geneticist has to make the best judgement about recurrence risk.

Conditions Showing Causal Heterogeneity

Many referrals to genetic clinics relate to a clinical phenotype rather than to a precise underlying diagnosis (Table 22.8). In these situations, great care must be taken

Table 22.7 Empiric Recurrence Risks for Common Multifactorial Disorders

Disorder	Incidence (Per 1000)	Sex Ratio (M:F)	Unaffected Parents Having a Second Affected Child (%)	Affected Parents Having an Affected Child (%)
Cleft lip ± cleft palate	1–2	3:2	4	4
Clubfoot (talipes)	1–2	2:1	3	3
Congenital heart defect	8	1:1	1–4	2 (father affected) 6 (mother affected)
Congenital dislocation of the hip	1	1:6	6	12
Hypospadias (in males)	2	—	10	10
Manic depression	4	2:3	10–15	10–15
Neural tube defect				
Anencephaly	1.5	1:2	4–5	—
Spina bifida	2.5	2:3	4–5	4
Pyloric stenosis				
Male index	2.5	—	2	4
Female index	0.5	—	10	17
Schizophrenia	10	1:1	10	14

Table 22.8 Empiric Recurrence Risks for Conditions Showing Causal Heterogeneity

Disorder	Incidence (Per 1000)	Sex Ratio (M:F)	Unaffected Parents Having a Second Affected Child (%)	Affected Parents Having an Affected Child (%)
Autism	1–2	4:1	2–3	—
Epilepsy (idiopathic)	5	1:1	5	5
Hydrocephalus	0.5	1:1	3	—
Mental retardation (idiopathic)	3	1:1	3–5	10
Profound childhood sensorineural hearing loss	1	1:1	10–15	5–10

to ensure that all appropriate diagnostic investigations have been undertaken before resorting to the use of empiric risk data (p. 346).

It is worth emphasizing that the use of empiric risks for conditions such as sensorineural hearing loss in childhood is at best a compromise, as the figure quoted to an individual family will rarely be the correct one for their particular diagnosis. Severe sensorineural hearing loss in a young child is usually caused either by single-gene inheritance, most commonly autosomal recessive, but occasionally autosomal dominant or sex-linked recessive, or by an environmental condition such as rubella embryopathy. Therefore, for most families the correct risk of recurrence will be either 25% or 0%. In practice, it is often not possible to establish the precise cause, so that the only option available is to offer the family an empiric or 'average' risk.

FURTHER READING

Bayes T 1958 An essay towards solving a problem in the doctrine of chances. Biometrika 45:296–315

A reproduction of the Reverend Bayes' original essay on probability theory that was first published, posthumously, in 1763.

Emery AEH 1986 Methodology in medical genetics, 2nd edn. Edinburgh, UK: Churchill Livingstone

An introduction to statistical methods of analysis in human and medical genetics.

Murphy EA, Chase GA 1975 Principles of genetic counseling. Chicago: Year Book Medical

A very thorough explanation of the use of Bayes' theorem in genetic counseling.

Young ID 1999 Introduction to risk calculation in genetic counselling, 2nd edn. Oxford, UK: Oxford University Press

A short introductory guide to all aspects of risk calculation in genetic counseling. Highly recommended.

ELEMENTS

1. Risk calculation in genetic counseling requires a knowledge and understanding of basic probability theory. Bayes' theorem enables initial background 'prior' risks to be modified by 'conditional' information to give an overall probability or risk for a particular event such as carrier status.
2. For disorders showing autosomal dominant inheritance it is often necessary to consider factors such as reduced penetrance and delayed age of onset. For disorders showing autosomal recessive inheritance, risks to offspring are determined by calculating the probability that each parent is a carrier and then multiplying the product of these probabilities by 1/4.
3. In sex-linked recessive inheritance, a particular problem arises when only one male in a family is affected. The results of carrier tests that show overlap between carriers and non-carriers can be incorporated in a bayesian calculation.
4. Polymorphic DNA markers linked to the disease locus can be used in many single-gene disorders for carrier detection, preclinical diagnosis, and prenatal diagnosis.
5. Empiric (observed) risks are available for multifactorial disorders and for etiologically heterogeneous conditions such as non-syndromal sensorineural hearing loss.

CHAPTER 23

Treatment of Genetic Disease

So little done. So much to do.
ALEXANDER GRAHAM BELL

Many genetic disorders are characterized by progressive disability or chronic ill health for which there is, at present, no effective treatment. Consequently, one of the most exciting aspects of the developments in biotechnology is the prospect of new treatments mediated through gene transfer, RNA modification, or stem cell therapy. It is important, however, to keep a perspective on the limitations of these approaches for the immediate future and to consider, in the first instance, conventional approaches to the treatment of genetic disease.

Conventional Approaches to Treatment of Genetic Disease

Most genetic disorders cannot be cured or even ameliorated using conventional methods of treatment. Sometimes this is because the underlying gene and gene product have not been identified so that there is little, if any, understanding of the basic metabolic or molecular defect. If, however, this is understood then dietary restriction, as in phenylketonuria (p. 167), or hormone replacement, as in congenital adrenal hyperplasia (p. 174), can be used very successfully in the treatment of the disorder. In a few disorders, such as homocystinuria (p. 172) and some of the organic acidurias (p. 183), supplementation with a vitamin or co-enzyme can increase the activity of the defective enzyme with beneficial effect (Table 23.1).

Protein/Enzyme Replacement

If a genetic disorder is found to be the result of a deficiency of or an abnormality in a specific enzyme or protein, treatment could, in theory, involve replacement of the deficient or defective enzyme or protein. An obviously successful example of this is the use of factor VIII concentrate in the treatment of hemophilia A (p. 309).

For most of the inborn errors of metabolism in which an enzyme deficiency has been identified, recombinant DNA techniques may be used to biosynthesize the missing or defective gene product; however, injection of the enzyme or protein may not be successful if the metabolic processes involved are carried out within cells and the protein or enzyme is not normally transported into the cell. Modifications in β-glucocerebrosidase as used in the treatment of Gaucher disease enable it to enter the lysosomes, resulting in an effective form of treatment (p. 178). Another example is the modification of adenosine deaminase (ADA) by an inert polymer, polyethylene glycol (PEG), to generate a replacement enzyme that is less immunogenic and has an extended half-life.

Drug Treatment

In some genetic disorders, drug therapy is possible; for example, statins can help to lower cholesterol levels in familial hypercholesterolemia (p. 175). Statins function indirectly through the low-density lipoprotein (LDL) receptor by inhibiting endogenous cholesterol biosynthesis at the rate-limiting step that is mediated by hydroxymethyl glutaryl co-enzyme A (HMG-CoA) reductase. This leads to upregulation of the LDL receptor and increased LDL clearance from plasma.

In others, avoidance of certain drugs or foods can prevent the manifestation of the disorder, for example sulfonamides in glucose-6-phosphate dehydrogenase (G6PD) deficiency (p. 187). Drug therapy might also be directed at a subset of patients according to their molecular defect. An example is a trial in which gentamicin was administered via nasal drops to patients with cystic fibrosis. Aminoglycoside antibiotics such as gentamicin or PTC124 cause read-through of premature stop codons in vitro and only patients with nonsense mutations (p. 25) showed evidence of expression of full-length cystic fibrosis transmembrane conductance regulator (CFTR) protein in the nasal epithelium. However, although gentamicin and PTC124 were effective in the *mdx* mouse model of Duchenne muscular dystrophy (DMD), a clinical trial with 174 patients failed to show convincing functional improvement after daily treatment for 48 weeks.

Tissue Transplantation

Replacement of diseased tissue has been a further option since the advent of tissue typing (p. 387). An example is renal transplantation in adult polycystic kidney disease or lung transplantation in patients with cystic fibrosis.

Islet transplantation for treating type 1 diabetes mellitus was transformed in 2000 with development of the 'Edmonton' protocol. Islet cells are prepared from donated pancreases (usually two per patient) and injected into the liver of the recipient: at 3 years post-transplant more than 80% of patients are still producing their own insulin.

Table 23.1 Examples of Various Methods for Treating Genetic Disease

Treatment	Disorder
Enzyme Induction by Drugs	
Phenobarbitone	Congenital non-hemolytic jaundice
Replacement of Deficient Enzyme/Protein	
Blood transfusion	SCID resulting from adenosine deaminase deficiency
Bone marrow transplantation	Mucopolysaccharidoses
Enzyme/Protein Preparations	
Trypsin	Trypsinogen deficiency
α_1-Antitrypsin	α_1-Antitrypsin deficiency
Cryoprecipitate/factor VIII	Hemophilia A
β-Glucosidase	Gaucher disease
α-Galactosidase	Fabry disease
Replacement of Deficient Vitamin or Coenzyme	
B_6	Homocystinuria
B_{12}	Methylmalonic acidemia
Biotin	Propionic acidemia
D	Vitamin D–resistant rickets
Replacement of Deficient Product	
Cortisone	Congenital adrenal hyperplasia
Thyroxine	Congenital hypothyroidism
Substrate Restriction in Diet	
AMINO ACIDS	
Phenylalanine	Phenylketonuria
Leucine, isoleucine, valine	Maple syrup urine disease
CARBOHYDRATE	
Galactose	Galactosemia
LIPID	
Cholesterol	Familial hypercholesterolemia
Protein	Urea cycle disorders
Drug Therapy	
Aminocaproic acid	Angioneurotic edema
Dantrolene	Malignant hyperthermia
Cholestyramine	Familial hypercholesterolemia
Pancreatic enzymes	Cystic fibrosis
Penicillamine	Wilson disease, cystinuria
Drug/Dietary Avoidance	
Sulfonamides	G6PD deficiency
Barbiturates	Porphyria
Replacement of Diseased Tissue	
Kidney transplantation	Adult-onset polycystic kidney disease, Fabry disease
Bone marrow transplantation	X-linked SCID, Wiskott-Aldrich syndrome
Removal of Diseased Tissue	
Colectomy	Familial adenomatous polyposis
Splenectomy	Hereditary spherocytosis

SCID, Severe combined immunodeficiency.

Therapeutic Applications of Recombinant DNA Technology

The advent of recombinant DNA technology has also led to rapid progress in the availability of biosynthetic gene products for the treatment of certain inherited diseases.

Biosynthesis of Gene Products

Insulin used in the treatment of diabetes mellitus was previously obtained from pig pancreases. This had to be purified for use very carefully, and even then it occasionally produced sensitivity reactions in patients. However, with recombinant DNA technology, microorganisms can be used to synthesize insulin from the human insulin gene. This is inserted, along with appropriate sequences to ensure efficient transcription and translation, into a recombinant DNA vector such as a plasmid and cloned in a microorganism such as *Escherichia coli*. In this way, large quantities of insulin can be made. An artificial gene that is not identical to the natural gene needs to be constructed for this purpose. However, synthetically produced genes cannot contain the non-coding intervening sequences, or introns (p. 17), found in the majority of structural genes in eukaryotic organisms, as microorganisms such as *E. coli* do not possess a means for splicing of the messenger RNA (mRNA) after transcription.

Recombinant DNA technology is being employed in the production of a number of other biosynthetic products (Table 23.2). The biosynthesis of medically important peptides in this way is usually more expensive than obtaining the product from conventional sources because of the research and development involved. For example, the cost of treating one patient with Gaucher disease can exceed £100,000 per year. However, biosynthetically derived products have the dual advantages of providing a pure product that is unlikely to induce a sensitivity reaction and one that is free of the risk of chemical or biological contamination. In the past, the use of growth hormone from human cadaver pituitaries has been associated with the transmission of Creutzfeldt-Jakob disease, and human immunodeficiency virus (HIV) has been a contaminant in cryoprecipitate containing factor VIII used in the treatment of hemophilia A (p. 309).

Gene Therapy

Gene therapy has been defined by the UK Gene Therapy Advisory Committee (GTAC) as 'the deliberate introduc-

Table 23.2 Proteins Produced Biosynthetically Using Recombinant DNA Technology

Protein	Disease
Insulin	Diabetes mellitus
Growth hormone	Short stature resulting from growth hormone deficiency
Factor VIII	Hemophilia A
Factor IX	Hemophilia B
Erythropoietin	Anemia
α-Galactosidase A	Fabry disease (X-linked lysosomal storage disorder)
β-Interferon	Multiple sclerosis

tion of genetic material into human somatic cells for therapeutic, prophylactic, or diagnostic purposes'. It includes techniques for delivering synthetic or recombinant nucleic acids into humans; genetically modified biological vectors (such as viruses or plasmids), genetically modified stem cells, oncolytic viruses, nucleic acids associated with delivery vehicles, naked nucleic acids, antisense techniques (e.g., gene silencing, gene correction or gene modification), genetic vaccines, DNA or RNA technologies such as RNA interference, and xenotransplantation of animal cells (but not solid organs).

Advances in molecular biology leading to the identification of many important human disease genes and their protein products have raised the prospect of gene therapy for many genetic and non-genetic disorders. The first human gene therapy trial began in 1990, but it is important to emphasize that, although it is often presented as the new panacea in medicine, progress to date has been limited, and there are many practical difficulties to overcome before gene therapy can deliver its promise.

Regulatory Requirements

There has been much publicity about the potential uses and abuses of gene therapy. Regulatory bodies have been established in several countries to oversee the technical, therapeutic, and safety aspects of gene therapy programs (p. 368). There is universal agreement that **germline gene therapy**, in which genetic changes could be distributed to both somatic and germ cells, and thereby be transmitted to future generations, is morally and ethically unacceptable. Therefore all programs are focusing only on **somatic cell gene therapy**, in which the alteration in genetic information is targeted to specific cells, tissues or organs in which the disorder is manifest.

In the USA the Human Gene Therapy Subcommittee of the National Institutes of Health has produced guidelines for protocols of trials of gene therapy that must be submitted for approval to both the Food and Drug Administration and the Recombinant DNA Advisory Committee, along with their institutional review boards. In the United Kingdom, the GTAC provides ethical oversight of proposals to conduct clinical trials involving gene or stem cell therapies in humans, taking account of the scientific merits, and the potential benefits and risks.

More than 1500 clinical trials of gene therapy have been approved for children and adults for a variety of genetic and non-genetic disorders. For the most part, these appear to be proceeding without event, although the unexpected death of a patient in one trial in 1999 and the development of leukemia in 3 of 20 children who received gene therapy for X-linked severe combined immunodeficiency (XL-SCID) (p. 201) has highlighted the risks of gene therapy.

Technical Aspects

Before a gene therapy trial is possible, there are a number of technical aspects that must be addressed.

Gene Characterization

One of the basic prerequisites of gene therapy is that the gene involved should have been cloned. This should include not only the structural gene, but also the DNA sequences involved in the control and regulation of expression of that gene.

Target Cells, Tissue, and Organ

The specific cells, tissue, or organ affected by the disease process must be identified and accessible before treatment options can be considered. Again, this seems obvious. Some of the early attempts at treating the inherited disorders of hemoglobin, such as β-thalassemia, involved removing bone marrow from affected individuals, treating it in vitro, and then returning it to the patient by transfusion. Although in principle this could have worked, to have any likelihood of success the particular cells that needed to be targeted were the small number of bone marrow stem cells from which the immature red blood cells, or reticulocytes, develop.

Vector System

The means by which a foreign gene is introduced need to be both efficient and safe. If gene therapy is to be considered as a realistic alternative to conventional treatments, there should be unequivocal evidence from trials of gene therapy carried out in animal models that the inserted gene functions adequately with appropriate regulatory, promoter, and enhancer sequences. In addition, it needs to be shown that the treated tissue or cell population has a reasonable lifespan, that the gene product continues to be expressed, and that the body does not react adversely to the gene product, for instance by producing antibodies to the protein product. Last, it is essential to demonstrate that introduction of the foreign gene or DNA sequence has no deleterious effects, such as inadvertently leading to a malignancy or a mutagenic effect on either the somatic or the germ-cell lines, for example through mistakes arising as a result of the insertion of the gene or DNA sequence into the host DNA, or what is known as **insertional mutagenesis**. In two patients who developed leukemia after gene therapy for XL-SCID, the retrovirus used to deliver the γ-c *(IL2RG)* gene was shown to have inserted into the *LMO-2* oncogene, which plays a role in some forms of childhood leukemia, on chromosome 11.

Animal Models

One of the basic prerequisites for assessing the suitability of gene therapy trials in humans is the existence of an animal model. Although there are naturally occurring animal models for some inherited human diseases, for most there is no animal counterpart. The techniques used to generate animal models for human disease are outside the scope of this book, but much effort has focused on the production of animal models that faithfully recreate disease phenotypes. Animal models for cystic fibrosis, DMD, Huntington disease, and Friedreich ataxia have been generated and provide just a few examples that may be used to evaluate gene therapy before trials in humans.

In Utero Fetal Gene Therapy

The report of successful adenovirus vector-mediated in utero gene therapy in a cystic fibrosis mouse model in 1997 means that fetal gene therapy in utero may be possible in humans. At present it is considered unacceptable because of the possibility of inadvertent germ-cell modification. The use of stem cells genetically modified ex vivo should reduce this risk. However, in utero stem-cell transplantation without genetic modification offers the best prospects for the successful treatment of serious neurodegenerative disorders with a very early onset, such as Krabbe disease or Hurler syndrome (p. 177).

Gene Transfer

Gene transfer can be carried out either ex vivo by treatment of cells or tissue from an affected individual in culture, with reintroduction into the affected individual, or in vivo if cells cannot be cultured or be replaced in the affected individual (Figure 23.1). The ex vivo approach is limited to disorders in which the relevant cell population can be removed from the affected individual, modified genetically, and then replaced. The in vivo approach is the most direct strategy for gene transfer and can theoretically be used to treat many hereditary disorders.

Target Organs

In many instances, gene therapy will need to be, and should be, directed or limited to a particular organ, tissue or body system.

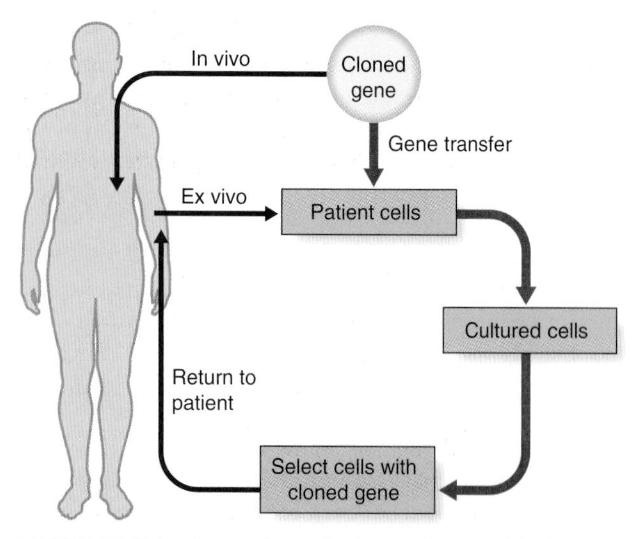

FIGURE 23.1 In vivo and ex vivo gene therapy. In vivo gene therapy delivers genetically modified cells directly to the patient. An example is *CFTR* gene therapy using liposomes or adenovirus via nasal sprays. Ex vivo gene therapy removes cells from the patient, modifies them in vitro and then returns them to the patient. An example is the treatment of fibroblasts from patients with hemophilia B by the addition of the factor IX gene. Modified fibroblasts are then injected into the stomach cavity.

Liver

Liver cells are susceptible to transfection by retroviruses in vitro. Cells removed from the liver by partial hepatectomy can be treated in vitro and then reinjected via the portal venous system, from which they seed in the liver. Hypercholesterolemia is a major cause of cardiovascular disease in the Western world. The most severe form, autosomal recessive familial hypercholesterolemia, is caused by mutations in the low-density lipoprotein receptor (LDLR) gene. Patients do not respond to statin therapy, require maintenance therapy with highly invasive LDL apheresis, and typically die of myocardial infarction in their third decade of life. Gene therapy for lipid disorders has a high potential for success, but studies to date based on viral-mediated overexpression of LDLR cDNA have been unsuccessful, probably because vectors have lacked the sterol response elements that are required for regulated transcription.

Central Nervous System

Gene therapy for central nervous system disorders, such as Parkinson and Alzheimer diseases, requires delivery to the brain. Lentiviral vectors are particularly suitable because they integrate into the host genome of non-dividing cells and can potentially act as a delivery system for stable expression. After a study in a non-human primate model of Parkinson disease that reported motor deficit improvements following the use of a lentiviral vector to restore extracellular dopamine levels, several clinical trials have been undertaken to evaluate the effectiveness of this approach in humans. Encouraging results came from a study in 2009 in which the neurotransmitter GABA was delivered to one side of the brain in 12 patients with Parkinson disease and improvement demonstrated by positron emission tomography scanning. Trials for Alzheimer disease are at an earlier stage, but results of a phase II trial of 50 patients with Parkinson disease to deliver nerve growth factor into the brain are eagerly awaited.

Muscle

Unlike other tissues, direct injection of foreign DNA into muscle has met with some success in terms of retention and expression of the foreign gene in the treated muscle. However, this will not provide a long-term response. A clinical trial is underway to introduce a 'microdystrophin' molecule, so called because it contains only the bare essential domains of the dystrophin gene, via an adeno-associated viral (AAV) vector into the arm muscles of patients with DMD.

Bone Marrow

In the treatment of disorders affecting the bone marrow, problems arise from the small numbers of stem cells, which need to be immortalized. Ex vivo gene therapy has been reported in two young boys with X-linked adrenoleukodystrophy for whom suitable bone marrow donors were not available. Stem cells were removed from their bone marrow, genetically corrected with a lentiviral vector encoding the normal *ABCD1* gene, and then re-infused after the patients received myeloablative treatment. A year later, the progressive cerebral demyelination had stopped, a clinical outcome comparable to that achieved by allogeneic bone marrow transplant.

There are two main methods for delivering gene transfer, viral and non-viral.

Viral Agents

A number of different viruses can be used to transport foreign genetic material into cells and the most successful viral agents are described in the following sections.

Lentiviruses

The lentivirus family includes HIV. Lentiviruses are complex viruses that infect macrophages and lymphocytes, but their main advantage is that they can be integrated into non-dividing cells. They may, therefore, be useful in the treatment of neurological conditions.

Adenoviruses

Adenoviruses can be used as vectors in gene therapy as they infect a wide variety of cell types. They are stable, can infect non-dividing cells and carry up to 36 kb of foreign DNA. In addition, they are suitable for targeted treatment of specific tissues such as the respiratory tract, and have been extensively used in gene therapy trials for the treatment of cystic fibrosis.

Adenoviruses do not integrate into the host genome, thereby avoiding the possibility of insertional mutagenesis but having the disadvantage that expression of the introduced gene is usually unstable and often transient. They also contain genes known to be involved in the process of malignant transformation, so there is a potential risk that they could inadvertently induce malignancy. By virtue of their infectivity, they can produce adverse effects secondary to infection and by stimulating the host immune response. This was demonstrated by a vector-related death following intravascular administration of high doses (3.8×10^{13}) of adenovirus particles to a patient with ornithine transcarbamylase deficiency.

Adeno-Associated Viruses

Adeno-associated viruses are non-pathogenic parvoviruses in humans that require co-infection with helper adenoviruses or certain members of the herpes virus family to achieve infection. In the absence of the helper virus, the adeno-associated virus DNA integrates into chromosomal DNA at a specific site on the long arm of chromosome 19 (19q13.3-qter). Subsequent infection with an adenovirus activates the integrated adeno-associated viral DNA-producing virions. They have the advantages of being able to infect a wide variety of cell types, exhibiting long-term gene expression and not generating an immune response to transduced cells. The safety of adeno-associated viruses as vectors occurs by virtue of their site-specific integration but, unfortunately, this is often impaired with the inclusion of foreign DNA in the virus. The disadvantages of adeno-associated viruses include the fact that they can be activated by any adenovirus infection and that, although 95% of the vector genome is removed, they can take inserts of foreign DNA of only up to 5 kb in size.

Non-Viral Methods

There are a number of different non-viral methods of gene therapy but the most popular is liposome-mediated DNA transfer. This has the theoretical advantage of not eliciting an immune response, being safer and simpler to use as well as allowing large-scale production, but efficacy is limited.

Liposomes

Liposomes are lipid bilayers surrounding an aqueous vesicle that can facilitate the introduction of foreign DNA into a target cell (Figure 23.2). A disadvantage of liposomes is that they are not very efficient in gene transfer and the expression of the foreign gene is transient, so that the treatment has to be repeated. An advantage of liposome-mediated

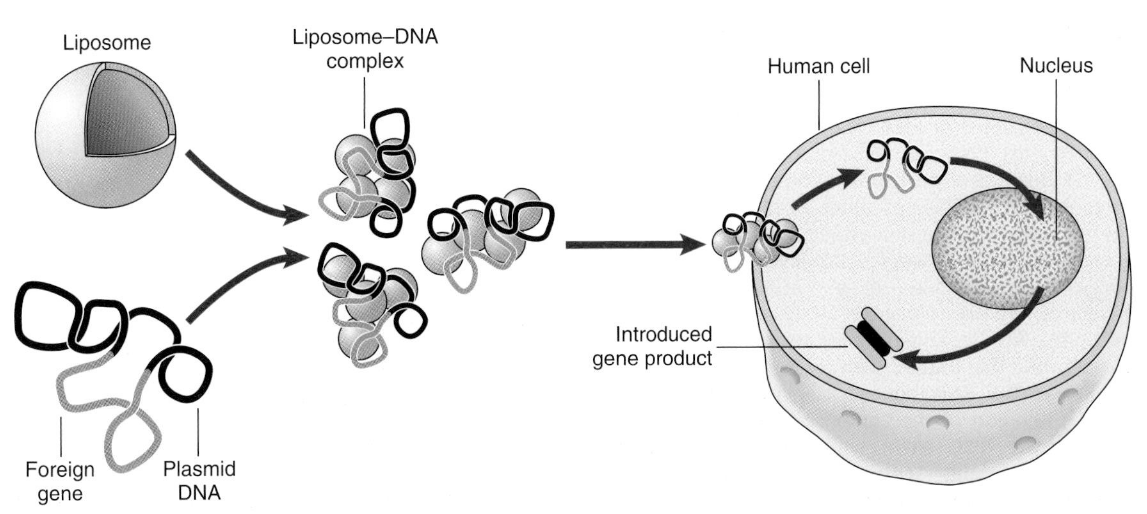

FIGURE 23.2 Diagrammatic representation of liposome-mediated gene therapy.

gene transfer is that a much larger DNA sequence can be introduced into the target cells or tissues than with viral vector systems.

RNA Modification

RNA modification therapy targets mRNA, either by suppressing mRNA levels or by correcting/adding function to the mRNA.

Antisense Oligonucleotides

Antisense therapy may be used to modulate the expression of genes associated with malignancies and other genetic disorders. The principle of antisense technology is the sequence-specific binding of an antisense oligonucleotide (typically 18 to 30 bases in length) to a target mRNA that results in inhibition of gene expression at the protein level.

The identification of exon-splicing enhancer (ESE) sequences has increased our understanding of the process of exon splicing. If an ESE is mutated, the exon is more likely to be spliced out. Some proteins with in-frame whole-exon deletions retain some residual activity; for example, dystrophin mutations in Becker muscular dystrophy (p. 307). Blocking an ESE using an antisense oligonucleotide in patients with DMD aims to restore the reading frame and ameliorate the phenotype to that of Becker muscular dystrophy. However, a specific antisense oligonucleotide is required to target each different exon. The potential for antisense therapy is perhaps greater in spinal muscular atrophy (p. 306) where the non-expressed *SMN2* gene could be converted to generate functional SMN1 protein in virtually all patients.

RNA Interference

This technique also has broad therapeutic application, as any gene may be a potential target for silencing by RNA interference. In contrast to antisense oligonucleotide therapy where the target mRNA is bound, as a result of RNA interference the target mRNA is cleaved and it is estimated to be up to 1000-fold more active. RNA interference works through the targeted degradation of mRNAs containing homologous sequences to synthetic double-stranded RNA molecules known as **small interfering RNAs (siRNAs)** (Figure 23.3). The siRNAs may be delivered in drug form using strategies developed to stabilize antisense oligonucleotides, or from plasmids or viral vectors. The first of these drugs is bevasiranib, an siRNA therapy designed to silence the genes that produce vascular endothelial growth factor, believed to be responsible for the vision loss in patients with the "wet" form of age-related macular degeneration.

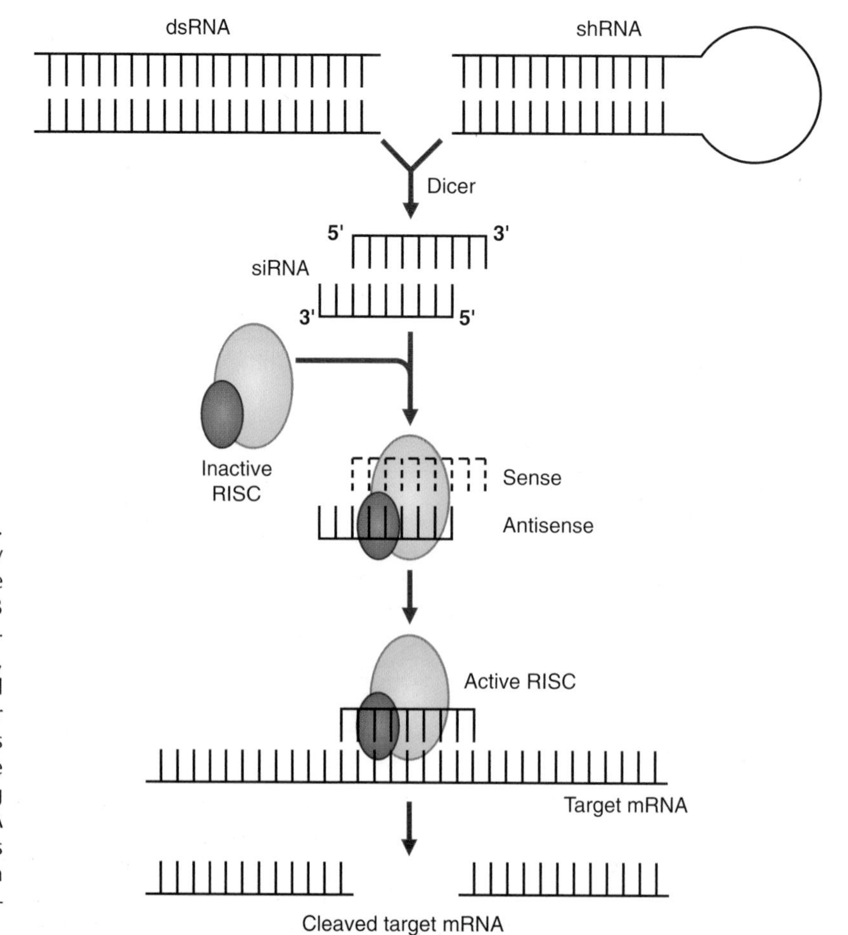

FIGURE 23.3 Mechanism of RNA interference. Double-stranded (ds) RNAs are processed by Dicer, in an ATP-dependent process, to produce small interfering RNAs (siRNA) of about 21–23 nucleotides in length with two-nucleotide overhangs at each end. Short hairpin (sh) RNAs, either produced endogenously or expressed from viral vectors, are also processed by Dicer into siRNA. An ATP-dependent helicase is required to unwind the dsRNA, allowing one strand to bind to the RNA-induced silencing complex (RISC). Binding of the antisense RNA strand activates the RISC to cleave mRNAs containing a homologous sequence. (From Lieberman, et al 2003 Trends Mol Med 9:397–403, with permission.)

Targeted Gene Correction

A promising new approach is to repair genes in situ through the cellular DNA repair machinery (p. 27). Proof of principle has been demonstrated in an animal model of Pompe disease. The point mutation was targeted by chimeric double-stranded DNA-RNA oligonucleotides containing the correct nucleotide sequence. Repair was demonstrated at the DNA level and normal enzyme activity was restored.

The latest strategy uses engineered zinc-finger nucleases (ZFNs) to stimulate homologous recombination. Targeted cleavage of DNA is achieved by zinc-finger proteins designed to recognize unique chromosomal sites and fused to the non-specific DNA cleavage domain of a restriction enzyme. A double-strand break induced by the resulting ZFNs can create specific changes in the genome by stimulating homology-directed DNA repair between the locus of interest and an extrachromosomal molecule. One possible application is to manufacture HIV resistant T cells through disruption of the HIV receptor CCR5 by zinc-finger nucleases.

Diseases Suitable for Treatment Using Gene Therapy

Disorders that are possible candidates for gene therapy include both genetic and non-genetic diseases (Table 23.3).

Genetic Disorders

There are a number of single-gene diseases that have been the focus of gene therapy attempts.

Adenosine Deaminase Deficiency

One of the first diseases for which gene therapy was attempted in humans is the inherited immunodeficiency disorder caused by adenosine deaminase (ADA) deficiency (p. 179). The most successful conventional treatment for ADA deficiency is bone marrow transplantation but, in the absence of a compatible donor, patients may be treated with PEG-conjugated ADA.

In 1990 the first gene therapy trial enrolled 10 patients with ADA-SCID. Although no adverse events were reported, none of the patients was 'cured', probably because of the low efficiency of gene transfer from the retroviral vector. More recently, an improved gene transfer protocol in bone-marrow CD34+ cells combined with low-dose chemotherapy resulted in multi-lineage, stable engraftment of transduced progenitors at substantial levels, restoration of immune functions, correction of the ADA metabolic defect, and proven clinical benefit, in the absence of PEG-ADA.

Hemoglobinopathies

Attempts at treating β-thalassemia and sickle-cell disease by gene therapy have not been effective as yet, primarily because the numbers of α- and β-globin chains must be equal (p. 157). Gene therapy must, therefore, be dose-specific, and this is not possible at the present time.

Cystic Fibrosis

In contrast to the hemoglobinopathies, cystic fibrosis should be more amenable to gene therapy as the level of functional protein sufficient to produce a clinical response may be as low as 5% to 10% and the lung is a relatively accessible tissue.

However, progress to date has been slow and, although gene therapy can correct the primary and secondary defects associated with cystic fibrosis, the extent and duration of gene expression has been inadequate, owing to the rapid turnover of lung epithelial cells. There have also been concerns about the safety of some current delivery systems, especially following the adenovirus-triggered death of one patient. For viral vectors, the main challenges are access to target cells and host immunity, which prevents efficient readministration. Liposomes have been used in animal models and clinical trials but are also relatively inefficient.

Hemophilia A and B

Hemophilia A and B were thought to be excellent candidates for gene therapy as a modest increase in the level of factor VIII or IX, respectively, will be of major clinical benefit. Trials have used direct intramuscular injection of adeno-associated virus expressing factor VIII or ex-vivo treatment of fibroblasts with plasmid-borne factor IX followed by injection into the stomach cavity. Although initial results were encouraging, the transient rise in factor VIII levels was modest (0.5% to 4% of normal) and these clinical trials were halted. A new way of targeting gene therapy to mouse liver sinusoidal endothelial cells (the cells that are the main source of factor VIII) has been developed in which nanoparticles are coated with hyaluron so that they target these cells. Even 50 weeks after hemophilia A mice were injected with these nanoparticles, levels of factor VIII in the blood were the same as in the blood of normal mice and bleeding times were also similar to those of normal mice.

Duchenne Muscular Dystrophy

The main difficulty with gene therapy for DMD is the sheer size of the dystrophin gene—the complementary DNA (cDNA) is 14 kb. Current trials use a 'microdystrophin' molecule that includes just the bare essential domains within an adeno-associated vector. The vector is delivered by multiple injections into the arm muscle.

An alternative strategy is to use antisense oligonucleotides to force exon skipping and convert out-of-frame deletions that cause DMD to in-frame deletions usually associated with the milder Becker muscular dystrophy phenotype. This approach could be successful for up to 80% of patients with DMD. Proof of concept has been demonstrated in cultured patient cells and the *mdx* mouse model where dystrophin re-expression was demonstrated. The first clinical trial involved four patients who underwent intramuscular injection of an antisense oligonucleotide to

Table 23.3 Diseases that Can Potentially Be Treated by Gene Therapy

Disorder	Defect
Immune deficiency	Adenosine deaminase deficiency Purine nucleoside phosphorylase deficiency Chronic granulomatous disease
Hypercholesterolemia	Low-density lipoprotein receptor abnormalities
Hemophilia	Factor VIII deficiency (A) Factor IX deficiency (B)
Gaucher disease	Glucocerebrosidase deficiency
Mucopolysaccharidosis VII	β-Glucuronidase deficiency
Emphysema	α_1-Antitrypsin deficiency
Cystic fibrosis	*CFTR* mutations
Phenylketonuria	Phenylalanine hydroxylase deficiency
Hyperammonemia	Ornithine transcarbamylase deficiency
Citrullinemia	Argininosuccinate synthetase deficiency
Muscular dystrophy	Dystrophin mutations
Spinal muscular atrophy	*SMN1* gene deletion
Thalassemia/sickle cell anemia	α- and β-globin mutations
Malignant melanoma	
Ovarian cancer	
Brain tumors	
Neuroblastoma	
Renal cancer	
Lung cancer	
AIDS	
Cardiovascular diseases	
Rheumatoid arthritis	

target exon 51. Dystrophin was restored in the vast majority of muscle fibres at levels between 17% and 35%, without any adverse effects. However, intramuscular injection of individual muscles is not feasible as a treatment and various chemistries are under investigation in order to deliver antisense oligonucleotides to muscles throughout the body.

One key hurdle in the use of antisense oligonucleotide therapy is the fact that each different antisense is considered a new drug and requires separate regulatory approval. This makes their development more expensive and not feasible for low prevalence mutations for which there would be insufficient patients for clinical trials.

There is also the possibility of upregulating a dystrophin homolog, utrophin. Immune rejection is not a problem and studies in the *mdx* mouse have shown significant improvement in muscle function. Pharmacological compounds that enhance utrophin expression in animal models and cultured patient cells have been identified and clinical trials will start shortly.

Leber's Congenital Amaurosis

This autosomal recessive disorder is caused by mutations in the *RPE65* gene and characterized by poor vision at birth with complete loss of vision in early adulthood. Ten years ago, studies in a naturally occurring dog model (the Briard dog) showed that gene therapy by means of a single operation involving sub-retinal injection of an adeno-associated vector carrying the full length *RPE65* gene sequence was both safe and effective. Recent studies in a small number of young adults have replicated these findings and there is much excitement regarding the potential benefits for younger children whose loss of vision is not so far advanced. One obvious advantage for measuring the success of gene therapy in this condition is that a single eye can be treated whilst the other eye serves as a control. This research provides proof of principle that genetic forms of blindness may be reversed.

Stem Cell Therapy

Stem cells are unspecialized cells that are defined by their capacity for self-renewal and the ability to differentiate into specialized cells along many lineages. Embryonic stem cells are pluripotent, which means they can give rise to derivatives of all three germ layers (i.e., all cell types that are found in the adult organism). Somatic stem cells can only differentiate into the cell types found in the tissue from which they are derived (Figure 23.4), but can be isolated from any human, whatever their age. Nowadays the term **induced pluripotent stem cell** (**iPS**) is used rather than **somatic** or **adult stem cell**.

Bone-marrow transplantation is a form of somatic stem cell therapy that has been used for more than 40 years. During the past 5 years, cord blood stem cells have emerged as an alternative source. Although these transplants can be an effective treatment for a number of genetic disorders, including ADA deficiency, SCID,X-linked adrenoleukodystrophy, lysosomal storage diseases and Fanconi anemia, the associated risks of infection due to immunosuppression and graft-versus-host disease are high. The main limitation is the lack of a suitable bone-marrow donor or availability of matched cord blood stem cells.

Transplantation of stem cells (e.g., pluripotent hematopoietic stem cells) in utero offers the prospect of a novel mode of treatment for genetic disorders with a congenital onset. The immaturity of the fetal immune system means that the fetus will be tolerant of foreign cells so that there is no need to match the donor cells with those of the fetus. A small number of trials for have been performed but engraftment has so far only been successful in cases of SCID.

Embryonic Stem Cell Therapy

Teratomas (benign) and teratocarcinomas (malignant) are tumors that are found most commonly in the gonads. Their name is derived from the Greek word 'teratos' (monster); it describes their appearance well, as these tumors contain teeth, pieces of bone, muscles, skin, and hair. A key experiment demonstrated that if a single cell is removed from one of these tumors and injected intraperitoneally,

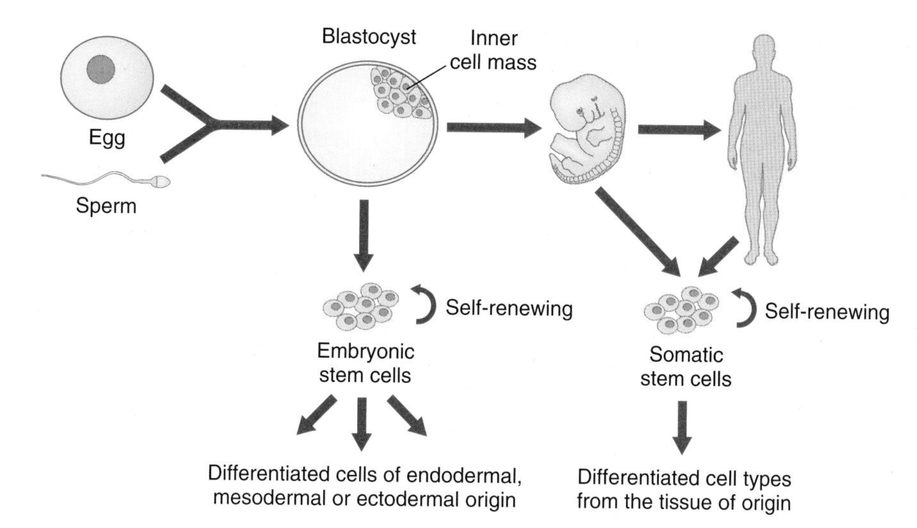

FIGURE 23.4 Generation of embryonic and somatic stem cells. The fusion of the sperm and egg during fertilization establishes a diploid zygote that divides to create the blastocyst. Embryonic stem cells (ESCs) are derived from the inner cell mass of the blastocyst. ESCs in culture are capable of self-renewal without differentiation and are able to differentiate into all cell types of the endoderm, mesoderm, and ectoderm lineages using appropriate signals. Somatic stem cells are also capable of self-renewal and, with appropriate signals, differentiate into various cell types from the tissue from which they are derived.

it acts as a stem cell by producing all the cell types found in a teratocarcinoma.

Mouse embryonic stem cells were first isolated and cultured 25 years ago. Studies of human embryonic stem cells have lagged behind, but the pace of research increased exponentially following the achievement in 1998 of the first cultured human embryonic stem cells.

Embryonic Stem Cells for Transplantation

The ability of an embryonic stem cell (ESC) to differentiate into any type of cell means that the potential applications of ESC therapy are vast. One approach involves the differentiation of ESCs in vitro to provide specialized cells for transplantation. For example, it is possible to culture mouse ESCs to generate dopamine-producing neurons. When these neural cells were transplanted into a mouse model for Parkinson disease, the dopamine-producing neurons showed long-term survival and ultimately corrected the phenotype. This 'therapeutic cloning' strategy has been proposed as a future therapy for other brain disorders such as stroke and neurodegenerative diseases. However, after many encouraging small studies of fetal cell transplantation for Parkinson disease, three randomized, double-blind, placebo-controlled studies found no net benefit. Also, patients in two of the studies developed dyskinesias that persisted despite reductions in medication. Further research is needed to understand and overcome the dual problems of unpredictable benefit and troublesome dyskinesias after dopaminergic cell transplantation. In addition, post mortem analysis of patients who received fetal brain cell transplantation revealed that implanted cells are prone to degeneration just like endogenous neurons in the same pathological area, indicating that long-term efficacy of cell therapy of Parkinson disease needs to overcome the degenerating environment in the brain.

Gene Therapy Using Embryonic Stem Cells

An alternate strategy is to use ESCs as delivery vehicles for genes that mediate phenotype correction through gene-transfer technology. One potential barrier to using human ESCs to treat genetic disorders is immunorejection of the transplanted cells by the host. This obstacle might be overcome by using gene transfer with the relevant normal gene to autologous cells (such as cultured skin fibroblasts), transfer of the corrected nucleus to an enucleated egg from an unrelated donor, development of 'corrected' ESCs and, finally, differentiation and transplantation of the corrected relevant cells to the same patient (Figure 23.5).

A crucial component of future clinical applications of this strategy is the ability to derive 'personalized' human ESC lines using the nuclear transfer technique. Although research on this technology has been controversial, the efficient transfer of somatic cell nuclei to enucleated oocytes from unrelated donors, and the subsequent derivation of human ESC lines from the resulting blastocysts, is a technical hurdle that has recently been overcome.

There has been much debate around the ethical issues of using ESCs and it seems that embryonic stem cells may not be an essential prerequisite, as iPS cells have been found in many more tissues than was once thought possible. Hence iPS cells might be used be used for transplantation.

Induced Pluripotent Stem Cell Therapy

Certain kinds of somatic stem cell seem to have the ability to differentiate into a number of different cell types, given the right conditions. Recent progress in stem cell biology has shown that iPS-derived cells can be used to successfully treat rodent Parkinson disease models, thus solving the problem of immunorejection and paving the way for future autologous transplantations for treating this disease and others.

Mesenchymal Stem Cells

Mesenchymal stem cell (MSC) therapy, through its promise of repair and regeneration of cardiac tissue, represents an exciting avenue of treatment for a range of cardiovascular

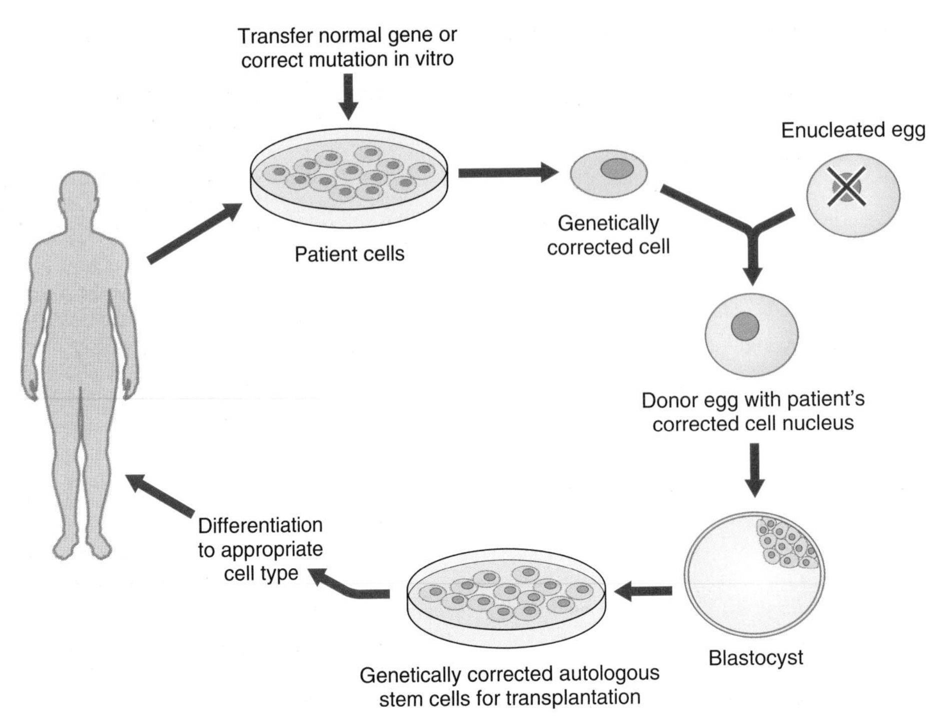

FIGURE 23.5 Embryonic stem cells for gene therapy. The strategy depicted starts with removing cells (e.g., fibroblasts) from a patient with a monogenic disorder and then transferring the normal gene using a vector (or perhaps by correcting the mutation in vitro). The nucleus from a corrected cell is then transferred to an enucleated egg obtained from an unrelated donor by somatic cell nuclear transfer. The egg, now containing the genetically corrected genome of the patient, is activated to develop into a blastocyst in vitro, and corrected autologous stem cells are derived from the inner cell mass. The stem cells are then directed to differentiate into a specific cell type and transferred to the patient, thereby correcting the disorder.

diseases. Cardiovascular disease is the leading cause of death in developed countries. Although cardiomyocytes retain limited plasticity following maturation, the heart is grossly unable to recover from structural damage.

MSCs are relatively immunopriviledged, lacking both major histocompatibility II and T cell co-stimulatory signal expression, and possess the unique ability to home into sites of myocardial damage when delivered systemically. They are obtained either from the bone marrow of healthy adult volunteers or from the patients themselves, and cultured in vitro with appropriate factors before being delivered to the damaged heart. Animal studies have shown therapeutic benefit via several distinct mechanisms, the most important of which appears to be the abundant secretion of paracrine factors that promote local regeneration. Phase I clinical trials have shown that this approach is safe and the results of phase II trials are eagerly awaited to see if there will be clear clinical benefit.

The genetic disorder retinitis pigmentosa (p. 182) results in the loss of photoreceptors, leading to visual symptoms in the teens and blindness by 40 to 50 years of age. Recently, systemic administration of pluripotent bone marrow–derived MSCs in a rat model has demonstrated improved visual function. This is a potentially exciting development for the future treatment of other forms of retinal degeneration and other ocular vascular diseases such as diabetic retinopathy.

A third application of MSC therapy is in bone repair and metabolic bone diseases such as osteogenesis imperfecta (p. 121) and hypophosphatasia, because MSCs can also differentiate to form bone and cartilage.

Limbal Stem Cells

The corneal limbus harbors corneal epithelial stem cells known as **limbal stem cells** (LSCs). Corneal conditions, such as infections, tumors, immunological disorders, trauma, and chemical burns, often lead to the deficiency of the corneal stem cells, and subsequent vision loss. Treatment of limbal stem cell deficiency (LSCD) has recently been achieved in eight patients who had complete LSCD in one eye. A small sample of the limbal epithelium of the patient's healthy eye was removed and grown in cell culture using the patient's own serum and donated amniotic cells to provide the required conditioning medium. Twelve days later, the LSCs were transplanted onto the patients' unhealthy eye and the group was followed for around 18 months. Overall, all patients had a decrease in pain and an increase in visual acuity.

Stem cell therapy has now progressed from preclinical (animal studies) to early clinical trials (in humans) for a variety of disorders. In general, these studies have shown enormous potential in the animal models but more limited success in humans so far. Aside from participation in regulated trials, patients should be advised that stem cell therapy is at an early stage and discouraged from undergoing forms of treatment whose safety and efficacy is not yet proven. An unwanted spin-off from stem cell research has been the development of so called stem cell tourism. Patients have travelled to countries where stem cell–based treatment is not regulated to receive expensive treatments that are scientifically unproven These treatments are at best, ineffective and at worst, dangerous.

FURTHER READING

Anderson WF 1992 Human gene therapy. Science 256:808–813

A consideration of gene therapy by one of its main proponents.

Aartsma-Rus A, van Ommen G-JB 2010 Progress in therapeutic antisense applications for neuromuscular disease. Eur J Hum Genet 18:146–153

A review of antisense oligonucleotide therapy for Duchenne muscular dystrophy, spinal muscular atrophy, and myotonic dystrophy.

Belmonte JCI, Ellis J, Hochedlinger K, Yamanaka S 2009 Induced pluripotent stem cells and reprogramming: seeing the science through the hype. Nat Rev Genet 10:878–883

An overview of the advantages and disadvantages of embryonic stem cell vs induced pluripotent stem cell therapy.

Brown BD, Naldini L 2009 Exploiting and antagonizing microRNA regulation for therapeutic and experimental applications. Nat Rev Genet 10:578–585

Article describing the potential applications of RNA interference.

Graw J, Brackmann HH, Oldenburg J, et al 2006 Haemophilia A: from mutation analysis to new therapies. Nat Rev Genet 6:488–501

Review of hemophilia A genetics and gene therapy.

Solter D 2006 From teratocarcinomas to embryonic stem cells and beyond: a history of embryonic stem cell research. Nat Rev Genet 7:319–327

The fascinating history of stem cell research with insights into the future applications of embryonic stem cells.

ELEMENTS

1 Treatment of genetic disease by conventional means requires identification of the gene product and an understanding of the pathophysiology of the disease process. Therapeutic options can include dietary restriction or supplementation, drug therapy, replacement of an abnormal or deficient protein or enzyme, and replacement or removal of an abnormal tissue.

2 Recombinant DNA technology has enabled human-derived biosynthetic gene products such as human insulin and growth hormone to be produced for the treatment of human disease.

3 Before a trial of gene therapy is carried out in humans, the gene involved must be characterized, the particular cell type or tissue to be targeted must be identified, an efficient, reliable, and safe vector system that results in stable continued expression of the introduced gene has to be developed, and the safety and effectiveness of the particular modality of gene therapy has to be demonstrated in an animal model.

4 Germline gene therapy is universally viewed as ethically unacceptable, whereas somatic cell gene therapy is generally viewed as being acceptable, because this is seen as similar to existing treatments such as organ transplantation.

5 Embryonic stem cells might be used therapeutically in a regenerative approach in which they are differentiated in vitro to specialized cell types (or progenitors of the target specialized cells), and then transplanted in vivo to replace diseased cells or tissues. Alternatively they could be used as delivery vehicles for gene-transfer technology.

CHAPTER 24

Ethical and Legal Issues in Medical Genetics

The mere existence of the complete reference map and DNA sequence down to the last nucleotide may lead to the absurdity of reductionism—the misconception that we know everything it means to be human; or to the absurdity of determinism—that what we are is a direct and inevitable consequence of what our genome is.

VICTOR MCKUSICK (1991)

Ethics is the branch of knowledge that deals with moral principles, which in turn relate to principles of right, wrong, justice, and standards of behavior. Traditionally, the reference points are based on a synthesis of the philosophical and religious views of well-informed, respected, thinking members of society. In this way, a code of practice evolves that is seen as reasonable and acceptable by a majority, which often forms the basis for professional guidelines or regulations. It might be argued that there are no 'absolutes' in ethical and moral debates. In complex scenarios, in which there may be competing and conflicting claims to an ethical principle, practical decisions and actions often have to be based on a balancing of duties, responsibilities, and rights. Ethics, like science, is not static but moves on, and in fact the development of the two disciplines is closely intertwined.

Ethical issues arise in all branches of medicine, but human genetics poses particular challenges because genetic identity impinges not just on an individual, but also on close relatives and the extended family, as well as society in general. In the minds of the general public, clinical genetics and genetic counseling can easily be confused with eugenics—defined as the science of 'improving' a species through breeding. It is important to stress that the modern specialty of clinical genetics has absolutely nothing in common with the appalling eugenic philosophies that were practiced in Nazi Germany and, to a much lesser extent, elsewhere in Europe and the United States between the two world wars. Emphasis has already been placed on the fundamental principle that genetic counseling is a *non-directive* and *non-judgmental* communication process whereby factual knowledge is imparted to facilitate informed personal choice (see Chapter 17). Indeed, clinical geneticists have been pioneers in recent times in practicing and promoting non-paternalism in medicine, and 5% of the original budget for the Human Genome Project was set aside for funding studies into the ethical and social implications of the knowledge gained from the project. Coercion and eugenics certainly have no place in modern medical genetics.

Nevertheless, this subject lends itself to ethical debate, not least because of the new challenges and opportunities provided by discoveries and new technologies in molecular genetics. In this chapter, some of the more controversial and difficult areas are considered. It soon becomes apparent that for many of these issues there is no clearly right or wrong approach and that individual views will vary widely. Sometimes in a clinical setting the best that can be hoped for is to arrive at a mutually acceptable compromise, with an explicit agreement that opposing views are respected and, personal conscience permitting, a patient's expressed wishes are carried out.

As genetic testing and DNA technologies enter the mainstream of medicine, and awareness of the ethical issues grows and impacts society, so there is a need for some restrictions and protections to be enshrined in law. This chapter therefore touches on some developments in this area. The Western world is becoming increasingly familiar with courts of law making final decisions—for example, in relation to contentious end-of-life issues—and this trend is likely to continue.

General Principles

The time-honored four principles of medical ethics that command wide consensus are listed in Box 24.1. Developed and championed by the American ethicists Tom Beauchamp and James Childress, these principles provide an acceptable framework, although close scrutiny of many difficult dilemmas highlights limitations in these principles and apparent conflicts between them. Everyone involved in clinical genetics will sooner or later be confronted by complex and challenging ethical situations, some of which pose particularly difficult problems with no obvious solution, and certainly no perfect one. Just as patients need to balance risks when making a decision about a treatment option, so the

Box 24.1 Fundamental Ethical Principles

Autonomy—incorporating respect for the individual, privacy, the importance of informed consent, and confidentiality
Beneficence—the principle of seeking to do good and therefore acting in the best interests of the patient
Non-maleficence—the principle of seeking, overall, not to harm (i.e., not to leave the patient in a worse condition than before treatment)
Justice—incorporating fairness for the patient in the context of the resources available, equity of access, and opportunity

clinician/counselor may need to balance these principles one against the other. A particular difficulty in medical genetics can be the principle of *autonomy*, given that we all share our genes with our biological relatives. Individual autonomy needs sometimes to be weighed against the principle of doing good, and doing no harm, to close family members.

The **Beauchamp and Childress framework** of ethical principles is, unsurprisingly, not the only one in use and others have developed them into practical approaches. These include the Jonsen framework (Box 24.2) and the more detailed scheme developed by Mike Parker of Oxford's Ethox Centre (Box 24.3), which builds on previous proposals. Taken together, these provide a practical approach to clinical ethics, which is an expanding discipline in health care.

In practice, the issues that commonly arise in the genetics clinic during any patient contact are outlined below.

Autonomy

It is the patient who should be empowered and in charge when it comes to decisions that have to be made. The degree to which this is possible is a function of the quality of information given. Sometimes patients are still seeking some form of guidance to give them confidence in the decision they reach, and it will require the judgment of the clinician/counselor as to how much guidance is appropriate in a given situation. The patient should feel comfortable to proceed no further, and opt out freely at any stage of the process; this applies particularly in the context of predictive genetic testing.

Box 24.2 The Jonson Framework: A Practical Approach to Clinical Ethics

Indications for medical intervention—Establish a diagnosis. Determine the options for treatment and the prognoses for each of the options.
Preferences of patient—Is the patient competent? If so, what does he or she want? If not competent, what is in the patient's best interest?
Quality of life—Will the proposed treatment improve the patient's quality of life?
Contextual features—Do religious, cultural, or legal factors have an impact on the decision?

Box 24.3 The Ethox Centre Clinical Ethics Framework (Mike Parker)

1. What are the relevant clinical and other facts (e.g., family dynamics, general practitioner support)?
2. What would constitute an appropriate decision-making process?
   - Who is to be held responsible?
   - When does the decision have to be made?
   - Who should be involved?
   - What are the procedural rules (e.g., confidentiality)?
3. List the available options.
4. What are the morally significant features of each option; for example:
   - What does the patient want to happen?
   - Is the patient competent?
   - If the patient is not competent, what is in his or her 'best interests'?
   - What are the foreseeable consequences of each option?
5. What does the law/guidance say about each of these options?
6. For each realistic option, identify the moral arguments in favor and against.
7. Choose an option based on judgment of the relative merits of these arguments:
   - How does this case compare with others?
   - Are there any key terms for which the meaning needs to be agreed (e.g., 'best interest', 'person')?
   - Are the arguments 'valid'?
   - Consider the foreseeable consequences (local and more broad).
   - Do the options 'respect persons'?
   - What would be the implications of this decision applied as a general rule?
8. Identify the strongest counterargument to the option you have chosen.
9. Can you rebut this argument? What are your reasons?
10. Make a decision.
11. Review this decision in the light of what actually happens, and learn from it.

Informed Choice

The patient is entitled to full information about all options available in a given situation, including the option of not participating. Potential consequences of each decision option should be discussed. No duress should be applied and the clinician/counselor should not have a vested interest in the patient pursuing any particular course of action.

Informed Consent

A patient is entitled to an honest and full explanation before any procedure or test is undertaken. Information should include details of the risks, limitations, implications, and possible outcomes of each procedure. In the current climate, with respect to full information and the doctor-patient contract, some form of *signed* consent is increasingly being obtained for every action that exposes the patient—access to medical records, clinical photography, genetic testing, and storage of DNA. In fact, there is no legal requirement to obtain signed consent for taking a blood test from which DNA is extracted and stored. The issue was addressed by the UK Human Tissue Act 2004. According to the act, DNA does not constitute 'human tissue' in the same way as biopsy samples or cellular material, for which formal consent is required, whether the tissue is from the living or the dead. The act does require that consent is formally obtained where cellular material is used to obtain genetic information *for another person*. In a clinical setting, this must be clearly discussed and documented.

In clinical genetics, many patients who are candidates for clinical examination and genetic testing are children or individuals with learning difficulties who may lack capacity to grant informed consent. Furthermore, the result of any examination or test may have only a small chance of directly benefiting the patient but is potentially very important for family members. Here the law is important. In England and Wales, the Mental Capacity Act of 2005 came into effect in 2007 and applies to adults aged 16 and older. It replaced case law for health (and social) care and there is a legal duty to use the legislation and apply the 'Test for Capacity' (Box 24.4) for any relevant decision for people who lack capacity. Decisions must take into account the 'best interests' of the patient, but can also embrace the wider interests that relate to the family. In England and Wales, the law allows for an appropriate person appointed by the Court of Protection to act on their behalf, whereas in Scotland it is legally permitted for certain designated adults, including family members, to give consent (or refuse) on behalf of a person lacking capacity.

Box 24.4 Mental Capacity Act, 2005, England and Wales (Outline)—Principles, Definition, and Test for Capacity

Principles:
- A person must be assumed to have capacity unless proved otherwise
- A decision taken for someone lacking capacity must be in the person's best interests
- Practical steps must be taken to help someone make a decision
- If the test of capacity is passed the decision taken must be respected

Definition of Capacity:
'... a person lacks capacity in relation to a matter if at the material time he is unable to make a decision for himself in relation to the matter because of an impairment of, or a disturbance in the functioning of, the mind or brain.'
In relation to any decision, it is therefore:
- Time specific (a person's capacity may change)
- Decision specific (capacity varies, depending on the decision)

Test for Capacity:
At a specific time and for a specific decision, the person should:
- Understand the information relevant to the decision
- Retain the information
- Weigh the information as part of decision making
- Communicate the decision

Confidentiality

A patient has a right to complete confidentiality, and there are clearly many issues relating to genetic disease that a patient, or a couple, would wish to keep totally private. Stigmatization and guilt may still accompany the concept of hereditary illness. Traditionally, confidentiality should be breached only under extreme circumstances; for example, when it is deemed that an individual's behavior could convey a high risk of harm to self or to others. In trying to help some patients in the genetics clinic, however, it may be desirable to have a sample of DNA from a key family member, necessitating at least some disclosure of detail. There is also the difficult area of sharing information and results between different regional genetic services. This is a complex and much debated area in the context of genetic and hereditary disease but the principle of patient consent for release and/or sharing of information should be the norm.

Universality

Much of traditional medical ethical thinking has upheld the autonomy of the *individual* as paramount. Growing appreciation of the ethical challenges posed by genetics has led to calls for a new pragmatism in bioethics, built on the concept that the human genome is fundamentally common to all humankind, and can—and indeed should—be considered a shared resource because we have a shared identity at this level. What we learn from one individual's genome, from a family's genome, or a population's genome, carries potential benefits far beyond the immediate relevance and impact for that individual or family. From this it is a direct and natural step to consider how best the genetic information is exchanged, for the medical benefits may be far reaching. This ethical *attitude* therefore leads on to a realization of mutual respect, reciprocity, and world citizenry in the context of human genetics. It prompts the individual to consider his or her responsibility toward others, as well as to society, both in the present and in the future.

Meanwhile, however, very real ethical problems have to be faced and dealt with in some way, and it is to a few of these that we now turn.

Ethical Dilemmas in the Genetic Clinic

Prenatal Diagnosis

Many methods are now widely available for diagnosing structural abnormalities and genetic disorders during the first and second trimesters (see Chapter 21). The past 35 to 40 years have seen the first real availability of *choice* in the context of pregnancy in human history. Not surprisingly, the issue of prenatal diagnosis and subsequent offer of termination of pregnancy raises many difficult issues for individuals and families, and raises serious questions about the way in which society views and cares for both children and adults with disability. In the United Kingdom, termination of pregnancy is permitted up to and beyond 24 weeks' gestation if the fetus has a lethal condition such as anencephaly, or if there is a serious risk of major physical or mental handicap. For good reason, terms such as 'serious' are not defined in the relevant legislation, but this can inevitably lead to controversy over interpretation.

The difficulties surrounding prenatal diagnosis can be illustrated by considering some of the general principles that have already been discussed. At the top of the list comes informed consent. In the United Kingdom, approximately 70% of all pregnancies are monitored for the presence of a neural tube defect by measurement of α-fetoprotein in maternal serum at approximately 16 weeks' gestation (p. 328). In theory, all women undergoing this test should have a full understanding of its potential implications. This also applies to every woman who is offered a detailed ultrasonographic scan to assess fetal anatomy at around 18 to 20 weeks' gestation (p. 326). For fully informed consent to be obtained in these situations, it is essential that pregnant women should have access to detailed counseling by unhurried staff members who are knowledgeable, experienced, and sympathetic. In practice this may not always be so; indeed, there is evidence that the quality of information provided varies widely.

The most difficult problems in prenatal diagnosis are those involving autonomy and individual choice. This relates particularly to disease severity and who should make the decision that termination is justified. This can be illustrated by considering the following situations. In the first situation, parents whose first child, a boy, has autism are expecting another baby. They have read that autism is more common in boys than girls, so they request sexing of the fetus with a view to terminating a male fetus but continuing if the sex is female. Overall, however, the risk of having another child with autism is only about 5%. Such a request presents the clinician and counselor with a challenge. There is general agreement that sex selection for purely social reasons is not justified as grounds for termination of pregnancy, nor indeed for embryo selection by preimplantation genetic diagnosis (PGD), although in the United States, it is permissible to perform sex selection by PGD for 'family balancing'. In the United Kingdom, the general public, through a public consultation process overseen by the Human Fertilization and Embryology Authority (HFEA), has overwhelmingly expressed the view that sex selection for social reasons and family balancing is not acceptable: children should be considered as gifts, not consumer commodities. But what about this situation, when the risk of a second child having autism is low and it cannot be guaranteed that a daughter would not be affected?

Next, consider the unusual but not unprecedented dilemma that arises when parents with an inherited condition indicate that they wish to *continue with a pregnancy* only if tests show that their unborn baby *is also affected*. Examples of conditions that could generate a request of this nature include achondroplasia and congenital sensorineural hearing loss. If the family's autonomy and right to choose is to be respected, then their request should be granted. Many readers of this chapter will be uncomfortable with the suggestion that an unaffected pregnancy should be terminated. This particular scenario illustrates the difficulty of interpretation and defining what is normal.

The issue of autonomy and individual choice can also arise when a fetus is found to have a relatively mild abnormality, such as a non-syndromal cleft lip and palate, for which surgical correction usually achieves an excellent outcome. For some parents, particularly those who themselves have had an unhappy childhood because of being stigmatized for a similar problem, the prospect of having a similarly affected child can be unacceptable. Understandably, medical and nursing staff may feel very uneasy about complying with a request for termination of pregnancy in such situations.

It is inevitable that a subject as emotive as termination of pregnancy will generate controversy, and the ethical dilemmas that arise are not easily resolved. Proponents of choice argue that selective termination should be available, particularly if the alternative involves a lifetime of pain and suffering. More often than not, prenatal diagnostic techniques provide reassurance, and the fact that tests are available provides many couples with the necessary confidence to embark on a pregnancy. Without the option of prenatal tests, these couples might decide against trying to have further children. When viewed in the context of abortion in general, termination on the grounds of fetal abnormality constitutes less than 2% of the total of approximately 200,000 abortions carried out each year in the United Kingdom.

Those who hold opposing views argue on religious, moral or ethical grounds that selective termination is little less than legalized infanticide. Key to the ethical issue here are views on the status and rights of the embryo and fetus. For those who believe that the fertilized egg constitutes full human status, PGD and embryo research are unacceptable. Indeed, logically, for people who hold this belief all in-vitro fertilization (IVF) programs are unacceptable by virtue of generating thousands of spare human embryos to be kept in freezers, most never used. There is also concern that prenatal diagnostic screening programs could lead to a devaluing of the 'disabled' and 'abnormal' in society (notwithstanding that these terms are difficult to define and all too often used pejoratively), with a possible shift of resources away from their care to the funding of programs aimed at 'preventing' their birth. The debate about the ethics of prenatal diagnosis is a fierce one that will become even more difficult when genes are identified for common multifactorial disorders such as depression and schizophrenia. Mutations or polymorphisms in such genes are likely to confer a *risk* that the fetus will develop the condition as an adult, not that the individual will definitely be affected. The arrival of microarray-CGH technology, and in the future rapid automated genome sequencing, is raising the specter of wide-ranging, affordable antenatal genetic screening, well before we know how to identify all copy number polymorphisms and variants of unknown significance.

The results of public consultation exercises conducted by the Advisory Committee on Genetic Testing (subsumed into the Human Genetics Commission which was abolished in 2010) and the HFEA are reasonably reassuring. The views expressed support the applications of genetics in prenatal testing for *serious* disorders but concern over wider applications of the techniques. Similarly, research published by the British Social Attitudes survey, in the context of genetics research and gene manipulation for the detection of disease, suggested that the public supports these activities in general but expressed deep reservations for application of the technologies for **genetic enhancement**. Genetic enhancement, through manipulation of embryos or gametes, strikes at the very heart of what it means to have one's own identity through natural laws of chance. This, it seems, is a powerful undercurrent in the understanding of who we are as individuals and as a species.

Predictive Testing In Childhood

Understandably, parents sometimes wish to know whether or not a child has inherited the gene for an adult-onset autosomal dominant disorder that runs in the family. It could be argued that this knowledge will help the parents guide their child toward the most appropriate educational and career opportunities and that to refuse their request is a denial of their rights as parents. Similarly, parents may request testing to clarify the status of young healthy children at risk of being carriers of a recessive disorder such as cystic fibrosis. Sometimes this information will have become available as a result of prenatal diagnostic testing.

The problem with agreeing to such a request is that it infringes the child's own future autonomy. Increasingly, it is felt that testing should be delayed until the child reaches an age at which he or she can make his or her own informed decision. There is also concern about the possible deleterious effects on a child of growing up with the certain knowledge of developing a serious adult-onset hereditary disorder or being a carrier of a recessive disorder, particularly if the tests have proved negative in the child's other siblings. Such a situation could raise a very real possibility of stigmatization. However, although there is consensus among geneticists that children should not be tested for carrier status, the evidence that such testing causes emotional or psychological harm is weak.

The situation is very different if predictive testing could directly benefit the child by identifying the need for a medical or surgical intervention *in childhood*. This applies to conditions such as familial hypercholesterolemia (p. 175), for which early dietary management can be introduced, and also to some of the familial cancer-predisposing syndromes (p. 225) for which early screening, and sometimes prophylactic surgery, is indicated. Generally, it is thought that in these situations genetic testing is acceptable at around the time when other screening tests or preventive measures would be initiated.

One of the arguments for not testing children for adult-onset disorders is that parents might view their child differently, perhaps prejudicially, in some way. This type of argument has been voiced in relation to the PGD cases that have selected embryos not only for their negative affection status for Fanconi anemia but also in order to be a potential stem-cell donor for their affected child—so-called 'savior siblings'. Those objecting to this use of technology cite a *utilitarian*, or *instrumental*, attitude toward the child created in this way. Furthermore, the child so created has no choice about whether to be a tissue-matched donor for the sick sibling. Will the child eventually feel 'used' by the parents, and how might he or she feel if the treatment fails and the sick sibling dies? At present these questions are imponderables because most children created for this purpose are too young to tell how they feel—the first successful case was in the United States in 2000. The numbers of such children will be very small for the foreseeable future.

Implications for the Immediate Family (Inadvertent Testing or Testing by 'Proxy')

A positive test result in an individual can have major implications for close antecedent relatives who themselves may not wish to be informed of their disease status. For example, consider Huntington disease (HD), for which direct mutation analysis is available. A young man age 20 years requests predictive testing before starting his family; his fears are based on a confirmed diagnosis in his 65-year-old paternal grandfather. Predictive testing would be relatively straightforward were it not for the fact that his father, who is obviously at a prior risk of 1 in 2, specifically does not wish to know whether he will develop the disease.

Thus the young man has raised the difficult question of how to comply with his request without inadvertently carrying out a predictive test on his father. A negative result in the young man leaves the situation unchanged for his father. However, a positive result in the son might be difficult to conceal from an observant father; and the son will know that his father will develop the disease if he has not done so already.

There is no easy solution to this particular problem. In the guidelines drawn up in 1994 for predictive testing in HD it was concluded that 'every effort should be made by the counselors and the persons concerned to come to a satisfactory solution', with the rider that 'if no consensus can be reached the right of the adult child to know should have priority over the right of the parent not to know'.

Implications for the Extended Family

It is widely agreed that the identification of a condition that could have implications for other family members should lead to the offer of tests for the extended family. This applies particularly to balanced translocations and serious X-linked recessive disorders. In the case of translocations, this is sometimes referred to as **translocation chasing**. For an autosomal recessive disorder such as cystic fibrosis, the term 'cascade screening' is applied (p. 304).

The main ethical problem that arises here is that of confidentiality. A carrier of a translocation or serious X-linked recessive disorder is usually urged to alert close family relatives to the possibility that they could also be carriers and therefore at risk of having affected children. Alternatively, permission can be sought for members of the genetics team to make these approaches. Occasionally a patient, for whatever reason, will refuse to allow this information to be disseminated.

Faced with this situation, what should the clinical geneticist do? In practice most clinical geneticists would try to convince their patient of the importance of offering information and tests to relatives, possibly by providing an explanation of the consequences and ill-feeling that could arise in the future if a relative was to have an affected child whose birth could have been avoided. In most cases, skilled and sensitive counseling will lead to a satisfactory solution. Ultimately, however, many clinical geneticists would opt to respect their patient's confidentiality rather than break the trust that forms a cornerstone of the traditional doctor-patient relationship. Not all would agree, and where the application of this standard could result in damage or morbidity to other family members, the clinician might seek to *persuade* the individual to disclose the medical/genetic information. This view is backed up by the statements of authoritative working parties, such as the Nuffield Council on Bioethics. Sometimes it is possible to draw the issue to the attention of the general practitioner of the family member at risk—he or she might be well placed to open the issue up in a sensitive way.

Informed Consent in Genetic Research

All individuals who agree to undergo genetic testing in a service context are obviously entitled to a full and clear explanation of what the test involves and how the results could have implications for both themselves and other family members. Vigorous efforts are usually made to ensure that these basic principles are adhered to, particularly when predictive testing for serious late-onset genetic disorders is being undertaken.

The issues relating to informed consent when participating in genetic research are just as complex. Many people are perfectly willing to hold out their arm for a blood test which might 'help others', particularly if they have personal experience of a serious disorder in their own family. However, few will have given any serious thought to the possible ramifications of their simple act of altruism. For example, it is unlikely that they will ever have considered whether their sample will be tested anonymously, who will be informed of the result, or whether other tests will be carried out on stored DNA in the future as new techniques are developed. These concerns, among others (Box 24.5), have prompted the US National Institutes of Health Office of Protection from Research Risks to draw up proposals on the steps that should be taken to try to ensure that all aspects of informed consent are addressed when samples are collected for genetic research. Just as signed consent for genetic testing and storage of DNA has become routine in the service setting (although not a legal requirement under the UK Human Tissue Act 2004), similar procedures should be adhered to in a research setting.

Box 24.5 Issues of Disclosure and Consent in Genetic Research—The Nature of the Study

- Who is doing the study and where is it being carried out?
- Availability of results and their implications for the individual and extended family regarding health, employment, and insurance
- Anonymity of testing and confidentiality of results
- Long-term storage of DNA and its possible use in other research projects
- Potential commercial applications and profit

Ethical Dilemmas and the Public Interest

Recent progress in genetics, most notably in the area of molecular testing, has brought the ethical debate into a much wider public arena. Topics such as insurance, forensic science and DNA databases, patenting, gene therapy, population screening, cloning, stem-cell research, and hybrids, are now rightly viewed as being of major societal, commercial, and political importance, and perhaps not surprisingly they feature prominently in media discussion. All of these subjects affect the specialty of medical genetics and each of these will now be considered in turn.

Genetics and Insurance

The availability of predictive tests for disorders of adult onset that convey a risk for chronic ill health, and possibly reduced life expectancy, has led to concern about the extent to which the results of these tests should be revealed to outside agencies, especially insurance companies providing life cover, private health care, and critical illness and disability income. For insurance arranged through an employer, in theory adverse genetic tests might compromise career prospects.

The life insurance industry is competitive and profit driven. Private insurance is based on 'mutuality', whereby risks are pooled for individuals in similar circumstances. In contrast, public health services are based on the principle of 'solidarity', whereby health provision for everyone is funded from general taxation. It is understandable that the life insurance industry is concerned that individuals who receive a positive predictive test result will take out large policies without revealing their true risk status. This is sometimes referred to as 'antiselection' or 'adverse selection'. On the other hand, the genetics community is concerned that individuals who test positive will become victims of discrimination, and perhaps uninsurable. This concern extends to those with a family history of a late-onset disorder, who might be refused insurance unless they undergo predictive testing.

The possibility that DNA testing will create an uninsurable 'genetic underclass' led to the introduction of legislation in parts of the United States aimed at limiting the use of genetic information by health insurers. In 1996 this culminated in President Clinton signing The Health Insurance Portability and Accountability Act, which expressly prevented employer-based health plans from refusing coverage on genetic grounds when a person changes employment. In the United Kingdom, this whole arena was considered in 1995 by the House of Commons Science and Technology Committee, which recommended that a Human Genetics Advisory Commission be established to overview developments in human genetics. In 1997 this Advisory Commission (subsumed into the Human Genetics Commission which was abolished in 2010) recommended that applicants for life insurance should not have to disclose the results of any genetic test to a prospective insurer and that a moratorium on disclosure of results should last for at least 2 years until genetic testing had been carefully evaluated.

Inevitably, the Association of British Insurers (ABI) had a view. In the 1999 revision of its Code of Practice, the ABI reiterated its view that applicants should not be asked to undergo genetic testing and that existing genetic test results need not be disclosed in applications for mortgage-related life assurance up to a total of £100,000. In 2005 the UK government negotiated an agreement with the ABI to extend restriction on the use of predictive genetic tests by insurers to November 2011. The document, entitled 'Concordat and Moratorium on Genetics and Insurance' (Box 24.6), stated that no one will be required to disclose the result of a predictive genetic test unless first approved by the government's Genetics and Insurance Committee (GAIC). The GAIC approved only one application—for HD for amounts greater than £500,000—and the body was abolished in 2009.

These issues are likely to come under renewed scrutiny in the future. Where various combinations of polymorphisms convey susceptibility to common disorders of adult life, large swathes of the population could find themselves at the mercy of a profit-driven, commercially-focused insurance industry. The medical genetics community therefore has an advocacy role to ensure that the genetically disadvantaged, through no fault of their own, do not face discrimination when seeking health care or long-term life insurance. These are powerful arguments favoring retention of the principles of the UK National Health Service.

Box 24.6 Key Points in the 'Concordat and Moratorium on Genetics and Insurance' Negotiated between the UK Government and the Association of British Insurers (ABI), 2005

- Applicants must not be asked to undergo predictive genetic testing.*
- Results of genetic tests that have been undertaken should be disclosed up to a certain limit.
- Only the results of tests that have been validated should be taken into account.
- Existing genetic test results need not be disclosed in applications for life insurance policies up to a total of £500,000, and for critical illness/income protection insurance up to £300,000.
- The results of genetic tests undertaken by an applicant will not be taken into account when assessing an application from another individual, and vice versa.
- Genetic tests taken as part of a research study do not have to be disclosed to insurers.

*A genetic test is defined as 'an examination of the chromosome, DNA or RNA to find out if there is an otherwise undetectable related genotype, which may indicate an increased chance of that individual developing a specific disease in the future'.

Forensic Science and DNA Databases

The existence of the police-controlled **National DNA Database** in the UK has been hotly debated as an issue of personal privacy. The use of DNA fingerprinting in criminal investigations, to the tune of approximately 10,000 cases per annum, is now so sophisticated that there is a natural desire on the part of law enforcers to be able to identify the DNA fingerprint for anyone in the general population. Furthermore, techniques have been developed, called **DNA boost**, to generate profiles of individuals from samples where the DNA of two or more persons is mixed. The UK's National DNA Database, which once contained material solely from sentenced offenders, has expanded rapidly and is now the largest of any country, containing information on more than 5 million people (0.8% of the population), including an estimated 1 million with no criminal conviction; this compares with 0.5% of the population in the United States. For certain types of crime, whole sections of a community are invited to come forward to give a sample of DNA so that they can be eliminated from enquiries. However, with significant numbers of DNA samples from children on the database, including that of at least one infant, the police came under political pressure in 2009 to scrap 'innocent' profiles after a declaration by the European Court of Human Rights that to hold the profiles of innocents indefinitely was a breach of privacy.

The National DNA Database is huge, but so too are the collections for big population studies, such as ALSPAC (Avon Longitudinal Study of Parents and Children) or the UK Biobank project. As research, these samples will have been rigorously consented, but it is essential for safeguards to be built into the use that is made of DNA collections such as these, and access to them. Debate will certainly continue on the use and misuse of personal genetic data.

Gene Patenting and the Human Genome Project

The controversy surrounding the patenting of naturally occurring human DNA sequences, whether complete genes or **expressed sequence tags**, neatly encapsulates the conflict between harsh commercial realism and altruistic academic idealism. As illustrations of the levels of investment, it is noteworthy that the rights to one gene associated with obesity were sold in 1995 for $70 million, whereas in 1997 DeCODE, the Icelandic genomics company at the center of the controversy regarding national assent, sold the potential rights to 12 genes, possibly associated with common complex diseases, to Hoffman-La Roche for $200 million. On the one hand, biotechnology companies that have invested heavily in molecular research can reasonably argue that they and their shareholders are entitled to benefit from the fruits of their labors. Biotechnology research is indeed expensive, and the realistic view is that commercial companies must at least cover their costs (otherwise they cease to exist) but preferably make a fair return on investments. The idealistic view argues that the human genome represents humankind's 'common heritage' and that information gained through the Human Genome Project, or other molecular research, should be freely available for all to benefit, thus maintaining the ethical principle of equity of access. Proponents of the latter cite alleged exploitation of patients and communities who have donated their blood samples for research, little realizing that their generosity could be exploited for financial gain. This is amply illustrated by the furor surrounding the proposed use (or abuse) of a centralized medical database of the entire Icelandic population to help identify 'polygenes' for potential commercial gain. There have been some high-profile court cases over the issue in the United States.

The legal issues are complex and, not surprisingly, the international community has struggled to identify satisfactory solutions. With the exception of the United States, most national regulatory bodies prohibit payment for the procurement of human genetic material, but their views on patenting are much less well defined. In recent years many important human genes have been patented, and companies such as Myriad Genetics in the United States sought to impose their exclusive licence for genetic testing for *BRCA1* and *BRCA2* (p. 224). In fact, in 2004 the European Patent Office revoked the patent, denying Myriad a license fee from every *BRCA* test undertaken in Europe, and thereby setting a precedent for other contentious cases. Gene patents may concern single-gene disorders, such as the *BRCA* genes, but increasingly commercial companies are offering a range of 'direct to consumer' tests, consisting of the analysis of a range of polymorphisms, with the promise of predicting the future risk of ill health from common disorders. These companies are often less than candid about the precise tests offered, and their validation, all of which runs counter to the spirit of scientific enquiry and evidence-based medicine.

Gene Therapy

One of the most exciting aspects of recent progress in molecular biology is the prospect of successful gene therapy (p. 350), although it is obviously disappointing that this potential has not yet been realized. It is understandable that both the general public and the health care professions should be concerned about the possible side effects and abuse of gene therapy. Frequent reference has been made to the 'slippery slope' argument, whereby to take the first step leads incrementally and inevitably to uncontrolled experimentation. To address these anxieties, advisory or regulatory committees have been established in several countries to assess the practical and ethical aspects of gene therapy research programs.

Concern centers around two fundamental issues. The first relates to the practical aspects of ensuring informed consent on the part of patients who wish to participate in gene therapy research. Adult patients and parents of children affected by otherwise incurable conditions may be desperate to participate in gene therapy research. Consequently they may disregard the possible hazards of what is essentially a new, untried, and unproven therapeutic approach. In the United Kingdom, the Committee on the Ethics of Gene Therapy recommended that, until shown to be safe, all gene therapy programs should be subjected to careful scrutiny by research ethics committees. In addition, a national supervisory body, the Gene Therapy Advisory Committee (GTAC), was established to review all proposals to conduct gene therapy in humans and to monitor ongoing trials, thus safeguarding patients' rights and confidentiality.

Second, gene therapy generates concern about potential eugenic applications. The GTAC recommended that genetic modification involving the germline should be prohibited, and limited to somatic cells, to prevent the possibility of newly modified genes being transmitted to future generations. Somatic cell gene therapy should be used only to try to treat serious diseases, and not to alter human characteristics, such as intelligence or athletic prowess for example.

Population Screening

Population screening programs offering carrier detection for common autosomal recessive disorders, have been in operation for many years (p. 318), and in many cases well received (e.g., thalassemia and Tay-Sachs disease). This was not so with respect to neonatal screening for α_1 antitrypsin deficiency in Scandinavia, which was abandoned because it proved stressful. Such is the progress in comparative genome analysis that it will be technically possible to compare the genome of the unborn baby with that of its parents; de novo differences may represent mutations that would lead to serious disease and it has been proposed that the technique is offered to couples undergoing PGD. Chorionic villus tissue could also be analyzed in this way, and even free fetal DNA from maternal blood eventually. The ethical problem relates to the range of diseases that might be tested this way, and the choices couples might face.

With respect to screening programs that detect carrier status for disease, the issues are slightly different. Early efforts to introduce sickle-cell carrier detection in North America were largely unsuccessful because of misinformation, discrimination, and stigmatization. Also, pilot studies assessing the responses to cystic fibrosis (CF) carrier screening in white populations yielded conflicting results (p. 321). These experiences illustrate the importance of informed consent and the difficulties of ensuring both autonomy and informed choice. For example, CF screening in the United Kingdom has been implemented as part of neonatal screening. Although aimed at identifying babies with CF, the screening detects a proportion who are simply carriers, but obviously newborn infants cannot make an informed choice. Consequently, some pilot studies have focused on adults and their responses to the offer of carrier testing, either from their general practitioner or at the antenatal clinic. This has raised the vexed question of whether an offer from a respected family doctor could be interpreted as an implicit recommendation to participate, and such an approach yields a higher acceptance rate than a casual written invitation to attend for screening at a future date. This begs the question as to whether, depending on the approach, individuals feel pressured to undergo a test that they do not necessarily want.

It is, therefore, important that even the most well-intended offer of carrier detection should be worded carefully to ensure that participation is entirely voluntary. Full counseling in the event of a positive result is also essential to minimize the risk of any feeling of stigmatization or genetic inferiority.

In population screening, confidentiality is also important. Many will not wish their carrier status to be known by classmates or colleagues at work. The issue of confidentiality will be particularly difficult for individuals found by genetic testing to be susceptible to a medical problem through environmental industrial hazards, which could lead to employment discrimination (p. 367). These anxieties led in 1995 to the Equal Employment Opportunity Commission in the United States issuing a guideline that allows for anyone denied employment because of disease susceptibility to claim protection under the Americans with Disabilities Act.

Cloning and Stem Cell Research

Dolly the sheep, born in July 1996 at Roslin, near Edinburgh, was the first mammal to be cloned from an adult cell, and when her existence was announced about 6 months later, the world suddenly became intensely interested in cloning. Dolly was 'conceived' by fusing individual mammary gland cells with unfertilized eggs from which the nucleus had been removed; 277 attempts failed before a successful pregnancy ensued. It was immediately assumed that the technology would sooner or later lead to a cloned human being and there have been some unsubstantiated bogus claims to this effect. In fact, there has been widespread rejection of any move toward human *reproductive* cloning, with strong statements emanating from politicians, religious leaders, and scientists. Experiments with animals have continued to have a very poor success rate, and for this reason alone no rational person is advocating 'experiments' in humans. In some cloned animals, the features have suggested possible defects in genomic imprinting. Dolly died prematurely from lung disease in February 2003 and she had a number of characteristics suggesting she was not biologically normal.

Lessons have been learned from Dolly in relation to cell nuclear replacement technology and this has, potentially, opened the door to understanding more about cell differentiation. The focus has therefore shifted to *therapeutic* cloning using stem cells, and to the prospects this holds with respect to human disease. If stem cells were subjected to nuclear transfer from a patient in need, they might be stimulated to grow into any tissue type, perhaps in unlimited quantities and genetically identical to the patient, thus avoiding rejection. To date there have been a number of successes and the potential possibilities are legion for degenerative disease and repair of damage from trauma and burns.

The main ethical difficulty arises in relation to the source of stem cells. No one voices serious ethical difficulties in relation to stem cells harvested from the fully formed person, whether taken from the umbilical cord or the mature adult, and there have been significant advances using these sources. But a strong school of scientific opinion maintains that there is no substitute for studying *embryonic* stem cells to understand how cells differentiate from primitive into more complex types. In 2005 the UK Parliament moved swiftly to approve an extension to research on early human embryos for this purpose. Research on human embryos up to 14 days of age was already permitted under the Human Fertilization and Embryology Act 1990. The United Kingdom therefore became one of the most attractive places to work in stem-cell research because, although regulated, it is legal. Publicly funded research of this kind was not permitted in the United States until a change of political direction in 2009. Progress has been painfully slow for those engaged in this work, and the focus shifted to the creation of animal-human ('human-admixed') hybrids and chimeras because of the poor supply and quality of human oocytes (usually 'leftovers' from infertility treatment) for use in nuclear cell transfer. In the United Kingdom, Newcastle was granted a license to collect fresh eggs for stem-cell research from egg donors in return for a reduction in the cost of IVF treatment, a decision greeted with alarm in some quarters. This group was also the first, in 2005, to create a human blastocyst after nuclear transfer.

Those who object to the use of embryonic stem cells believe it is not only treating the human embryo with disrespect and tampering with the sanctity of life, but also could lead eventually to reproductive cloning. In fact,

Box 24.7 The Key 2008 Amendments to the Human Fertilization and Embryology Act (HFEA) 1990

- Ensure that all human embryos outside the body—whatever the process used in their creation—are subject to regulation.
- Ensure regulation of 'human-admixed' embryos created from a combination of human and animal genetic material for research.
- Ban sex selection of offspring for non-medical reasons. This puts into statute a ban on non-medical sex selection currently in place as a matter of HFEA policy. Sex selection is allowed for medical reasons—for example, to avoid a serious disease that affects only boys.
- Recognize same-sex couples as legal parents of children conceived through the use of donated sperm, eggs, or embryos. These provisions enable, for example, the civil partner of a woman who carries a child via IVF to be recognized as the child's legal parent.
- Retain a duty to take account of the welfare of the child in providing fertility treatment, but replace the reference to 'the need for a father' with 'the need for supportive parenting'—hence valuing the role of all parents.
- Alter the restrictions on the use of HFEA-collected data to help enable follow-up research of infertility treatment.

the Human Fertilization and Embryology Act of 1990 permits the creation of human embryos for research, but very few have been created since the HFEA began granting licenses. This Act of Parliament has been reviewed and updated to accommodate new developments, and came into effect in 2009. The main provisions are listed in Box 24.7 and will continue to generate contentious ethical debate.

Conclusion

Advances in human molecular genetics and cell biology generate major issues in medical ethics. Each new discovery brings new challenges and raises new dilemmas for which there are usually no easy answers. On a global scale the computerization of medical records, together with the widespread introduction of genetic testing, makes it essential that safeguards are introduced to ensure that fundamental principles such as privacy and confidentiality are maintained. Members of the medical genetics community will continue to play a pivotal role in trying to balance the needs of their patients and families with the demands of an increasingly cost-conscious society and a commercially driven biotechnology industry. Cost-benefit arguments can be persuasive in cold financial terms but take no account of the fundamental human and social issues that are often involved. The medical genetics community must take an advocacy role to ensure that the interests of their patients and families take precedence, and toward that end it is hoped that this chapter, and indeed the rest of this book, can make a positive contribution.

FURTHER READING

American Society of Human Genetics Report 1996 Statement on informed consent for genetic research. Am J Hum Genet 59:471–474

The statement of the American Society of Human Genetics Board of Directors on the issues relating to informed consent in genetic research.

Association of British Insurers 1999 Genetic testing. ABI Code of Practice. London, UK: ABI, London

A formal statement of the principles and practice adopted by the British insurance industry with regard to genetic testing.

Baily MA, Murray TH (eds) 2009 Ethics and newborn genetic screening: new technologies, new challenges. Baltimore, MD: Johns Hopkins University Press, Baltimore

A multiauthor volume with a focus on the health economics of newborn screening and distributive justice.

British Medical Association 1998 Human genetics. Choice and responsibility. Oxford, UK: Oxford University Press

A comprehensive, wide-ranging report produced by a BMA medical ethics committee steering group on the ethical issues raised by genetics in clinical practice.

Bryant J, Baggott la Velle L, Searle J (eds) 2002 Bioethics for scientists. Chichester, UK: John Wiley

A multiauthor text of wide scope with many contributions relevant to medical genetics.

Buchanan A, Daniels N, Wikler D, Brock DW 2000 From chance to choice: genetics and justice. Cambridge, UK: Cambridge University Press

An acclaimed book, intellectually rigorous and wide-ranging, addressing issues related to our knowledge of the human genome.

Clarke A (ed) 1997 The genetic testing of children. Oxford, UK: Bios Scientific

A comprehensive multiauthor text dealing with this important subject.

Clothier Committee 1992 Report of the Committee on the Ethics of Gene Therapy. London, UK: HMSO

Recommendations of the committee chaired by Sir Cecil Clothier on the ethical aspects of somatic cell and germline gene therapy.

Collins FS 1999 Shattuck lecture—medical and societal consequences of the human genome project. N Engl J Med 341:28–37

A contemporary overview of the Human Genome Project with emphasis on its possible ethical and social implications.

Harper PS, Clarke AJ 1997 Genetics society and clinical practice. Oxford, UK: Bios Scientific

A thoughtful account of the important ethical and social aspects of recent developments in clinical genetics.

Human Genetics Commission 2002 Inside information: balancing interests in the use of personal genetic data. London, UK: Department of Health

A detailed working party report by the Human Genetics Commission covering the use and abuse of personal genetic information.

Jonsen AR, Siegler M, Winslade WJ 1992 Clinical ethics: a practical approach to ethical decisions in clinical medicine, 3rd edn. New York: McGraw-Hill

The key reference that outlines the Jonsen framework for decision-making in clinical ethics.

Knoppers BM 1999 Status, sale and patenting of human genetic material: an international survey. Nat Genet 22:23–26

An article written in the light of a landmark legal and social policy document, the 'Directive on the Legal Protection of Biotechnology Inventions', from the European Parliament, 1998.

Knoppers BM, Chadwick R 2005 Human genetic research: emerging trends in ethics. Nat Rev Genet 6:75–79

An overview of current international policies on gene patenting.

McInnis MG 1999 The assent of a nation: genetics and Iceland. Clin Genet 55:234–239

A critical review of the complex ethical issues raised by the decision of the Icelandic government to collaborate in genetic research with a biotechnology company.

Nuffield Council on Bioethics 1993 Genetic screening: ethical issues. London, UK: Nuffield Council on Bioethics

A very helpful document for professional guidance.

Nuffield Council on Bioethics 1998 Mental disorders and genetics: the ethical context. London, UK: Nuffield Council on Bioethics

A further detailed document dealing with genetic issues in the context of mental health.

Pokorski RJ 1997 Insurance underwriting in the genetic era. Am J Hum Genet 60:205–216

A detailed account of the issues surrounding the use of genetic tests by the insurance industry.

Royal College of Physicians, Royal College of Pathologists, British Society of Human Genetics: Consent and Confidentiality in Genetic Practice 2006 Guidance on genetic testing and sharing genetic information. Report of the Joint Committee on Medical Genetics. London, UK: RCP, RCPath, BSHG

A detailed working party report that considers confidentiality issues, especially in the context of the Human Tissue Act 2004.

ELEMENTS

1 Ethical considerations impinge on almost every aspect of clinical genetics. In a wider context, developments in molecular genetics have important ethical implications for society at large.

2 Particularly difficult problems in clinical genetics include prenatal diagnosis, predictive testing in childhood, genetic testing in the extended family, confidentiality, consent, privacy, and disclosure of information.

3 Possible applications of molecular genetics, of ethical importance on a wider scale, include population screening, the storage of large amounts of genetic information in databases, the use of genetic test results by the insurance industry, gene patenting, and gene therapy.

4 There are no easy or correct solutions for many of the difficult ethical problems that arise in medical genetics. Guidelines, codes of practice, and sometimes regulations have an important role in establishing and maintaining standards, and preserving respect for the individual, the family, and wider societal needs.

APPENDIX

Websites and Clinical Databases

The rate of generation of information about human, medical, and clinical genetics means that access to current information is vital to both the student and the doctor, particularly as patients and families often come to the clinic armed with the same information!

There are several general websites that students will find useful as entry points, with a wealth of links to other sites for 'surfing'. A number of educational websites is now available; many include animated diagrams to assist the student.

Clinical geneticists regularly use a number of expert databases to assist in the diagnosis of genetic disorders and diseases, some of which are listed. Other specialized websites include mutation databases, information on nucleotide and protein sequences and current projects such as HapMap (p. 148).

Last, students may find it of interest to look at the professional societies' websites as they contain many useful links.

General Genetic Websites

Online Mendelian Inheritance in Man (OMIM)
http://www.ncbi.nlm.nih.gov/omim/
Online access to McKusick's catalogue, an invaluable resource for clinical genetic information with a wealth of links to many other resources.

Genetic Alliance UK
http://www.geneticalliance.org.uk/
Website for alliance of organizations supporting people affected with genetic disorders.

Gene tests
http://www.genetests.org/
Includes useful reviews of genetic disorders.

Orphanet
http://www.orpha.net/
A website with information about rare diseases, including many genetic disorders.

Human Genome Websites

Policy, Legal, and Ethical Issues in Genetic Research
http://www.nhgri.nih.gov./PolicyEthics/

Ensembl Genome Browser
http://www.ensembl.org/
Joint project between the European Bioinformatics Society and the Wellcome Trust Sanger Institute to provide annotated eukaryotic genomes.

UCSC Genome Bioinformatics
http://genome.ucsc.edu/
University of California at Santa Cruz genome browser.

Human Genome Organization
http://www.hugo-international.org/
The website for HUGO, the Human Genome Organization, which was set up as a "U.N. for the human genome".

International HapMap Project
http://www.hapmap.org/
The website of the project to map common DNA variants.

1000 Genomes Project
http://1000genomes.org/
A deep catalog of human genetic variation.

Molecular Genetics Websites

Human Gene Mutation Database
http://www.hgmd.cf.ac.uk/ac/index.php
A database of the reported mutations in human genes.

BROAD Institute
http://www.broad.mit.edu/
Human gene map, sequencing, and software programs.

Mammalian Genetics Unit and Mouse Genome Centre
http://www.mgu.har.mrc.ac.uk/
Mouse genome site.

Drosophila melanogaster Genome Database
http://flybase.org/
A comprehensive database for information on the genetics and molecular biology of D. melanogaster, including the genome sequence.

***Caenorhabditis elegans* Genetics and Genomics**
http://elegans.swmed.edu/genome.shtml
C. elegans *genome project information.*

Yeast Genome Project
http://mips.gsf.de/genre/proj/yeast/index.jsp
Yeast genome project information.

Cytogenetics Websites

Decipher Website
http://www.decipher.ac.uk/
A database of submicroscopic chromosome imbalance that includes phenotypic data.

Educational Human Genetics Websites

The National Genetics Education and Development Centre
http://www.geneticseducation.nhs.uk/
Supporting education in genetics and genomics for health.

Dolan DNA Learning Center at Cold Spring Harbor Laboratory
http://www.dnalc.org/
Information about genes in education.

University of Kansas Medical Center
http://www.kumc.edu/gec/
For educators interested in human genetics and the Human Genome Project.

Human Genetics Societies

American Society of Human Genetics
http://www.ashg.org/

British Society for Human Genetics
http://www.bshg.org.uk/

European Society of Human Genetics
http://www.eshg.org/

Human Genetics Society of Australasia
http://www.hgsa.com.au/

Clinical Databases

London Medical Databases
http://www.lmdatabases.com/
Includes the Winter-Baraitser Dysmorphology Database, the Baraitser-Winter Neurogenetics Database, and the London Ophthalmic Genetics Database.

Glossary

A. Abbreviation for adenine.
Acentric. Lacking a centromere.
Acetylation. The introduction of an acetyl group into a molecule; often used by the body to help eliminate substances by the liver.
Acrocentric. Term used to describe a chromosome where the centromere is near one end and the short arm usually consists of satellite material.
Acute-phase proteins. Proteins involved in innate immunity produced in reaction to infection, including C-reactive protein, mannose-binding protein, and serum amyloid P component.
Adenine. A purine base in DNA and RNA.
Adenomatous polyposis coli (APC). See Familial adenomatous polyposis.
AIDS. Acquired immune deficiency syndrome.
Allele (= allelomorph). Alternative form of a gene found at the same locus on homologous chromosomes.
Allograft. A tissue graft between non-identical individuals.
Allotypes. Genetically determined variants of antibodies.
Alpha (α)-thalassemia. Inherited disorder of hemoglobin involving underproduction of the α-globin chains occurring most commonly in people from South-East Asia.
Alternative pathway. One of the two pathways of the activation of complement that, in this instance, involves cell membranes of microorganisms.
Alu repeat. Short repeated DNA sequences that appear to have homology with transposable elements in other organisms.
Am. The group of genetic variants associated with the immunoglobulin (Ig) A heavy chain.
Amino acid. An organic compound containing both carboxyl (–COOH) and amino ($-NH_2$) groups.
Amniocentesis. Procedure for obtaining amniotic fluid and cells for prenatal diagnosis.
Amorph. A mutation that leads to complete loss of function.
Anaphase. The stage of cell division when the chromosomes leave the equatorial plate and migrate to opposite poles of the spindle.
Anaphase lag. Loss of a chromosome as it moves to the pole of the cell during anaphase; can lead to monosomy.
Aneuploid. A chromosome number that is not an exact multiple of the haploid number (i.e., $2N - 1$ or $2N + 1$, where N is the haploid number of chromosomes).
Anterior information. Information previously known that leads to the prior probability.
Antibody (= immunoglobulin). A serum protein formed in response to an antigenic stimulus and reacts specifically with that antigen.
Anticipation. The tendency for some autosomal dominant diseases to manifest at an earlier age and/or to increase in severity with each succeeding generation.
Anticodon. The complementary triplet of the transfer RNA (tRNA) molecule that binds to it with a particular amino acid.
Antigen. A substance that elicits the synthesis of antibody with which it specifically reacts.
Antigen binding fragment (Fab). The fragment of the antibody molecule produced by papain digestion responsible for antigen binding.
Antiparallel. Opposite orientation of the two strands of a DNA duplex; one runs in the 3′ to 5′ direction, the other in the 5′ to 3′ direction.
Antisense oligonucleotide. A short oligonucleotide synthesized to bind to a particular RNA or DNA sequence to block its expression.
Antisense strand. The template strand of DNA.
Apical ectodermal ridge. Area of ectoderm in the developing limb bud that produces growth factors.
Apolipoproteins. Proteins involved in lipid transportation in the circulation.
Apoptosis. Programmed involution or cell death of a developing tissue or organ of the body.
Artificial insemination by donor (AID). Use of semen from a male donor as a reproductive option for couples at high risk of transmitting a genetic disorder.
Ascertainment. The finding and selection of families with a hereditary disorder.
Association. The occurrence of a particular allele in a group of patients more often than can be accounted for by chance.
Assortative mating (= non-random mating). The preferential selection of a spouse with a particular phenotype.
Atherosclerosis. Fatty degenerative plaque that accumulates in the intimal wall of blood vessels.
Autoimmune diseases. Diseases believed to be caused by the body not recognizing its own antigens.
Autonomous replication sequences. DNA sequences that are necessary for accurate replication within yeast.
Autoradiography. Detection of radioactively labeled molecules on an X-ray film.
Autosomal dominant. A gene on one of the non-sex chromosomes that manifests in the heterozygous state.
Autosomal inheritance. The pattern of inheritance shown by a disorder or trait determined by a gene on one of the non-sex chromosomes.
Autosomal recessive. A gene located on one of the non-sex chromosomes that manifests in the homozygous state.
Autosome. Any of the 22 non-sex chromosomes.
Autozygosity. Homozygosity as a result of identity by descent from a common ancestor.
Azoospermia. Absence of sperm in semen.
B lymphocytes. Antibody-producing lymphocytes involved in humoral immunity.
Bacterial artificial chromosome (BAC). An artificial chromosome created from modification of the fertility factor of plasmids that allows incorporation of up to 330 kb of foreign DNA.
Bacteriophage (= phage). A virus that infects bacteria.
Balanced polymorphism. Two different genetic variants that are stably present in a population (i.e., selective advantages and disadvantages cancel each other out).
Balanced translocation. See Reciprocal translocation.
Bare lymphocyte syndrome. A rare autosomal recessive form of severe combined immunodeficiency resulting from absence of the class II molecules of the major histocompatibility complex.
Barr body. The condensation of the inactive X chromosome seen in the nucleus of certain types of cells from females. See Sex chromatin.
Base. Short for the nitrogenous bases in nucleic acid molecules (A, adenine; T, thymine; U, uracil; C, cytosine, G, guanine).
Base pair (bp). A pair of complementary bases in DNA (A with T, G with C).

Bayes' theorem. Combining the prior and conditional probabilities of certain events or the results of specific tests to give a joint probability to derive the posterior or relative probability.

Bence-Jones protein. The antibody of a single species produced in large amounts by a person with multiple myeloma, a tumor of antibody-producing plasma cells.

Beta (β)-thalassemia. Inherited disorder of hemoglobin involving underproduction of the β-globin chain, occurring most commonly in people from the Mediterranean region and Indian subcontinent.

Bias of ascertainment. An artifact that must be taken into account in family studies when looking at segregation ratios, caused by families coming to attention because they have affected individual(s).

Biochemical disorder. An inherited disorder involving a metabolic pathway (i.e., an inborn error of metabolism).

Biological or genetic determinism. The premise that our genetic makeup is the only factor determining all aspects of our health and disease.

Biosynthesis. Use of recombinant DNA techniques to produce molecules of biological and medical importance in the laboratory or commercially.

Bipolar illness. Affective manic–depressive illness.

Bivalent. A pair of synapsed homologous chromosomes.

Blastocyst. Early embryo consisting of embryoblast and trophoblast.

Blastomere. A single cell of the early fertilized conceptus.

Blood chimera. A mixture of cells of different genetic origin present in twins as a result of an exchange of cells via the placenta between non-identical twins in utero.

Boundary elements. Short sequences of DNA, usually from 500 bp to 3 kb in size, that block or inhibit the influence of regulatory elements of adjacent genes.

Break-point cluster (bcr). Region of chromosome 22 involved in the translocation seen in the majority of people with chronic myeloid leukemia.

C. Abbreviation for cytosine.

CAAT box. A conserved, non-coding, so-called promoter sequence about 80 bp upstream from the start of transcription.

Cancer family syndrome. Clustering in certain families of particular types of cancers, in which it has been proposed that the different types of malignancy could be due to a single dominant gene, specifically Lynch type II.

Cancer genetics. The study of the genetic causes of cancer.

Candidate gene. A gene whose function or location suggests that it is likely to be responsible for a particular genetic disease or disorder.

5′ Cap. Modification of the nascent mRNA by the addition of a methylated guanine nucleotide to the 5′ end of the molecule by an unusual 5′ to 5′ triphosphate linkage.

CA repeat. A short dinucleotide sequence present as tandem repeats at multiple sites in the human genome, producing microsatellite polymorphisms.

Carrier. Person heterozygous for a recessive gene; male or female for autosomal genes or female for X-linked genes.

Cascade screening. Identification within a family of carriers for an autosomal recessive disorder or people with an autosomal dominant gene after ascertainment of an index case.

Cell-mediated immunity. Immunity that involves the T lymphocytes in fighting intracellular infection; is also involved in transplantation rejection and delayed hypersensitivity.

Cellular oncogene. See Proto-oncogene.

Centimorgan (cM). Unit used to measure map distances, equivalent to a 1% chance of recombination (crossing over).

Central dogma. The concept that genetic information is usually transmitted only from DNA to RNA to protein.

Centric fusion. The fusion of the centromeres of two acrocentric chromosomes to form a robertsonian translocation.

Centriole. The cellular structure from which microtubules radiate in the mitotic spindle involved in the separation of chromosomes in mitosis.

Centromere (= kinetochore). The point at which the two chromatids of a chromosome are joined, and the region of the chromosome that becomes attached to the spindle during cell division.

Chemotaxis. The attraction of phagocytes to the site of infection by components of complement.

Chiasmata. Crossovers between chromosomes in meiosis.

Chimera. An individual composed of two populations of cells with different genotypes.

Chorion. Layer of cells covering a fertilized ovum, some of which (the chorion frondosum) will later form the placenta.

Chorionic villus sampling. Procedure using ultrasonographic guidance to obtain chorionic villi from the chorion frondosum for prenatal diagnosis.

Chromatid. During cell division, each chromosome divides longitudinally into two strands, or chromatids, which are held together by the centromere.

Chromatin. The tertiary coiling of the nucleosomes of the chromosomes with associated proteins.

Chromatin fiber FISH. Use of extended chromatin or DNA fibers with fluorescent in situ hybridization (FISH) to order physically DNA clones or sequences.

Chromosomal analysis. The process of counting and analyzing the banding pattern of the chromosomes of an individual.

Chromosomal fragments. Acentric chromosomes that can arise as a result of segregation of a paracentric inversion and that are usually incapable of replication.

Chromosome instability. The presence of breaks and gaps in the chromosomes from people with a number of disorders associated with an increased risk of neoplasia.

Chromosome mapping. Assigning a gene or DNA sequence to a specific chromosome or a particular region of a chromosome.

Chromosome-mediated gene transfer. The technique of transferring chromosomes or parts of chromosomes to somatic cell hybrids to enable more detailed chromosome mapping.

Chromosome painting. The hybridization in situ of fluorescent-labeled probes to a chromosome preparation to allow identification of a particular chromosome(s).

Chromosome walking. Using an ordered assembly of clones to extend from a known start point.

Chromosomes. Thread-like, darkly staining bodies within the nucleus composed of DNA and chromatin that carry the genetic information.

***Cis*-acting.** Regulatory elements in the promoter region that act on genes on the same chromosome.

Class switching. The normal change in antibody class from IgM to IgG in the immune response.

Classic gene families. Multigene families that show a high degree of sequence homology.

Classic pathway. One of the two ways of activation of complement, in this instance involving antigen–antibody complexes.

Clone. A group of cells, all of which are derived from a single cell by repeated mitoses and all having the same genetic information.

Clone contigs. Assembly of clones that have been mapped and ordered to produce an overlapping array.

Cloning in silico. The use of a number of computer programs that can search genomic DNA sequence databases for sequence homology to known genes, as well as DNA sequences specific to all genes such as the conserved intron/exon splice junctions, promoter sequences, polyadenylation sites and stretches of open-reading frames (ORFs) to identify novel genes.

cM. Abbreviation for centimorgan.

Co-dominance. When both alleles are expressed in the heterozygote.

Codon. A sequence of three adjacent nucleotides that codes for one amino acid or chain termination.

Common cancers. The cancers that occur commonly in humans, such as bowel and breast cancer.

Common diseases. The diseases that occur commonly in humans (e.g., cancer, coronary artery disease, diabetes).

Community genetics. The branch of medical genetics concerned with screening and the prevention of genetic diseases on a population basis.

Comparative genomics. The identification of orthologous genes in different species.

Competent. Making bacterial cell membrane permeable to DNA by a variety of different methods, including exposure to certain salts or high voltage.

Complement. A series of at least 10 serum proteins in humans (and other vertebrates) that can be activated by either the 'classic' or the 'alternative' pathway and that interact in sequence to bring about the destruction of cellular antigens.

Complementary DNA (cDNA). DNA synthesized from mRNA by the enzyme reverse transcriptase.

Complementary strands. The specific pairing of the bases in the DNA of the purines adenine and guanine with thymine and cytosine.

Complete ascertainment. A term used in segregation analysis for a type of study that identifies all affected individuals in a population.

Compound heterozygote. An individual who is affected with an autosomal recessive disorder having two different mutations in homologous genes.

Concordance. When both members of a pair of twins exhibit the same trait, they are said to be concordant. If only one twin has the trait, the twins are said to be discordant.

Conditional knockout. A mutation that is expressed only under certain conditions (e.g., raised temperature).

Conditional probability. Observations or tests that can be used to modify prior probabilities using bayesian calculation in risk estimations.

Conditionally toxic or suicide gene. Genes that are introduced in gene therapy and that, under certain conditions or after the introduction of a certain substance, will kill the cell.

Confined placental mosaicism. The occurrence of a chromosomal abnormality in chorionic villus samples obtained for first-trimester prenatal diagnosis in which the fetus has a normal chromosomal complement.

Congenital. Any abnormality, whether genetic or not, that is present at birth.

Congenital hypertrophy of the retinal pigment epithelium (CHRPE). Abnormal retinal pigmentation that, when present in people at risk for familial adenomatous polyposis, is evidence of the heterozygous state.

Conjugation. A chemical process in which two molecules are joined, often used to describe the process by which certain drugs or chemicals can then be excreted by the body (e.g., acetylation of isoniazid by the liver).

Consanguineous marriage. A marriage between 'blood relatives'; that is, between people who have one or more ancestors in common, most frequently between first cousins.

Consanguinity. Relationship between blood relatives.

Consensus sequence. A GGGCGGG sequence promoter element to the 5′ side of genes in eukaryotes involved in the control of gene expression.

Conservative substitution. Single base-pair substitution that, although resulting in the replacement by a different amino acid, if chemically similar, has no functional effect.

Constant region. The portion of the light and heavy chains of antibodies in which the amino acid sequence is relatively constant from molecule to molecule.

Constitutional. Present in the fertilized gamete.

Constitutional heterozygosity. The presence in an individual at the time of conception of obligate heterozygosity at a locus when the parents are homozygous at that locus for different alleles.

Consultand. The person presenting for genetic advice.

Contigs. Contiguous or overlapping DNA clones.

Contiguous gene syndrome. Disorder resulting from deletion of adjacent genes.

Continuous trait. A trait, such as height, for which there is a range of observations or findings, in contrast to traits that are all or none (see Discontinuous trait), such as cleft lip and palate.

Control gene. A gene that can turn other genes on or off (i.e., regulate).

Cordocentesis. The procedure of obtaining fetal blood samples for prenatal diagnosis.

Corona radiata. Cellular layer surrounding the mature oocyte.

Cor pulmonale. Right-sided heart failure that can occur after serious lung disease, such as in people with cystic fibrosis.

Correlation. Statistical measure of the degree of association or resemblance between two parameters.

Cosmid. A plasmid that has had the maximum DNA removed to allow the largest possible insert for cloning but still has the DNA sequences necessary for in vitro packaging into an infective phage particle.

Co-twins. Both members of a twin pair, whether dizygotic or monozygotic.

Counselee. Person receiving genetic counseling.

Coupling. When a certain allele at a particular locus is on the same chromosome with a specific allele at a closely linked locus.

CpG dinucleotides. The occurrence of the nucleotides cytosine and guanine together in genomic DNA, which is frequently methylated and associated with spontaneous deamination of cytosine converting it to thymine as a mechanism of mutation.

CpG islands. Clusters of unmethylated CpGs occur near the transcription sites of many genes.

Crossover (= recombination). The exchange of genetic material between homologous chromosomes in meiosis.

Cross-reacting material (CRM). Immunologically detected protein or enzyme that is functionally inactive.

Cryptic splice site. A mutation in a gene leading to the creation of the sequence of a splice site that results in abnormal splicing of the mRNA.

Cystic fibrosis transmembrane conductance regulator (CFTR). The gene product of the cystic fibrosis gene responsible for chloride transport and mucin secretion.

Cytogenetics. The branch of genetics concerned principally with the study of chromosomes.

Cytokinesis. Division of the cytoplasm to form two daughter cells in meiosis and mitosis.

Cytoplasm. The ground substance of the cell, in which are situated the nucleus, endoplasmic reticulum, or mitochondria.

Cytoplasmic inheritance. See Mitochondrial inheritance.

Cytosine. A pyrimidine base in DNA and RNA.

Cytosol. The semi-soluble contents of the cytoplasm.

Cytotoxic T cells. A subclass of T lymphocytes sensitized to destroy cells bearing certain antigens.

Cytotoxic T lymphocytes (= killer). A group of T cells that specifically kill foreign or virus-infected vertebrate cells.

Daltonism. A term given formerly to X-linked inheritance, after John Dalton, who noted this pattern of inheritance in color blindness.

Deformation. A birth defect that results from an abnormal mechanical force which distorts an otherwise normal structure.

Degeneracy. Certain amino acids being coded for by more than one triplet codon of the genetic code.

Deleted in colorectal carcinoma (DCC). A region on the long arm of chromosome 18 often found to be deleted in colorectal carcinomas.

Deletion. A type of chromosomal aberration or mutation at the DNA level in which there is loss of part of a chromosome or of one or more nucleotides.
Delta–beta (δβ)-thalassemia. A form of thalassemia in which there is reduced production of both the δ- and β-globin chains.
De novo. Literally 'from new', as opposed to inherited.
Deoxyribonucleic acid. See DNA.
Desert hedgehog. One of three mammalian homologs of the segment polarity hedgehog genes.
Dicentric. Possessing two centromeres.
Dictyotene. The stage in meiosis I in which primary oocytes are arrested in females until the time of ovulation.
Digenic inheritance. An inheritance mechanism resulting from the interaction of two non-homologous genes.
Diploid. The condition in which the cell contains two sets of chromosomes. Normal state of somatic cells in humans where the diploid number (2N) is 46.
Discontinuous trait. A trait that is all or none (e.g., cleft lip and palate), in contrast to continuous traits such as height.
Discordant. Differing phenotypic features between individuals, classically used in twin pairs.
Disomy. The normal state of an individual having two homologous chromosomes.
Dispermic chimera. Two separate sperm fertilize two separate ova and the resulting two zygotes fuse to form one embryo.
Dispermy. Fertilization of an oocyte by two sperm.
Disruption. An abnormal structure of an organ or tissue as a result of external factors disturbing the normal developmental process.
Diversity region. DNA sequences coding for the segments of the hypervariable regions of antibodies.
Dizygotic twins (= fraternal). Type of twins produced by fertilization of two ova by two sperm.
DNA (= deoxyribonucleic acid). The nucleic acid in chromosomes in which genetic information is coded.
DNA chip. DNA microarrays that, with the appropriate computerized software allow rapid, automated, high-throughput DNA sequencing and mutation detection.
DNA fingerprint. Pattern of hypervariable tandem DNA repeats of a core sequence that is unique to an individual.
DNA haplotype. The pattern of DNA sequence polymorphisms flanking a DNA sequence or gene of interest.
DNA library. A collection of recombinant DNA molecules from a particular source, such as genomic or cDNA.
DNA ligase. An enzyme that catalyzes the formation of a phosphodiester bond between a 3′-hydroxyl and a 5′-phosphate group in DNA, thereby joining two DNA fragments.
DNA mapping. The physical relationships of flanking DNA sequence, polymorphisms, and the detailed structure of a gene.
DNA polymorphisms. Inherited variation in the nucleotide sequence, usually of non-coding DNA.
DNA probes. A DNA sequence that is labeled, usually radioactively or fluorescently, and used to identify a gene or DNA sequence (e.g., a cDNA or genomic probe).
DNA repair. DNA damaged through a variety of mechanisms can be removed and repaired by a complex set of processes.
DNA replication. The process of copying the nucleotide sequence of the genome from one generation to the next.
DNA sequence amplification. See Polymerase chain reaction.
DNA sequence variants. See DNA polymorphisms.
DNA sequencing. Analysis of the nucleotide sequence of a gene or DNA fragment.
Dominant. A trait expressed in individuals who are heterozygous for a particular allele.
Dominant-negative mutation. A mutant allele in the heterozygous state that results in the loss of activity or function of its mutant gene product as well as interfering with the function of the normal gene product of the corresponding allele.
Dosage compensation. The phenomenon in women who have two copies of genes on the X chromosome having the same level of the products of those genes as males who have a single X chromosome.
Dosimetry. The measurement of radiation exposure.
Double heterozygote. An individual who is heterozygous at two different loci.
Double-minute chromosomes. Amplified sequences of DNA in tumor cells that can occur as small extra chromosomes, as in neuroblastoma.
Drift (= random genetic drift). Fluctuations in gene frequencies that tend to occur in small isolated populations.
Dynamic mutation. See Unstable mutation.
Dysmorphology. The study of the definition, recognition, and etiology of multiple malformation syndromes.
Dysplasia. An abnormal organization of cells into tissue.
Dystrophin. The product of the Duchenne muscular dystrophy gene.
Ecogenetics. The study of genetically determined differences in susceptibility to the action of physical, chemical and infectious agents in the environment.
Em. The group of genetic variants of the IgE heavy chain of immunoglobulins.
Embryoblast. Cell layer of the blastocyst which forms the embryo.
Embryonic stem cells. A cell in the early embryo that is totipotent in terms of cellular fate.
Empiric risks. Advice given in recurrence risk counseling for multifactorially determined disorders based on observation and experience, in which the inherited contribution is due to a number of genes (i.e., polygenic).
Endoplasmic reticulum. A system of minute tubules within the cell involved in the biosynthesis of macromolecules.
Endoreduplication. Duplication of a haploid sperm chromosome set.
Enhancer. DNA sequence that increases transcription of a related gene.
Enzyme. A protein that acts as a catalyst in biological systems.
Epidermal growth factor (EGF). A growth factor that stimulates a variety of cell types including epidermal cells.
Epigenetic. Heritable changes to gene expression that are *not* due to differences in the genetic code.
Epistasis. Interaction between non-allelic genes.
Erythroblastosis fetalis. See Hemolytic disease of the newborn.
Essential hypertension. Increased blood pressure for which there is no recognized primary cause.
Euchromatin. Genetically active regions of the chromosomes.
Eugenics. The 'science' that promotes the improvement of the hereditary qualities of a race or a species.
Eukaryote. Higher organism with a well defined nucleus.
Exon (= expressed sequence). Region of a gene that is not excised during transcription, forming part of the mature mRNA, and therefore specifying part of the primary structure of the gene product.
Exon trapping. A process by which a recombinant DNA vector that contains the DNA sequences of the splice-site junctions is used to clone coding sequences or exons.
Expansion. Refers to the increase in the number of triplet repeat sequences in the various disorders due to dynamic or unstable mutations.
Expressed sequence tags. Sequence-specific primers from cDNA clones designed to identify sequences of expressed genes in the genome.
Expressivity. Variation in the severity of the phenotypic features of a particular gene.
Extinguished. Loss of one allelic variant at a locus resulting from random genetic drift.
Extrinsic malformation. Term previously used for disruption.

Fab. The two antigen-binding fragments of an antibody molecule produced by digestion with the proteolytic enzyme papain.

False negative. Affected cases missed by a diagnostic or screening test.

False positive. Unaffected cases incorrectly diagnosed as affected by a screening or diagnostic test.

Familial adenomatous polyposis. A dominantly inherited cancer-predisposing syndrome characterized by the presence of a large number of polyps of the large bowel with a high risk of developing malignant changes.

Familial cancer-predisposing syndrome. One of a number of syndromes in which people are at risk of developing one or more types of cancer.

Favism. A hemolytic crisis resulting from glucose 6-phosphate dehydrogenase (G6PD) deficiency occurring after eating fava beans.

Fc. The complement binding fragment of an antibody molecule produced by digestion with the proteolytic enzyme papain.

Fetoscopy. Procedure used to visualize the fetus and often to take skin and/or blood samples from the fetus for prenatal diagnosis.

Fetus. Unborn infant during the final stage of in utero development, usually from 12 weeks' gestation to term.

Filial. Relating to offspring.

First-degree relatives. Closest relatives (i.e., parents, offspring, siblings), sharing on average 50% of their genes.

Fitness (= biological fitness). The number of offspring who reach reproductive age.

Five-prime (5′) end. The end of a DNA or RNA strand with a free 5′ phosphate group.

Fixed. The establishment of a single allelic variant at a locus from random genetic drift.

Fixed mutation. See Stable mutation.

Flanking DNA. Nucleotide sequence adjacent to the DNA sequence being considered.

Flanking markers. Polymorphic markers that are located adjacent to a gene or DNA sequence of interest.

Flow cytometry. See Fluorescence-activated cell sorting.

Flow karyotype. A distribution histogram of chromosome size obtained using a fluorescence-activated cell sorter.

Fluorescence-activated cell sorting (FACS). A technique in which chromosomes are stained with a fluorescent dye that binds selectively to DNA; the differences in fluorescence of the various chromosomes allow them to be physically separated by a special laser.

Fluorescent in situ hybridization (FISH). Use of a single-stranded DNA sequence with a fluorescent label to hybridize with its complementary target sequence in the chromosomes, allowing it to be visualized under ultraviolet light.

Foreign DNA. A source of DNA incorporated into a vector in producing recombinant DNA molecules.

Founder effect. Certain genetic disorders can be relatively common in particular populations through all individuals being descended from a relatively small number of ancestors, one or a few of whom had a particular disorder.

Fragile site. A non-staining gap in a chromatid where breakage is liable to occur.

Frameshift mutations. Mutations, such as insertions or deletions, that change the reading frame of the codon triplets.

Framework map. A set of markers distributed at defined approximately evenly spaced intervals along the chromosomes in the human genome.

Framework region. Parts of the variable regions of antibodies that are not hypervariable.

Fraternal twins. Non-identical twins.

Freemartin. A chromosomally female twin calf with ambiguous genitalia resulting from gonadal chimerism.

Frequency. The number of times an event occurs in a period (e.g., 1,000 cases per year).

Full ascertainment. See Complete ascertainment.

Functional cloning. Identification of a gene through its function (e.g., isolation of cDNAs expressed in a particular tissue in which a disease or disorder is manifest).

Functional genomics. The normal pattern of expression of genes in development and differentiation and the function of their protein products in normal development as well as their dysfunction in inherited disorders.

Fusion polypeptide. Genes that are physically near to one another and have DNA sequence homology can undergo a crossover, leading to formation of a protein that has an amino acid sequence derived from both of the genes involved.

G. Abbreviation for the nucleotide guanine.

Gain-of-function. Mutations that, in the heterozygote, result in new functions.

Gap mutant. Developmental genes identified in *Drosophila* that delete groups of adjacent segments.

Gastrulation. The formation of the bi- then tri-laminar disc of the inner cell mass that becomes the early embryo.

Gene. A part of the DNA molecule of a chromosome that directs the synthesis of a specific polypeptide chain.

Gene amplification. Process in tumor cells of the production of multiple copies of certain genes, the visible evidence of which are homogeneously staining regions and double-minute chromosomes.

Gene flow. Differences in allele frequencies between populations that reflect migration or contact between them.

Gene superfamilies. Multigene families that have limited sequence homology but are functionally related.

Gene targeting. The introduction of specific mutations into genes by homologous recombination in embryonic stem cells.

Gene therapy. Treatment of inherited disease by addition, insertion or replacement of a normal gene or genes.

Genetic code. The triplets of DNA nucleotides that code for the various amino acids of proteins.

Genetic counseling. The process of providing information about a genetic disorder that includes details about the diagnosis, cause, risk of recurrence, and options available for prevention.

Genetic heterogeneity. The phenomenon that a disorder can be caused by different allelic or non-allelic mutations.

Genetic isolates. Groups isolated for geographical, religious, or ethnic reasons that often show differences in allele frequencies.

Genetic load. The total of all kinds of harmful alleles in a population.

Genetic register. A list of families and individuals who are either affected by or at risk of developing a serious hereditary disorder.

Genetic susceptibility. An inherited predisposition to a disease or disorder that is not due to a single-gene cause and is usually the result of a complex interaction of the effects of multiple different genes (i.e., polygenic inheritance).

Genocopy. The same phenotype but from different genetic causes.

Genome. All the genes carried by a cell.

Genomic DNA. The total DNA content of the chromosomes.

Genomic imprinting. Differing expression of genetic material dependent on the sex of the transmitting parent.

Genotype. The genetic constitution of an individual.

Genotype–phenotype correlation. Correlation of certain mutations with particular phenotypic features.

Germ cells. The cells of the body that transmit genetic information to the next generation.

Germline gene therapy. The alteration or insertion of genetic material in the gametes.

Germline mosaicism. The presence in the germline or gonadal tissue of two populations of cells that differ genetically.

Germline mutation. A mutation in a gamete.

Gestational diabetes. Onset of an abnormal glucose tolerance in pregnancy that usually reverts to normal after delivery.

Gm. Genetic variants of the heavy chain of IgG immunoglobulins.

Goldberg–Hogness box. See CAAT box.

Gonad dose. Radiation dosimetry term that describes the radiation exposure of an individual to a particular radiological investigation or exposure.

Gonadal mosaicism. See Germline mosaicism.

Gray (Gy). Equivalent to 100 rad.

Growth factor. A substance that must be present in culture medium to permit cell multiplication, or involved in promoting the growth of certain cell types, tissues, or parts of the body in development (e.g., fibroblast growth factor).

Growth factor receptors. Receptors on the surfaces of cells for a growth factor.

Guanine. A purine base in DNA and RNA.

Haploid. The condition in which the cell contains one set of chromosomes (i.e., 23). This is the chromosome number in a normal gamete.

Haploinsufficiency. Mutations in the heterozygous state that result in half normal levels of the gene product leading to phenotypic effects (i.e., are sensitive to gene dosage).

Haplotype. Conventionally used to refer to the particular alleles present at the four genes of the HLA complex on chromosome 6. The term is also used to describe DNA sequence variants on a particular chromosome adjacent to or closely flanking a locus of interest.

Hardy-Weinberg equilibrium. The maintenance of allele frequencies in a population with random mating and absence of selection.

Hardy-Weinberg formula. A simple binomial equation in population genetics that can be used to determine the frequency of the different genotypes from one of the phenotypes.

Hardy-Weinberg principle. The relative proportions of the different genotypes remain constant from one generation to the next.

Hb Barts. The tetramer of γ-globin chains found in the severe form of α-thalassemia, which causes hydrops fetalis.

Hb H. Tetramer of the β-globin chains found in the less severe form of thalassemia.

Hedgehog. A group of morphogens produced by segment polarity genes.

Helix-loop-helix. DNA-binding motif that controls gene expression.

Helix-turn-helix proteins. Proteins made up of two a helices connected by a short chain of amino acids that make up a 'turn.'

Helper lymphocytes. A subclass of T lymphocytes necessary for the production of antibodies by B lymphocytes.

Helper virus. A retroviral provirus engineered to remove all but the sequences necessary to produce copies of the viral RNA sequences along with the sequences necessary for packaging of the viral genomic RNA in retrovirus-mediated gene therapy.

Heme. The iron-containing group of hemoglobin.

Hemizygous. A term used when describing the genotype of a male with regard to an X-linked trait, as males have only one set of X-linked genes.

Hemoglobinopathy. An inherited disorder of hemoglobin.

Hemolytic disease of the newborn. Anemia resulting from an antibody produced by an Rh-negative mother to the Rh-positive blood group of the fetus crossing the placenta and causing hemolysis. If this hemolytic process is severe, it can cause death of the fetus from heart failure because of the anemia, or what is known as hemolytic disease of the newborn.

Hereditary non-polyposis colorectal cancer (HNPCC). A form of familial cancer in which people are at risk of developing bowel cancer; not associated with a large number of polyps as in familial polyposis coli, in which the bowel cancer is usually proximal and right-sided.

Hereditary persistence of fetal Hb (HPFH). Persistence of the production of fetal hemoglobin into childhood and adult life.

Heritability. The proportion of the total variation of a character attributable to genetic as opposed to environmental factors.

Hermaphrodite. An individual with both male and female gonads, often in association with ambiguous external genitalia.

Heterochromatin. Genetically inert or inactive regions of the chromosomes.

Heterogeneity. The phenomenon of there being more than a single cause for what appears to be a single entity. See Genetic heterogeneity.

Heteromorphism. An inherited structural polymorphism of a chromosome.

Heteroplasmy. The mitochondria of an individual consisting of more than one population.

Heteropyknotic. Condensed darkly staining chromosomal material (e.g., the inactivated X chromosome in females).

Heterozygote (= carrier). An individual who possesses two different alleles at one particular locus on a pair of homologous chromosomes.

Heterozygote advantage. An increase in biological fitness seen in unaffected heterozygotes compared with unaffected homozygotes (e.g., sickle cell trait and resistance to infection by the malarial parasite).

Heterozygous. The state of having different alleles at a locus on homologous chromosomes.

High-resolution DNA mapping. Detailed physical mapping at the level of restriction site polymorphisms, expressed sequence tags, and so on.

Histocompatibility. Antigenic similarity of donor and recipient in organ transplantation.

Histone. Type of protein rich in lysine and arginine found in association with DNA in chromosomes.

HIV. Human immunodeficiency virus.

HLA (human leukocyte antigen). Antigens present on the cell surfaces of various tissues, including leukocytes.

HLA complex. The genes on chromosome 6 responsible for determining the cell-surface antigens important in organ transplantation.

Hogness box (= TATA box). A conserved, non-coding, so-called promoter sequence about 30 bp upstream from the start of transcription.

Holandric inheritance. The pattern of inheritance of genes on the Y chromosome; only males are affected and the trait is transmitted by affected males to their sons but to none of their daughters.

Homeobox. A stretch of approximately 180 bp conserved in different homeotic genes.

Homeotic gene. Genes that are involved in controlling the development of a region or compartment of an organism producing proteins or factors that regulate gene expression by binding particular DNA sequences.

Homogeneously staining regions (HSRs). Amplification of DNA sequences in tumor cells that can appear as extra or expanded areas of the chromosomes, which stain evenly.

Homograft. Graft between individuals of the same species but with different genotypes.

Homologous chromosomes. Chromosomes that pair during meiosis and contain identical loci.

Homologous recombination. The process by which a DNA sequence can be replaced by one with a similar sequence to determine the effect of changes in DNA sequence in the process of site-directed mutagenesis.

Homoplasmy. The mitochondria of an individual consisting of a single population.
Homozygote. An individual who possesses two identical alleles at one particular locus on a pair of homologous chromosomes.
Homozygous. The presence of two identical alleles at a particular locus on a pair of homologous chromosomes.
Hormone nuclear receptors. Intracellular receptors involved in the control of transcription.
Housekeeping genes. Genes that express proteins common to all cells (e.g., ribosomal, chromosomal, and cytoskeletal proteins).
HTF islands. Methylation-free clusters of CpG dinucleotides found near transcription initiation sites at the 5′ end of many eukaryotic genes; can be detected by cutting with the restriction enzyme *Hpa*II, producing *t*iny DNA *f*ragments.
Human Genome Project. A major international collaborative effort to map and sequence the entire human genome.
Humoral immunity. Immunity that is due to circulating antibodies in the blood and other bodily fluids.
Huntingtin. The protein product of the Huntington disease gene.
H-Y antigen. A histocompatibility antigen originally detected in the mouse and thought to be located on the Y chromosome.
Hydatidiform mole. An abnormal conceptus that consists of abnormal tissues. A complete mole contains no fetus, but can undergo malignant change and receives both sets of chromosomes from the father; a partial mole contains a chromosomally abnormal fetus with triploidy.
Hydrops fetalis. The most severe form of α-thalassemia, resulting in death of the fetus in utero from heart failure secondary to the severe anemia caused by hemolysis of the red cells.
Hypervariable DNA length polymorphisms. Different types of variation in DNA sequence that are highly polymorphic (e.g., variable number tandem repeats, mini- and microsatellites).
Hypervariable minisatellite DNA. Highly polymorphic DNA consisting of a 9- to 24-bp sequence often located near the telomeres.
Hypervariable region. Small regions present in the variable regions of the light and heavy chains of antibodies in which the majority of the variability in antibody sequence occurs.
Hypomorph. Loss-of-function mutations that result in either reduced activity or complete loss of the gene product from either reduced activity or to decreased stability of the gene product.
Identical twins. See Monozygotic twins.
Idiogram. An idealized representation of an object (e.g., an idiogram of a karyotype).
Immunoglobulin. See Antibody.
Immunoglobulin allotypes. Genetically determined variants of the various antibody classes (e.g., the Gm system associated with the heavy chain of IgG).
Immunoglobulin superfamily. The multigene families primarily involved in the immune response with structural and DNA sequence homology.
Immunological memory. The ability of the immune system to 'remember' previous exposure to a foreign antigen or infectious agents, leading to the enhanced secondary immune response on re-exposure.
Imprinting. The phenomenon of a gene or region of a chromosome showing different expression depending on the parent of origin.
Inborn error of metabolism. An inherited metabolic defect that results in deficient production or synthesis of an abnormal enzyme.
Incest. Union between first-degree relatives.
Incestuous. Description of a relationship between first-degree relatives.
Incidence. The rate at which new cases occur; for example, 2 in 1,000 births are affected by neural tube defects.
Incompatibility. A donor and host are incompatible if the latter rejects a graft from the former.
Incomplete ascertainment. A term used in segregation analysis to describe family studies in which complete ascertainment is not possible.
Index case. See Proband.
Index map. See Framework map.
Indian hedgehog. One of three mammalian homologs of the segment polarity hedgehog genes.
Inducer. Small molecule that interacts with a regulator protein and triggers gene transcription.
Informative. Variation in a marker system in a family that enables a gene or inherited disease to be followed in that family.
Innate immunity. A number of non-specific systems involved in immunity that do not require or involve prior contact with the infectious agent.
Insertion. Addition of chromosomal material or DNA sequence of one or more nucleotides within the genome.
Insertional mutagenesis. The introduction of mutations at specific sites to determine the effects of these changes.
In situ hybridization. Hybridization with a DNA probe carried out directly on a chromosome preparation or histological section.
Insulin-dependent diabetes mellitus. Diabetes requiring the use of insulin, usually of juvenile onset, now known as type 1 diabetes.
Intermediate inheritance. See Co-dominance.
Interphase. The stage between two successive cell divisions during which DNA replication occurs.
Interphase cytogenetics. The study of chromosomes during interphase, usually by FISH.
Intersex. An individual with external genitalia not clearly male or female.
Interval cancer. Developing cancer in the interval between repeated screening procedures.
Intra-cytoplasmic sperm injection (ICSI). A technique whereby a secondary spermatocyte or spermatozoon is removed from the testis and used to fertilize an egg.
Intrinsic malformation. A malformation resulting from an inherent abnormality in development.
Intron (= intervening sequence). Region of DNA that generates the part of precursor RNA that is spliced out during transcription and does not form mature mRNA and therefore does not specify the primary structure of the gene product.
Inv. Genetic variants of the κ light chains of immunoglobulins.
Inversion. A type of chromosomal aberration or mutation in which part of a chromosome or sequence of DNA is reversed in its order.
Inversion loop. The structure formed in meiosis I by a chromosome with either a paracentric or pericentric inversion.
In vitro. In the laboratory—literally 'in glass.'
In vivo. In the normal cell—literally 'in the living organism.'
Ionizing radiation. Electromagnetic waves of very short wavelength (X-rays and γ-rays), and high-energy particles (α particles, β particles, and neutrons).
Isochromosome. A type of chromosomal aberration in which one of the arms of a particular chromosome is duplicated because the centromere divides transversely and not longitudinally as normal during cell division. The two arms of an isochromosome are therefore of equal length and contain the same set of genes.
Isolate. A term used to describe a population or group of individuals that for religious, cultural, or geographical reasons has remained separate from other groups of people.
Isozymes. Enzymes that exist in multiple molecular forms which can be distinguished by biochemical methods.
Joining region. Short, conserved sequence of nucleotides involved in somatic recombinational events in the production of antibody diversity.

Joint probability. The product of the prior and conditional probability for two events.

Karyogram. Photomicrograph of chromosomes arranged in descending order of size.

Karyotype. The number, size, and shape of the chromosomes of an individual. Also used for the photomicrograph of an individual's chromosomes arranged in a standard manner.

Kb. Abbreviation for kilobase.

Killer lymphocytes. See Cytotoxic T lymphocytes.

Kilobase. 1,000 base pairs (bp).

Km. Genetic variants of the κ light chain of immunoglobulins.

Knockout mutation. Complete loss of function of a gene.

Lagging strand. One of the two strands created in DNA replication which is synthesized in the 3′ to 5′ direction made up of pieces synthesized in the 5′ to 3′ direction, which are then joined together as a continuous strand by the enzyme DNA ligase.

Law of addition. If two or more events are mutually exclusive then the probability that either one or the other will occur equals the sum of their individual probabilities.

Law of independent assortment. Members of different gene pairs segregate to offspring independently of one another.

Law of multiplication. If two or more events or outcomes are independent, the probability that both the first and the second will occur equals the product of their individual probabilities.

Law of segregation. Each individual possesses two genes for a particular characteristic, only one of which can be transmitted at any one time.

Law of uniformity. When two homozygotes with different alleles are crossed, all of the offspring in the F1 generation are identical and heterozygous (i.e., the characteristics do not blend and can reappear in later generations).

Leading strand. The synthesis of one of the DNA strands created in DNA replication; occurs in the 5′ to 3′ direction as a continuous process.

Lethal mutation. A mutation that leads to the premature death of an individual or organism.

Leucine zipper. A DNA-binding motif controlling gene expression.

Liability. A concept used in disorders that are determined multifactorially to take into account all possible causative factors.

Library. Set of cloned DNA fragments derived from a particular DNA source (e.g., a cDNA library from the transcript of particular tissue, or a genomic library.)

Ligase. Enzyme used to join DNA molecules.

Ligation. Formation of phosphodiester bonds to link two nucleic acid molecules.

Linkage. Two loci situated close together on the same chromosome, the alleles at which are usually transmitted together in meiosis in gamete formation.

Linkage disequilibrium. The occurrence together of two or more alleles at closely linked loci more frequently than would be expected by chance.

Liposomes. Artificially prepared cell-like structures in which one or more bimolecular layers of phospholipid enclose one or more aqueous compartments, which can include proteins.

Localization sequences. Certain short amino acid sequences in newly synthesized proteins that result in their transport to specific cellular locations, such as the nucleus, or their secretion.

Location score. Diagrammatic representation of likelihood ratios used in multipoint linkage analysis.

Locus. The site of a gene on a chromosome.

Locus control region (LCR). A region near the β-like globin genes involved in the timing and tissue specificity of their expression in development.

Locus heterogeneity. The phenomenon of a disorder being due to mutations in more than one gene or locus.

Lod score. A mathematical score of the relative likelihood of two loci being linked.

Long interspersed nuclear elements (LINEs). 50,000 to 100,000 copies of a DNA sequence of approximately 6,000 bp that occurs approximately once every 50 kb and encodes a reverse transcriptase.

Long terminal repeat (LTR). One of two long sections of double-stranded DNA synthesized by reverse transcriptase from the RNA of a retrovirus involved in regulating viral expression.

Loss of constitutional heterozygosity (LOCH). Loss of an allele inherited from a parent; frequently seen as evidence of a 'second hit' in tumorigenesis.

Loss-of-function mutation. Phenotypic features of a disorder due to reduced or absent activity of the gene.

Low-copy repeats (LCRs). Homologous sequences of DNA (more than 95% sequence identity) interspersed throughout the genome, predisposing to unequal recombination.

Low-resolution mapping. See Chromosome mapping.

Lymphokines. Glycoproteins released from T lymphocytes after contact with an antigen that act on other cells of the host immune system.

Lyonization. The process of inactivation of one of the X chromosomes in females, originally proposed by the geneticist Mary Lyon.

Major histocompatibility complex (MHC). A multigene locus that codes for the histocompatibility antigens involved in organ transplantation.

Malformation. A primary structural defect of an organ or part of an organ that results from an inherent abnormality in development.

Manifesting heterozygote or carrier. The phenomenon of a female carrier for an X-linked disorder having symptoms or signs of that disorder due to non-random X-inactivation (e.g., muscular weakness in a carrier for Duchenne muscular dystrophy).

Map unit. See Centimorgan.

Marker. A loose term used for a blood group, biochemical, or DNA polymorphism that, if shown to be linked to a disease locus of interest, can be used in presymptomatic diagnosis, determining carrier status and prenatal diagnosis.

Marker chromosome. A small, extra, structurally abnormal chromosome.

Maternal (matrilineal) inheritance. Transmission of a disorder through females.

Matrilineal inheritance. See Maternal inheritance.

Maximum likelihood method. The calculation of the Lod score for various values of the recombination fraction θ to determine the best estimate of the recombination fraction.

Meconium ileus. Blockage of the small bowel in the newborn period resulting from inspissated meconium, a presenting feature of cystic fibrosis.

Meiosis. The type of cell division that occurs in gamete formation with halving of the somatic number of chromosomes, with the result that each gamete is haploid.

Meiotic drive. Preferential transmission of one of a pair of alleles during meiosis.

Mendelian inheritance. Inheritance that follows the laws of segregation and independent assortment as proposed by Mendel.

Merlin. The protein product of the neurofibromatosis type II gene.

Messenger RNA (mRNA). A single-stranded molecule complementary to one of the strands of double-stranded DNA that is synthesized during transcription and transmits the genetic information in the DNA to the ribosomes for protein synthesis.

Metabolic disorder. An inherited disorder involving a biochemical pathway (i.e., an inborn error of metabolism).

Metacentric. Term used to describe chromosomes in which the centromere is central with both arms being of approximately equal length.

Metaphase. The stage of cell division at which the chromosomes line up on the equatorial plate and the nuclear membrane disappears.

Metaphase spreads. The preparation of chromosomes during the metaphase stage of mitosis in which they are condensed.

Methemoglobin. A hemoglobin molecule in which the iron is oxidized.

Methylation. The chemical imprint applied to certain DNA sequences in their passage through gametogenesis (applying to a small proportion of the human genome).

Microdeletion. A small chromosomal deletion detectable by high-resolution prometaphase chromosomal analysis or FISH.

Microdeletion syndrome. The pattern of abnormalities caused by a chromosome microdeletion.

Microsatellite DNA. Polymorphic variation in DNA sequences resulting from a variable number of tandem repeats of the dinucleotide CA, trinucleotides, or tetranucleotides.

Microsatellite instability. The alteration of the size of microsatellite polymorphic markers compared with the constitutional markers of an individual with hereditary non-polyposis colorectal cancer from mutations in the genes for the mismatch repair enzymes.

Microtubules. Long cylindrical tubes composed of bundles of small filaments that are an important part of the cytoskeleton.

Minichromosomes. Artificially constructed chromosomes containing centromeric and telomeric elements that allow replication of foreign DNA as a separate entity.

Minidystrophin. A modified dystrophin gene in which a large amount of the gene has been deleted, but that still has relatively normal function.

Minigene. A construct of a gene with the majority of the sequence removed that still remains functional (e.g., a dystrophin minigene).

Minisatellite. Polymorphic variation in DNA sequences from a variable number of tandem repeats of a short DNA sequence.

Mismatch repair genes. Genes for the DNA proof-reading enzymes, mutations in which cause hereditary non-polyposis colorectal cancer.

Missense mutation. A point mutation that results in a change in an amino acid-specifying codon.

Mitochondria. Minute structures situated within the cytoplasm that are concerned with cell respiration.

Mitochondrial DNA (mtDNA). Mitochondria possess their own genetic material that codes for enzymes involved in energy-yielding reactions, mutations of which are associated with certain diseases in humans.

Mitochondrial inheritance. Transmission of a mitochondrial trait exclusively through maternal relatives.

Mitosis. The type of cell division that occurs in replication of somatic cells.

Mixoploidy. The presence of cell lines with a different genetic constitution in an individual.

Modifier gene. Phenotypic variability from the consequence of interactions with other genes.

Monosomy. Loss of one member of a homologous pair of chromosomes so that there is one less than the diploid number of chromosomes (2N – 1).

Monozygotic twins (= identical). Type of twins derived from a single fertilized ovum.

Morphogen. A chemical or substance that determines a developmental process.

Morphogenesis. The evolution and development of form and shape.

Morula. The 12- to 16-cell stage of the early embryo at 3 days after conception.

Mosaic. An individual with two different cell lines derived from a single zygote.

mRNA splicing. The excision of intervening non-coding sequences or introns in the primary mRNA resulting in the non-contiguous exons being spliced together to form a shorter mature mRNA before its transportation to the ribosomes in the cytoplasm for translation.

Mucoviscidosis. An older term used for cystic fibrosis.

Multifactorial inheritance. Inheritance controlled by many genes with small additive effects (polygenic) plus the effects of the environment.

Multigene families. Genes with functional and/or sequence similarity.

Multiple alleles. The existence of more than two alleles at a particular locus in a population.

Multiple myeloma. A cancer of antibody-producing B cells that leads to the production of a single species of an antibody in large quantities.

Multipoint linkage analysis. Analysis of the segregation of alleles at a number of closely adjacent loci.

Mutagen. Natural or artificial ionizing radiation, chemical or physical agents that can induce alterations in DNA.

Mutant. A gene that has undergone a change or mutation.

Mutated in colorectal carcinoma (MCC). A gene involved in colorectal cancer as evidenced by loss of constitutional heterozygosity at 5q21.

Mutation. A change in genetic material, either of a single gene or in the number or structure of the chromosomes. A mutation that occurs in the gametes is inherited; a mutation in the somatic cells (somatic mutation) is not inherited.

Mutation rate. The number of mutations at any one particular locus that occur per gamete per generation.

Mutational heterogeneity. The occurrence of more than one mutation in a particular single-gene disorder.

Mutator genes. The equivalent in yeast to the DNA proof-reading enzymes that cause hereditary non-polyposis colorectal cancer.

Myeloma. A tumor of the plasma or antibody-producing cells.

Natural killer cells. Large granular lymphocytes with carbohydrate-binding receptors on their cell surface that recognize high-molecular-weight glycoproteins expressed on the cell surface of the infected cell as a result of the virus taking over the cellular replicative functions.

Neurofibromin. The protein product of the neurofibromatosis type I gene.

Neutral gene. A gene that appears to have no obvious effect on the likelihood of an individual's ability to survive.

New mutation. The occurrence of a change in a gene arising as a new event.

Non-conservative substitution. A mutation that codes for an amino acid which is chemically dissimilar (e.g., a different charge) will result in a protein with an altered structure.

Non-disjunction. The failure of two members of a homologous chromosome pair to separate during cell division so that both pass to the same daughter cell.

Non-identical twins. See Dizygotic twins.

Non-insulin-dependent diabetes mellitus. Diabetes that can often be treated with diet and/or oral medication, now known as type 2 diabetes.

Non-paternity. The biological father is not as stated or believed.

Non-penetrance. The occurrence of an individual being heterozygous for an autosomal dominant gene but showing no signs of it.

Non-random mating. See Assortative mating.

Nonsense mutation. A mutation that results in one of the termination codons, thereby leading to premature termination of translation of a protein.

Non-synonymous mutation. A mutation that leads to an alteration in the encoded polypeptide.

Northern blotting. Electrophoretic separation of mRNA with subsequent transfer to a filter and localization with a radio-labeled probe.
Nuclear envelope. The membrane around the nucleus, separating it from the cytoplasm.
Nuclear pores. Gaps in the nuclear envelope that allow substances to pass from the nucleus to the cytoplasm and vice versa.
Nucleolus. A structure within the nucleus that contains high levels of RNA.
Nucleosome. DNA-histone subunit of a chromosome.
Nucleotide. Nucleic acid is made up of many nucleotides, each of which consists of a nitrogenous base, a pentose sugar and a phosphate group.
Nucleus. A structure within the cell that contains the chromosomes and nucleolus.
Null allele. See Amorph.
Nullisomy. Loss of both members of a homologous pair of chromosomes.
Obligate carrier. An individual who, by pedigree analysis, must carry a particular gene (e.g., parents of a child with an autosomal recessive disorder).
Oligogene. One of a relatively small number of genes that contribute to a disease phenotype.
Oligonucleotide. A chain of, literally, a few nucleotides.
Oncogene. A gene affecting cell growth or development that can cause cancer.
Oncogenic. Literally, 'cancer causing.'
Opsonization. The 'making ready' of an infectious agent in the production of an immune response.
Origins of replication. The points at which DNA replication commences.
Orthologous. Conserved genes or sequences between species.
Ova. Mature haploid female gametes.
Oz. The group of genetic variants of the λ light-chain immunoglobulins.
P1-derived artificial chromosomes (PACs). Combination of the P1- and F-factor systems to incorporate foreign DNA inserts up to 150 kb.
Pachytene quadrivalent. The arrangement adopted by the two pairs of chromosomes involved in a reciprocal translocation when undergoing segregation in meiosis I.
Packaging cell line. A cell line that has been infected with a retrovirus in which the provirus is genetically engineered to lack the packaging sequence of the proviral DNA necessary to produce infectious viruses.
Packaging sequence. The DNA sequence of the proviral DNA of a retrovirus necessary for packaging of the retroviral RNA into an infectious virus.
Paint. Use of fluorescently labeled probes derived from a chromosome or region of a chromosome to hybridize with a chromosome in a metaphase spread.
Pair-rule mutant. Developmental genes identified in *Drosophila* that cause pattern deletions in alternating segments.
Panmixis. See Random mating.
Paracentric inversion. A chromosomal inversion that does not include the centromere.
Paralogous. Close resemblance of genes from different clusters (e.g., *HOXA13* and *HOXD13*).
Parthenogenesis. The development of an organism from an unfertilized oocyte.
Partial sex-linkage. A term used to describe genes on the homologous or pseudoautosomal portion of the X and Y chromosomes.
Penetrance. The proportion of heterozygotes for a dominant gene who express a trait, even if mildly.
Peptide. An amino acid, a portion of a protein.
Pericentric inversion. A chromosomal inversion that includes the centromere.
Permissible dose. An arbitrary safety limit that is probably much lower than that which would cause any significant effect on the frequency of harmful mutations within the population.
Phage. Abbreviation for bacteriophage.
Pharmacogenetics. The study of genetically determined variation in drug metabolism.
Phase. The relation of two or more alleles (DNA 'markers') at two linked genetic loci. If the alleles are located on the same physical chromosome they are 'in phase,' or 'coupled.'
Phenocopy. A condition that is due to environmental factors but resembles one that is genetic.
Phenol-enhanced reassociation technique (pERT). Use of the chemical phenol to facilitate rehybridization of slightly differing sources of double-stranded DNA to enable isolation of sequences that are absent from one of the two sources.
Phenotype. The appearance (physical, biochemical, and physiological) of an individual that results from the interaction of the environment and the genotype.
Pink-eyed dilution. Human homolog to mouse pink-eye gene for albinism.
Plasma cells. Mature antibody-producing B lymphocytes.
Plasmid. Small, circular DNA duplex capable of autonomous replication within a bacterium.
Platelet-derived growth factor (PDGF). A substance derived from platelets that stimulates the growth of certain cell types.
Pleiotropy. The multiple effects of a gene.
Point mutation. A single base-pair change.
Polar body. The daughter cell of gamete division in the female in meiosis I and II that does not go on to become a mature gamete.
Poly(A) tail. A sequence of 20 to 200 adenylic acid residues that is added to the 3′ end of most eukaryotic mRNAs, increasing its stability by making it resistant to nuclease digestion.
Polygenes. Genes that make a small additive contribution to a polygenic trait.
Polygenic inheritance. The genetic contribution to the etiology of disorders in which there are both environmental and genetic causative factors.
Polymerase chain reaction (PCR). The repeated serial reaction involving the use of oligonucleotide primers and DNA polymerase that is used to amplify a particular DNA sequence of interest.
Polymorphic information content (PIC). The amount of variation at a particular site in the DNA.
Polymorphism. The occurrence in a population of two or more genetically determined forms in such frequencies that the rarest of them could not be maintained by mutation alone.
Polypeptide. An organic compound consisting of three or more amino acids.
Polyploid. Any multiple of the haploid number of chromosomes (3N, 4N, etc.).
Polyribosome. See Polysome.
Polysome (= polyribosome). A group of ribosomes associated with the same molecule of mRNA.
Population genetics. The study of the distribution of alleles in populations.
Positional candidate gene. A gene located within a chromosome region believed to harbor the gene responsible for a disease or phenotype under study. It is a candidate because it is positioned within the critical chromosomal region.
Positional cloning. The mapping of a disorder to a particular region of a chromosome and leading to identification of the gene responsible.
Posterior information. Information available for risk calculation from the results of tests or analysis of offspring in pedigrees.
Posterior probability. The joint probability for a particular event divided by the sum of all possible joint probabilities.
Post-genomic genomics. See Functional genomics.

Post-translational processing. Various modifications of protein that occur after their synthesis (e.g., the addition of carbohydrate moieties).

Predictive testing. Presymptomatic testing, often used in relation to testing of people at risk for Huntington disease.

Preimplantation genetic diagnosis. The ability to detect the presence of an inherited disorder in an in-vitro fertilized conceptus before reimplantation.

Premutation. The existence of a gene in an unstable form that can undergo a further mutational event to cause a disease.

Prenatal diagnosis. The use of tests during a pregnancy to determine whether an unborn child is affected by a particular disorder.

Presymptomatic diagnosis. The use of tests to determine whether a person has inherited a gene for a disorder before he or she has any symptoms or signs.

Prevalence. At a point in time, the proportion of people in a given population with a disorder or trait.

Primary response. The response to an infectious agent with an initial production of IgM, then subsequently IgG.

Prion. A proteinaceous infectious particle implicated in the cause of several rare neurodegenerative diseases.

Prior probability. The initial probability of an event.

Probability. The proportion of times an outcome occurs in a large series of events.

Proband (= index case). An affected individual (irrespective of sex) through whom a family comes to the attention of an investigator. Propositus if male; proposita if female.

Probe. A labeled, single-stranded DNA fragment that hybridizes with, and thereby detects and locates, complementary sequences among DNA fragments on, for example, a nitrocellulose filter.

Processing. Alterations of mRNA that occur during transcription including splicing, capping, and polyadenylation.

Progress zone. The area of growth beneath the apical ectodermal ridge in the developing limb bud.

Prokaryotes. Lower organisms with no well defined nucleus (e.g., bacteria).

Prometaphase. The stage of cell division when the nuclear membrane begins to disintegrate, allowing the chromosomes to spread, with each chromosome attached at its centromere to a microtubule of the mitotic spindle.

Promoter. Recognition sequence for the binding of RNA polymerase.

Promoter elements. DNA sequences that include the GGGCGGG consensus sequence, the AT-rich TATA or Hogness box, and the CAAT box, in a 100- to 300-bp region located 5′ or upstream to the coding sequence of many structural genes in eukaryotic organisms and which control individual gene expression.

Pronuclei. The stage just after fertilization of the oocyte with the nucleus of the oocyte and sperm present.

Prophase. The first visible stage of cell division when the chromosomes are contracted.

Proposita. A female individual as the presenting person in a family.

Propositus. A male individual as the presenting person in a family.

Protein. A complex organic compound composed of hundreds or thousands of amino acids.

Proto-oncogene. A gene that can be converted to an oncogene by an activating mutation. The term 'oncogene' is now commonly used for both the normal and activated gene forms. The DNA genomic sequence shows homology to viral oncogenes.

Pseudoautosomal. Genes that behave like autosomal genes as a result of being located on the homologous portions of the X and Y chromosomes.

Pseudodominance. The apparent dominant transmission of a disorder when an individual homozygous for a recessive gene has affected offspring through having children with an individual who is also a carrier.

Pseudogene. DNA sequence homologous with a known gene but non-functional.

Pseudohermaphrodite. An individual with ambiguous genitalia or external genitalia opposite to the chromosomal sex in which there is gonadal tissue of only one type.

Pseudohypertrophy. Literally, false enlargement. Seen in the calf muscles of boys with Duchenne muscular dystrophy.

Pseudomosaicism. False mosaicism seen occasionally as an artifact with cells in culture.

Pulsed-field gel electrophoresis (PFGE). A technique of DNA analysis using electrophoretic methods to separate large DNA fragments, up to 2 million base pairs in size, produced by digesting DNA with restriction enzymes with relatively long DNA recognition sequences that, as a consequence, cut DNA relatively infrequently.

Purine. A nitrogenous base with fused five- and six-member rings (adenine and guanine).

Pyrimidine. A nitrogenous base with a six-membered ring (cytosine, uracil, thymine).

Quantitative inheritance. See Polygenic inheritance.

Radiation absorbed dose (rad). A measure of the amount of any ionizing radiation that is absorbed by the tissues; 1 rad is equivalent to 100 erg of energy absorbed per gram of tissue.

Radiation hybrid. An abnormal cell containing numerous small fragments of human chromosomes, brought about by fusion with a lethally irradiated human cell. These cells have a very useful role in physical gene mapping.

Random genetic drift. The chance variation of allele frequencies from one generation to the next.

Random mating (= panmixis). Selection of a spouse regardless of the spouse's genotype.

Reading frame. The order of the triplets of nucleotides in the codons of a gene that are translated into the amino acids of the protein.

Recessive. A trait expressed in individuals who are homozygous for a particular allele but not in those who are heterozygous.

Reciprocal translocation. A structural rearrangement of the chromosomes in which material is exchanged between one homolog of each of two pairs of chromosomes. The rearrangement is balanced if there is no loss or gain of chromosome material.

Recombinant DNA molecule. A union of two different DNA sequences from two different sources (e.g., a vector containing a 'foreign' DNA sequence).

Recombination. Cross-over between two linked loci.

Recombination fraction (θ). A measure of the distance separating two loci determined by the likelihood that a crossover will occur between them.

Reduced penetrance. A dominant gene that does not manifest itself in a proportion of heterozygotes.

Relative probability. See Posterior probability.

Relative risk. The frequency with which a disease occurs in an individual with a specific marker compared with that in those without the marker in the general population.

Repetitive DNA. DNA sequences of variable length that are repeated up to 100,000 (middle repetitive) or more than 100,000 (highly repetitive) copies per genome.

Replication. The process of copying the double-stranded DNA of the chromosomes.

Replication bubble. The structure formed by coalescence of two adjacent replication forks in copying the DNA molecule of a chromosome.

Replication error. The phenomenon of microsatellite instability seen in hereditary non-polyposis colorectal cancer from a mutation in one of the DNA proof-reading enzymes.

Replication fork. The structure formed at the site(s) of origin of replication of the double-stranded DNA molecule of chromosomes.

Replication units. Clusters of 20 to 80 sites of origin of DNA replication.

Replicons. A generic term for DNA vectors such as plasmids, phages, and cosmids that replicate in host bacterial cells.

Repressor. The product of the regulator gene of an operon that inhibits the operator gene.

Repulsion. When a particular allele at a locus is on the homologous chromosome for a specific allele at a closely linked locus.

Response elements. Regulatory sequences in the DNA to which signaling molecules bind, resulting in control of transcription.

Restriction endonucleases or enzymes. Group of enzymes each of which cleaves double-stranded DNA at a specific nucleotide sequence and so produces fragments of DNA of different lengths.

Restriction fragment. DNA fragment produced by a restriction endonuclease.

Restriction fragment length polymorphism (RFLP). Polymorphism resulting from the presence or absence of a particular restriction site.

Restriction map. Linear arrangement of restriction enzyme sites.

Restriction site. Base sequence recognized by a restriction endonuclease.

Reticulocytes. Immature red blood cells that still contain mRNA.

Retrovirus. RNA virus that replicates via conversion into a DNA provirus.

Reverse genetics. The process of identifying a protein or enzyme through its gene product.

Reverse painting. Amplification using PCR of an unidentified portion of chromosomal material, such as a small duplication or marker chromosome, which is then used as a probe for hybridization to a normal metaphase spread to identify its source of origin.

Reverse transcriptase. An enzyme that catalyzes the synthesis of DNA from RNA.

Reverse transcriptase–PCR (RT-PCR). Using a special primer that contains a promoter and translation initiator from mRNA (for PCR) to make cDNA.

Ribonucleic acid (RNA). See RNA.

Ribosomes. Minute spherical structures in the cytoplasm, rich in RNA; the location of protein synthesis.

Ring chromosome. An abnormal chromosome caused by a break in both arms of the chromosome, the ends of which unite leading to the formation of a ring.

RNA (= ribonucleic acid). The nucleic acid found mainly in the nucleolus and ribosomes. Messenger RNA transfers genetic information from the nucleus to the ribosomes in the cytoplasm and also acts as a template for the synthesis of polypeptides. Transfer RNA transfers activated amino acids from the cytoplasm to mRNA.

RNA-directed DNA synthesis. An exception to the central dogma—a process used by many RNA viruses to produce DNA that can integrate with the host genome.

Robertsonian translocation. A translocation between two acrocentric chromosomes with loss of satellite material from their short arms.

Roentgen equivalent for man (rem). The dose of any radiation that has the same biological effect as 1 rad of X-rays.

Satellite. A distal portion of the chromosome separated from the remainder of the chromosome by a narrowed segment or stalk.

Satellite DNA. A class of DNA sequences that separates out on density gradient centrifugation as a shoulder or 'satellite' to the main peak of DNA and corresponds to 10% to 15% of the DNA of the human genome, consisting of short, tandemly repeated, DNA sequences that code for ribosomal and transfer RNAs.

Screening. The identification of people from a population with a particular disorder, or who carry a gene for a particular disorder.

Secondary hypertension. Increased blood pressure that occurs as a result of another primary cause.

Secondary oocyte or spermatocyte. The intermediate stage of a female or male gamete in which the homologous duplicated chromosome pairs have separated.

Secondary response. The enhanced immune response seen after repeated exposure to an infectious organism or foreign antigen.

Secretor locus. A gene in humans that results in the secretion of the ABO blood group antigens in saliva and other body fluids.

Secretor status. The presence or absence of excretion of the ABO blood group antigens into various body fluids (e.g., saliva).

Segment polarity mutants. Developmental genes identified in *Drosophila* that cause pattern deletions in every segment.

Segmental. Limited area of involvement (e.g., a somatic mutation limited to one area of embryonic development).

Segregation. The separation of alleles during meiosis so that each gamete contains only one member of each pair of alleles.

Segregation analysis. Study of the way in which a disorder is transmitted in families to establish the mode of inheritance.

Segregation ratio. The proportion of affected to unaffected individuals in family studies.

Selection. The forces that affect biological fitness and therefore the frequency of a particular condition within a given population.

Selfish DNA. DNA sequences that appear to have little function and that, it has been proposed, preserve themselves as a result of selection within the genome.

Semi-conservative. The process in DNA replication by which only one strand of each resultant daughter molecule is newly synthesized.

Sense strand. Strand of genomic DNA to which the mRNA is identical.

Sensitivity. Refers to the proportion of cases that are detected. A measure of sensitivity can be made by determining the proportion of false-negative results (i.e., how many cases are missed).

Sequence. A stretch of DNA nucleotides. Also used in relation to birth defects or congenital abnormalities that occur as a consequence of a cascade of events initiated by a single primary factor (e.g., Potter's sequence), which occurs as a consequence of renal agenesis.

Severe combined immunodeficiency. A genetically heterogeneous lethal form of inherited immunodeficiency with abnormal B- and T-cell function leading to increased susceptibility to both viral and bacterial infections.

Sex chromatin (= Barr body). A darkly staining mass situated at the periphery of the nucleus during interphase which represents a single, inactive, condensed X chromosome. The number of sex chromatin masses is one less than the number of X chromosomes (e.g., none in normal males and 45,X females, one in normal females and XXY males).

Sex chromosomes. The chromosomes responsible for sex determination (XX in women, XY in men).

Sex-determining region of the Y (SRY). The part of the Y chromosome that contains the testis-determining gene.

Sex influence. When a genetic trait is expressed more frequently in one sex than another. In the extreme, when only one sex is affected, this is called sex limitation.

Sex limitation. When a trait is manifest only in individuals of one sex.

Sex linkage. The pattern of inheritance shown by genes carried on the sex chromosomes. Because there are very few mendelizing genes on the Y chromosome, the term is often used synonymously for X-linkage.

Sex-linked inheritance. A disorder determined by a gene on one of the sex chromosomes.

Sex ratio. The number of male births divided by the number of female births.

Short interspersed nuclear elements (SINEs). Five percent of the human genome consists of some 750,000 copies of DNA sequences of approximately 300 bp that have sequence similarity to a signal recognition particle involved in protein synthesis.

Siamese twins. Conjoined identical twins.

Sib (= sibling). Brother or sister.

Sickle cell crisis. An acute hemolytic episode in people with sickle cell disease associated with a sudden onset of chest, back or limb pain, fever and dark urine from the presence of free hemoglobin in the urine.

Sickle cell disease. The homozygous state for hemoglobin S associated with anemia and the risk of sickle cell crises.

Sickle cell trait. The heterozygous state for hemoglobin S which is not associated with any significant medical risks under ordinary conditions.

Sickling. The process of distortion of red blood cell morphology under low oxygenation conditions in people with sickle cell disease.

Sievert (Sv). Equivalent to 100 rem.

Signal transduction. A complex multistep pathway from the cell membrane, through the cytoplasm to the nucleus, with positive and negative feedback loops for accurate cell proliferation and differentiation.

Silencers. A negative 'enhancer,' the normal action of which is to repress gene expression.

Silent mutation. A point mutation in a codon that, because of the degeneracy of the genetic code, still results in the same amino acid in the protein.

Single-nucleotide polymorphisms (SNPs). Single-nucleotide DNA sequence variation that is polymorphic, occurring every 1/500 to 1/2,000 base pairs.

Single-stranded conformational polymorphism (SSCP). A mutation detection system in which differences in the three-dimensional structure of single-stranded DNA result in differential gel electrophoresis mobility under special conditions.

Sister chromatids. Identical daughter chromatids derived from a single chromosome.

Sister chromatid exchange. Exchange (crossing over) of genetic material between two chromatids of any particular chromosome in mitosis.

Site-directed mutagenesis. The ability to alter or modify DNA sequences or genes in a directed fashion by processes such as insertional mutagenesis or homologous recombination to determine the effect of these changes on their function.

Skeleton map. See Framework map.

Skewed X-inactivation. A non-random pattern of inactivation of one of the X chromosomes in a female that can arise through a variety of mechanisms (e.g., an X-autosome translocation).

Slipped strand mispairing. Incorrect pairing of the tandem repeats of the two complementary DNA strands during DNA replication that is thought to lead to variation in DNA microsatellite repeat number.

Small nuclear RNA molecules. RNA molecules involved in RNA splicing.

Soft markers. Minor structural ultrasound findings associated with the possibility of fetal abnormality.

Solenoid model. The complex model of the quaternary structure of chromosomes.

Somatic cell gene therapy. The alteration or replacement of a gene limited to the non-germ cells.

Somatic cell hybrid. A technique involving the fusion of cells from two different species that results in the loss of chromosomes from one of the cell types and is used in assigning genes to particular chromosomes.

Somatic cells. The non-germline cells of the body.

Somatic mosaicism. The occurrence of two different cell lines in a particular tissue or tissues that differ genetically.

Somatic mutation. A mutation limited to the non-germ cells.

Sonic hedgehog. One of three mammalian homologs of the segment polarity hedgehog genes.

Southern blot. Technique for transferring DNA fragments from an agarose gel to a nitrocellulose filter on which they can be hybridized to a radiolabeled single-stranded complementary DNA sequence or probe.

Specific acquired or adaptive immunity. A tailor-made immune response that occurs after exposure to an infectious agent.

Specificity. The extent to which a test detects only affected individuals. If unaffected people are detected, these are referred to as false positives.

Spermatid. Mature haploid male gamete.

Spindle. A structure responsible for the movement of the chromosomes during cell division.

Splicing. The removal of the introns and joining of exons in RNA during transcription, with introns being spliced out and exons being spliced together.

Splicing branch site. Intronic sequence involved in splicing of mRNA.

Splicing consensus sequences. DNA sequences surrounding splice sites.

Spontaneous mutation. A mutation that arises de novo, apparently not from environmental factors such as mutagens.

Sporadic. When a disorder affects a single individual in a family.

SRY (sex-reversed Y). The sex-determining region of the Y chromosome that contains the testis-determining gene.

Stable mutation. A mutation that is transmitted unchanged.

Stop codons. One of three codons (UAG, UAA, and UGA) that cause termination of protein synthesis.

Subchromosomal mapping. Mapping of a gene or DNA sequence of interest to a region of a chromosome.

Submetacentric. Chromosomes in which the centromere is slightly off center.

Substitution. A single base pair replaced by another nucleotide.

Suppressor lymphocytes. A subclass of T lymphocytes that regulate immune responses, particularly suppressing an immune response to 'self'.

Switching. Change in the type of β- or α-like globin chains produced in embryonic and fetal development.

Synapsis. The pairing of homologous chromosomes during meiosis.

Synaptonemal complex. A complex protein structure that forms between two homologous chromosomes which pair during meiosis.

Syndrome. The complex of symptoms and signs that occur together in any particular disorder.

Synonymous mutation. See Silent mutation.

Syntenic genes. Two genes at different loci on the same chromosome.

T. Abbreviation for thymine.

TATA (Hogness) box. See Hogness box.

T cell surface antigen receptor. Antigenic receptor on the cell surface of T lymphocytes.

T lymphocytes. Lymphocytes involved in the cellular immune response—'thymus'-derived.

Tandemly repeated DNA sequences. DNA consisting of blocks of tandem repeats of non-coding DNA that can be either highly dispersed or restricted in their location in the genome.

Target DNA. The carrier or vector DNA to which foreign DNA is incorporated or attached to produce recombinant DNA.

Telomere. The distal portion of a chromosome arm.

Telomeric DNA. The terminal portion of the telomeres of the chromosomes contains 10 to 15 kb of tandem repeats of a 6 base-pair DNA sequence.

Telophase. The stage of cell division when the chromosomes have separated completely into two groups and each group has become invested in a nuclear membrane.

Template strand. The strand of the DNA double helix that is transcribed into mRNA.

Teratogen. An agent that causes congenital abnormalities in the developing embryo or fetus.

Teratogene. A gene that can mutate to form a developmental abnormality.

Terminator. A sequence of nucleotides in DNA that codes for the termination of translation of mRNA.

Tertiary trisomy. The outcome when 3 to 1 segregation of a balanced reciprocal translocation results in the presence of an additional derivative chromosome.

Tetraploidy. Twice the normal diploid number of chromosomes (4N).

Thalassemia intermedia. A less severe form of β-thalassemia that requires less frequent transfusions.

Thalassemia major. An inherited disorder of human hemoglobin that is due to underproduction of one of the globin chains.

Thalassemia minor. See Thalassemia trait.

Thalassemia trait. The heterozygous state for β-thalassemia, associated with an asymptomatic, mild, microcytic, hypochromic anemia.

Three-prime (3′) end. The end of a DNA or RNA strand with a free 3′ hydroxyl group.

Threshold. A concept used in disorders that exhibit multifactorial inheritance to explain a discontinuous phenotype in a process or trait that is continuous (e.g., cleft lip as a result of disturbances in the process of facial development).

Thymine. A pyrimidine base in DNA.

Tissue typing. Cellular, serological and DNA testing to determine histocompatibility for organ transplantation.

Trait. Any detectable phenotypic property or character.

***Trans*-acting.** Transcription factors that act on genes at a distance, usually on both copies of a gene on each chromosome.

Transcription. The process whereby genetic information is transmitted from the DNA in the chromosomes to mRNA.

Transcription factors. Genes, including the *Hox*, *Pax*, and zinc finger-containing genes, that control RNA transcription by binding to specific DNA regulatory sequences and forming complexes that initiate transcription by RNA polymerase.

Transfection. The transformation of bacterial cells by infection with phage to produce infectious phage particles. Also the introduction of foreign DNA into eukaryotic cells in culture.

Transfer RNA. RNA molecule involved in transfer of amino acids in the process of translation.

Transformation. Genetic recombination in bacteria in which foreign DNA introduced into the bacterium is incorporated into the chromosome of the recipient bacterium. Also the change of a normal cell into a malignant cell; for example, as results from infection of normal cells by oncogenic viruses.

Transgenic animal model. Use of techniques such as targeted gene replacement to introduce mutations into a particular gene in another animal species to study an inherited disorder in humans.

Transient polymorphism. Two different allelic variants present in a population whose relative frequencies are altering due to either selective advantage or disadvantage of one or the other.

Transition. A substitution involving replacement by the same type of nucleotide (i.e., a pyrimidine for a pyrimidine [C for T, or vice versa] or a purine for a purine [A for G, or vice versa]).

Translation. The process whereby genetic information from mRNA is translated into protein.

Translocation. The transfer of genetic material from one chromosome to another chromosome. If there is an exchange of genetic material between two chromosomes then this is referred to as a reciprocal translocation. A translocation between two acrocentric chromosomes by fusion at the centromeres is referred to as a robertsonian translocation.

Transposon. Mobile genetic element able to replicate and insert a copy of itself at a new location in the genome.

Transversion. Substitution of a pyrimidine by a purine, or vice versa.

Triple test. The test that gives a risk for having a fetus with Down syndrome in mid-trimester as a function of age, serum α-fetoprotein, estriol, and human chorionic gonadotropin levels.

Triplet amplification or expansion. Increase in the number of copies of triplet repeat sequences responsible for mutations in a number of single-gene disorders.

Triplet code. A series of three bases in the DNA or RNA molecule that codes for a specific amino acid.

Triploid. A cell with three times the haploid number of chromosomes (i.e., 3N).

Trisomy. The presence of a chromosome additional to the normal complement (i.e., 2N + 1), so that in each somatic nucleus one particular chromosome is represented three times rather than twice.

Trophoblast. The outer cell mass of the early embryo that gives rise to the placenta.

Truncate ascertainment. See Incomplete ascertainment.

Tumor suppressor gene. Genes that appear to prevent the development of certain types of tumor.

Tyrosinase-negative albinism. Form of oculocutaneous albinism with no melanin production that can be tested for in vitro.

Tyrosinase-positive albinism. Form of oculocutaneous albinism with some melanin production that can be tested for in vitro.

U. Abbreviation for uracil.

Ultrasonography. Use of ultrasonic sound waves to image objects at a distance (e.g., the developing fetus in utero).

Unbalanced translocation. A translocation in which there is an overall loss or gain of chromosomal material (e.g., partial monosomy of one of the portions involved and partial trisomy of the other portion involved).

Unifactorial (= mendelizing). Inheritance controlled by a single locus.

Uniparental disomy. When an individual inherits both chromosomes of a homologs pair from one parent.

Uniparental heterodisomy. Uniparental disomy resulting from inheritance of the two different homologs from one parent.

Uniparental isodisomy. Uniparental disomy resulting from inheritance of two copies of a single chromosome of a homologous pair from one parent.

Unipolar illness. Affective depressive illness.

Universal donor. A person of blood group O, Rh-negative, who can donate blood to any person irrespective of their blood group.

Universal recipient. A person of blood group AB, Rh-positive, who can receive blood from any donor irrespective of their blood group.

Unstable mutation. A mutation that, when transmitted, is passed on in altered form (e.g., triplet repeat mutations).

Uracil. A pyrimidine base in RNA.

Utrophin. A gene on chromosome 6 with homology to the dystrophin gene.

Variable expressivity. The variation in the severity of phenotypic features seen in people with autosomal dominant disorders (e.g., variable number of café-au-lait spots or neurofibromata in neurofibromatosis type I).

Variable region. The portion of the light and heavy chains of immunoglobulins that differs between molecules and helps to determine antibody specificity.

Variants. Alleles that occur less frequently than in 1% of the population.
Vector. A plasmid, phage or cosmid into which foreign DNA can be inserted for cloning.
Virions. Infectious viral particles.
Virus. A protein-covered DNA- or RNA-containing organism that is capable of replication only within bacterial or eukaryotic cells.
Wingless. A group of morphogens produced by segment polarity genes.
X-chromatin. See Barr body or sex chromatin.
X-inactivation. See Lyonization.
X-inactivation center. The part of the X chromosome responsible for the process of X-inactivation.
X-linkage. Genes carried on the X chromosome.
X-linked dominant. Genes on the X chromosome that manifest in heterozygous females.
X-linked dominant lethals. Disorders that are seen only in females and not in males, as they are thought to be incompatible with survival of the early embryo in hemizygous males (e.g., incontinentia pigmenti).
X-linked recessive. Genes that are carried by females and expressed in hemizygous males.
Xanthomata. Subcutaneous depositions of lipid, often around tendons; a physical sign associated with disordered lipid metabolism.
Yeast artificial chromosome (YAC). A plasmid-cloning vector that contains the DNA sequences for the centromere, telomere and autonomous chromosome replication sites that enable cloning of large DNA fragments up to 2 to 3 million base pairs in length.
Y-linked inheritance. See Holandric inheritance.
Zinc finger. A finger-like projection formed by amino acids, positioned between two separated cysteine residues, which is stabilized by forming a complex with a zinc ion and can then bind specifically to DNA sequences; they are commonly found in transcription factors.
Zona pellucida. Cellular layer surrounding the mature unfertilized oocyte.
Zone of polarizing activity. An area on the posterior margin of the developing limb bud that determines the anteroposterior axis.
Zoo blot. A Southern blot of DNA from a number of different species used to look for evidence of DNA sequences conserved during evolution.
Zygote. The fertilized ovum.

Multiple-Choice Questions

There may be more than one correct answer per question.

CHAPTER 2: The Cellular and Molecular Basis of Inheritance

1. Base substitutions:

a. May result in nonsense mutations
b. Can affect splicing
c. Are always pathogenic
d. Can affect gene expression
e. Result in frameshift mutations

2. Transcription:

a. Describes the production of polypeptides from the mRNA template
b. Occurs in the nucleus
c. Produces single-stranded mRNA using the antisense DNA strand as a template
d. Is regulated by transcription factors that bind to the 3′ UTR
e. Precedes 5′ capping and polyadenylation

3. The following are directly involved in DNA repair:

a. Glycosylases
b. DNA polymerases
c. Ligases
d. Splicing
e. Ribosomes

4. During DNA replication:

a. DNA helicase separates the double-stranded DNA
b. DNA is synthesized in one direction
c. Okazaki fragments are synthesized
d. DNA is synthesized in a conservative manner
e. Uracil is inserted to pair with adenine

CHAPTER 3: Chromosomes and Cell Division

1. Meiosis differs from mitosis in the following ways:

a. Daughter cells are haploid, not diploid
b. Meiosis is restricted to the gametes and mitosis occurs only in somatic cells
c. In mitosis, there is only one division
d. Meiosis generates genetic diversity
e. The prophase stage of mitosis is one step; in meiosis I, there are four stages

2. Chromosome abnormalities reliably detected by light microscopy include:

a. Trisomy
b. Monosomy
c. Reciprocal translocation
d. Interstitial deletion
e. Robertsonian translocation

3. Fluorescent in situ hybridization using whole-chromosome (painting) or specific locus probes enables routine detection of:

a. Gene amplification
b. Subtelomeric deletion
c. Trisomy
d. Supernumerary marker chromosomes
e. Reciprocal translocation

4. Chemicals used in the preparation of metaphase chromosomes for analysis by light microscopy include:

a. Colchicine
b. Phytohemagglutinin
c. Giemsa
d. Quinacrine
e. Hypotonic saline

CHAPTER 4: DNA Technology and Applications

1. The following statements apply to restriction enzymes:

a. They can generate DNA fragments with ‘sticky’ ends
b. They are viral in origin
c. They are used to detect point mutations
d. They are used in Southern blotting
e. They are also called restriction exonucleases

2. The following describe the polymerase chain reaction (PCR):

a. A type of cell-free cloning
b. A process that uses a heat-labile DNA polymerase
c. A very sensitive method of amplifying DNA that can be prone to contamination
d. A technique that can routinely amplify up to 100 kb of DNA
e. A method of amplifying genes that requires no prior sequence knowledge

3. Types of nucleic acid hybridization include:
a. Southern blotting
b. Microarray
c. Western blotting
d. Northern blotting
e. DNA fingerprinting

4. The following techniques can be used to screen genes for unknown mutations:
a. Sequencing
b. Single-stranded conformational polymorphism (SSCP)
c. Denaturing high-performance liquid chromatography (DHPLC)
d. Oligonucleotide ligation assay (OLA)
e. Real-time PCR

CHAPTER 5: Mapping and Identifying Genes for Monogenic Disorders

1. Positional cloning uses:
a. Genetic databases
b. Knowledge of orthologous genes
c. Patients with chromosomal abnormalities
d. Candidate genes selected by biological knowledge
e. Microsatellite markers

2. A candidate gene is likely to be a disease-associated gene if:
a. A loss-of-function mutation causes the phenotype
b. An animal model with a mutation in the orthologous gene has the same phenotype
c. Multiple different mutations cause the phenotype
d. The pattern of expression of the gene is consistent with the phenotype
e. It is a pseudogene

3. Achievements of the Human Genome Project include:
a. Draft sequence published in 2000
b. Sequencing completed in 2003
c. Development of bioinformatics tools
d. Identification of all disease-causing genes
e. Studies of ethical, legal, and social issues

CHAPTER 6: Developmental Genetics

1. In development, *HOX* genes:
a. Function as transcription factors
b. When mutated have been shown to be associated with many malformation syndromes
c. Show very divergent structures across different species
d. Are functionally redundant in postnatal life
e. Individually can be important in the normal development of widely different body systems

2. In the embryo and fetus:
a. Gastrulation is the process leading to the formation of the 16-cell early embryo 3 days after fertilization
b. Organogenesis takes place at between 8 and 12 weeks' gestation
c. The notch signaling and sonic hedgehog pathways are important for ensuring normal development in diverse organs and tissues
d. Somites form in a caudo-rostral direction from the presomitic mesoderm
e. *TBX* genes appear to be crucial to normal limb development

3. Concerning developmental pathways and processes:
a. In mammalian development, the jaw is formed from the second pharyngeal arch
b. Pharyngeal arch arteries ultimately become the great vessels around the heart
c. *TBX1* is a key gene in the defects associated with DiGeorge syndrome
d. Achondroplasia can be caused by a wide variety of mutations in the *FGFR3* gene
e. Loss-of-function mutations and gain-of-function mutations usually cause similar defects

4. Regarding the X-chromosome:
a. In most males with a karyotype of 46,XX, the *SRY* gene is present and found on one of the X chromosomes
b. In lyonization, or X-chromosome inactivation, all the genes of one X chromosome are switched off
c. As a result of lyonization, all females are X-chromosome mosaics
d. Male fetal development is solely dependent on the *SRY* gene having normal functioning
e. X-chromosome inactivation may be linked in some way to the monozygotic twinning process

5. Transcription factors:
a. Are RNA sequences that interfere with translation in the ribosomes
b. Their only function is to switch off genes in development
c. When mutated in *Drosophila* body segments may be completely reorganized
d. Are not involved in defects of laterality
e. Include genes that have a zinc finger motif

CHAPTER 7: Patterns of Inheritance

1. Concerning autosomal recessive inheritance:
a. Females are more likely to be affected than males
b. If both parents are carriers, the risk at conception that any child might be a carrier is ¾
c. Diseases following this pattern of inheritance are more common in societies where cousin marriage is normal
d. Usually only a single generation has affected individuals
e. Angelman syndrome follows this pattern

2. Concerning X-linked inheritance:
a. The condition cannot be passed from an affected father to his son
b. When recessive, an affected man will not see the condition in his children but it may appear in his grandchildren
c. When dominant, females are usually as severely affected as males
d. When dominant, there are usually more affected females than affected males in a family
e. The risk of germline mosaicism does not need to be considered

3. In mitochondrial genetics:
a. Heteroplasmy refers to the presence of more than one mutation in mitochondria
b. Mitochondrial genes mutate less often than nuclear genes
c. Mitochondrial conditions affect only muscle and nerve tissue
d. The risk of passing on a mitochondrial condition to the next generation may be as high as 100%
e. Mitochondrial diseases have nothing to do with nuclear genes

4. Concerning terminology:
a. Locus heterogeneity means that the same disease can be caused by genes on different chromosomes
b. Pseudo-dominance refers to the risk to the offspring when both parents have the same dominantly inherited condition
c. If a condition demonstrates reduced penetrance it may skip generations
d. Variable expression characterizes diseases that demonstrate anticipation
e. Pleiotropy is simply a more striking form of variable expression

5. In inheritance:
a. An autosomal recessive condition can occasionally arise through uniparental disomy
b. Imprinted genes can be unmasked through uniparental disomy
c. Digenic inheritance is simply another way of referring to uniparental disomy
d. Hormonal factors may account for conditions demonstrating sex influence
e. Most of the human genome is subject to imprinting

CHAPTER 8: Mathematical and Population Genetics

1. In applying the Hardy-Weinberg equilibrium, the following assumptions are made:
a. The population is small
b. There is no consanguinity
c. New mutations do not occur
d. No babies are born by donor insemination
e. There is no significant movement of population

2. If the population incidence of a recessive disease is 1 in 10,000, the carrier frequency in the population is:
a. 1 in 100
b. 1 in 200
c. 1 in 25
d. 1 in 50
e. 1 in 500

3. Heterozygote advantage:
a. May lead to an increased incidence of autosomal dominant disorders
b. Does not mean that biological fitness is increased in the homozygous state
c. May explain the worldwide distribution of sickle cell disease and malaria
d. May lead to distortion of the Hardy-Weinberg equilibrium
e. Is very unlikely to be traced to a founder effect

4. Polymorphic loci:
a. Are defined as those loci at which there are at least two alleles, each with frequencies greater than 10%
b. Have been crucial to gene discoveries
c. Can be helpful in determining someone's genetic status in a family
d. Have nothing to do with calculating LOD scores
e. By themselves have no consequence for genetically determined disease

5. In population genetics:

a. To calculate the mutation rate for a disorder, it is necessary only to know the biological fitness for the condition
b. If medical treatment can improve biological fitness, the frequency of an autosomal dominant condition will increase far more rapidly than that of an autosomal recessive condition
c. Even when a large number of families is studied, the calculated segregation ratio for a disorder might not yield the expected figures for a given pattern of inheritance
d. For X-linked disorders, the frequency of affected males equals the frequency of the mutated allele
e. Autozygosity mapping is a useful strategy to look for the gene in any autosomal recessive condition

CHAPTER 9: Polygenic and Multifactorial Inheritance

1. Concerning autism:

a. It is best classified as an inborn error of metabolism
b. The concordance rate in dizygotic twins is approximately 50%
c. Fragile X syndrome is a major cause
d. The risk to the siblings of an affected person is approximately 5%
e. Girls are more frequently affected than boys

2. Linkage analysis is more difficult in multifactorial conditions than in single-gene disorders because:

a. Variants in more than one gene are likely to contribute to the disorder
b. The number of affected persons within in a family is likely to be fewer than for a single-gene disorder
c. The mode of inheritance is usually uncertain
d. Some multifactorial disorders are likely to have more than one etiology
e. Many multifactorial conditions have a late age of onset

3. Association studies:

a. Can give false-positive results because of population stratification
b. Include the transmission disequilibrium test (TDT)
c. Positive association studies should be replicated
d. Are used to map genes in multifactorial disorders
e. Require closely matched control and patient groups

4. Variants in genes that confer susceptibility to type 2 diabetes (T2DM) have been found:

a. By linkage analysis using affected sibling pairs
b. Using animal models
c. By candidate gene studies from monogenic subtypes of diabetes
d. Through the study of biological candidates
e. In isolated populations

5. Variants in the *NOD2/CARD15* gene:

a. Are associated with Crohn disease and ulcerative colitis
b. Can result in a 40-fold increased risk of disease
c. Were identified after the gene was mapped to chromosome 16p12 by positional cloning
d. Has led to novel therapies
e. Are very rare in the general population

CHAPTER 10: Hemoglobin and the Hemoglobinopathies

1. For different hemoglobins:

a. The fetal hemoglobin chain, γ, resembles the adult β chain
b. The Hb chains, α, β and γ are all expressed throughout fetal life
c. In α-thalassemia, there are too many α chains
d. Hb Barts is a form of β-thalassemia
e. Carriers of β-thalassemia frequently suffer from symptomatic anemia

2. Regarding sickle cell disease:

a. The sickling effect of red blood corpuscles is the result of abnormal Hb binding with the red blood cell membrane
b. Life-threatening thrombosis can occur
c. Hb S differs from normal Hb A by a single amino-acid substitution
d. Splenic infarction may occur but this has little clinical consequence
e. Point (missense) mutations are the usual cause of abnormal Hb in the sickling disorders

3. Concerning hemoglobin variants:

a. Many Hb variants are harmless
b. The types of mutation occurring in the hemoglobinopathies are very limited
c. In the thalassemias, hypoplasia of the bone marrow occurs
d. In the thalassemias, Hb demonstrates abnormal oxygen affinity
e. In some thalassemias, increased red cell hemolysis occurs

4. Regarding hemoglobins during life:

a. Persistence of fetal Hb into adult life is an acquired disorder
b. Throughout fetal life, it is the liver that produces most of the body's Hb
c. The bone marrow is not involved in Hb production before birth
d. The liver continues to produce Hb into the second year of postnatal life
e. Persistence of fetal Hb into adult life is a benign condition

CHAPTER 11: Biochemical Genetics

1. In congenital adrenal hyperplasia (CAH):
- a. Females may show virilization and ambiguous genitalia
- b. Males may show undermasculinization and ambiguous genitalia
- c. Mineralocorticoid deficiency can be life threatening
- d. Treatment is required during childhood but not usually in adult life
- e. In affected females, fertility is basically unaffected

2. Phenylketonuria:
- a. Is the only cause of a raised phenylalanine level in the neonatal period
- b. Requires lifelong treatment
- c. Is a cause of epilepsy and eczema
- d. Results in reduced levels of melanin
- e. Is part of the same pathway as cholesterol production

3. Hepatomegaly is an important feature of:
- a. Hurler syndrome
- b. Glycogen storage disorders
- c. Abnormalities of porphyrin metabolism
- d. Niemann-Pick disease
- e. Galactosemia

4. Concerning mitochondrial disorders:
- a. All follow matrilinear inheritance
- b. Retinal pigmentation and diabetes can both be features
- c. There are fewer than 50 gene products from the mitochondrial genome
- d. Leigh disease is always caused by the same point mutation
- e. The gene for Barth syndrome is known but the metabolic pathway is uncertain

5. Regarding metabolic conditions:
- a. The carnitine cycle and long-chain fatty acids are linked
- b. A single point mutation explains most cases of MCAD (medium-chain acyl-CoA dehydrogenase) deficiency
- c. Peroxisomal disorders include Menkes disease and Wilson disease
- d. Inborn errors of metabolism may present with hypotonia and acidosis alone
- e. X-rays are of no value in making a diagnosis of inborn errors of metabolism

CHAPTER 12: Pharmacogenetics

1. Thiopurine drugs used to treat leukemia:
- a. Include 6-mercaptopurine, 6-thioguanine, and azathioprine
- b. Are also used to suppress the immune system
- c. May be toxic in 1–2% of patients
- d. Can have serious side-effects
- e. Are metabolized by thiopurine methyltransferase (TPMT)

2. Liver enzymes that show genetic variation of expression and hence influence the response to drugs include:
- a. UDP-glucuronosyltransferase
- b. O-acetyltransferase
- c. Alcohol dehydrogenase
- d. CYP2D6
- e. CYP2C9

3. The following have an increased risk of complications from anesthesia:
- a. First-degree relatives of patients affected with malignant hyperthermia
- b. Patients with *RYR1* mutations
- c. Patients with *G6DP* deficiency
- d. Patients with *CHE1* mutations
- e. Patients with raised pseudocholinesterase levels

4. Examples of diseases in which treatment may be influenced by pharmacogenetics include:
- a. Maturity-onset diabetes of the young (MODY), subtype glucokinase
- b. Maturity-onset diabetes of the young (MODY), subtype HNF-1α
- c. HIV infection
- d. Epilepsy
- e. Tuberculosis

CHAPTER 13: Immunogenetics

1. Concerning complement:
- a. The complement cascade can be activated only by the binding of antibody and antigen
- b. C1-inhibitor deficiency can result in complement activation through the classic pathway
- c. C3 levels are reduced in hereditary angioneurotic edema
- d. Complement helps directly in the attack on microorganisms
- e. Complement is found mainly in the intracellular matrix

2.

a. The immunoglobulin molecule is made up of six polypeptide chains
b. The genes for the various light and heavy immunoglobulin chains are found close together in the human genome
c. Close relatives make the best organ donors because they are likely to share the same complement haplotypes
d. The DNA encoding the κ light chain contains four distinct regions
e. The diversity of T-cell surface antigen receptor can be compared with the process of immunoglobulin diversity

3.

a. Maternal transplacental mobility of antibodies gives infants protection for about 12 months
b. X-linked severe combined immunodeficiency (SCID) accounts for about 5–10% of the total of SCID
c. SCID, despite its name, is not always a severe condition
d. There is always a T-cell abnormality in the different forms of SCID
e. Chronic granulomatous disease (CGD) is a disorder of humoral immunity

4.

a. DiGeorge/Sedláčková syndrome is a primary disorder of immune function
b. Severe opportunistic bacterial infections are uncommon in DiGeorge syndrome
c. Genetic prenatal diagnosis is possible for common variable immunodeficiency (CVID)
d. Autoimmune disorders follow autosomal dominant inheritance
e. Investigation of immune function should be considered in any child with failure to thrive (FTT)

CHAPTER 14: Cancer Genetics

1.

a. Chromosome translocations can lead to cancer through modification of oncogene activity
b. Oncogenes are the most common forms of genes predisposing to hereditary cancer syndromes
c. Defective apoptosis may lead to tumorigenesis
d. Loss of heterozygosity (LOH) is another term for a mutational event in an oncogene
e. A mutation in the *APC* gene is sufficient to cause colorectal cancer

2.

a. The two-hit hypothesis predicted that a tumor would develop when both copies of a critical gene were mutated
b. *TP53* mutations are found only in Li-Fraumeni syndrome
c. The *RET* proto-oncogene is implicated in all forms of multiple endocrine neoplasia (MEN)
d. Individuals with familial adenomatous polyposis (FAP) should have screening of the upper gastrointestinal tract
e. Endometrial cancer is a feature of hereditary non-polyposis colon cancer (HNPCC)

3.

a. Thyroid cancer is a risk in Bannayan-Riley-Ruvalcaba syndrome
b. Men with a germline mutation in *BRCA2* are at increased risk of prostate cancer
c. The genetic basis of all familial breast cancer is now well established
d. Familial breast cancer is usually fully penetrant
e. For men with prostate cancer, 3% of male first-degree relatives are similarly affected

4.

a. Medulloblastoma is a common tumor in von Hippel-Lindau (VHL) disease
b. Pheochromocytoma is frequently seen in Gorlin syndrome
c. There is a risk of ovarian cancer in Peutz-Jegher syndrome and HNPCC
d. Cutaneous manifestations occur in Peutz-Jegher syndrome, Gorlin syndrome, and HNPCC
e. In two-thirds of HNPCC cases the predisposing gene is unknown

5.

a. Screening for renal cancer in VHL is recommended
b. Mammography detects breast cancer more easily in premenopausal than postmenopausal women
c. Screening for retinoblastoma should begin in the second year of life
d. Colorectal cancer in two close relatives is sufficient to indicate the need for colonoscopy screening in other family members
e. Preventive surgery is strongly indicated in FAP and women positive for *BRCA1* mutations

CHAPTER 15: Genetic Factors in Common Diseases

1.

a. A genetic factor is thought to be present in 10% of epilepsy
b. Mutations in potassium channel genes predispose to epilepsy
c. Juvenile myoclonic epilepsy follows autosomal dominant inheritance
d. Epilepsy is common in deletion 1p36 syndrome
e. Unverricht-Lundborg disease is a benign form of mendelian epilepsy

2.

a. There is a high risk of bipolar depressive illness in DiGeorge syndrome
b. The concordance rate for schizophrenia in monozygotic twins is approximately 80%
c. No mutations have so far been found in any gene to provide a cause for schizophrenia
d. Twin studies suggest a similar heritability for bipolar and unipolar illness
e. Analysis of APOE ε4 is recommended as a screening procedure for Alzheimer disease

3.

a. Susceptibility for Creutzfeldt-Jakob disease is conferred by homozygosity for a prion protein (PRNP) polymorphism
b. So far, three genes are known to cause autosomal dominant Alzheimer disease
c. Presenilin-1 is a low-penetrance gene for early-onset Alzheimer disease
d. Linkage studies are more important than twin studies in understanding the genetic contribution to bipolar illness
e. Persons with Down syndrome are at significant risk of Alzheimer disease

4. Hemochromatosis:

a. Is genetically heterogeneous
b. Affects more women than men
c. Can be treated by phlebotomy
d. Is fully penetrant
e. Is usually caused by a homozygous H63D mutation in the *HFE* gene

5. The following statements about venous thrombosis are true:

a. Factor V Leiden and the prothrombin G20210A variant are common causes of venous thrombosis
b. Testing for factor V Leiden and the prothrombin variant will change the management of a patient with venous thrombosis
c. The factor V Leiden mutation causes reduced expression of the factor V gene
d. The prothrombin variant causes increased expression of the prothrombin gene
e. Compound heterozygotes for factor V Leiden and the prothrombin variant have a 20-fold increased risk of venous thrombosis

CHAPTER 16: Congenital Abnormalities and Dysmorphic Syndromes

1.

a. Around 5% of all infant deaths are due to congenital abnormalities
b. At least half of all spontaneous miscarriages have a genetic basis
c. A major congenital abnormality affects approximately one newborn baby in every 100
d. Positional talipes is an example of a disruption to normal intrauterine development
e. Multiple abnormalities are sometimes the result of a sequence

2.

a. Down syndrome should more accurately be termed 'Down association'
b. Sotos syndrome, as with Down syndrome, is due to a chromosomal abnormality
c. Spina bifida affects approximately 2 per 1000 births
d. Infantile polycystic kidney disease is an example of a condition with different patterns of inheritance
e. Holoprosencephaly is an example of a condition with different patterns of inheritance

3.

a. Thalidomide embryopathy was an example of a disruption to normal intrauterine development
b. Talipes may be a consequence of renal agenesis
c. Limb defects are not caused by fetal exposure to sodium valproate
d. Symmetrical defects tend to feature in a dysplasia
e. Birth defects are unexplained in 20% of cases

4.

a. Congenital infection could lead to someone being both blind and deaf
b. The mid-trimester is the most dangerous time for a fetus to be exposed to a maternal infection
c. Vertebral body defects can be a consequence of poorly treated diabetes mellitus in the first trimester
d. A polymorphism in the methylenetetrahydrofolate reductase *(MTHFR)* gene is always associated with an increased risk of neural tube defect
e. Pulmonary stenosis is a feature of Noonan syndrome and congenital rubella

5.

a. Cleft lip–palate occurs more frequently than 1 in 1000 births
b. Associations generally have a high recurrence risk
c. The recurrence risk for a multifactorial condition can usually be determined by looking at patient's family pedigree
d. One cause of holoprosencephaly is a metabolic defect
e. Congenital heart disease affects 1 in 1000 babies

CHAPTER 17: Genetic Counseling

1.

a. The individual who seeks genetic counseling is the proband
b. Retinitis pigmentosa is not genetically heterogeneous
c. Genetic counseling is all about recurrence risks
d. The counselor's own opinion about a difficult choice is always helpful
e. Good counseling should not be measured by the patient's/client's ability to remember genetic risks

2.

a. First-cousin partnerships are 10 times more likely to have babies with congenital abnormalities than the general population
b. On average, a grandparent and grandchild share ¼ of their genes
c. Incestuous relationships virtually always result in severe learning difficulties in the offspring
d. Consanguinity should be regarded as extremely abnormal
e. Consanguinity refers exclusively to cousin marriages/partnerships

3.

a. Genetic disorders are accidents of nature, so guilt feelings are rare
b. Clear genetic counseling changes patients' reproductive decisions in virtually all cases
c. The chance of first cousins having their first child affected with an autosomal recessive condition due to a deleterious gene inherited from a grandparent is 1 in 32
d. Far more genetic testing of children for adoption takes place than for children reared by their birth parents
e. Patient support groups have little value given that modern medical genetics is so technically complex

CHAPTER 18: Chromosome Disorders

1.

a. The chromosome number in humans was discovered after the structure of DNA
b. The Turner syndrome karyotype is the most common single chromosome abnormality in spontaneous abortuses
c. The rate of miscarriage in Down syndrome is similar to the rate in karyotypically normal fetuses
d. Most babies with Down syndrome are born to mothers age younger than 30 years
e. All children with Down syndrome have to go to special school

2.

a. The life expectancy of children with trisomy 18 (Edwards syndrome) is about 2 years
b. 47,XYY males are fertile
c. The origin of Turner syndrome (45,X) can be in paternal meiosis
d. All persons with Angelman syndrome have a cytogenetically visible deletion on chromosome 15q
e. DiGeorge syndrome results from misaligned homologous recombination between flanking repeat gene clusters

3.

a. Premature vascular problems occur in adults with Williams syndrome
b. Congenital heart disease is a feature of Prader-Willi and Smith-Magenis syndromes
c. The Wilms tumor locus is on chromosome 13
d. Aniridia may be caused by either a gene mutation or a chromosome microdeletion
e. A child's behavior may help to make a diagnosis of a malformation syndrome

4.

a. Klinefelter syndrome affects approximately 1 in 10,000 male live births
b. Learning difficulties are common in Klinefelter syndrome
c. Chromosome mosaicism is commonly seen in Turner syndrome
d. Females with a karyotype 47,XXX are infertile
e. Chromosome breakage syndromes can cause cancer

5.

a. In fragile X syndrome, the triplet repeat does not change in size significantly when passed from father to daughter
b. Fragile X syndrome is a single, well-defined, condition
c. Girls with bilateral inguinal herniae should have their chromosomes tested
d. Normal karyotyping is a good way of diagnosing fragile X syndrome in girls
e. FISH analysis using multi-telomeric probes diagnoses about 25% of non-specific learning difficulties

CHAPTER 19: Single-Gene Disorders

1.

a. In Huntington disease (HD), an earlier age of onset in the offspring is more likely if the gene is passed from an affected mother rather than an affected father
b. In HD, those homozygous for the mutation are no more severely affected than those who are heterozygous
c. From the onset of HD, the average duration of the illness until a terminal event is 25–30 years
d. In HD, non-penetrance of the disease may be associated with low repeat abnormal alleles
e. Anticipation is a feature of the inheritance of myotonic dystrophy

2.

a. Insomnia is a feature of myotonic dystrophy
b. Myotonic dystrophy is a cause of neonatal hypertonia
c. The clinical effects of myotonic dystrophy are mediated through RNA
d. Cardiac conduction defects are a feature of myotonic dystrophy and ion channelopathies
e. The diagnosis of cystic fibrosis (CF) may come to light only in the infertility clinic

3.

a. In CF the R117H mutation is the most common one in northern Europe
b. In the *CFTR* gene a modifying intragenic polymorphism affects the phenotype
c. Hypertrophic cardiomyopathies are mostly due to mutations in ion channelopathy genes
d. Many different inherited muscular dystrophies can be linked to the complex that includes dystrophin (mutated in Duchenne and Becker muscular dystrophies)
e. Learning difficulties are part of spinal muscular atrophy

4.

a. Cystic fibrosis and hemophilia are unlikely candidates for gene therapy
b. An abnormal span : height ratio alone is a major feature of Marfan syndrome
c. Neurofibromatosis type 1 (NF1) sometimes 'skips generations'
d. Scoliosis can be a feature of both NF1 and Marfan syndrome
e. Cataracts can be a feature of NF1 but not of NF2

5.

a. HMSN types I and II refer to a genetic classification
b. HMSN can follow all major patterns of inheritance
c. It is the nerve sheath, rather than the nerve itself, that is altered in the most common form of HMSN
d. Estimation of the creatine kinase level and factor VIII level is good for identifying carriers of Duchenne dystrophy and hemophilia, respectively
e. Brugada syndrome is one of the varieties of spinal muscular atrophy

CHAPTER 20: Screening for Genetic Disease

1.

a. X-inactivation studies provide a useful means of identifying female carriers of some X-linked disorders
b. Reliable clinical signs to detect most carriers of X-linked disorders are lacking
c. DNA sequence variants are useful in targeted screening as long as they are not polymorphic
d. DNA markers are useful in targeted screening for retinitis pigmentosa
e. For the purposes of screening family members, opportunities should be taken for the banking of DNA from probands with lethal conditions

2.

a. Patients with presymptomatic tuberous sclerosis always have a characteristic facial rash
b. It is always possible to diagnose neurofibromatosis type 1 by age 2 years because it is a fully penetrant condition
c. Biochemical tests should not be considered as diagnostic genetic tests
d. Magnetic resonance imaging of the lumbar spine may be useful in diagnosing Marfan syndrome
e. Predictive genetic testing must always be done by direct gene analysis

3.

a. Population screening programs should be legally enforced
b. Population screening programs should be offered if some form of treatment or prevention is available
c. The sensitivity of a test refers to the extent to which the test detects only affected individuals
d. The positive predictive value of a screening test refers to the proportion of positive tests that are true positives
e. If there is no effective treatment for a late-onset condition, predictive genetic testing should be undertaken with great care

4.

a. A high proportion of people who undergo carrier testing cannot remember their result properly
b. Carrier screening for cystic fibrosis is the most useful program among Greek Cypriots
c. The possibility of a screening test leading to employment discrimination is not a major concern
d. Neonatal screening for Duchenne muscular dystrophy improves life expectancy
e. Neonatal screening for cystic fibrosis is a DNA-based test

5.

a. Newborn screening for hemochromatosis, the most common mutated gene in European populations, is a nationally managed program in the United Kingdom
b. The presymptomatic screening of children for adult-onset genetic disease is a decision made by the parents
c. Neonatal screening for phenylketonuria and congenital hypothyroidism are the longest-running screening programs
d. Screening for MCAD (medium-chain acyl-CoA dehydrogenase) deficiency is integrated into neonatal population screening
e. Genetic registers are mainly for research

CHAPTER 21: Prenatal Testing and Reproductive Genetics

1. In prenatal testing:

a. Amniocentesis is being routinely practiced earlier and earlier in pregnancy
b. The cells grown from amniocentesis originate purely from fetal skin
c. The risk of miscarriage is higher from chorionic villus sampling (CVS) than from amniocentesis
d. CVS is a safe procedure at 9 weeks' gestation
e. Fetal skin disorders can be diagnosed by ultrasonography

2. Regarding prenatal markers:

a. In Down syndrome pregnancies, maternal serum human chorionic gonadotropin (hCG) levels are usually raised
b. In Down syndrome pregnancies, maternal serum α-fetoprotein (αFP) concentration is usually reduced
c. In trisomy 18 pregnancies, maternal serum markers behave in just the same way as in Down syndrome pregnancies
d. About 95% of Down syndrome pregnancies are picked up by determining maternal age, serum αFP and hCG levels, and fetal nuchal translucency
e. Twin pregnancy is a cause of increased maternal serum αFP levels

3.

a. Chromosome mosaicism is detected in about 5% of CVS samples
b. Chromosome disorders are the main cause of abnormal nuchal translucency
c. Echogenic fetal bowel on ultrasonography is a risk factor for cystic fibrosis
d. For a couple who have had one child with Down syndrome, the risk in the next pregnancy is usually not greatly increased
e. Familial marker chromosomes are usually not clinically significant

4.

a. Donor insemination is a procedure not requiring a license from the HFEA
b. Surrogacy is illegal in the United Kingdom
c. For preimplantation genetic diagnosis (PGD), fertilization of the egg is achieved by intracytoplasmic sperm injection (ICSI)
d. The success rate from IVF, in terms of taking home a baby, is only about 50%
e. The largest group of diseases being tested in PGD is single-gene conditions

5.

a. There is an increased risk of genetic conditions in the children conceived by ICSI
b. The sperm of one donor may be used only five times
c. Children conceived by donor insemination are entitled to as much information as adopted children about their biological parents
d. If prenatal diagnosis could be achieved by analyzing fetal cells in the maternal circulation, the couple would still have to consider termination of pregnancy
e. Infertility affects about 1 in 20 couples

CHAPTER 22: Risk Calculation

1.

a. A probability of 0.5 is the same as a 50% risk
b. The probability of an event never exceeds unity
c. In a dizygotic twin pregnancy, the probability that the babies will be the same sex equals 0.5
d. Bayes' theorem takes account of both prior probability and conditional information
e. In an autosomal dominant condition, a penetrance of 0.7 means that 30% of heterozygotes will not manifest the disorder

2. For an autosomal recessive condition, the chance that the first cousin of an affected individual is a carrier is:

a. 1 in 8
b. 1 in 2
c. 1 in 4
d. 1 in 10
e. 1 in 6

3. In X-linked recessive inheritance:

a. The sons of a female carrier have a 1 in 4 chance of being affected
b. The mother of an affected male is an obligate carrier
c. The gonadal mosaicism risk in Duchenne muscular dystrophy may be as high as 10%
d. For a woman who has an affected son, her chance of being a carrier is reduced if she goes on to have three unaffected sons
e. A dummy consultand refers to an individual in a pedigree who is ignored when it comes to calculating risk

4. In autosomal recessive inheritance, the risk that the nephew of an affected individual, born to the affected individual's healthy sibling, is a carrier is:

a. 1 in 2
b. 1 in 4
c. 2 in 3
d. 1 in 3
e. 1 in 6

5.

a. In calculating risk, conditional information can include negative DNA data
b. In delayed onset of a dominantly inherited condition, calculation of heterozygote risk requires clinical expression data
c. Calculating odds ratios does not require information about prior probabilities
d. Empiric risks derived from epidemiological studies have limited application to a particular situation
e. When using DNA marker data to predict risk, the recombination fraction does not really matter

CHAPTER 23: Treatment of Genetic Disease

1. Methods currently used to treat genetic disease include:

a. Germ-cell gene therapy
b. Stem-cell transplantation
c. Enzyme/protein replacement
d. Dietary restriction
e. In situ repair of mutations by cellular DNA repair mechanism

2. Gene therapy may be delivered by:

a. Liposomes
b. Adeno-associated viruses
c. Antisense oligonucleotides
d. Lentiviruses
e. Injection of plasmid DNA

3. Gene therapy has been used successfully to treat patients with the following diseases:

a. Cystic fibrosis
b. Severe combined immunodeficiency (XL-SCID)
c. Sickle cell disease
d. Hemophilia
e. Adenosine deaminase deficiency

4. Potential gene therapy methods for cancer include:

a. Inhibition of fusion proteins
b. Stimulation of the immune system
c. Increased expression of the angiogenic factors
d. RNA interference
e. Antisense oligonucleotides

Case-Based Questions

CHAPTER 6: Developmental Genetics

Case 1

A 2-year-old child is referred to geneticists because of a large head circumference above the 97th centile, although it is growing parallel to the centile lines. The parents would like to have another child and are asking about the recurrence risk. The cerebral ventricles are dilated and there has been much discussion with the neurosurgeons about possible ventriculoperitoneal shunting. On taking a full family history, it emerges that the paternal grandmother is under review by dermatologists for skin lesions, some of which have been removed, and a paternal uncle has had some teeth cysts removed by a hospital dentist.

1. *Is there a diagnosis that embraces the various features in these different family members?*
2. *What investigations would be appropriate for the child's father, and what is the answer to the couple's question about recurrence risk?*

Case 2

A 4-year-old girl is brought to a pediatrician because of behavioral difficulties, including problems with potty training. The pediatrician decides to test the child's chromosomes because he has previously seen a case of 47,XXX (triple X) syndrome in which the girl had oppositional behavior. Somewhat to his surprise the chromosome result is 47,XY—i.e., the 'girl' is genetically 'male'.

1. *What are the most important causes of sex reversal in a 4-year-old child who is phenotypically female and otherwise physically healthy?*
2. *What should the pediatrician tell the parents, and which investigations should be performed?*

CHAPTER 7: Patterns of Inheritance

Case 1

A 34-year-old man has developed some spasticity of his legs in the past few years and his family has noted some memory problems and alteration in behavior. He has very brisk peripheral reflexes. He is seen with his mother in the genetic clinic and she is found to have significantly brisk peripheral reflexes on examination but has no health complaints. It emerges that her own father probably had problems similar to her son's in his thirties, but he was killed in the war.

1. *Which patterns of inheritance need to be considered in this scenario?*
2. *What diagnostic possibilities should be considered?*

Case 2

A couple has a child who suffers a number of bone fractures during early childhood after minor trauma and is told that this is probably a mild form of osteogenesis imperfecta. The parents did not suffer childhood fractures themselves, and when they have another child who also develops fractures they are told the inheritance is autosomal recessive. This includes an explanation that the affected children should not see the condition occur in their offspring in the future.

1. *Is the information given to the parents correct?*
2. *If not, what is the most likely pattern of inheritance and explanation for the sibling recurrence of fractures?*

CHAPTER 8: Population and Mathematical Genetics

Case 1

The incidence of a certain autosomal recessive disorder in population A is well established at approximately 1 in 10,000, whereas in population B the incidence of the same disorder is much higher at approximately 1 in 900. A man from the first population group and a woman from the second population group are planning to marry and start a family. Being aware of the relatively high incidence of the disorder in population B, they seek genetic counseling.

1. *What essential question must be asked of each individual?*
2. *What is the risk of the disorder occurring in their first pregnancy, based on application of the Hardy-Weinberg equilibrium?*

Case 2

Neurofibromatosis type 1 is a relatively common mendelian condition. In a population survey of 50,000 people in one town, 12 cases are identified, of which 8 all belong to one large affected family.

1. *Based on these figures, what is the mutation rate in the neurofibromin gene?*
2. *Name some limitations to the validity of calculating the mutation rate from a survey such as this.*

CHAPTER 9: Polygenic and Multifactorial Inheritance

Case 1

A 16-year-old requests oral contraceptives from her general practitioner. On taking a family history, it emerges that her mother had a deep vein thrombosis at the age of 40 years and died after a pulmonary embolism at age 55 years. There is no other relevant family history.

1. *What genetic testing is appropriate?*
2. *What are the limitations of testing in this situation?*

Case 2

A 35-year-old woman is diagnosed with diabetes and started on insulin treatment. She and her 29-year-old brother were adopted and have no contact with their birth parents. Her brother has no symptoms of hyperglycemia. Both have normal hearing and no other significant findings.

1. *What possible subtypes of diabetes might she have and what are the modes of inheritance of these subtypes?*
2. *For each of these subtypes, what is the risk of her brother developing diabetes?*

CHAPTER 10: Hemoglobin and the Hemoglobinopathies

Case 1

A Chinese couple residing in the United Kingdom has had two pregnancies and the outcome in both was a stillborn edematous baby (hydrops fetalis). These pregnancies occurred when they lived in Asia and they have no living children. They seek some genetic advice about the chances of this happening again, but no medical records are available for the pregnancies.

1. *What diagnostic possibilities should be considered?*
2. *What investigations are appropriate to this situation?*

Case 2

A young adolescent whose parents are of West Indian origin is admitted from accident and emergency after presenting with severe abdominal pain and some fever. An acute abdomen is suspected and the patient undergoes laparotomy for possible appendicitis. However, no surgical pathology is identified. Subsequently the urine appears dark.

1. *What other investigations might be appropriate at this stage?*
2. *What form of follow-up is appropriate?*

CHAPTER 11: Biochemical Genetics

Case 1

A 2-year-old boy, who has a baby sister age 4 months, is admitted to hospital with a vomiting illness and drowsiness. Despite his vomiting symptoms improving quite quickly with intravenous fluid support, his blood glucose remains low and intravenous fluids are required longer than might normally be expected. The parents say that something like this happened before, although he recovered without seeing a doctor.

1. *What does this history suggest?*
2. *What investigations are appropriate?*

Case 2

A 28-year-old woman has become aware over several years that she does not have the same energy as she did at the age of 20 years. She tires relatively easily on exertion and family members have noticed that she has developed slightly droopy eyelids as well as wondering whether her hearing is deteriorating, which she vigorously denies.

1. *How might a detailed family history help towards a diagnosis in this case?*
2. *What investigations should be performed?*

CHAPTER 13: Immunogenetics

Case 1

A 32-year-old man has had low back pain and stiffness for 2 years and recently developed some irritation in his eyes. Radiography is performed and a diagnosis of ankylosing spondylitis made. He remembers his maternal grandfather having similar back problems as well as arthritis in other joints. He has three young children.

1. *Is it likely that his grandfather also had ankylosing spondylitis?*
2. *What is the risk of passing the condition to his three children?*

Case 2

A 4-year-old girl suffers frequent upper respiratory infections with chest involvement, and each episode lasts longer than in her preschool peers. Doctors have always assumed this is somehow a consequence of her stormy early months, when she had major heart surgery for tetralogy of Fallot. She also has nasal speech and in her neonatal record a radiologist commented on the small size of the thymus gland.

1. *Is there another diagnosis that could explain her frequent and prolonged upper respiratory infections?*
2. *What further management of the family is indicated?*

CHAPTER 14: Cancer Genetics

Case 1

A 38-year-old woman, who recently had a mastectomy for breast cancer, requests a referral to the genetic service. Her father had some bowel polyps removed in his fifties and a cousin on the same side of the family had some form of thyroid cancer in her forties. The general practitioner consults a set of guidelines that suggest a familial form of breast cancer is unlikely because she is the only one affected, even though quite young. He is reluctant to refer her.

1. *Could the history suggest another familial condition? If so, which one?*
2. *What other clinical features might give a clue to the diagnosis?*

Case 2

A 30-year-old is referred for genetic counseling because she is concerned about her risk of developing breast cancer. The consultand's mother has recently been diagnosed with breast cancer at 55 years of age. Her brother's daughter (the consultand's cousin) had bilateral breast cancer diagnosed at age 38 years and died 5 years ago from metastatic disease. The cousin had participated in a research study that identified a *BRCA2* gene mutation. The clinical geneticist suggests that the consultand's mother should be tested before predictive testing is offered to her daughter. They are surprised when a negative result is reported by the laboratory.

1. *What are the possible explanations for this result?*
2. *What is the risk of the consultand's uncle developing breast cancer?*

CHAPTER 15: Genetic Factors in Common Diseases

Case 1

A 2-year-old girl presents with partial seizures. The episode is brief and unaccompanied by fever. Because the child is well with no neurological deficit, a decision is made not to treat with an anticonvulsant drug. A year later she suffers a generalized seizure, again without fever. On this occasion, her 30-year-old mother asks whether this might have anything to do with her own seizures that began at the age of 15 years, although she has had only two episodes since. She had undergone computed tomography of the brain and the doctors mentioned a condition whose name she could not remember. Magnetic resonance imaging of the child's brain shows uncalcified nodules on the lateral ventricular walls.

1. *The mother asks whether the epilepsy is genetic and whether it could happen again if she has another child. What can she be told?*
2. *What diagnoses should be considered and can genetic testing be offered?*

Case 2

A 5-year-old child is admitted to hospital with an unexplained fever and found to have a raised blood glucose level. He makes a good recovery, but 2 weeks later his fasting blood glucose level is shown to be increased at 7 mmol/L. There is a strong family history of diabetes on his mother's side, with his mother, maternal uncle, and maternal grandfather all affected. His father has no symptoms of diabetes, but his sister had gestational diabetes during her recent pregnancy. Molecular genetic testing identifies a heterozygous glucokinase gene mutation in the child.

1. *The parents believe that their son's hyperglycemia is inherited from the mother's side of the family. Is this correct?*
2. *What are the consequences of finding a glucokinase gene mutation for this family?*

CHAPTER 16: Congenital Abnormalities and Dysmorphic Syndromes

Case 1

A young couple has just lost their first pregnancy through fetal abnormality. Polyhydramnios was diagnosed on ultrasonography as well as a small fetal kidney on one side. Amniocentesis was performed and the karyotype showed a normal 46,XY pattern. The couple was unsure what to do, but eventually elected for a termination of pregnancy at 21 weeks. They were very upset and did not want any further investigations performed, including an autopsy. They did agree to a whole-body radiography of the fetus and some of the upper thoracic vertebrae were misshapen.

1. *The couple asks whether such a problem could recur—they do not feel they can go through this again. What can they be told?*
2. *What further investigations might have helped to inform the genetic risk?*

Case 2

On routine neonatal examination on the second day, a baby is found to have a cleft palate. The pregnancy was uneventful with no exposure to potential teratogens, and the family history is negative. The pediatric registrar also wonders whether the limbs are slightly short. The baby's birth weight is on the 25th percentile, with length on the 2nd percentile.

1. *What diagnoses might be considered?*
2. *What are the management issues in a case like this?*

CHAPTER 17: Genetic Counseling

Case 1

A couple has a son with dysmorphic features, short stature, and moderately severe developmental delay. On the second occasion when his karyotype is analyzed, he is found to have a subtle chromosome translocation that was missed first time. The father is found to be a balanced translocation carrier. His family has always blamed the mother for the child's condition because of her history of drug abuse, with the result that the couple no longer talks to his wider family. However, through friends he has learned that his sister is trying to start a family.

1. *What are the important genetic issues?*
2. *What other issues does this case raise?*

Case 2

A couple has a child who is diagnosed with cystic fibrosis (CF) at age 18 months. The child is homozygous for the common ΔF508 mutation. They request prenatal diagnosis in the next pregnancy, but DNA analysis shows that the father is not a carrier of ΔF508. It must be assumed he is not the biological father of the child with CF, and this is confirmed when further analysis shows that the child does not have a haplotype in common with him.

1. *What medical issue does this information raise?*
2. *What counseling issues are raised by these results?*

CHAPTER 18: Chromosome Disorders

Case 1

A newborn baby girl looks somewhat dysmorphic, is diagnosed with an atrioventricular septal defect (AVSD) congenital heart, and the pediatricians think this may be Down syndrome. This is discussed with the parents and a karyotype performed. The result comes back as normal: 46,XX. The baby is very 'good' during infancy with very little crying, and no further investigations are done. Subsequently the child shows global developmental delay, head-banging, wakes every night for about 3 hours, and has mild brachydactyly. The pediatricians refer her to a geneticist for an opinion.

1. *Does the history suggest a diagnosis?*
2. *What investigation should be requested?*

Case 2

The parents of a 10-year-old girl seek a follow-up appointment in the genetics clinic. At the age of 4 years, she had some behavioral problems and chromosomes were tested from a blood sample. The result came back as 47,XXX, and it was explained that these girls sometimes have behavioral problems, are usually tall, fertility is normal, and 'everything would be alright'. However, by age 10 years, she is the smallest girl in the class and still has the slightly webbed neck that was present at birth.

1. *What diagnosis should be considered and what investigation should now be offered?*
2. *How are the genetic counseling and future management modified by the new diagnosis?*

CHAPTER 19: Single-Gene Disorders

Case 1

A 31-year-old woman would like to start a family but is worried because her 39-year-old brother was diagnosed as having Becker muscular dystrophy nearly 30 years ago and she remembers having being told that the condition affects boys but the women pass it on. Her brother is still living, but is now quite disabled by his condition. There is no wider family history of muscular dystrophy.

1. *Is the original diagnosis reliable?*
2. *What are the next steps in managing this situation?*

Case 2

A middle-aged couple is devastated when their 21-year-old daughter collapses at a dance and cannot be resuscitated. At post-mortem examination, all toxicology tests are negative and no cause of death is found. The mother recalls that her father died suddenly in his fifties from what was presumed at the time to be a cardiac cause, and her sister has had some dizzy spells but has not thought it necessary to see her doctor. The couple has three other children who are young, sport-loving adults and wonder whether this could happen again.

1. *What investigations are appropriate?*
2. *What advice should the family be given?*

CHAPTER 20: Screening for Genetic Disease

Case 1

A 32-year-old man is tall and thin, and 20 years ago his father died suddenly at age 50 years. The general practitioner wonders whether his patient has Marfan syndrome and refers him to a genetics clinic. He has some features of Marfan syndrome but, strictly speaking, would meet the accepted criteria only if the family history was definitely positive for the disorder. He has a brother of average height and three young children who are in good health.

1. *In terms of genetic testing, what are the limitations to screening if the diagnosis is Marfan syndrome?*
2. *What are the screening issues for the family?*

Case 2

A screening test for cystic fibrosis (CF) is being evaluated on a population of 100,000 newborn babies. The test is positive in 805 babies, of whom 45 are eventually shown to have CF by a combination of DNA analysis and sweat testing. Of those babies whose screening test is negative, five subsequently develop symptoms and are diagnosed with CF.

1. *What is the sensitivity and specificity of this screening test?*
2. *What is the positive predictive value of the screening test?*

CHAPTER 21: Prenatal Testing and Reproductive Genetics

Case 1

A 36-year-old pregnant woman elects to undergo prenatal testing by chorionic villus biopsy after the finding of increased nuchal translucency on ultrasonography. The initial result, using FISH probes, is good news—there is no evidence for trisomy 21—and the woman is greatly relieved.

However, on the cultured cells more than 2 weeks later, it emerges that there is mosaicism for trisomy 20. She undergoes amniocentesis a week later, and 3 weeks after that the result also shows some cells with trisomy 20.

1. *Why was an amniocentesis performed in addition to the chorionic villus biopsy?*
2. *What else can be done following the amniocentesis result?*

Case 2

A couple has two autistic sons and would very much like to have another child. They are prepared to do anything to ensure that the problem does not recur. They acquire a lot of information from websites and learn that boys are more commonly affected—the male: female sex ratio is approximately 4:1. As they see it, the simple solution to their problem is sex selection by preimplantation genetic diagnosis (PGD).

1. *What investigations might be performed on the autistic sons?*
2. *If tests on the sons fail to identify a diagnosis, can the request of the couple for sex selection by PGD be supported by the geneticist?*

CHAPTER 22: Risk Calculation

Case 1

In the pedigree shown below, two cousins have married and would like to start a family. However, their uncle died many years ago from Hurler syndrome, one of the mucopolysaccharidoses, an inborn error of metabolism following autosomal recessive inheritance. No tissue samples are available for genetic studies.

1. *What is the risk that the couple's first child will be affected by Hurler syndrome?*
2. *Can the couple be offered anything more than a risk figure?*

Case 2

A woman has a brother and a maternal uncle affected by hemophilia A. She herself has had two unaffected sons and would like more children. She is referred to a genetics clinic to discuss the risk and the options.

1. *Purely on the basis of the information given, what is the woman's carrier risk for hemophilia A?*
2. *Can anything be done to modify her risk?*

Multiple-Choice Answers

CHAPTER 2: The Cellular and Molecular Basis of Inheritance

1. Base substitutions:

a. **True.** When a stop codon replaces an amino acid
b. **True.** For example, by mutation of conserved splice donor and acceptor sites
c. **False.** Silent mutations or substitutions in non-coding regions may not be pathogenic
d. **True.** For example promoter mutations may affect binding of transcription factors
e. **False.** Frameshifts are caused by the insertion or deletion of nucleotides

2. Transcription:

a. **False.** During transcription mRNA is produced from the DNA template
b. **True.** The mRNA product is then translocated to the cytoplasm
c. **True.** The mRNA is complementary to the antisense strand
d. **False.** Transcription factors bind to regulatory sequences within the promoter
e. **True.** The addition of the 5′ cap and 3′ poly(A) tail facilitates transport to the cytoplasm

3. The following are directly involved in DNA repair:

a. **True.** The DNA glycosylase MYH is involved in base excision repair (BER)
b. **True.** They incorporate the correct bases
c. **True.** They seal gaps after abnormal base excision and correct base insertion
d. **False.** Splicing removes introns during mRNA production
e. **False.** Ribosomes are involved in translation

4. During DNA replication:

a. **True.** It unwinds the DNA helix
b. **False.** Replication occurs in both directions
c. **True.** These fragments are joined by DNA ligase to form the lagging strand
d. **False.** DNA replication is semiconservative as only one strand is newly synthesized
e. **False.** Uracil is incorporated in mRNA, thymine in DNA

CHAPTER 3: Chromosomes and Cell Division

1. Meiosis differs from mitosis in the following ways:

a. **True.** During human meiosis, the number of chromosomes is reduced from 46 to 23
b. **False.** Early cell divisions in gametogenesis are mitotic; meiosis occurs only at the final division
c. **True.** In meiosis, the two divisions are known as meiosis I and II
d. **True.** The bivalents separate independently during meiosis I, and crossovers (chiasmata) occur between homologous chromosomes
e. **False.** The five stages of meiosis I prophase are leptotene, zygotene, pachytene, diplotene, and diakinesis

2. Chromosome abnormalities reliably detected by light microscopy include:

a. **True.** An extra chromosome (e.g., chromosome 21 in Down syndrome) is easily seen
b. **True.** A missing chromosome (e.g., Turner syndrome in females with a single X) is easily seen
c. **False.** A subtle translocation may not be visible
d. **False.** A small deletion may not be visible
e. **True.** Centric fusion of the long arms of two acrocentric chromosomes is readily detected

3. Fluorescent in situ hybridization using whole-chromosome (painting) or specific locus probes enables routine detection of:

a. **False.** Changes in gene dosage may be identified by comparative genomic hybridization (CGH)
b. **True.** Subtelomeric probes are useful in the investigation of non-specific learning difficulties
c. **True.** Trisomies can be detected in interphase cells
d. **True.** The origin of marker chromosomes can be determined by chromosome painting
e. **True.** Subtle rearrangements can be detected by chromosome painting

4. Chemicals used in the preparation of metaphase chromosomes for analysis by light microscopy include:

a. **True.** Colchicine inhibits spindle formation, thus arresting cells at metaphase
b. **True.** Phytohemagglutinin stimulates cell division of T lymphocytes
c. **True.** Giemsa is used to stain chromosomes a pink/purple color
d. **False.** Quinacrine is a fluorescent stain not visible by light microscopy
e. **True.** Hypotonic saline swells the cells, causing cell lysis and spreading of chromosomes

CHAPTER 4: DNA Technology and Applications

1. The following statements apply to restriction enzymes:

a. **True.** Double-stranded DNA can be digested to give overhanging (sticky) ends or blunt ends
b. **False.** More than 300 restriction enzymes have been isolated from various bacteria
c. **True.** If the mutation creates or destroys a recognition site
d. **True.** DNA digestion by a restriction enzyme is the first step in Southern blotting
e. **False.** They are endonucleases as they digest DNA fragments internally, as opposed to exonuclease digestion from the 5′ or 3′ ends of DNA fragments

2. The following describe polymerase chain reaction (PCR):

a. **True.** Millions of copies of DNA can be produced from one template without using cloning vectors
b. **False.** PCR uses the heat-stable Taq polymerase, because a high denaturing temperature (around 95°C) is required to separate double-stranded products at the start of each cycle
c. **True.** PCR may be used to amplify DNA from single cells (e.g., in preimplantation genetic diagnosis); therefore, appropriate control measures are important to avoid contamination
d. **False.** PCR routinely amplifies targets of up to 1 kb and long-range PCR is limited to around 40 kb
e. **False.** Knowledge of the sequence is required to design primers to flank the region of interest

3. Types of nucleic acid hybridization include:

a. **True.** Southern blotting describes the hybridization of a radioactively labeled probe with DNA fragments separated by electrophoresis
b. **True.** Hybridization between the target and probe DNA takes place on a glass slide
c. **False.** Western blotting is used to analyze protein expression using antibody detection methods
d. **True.** Northern blotting is used to examine RNA expression
e. **True.** DNA fingerprinting employs a minisatellite DNA probe to hybridize to hypervariable DNA fragments

4. The following techniques can be used to screen genes for unknown mutations:

a. **True.** Sequencing can be used to detect known or unknown mutations and will characterize an unknown mutation
b. **True.** SSCP is an inexpensive method for mutation screening although its sensitivity is limited
c. **True.** DHPLC is an efficient method for detecting heterozygous mutations
d. **False.** Oligonucleotide ligation assay is used to detect known mutations as the probe design is mutation specific
e. **False.** Real-time PCR is also used to detect known mutations as the probe design is mutation specific

CHAPTER 5: Mapping and Identifying Genes for Monogenic Disorders

1. Positional cloning uses:

a. **True.** Now that the human genome sequence is complete, it is possible to identify a disease-associated gene in silico
b. **True.** After a gene has been mapped to a region, it can be helpful to check for syntenic regions in animal models
c. **True.** Many genes have been identified through mapping of translocation or deletion break-points
d. **False.** Positional cloning describes the search for genes based on their chromosomal location
e. **True.** A genome-wide scan uses microsatellite markers located throughout the genome for linkage mapping

2. A candidate gene is likely to be a disease associated gene if:

a. **True.** This implies causality
b. **True.** This is strong evidence
c. **True.** This excludes the possibility that a single variant is a marker in linkage disequilibrium rather than a pathogenic mutation
d. **True.** For example, a gene associated with blindness might be expected to be expressed in the eye
e. **False.** A pseudogene does not encode a functional protein and mutations are therefore unlikely to be pathogenic

3. Achievements of the Human Genome Project include:

a. **False.** The draft sequence was completed in 2000, but its publication date was February 2001
b. **True.** Sequencing was finished 2 years ahead of the original schedule
c. **True.** Annotation tools such as Ensembl were developed to assist users
d. **False.** Nearly 1500 have been identified to date
e. **True.** Around 5% of the US budget for the Human Genome Project was devoted to studying these issues

CHAPTER 6: Developmental Genetics

1. In development, *HOX* genes:

a. **True.** They are important in spatial determination and patterning
b. **False.** Only a few malformation syndromes can be directly attributed to *HOX* gene mutations at present, probably because of paralogous compensation
c. **False.** They contain an important conserved homeobox of 180 bp
d. **True.** They are probably important only in early development
e. **True.** Where malformation-causing mutations have been identified, different organ systems may be involved, e.g., the hand–foot–genital syndrome (HOXA13)

2. In the embryo and fetus:

a. **False.** This occurs later and is the process of laying down the primary body axis in the second and third weeks
b. **False.** Organogenesis takes places mainly between 4 and 18 weeks' gestation
c. **True.** The genes in these pathways are expressed widely throughout the body
d. **False.** Somites form in a rostro-caudal direction
e. **True.** When mutated these genes lead to the ulnar–mammary syndrome and Holt-Oram syndrome

3. Concerning development pathways and processes:

a. **False.** It is formed from the first pharyngeal (branchial) arch
b. **True.** Remodeling occurs so that these vessels become the great arteries
c. **True.** This has been established in animals and is proving to be highly likely in humans
d. **False.** Achondroplasia is associated with very specific *FGFR3* mutations affecting the membrane-bound part of the protein
e. **False.** These different types of mutation usually cause widely differing phenotypes (e.g., the *RET* gene)

4. Regarding the X-chromosome:

a. **True.** Sometimes the *SRY* gene is involved in recombination with the pseudoautosomal regions of X and Y
b. **False.** Not all regions of the X are switched off; otherwise, there would presumably be no phenotypic effects in Turner syndrome
c. **True.** However, only when there is a mutated gene on one X chromosome does this have any consequences
d. **False.** *SRY* has an important initiating function, but other genes are very important
e. **True.** Some unusual phenomena occur in twins leading to the conclusion that these processes are linked

5. Transcription factors:

a. **False.** They are usually proteins that bind to specific regulatory DNA sequences
b. **False.** They also switch genes on
c. **True.** For example, a leg might develop in place of an antenna
d. **False.** Transcription factors are crucial to normal laterality
e. **True.** The zinc finger motif encodes a finger-like projection of amino acids that forms a complex with a zinc ion

CHAPTER 7: Patterns of Inheritance

1. Concerning autosomal recessive inheritance:

a. **False.** Sex ratio is equal
b. **False.** The risk at the time of conception is ½
c. **True.** All people carry mutated genes; cousins are more likely to share a mutated gene inherited from a common grandparent
d. **True.** Affected individuals would have to partner a carrier or another affected person for their offspring to be affected as well
e. **False.** The mechanisms causing Angelman syndrome are varied but autosomal recessive inheritance is not one of them

2. Concerning X-linked inheritance:

a. **True.** A father passes his Y chromosome to his son
b. **True.** He might have affected grandsons through his daughters who are carriers
c. **False.** Although the condition affects females, the males are more severely affected because the female has a normal copy of the gene on her other X chromosome, and X-inactivation means that the normal copy is expressed in about half of her tissues
d. **True.** All daughters of an affected man will be affected, but none of his sons
e. **False.** Germline mosaicism always needs to be considered when an isolated case of an X-linked condition occurs

3. In mitochondrial genetics:

a. **False.** This refers to two populations of mitochondrial DNA, one normal, one mutated
b. **False.** The opposite, probably because they replicate more frequently
c. **False.** Any tissue with mitochondria can be affected
d. **True.** If the affected woman's oocytes contain only mutated mitochondria
e. **False.** Many mitochondrial proteins are encoded by nuclear genes

4. Concerning terminology:

a. **False.** The same disease caused by different genes—but not necessarily on different chromosomes
b. **False.** The basic pattern of inheritance in pseudodominance is autosomal recessive
c. **True.** A proportion of individuals with the mutated gene show no signs or symptoms
d. **True.** Diseases showing anticipation demonstrate increased severity, and earlier age of onset, with succeeding generations
e. **False.** Not a variation in severity (variable expression), but two or more apparently unrelated effects from the same gene

5. In inheritance:

a. **True.** Both copies of a mutated gene can be passed to a child this way
b. **True.** This explains a proportion of cases of Prader-Willi and Angelman syndromes
c. **False.** Digenic inheritance refers to heterozygosity for two different genes causing a phenotype
d. **True.** This explains presenile baldness and gout
e. **False.** Only a small proportion

CHAPTER 8: Mathematical and Population Genetics

1. In applying the Hardy-Weinberg equilibrium, the following assumptions are made:

a. **False.** The population should be large to increase the likelihood of non-random mating
b. **True.** Consanguinity is a form of non-random mating
c. **True.** The introduction of new alleles introduces variables
d. **True.** In theory, if the population is small and donors are used many times this would introduce a form of non-random mating
e. **True.** Migration introduces new alleles

2. If the population incidence of a recessive disease is 1 in 10,000, the carrier frequency in the population is:

a. **False.**
b. **False.**
c. **False.**
d. **True.** The carrier frequency is double the square root of the incidence
e. **False.**

3. Heterozygote advantage:

a. **False.** It refers to conditions that follow autosomal recessive inheritance
b. **True.** The homozygote may show markedly reduced biological fitness (e.g., cystic fibrosis)
c. **True.** People with sickle-cell trait are more able to remove parasitized cells from the circulation
d. **True.** A process of selective advantage may be at work
e. **False.** The presence of the allele in a population may indeed be a founder effect

4. Polymorphic loci:

a. **False.** The alleles need have frequencies of only 1%
b. **True.** They are crucial to gene mapping by virtue of their co-segregation with disease
c. **True.** Linkage analysis using polymorphic loci may be the only way to determine genetic status in presymptomatic diagnosis and prenatal testing
d. **False.** The association of polymorphic loci with disease segregation is key to calculating a logarithmic of the odds score
e. **False.** They may be important (e.g., blood groups)

5. In population genetics:

a. **False.** The incidence of the disease must also be known
b. **True.** In autosomal recessive disease most of the genes in the population are present in unaffected heterozygotes
c. **True.** In recessive conditions unaffected sibships will not be ascertained
d. **True.** Basically, only males are affected and they always manifest the condition
e. **False.** It is useful only when there is a common ancestor from both sides of the family, (i.e., inbreeding)

CHAPTER 9: Polygenic and Multifactorial Inheritance

1. Concerning autism:

a. **False.** It is a neurodevelopmental disorder and no metabolic abnormalities are found
b. **False.** This would imply autosomal dominant inheritance. The rate is about 20%
c. **False.** Although autism occurs in fragile X syndrome the vast majority of affected individuals do not have this condition
d. **True.** The figure is nearly 3% for full-blown autism and a further 3% for milder features—autistic spectrum disorder
e. **False.** The male:female ratio is approximately 4:1

2. Linkage analysis is more difficult in multifactorial conditions than in single-gene disorders because:

a. **True.** Detection of polygenes with small effects is very difficult
b. **True.** In a fully penetrant single-gene disorder, it is easier to find families with sufficient informative meioses
c. **True.** Parametric linkage analysis requires that the mode of inheritance is known
d. **True.** Different genetic and environmental factors may be involved
e. **True.** The late age of onset means that affection status may be uncertain

3. Association studies:

a. **True.** The disease and variant tested may be common in a population subset but there is no causal relationship
b. **True.** The TDT test uses family controls and thus avoids population stratification effects
c. **True.** Replication of positive studies in different populations will increase the evidence for an association
d. **False.** Association studies are used to test variants identified by gene mapping techniques, including affected sibling-pair analysis
e. **True.** Variants with small effects may be missed if the patients and controls are not closely matched

4. Variants in genes that confer susceptibility to type 2 diabetes (T2DM) have been found:

a. **True.** The calpain-10 gene was identified by positional cloning in Mexican-American sibling pairs
b. **False.** No confirmed T2DM susceptibility genes have been identified by this approach
c. **True.** Examples include two subtypes of maturity-onset diabetes of the young (MODY)
d. **True.** The genes encoding the potassium channel subunits in the pancreatic β-cell were biological candidates
e. **True.** For example, the HNF-1A variant G319S has been reported only in the Oji Cree population

5. Variants in the *NOD2/CARD15* gene:

a. **False.** Evidence to date supports a role in Crohn disease, but not ulcerative colitis
b. **True.** Increased risk is estimated at 40-fold for homozygotes and 2.5-fold for heterozygotes
c. **True.** A genome-wide scan for inflammatory bowel disease (IBD) initially identified the 16p12 region
d. **False.** The *NOD2/CARD15* gene activates NF-κB, but this complex is already targeted by the most effective drugs used to treat Crohn disease
e. **False.** The three reported variants are found at a frequency of 5% in the general population, compared with 15% in patients with Crohn disease

CHAPTER 10: Hemoglobin and the Hemoglobinopathies

1. For different hemoglobins:

a. **True.** The population should be large to increase the likelihood of non-random mating
b. **False.** This is true for the α and γ chains only; the β chain appears toward the end of fetal life
c. **False.** There are too few α chains, which are replaced by β chains
d. **False.** It is a form of α-thalassemia
e. **False.** They have a mild anemia and clinical symptoms are rare

2. Regarding sickle-cell disease:

a. **False.** The effect is due to reduced solubility and polymerization
b. **True.** Obstruction of arteries can be the result of sickling crises
c. **True.** A valine residue is substituted for a glutamic acid residue
d. **False.** Life-threatening sepsis can result from splenic infarction
e. **True.** These mutations give rise to an amino acid substitution

3. Concerning hemoglobin variants:

a. **True.** This applies to the majority of those known
b. **False.** All types of mutation are known
c. **False.** Bone marrow hyperplasia occurs, which leads to physical changes such as a thickened calvarium
d. **True.** Oxygen is not released so readily to tissues
e. **True.** Hb H, for example, is unstable

4. Regarding hemoglobins during life:

a. **False.** It is a hereditary condition
b. **False.** This is true only between about 2 and 7 months' gestation
c. **False.** The bone marrow starts producing Hb from 6 to 7 months of fetal life
d. **False.** Production ceases from 2 to 3 months of postnatal life
e. **True.** It gives rise to no symptoms—the Hb chains produced are normal

CHAPTER 11: Biochemical Genetics

1. In congenital adrenal hyperplasia (CAH):

a. **True.** The most common enzyme defect is 21-hydroxylase deficiency
b. **True.** This occurs in the rare forms: 3β-dehydrogenase, 5α-reductase, and desmolase deficiencies
c. **True.** Hyponatremia and hyperkalemia may be severe and lead to circulatory collapse
d. **False.** Cortisol and fludrocortisone are required lifelong in salt-losing CAH
e. **False.** Fertility is reduced in the salt-losing form

2. Phenylketonuria:

a. **False.** There is a benign form as well as abnormalities of cofactor synthesis
b. **False.** Dietary restriction of phenylalanine is necessary only during childhood and pregnancy
c. **True.** These are features if untreated
d. **True.** Affected individuals have reduced pigment and are fair
e. **False.** A different pathway

3. Hepatomegaly is an important feature of:

a. **True.** Hepatomegaly is a feature of most of the mucopolysaccharidoses
b. **True.** Hepatomegaly is a feature of most of the glycogen storage disorders, although not all
c. **False.** This is not a feature, even in the so-called hepatic porphyrias
d. **True.** This is one of the sphingolipidoses—lipid storage diseases
e. **False.** Cirrhosis can occur in the untreated

4. Concerning mitochondrial disorders:

a. **False.** The main patterns of inheritance also apply where mitochondrial proteins are encoded by nuclear genes
b. **True.** Especially in NARP and MIDD, respectively
c. **True.** There are 37 gene products
d. **False.** Leigh disease is genetically heterogeneous
e. **True.** The *G4.5* gene is mutated, urinary 3-methyglutaconic acid is raised, but the link remains to be elucidated

5. Regarding metabolic conditions:

a. **True.** The carnitine cycle is important for long-chain fatty acid transport into mitochondria
b. **True.** 90% of alleles are due to the same mutation and neonatal population screening has been suggested
c. **False.** These are inborn errors of copper transport metabolism
d. **True.** These features should prompt investigation for organic acidurias and mitochondrial disorders, among others
e. **False.** Important radiological features may be seen in peroxisomal and storage disorders

CHAPTER 12: Pharmacogenetics

1. Thiopurine drugs used to treat leukemia:

a. **True.**
b. **True.** They are used to treat autoimmune disorders and to prevent rejection of organ transplants
c. **False.** They can be toxic in 10% to 15% of patients
d. **True.** These include leukopenia and severe liver damage
e. **True.** Variants in the TPMT gene are associated with thiopurine toxicity

2. Liver enzymes that show genetic variation of expression and hence influence the response to drugs include:

a. **True.** Complete deficiency of this enzyme causes type 1 Crigler-Najjar disease
b. **False.** N-acetyltransferase (NAT2) variation influences the metabolism of isoniazid
c. **False.** Absence of ALDH2 (acetaldehyde dehydrogenase) is associated with an acute flushing response to alcohol
d. **True.** Approximately 5% to 10% of the European population is poor metabolizers of debrisoquine because of a homozygous variant in the CYP2D6 gene
e. **True.** CYP2C9 variants are associated with decreased metabolism of warfarin

3. The following have an increased risk of complications from anesthesia:
- a. **True.** Malignant hyperthermia is dominantly inherited
- b. **True.** RYR1 mutations are associated with malignant hyperthermia
- c. **False.** G6DP deficiency results in sensitivity to certain drugs and fava beans
- d. **True.** Homozygous CHE1 mutations cause suxamethonium sensitivity
- e. **False.** Pseudocholinesterase metabolizes suxamethonium; therefore, decreased levels are associated with suxamethonium sensitivity

4. Examples of diseases in which treatment may be influenced by pharmacogenetics include:
- a. **False.** Patients with glucokinase mutations are usually treated with diet alone
- b. **True.** Patients with HNF1A mutations are sensitive to sulfonylureas
- c. **True.** Abacavir is an effective drug but approximately 5% of patients show potentially fatal hypersensitivity
- d. **True.** Some patients show adverse reactions to the drug felbamate
- e. **True.** Slow inactivators of isoniazid are more likely to suffer side effects

CHAPTER 13: Immunogenetics

1. Concerning complement:
- a. **False.** The cascade can also be activated by the alternative pathway
- b. **True.** C4 levels are reduced and production of C2b is uncontrolled
- c. **False.** C3 levels are normal, C4 levels are reduced
- d. **True.** C3b adheres to the surface of microorganisms
- e. **False.** Complement is a series of at least 20 interacting plasma proteins

2.
- a. **False.** It is made up of four polypeptide chains—two 'light' and two 'heavy'
- b. **False.** They are distributed around different chromosomes
- c. **False.** Donors are likely to share HLA haplotypes, which are crucial to tissue compatibility
- d. **True.** These are variable, diversity, junctional and constant regions
- e. **True.** Antigen receptors contain two immunoglobulin-like domains

3.
- a. **False.** They are protected for only 3 to 6 months
- b. **False.** X-linked SCID is 50% to 60% of the total
- c. **True.** B-cell positive SCID due to JAK3 deficiency can be subclinical
- d. **True.** A defect in either T-cell function or development
- e. **False.** CGD is an X-linked disorder of cell-mediated immunity

4.
- a. **False.** It is classed as a secondary or associated immunodeficiency
- b. **True.** Immunodeficiency is usually mild and the immune system improves with age as the thymus grows; there is a proneness to viral infections in childhood
- c. **False.** The causes of common invariable immunodeficiency are not known and it is often a disorder of adult life
- d. **False.** The risk to first-degree relatives is increased but the pedigree pattern is more suggestive of multifactorial conditions
- e. **True.** FTT may be the only clue to an immunodeficiency disorder

CHAPTER 14: Cancer Genetics

1.
- a. **True.** The best known example is chronic myeloid leukemia and the Philadelphia chromosome
- b. **False.** Tumor suppressor genes are more common than oncogenes
- c. **True.** Apoptosis is normal programmed cell death
- d. **False.** LOH refers to the presence of two defective alleles in a tumor suppressor gene
- e. **False.** Although important, APC mutations are part of a sequence of genetic changes leading to colonic cancer

2.
- a. **True.** The paradigm was retinoblastoma and the hypothesis was subsequently proved to be correct
- b. **False.** Mutations in *TP53* are found in many cancers, but are germline in Li-Fraumeni syndrome
- c. **False.** It is implicated in MEN-2, but not MEN-1
- d. **True.** There is a significant risk of small bowel polyps and duodenal cancer
- e. **True.** Women with this condition have a lifetime risk of up to 50%

3.

a. **True.** This syndrome is allelic with Cowden disease, in which papillary thyroid cancer can occur
b. **True.** The lifetime risk may be in the region of 16%
c. **False.** The *BRCA1* and *BRCA2* genes do not account for all familial breast cancer
d. **False.** The lifetime risk of breast cancer for female *BRCA1* or *BRCA2* carriers is 60% to 85%
e. **False.** The figure is approximately 15%

4.

a. **False.** Cerebellar hemangioblastoma is a common tumor in VHL
b. **False.** This tumor is seen in MEN-2 and VHL disease
c. **True.** There is also an increased risk in familial breast cancer
d. **True.** Melanin spots in Peutz-Jegher syndrome, basal cell carcinomas in Gorlin syndrome, and skin tumors in the Muir-Torre form of HNPCC
e. **False.** The figure is approximately one-third

5.

a. **True.** Clear cell renal carcinoma is a significant risk in VHL
b. **False.** It is easier to detect breast cancer by mammography in postmenopausal women
c. **False.** It should begin at birth
d. **False.** This does not meet the Amsterdam criteria for HNPCC
e. **False.** It is strongly indicated in FAP, but not in women positive for *BRCA1* mutations

CHAPTER 15: Genetic Factors in Common Diseases

1.

a. **False.** The figure is approximately 40%
b. **True.** One form of epilepsy is autosomal dominant, benign, familial neonatal convulsions
c. **False.** This epilepsy is non-mendelian
d. **True.** A large proportion of patients have epilepsy of different kinds, including infantile spasms
e. **False.** It is mendelian (autosomal recessive) but is progressive with neurodegeneration

2.

a. **False.** The high risk is for schizophrenia
b. **False.** The risk lies between 40% and 50%
c. **True.** This is the case despite many studies suggesting significant linkage
d. **True.** The concordance rates in monozygotic and dizygotic twins are 67% and 20%, respectively
e. **False.** Although present in about 40% of patients with Alzheimer disease, it is not discriminatory enough for screening

3.

a. **True.** Conversion of the prion protein to the abnormal isoform, PrP^{Sc}, is more likely
b. **True.** The genes are APP, Presenilin-1, and Presenilin-2
c. **False.** The penetrance is very high for all mendelian Alzheimer disease so far reported
d. **False.** Epidemiological studies using twins have proved more important so far in answering this question
e. **True.** However, the reasons are not entirely clear, despite the APP gene being located on chromosome 21

4. Hemochromatosis:

a. **True.** Mutations in at least five genes can cause hemochromatosis
b. **False.** Males are five times more likely to be affected
c. **True.** Regular phlebotomy is an effective treatment
d. **False.** Penetrance may be as low as 1%
e. **False.** The most common cause is homozygosity for the C282Y HFE mutation

5. The following statements about venous thrombosis are true:

a. **True.** They account for more than 50% of cases
b. **False.** Unlikely, but their first-degree relatives can be offered testing and prophylactic treatment if necessary
c. **False.** The factor V Leiden mutation renders the factor V protein resistant to cleavage by activated protein C
d. **True.** Raised serum prothrombin levels result from the G20210A variant
e. **True.** The risks are multiplicative

CHAPTER 16: Congenital Abnormalities and Dysmorphic Syndromes

1.

a. **False.** The figure is approximately 25%
b. **True.** This is the figure from chromosome studies. It might be much higher if all lethal single-gene abnormalities could be included
c. **False.** The figure is ~3%
d. **False.** This is an example of deformation
e. **True.** 'Sequence' implies a cascade of events traced to a single abnormality

2.

a. **False.** Syndrome is correct because of the highly recognizable nature of the condition
b. **False.** It has been found to be due to mutations in a single gene
c. **True.** The figure varies between populations and is lowered by periconceptional folic acid
d. **False.** This well defined entity is an autosomal recessive condition
e. **True.** It may be chromosomal, autosomal dominant, and autosomal recessive

3.

a. **True.** A teratogen represents a chemical or toxic disruption
b. **True.** Renal agenesis causes oligohydramnios, which leads to talipes through deformation
c. **False.** A variety of limb defects can occur
d. **True.** There is a generalized effect on a particular tissue, such as bone or skin
e. **False.** The figure is much higher, at around 50%

4.

a. **True.** Deafness and various visual defects are features
b. **False.** The first trimester is much more dangerous
c. **True.** Vertebral defects at any level are possible, including sacral agenesis
d. **False.** This is true for some populations, not all
e. **True.** Peripheral pulmonary artery stenosis in the case of congenital rubella

5.

a. **True.** The incidence is between 1 in 500 and 1 in 1000
b. **False.** Low recurrence risk because they are thought not to be genetic
c. **False.** Large studies of many families are required
d. **True.** Smith-Lemli-Opitz syndrome is a defect of cholesterol metabolism, affecting the sonic hedgehog pathway
e. **False.** The figure is much nearer 1 in 100

CHAPTER 17: Genetic Counseling

1.

a. **False.** This is the consultand; the proband is the affected individual
b. **False.** Retinitis pigmentosa can follow all the main patterns of inheritance
c. **False.** It is far more—transfer of relevant information, presentation of options, and facilitation of decision making in the face of difficult choices
d. **False.** Non-directive counseling is the aim because patients/clients should be making their own decisions
e. **True.** Patients do not remember risk information accurately and there are other important measures of patient satisfaction

2.

a. **False.** The risk is approximately twice the background risk
b. **True.** This is a second-degree relationship
c. **False.** The risk is roughly 25%
d. **False.** It is perfectly normal in many societies
e. **False.** It refers to anything from, for example, uncle–niece partnerships (second degree) to third cousins (seventh degree)

3.

a. **False.** Guilt feelings from parents and grandparents are common when a genetic disease is first diagnosed in a child
b. **False.** Many patients make the choice they would have made before genetic counseling—but after the counseling they should be much better informed
c. **True.** The risk from each grandparent is 1 in 64
d. **False.** Such a practice is strongly discouraged and the indications for genetic testing should be the same
e. **False.** Good patient support groups have a huge role, and the patients/families themselves become the experts for their condition

CHAPTER 18: Chromosome Disorders

1.

a. **True.** Chromosome number was identified in 1956, DNA structure in 1953
b. **True.** A wide variety of abnormal karyotypes occur in spontaneous abortuses but 45,X is the single most common one
c. **False.** It is estimated that 80% of all Down syndrome fetuses are lost spontaneously
d. **True.** Although the risk of Down syndrome increases with maternal age, the large proportion of babies born to younger mothers means that most Down syndrome babies are born to this group
e. **False.** A small proportion has an IQ at the lower end of the normal range

2.

a. **False.** Such children usually die within days or weeks of birth
b. **True.** Males with Klinefelter syndrome (47,XXY) are usually infertile
c. **True.** This accounts for a substantial proportion of cases
d. **False.** This is not seen in either uniparental disomy or imprinting center defect cases
e. **True.** The deletion on 22q11.2 is a 3-Mb region flanked by very similar DNA sequences

3.

a. **True.** Probably because of haploinsufficiency for elastin
b. **False.** Congenital heart disease is not a recognized feature of Prader-Willi syndrome
c. **False.** Chromosome 11p13, and may be a feature of WAGR and Beckwith-Wiedemann syndrome
d. **True.** A mutation in PAX6 or a deletion encompassing this locus at 11p15
e. **True.** Behavioral phenotypes can be very informative (e.g., Smith-Magenis syndrome)

4.

a. **False.** The figure is approximately 1 in 1000
b. **False.** IQ is reduced by 10 to 20 points but learning difficulties are not a feature
c. **True.** The other cell line may be normal but could also contain Y-chromosome material
d. **False.** They have normal fertility
e. **True.** This occurs because of DNA instability

5.

a. **True.** The mutation passes from a normal transmitting male to his daughters essentially unchanged
b. **False.** In addition to FRAXA, there is also FRAXE and FRAXF
c. **True.** Androgen insensitivity syndrome can present in this way
d. **False.** This is unreliable. DNA analysis is necessary
e. **False.** The figure is around 5%

CHAPTER 19: Single-Gene Disorders

1.

a. **False.** Meiotic instability is greater in spermatogenesis than in oogenesis
b. **True.** This has been shown from studies in Venezuela
c. **False.** The duration is approximately 10 to 15 years
d. **True.** This is so for the reduced penetrance alleles of 36 to 39 repeats
e. **True.** The condition is due to a very large triplet repeat expansion, as in fragile X syndrome

2.

a. **False.** Somnolence is common
b. **False.** Neonatal hypotonia
c. **True.** Through a CUG RNA-binding protein, which interferes with a variety of genes
d. **True.** An important feature of myotonic dystrophy and the defining abnormality of many channelopathies
e. **True.** Males with mild or subclinical CF have congenital bilateral absence of the vas deferens

3.

a. **False.** The DF508 mutation is the most common
b. **True.** The polythymidine tract—5T, 7T, and 9T—can be crudely correlated with different CF phenotypes
c. **False.** This is true for most of the inherited cardiac arrhythmias; cardiomyopathies are often from defects in cardiac muscle proteins
d. **True.** This glycoprotein complex in the muscle membrane contains a variety of units; defects in these cause various limb-girdle dystrophies
e. **False.** These patients have normal intelligence

4.

a. **False.** They are good candidates according to current thinking
b. **False.** It is only a component of the skeletal system criteria
c. **False.** It is thought to be a fully penetrant disorder
d. **True.** This is not usually severe but is a recognized feature
e. **False.** The opposite is the case

5.

a. **False.** This is a neurophysiological classification
b. **True.** Autosomal dominant, autosomal recessive and X-linked
c. **True.** Mutations in the peripheral myelin protein affect Schwann cells
d. **False.** They are not good discriminatory tests and DNA analysis is much preferred
e. **False.** It is an inherited cardiac arrhythmia

CHAPTER 20: Screening for Genetic Disease

1.

a. **True.** By looking for evidence of two populations of cells
b. **True.** Firm clinical signs are the exception rather than the rule
c. **False.** DNA sequence variants must be polymorphic to be useful
d. **False.** There is far too much locus heterogeneity in this condition
e. **True.** As a general rule this may be vital, but should be undertaken with informed consent

2.

a. **False.** The facial rash of angiokeratoma (adenoma sebaceum) is often not present
b. **False.** There may not be sufficient numbers of café-au-lait spots until age 5 to 6 years
c. **False.** They may be fully informative of an individual's genetic status
d. **True.** Dural ectasia of the lumbar spine is one of the major criteria
e. **False.** Linked DNA markers, and sometimes biochemical tests, may be the best modality available

3.

a. **False.** Participation should, in principle, be voluntary
b. **True.** The outcome of population screening programs should be an improvement in health benefit
c. **False.** This is the specificity of a test
d. **True.** This is different from the sensitivity, which refers to the proportion of affected cases that are detected (i.e., there may be some false negatives)
e. **True.** Adequate expert counseling should be part of the predictive test program

4.

a. **True.** Recall of the result itself, or the interpretation, is frequently inaccurate
b. **False.** The highest incidence for a serious disease is that in β-thalassemia: 1 in 8 is carriers
c. **False.** This has happened before and should be a major concern
d. **False.** The benefit lies in informing the family for subsequent reproductive decisions
e. **False.** The first assay is biochemical, a measure of immunoreactive trypsin

5.

a. **False.** Although the carrier frequency is about 1 in 10, no population screening is undertaken in the United Kingdom
b. **False.** In general, unless a beneficial medical intervention can be offered, such testing should be deferred until the child is old enough to make the decision
c. **True.** They have been operational for about 30 years
d. **False.** This is only at the research stage
e. **False.** Their prime function in a service department is for clinical management of patients and families

CHAPTER 21: Prenatal Testing and Reproductive Genetics

1.

a. **False.** It is still mainly performed around 16 weeks' gestation
b. **False.** They also derive from the amnion and fetal urinary tract epithelium
c. **True.** The risk is 1% to 2%, compared with 0.5% to 1%
d. **False.** There is a small risk of causing limb abnormalities; CVS should not be performed before 11 weeks' gestation
e. **False.** Fetoscopy is required for this if a DNA test is not available

2.

a. **True.** This forms part of the triple test
b. **True.** This forms part of the triple test
c. **False.** In trisomy 18 all maternal serum markers are low
d. **False.** The best figure achieved is around 86%
e. **True.** There are two fetuses rather than one

3.

a. **False.** The figure is about 1%
b. **True.** Especially aneuploidies
c. **True.** Probably because of the presence of inspissated meconium
d. **True.** Most cases of Down syndrome are due to meiotic non-disjunction
e. **True.** They are unlikely to have different clinical effects in different members of the same family

4.

a. **False.** A license from the HFEA is required
b. **False.** It is not illegal but does require a HFEA license
c. **True.** This is undertaken to avoid the presence of extraneous sperm
d. **False.** The figure is about 25%
e. **False.** Chromosome disorders are the largest group

5.

a. **True.** Chromosome abnormalities are present in 10% to 12% of men with azoospermia or severe oligospermia, some of them heritable
b. **False.** The rule is that no more than 10 pregnancies may result from one donor
c. **False.** They are currently not entitled to know the identity of their biological parents
d. **True.** Analyzing fetal cells in the maternal circulation only removes the miscarriage risk
e. **False.** The figure is approximately 1 in 7

CHAPTER 22: Risk Calculation

1.

a. **True.** These are two ways of expressing the same likelihood
b. **True.** A probability of 1 means that the event will happen 100% of the time
c. **True.** The probability that both will be boys is $\frac{1}{2} \times \frac{1}{2} = \frac{1}{4}$, for girls the same; therefore the chance of being the same sex is $\frac{1}{4} + \frac{1}{4} = \frac{1}{2}$ (0.5)
d. **True.** These two approaches to a probability calculation are essential
e. **True.** 70% of heterozygotes will manifest the condition

2. For an autosomal recessive condition the chance that the first cousin of an affected individual is a carrier is:

a. **False.**
b. **False.**
c. **True.** The affected individual's parents are obligates carriers, aunts and uncles have a 1 in 2 risk, cousins a 1 in 4 risk
d. **False.**
e. **False.**

3. In X-linked recessive inheritance:

a. **False.** The risk is 1 in 2 if the sex of the fetus is known to be male
b. **False.** The male might be affected because a new mutation has occurred
c. **True.** This is significant and has to be allowed for in risk calculation and counseling
d. **True.** This is conditional information that can be built into a Bayes' calculation
e. **False.** This is a key individual whose risk must be calculated before the consultand's risk

4. In autosomal recessive inheritance the risk that the nephew of an affected individual, born to the affected individual's healthy sibling, is a carrier is:

a. **False.**
b. **False.**
c. **False.**
d. **True.** The healthy sibling of the affected individual has a 2 in 3 chance of being a carrier; this person's son has a risk which is half that
e. **False.**

5.

a. **True.** For example, negative mutation findings when testing for cystic fibrosis
b. **True.** Age of onset (clinical expression) data must be derived from large family studies
c. **False.** Without this information huge errors will be made
d. **True.** An empiric risk is really a compromise figure
e. **False.** It may matter a lot because it is a measure of the likelihood that a meiotic recombination event will take place between the marker and the gene mutation causing the disease

CHAPTER 23: Treatment of Genetic Disease

1. Methods currently used to treat genetic disease include:

a. **False.** Germ-cell gene therapy is considered unacceptable because of the risk of transmitting genetic changes to future generations
b. **True.** For example, bone marrow transplantation is used to treat various inherited immunodeficiencies
c. **True.** Examples include the replacement of factor VIII or IX in patients with hemophilia
d. **True.** For example, restricted phenylalanine in patients with phenylketonuria
e. **False.** This potential treatment has been tested in animal models

2. Gene therapy may be delivered by:

a. **True.** Liposomes are widely used as they are safe and can facilitate transfer of large genes
b. **True.** CFTR gene therapy trials have used adeno-associated viral vectors
c. **False.** Antisense oligonucleotides need to be delivered to the target cells
d. **True.** Lentiviruses may be useful for delivery of genes to non-dividing cells
e. **True.** An example is the injection of plasmid-borne factor IX into fibroblasts from patients with hemophilia B

3. Gene therapy has been used successfully to treat patients with the following diseases:

a. **False.** Trials have shown safe delivery of the *CFTR* gene to the nasal passages but effective treatment of cystic fibrosis is not possible at present
b. **True.** A number of patients have been treated successfully, although concern was raised when two boys developed leukemia
c. **False.** This will be difficult because the number of α- and β-globin chains must be equal or a thalassemia phenotype might result
d. **True.** Some patients have been able to reduce their exogenous clotting factors
e. **True.** Although early attempts were unsuccessful, two patients have now been treated successfully by ex vivo gene transfer

4. Potential gene therapy methods for cancer include:

a. **True.** An example is the protein kinase inhibitor used to treat chronic myeloid leukemia
b. **True.** Perhaps through overexpression of interleukins
c. **False.** Anti-angiogenic factors might be used to reduce blood supply to tumors
d. **True.** RNA interference is a promising new technique that can used to target overexpressed genes associated with cancers
e. **True.** A number of trials are ongoing to determine the utility of this technique

Case-Based Answers

CHAPTER 6: Developmental Genetics

Case 1

1. The combination of macrocephaly, odontogenic keratocysts, and basal cell carcinomas occurs in Gorlin (basal cell nevoid carcinoma) syndrome. This condition should be in the differential diagnosis of a child with macrocephaly, with appropriate exploration of the family history. It is understandable that hydrocephalus would be the main concern, but true hydrocephalus is unusual in Gorlin syndrome.
2. The child's father is an obligate carrier for the *PTCH* gene mutation causing Gorlin syndrome in the family. He should be screened regularly (at least annually) by radiography for odontogenic keratocysts, and be under regular surveillance by a dermatologist for basal cell carcinomas. *PTCH* gene mutation analysis is possible for any affected family member.

Case 2

1. The two most likely causes of sex reversal in a young 'girl' are androgen insensitivity syndrome (AIS), which is X-linked and results from mutations in the androgen receptor gene, and mutations in the *SRY* gene on the Y chromosome.
2. Mutation analysis in both the androgen receptor and *SRY* genes can be performed to determine the genetic basis of the sex reversal. It is very important to investigate and locate, if present, remnants of gonadal tissue because this will have to be removed to avoid the risk of malignant change. Because of this, the parents should be given a full explanation, but the phenotypic sex of the child should be affirmed as female.

CHAPTER 7: Patterns of Inheritance

Case 1

1. It is possible that the problems described in family members are unrelated, but this is unlikely. If the condition has passed from the grandfather, mitochondrial inheritance is very unlikely. The condition is either autosomal dominant with variability, or X-linked.
2. The spinocerebellar ataxias are a genetically heterogeneous group of conditions that usually follow autosomal dominant inheritance and could present in this way. A form of hereditary spastic paraparesis is possible, also genetically heterogeneous and usually after autosomal dominant inheritance, although recessive and X-linked forms are described. Apart from these, X-linked adrenoleukodystrophy must be considered, especially because the man has signs of cognitive and behavioral problems. This is very important, not only because it can present early in life but also because of the potential for adrenal insufficiency.

Case 2

1. The information *may* be correct but is probably not and other possibilities must be explained to them.
2. Most forms of osteogenesis imperfecta (brittle bone disease) follow autosomal dominant inheritance. Sibling recurrence, when neither parent has signs or symptoms, can be explained by somatic and/or germline mosaicism in one of the parents. The risk to the offspring of those affected would then be 50% (i.e., high). In this case history, the possibility of a non-genetic diagnosis must be considered, namely non-accidental injury. It is therefore important to try to confirm the diagnosis.

CHAPTER 8: Mathematical and Population Genetics

Case 1

1. Clearly, it is essential to know whether the condition in question has ever knowingly occurred in the families of either of the two consultands. If this had occurred, it would potentially modify the carrier risk for one of the consultands regardless of the frequency of the disease in the population.
2. Assuming the disorder in question has not occurred previously in the family, the carrier frequency in the population A is 1 in 50, and 1 in 15 in the population B. The risk in the first pregnancy is therefore $\frac{1}{50} \times \frac{1}{15} \times \frac{1}{4} = \frac{1}{3000}$.

Case 2

1. From the figures given, four cases in the town appear to be new mutations, i.e., four new mutations per 100,000 genes inherited. The mutation rate is therefore 1 per 25,000 gametes.
2. The population sample is small and may not therefore be representative of the larger, wider population. For example, if there is a bias toward an older, retired

population, the reproducing subpopulation may be smaller and the figures distorted by the migration of younger people away from the town. In addition, the four 'new mutation' cases should be verified by proper examination of the parents.

CHAPTER 9: Polygenic and Multifactorial Inheritance

Case 1

1. Testing for factor V Leiden and the prothrombin G20210A variant is appropriate. A positive result would provide a more accurate risk of her developing thromboembolism and would inform her choice of contraception. Heterozygosity for factor V Leiden or the prothrombin G20210A variant would increase her risk by four- to fivefold. Homozygosity or compound heterozygosity would increase her risk by up to 80-fold.
2. Negative results for factor V Leiden and the prothrombin 20210A variant in the consultand should be interpreted with caution as up to 50% of cases of venous thrombosis are not associated with these genetic risk factors.

Case 2

1. The proband might have type 1 diabetes (T1DM), type 2 diabetes (T2DM), or maturity-onset diabetes of the young (MODY). Because both have normal hearing, a diagnosis of maternally inherited diabetes and deafness (MIDD) is unlikely. T1DM and T2DM show multifactorial inheritance with environmental factors playing a role in addition to predisposing genetic susceptibility factors. MODY shows autosomal dominant inheritance.
2. The brother's risk of developing diabetes is 6%, 35%, or 50%, respectively. If his sister is found to have a mutation in one of the genes causing MODY, then he could have predictive genetic testing. A negative test would reduce his risk to that of the population. A positive test would allow regular monitoring in order to make an early diagnosis of diabetes and avoid diabetic complications from long-standing undiagnosed diabetes.

CHAPTER 10: Hemoglobin and the Hemoglobinopathies

Case 1

1. The ethnic origin of the couple and the limited information should suggest the possibility of a hematological disorder. α-Thalassemia is the likely cause of stillbirth, hydrops being secondary to heart failure, which in turn is secondary to anemia. Rhesus isoimmunization and glucose-6-phosphate dehydrogenase deficiency are other possibilities. Severe forms of congenital heart disease are frequently associated with hydrops, but the chance of a sibling recurrence (which occurred in the case history) is low. However, there are many other causes of hydrops and these would need to be considered. Among those that are genetic with a chance of recurrence are lethal forms of rare skeletal dysplasias and a wide range of metabolic disease.
2. A full blood count, blood groups, Hb electrophoresis, and maternal autoantibody and glucose-6-phosphate dehydrogenase deficiency screens should be performed for the couple. DNA analysis may detect the common mutation seen in Southeast Asia, which would then make it possible to offer genetic prenatal diagnosis by chorionic villus sampling. If no disorder is identified by these investigations it is unlikely that further diagnostic progress will be made unless the couple has another affected pregnancy that can be fully investigated by examination of the fetus.

Case 2

1. This presentation is consistent with acute intermittent porphyria and hemolytic uremic syndrome. However, the ethnic origin should suggest the possibility of sickle cell disease. The contents of the dark urine, and specific tests for porphyria, will help to differentiate these, and a sickle cell test should be performed.
2. If the diagnosis is sickle cell disease there are various agents that can be tried to reduce the frequency of sickling crises—hydroxyurea in particular. Prophylactic penicillin is important for reducing the risk of serious pneumococcal infections, and the family should be offered genetic counseling and cascade screening of relatives.

CHAPTER 11: Biochemical Genetics

Case 1

1. Hypoglycemia can be part of severe illness in young children, but in this case the intercurrent problem appears relatively minor, suggesting that the child's metabolic capacity to cope with stress is compromised. This history should prompt investigations for a possible inborn error of metabolism and, if a diagnosis is made, the younger sibling should be tested.
2. Hypoglycemia is a common consequence of a number of inborn errors of amino acid and organic acid metabolism. Investigation should begin with analysis of urinary organic acids, and plasma amino acids, ammonia and liver function tests.

Case 2

1. If there is a family history of similar symptoms, it might demonstrate matrilinear inheritance with all the offspring of affected males being normal. If this person is the only affected individual, a family history will not be informative with respect to the diagnosis.

2. All causes of myopathy need to be considered, but the combination of features is suggestive of a mitochondrial cytopathy. This would explain the muscular symptoms, ptosis, and hearing impairment—and there might also be evidence of a cardiomyopathy, neurological disturbance, retinitis pigmentosa, and diabetes mellitus. Mitochondrial DNA analysis on peripheral lymphocytes might identify a mutation, although a negative result would not rule out the diagnosis. A muscle biopsy might show ragged red fibers, and DNA analysis on this tissue might be more informative than lymphocytes.

CHAPTER 13: Immunogenetics

Case 1

1. The nature of his grandfather's symptoms are rather non-specific—back pain and arthritis are both very common in the general population. However, it is certainly possible that he also had ankylosing spondylitis, a form of enthesitis (inflammation at the site of insertion of a ligament or tendon into bone) with involvement of synovial joints, as the heritability is greater than 90%.
2. About 95% of patients with ankylosing spondylitis are positive for the HLA-B27 antigen; however, in the general population this test has only a low positive predictive value. His children have a 50% chance of being HLA-B27 positive; if positive, the risk of developing clinical ankylosing spondylitis is about 9%; if negative, the risk is less than 1%.

Case 2

1. This history points strongly towards a diagnosis of deletion 22q11 (DiGeorge/Sedláčková) syndrome, which can easily be confirmed by specific FISH testing. Immunity is impaired but usually only mildly, and gradual improvement occurs through childhood and adolescence.
2. Deletion 22q11 syndrome can be familial and does not always give rise to congenital heart disease. If confirmed in the child, both parents should be offered FISH testing, and other family members as appropriate. Genetic counseling for the child will be important when she is older.

CHAPTER 14: Cancer Genetics

Case 1

1. The family history should first of all be confirmed with the consent of the affected individuals. If the thyroid cancer in the cousin was papillary in type, and the polyps in her father hamartomatous, the pattern would be very suspicious for Cowden disease. This is also known as multiple hamartoma syndrome, which is autosomal dominant and often the result of mutations in *PTEN*; the risk of breast cancer is approximately 50% in females.
2. Macrocephaly, a cobblestone appearance of the oral mucosa, and generalized lipomas are other features to look for in patients with this unusual history.

Case 2

1. If the *BRCA2* mutation has not been confirmed in another family member or by testing another sample from the deceased cousin (e.g., a tissue section embedded in paraffin), the possibility of a sample mix-up in the research laboratory cannot be excluded. If, however, the uncle tests positive for the mutation, the consultand's mother is a phenocopy. Alternatively the mutation may have been inherited from the cousin's mother.
2. If the uncle tests positive for the *BRCA2* mutation, then his lifetime risk of developing breast cancer is approximately 6%, more than 100-fold higher than that in the general population.

CHAPTER 15: Genetic Factors in Common Diseases

Case 1

1. Generally, the risk of epilepsy to first-degree relatives is around 4%. However, mother and daughter are affected here, which suggests the possibility of a mendelian form of epilepsy. Furthermore, it seems that both have an abnormal finding on brain imaging and the mother's computed tomograms should be located and reviewed. At this stage, an explanation of both autosomal dominant and X-linked inheritance should be given, as well as the possibility that the two cases of epilepsy are coincidental.
2. The condition that the mother's doctors mentioned would almost certainly have been tuberous sclerosis (TS), which follows autosomal dominant inheritance. Further evaluation of both mother and daughter looking for clinical features of TS might be indicated and genetic testing is available. However, the nodules on the lateral ventricle walls may be pathognomonic of bilateral periventricular nodular heterotopia (BPVNH) and the images should be reviewed by someone who can recognize this. BPVNH is an abnormality of neuronal migration and is inherited as an X-linked dominant condition, caused by mutations in the filamin-1 (*FLNA*) gene. Testing is available. In general, mendelian forms of epilepsy are rare.

Case 2

1. Not necessarily. Many people with glucokinase gene mutations are asymptomatic and their mild hyperglycemia is detected only upon screening (routine medicals, during pregnancy or intercurrent illness). Gestational diabetes in the father's sister suggests that the mutation could have been inherited from his side of the family.

2. Identification of a glucokinase gene is 'good news' because the mild hyperglycemia is likely to be stable throughout life, treated by diet alone (except during pregnancy) and unlikely to result in diabetic complications. Cascade testing can be offered to other relatives. If the mutation has been inherited from the father, his sister and her child may be tested. The sister might avoid the anxiety of having a young child diagnosed with unexplained hyperglycemia.

CHAPTER 16: Congenital Abnormalities and Dysmorphic Syndromes

Case 1

1. This is not an unusual scenario. The karyotype on amniocentesis was normal and polyhydramnios suggests the possibility of a gastrointestinal obstruction such as esophageal atresia. The abnormalities are more likely to represent an 'association' rather than a syndrome or mendelian condition. The empiric recurrence risk is low, and all that can be offered is ultrasonography in subsequent pregnancies.
2. A fetal autopsy is highly desirable in this situation to know the full extent of internal organ anomalies. A repeat karyotype on fetal skin might have shown something that was not detected on amniocentesis, and DNA should be stored for possible future use. Maternal diabetes should be excluded. As a last resort, the parents' chromosomes could be tested, including telomere screens to look for the possibility of a cryptic translocation.

Case 2

1. Isolated, non-syndromic cleft palate is statistically the most likely diagnosis, but the mild short stature might be significant. Syndromic possibilities include spondyloepiphyseal dysplasia congenita (SEDC)—although there are many rare syndromes with more severe short stature and other features.
2. The short stature appears mild; it is, therefore, important to try to determine whether this might be familial—the parents need to be assessed. Follow-up of the baby is indicated, including a radiological skeletal survey to see whether there is an identifiable skeletal dysplasia. SEDC may be accompanied by myopia and sensorineural hearing impairment; therefore hearing and vision assessment is important. However, the child has cleft palate and is at risk of conductive hearing problems as a result. The cleft palate team needs to be involved from the beginning.

CHAPTER 17: Genetic Counseling

Case 1

1. The couple is at risk of having further affected children and prenatal diagnosis can be offered. The father may have inherited the balanced translocation from one of his parents and his sister may also be a carrier. Carrier testing should be offered to his family, especially as his sister is trying to get pregnant.
2. The father's wider family needs to be made aware of the child's diagnosis, but have fixed misconceptions and it might be very difficult to accept that the child's problems have their origin on their side of the family. There is a severe communication problem but a way needs to be found to inform the father's wider family of the genetic risk. It might be necessary to involve their general practitioners.

Case 2

1. There is now no need for the woman to undergo an invasive prenatal test in future pregnancies; this would be a waste of resources and place the pregnancy at a small but unnecessary risk of miscarriage.
2. There is the difficulty of communicating the fact that a prenatal test is not necessary, but disclosure of non-paternity may have far-reaching consequences for the marriage. The counselors do not know whether the 'father' suspects non-paternity, and the mother may think he is the biological father of the child.

CHAPTER 18: Chromosome Disorders

Case 1

1. Head-banging is not rare in early childhood, especially in children with developmental delay, and it is not necessarily a helpful feature in making a diagnosis. However, combined with the persistently disturbed sleep pattern, the diagnosis of Smith-Magenis syndrome should be considered. These children can be quiet as babies and have congenital heart disease.
2. Smith-Magenis syndrome is usually due to a microdeletion at 17p11.2, for which a FISH probe test is available. They can also exhibit self-hugging behavior and develop scoliosis. Melatonin has proved a very effective treatment for sleep disturbance.

Case 2

1. The previous counseling given naturally assumed the girl was *pure* 47,XXX. However, the subsequent course raises the possibility that she has chromosome mosaicism, and in particular she might be mosaic for Turner syndrome (45,X). A buccal smear and/or skin biopsy should be offered to look at chromosomes in a tissue other than blood. If this is normal, other causes of short stature would need to be considered.
2. If the child is indeed found to be a 45,X/47,XXX mosaic, she needs to be investigated for the complications of Turner syndrome—congenital heart disease and horseshoe kidney. In addition, her fertility is in question and she would need to be referred to a pediatric

endocrinologist, who would also assess her for possible growth hormone treatment.

CHAPTER 19: Single-Gene Disorders

Case 1

1. The history in the brother is consistent with his having Becker muscular dystrophy (BMD) but is also consistent with other diagnostic possibilities, e.g., limb-girdle muscular dystrophy (LGMD). These two conditions have sometimes been difficult to distinguish and the inheritance is different, with quite different implications for the woman who wishes to start a family.
2. The medical records of the affected brother should be reviewed and he should be reassessed. Thirty years ago the tests for BMD were very basic but now dystrophin gene mutation analysis is available. A muscle biopsy subjected to specific dystrophin staining may be diagnostic, but if this is negative staining techniques for different forms of LGMD are available. If one of the LGMD group, the woman can be reassured because these follow autosomal recessive inheritance. If BMD, carrier testing for the consultand would be straightforward if only a specific mutation were found in her brother.

Case 2

1. The sudden, unexpected death of young adults, especially when no cause can be identified, is extremely shocking for a family. The focus of attention becomes the inherited arrhythmias and cardiomyopathies—sometimes the latter show no obvious features at post-mortem examination. All close family members are eligible for cardiac evaluation by echocardiography, ECG, and provocation tests, looking for evidence of the long QT and Brugada syndromes. Genetic testing is available but not guaranteed to identify a pathogenic mutation. Some forms of inherited arrhythmia/cardiomyopathy are amenable to prophylactic treatment; for others very little is available.
2. Management will depend on the outcome of investigations and genetic testing. However, if no specific findings are made it is very difficult to know how to advise families like this. High-intensity sports and swimming should probably be avoided in case such activity is a precipitating factor for a life-threatening arrhythmia.

CHAPTER 20: Screening for Genetic Disease

Case 1

1. Mutation analysis in the fibrillin-1 gene is possible for the consultand but not guaranteed to identify a mutation even if the clinical diagnosis is confident. Linkage analysis at the fibrillin-1 locus on chromosome 15 would be possible using DNA from the consultand's mother and inferring the haplotype of the father, whose affected status would have to be assumed. In a small family, however, linkage may be consistent with segregation of the condition by chance. Furthermore, in unusual cases Marfan syndrome is due to a mutated gene at a separate locus on chromosome 3. There are serious pitfalls to genetic testing in this situation.
2. The important life-threatening complication of Marfan syndrome is progressive aortic root dilatation carrying a risk of dissection. Those with a firm diagnosis must be followed until at least the age of 30 years. If there is doubt about the diagnosis, regular cardiac screening is probably a sensible precaution for all those at risk until their mid-twenties.

Case 2

1. The sensitivity is the proportion of true positives detected by the test, i.e., $^{45}/_{45} + 5 = 90\%$. The specificity is the proportion of true negatives detected by the test, i.e., 99,190 (the unaffected cases who test negative)/99,190 + 760 (the unaffected cases who test positive) = 99.2%.
2. The positive predictive value is the proportion of cases with a positive test who truly have the disease, i.e., $^{45}/_{805}$ = 5.6%.

CHAPTER 21: Prenatal Testing and Reproductive Genetics

Case 1

1. The finding of mosaicism for trisomy 20 in chorionic villus tissue might have been a case of confined placental mosaicism. The latter is not a rare event for a wide variety of chromosome aberrations but, as long as it is confined, there are no serious consequences for the pregnancy. The problem with going on to perform an amniocentesis is in interpretation of the result. If no abnormal cells are found, this does not completely rule out chromosome mosaicism in the fetus. If abnormal cells are found, the clinical implications are very difficult, if not impossible, to predict.
2. This case illustrates the rollercoaster of emotions and experiences that some women and couples have to cope with as a result of different forms of prenatal tests and their interpretation. In fact, trisomy 20 mosaicism is unlikely to be of great clinical significance—but it is very difficult to be sure. Renal abnormalities have been reported, and detailed fetal anomaly scanning can be offered for the remainder of the pregnancy. However, what might have been an enjoyable pregnancy will probably continue to be an anxious one.

Case 2

1. In the large majority of autism cases, no specific diagnosis is reached. Chromosome analysis, including a multitelomere screen, fragile X syndrome, a metabolic screen,

and examination for neurocutaneous disorders, should all be performed.

2. This is a very difficult situation. However, there is no proof in this case that autism is X-linked, and therefore no guarantee that any daughter will be unaffected. It would, therefore, be very difficult to support this request in the United Kingdom where PGD is regulated by the Human Fertilization and Embryology Authority, and sex selection for anything other than clearly X-linked conditions is not licensed. In other countries, where these techniques are not regulated, the couple might find clinicians who acquiesce to their request.

CHAPTER 22: Risk Calculation

Case 1

1. Each of the siblings of the affected aunt has a chance of being a carrier; therefore, each of the cousins has a chance of being a carrier. The chance of the couple's first baby being affected is $\frac{1}{3} \times \frac{1}{3} \times \frac{1}{4} = \frac{1}{36}$.
2. Even though genetic studies cannot be performed, biochemical prenatal testing can be offered for their pregnancies, although biochemical tests will not reliably determine whether they are carriers. If they elect for prenatal testing, it would also be worth testing for Hunter syndrome, which can easily be confused with Hurler syndrome clinically, and is X-linked.

Case 2

1. A simple Bayes' calculation can be performed, taking into account that she has had two normal sons (Table 1). She therefore has a $\frac{1}{5}$, or 20%, chance of being a carrier.
2. There is a good chance of identifying the factor VIII gene mutation in either her brother or uncle if either of them is still alive. If so, it should then be possible to determine her carrier status definitively. If not, tests of factor VIII levels and factor VIII–related antigen may give a result that can modify her risk, but this is not necessarily discriminatory. DNA linkage analysis would be much more reliable, provided DNA samples are available.

Table 1

Probability	Is a Carrier	Is Not a Carrier
Prior	$\frac{1}{2}$	$\frac{1}{2}$
Conditional (2 normal sons)	$\frac{1}{2} \times \frac{1}{2}$	1
Joint	$\frac{1}{8}$	$\frac{1}{2}$
Posterior	$\frac{1}{8} / \frac{1}{8} + \frac{1}{2} = \frac{1}{5}$	

Index

Page numbers followed by "f" indicate figures, "t" indicate tables, and "b" indicate boxes.

D